AF342677

Recent Progress in MANY-BODY THEORIES

VOLUME 3

Recent Progress in
MANY-BODY THEORIES

VOLUME 3

Edited by

T. L. Ainsworth
Texas A&M University
College Station, Texas

C. E. Campbell
University of Minnesota
Minneapolis, Minnesota

B. E. Clements
Texas A&M University
College Station, Texas

and

E. Krotscheck
Texas A&M University
College Station, Texas

Plenum Press • New York and London

Proceedings of the Seventh International Conference on Recent
Progress in Many-Body Theories, held August 26–31, 1991,
in Minneapolis, Minnesota

Library of Congress Catalog Card Number 88-645051

ISBN 0-306-44246-9

PREFACE

The present volume contains the texts of the invited talks delivered at the Seventh International Conference on Recent Progress in Many-Body Theories held at the University of Minnesota during the period August 26-31, 1991. The proceedings of the Fourth Conference (Oulu, Finland, 1987) and Fifth Conference (Arad, Israel, 1989) have been published by Plenum as the first two volumes of this series. Papers from the First Conference (Trieste, 1978) comprise Nuclear Physics volume A328, Nos. 1, 2. The Second Conference (Oaxtepec, Mexico, 1989) was published by Springer-Verlag as volume 142 of "Lecture Notes in Physics," entitled "Recent Progress in Many-Body Theories." Volume 198 of the same series contains the papers from the Third Conference (Altenberg, Germany, 1983).

These volumes are intended to cover a broad spectrum of current research topics in physics that benefit from the application of many-body theories for their elucidation. At the same time there is a focus on the development and refinement of many-body methods. One of the major aims of the conference series has been to foster the exchange of ideas among physicists working in such diverse areas as nucleon-nucleon interactions, nuclear physics, astronomy, atomic and molecular physics, quantum chemistry, quantum fluids, and condensed matter physics. The present volume contains contributions from all of these areas.

Several of the contributors to this volume were asked to provide a broad overview of a particular area, including a discussion of the challenges particular problems pose for the future. The contribution by A. D. Jackson on the nuclear many-body problem, and the contribution by B. Serot and J. D. Walecka on the relativistic nuclear many-body problem, provide excellent surveys of the present status and future challenges of nuclear many-body physics. The paper by I. Lindgren focuses on developments of relativistic many-body theory in atomic physics. Eugene Bashkin's contribution provides a picture of the present status of the theory of "ordinary" quantum fluids, while the article by D. P. Arovas and S. M. Girvin contends with the present and future of what might be dubbed "exotic" quantum fluids. Yu Lu accepted the daunting challenge of providing a review/preview/overview of high temperature superconductivity.

The remaining invited papers are generally focused more narrowly on recent research results. Most are on topics related to the overview papers described above, in which case they appear in the same section as the overview. There are a number of papers on the subject of large-scale computations, reflecting the fact that an entire day was set aside for the Symposium on Large-Scale Computations Applied to Many-Body Systems, held at the Minnesota Supercomputer Institute on August 28, 1991. M. H. Kalos presented the keynote talk for the Symposium, which is included herein.

The conference was the occasion for the presentation of the fourth Eugene Feenberg Medal in Many-Body Physics to Walter Kohn of the University of California at

Santa Barbara. The presentation was made by M. H. Kalos of Cornell University, recipient of the Third Eugene Feenberg Medal awarded at the Sixth Conference in Arad. The presentation remarks are included in this volume, along with Professor Kohn's gracious acceptance remarks. Also included is a brief outline of Kohn's review talk, together with citations to complete recent reviews by him on density functional theory.

A second high point of the conference was the delightful talk by A. A. Abrikosov, entitled "L. D. Landau — his life, achievements, and my own memories," on the occasion of the fiftieth anniversary of Landau's introduction of the roton into the theory of liquid helium. We regret that there is no manuscript for that talk that we could include in this volume. We encourage the reader to take any opportunity to listen to Dr. Abrikosov on this subject.

We would like to take this opportunity to acknowledge a generous contribution to the fund that supports the Feenberg Award by Professor Hermann Kümmel, and also to acknowledge his efforts and good taste as he oversaw the design and execution of the Feenberg Medal.

We would like to acknowledge support for the Seventh International Conference on Recent Progress in Many-Body Theories by the Theoretical Physics Institute, the School of Physics and Astronomy, the Institute of Technology, the Minnesota Supercomputer Institute, the Army High Performance Computing Research Center, the Graduate School and the Office of the Provost, University of Minnesota, and by the Department of Physics, Texas A&M University.

We would also like to acknowledge and thank Sandra Smith for her Herculean efforts in preparation for and administration during and after the conference, Michael Olesen for his excellent facilitation of the Symposium on Large-Scale Computations Applied to Many-Body Systems, and Friedlinde Krotscheck for her fine arrangements of the reception and coffee breaks. We would also like to express our appreciation to the many-body theory graduate students from Texas A&M, Washington University, and the University of Minnesota for their assistance during the conference.

T. L. Ainsworth
C. E. Campbell
B. E. Clements
E. Krotscheck

CONTENTS

THE ONCE AND FUTURE NUCLEAR MANY–BODY PROBLEM

Andrew D. Jackson

Department of Physics
The State University of New York
Stony Brook, New York 11794, USA

INTRODUCTION

A friend of mine — an anthropologist in Copenhagen — described a seminar given there by a French colleague under the title of *The Greenland Eskimo: His Past, His Present, His Future.* The seminar was short, and I would like to share it with you. 'Nothing is known about the past of the Greenland Eskimo. You,' he said with an all-encompassing sweep of his arm 'are the world's experts about His present. And who, my Friends, can predict the future?'

I feel a certain sympathy with the speaker. I have worked on the nuclear many–body problem in the past. I shall return to it in the future. But you are the world's experts about the present status of this problem. It was thus with some trepidation that I agreed to give this talk. Nuclear physics is a field in transition. It is moving from the study of low–energy nuclear structure with a simple model for every nucleus to something much more closely related to elementary particle physics. There can be no doubt that the microscopic nuclear many-body problem has played an important role in our understanding of traditional nuclear structure questions. At the very least, it has provide some theoretical justification for the many of the *ad hoc* models employed. That role is far from finished and, as I hope to convince you, many interesting questions remain open. I am equally convinced that natural extensions of the traditional many-body problem will be central in exploring the new nuclear physics of the coming decade.

THE TRADITIONAL NUCLEAR MANY–BODY PROBLEM

For almost thirty years the central question of the nuclear many–body problem has been: To what extent can the properties of nuclei be described by microscopic calculations using a two–body interaction which describes two nucleon scattering data?[1] The problem has been a difficult one, and most people have concentrated on those bulk properties of infinite nuclear matter (*i.e.,* binding energy and equilibrium density) which can be determined empirically with some confidence. To a lesser extent, microscopic calculations have been performed to construct the Fermi liquid parameters for nuclear matter. As I shall note later, it has been far more difficult to get a reliable empirical handle on these numbers. I think we now have a reasonable understanding of the more restricted nuclear matter problem.

Before turning to a summary of these developments, I think it is valuable to understand in very simple terms why the nuclear matter problem appears easier than the problem of atomic liquids and why it is, in reality, far more difficult. Nuclei (or nuclear matter) are bound and saturating systems. Their average density can be expressed in terms of a nuclear radius, $R \sim r_o A^{1/3}$, where $r_o \sim 1.1$ fm. The two–body interaction can be characterized by hard core repulsion when the interparticle spacing is less than $c \sim 0.4$ fm. The ratio of the excluded volume to the average volume particle is evidently very much less than 1. Historically, this fact has been used to argue for the perturbative treatment of particle–hole excitations in the nuclear medium. Clearly, the situation is rather different in atomic liquids (*e.g.,* liquid ^{3}He) where $c \sim 2.55$ Å and $r_o \sim 2.44$ Å. It is more instructive to look at the maximum value of r_o which is consistent with binding. (The minimum value, corresponding to the close packing of hard spheres, is simply $r_o \sim 0.552c$.) To do this roughly, assume that the saturation of nuclear matter is due solely to the hard core. (It is not.) Familiar arguments from hard sphere gases suggest that the kinetic energy per particle, including the effects of the hard core, is given approximately as

$$\epsilon_{kin} \sim \frac{3}{5}\epsilon_F \left(1 - 0.8\frac{c}{r_o}\right)^{-2}. \tag{1}$$

(This expression is quantitatively reliable for r_o greater than $2c$.) Let us assume that the potential energy per particle, ϵ_{pot}, is proportional to $r_o{}^{-3}$ but otherwise unspecified. We now minimize the total energy with respect to r_o and consider the limiting case where the resulting energy is precisely 0. This will lead us to the maximum value of r_o. We find

$$\epsilon_{kin} + \epsilon_{pot} = 0 \tag{2}$$

and

$$\frac{\partial}{\partial r_o}(\epsilon_{kin} + \epsilon_{pot}) = 0. \tag{3}$$

These equations are readily solved to yield

$$\frac{r_o}{\epsilon_{kin}}\frac{\partial \epsilon_{kin}}{\partial r_o} = -3 \tag{4}$$

which yields $0.552c \leq r_o \leq 2.4c$. Qualitatively, this suggests nuclear matter, where $r_o \sim 2.75c$ is at — or beyond — the very limit of binding. Liquid helium, on the other hand, has $r_o \sim 0.96c$. According to these arguments it should be comfortably bound.[2] While one can question the ingredients in this crude estimate, the message is clear: It is a delicate matter to bind nuclei at all, and the properties of nuclear matter are very likely to emerge from fine cancellations of relatively large effects. The problem is intrinsically difficult.

Initial progress was hampered by our double–barreled ignorance. Ignorance of the NN interaction itself and of reliable microscopic techniques for the description of strongly interacting quantum liquids. Each bit of ignorance could be (and was) used to justify unreasonable assumptions about the other. For some time things floated in a sea of self–consistent if not self–contented vagueness. It was known (through, for example, the study of phase shift equivalent potentials) that the bulk binding energy and equilibrium density of nuclear matter were likely to be rather sensitive to off–shell properties of the two body interaction. Until such ambiguities were resolved there seemed little point in even bothering with good many–body techniques. To my mind, the situation became better focussed in the early 1970s when particle physics techniques from the 1960s were used to describe the NN interaction in terms of the exchange of mesons. Pion exchange had long been known to account for the longest–range part of the NN interaction even though its spin– and isospin–dependence rendered OPEP an unsuitable mechanism for providing the bulk of the central attraction needed to hold nuclei together. Vector (*i.e.,* ω) meson exchange was found to be largely successful in providing the short–range repulsion required for saturation.[1] Two pion exchange potentials were constructed (using dispersion relations) from "empirical" pseudophysical $N\bar{N} \to \pi\pi$ scattering amplitudes obtained from a combination of (analytically continued) πN scattering data and empirical $\pi\pi$ phase shifts. Despite serious uncertainties in both the data and its analytic continuation (which persist to this day), the resulting (dominantly central) potential was of roughly the right range and strength to hold nuclei together.[2] What emerged could not be claimed to be a "theory" of the NN interaction,[3] but rather a self–consistent picture in which dominant t– and u–channel processes were found to give a better–than–qualitative description of the s–channel NN interaction. With fine tuning of coupling constants and a not unappreciable amount of phenomenology at short distances, it was possible to make precision descriptions of NN phase shifts.[4] Using such interactions and Brueckner calculations, a simple but somewhat delicate qualitative picture emerged for nuclear matter. Intermediate–range attraction wins slightly over short–range repulsion to provide a net attraction of roughly 30 MeV per particle which is largely cancelled by the Fermi energy of nuclear matter over a wide range of nuclear densities. The final binding is provided by the central component of the iterated OPE potential, and saturation is provided by the reduction of this attraction by the Pauli Principle at higher densities.[5]

With the conviction (right or wrong) that off–shell ambiguities had been minimized with the adoption of physically dictated forms in the phenomenology of the NN interactions, it seemed reasonable to take a more critical look at the ubiquitous use of Brueckner theory. Fortunately, microscopic calculations of liquid helium had excellent advice to offer. Some were quick to adopt Jastrow trial wave functions and the hypernetted chain approximation. Others stuck to the diagrammatic language of BHF theory and its various perturbative improvements. The situation with differing approaches to the many–body problems was, if anything, worse than the previous spectrum of NN interactions.[6] To my (prejudiced) mind, it was important that some people wanted to determine the Jastrow correlation function by minimizing the HNC energy and not merely by parameterizing it. Once suitable (*e.g.,* Jackson–Feenberg) forms of the kinetic energy were adopted, this task became straightforward. The result was a far more direct comparison of "variational" and diagrammatic approaches to the nuclear many–body problem.

Specifically, it was possible to show that HNC "variational"[7] descriptions of bose

---

[1]It is interesting to note that the existence of vector mesons of roughly the correct mass was predicted by Breit years before their discovery as a way of understanding the microscopic origin of the spin–orbit force so essential in the traditional nuclear shell model.

[2]There is not and never has been evidence for the low mass σ–meson which was a mainstay of so many one boson exchange descriptions of the NN interaction.

[3]Nor was it offered as a "theory" by even its most enthusiastic proponents.

[4]Indeed, this is the approach was is still adopted in the construction of virtually all quantitative NN potentials.

[5]This picture of saturation being dominated by the iterated one–pion–exchange potential is *not* consistent with the assumption adopted in the crude estimate above the ϵ_{pot} is proportional to $1/r_o{}^3$. Indeed, if OPEP is turned off, most nuclear systems do not bind at all.

[6]Hence, Bethe's famous homework problem.

[7]I put the word variational in quotes because these calculations do not provide an upper bound on the ground state energy. Even using the relatively safe Jackson–Feenberg

liquids are absolutely equivalent to the sum of "parquet" diagrams with well–defined approximations to the s– and t–channel propagators. This identification makes it easier to understand why both sets of calculations were successful. They represent the self–consistent summation of RPA diagrams (needed to describe long wave length correlations) and the particle–particle ladder diagrams (necessary for a description of short–range correlations). As a result, such calculations are capable of reproducing the spinodal point in bose systems and of providing convincing indications of solidification. Indeed, the results of optimized HNC calculations (or, equivalently, of parquet summations) were found to be in strikingly good agreement with GFMC calculations.

At the same time, new approaches were being developed to describe the Fermi liquid parameters of strongly interacting fluids.[3] Although such approaches were largely phenomenological in nature, there basic structure made it clear that they were really firmly within the parquet structure. Slowly, some measure of uniformity appeared to be emerging in the treatment of the many–body problem. This comparison also revealed some cautionary clouds on the horizon. While the topological form of parquet theory and the "induced interaction" were similar, the details of the approximate treatment of propagators were different. In the former case, one desired a good description of global properties (*e.g.*, bulk binding energy and equilibrium density). In the latter, one wanted to describe long wave length excitations (*i.e.*, Fermi liquid parameters) with care. It seemed (and still seems) that one computationally practical scheme is not going to fit all needs.

This fact can be emphasized by looking at two definitions of the four–point function, Γ, in terms of functional derivatives of the total energy:

$$\Gamma \sim \frac{\delta E}{\delta V} \sim \frac{\delta^2 E}{\delta G\, \delta G'}. \tag{5}$$

Of course, the two definitions should lead to the same Γ. Of course, they do not — either for parquet theory or for *any* other approximate theory. It is straightforward to show that the smallest set of diagrams which maintains the consistency of these definitions is the set of *all* diagrams.[4] Inconsistencies are an inevitable result of the construction of approximate many–body theories. Approximations must be made on a case–by–case basis.

The theories upon which most variational calculations of nuclear matter are based were developed to describe bose systems. Their "natural" extension to fermi systems has been somewhat uneasy and *ad hoc*. For a variety of reasons, this is not such a serious problem in going from liquid ^{4}He to liquid ^{3}He. There, the interaction is largely spin–independent, the Fermi energy is small compared to the single particle mass, and the dynamic repulsion at short distances is sufficiently strong to render calculations of bulk properties rather insensitive to Pauli Principle effects. This is not the case in nuclear matter where, for example, the tensor structure of the interaction is enormously rich. To date, there is no consensus regarding the treatment of this richer operator structure in the NN problem. Diagrammatic theories can provide some guidance if one is prepared to assume that the same diagrammatic content is dominant and one merely needs to evaluate the corresponding diagrams for fermions. One can, for example, construct a hybrid parquet theory aimed at describing bosons with spin and isospin 1/2 interacting with the full panoply of spin– and isospin–dependent forces. The resulting "crossing" matrices needed to take the interaction from s– to t–channels and back again are readily constructed, and (due to the equivalence between diagrammatic and "variational"

---

form, the kinetic energy is unbounded from below! However, a physically sensible local energy minimum exists. Other widely–used choices of the kinetic energy operator do not have a local energy minimum, and attempts to solve the related Euler equation lead monotonically to disaster.

theories noted above) some guidance can be obtained regarding the treatment of spin– and isospin–dependent forces in the nuclear problem.[5] This problem remains an open one, and I find it somewhat unfortunate that relatively little effort has been invested on it in recent years.

Variational necromancers were not idle during these diagrammatic exercises, and significant improvements (such as the method of Correlated Basis Functions) were made. I think it is fair to say, based largely on these calculations, that the primary question of the traditional nuclear many–body problem has been answered with regard to the bulk properties of nuclear matter. The result is that there is fairly satisfactory agreement between the calculated bulk binding energy and the corresponding "empirical" value. ("Satisfactory" both in view of the strong cancellation evident in the calculations and the inevitable uncertainties in extrapolating the binding energy from mass 208 to ∞.) The calculated equilibrium density is much less satisfactory being a factor of two larger than the empirical value.[8] This situation stands in rather sharp contrast to that in liquid ^{4}He. There the agreement in bulk energy (as obtained in GFMC calculations) is within the 0.5 K uncertainty expected from the choice of the interaction and the roughly 0.25 K which might be expected from genuine many–body forces. The equilibrium density agrees to a few percent. Further, approximate calculations (parquet, again) agree with GFMC results at the level of roughly 1% for the binding energy and 8% for the equilibrium density. Finally, the liquid structure function (which is not available for nuclear matter) is also in reasonable agreement.

Do such nuclear matter results represent victory or defeat? I suggest the former. Genuine many–body forces are expected to be relatively large in the nuclear matter problem. For example, the same tensor force which renders OPEP relatively impotent at the two–body level ensures important three–body forces of two–pion range.[9] Several things are clear. First, it will require *two* mechanisms to correct both the equilibrium density and maintain the bulk binding energy. Second, the *ad hoc* construction of such mechanisms suggests that significant changes in equilibrium can be obtained with a very small net change in energy.[6] (This serves as a reminder that the equilibrium density is a particularly — perhaps unreasonably — sensitive quantity to calculate.) Finally, I think it unlikely that we will be able to calculate such many–body forces with the confidence required in the near future. I believe that the most satisfactory approach is to add many–body forces at the level required, declare victory and move on to more interesting questions. The problem is that there is too little dirt and too much rug under which to sweep it. One needs to study a wider variety of properties and systems in order to pin down such many–body forces with confidence.[10]

One might be tempted to ignore repairs and simply look at other nuclear matter properties, such as Fermi liquid parameters, at the empirical equilibrium density. Unfortunately, this does not work due to the factor of two overestimate of the equilibrium density. Like most other quantum fluids, the compressibility of nuclear matter changes sign when the density is roughly two–thirds of the (calculated) equilibrium density. For lower densities homogeneous solutions do not exist. As a result, any attempt to extract nuclear matter properties at the empirical equilibrium density from existing calculations will either fail outright or be extremely sensitive to arbitrary assumptions. One is re-

---

[8]It is conventional to minimize the problem by plotting energy versus Fermi momentum where a 25% discrepancy seems more palatable.

[9]Such three–body forces can again be constructed from pseudophysical $N\bar{N} \to \pi\pi$ amplitudes using dispersion results and the results demonstrate the important role played by the $\Delta(1238)$ isobar.

[10]In this regard it has been profitable to study many–body forces simultaneously in the context of the three–, four– and many–body problems.

ally obliged to offer a physically sensible mechanism for moving the current calculated equilibrium density.

Of course, other resolutions to this discrepancy are not prohibited. In spite of the use of theoretical suggested forms, there may still be significant ambiguities in the choice of the NN interaction. Relativistic effects, which I have neglected throughout, might play an important role. To investigate this question properly, it is necessary to have a *consistent* relativistic treatment of the two– and many–body systems. To my mind, methods fine enough to address such relatively delicate problems do not currently exist.

But whether this exercise has been a success or a failure is quite beside the point. It had to be done and must be continued with equal vigor to the study of the finite nuclei. Weisskopf likes to tell about a tourist in Austria who asked why the Austrian Railways bothered to print time tables since the trains are always late. The answer was simple. Without timetables no one would know how late they really were. In coming years more attention will be paid to genuine many–body forces, to relativistic effects, to the explicit meson presence in nuclei, and to the quark substructure of nucleons. But we will not be able to claim genuine evidence for such effects until we know conclusively that the traditional nuclear many–body problem fails to describe them.

WHAT THE CURRENT STATUS OF THE NUCLEAR MANY–BODY PROBLEM OUGHT TO BE

Short of neutron stars, there is not much nuclear matter accessible in bulk. To find the analogue of the liquid structure function, nuclear physicists most turn to electron scattering and the form factors of finite nuclei. This has been a wonderfully rich problem in light nuclei (*i.e.,* $A = 2$–4) where the exact solution of the few–body problem is now largely a matter of routine. These studies provide an excellent suggestion of the differences between the nuclear many–body problem that of atomic liquids. One's naive expectation is that the electromagnetic form factor of the deuteron can be calculated as the product of the Fourier transform of the square of the deuteron wave function times the appropriate EM nucleon form factor. As such, this form factor should provide a useful check of the quality of the two–body interaction adopted and little else. Nothing could be farther from the truth. Such 'impulse approximation' calculations fail to describe the data. The zero in the calculated deuteron form factor occurs at far too large a momentum transfer.[11] The height of the empirical "secondary maximum" is grossly underestimated by impulse approximation calculations. The disagreement is due to the fact that the electromagnetic interaction is modified by hadronic interactions in the nuclear medium. The collaboration between experiment and theory in mapping out the role of mesonic corrections to electromagnetic has been one of the most satisfying and fruitful in nuclear physics.[12] The result is an *extraordinary* agreement between theory and experiment up to values of q^2 large enough that even the faithful might begin to think that the quark structure of nucleons had a role to play.[7] These experiments and their theoretical interpretations make a compelling case for an important mesonic presence in nuclei. Surprisingly little insight in this connection has come from heavier nuclei. This would appear to be a direct consequence of the general absence of reliable microscopic calculations for finite nuclei. Since all interactions involving pions vanish in the limit of zero momentum transfer, the most important mesonic presence in nuclei

---

[11]This zero is sometimes erroneously called a 'diffraction minimum'. It is rather a manifestation of the short–range repulsion in the NN interaction.

[12]The experimental role has by no means been trivial. The deuteron, spin 1, has three distinct electromagnetic form factors. The electric and magnetic form factors can be separated with relative ease by measurements with fixed q^2 at different angles. Separation of the electric form factor into its "charge" and "tensor" components has been a far more challenging task requiring the use of polarized deuterons.

can only be seen in regions where a careful treatment of correlations is also required.

For some time much of the best quality electron scattering data has come from SACLAY. There is every reason to believe that CEBAF will assume this role in coming years. Whatever one's opinion about the role of quarks in nuclei may be, the only quantitative description of electron scattering currently available is provided by quality solutions of the nuclear many–body problem supplemented by mesonic exchange currents. Until such descriptions break down (for reasons of principle and not for want of adequate implementation), it will be impossible to make any convincing case for quark structure in nuclei.

Fortunately, a variety of factors have again led to significant advances in many–body techniques applicable to finite and semi–finite systems. The problems of thin films of liquid He and the lovely interplay between bulk and surface modes which they display have led to new many–body techniques. Recent progress in the description of the properties of small rare gas droplets (involving as many as ~ 100 particles) should provide stimulation and encouragement for nuclear physics. For example, studies of the giant monopole resonance in helium droplets have led to the result that the standard semi–classical approach is surprisingly incapable of describing either the energy of these collective states or, far more dramatically, their transition densities.[8] Further, these studies suggest that these monopole states are really not sufficiently collective to ensure the faithful extraction of properties of the bulk liquid from their energy. While this failure of semi–classical techniques may well be a manifestation of the extreme strength of the short–range repulsion between argon atoms, no one really knows if similar surprises await us in nuclear giant monopole resonances. This is an important question for nuclear physics, and its answer would do more than satisfy idle curiosity.

Not all nuclear excited states are equally worthy of consideration. I include among the most interesting those relatively collective superpositions of particle–hole excitations which are the closest nuclear physicists can get to the low–lying bulk modes of nuclear matter — the giant resonances. Experimentalists have mapped–out the distribution of strength of giant monopole, dipole and quadrupole resonances across the periodic table. A great deal is known about the systematics of their energies. Next to nothing is known about their transition densities. Nuclear theorists — making heavy use of semi–classical guidance — have used the giant monopole (and quadrupole) resonances to determine the bulk compressibility of nuclear matter. But is this correct? Condensed matter experimentalists, with the luxury of making liquid helium by the bucket, simply measure the sound velocity. Of course, at low frequencies collisions are essential, local equilibrium is maintained, and the velocity of first–sound is measured. At high frequencies (when the collision frequency is small compared with the frequency of the signal), the velocity of zero sound results. The first sound velocity is directly related to the bulk compressibility. The zero sound velocity is not. Due to the accident that the Fermi liquid parameter, F_0, is extremely large for liquid He, these velocities are very similar. However, in nuclear systems where $F_0 \sim 0$, these velocities differ by a factor of $\sqrt{3}$. Which sound velocity (if any) can actually be extracted from these giant resonances? The standard description of nuclei as an ensemble of weakly interacting quasiparticles would suggest the zero sound velocity. Common practice assumes it is the first sound velocity. Since giant resonances represent our only real hope to determine the Fermi liquid parameters of nuclear matter, it is important to understand precisely what they determine. Quality nuclear many–body calculations would appear to play an absolutely central role in this discussion. To date they have not done so.

Every august National Committee convened during the past two decades to consider future directions for nuclear physics manages to advocate the use of the nucleus as a "laboratory" for the study of fundamental interactions. One can, for example, search the nuclear data sheets for felicitous accidents. There are, for example, parity doublets with accidentally tiny splitting which would be lovely candidates for the detailed study of parity–violating weak interactions. But what should have become a "chicken in the pot" for nuclear physics has remained "pie in the sky" due to the lack of quality calculations of the necessary nuclear wave functions. The case of the parity–violating weak interactions is a case in point. These interactions can be thought of, roughly, as including a long–range weak pion exchange piece and a short–range weak vector meson exchange piece. Various models of this interaction differ through the relative importance of these two components. Evidently, long–range NN correlations will be important in the accurate evaluation of the former while a good description of short–range correlations is required for the latter. Shell model wave functions or their elementary extensions simply will not serve to discriminate between models.

The current status of the nuclear many–body problem is not what it ought to be. Existing techniques for dealing with nuclear matter and for treating finite and semi–finite atomic systems have not yet been applied with sufficient vigor to the large number of interesting nuclear problems which merit more reliable microscopic description.

THE FUTURE OF THE NUCLEAR MANY–BODY PROBLEM

There is a significant body of nuclear theorists who appear to believe that the nuclear many–body problem has no future in anything like its present form. They suggest, with more or less conviction, that the future progress in nuclear physics will best be made by abandoning nucleons and mesons altogether and describing nuclei directly in terms of QCD. I suspect that rather fewer condensed matter physicists embrace the abandonment of atoms in favor of a pure QED description of quantum fluids with equal fervor. I do not endorse this point of view and, for many purposes in nuclear physics, see chromodynamics as a red herring. Almost a half century of experience with the nuclear many–body problem indicates that, although nucleons and mesons may not be "elementary" particles, they are the correct (*i.e.,* most efficient) effective degrees of freedom for most nuclear problems.

As noted above, there is a clear mesonic presence in nuclei. We know that pions interact strongly with nucleons and have profitted from our knowledge of the πN interaction. Why shouldn't the properties of pions and other mesons — and, hence, the NN interaction — be modified in a significant way by the presence of the nuclear medium even at zero temperature and normal density? In this regard, nuclear matter is likely to behave rather differently than atomic liquids. One can make simple models of such effects based, again, on the πN scattering amplitudes. In this case, by closing the nucleon line to describe the pion propagator or self–energy with an eye towards studying the nuclear density dependence. Indeed, these has become a small industry — both theoretical and experimental — with specific attention frequently focussed on the role of the $\Delta(1238)$ nucleon resonance in the intermediate state. Arguments can be made, for example, that this isobar plays a crucial role in the 20% "quenching" of the nucleon axial vector coupling constant, g_A, governing Gamow–Teller nuclear beta decays.[9] Of course, such predictions depend critically on nuclear wave functions. Inadequate nuclear wave functions make it impossible to identify the role of the Δ resonance conclusively. Some years ago theorists were fascinated by these attractive interactions might be sufficiently strong to lead to a pion condensate as the ground state of nuclear matter. While this idea is no longer popular, there are even more compelling reasons to hope that kaon condensates might play a role in the properties of neutron matter and, hence, neutron stars. To date such suggestions and calculations have fallen far short of the standards of many–body theory which have been regarded as necessary in the traditional nuclear many–body problem.

Minimally, one can use the best many–body techniques available to study NN interactions with a density–dependence suggested by models. While some studies of this nature have been carried out, they have consistently avoided using many–body techniques of high quality. It would be better to aim towards a description of nuclear matter in which (real) nucleons and (virtual) mesons are placed on a more equal footing. The properties of nuclear matter, the properties of the mesons and, hence, the nature of the NN interaction in medium should be determined in a self–consistent manner.

But, if the properties of meson can change in medium, what about the properties of the nucleons themselves? In this regard nuclear systems are very different from their atomic counterparts. the static structure function, $S(k)$, for liquid helium shows significant structure for momenta as up to ~ 4 Å^{-1}. The associated energy is some three orders of magnitude smaller than the energy of the excited states of the helium atom. It is surely a good approximation to neglect the effects of these excited states and to regard helium atoms as 'elementary particles' with their free space properties. This is not the case for nuclear matter where calculated structure functions show important contributions from momenta up to ~ 4 fm^{-1} which is essentially *equal* to the excitation energy of the nucleon isobar. The properties of single nucleons may well be modified significantly in medium.

Let me express this somewhat differently. In the usual hadronic description, the nucleon has a distributed baryon density — a point nucleon surrounded by a meson cloud. For every contribution to the NN coming from the emission of a meson by a bit of baryon density in one nucleon and its absorption from a bit of baryon density in a second, there should be a corresponding contribution to the self–energy of the nucleon from emission and absorption of the same meson by bits of baryon density in the same nucleon. Given the significant mesonic presence in nuclei, two things seem clear. First and as noted, *in medium* modifications of nucleon processes seem likely. Second, it may be difficult to disentangle interaction effects from those due modifications of the structure of single nucleons. This is, of course, one of the lessons of the widely disparate but equally successful means of analyzing the EMC effect. A simple estimate may be instructive. Models of nucleon structure (like the Skyrme model) predict a nucleon compressibility, defined as $r^2 d^2 E/dr^2$ where r is the nucleon radius, of roughly the nucleon mass. One might try to account for the empirical binding energy of nuclei in terms of "swollen" but non–interacting nucleons. This would require a change in the nucleon radius of $\sim 15\%$ — which is precisely the sort of number which emerges from estimates of the EMC effect.

Before facing the challenge of making models which are capable of describing single nucleon structure and nucleon–nucleon interactions simultaneously and self–consistently, it would be nice to know that there was some empirical indication of modified nucleon structure in nuclei. This is a difficult problem. For example, the 15% increase in nucleon radius suggested by the EMC effect would correspond to a 1% change in the charge radius of ^{16}O and a 0.2% change in the charge radius of ^{208}Pb. Nonetheless a serious investigations of nuclear electromagnetic form factors — based on the best many–body theories and including meson exchange current effects — might reveal the need for modified nucleon form factors.

At the moment there are a few tantalizing suggestions of modified nucleon properties in normal nuclei. One of these is a suggested resolution of the long–standing Nolen–Schiffer Anomaly.[10] This is a case where cursory inspection might suggest that accurate many–body calculations are not really necessary. One compares the energy of mirror nuclei (*i.e.,* isobaric analogue states) in two or more nuclei. Given the usual assumption of the charge independence of hadronic forces, the energy difference between these states should be due to the neutron–proton mass difference and the presumably calculable (and much larger) difference of Coulomb energies. Some 95% of this energy difference can be explained. The remaining 300–500 keV has remained a nagging problem across the periodic table — the Nolen–Schiffer Anomaly. Some authors have argued (unconvincingly, to my taste) that this provides evidence for charge symmetry breaking of the *in medium NN* interaction. Hatsuda, Prakash and Høgassen have suggested that the resolution is rather due to an *in medium* change in the neutron–proton mass difference. They argue that the neutron–proton mass difference is related to the difference in quark masses, $(m_d - m_u)$, which is in turn related to expectation values of quark condensates, $< \bar{q}q >$. Models suggest that these condensates should become smaller with increasing density as a prelude to the ultimate restoration of chiral symmetry. Using QCD Sum Rule techniques, they estimate that this density dependence is sufficient to explain the Nolen–Schiffer Anomaly.[13] This result would be even more exciting were it not for the fact that one ingredient in the empirical discrepancy is the *calculation* of a Coulomb energy difference to significantly better than 5% accuracy. To obtain this kind of reliability quality nuclear wave functions are mandatory. This is a case where a relatively straightforward calculation could reveal the presence of genuinely new physics coming from the nuclear medium.

Based on the theoretical conviction that nucleon properties must change in a nuclear medium, many people have considered models which place hadronic structure and hadronic interactions on a common footing. Since caution suggests that we should have a reasonable description of normal nuclear matter before proceeding to more exotic possibilities and since we know that normal nuclear matter is delicate, it seems reasonable to require that any such model should have the freedom to describe the NN interaction along more–or–less traditional lines (*i.e.*, boson exchange). Effective Lagrangian models (like the Skyrme model [11]) represent a natural choice since they are expressed purely in terms of the pion field and powers of the pion field carrying the quantum numbers of familiar mesons. The leading term in such models is the (chiral symmetric) non–linear sigma model which is well–known to reproduce low energy $\pi\pi$ scattering data. In such models nucleons — fermions which must now be constructed out of bosons — emerge as topological "knots" in the pion field with baryon number being identical to the winding number. I do not wish to dwell on the details of such models, but I think a number of points are useful in the present context.

First, such models are invariably treated classically so that the generic baryon emerges as a classical, topological soliton. A minimal quantal treatment is required to recover baryons with well–defined spin and isospin through the quantization of zero modes (which follows "cranking" techniques familiar from the treatment of rotational bands of deformed nuclei). What results is a description of the properties of single baryons accurate to roughly 30%.[12] Such accuracy is expected from the our loose understanding that classical results should emerge as a leading order term in a $1/N_c$ expansion. Indeed, many of the results obtained for the properties of single baryons are identical to those of the Bag Model in the limit of large N_c. A similar description of the NN interaction (*i.e.*, at the classical level) reveals many — but not all — of the dominant features of the NN interaction. Terms reminiscent of π–exchange, (vector coupled) ω–exchange and (tensor coupled) ρ–exchange are readily identified. The intermediate range central attraction does *not* appear at the classical level but does arise from one loop quantum corrections.[13] In short, these models and their essentially classical implementation seem to provide a crude but serviceable way of tying nucleon structure and NN interactions together.

---

[13]The operative notion here is the partial restoration of chiral symmetry and not its QCD realization. Similar effects can almost surely be obtained from the large scalar and vector potentials produced, for example, by the Walecka model.

These models have interesting things to say about nuclear matter (described either as a periodic array of classical skyrmions or as a single skyrmion in a compactified space). For low baryon densities, nucleons exist as isolated objects somewhat larger than they would be in free space. In this domain, the underlying chiral symmetry of the effective Lagrangian is spontaneously broken. The classical approximations preclude realization of the delicate cancellation between kinetic and potential energies noted in the first section, and the resulting binding energies are necessarily unreasonably large.[14] At higher densities, one encounters a phase transition (of second order or weakly first-order depending on the details of the calculation.) Above this density, nucleons lose their identity and the excitation spectrum reveals a parity doubling identical to that expected from the restoration of chiral symmetry. At high densities one finds that the energy density is proportional to the 4/3 power of the baryon density as it would be for a gas of deconfined quarks. In short, although such models were constructed to contain the essential features of low–energy hadron physics, they appear to contain many of the qualitative features expected from QCD at high density (and/or temperature). Given the immense computational difficulty of lattice gauge calculations and the virtually impossibility of performing a lattice gauge description of normal nuclear matter, these models offer a potentially fruitful scheme for extending the nuclear many–body problem into the realm of RHIC physics.

One might fear that the predictions of such models are inordinately sensitive to the details of the effective Lagrangian chosen. One would certainly feel more comfortable if one could derive the correct effective Lagrangian from QCD or — at least — rigorously demonstrate its existence. Neither has been done to date. Fortunately, however, it is rather easy to demonstrate that virtually all qualitative predictions of the Skyrme model are robust. Variations in the properties of single nucleons and the NN interaction due to the choice of the effective Lagrangian are smaller than the unavoidable $1/N_c$ errors. In dense matter one can show that (i) low–density matter must have spontaneously broken chiral symmetry (*i.e.*, matter as we know it), (ii) chiral symmetry must be restored at high densities (as it would be in a quark–gluon plasma), (iii) the classical energy can have no maxima at any finite density and must have exactly one minimum, and (iv) this minimum must occur in the domain where chiral symmetry is restored.[15] Further, the $SU(3)$ extension of the Skyrme model (which includes K–mesons) shows strong suggestions of kaon condensation at the chiral symmetry restoration density.

The dominant notion driving these results is chiral symmetry. Indeed, chiral symmetry has been the most important single organizing principle in low–energy hadron physics for many years. This continuous symmetry of the underlying hadronic Lagrangian is spontaneously broken in our normal density world with the result that pions should exist as massless Goldstone bosons. (The fact that the pion has a small but non–zero mass is an indication of additional dynamical symmetry breaking, but this has not really diminished the utility of chiral symmetry.) Since any reasonable hadronic model will respect chiral symmetry (either by design or through careful phenomenological fits), it is perfectly reasonable to expect that, if permitted by adequate implementations, such models will also reveal the restoration of chiral symmetry at sufficiently high temperature or density.

Of course, QCD is a theory which respects chiral symmetry. So long as quarks have zero mass (as they do at high temperature and density), chiral symmetry will remain unbroken. At lower temperature and density, quark condensates have a non–zero expectation value, the quarks acquire a mass and chiral symmetry is spontaneously

---

[14]Some skyrmists have attempted to make detailed comparisons between classical solitons of winding number greater than 1 and light nuclei. Given this delicacy of nuclear problems, I do not find such comparisons useful.

broken. (This fact has led to the simple suggestion [16] that it is useful to give *all* meson masses and coupling constants at an explicit T–dependence governed by the "melting" of the quark condensate. For example,

$$\frac{m(T)}{m(T=0)} = \left(1 - \frac{T^2}{T_{\chi SR}^{\;2}}\right)^{1/3} \tag{6}$$

where $T_{\chi SR}$, the temperature for chiral symmetry restoration, is estimated as being approximately 170 MeV.) Here, m is a generic meson mass. To the extent that chiral symmetry restoration is the really important element required for the description of relativistic heavy ion collisions, it may well make very little difference whether we do this using hadronic models or explicitly with QCD.

There is another possibility. At low temperature and normal density, the gluon condensate also has a non–zero expectation value. The expectation value of this condensate is proportional to the familiar volume energy term, B, in the MIT Bag Model. At sufficiently high temperature and density, this condensate also vanishes leading to deconfinement. While it is perfectly reasonable to expect hadronic models to know about chiral symmetry and its restoration, it is equally unreasonable to expect them to know about deconfinement. To the extent that the salient features of RHIC experiments are determined by the gluon condensate, it is likely that hadronic models will fail completely. There are some modest grounds for optimism that chiral symmetry restoration is more important. Campostrini and Di Giacomo [17] find little change in the gluon condensate associated with chiral symmetry restoration suggesting only a weak coupling between gluon and quark condensates. There are also suggestions (based on lattice gauge calculations) that the temperature at which the gluon condensate vanishes is somewhat larger than $T_{\chi SR}$.

There are also purely practical reasons to hope that RHIC physics is dominated by a weakly first–order (or second–order) transition such as chiral symmetry restoration. Crudely put, a first–order phase transition occurs when one local minimum of the free energy (the quark–gluon plasma) drops below a another well–separated minimum (hadronic matter) at a temperature, T_c, which is a function of the chemical potential. The lower energy state is now thermodynamically favored but may not be reached quickly since the initial state remains a local minimum. Thus, the possibility of creating a superheated phase of hadronic matter exists until some higher temperature, T_x at which hadronic matter no longer corresponds to a minimum of the free energy. There is a window with $T_c \leq T \leq T_x$ where the system may exist as hadronic matter even though a quark–gluon plasma is thermodynamically favored. Ellis and Olive [18] have used phenomenological models this superheating window and find $T_x - T_c$ to be roughly 80 MeV over a wide range of chemical potentials. They offer no estimates of the lifetime of such a superheated phase which are likely to be very difficult to get. If this scenario is realized, insight gained from equilibrium calculations will be worthless and RHIC physics is very likely to come up dry.

There are many ways in which practitioners of the nuclear many–body problem can make significant contributions in providing substance for all of this hand waving. At the most immediate level, it would be of value to assume some specific temperature or density dependence for mesonic masses and coupling constants (as in eqn.(6) above) and calculate nuclear matter properties using the best available many–body techniques. One might also consider ways in which effective Lagrangian models for dense matter can be implemented beyond the classical level so that there is some hope of obtaining the various cancellations which must be awaiting us.

One might also consider the question of deriving an effective Lagrangian from QCD. There are some cases in $(1+1)$–dimensions where fermion theories are rigorously equivalent to boson theories. For example, the massive Thirring model is equivalent to the Sine–Gordon model. While one does not expect to find rigorously equivalent boson theories in $(3 + 1)$–dimensions in general or, specifically, a boson theory strictly equivalent to QCD, one should be able to find an equivalent theory in the large N_c limit. This equivalence is contained in the planar diagrams of QCD which have not received much attention in recent years. Since all parquet diagrams are planar (although not all planar diagrams are parquet), there is some possibility that traditional many–body techniques could be adapted to address this important question.

CONCLUSIONS

The nuclear many–body problem is reaching maturity. After many years of effort, its practitioners have a clear sense of the difficulties which must be overcome and a number of effective methods for doing so. What remains are some further refinements and a significant number of well–chosen applications. These applications fall into two relatively distinct categories. In a more traditional vein, we should turn our attention to the microscopic description of highly collective states in finite nuclei. In the best case, these states will provide the only handle we shall ever get on the Fermi liquid parameters of nuclear matter. I think it is of considerable importance to understand precisely what they have to tell us and to extract as much information as possible. It is equally important to perform quality calculations of, *e.g.*, electromagnetic properties of nuclei in the hope of (i) providing even more convincing evidence of the nature of the mesonic presence in nuclei and (ii) looking for possible indications that the properties of nucleons (and mesons) even at normal nuclear density and zero temperature differ in discernable ways from their free space values. This second goal is of some importance in creeping up on the kinds of challenges which relativistic heavy ion experiments will inevitably provide for nuclear many–body theorists. The most optimistic — and, in my opinion, the most likely — scenario as that RHIC experiments will provide some indication of chiral symmetry restoration at the high temperatures and densities which they will briefly explore. The mechanisms which lead to this symmetry restoration are likely to be the same as those which yield modified hadronic properties in ordinary matter.

In a more tentative direction, we should consider (chiral symmetric) models which place hadronic structure and hadronic interactions on the same footing and their self–consistent solutions both in normal matter and in matter under extreme conditions. Since much of the time evolution of a RHIC event will occur under relatively normal conditions and since, as we have seen, nuclear matter is a delicate system by its very nature, it is particularly important that such studies be performed with quality many–body theories. Such applications will require new work on the many–body problem. Our treatment of fermion systems must be improved. We must develop the capacities to work at non–zero temperature and with finite systems. We should also consider the consequences of chiral symmetry restoration in broader terms. Are there relatively simple signatures, for example, of the universal parity doubling which must accompany chiral symmetry restoration whether it is described by QCD or hadronic models?

At every step in this task we must be prepared to meet prophets of doom who remind us that nuclei are "really" made of quarks and glue and that we should "really" be doing lattice gauge calculations. We should continue nonetheless with the firm conviction that we will never know how wrong the traditional picture of nuclei built from nucleons and mesons is unless we try. Put another way, it will not be possible to claim any nuclear evidence for QCD unless the very best hadronic dynamics and the very best many–body techniques can be demonstrated to fail conclusively.

ACKNOWLEDGEMENTS

I would like to thank Ben Mottelson for several useful discussions. I gratefully acknowledge the hospitality of NORDITA where this talk was prepared. This work was supported in part by the US Department of Energy under Contract No. DE-FG02-88ER40388.

REFERENCES

[1] A.D. Jackson, Annual Review of Nuclear & Particle Science (1983) 105.
[2] B. Mottelson, private communication.
[3] S. Babu and G.E. Brown, Ann.Phys. (NY) **78** (1973) 1.
[4] A.D. Jackson and R.A. Smith, Phys.Rev. **36A** (1987) 2517.
[5] A.D. Jackson and R.A. Smith, Nucl.Phys. **A476** (1988) 448.
[6] A.D. Jackson, E. Krotscheck and M. Rho, Nucl.Phys. **A407** (1983) 495.
[7] B. Frois, Prog.Part. & Nucl.Phys., vol.24, Pergamon Press (1990) 1.
[8] S.A. Chin and E. Krotscheck, Phys.Rev.Lett. **66** (1990) 2658.
[9] M. Rho, Nucl.Phys. **A231** (1974) 493.
[10] T. Hatsuda, M. Prakash and H. Høgassen, Phys.Rev.Lett. **66** (1991) 2581.
[11] T.H.R. Skyrme, Nucl.Phys. **31** (1962) 556.
[12] G.S. Adkins, C.R. Nappi and E. Witten, Nucl.Phys. **B228** (1983) 552.
[13] A. Jackson, A.D. Jackson and V. Pasquier, Nucl.Phys. **A432** (1985) 567.
[14] L. Castillejo *et al.,* Nucl.Phys. **A501** (1989) 801.
[15] A.D. Jackson, C. Weiss and A. Wirzba, Nucl.Phys. **A529** (1991) 741.
[16] G.E. Brown, H.A. Bethe and P.M. Pizzochero, Phys.Lett. **263B** (1991) 337.
[17] M. Campostrini and A. DiGiacomo, Phys.Lett. **197B** (1987) 403.
[18] J. Ellis and K.A. Olive, Phys.Lett. **260B** (1991) 173.

FERMI LIQUID AND "QCD" ASPECTS OF THE NUCLEUS

W. H. Dickhoff

Department of Physics, Washington University
St. Louis, Missouri 63130, U.S.A.

ABSTRACT

Recent exclusive electron scattering results which involve the ejection of a proton from the nucleus, have established properties of the nucleus that can be brought into one-to-one correspondence with Fermi Liquid properties of infinite homogeneous systems. Theoretical calculations in finite nuclei are discussed which establish these "Fermi Liquid" properties by including the coupling of single-particle motion to more complicated states in an energy domain of about 100 MeV around the Fermi energy. Although sufficient to explain the shape of the experimental strength distributions, these calculations need to be supplemented by including the coupling to even higher energy states for quantitative agreement. Such calculations are reported for nuclear matter. These results clearly establish the importance of the coupling between low-lying single-particle states and very high-lying $2p1h$-states. This coupling is provided within the constraints of non-relativistic many-particle theory by the use of a strongly repulsive realistic interaction between nucleons. This feature is illustrated with numerical results. The simulation of QCD degrees of freedom in terms of short-range correlations turns out to be crucial for a quantitative understanding of $(e, e'p)$ results as well as other electron scattering data, as will be discussed for various examples.

INTRODUCTION

Experimental results from electron scattering facilities in the last decade have resulted in considerable progress in understanding the physics of the nucleus. The traditional picture of the nucleus in terms of the simple shell model with its strong single-particle (sp) spin-orbit potential[1,2] has on the one hand been confirmed, but on the other hand its limitations have also been elucidated. The confirmation has come from high-precision elastic electron scattering from ^{205}Tl and ^{206}Pb[3] which differ by one proton in the $3s\frac{1}{2}$ orbital. The resulting charge density difference of these nuclei corresponds in a striking way to the charge density of a $3s\frac{1}{2}$ proton. This result shows that the concept of a sp wave function for the description of nucleons in nuclei is of great relevance. The limitations of the shell model have been most clearly demonstrated by experimental work on the $(e, e'p)$ reaction. A typical result from the NIKHEF facility in Amsterdam for the $(e, e'p)$ reaction on ^{208}Pb is shown in Fig. 1[4,5] In this figure the distribution of sp strength that is obtained from the analysis of this experiment, is shown as a function of removal energy for various proton orbitals. It is useful to contrast this result with the simple mean-field or Hartree-Fock picture of the nucleus. In this picture the strength distribution would peak at the sp energy of each orbital and the corresponding sp strength at that energy would be 1. Fig. 1 clearly deviates from this simple picture in two important ways. First, only part of the mean-field single-particle (sp) strength for hole states is observed in the experimentally accessible energy region. Second, this strength, while concentrated at one energy for orbitals close to the mean-field Fermi energy, becomes more strongly fragmented when the orbital is further removed from the Fermi energy.

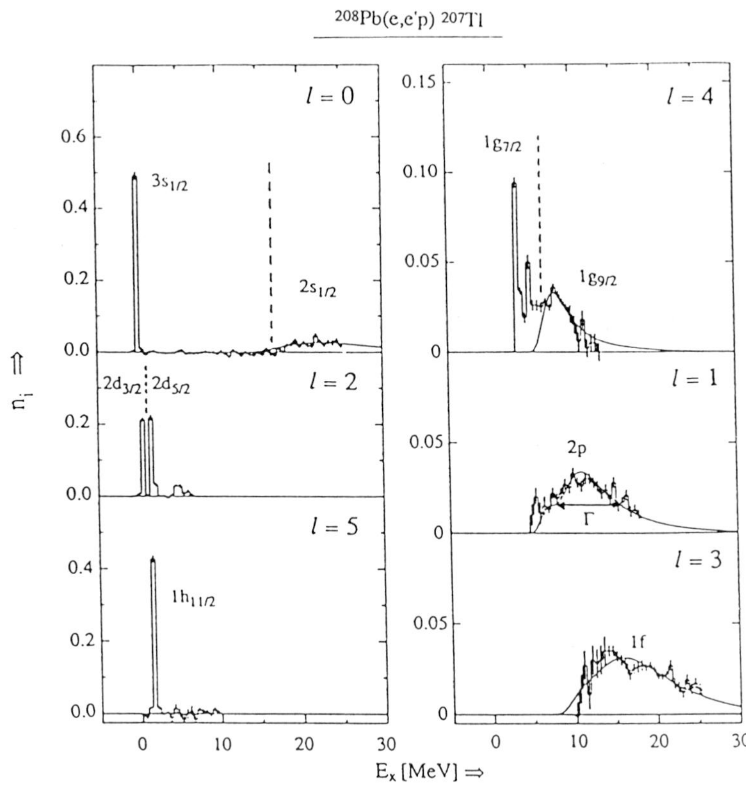

Figure 1. Distribution of sp strength for several proton orbitals below the Fermi energy in ^{208}Pb as a function of removal energy.

Occupation numbers for shell model orbitals can be obtained by combining (e, e') results for charge densities with these $(e, e'p)$ results using the so-called CERES-method.[6] Recently, including Coulomb distortion effects,[7] a spectroscopic factor of 0.65 for the $3s\frac{1}{2}$ proton with occupation number 0.75 has been obtained.[8]

It is the purpose of this paper to illustrate how one can obtain a detailed understanding of these results. This will include an analysis of the coupling of sp degrees of freedom as obtained from a mean-field picture, to more complicated states. It will be shown that this will automatically lead to a qualitative understanding of the fragmentation of the sp strength as it is observed. It will, however, also become evident that the energy domain in which this coupling can be studied in finite nuclei, is not sufficient to explain the experimental results quantitatively. Similar conclusions can be drawn from studies of the low-energy response functions of nuclei and this evidence will also be discussed in this paper. The energy domain that can presently be studied in nuclei, is related to the typical particle-hole (ph) energy difference and will be further discussed below. This energy scale results from the shell structure of the nucleus which reflects the finiteness of the system. This means that such a new energy scale is automatically introduced in the case of a finite many-particle system.

Higher energy scales relevant in nuclear physics are determined by the pion mass, the Δ-nucleon mass difference, and other, higher energy, mesonic and nucleonic excitations. Ultimately, at infinite energy, there are the colored quark and gluon states. A practical way to simulate the coupling of low-energy nucleons to these degrees of freedom at higher energy, is obtained by introducing a non-relativistic interaction between nucleons which is required to describe the nucleon-nucleon (NN) scattering data up to pion production threshold.

Important components of such so-called realistic interactions include the long-range pion exchange which determines most of the nuclear tensor force, an intermediate range attraction, and a short-range repulsion in all channels. In a non-relativistic framework, these features of the NN interaction require the solution of the Lippmann-Schwinger equation for the description of the NN scattering results. In diagrammatic language this means the inclusion of ladder diagrams. A non-relativistic many-particle description in turn will require a similar infinite summation to obtain physically meaningful results. A study of the sp properties in the nuclear medium will be discussed which includes the effect of these terms. The technically convenient nuclear matter system at normal density, will be employed for this study. It will be shown that the influence of the short-range and tensor components of the nuclear interaction on the sp properties of the nucleon are non-negligible and are of the right size to make a quantitative understanding of the $(e, e'p)$ results and low-energy response functions possible. Throughout this paper the language of the Green's function method will be used since it is particularly well suited to deal with the coupling of sp degrees of freedom to more complicated ones both at low and at high energy.

FERMI LIQUID ASPECTS OF NUCLEI

The distribution of sp strength as shown in Fig. 1 can be directly related to the sp propagator of the corresponding many-particle system which in the Lehmann-representation has the from

$$g(\alpha, \beta; \omega) = \sum_m \frac{\left\langle \Psi_0^A \middle| a_\alpha \middle| \Psi_m^{A+1} \right\rangle \left\langle \Psi_m^{A+1} \middle| a_\beta^\dagger \middle| \Psi_0^A \right\rangle}{\omega - (E_m^{A+1} - E_0^A) + i\eta}$$
$$+ \sum_n \frac{\left\langle \Psi_0^A \middle| a_\beta^\dagger \middle| \Psi_n^{A-1} \right\rangle \left\langle \Psi_n^{A-1} \middle| a_\alpha \middle| \Psi_0^A \right\rangle}{\omega - (E_0^A - E_n^{A-1}) - i\eta}. \tag{1}$$

The combined probability for removing a particle with quantum numbers α from the ground state while leaving the remaining $A-1$-system at an energy $E_n^{A-1} = E_0^A - \omega$ corresponds to the experimental results shown in Fig. 1. Theoretically, it is referred to as the "hole" part of the spectral function which is related to the single-particle propagator by

$$S_h(\alpha, \omega) = \frac{1}{\pi} Im\, g(\alpha, \alpha; \omega) \qquad\qquad \omega < \epsilon_F^-$$
$$= \sum_n \left| \left\langle \Psi_n^{A-1} \middle| a_\alpha \middle| \Psi_0^A \right\rangle \right|^2 \delta(\omega - (E_0^A - E_n^{A-1})). \tag{2}$$

A similar probability for the addition of a particle with quantumnumbers α leaving the $A + 1$-system at energy $E_m^{A+1} = E_0^A + \omega$ is obtained from

$$
\begin{aligned}
S_p(\alpha, \omega) &= -\frac{1}{\pi} Im \, g(\alpha, \alpha; \omega) \qquad\qquad\qquad\qquad \omega > \epsilon_F^+ \\
&= \sum_m \left| \langle \Psi_m^{A+1} | a_\alpha^\dagger | \Psi_0^A \rangle \right|^2 \delta(\omega - (E_m^{A+1} - E_0^A))
\end{aligned}
\tag{3}
$$

which is referred to as the particle spectral function. The corresponding Fermi energies are given by

$$
\begin{aligned}
\epsilon_F^- &= E_0^A - E_0^{A-1} \\
\epsilon_F^+ &= E_0^{A+1} - E_0^A.
\end{aligned}
\tag{4}
$$

The occupation number of single-particle state α can be obtained from the hole spectral function by using

$$
n(\alpha) = \int_{-\infty}^{\epsilon_F^-} d\omega \, S_h(\alpha, \omega).
\tag{5}
$$

In a similar way one can obtain the depletion number from the particle spectral function

$$
d(\alpha) = \int_{\epsilon_F^+}^{\infty} d\omega \, S_p(\alpha, \omega).
\tag{6}
$$

Naturally $n(\alpha) + d(\alpha) = 1$ but this relation can in practice serve as an excellent test of the numerical calculations.

This distribution between occupation and emptiness of a single-particle orbital in the correlated ground state is a sensitive measure of the strength of correlations provided a suitable single-particle basis is chosen. In order to treat correlation effects beyond the mean-field (Hartree-Fock) picture it is necessary to treat higher order contributions to the self-energy. The simplest extension is to include the second order contribution to the self-energy.

$$
\begin{aligned}
\Sigma^{(2)}(\gamma, \delta; \omega) = &\frac{1}{2} \sum_{\epsilon, \mu, \nu} \langle \gamma\mu | V | \epsilon\nu \rangle \, \langle \epsilon\nu | V | \delta\mu \rangle \\
&\left\{ \frac{\theta(F - \epsilon)\theta(F - \nu)\theta(\mu - F)}{\omega - (\epsilon_\epsilon + \epsilon_\nu - \epsilon_\mu) - i\eta} + \frac{\theta(\epsilon - F)\theta(\nu - F)\theta(F - \mu)}{\omega - (\epsilon_\epsilon + \epsilon_\nu - \epsilon_\mu) - i\eta} \right\}
\end{aligned}
\tag{7}
$$

Contrary to a mean-field self-energy, Eq. (7) displays an explicit energy-dependence. It is also very easy to interpret the two terms. The first one represents the coupling of the sp motion to 1particle-2hole ($1p2h$) intermediate states, the other represents the coupling to $2p1h$ intermediate states. It is also clear that this self-energy has a pole structure which has important consequences for the solution of the Dyson equation in which this self-energy is employed. The solution of the Dyson equation

$$
g(\alpha, \beta; \omega) = g^{(0)}(\alpha, \beta; \omega) + \sum_{\gamma, \delta} g^{(0)}(\alpha, \gamma; \omega)\Sigma(\gamma, \delta; \omega)g(\delta, \beta; \omega)
\tag{8}
$$

with an energy dependent self-energy results in a simple but interesting eigenvalue problem when a discrete sp basis is employed. Details of this procedure can be found in Ref. 9. The sp propagator $g^{(0)}$ in (8) refers to an appropriate mean-field propagator obtained from the non-interacting reference state in which all sp orbitals below the Fermi energy are occupied. The resulting eigenvalue equation has the form

$$\sum_{\delta} \left\{ \epsilon_{\delta} \delta_{\alpha,\delta} + \Sigma(\alpha,\delta;\epsilon_n^-) \right\} z_{\delta}^{n-} = \epsilon_n^- z_{\alpha}^{n-} \tag{9}$$

for the eigenvalues $\epsilon_n^- = E_0^A - E_n^{A-1}$ and eigenvectors $z_{\alpha}^{n-} = \langle \Psi_n^{A-1} | a_{\alpha} | \Psi_0^A \rangle$. Note that ϵ_n^- and $|z_{\alpha}^{n-}|^2$ can be directly compared to the type of experimental results shown in Fig. 1. A similar eigenvalue equation can be obtained for the $A+1$ particle states. The energy argument of the self-energy in (9) must coincide with with the actual eigenvalue on the right hand side of the equation. In practice one can choose an input energy to calculate the self-energy and then perform the diagonalization. This means that if one diagonalizes a finite matrix (sum over δ finite) at most one eigenvalue can coincide with this input energy. The pole structure in the self-energy (7) implies that between each pair of $2p1h$ (or $1p2h$) energies the self-energy goes from ∞ to $-\infty$. This implies that a solution of the eigenvalue equation must exist between such a pair of $2p1h$ (or $1p2h$) energies. One therefore does not obtain k eigenvalues for a k-dimensional sp space but instead as many eigenvalues as there are different poles in the self-energy plus the dimension of the self-energy matrix. It is important to properly normalize the eigenvectors, z_{α}^{n-}. To illustrate this, it is useful to consider a one-dimensional sp space. In this case the normalization condition reads[9]

$$|z_{\alpha}^{n\pm}|^2 = \left\{ 1 - \left. \frac{\partial \Sigma}{\partial \omega} \right|_{\epsilon_n^{\pm}} \right\}^{-1} \tag{10}$$

Since the derivative of Σ is always negative this results in spectroscopic factors, $|z|^2$, which are smaller than one. The distribution of single-particle strength depends naturally on the interaction used to calculate Σ but some qualitative statements can nevertheless be made.

In Fig. 2 results of a calculation[9] of the sp strength distribution for proton removal from ^{48}Ca for $l = 2$ states is compared to experimental data.[10] The normalization in the figure is such that if a single state carried all the strength the peak height at the corresponding energy would be $2j + 1$. The first peak corresponds to $d\frac{3}{2}$ removal whereas most of the rest of the strength corresponds to $d\frac{5}{2}$ removal.

The difference between the two fragmentation patterns already illustrates the basic features which are found in experiment. For an orbital which in mean field is very near the Fermi energy, like the $d\frac{3}{2}$ orbital, fragmentation of strength to all $2p1h$ and $1p2h$-like states takes place in such a way that the distribution is characterized by a single large fragment and hundreds of tiny contributions spread out over the whole energy domain covered by the $2p1h$ and $1p2h$ states. This result is related to the position of the $d\frac{3}{2}$ mean-field sp energy which lies in between the $2p1h$ and $1p2h$ states with correspondingly large energy denominators in (7) and large normalization (10) since the derivative of the self-energy has a weaker energy dependence there. In the example of Fig. 2, the theoretical strength in the peak corresponds to 0.73 (peak height divided by $2j + 1$), whereas experimentally it is 0.56. The total occupation number in the calculation is 0.89 with 0.80 residing in the experimentally accessible domain which implies a tail of about 10% of sp strength.

For an orbital which in mean field just resides between $1p2h$ states, like the $d\frac{5}{2}$ state, the situation is already very different. In this case, the energy denominators in (7) are much smaller for the energies near the $d\frac{5}{2}$ mean-field sp energy which solve (9). In addition, the derivative of the self-energy is typically large resulting in smaller fragments at those energies. As a result the $d\frac{5}{2}$ orbital in ^{48}Ca is already strongly

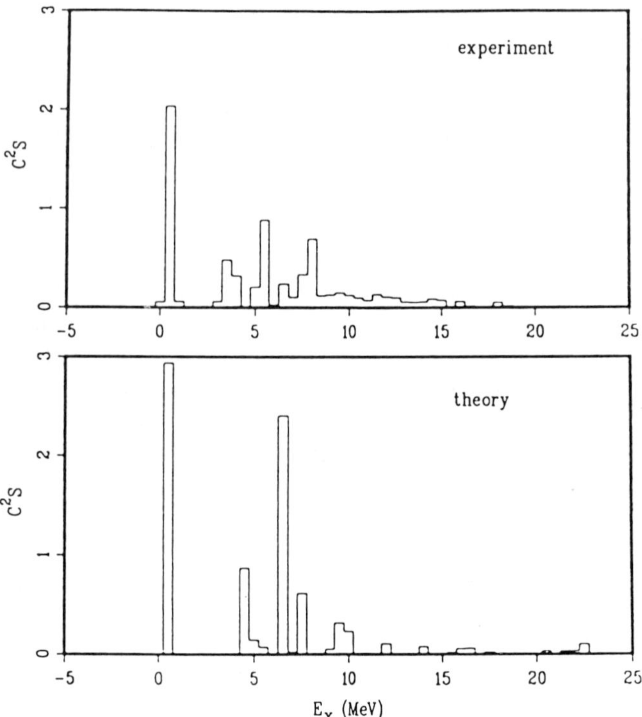

Figure 2. Distribution of *sp* strength for $l = 2$ proton removal from ^{48}Ca.

fragmented in accord with experiment. As discussed in detail in Ref. 9, more deeply bound *sp* orbitals acquire even more fragmentation and are correspondingly more spread in energy. This shows that calculations of the strength distribution using the Dyson equation for a finite nucleus, are capable of explaining all the features that are observed in the $(e, e'p)$ reaction as *e.g.* shown in Fig. 1.

It is possible to compare the characteristics of these *sp* strength distributions in nuclei with the corresponding distributions in a strongly correlated Fermi liquid. In the latter case, one observes for an orbital close to the Fermi energy, which in this case means momentum k close to the Fermi momentum k_F, also one large peak in the spectral distribution, better known as the quasiparticle peak. For orbitals which in mean field are far from the Fermi surface, *i.e.* k around 0, one expects a broad peak implying a much stronger fragmentation of the *sp* strength. This observation shows that the nucleus has features which are identical to those of a Fermi liquid. The only essential difference being the switch from a *sp* basis relevant for an infinite system with momentum as a good quantum number, to a *sp* basis which is relevant for the corresponding finite nucleus.

To understand the implications of the quantitative comparison of theory and experiment, it should be realized that the calculations of Ref. 9 used a *sp* space which includes all shells below the Fermi energy and about three major shells above. In this space one assumes all *sp* strength to reside. As a consequence, the spread of *sp* strength resulting from solving the Dyson equation corresponds roughly to one hundred MeV above and below the Fermi energy. The interaction that was used in Ref. 9 is a *G*-matrix type interaction which takes the effect of the short-range repulsion in the interaction into account at the two-body level. Note that the effect of short-range correlations at the sp level is not properly taken into account in the calculations of Ref. 9 since the coupling of the sp motion is truncated at about 100 MeV above the Fermi energy. As a general conclusion of this work on ^{48}Ca and ^{90}Zr, it is found that the total strength in the experimentally accessible domain in the $(e, e'p)$ reaction is overestimated by about 10-15% for these nuclei although the strength distribution is well described. The background contribution to the hole strength, $i.e.$ the strength which is not contained in the area of the main peak, is of the order of 10%. This is in agreement with the experimental result for the $3s\frac{1}{2}$ in ^{208}Pb discussed above. Stronger fragmentation at low energy in the theoretical calculation is also required which can be obtained by further correlating the intermediate $2p1h$ and $1p2h$ in the self-energy.[11]

Before discussing nuclear matter calculations which treat the effect of short-range and tensor correlations on the location of the sp strength, it is useful to point to the all pervasive quenching of nuclear excitations. Calculations within the shell model or RPA results systematically overestimate the experimentally observed magnetic and Gamow-Teller (GT) strength.[12,13] Detailed calculations of the low energy response beyond the RPA framework have been performed in which the coupling to $2p2h$ excitations up to about 100 MeV of excitation energy is included.[14] This coupling is obviously mandatory since already at the non-interacting ph level the particle and the hole individually display strong coupling to the nearby lying more complicated configurations as discussed above. The results of Ref. 14 indicate that it is possible to obtain good agreement with the shapes of the strength distributions for all multipoles including the electric, magnetic, and GT transitions. The only ingredient missing in these calculations is that all strength distributions at low energy are overestimated by the same amount of 20-30%. This means that quenching still persists when $2p2h$ excitations are included. However, the results in Ref. 14 show that quenching is not restricted to the magnetic and GT states when the response is considered in this extended RPA framework.

The conclusion that quenching is a much more general phenomenon than previously believed, suggests that an explanation is called for that is independent of the quantum numbers of a particular excitation mode. The calculations reported in Refs. 9 and 14 assume that all sp strength of the orbitals in the considered configuration space is available. It is therefore natural to seek an explanation for the remaining discrepancies focusing on the distribution of sp strength. This will be relevant for an explanation of both the 10-15% discrepancy between theoretical and experimental sp strength discussed in Ref. 9 and shown in Fig. 2, as well as for the additional quenching that is required for the description of low energy response functions.

FERMI LIQUID ASPECTS OF NUCLEAR MATTER

The study of the nucleon spectral function has been made for nuclear matter in order to accurately study the influence of the short-range repulsion and the tensor force. Calculations for the hole part of the spectral function have been performed in the context of the Correlated Basis Functions (CBF) method.[15] Both hole and particle part of the spectral function were studied in Ref. 16 by applying the Self Consistent Green's Function (SCGF) method to a semi-realistic central interaction. A discussion was given in Ref. 16 indicating that the impact of the short-range correlations is clearly seen in the particle spectral function although the presence of high momentum components in the ground state is reflected in the hole spectral function. Calculations of the particle spectral function for a realistic interaction including the effect of the nuclear tensor force are reported in Ref. 17.

The SCGF-method is $e.g.$ discussed in Refs. 18 and 16. The basic physical idea underlying this method is that the properties of the particle are determined by its interaction with the other particles. This interaction is then in first approximation determined by the the most dominant physical correlation in the system. In the nuclear case this is the short-range repulsion in the basic interaction. It is therefore necessary to sum the ladder diagrams into the interaction with the inclusion of hole-hole propagation. The hole-hole propagation is required since it provides the information for the calculation of the hole spectral function. The resulting interaction is used to calculate the self-energy which then is used to solve Dyson's equation which provides the spectral functions. The interaction should then be recalculated since the particles are dressed by their interactions with the medium. A non-linear formulation of the many-particle problem is thus obtained which requires the self-consistent determination of the sp propagator.

Results of calculations will now be disccused in which the sp spectrum is determined self-consistently from

$$\epsilon(k) = \frac{\int d\omega\, \omega S_h(k,\omega)}{\int d\omega\, S_h(k,\omega)} \qquad k < k_F \tag{11}$$

and

$$\epsilon(k) = \frac{\int d\omega\, \omega S_{QP}(k,\omega)}{\int d\omega\, S_{QP}(k,\omega)} \qquad k > k_F. \tag{12}$$

Once the spectrum given by equations (11) and (12) has been determined self-consistently, the final spectral functions are obtained from the Dyson equation. For the hole spectral function this yields the relation to the self-energy given by

$$S_h(k,\omega) = \frac{1}{\pi} \frac{Im\ \Sigma(k,\omega)}{\left(\omega - \frac{k^2}{2m} - Re\ \Sigma(k,\omega)\right)^2 + \left(Im\ \Sigma(k,\omega)\right)^2} \qquad \omega < \epsilon_F^- \tag{13}$$

For momenta above k_F, the quasiparticle contribution, S_{QP},[16] to the particle spectral function, $S_p(k,\omega)$, was used to determine the sp energy. This choice is dictated by the actual shape of S_p (discussed below) which would result in an unrealistically large gap at k_F. Equation (12) now, instead, generates the quasiparticle energy. Distinguishing between states below and above k_F introduces a gap in the sp energy which is necessary to avoid pairing instabilities and properly take into account the collective bound pair states that result from solving the ladder equation.[19−21] Any sensible choice of the sp spectrum will lead to very similar particle spectral functions. The specific choice given in (11) and (12) therefore will have consequences for the discussion of the particle spectral functions below. Its effect on the hole spectral functions might be somewhat more important although the characteristic features of the strength distribution are not changed. Ideally, the full spectral distribution should be determined self-consistently and in that case no sp spectrum needs to be chosen. Such calculations are presently under way.[22]

Results for the hole spectral function calculated for the Reid interaction[23] at normal nuclear matter density are shown in Fig. 3.[24] The three curves correspond to $k = 0.48, 0.79$, and $0.93 fm^{-1}$, respectively. As discussed above, the expected quasiparticle features of the strength distribution become more pronounced when the sp momentum is closer to the Fermi momentum. The background strength outside the peak of these spectral functions contains about 10% of the sp strength. This number is very similar to the results obtained in finite nuclei and the experimental result for the $3s\frac{1}{2}$ proton orbital in ^{208}Pb.

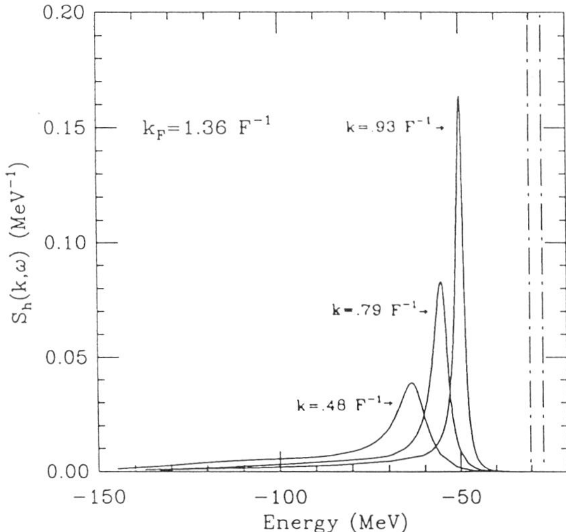

Figure 3. Hole spectral function for different momenta below k_F. As in a finite nucleus, one observes that the closer a *sp* state is to the mean-field Fermi energy, the sharper peaked the spectral distribution becomes.

The integral appearing in the denominator of (11) generates the occupation number of the *sp* state with momentum k,

$$n(k) = \int_{-\infty}^{\epsilon_F^-} d\omega\, S_h(k,\omega). \tag{14}$$

The result for the occupation number for $k = 0$ calculated according to (14) for the Reid soft core potential at $k_F = 1.36\,fm^{-1}$ is 0.83. Results from other methods such as Brueckner theory[25,26] and CBF theory[27,28] using other realistic interactions, for the occupation of $k = 0$ are very similar. In Refs. 25 and 26 0.82 is reported for the Paris potential.[29] Older CBF calculations[27] for the Urbana v_{14} interaction[30] give 0.87 whereas more recent CBF results[15,28] give 0.83. All recent calculations for different interactions using different methods give a strikingly similar result for $n(0)$. This should be contrasted to the result at k_F which is most easily expressed in terms of $z(k_F) = n(k_F^-) - n(k_F^+)$. For the Reid interaction $z(k_F) = 0.72$ is obtained from the present results. In Refs. 28 and 27 and $z(k_F) = 0.7$ is reported for the Urbana v_{14} interaction. However, for the Paris interaction $z(k_F) = 0.35$ has been obtained in Ref. 25 and 0.47 in Ref. 26. These results suggest that for an orbital which is far removed from the Fermi energy ($k = 0$) the depletion is rather uniquely pinned down and one has reason to expect a similar depletion for deeply bound orbitals in nuclei. In contrast to the $k = 0$ case, the depletion and occupation of momenta around k_F is very sensitive to the low-lying excitation modes of nuclear matter which are anyway very different from those of real nuclei where shell effects and collective low energy surface vibrations dominate the low energy excitation modes. Occupation numbers around k_F should therefore be considered less relevant in comparing with finite nuclei.[31]

The importance of the particle spectral function is to exhibit where the unoccupied *sp* strength is located in energy. It is obtained by solving the Dyson equation and is related to the self-energy by

$$S_p(k,\omega) = -\frac{1}{\pi} \frac{Im\ \Sigma(k,\omega)}{\left(\omega - \frac{k^2}{2m} - Re\ \Sigma(k,\omega)\right)^2 + \left(Im\ \Sigma(k,\omega)\right)^2} \qquad \omega > \epsilon_F^+. \qquad (15)$$

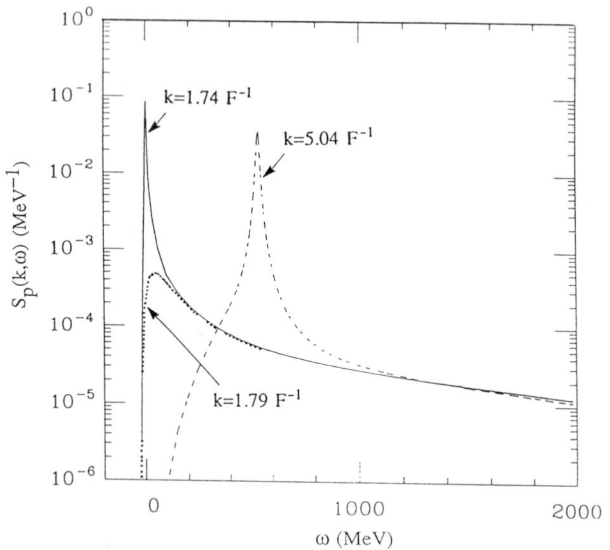

Figure 4. Particle spectral function for three different momenta, $k = 0.79$ (dotted), 1.74 (full), and 5.04 fm^{-1} (dashed). The high-energy tail is identical for all momenta below 5 fm^{-1} and is related to the short-range repulsion in the nucleon-nucleon interaction.

The striking feature of the particle spectral function is that is has a high-energy tail that can be directly related to the presence of repulsion at short distances as shown in Ref. 16. Results for the particle spectral functions are shown in Fig. 4 for for three different momenta, $k = 0.79, 1.74$, and $5.04 fm^{-1}$ as a function of energy. All momenta below k_F have the same high energy tail as the dotted curve for $k = 0.79 fm^{-1}$ in Fig. 4. For momenta larger than k_F a quasiparticle peak, which broadens with increasing momentum, can be observed on top of the same high energy tail. The results therefore display a common, essentially momentum independent, high-energy tail. The location of *sp* strength at high energy simply means that the interaction has sufficiently large matrix elements to compensate energy denominators encountered in the ladder equation. For this particular interaction a significant amount of strength is found at high energy. This result was of course already anticipated a long time ago.[32]

The integrated strength accounts for 17% of the *sp* strength for $k = 0.79 fm^{-1}$. This is in agreement with the sum rule since the integrated hole strength (see (14)) provides 83% of *sp* strength. Quantitative results for the location of the missing *sp* strength for this momentum indicate that the strength in the interval from 100 MeV above the Fermi energy to infinity amounts to 13% with 7% residing above 500 MeV. Additional study of the influence of tensor correlations[17] show that without tensor effects only a total of 10.5% depletion is obtained which should be regarded as resulting from pure short-range correlations. This is consistent with CBF calculations of the momentum distribution which show depletions of a similar size due to tensor correlations.[27,15,28]

DISCUSSION

The implications of these results for nuclear structure are manifold. Making the very plausible assumption that the role of short-range and tensor correlations will be identical in a heavy nucleus, one can expect that all *sp* states will have about 15% of their strength removed to high energy. This is precisely the amount necessary to bring the results of Ref. 9 into quantitative agreement with $(e, e'p)$ results for ^{48}Ca and ^{90}Zr. The present experimental information[8] on the $3s\frac{1}{2}$ proton in ^{208}Pb shows that this orbital is occupied for 75% (with an error of 9%) with 10% residing in the background. This is consistent with the notion that deep-lying orbitals in a heavy nucleus should have occupation numbers similar to those in nuclear matter *i.e.* about 80%. The theoretical results obtained here suggest that 15% of the missing 25% of *sp* strength for the $3s\frac{1}{2}$ orbital is removed to very high energy. The remaining 10% should then be found in the first 100 MeV above the Fermi energy in the particle domain. Calculations for lighter nuclei[9] show that this is indeed possible although this strength is expected to be highly fragmented as well. A satisfactory picture of the nucleus therefore emerges in which Landau's quasiparticle picture is valid[33] For the $3s\frac{1}{2}$ proton the strength of the quasihole pole, z_h, is 65%, 10% additional occupation is obtained as background, and 25% of the strength is depleted and spread to very high energies. This distribution is schematically shown in Fig. 5.

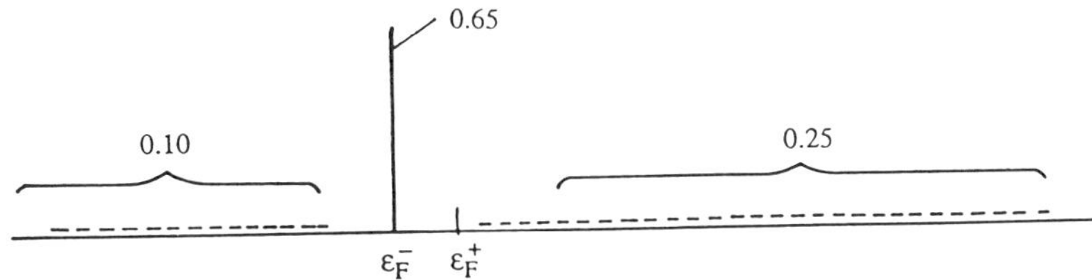

Figure 5. Spectral function for the $3s\frac{1}{2}$ proton orbital in ^{208}Pb. The strength in the peak and the occupation number (sum of peak and background strength) are experimental numbers. The distribution of the background strength as suggested in this figure is based on the theoretical work discussed in this paper.

The implications of these results to the low energy response are also important. Extending this picture to the mostly empty low-lying particle states one can expect a similar 65% strength of the quasiparticle pole, z_p, with an identical distribution of the background. It is then clear that for *ph* excitations at low energy for which the influence of the ph interactions is expected to be small, like $M12$ and $M14$ excitations in ^{208}Pb only the fraction $z_p * z_h$ of the the simple shell model estimate will be found experimentally[34] A similar conclusion holds also for the presence of low-lying GT strength and other excitation modes. The low-energy response functions can therefore also be understood quantitatively when the present results for particle spectral functions are taken into account[14] The presence of considerable sp strength at high energy due to short-range and tensor correlations, therefore provides a simple explanation for the appearance of quenching phenomena in nuclear physics.

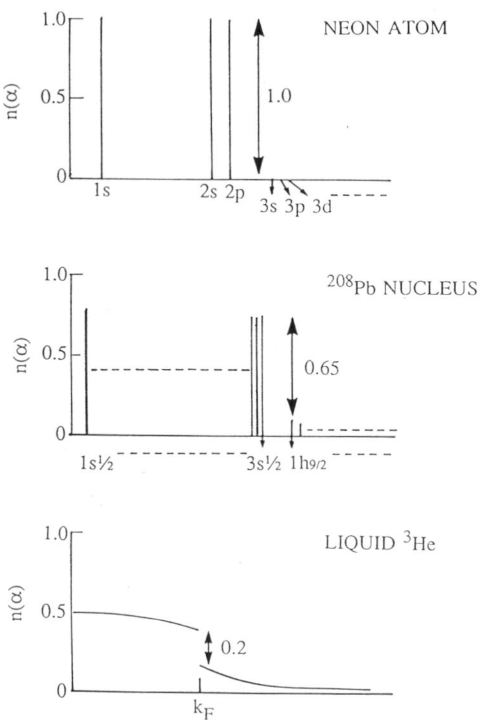

Figure 6. Comparison of occupation numbers in different physical systems characterized by the strength of the correlations. Weak correlations are found in atoms, strong correlations in nuclei, and very strong correlations in liquid ^3He.

With the results obtained in this work, it becomes also possible to compare the nucleus with other many-fermion systems. This is schematically done in Fig. 6, where occupation numbers are shown for three representative systems. In the top part, occupation numbers for the Ne atom are shown. The first three sp levels, $1s, 2s$, and $2p$, are basically full, whereas all the other orbitals have no occupation to speak of. This distribution reflects the relevance of the mean-field picture associated with a weak interaction between the particles which can be treated in Hartree-Fock approximation. This results in a jump in occupation of 1 at the Fermi energy. Occupation numbers for protons in ^{208}Pb are shown in the middle part of Fig. 6. The occupation of the $3s\frac{1}{2}$ orbital (0.75) and the jump in occupation (0.65) correspond to experimental numbers[8] whereas the occupation of the deeply bound orbitals is inferred from the nuclear matter calculations discussed above. Finally, in the bottom part the momentum distribution of liquid ^3He is shown indicating that this system is very strongly correlated with only a jump in occupation of about 0.2. The momentum distribution shown here is adapted from the results dicussed in Ref. 35.

At this point is useful to return to the discussion at the beginning of the paper where it was suggested that the coupling of low energy nucleons to high-lying mesonic and nucleonic or other QCD excitations is simulated by the use of a strongly repulsive NN interaction at short distance. In this paper it has been demonstrated on the one hand that the physics of the nucleus is very similar to that of a Fermi liquid with an appropriate choice of sp basis. On the other hand, the influence of the short-range part of the interaction has been studied in detail and found to be non-negligible. Its effect is to remove 10-15% of sp strength to energies where it is no longer available for low energy phenomena. This means that in a manner of speaking QCD-effects are "visible" in nuclei in terms of a global depletion of sp strength resulting in a host of quenching phenomena in nuclear physics.

ACKNOWLEDGEMENT

This work is supported by NSF Grant PHY-9002863. Some of the calculations reported in this work were performed at the Pittsburgh Supercomputing Center. The author would also like to acknowledge the fruitful collaboration with Klaas Allaart, Mario Brand, Arturo Polls, Angels Ramos, Gustl Rijsdijk, and Brian Vonderfecht, which has produced the results that were discussed in this paper and the permission from the NIKHEF group to use Fig. 1.

REFERENCES

1. O. Haxel, J. H. D. Jensen, and H. E. Suess, Phys. Rev. **75**, 1766 (1949).
2. M. Goeppert-Mayer, Phys. Rev. **75**, 1969 (1949).
3. J. M. Cavedon *et al.*, Phys. Rev. Lett. **49**, 978 (1982).
4. E. N. M. Quint, Thesis University of Amsterdam (1988).
5. P. K. A. de Witt Huberts, J. Phys. G: Nucl. Part. Phys. **16**, 507 (1990).
6. G. J. Wagner, AIP Conf. Proc. **142**,220 (1986); P. Grabmayr *et al.* Phys. Lett. **B164**, 15 (1985).
7. J. P. McDermott, Phys. Rev. Lett. **65**, 1991 (1990).
8. G. van der Steenhoven, Nucl. Phys. **A527**, 17c (1991).
9. M. G. E. Brand, G. A. Rijsdijk, F. A. Muller, K. Allaart, and W. H. Dickhoff, Nucl. Phys. **A** in press; poster this conference.
10. G. Kramer, Thesis University of Amsterdam (1990) and private communication.
11. G. A. Rijsdijk, work in progress; poster this conference.
12. A. Arima, K. Shimizu, W. Bentz, and H. Hyuga, Adv. Nucl. Phys. **18**, 1 (1987).
13. I. S. Towner, Phys. Reports **155**, 263 (1987).
14. M. G. E. Brand, K. Allaart, and W. H. Dickhoff, Nucl. Phys. **A509**, 1 (1990); poster this conference.
15. O. Benhar, A. Fabrocini, and S. Fantoni, Nucl. Phys. **A505**, 267 (1989).
16. A. Ramos, A. Polls, and W.H. Dickhoff, Nucl. Phys. **A503**, 1 (1989).

17. B. E. Vonderfecht, W. H. Dickhoff, A. Polls, and A. Ramos, Phys. Rev. C, in press; poster this conference.

18. A. Ramos, W. H. Dickhoff, and A. Polls, Phys. Lett. **B219**, 15 (1989).

19. W. H. Dickhoff, Phys. Lett. **B210**, 15 (1988).

20. W. H. Dickhoff, C. C. Gearhart, B. E. Vonderfecht, A. Polls, and A. Ramos, in *Recent Progress in Many-Body Theories*, Vol. 2, ed. Y. Avishai (Plenum, New York, 1990), p. 141.

21. B. E. Vonderfecht, C. C. Gearhart, W. H. Dickhoff, A. Polls, and A. Ramos, Phys. Lett. **B253**, 1 (1991).

22. C. C. Gearhart, work in progress.

23. R. V. Reid, Ann. Phys. **50**, 411 (1968).

24. B. E. Vonderfecht, Thesis Washington University (1991).

25. P. Grangé, J. Cugnon, and A. Lejeune, Nucl. Phys. **A473**, 365 (1987).

26. M. Baldo, I. Bombaci, G. Giansiracusa, U. Lombardo, C. Mahaux, and R. Sartor, Phys. Rev. **C41**, 1748 (1990).

27. S. Fantoni and V. R. Pandharipande, Nucl. Phys. **A427**, 473 (1984).

28. O. Benhar, A. Fabrocini, and S. Fantoni, Phys. Rev. **C41**, R24 (1990).

29. M. Lacombe *et al.*, Phys. Rev. **C21**, 861 (1980).

30. I. E. Lagaris and V.R. Pandharipande, Nucl. Phys. **A359**, 331 (1981).

31. V. R. Pandharipande, C. N. Papanicolas, and J. Wambach, Phys. Rev. Lett. **53**, 1133 (1984).

32. G. E. Brown, Comm. Nucl. Part. Phys. **3**, 136 (1969).

33. L. D. Landau, Soviet Phys. **8**, 70 (1958).

34. J. Lichtenstadt *et al.*, Phys. Rev. **C20**, 497 (1979).

35. S. Rosati, M. Viviani, and E. Buendia, in *Condensed Matter Theories*, Vol. 5, ed. V. C. Aguilera-Navarro (Plenum, New York, 1990), p. 119.

INCLUSIVE SCATTERING AND FSI IN NUCLEAR MATTER

Omar Benhar

INFN, Sezione Sanitá, Physics Laboratory
Istituto Superiore di Sanitá. I-00161 Roma, Italy

Adelchi Fabrocini

Dept. of Physics, University of Pisa and
INFN, Sezione di Pisa,I-56100 Pisa, Italy

Stefano Fantoni

International School for Advanced Studies (ISAS) and
INFN, Sezione di Trieste, I-34014 Trieste, Italy

Abstract: *The dynamical response and the cross section for inclusive scattering of electrons by nuclear matter are calculated at intermediate and high momentum transfers, using correlated basis function theory. Particular emphasis is placed on the effects of the final state interactions. By comparing the plane wave impulse approximation with the full response at $q \simeq 1\ GeV/c$, large contributions from final state interactions are found at the quasielastic peak, whereas, at higher momentum transfers, their contribution is located mainly in the low energy tails. For relativistic electrons, the inclusive cross section is computed by using a microscopic spectral function and final state interactions are introduced via correlated Glauber theory. The results are compared with recent empirical data on nuclear matter. The cross section obtained within plane wave impulse approximation is close to the observed data at large values of the energy loss, but it is too small at low values. After including final state interactions effects, the theory is in fair agreement with the data.*

1. Introduction

A longstanding problem in the study of scattering by strongly interacting many-

Recent Progress in Many-Body Theories, Vol. 3,
Edited by T.L. Ainsworth et al., Plenum Press, New York, 1992

body systems is to estimate the effects of the Final State Interactions (FSI). In the past large efforts have been devoted to the undertsanding of these effects in the large momentum transfer scattering of neutron by liquid Helium (both ^4He and ^3He) at low T [1]. The influence of FSI, particularly in ^4He, is by now quite well understood and the agreement of the theoretical results with experimental data is impressive.

The situation is less satisfactory in nuclei, where shell and surface effects may play an important role when comparing results from infinite systems with experiments. However recently *Day et al.* [2] have obtained a measure of the inclusive response $\Sigma_V(q,\omega)$ for scattering of electrons by nuclear matter, by fitting the experimental cross–section per nucleon $\Sigma_A(\mathbf{q},\omega)$ for $A \geq 12$ with the mass formula

$$\Sigma_A(\mathbf{q},\omega) \sim \Sigma_V(\mathbf{q},\omega) + \Sigma_S(\mathbf{q},\omega)A^{-1/3}, \qquad (1.1)$$

where $\Sigma_V(\mathbf{q},\omega)$ and $\Sigma_S(\mathbf{q},\omega)$ represent the volume and the surface terms respectively.

An effective, non relativistic approach to the nuclear many-body problem is provided by the *Correlated Basis Function* (CBF) theory [3,4]. CBF is based upon the following set of *correlated* states:

$$|N) = \frac{\mathcal{G}|N]}{[N|\mathcal{G}^\dagger\mathcal{G}|N]^{1/2}}, \qquad (1.2)$$

where $|N]$ is the generic eigenstate of an independent particle hamiltonian (Fermi gas hamiltonian in the case of nuclear matter) and $\mathcal{G}$ is a correlation operator which has the same structure as the N–N interaction. A realistic form of $\mathcal{G}$ is:

$$\mathcal{G} = S \prod_{j>i=1,A} F(i,j), \qquad (1.3)$$

$$F(i,j) = \sum_n f^n(r_{ij})O^n(i,j), \qquad (1.4)$$

where S is the symmetrization operator acting on $\prod_{i<j} F(i,j)$ and the operators $O^n(i,j)$ include the four central components $(1,\sigma_i\cdot\sigma_j,\tau_i\cdot\tau_j,\sigma_i\cdot\sigma_j\tau_i\cdot\tau_j)$ for $n = 1,4$ and both the isoscalar and the isovector tensor components S_{ij} and $S_{ij}\tau_i \cdot \tau_j$ for $n = 5,6$.

The correlated basis states are required to be as close as possible to the eigenstates of the nuclear hamiltonian H. This is achieved by variationally determining the correlation functions $f^n(r_{ij})$ via the minimization of the expectation value of H on the ground state.

The results obtained for different nuclear matter properties, like the equation of state [5], the optical potential [6] and the momentum distribution [7], by using

CBF theory are in good agreement with the available empirical data. Moreover, the theory has been recently developed to evaluate quantities measured in e–nucleus scattering experiments, such as the dynamical responses and the single particle Green functions [8,9].

Microscopic calculations of electron scattering observables in nuclear matter are mostly significant in the region of momentum transfers $q \geq 2k_F$, with k_F being the Fermi momentum ($k_F \sim 240 MeV/c$). In fact, in this region the electron mainly probes the short range behavior of the wave function and this is not affected by finite size effects. Furthermore, for momentum transfers in the GeV region nuclear matter is well suited to study the properties of the nucleon as a composite particle.

The plan of the paper is as follows: sect.2 is devoted to the analysis of the nuclear matter dynamical response in the region $q \sim 1\ GeV/c$, where relativistic effects are expected to be small and therefore a theoretical calculation can be carried out within CBF theory. FSI effects are studied by comparing the *exact* results with those obtained by using the *Plane Wave Impulse Approximation* (PWIA) and a microscopic hole spectral function computed within CBF. The inclusive cross–section at high momentum transfer (up to $q \sim 2\ GeV/c$) is discussed in sect.3. The main theoretical problem here arises from the need of a relativistic treatment for the struck nucleon. Using PWIA it is possible to decouple the motion of the relativistic struck nucleon from the spectator $(A-1)$–particle system, described in terms of the non relativistic hole spectral function. FSI effects turn out to be quite sizeable even at very high momentum transfers and are included via a correlated version of the Glauber theory (CGT) in the same section.

2. Dynamical response functions at intermediate momentum transfers

In this section we briefly outline the CBF theory of the longitudinal dynamical structure function $S_L(q,\omega)$, related to the longitudinal response $R_L(q,\omega)$ by

$$R_L(q,\omega) = G_L^2(q,\omega)S_L(\mathbf{q},\omega), \tag{2.1}$$

where $G_L(q,\omega)$ can be easily expressed in terms of the *Sachs* nucleon form factor. $S_L(q,\omega)$ is defined as

$$S_L(q,\omega) = \sum_{\bar{N}} |< \bar{0}|\rho_{L,\mathbf{q}}|\bar{N} >|^2 \delta(\omega - E_N + E_0), \tag{2.2}$$

where

$$\rho_{L,\mathbf{q}} = \sum_{i=1,A} e^{i\mathbf{q}\cdot\mathbf{r}_i}\{\frac{1+\tau_i^3}{2} - \mu_n\frac{q^2}{4(q^2+m^2)}\frac{1-\tau_i^3}{2}\}, \tag{2.3}$$

with μ_n being the neutron magnetic moment, and $|\bar{0} >$ and $|\bar{N} >$ exact ground state and intermediate state wave functions with energies E_0 and E_N respectively. The dynamical structure functions is calculated within non relativistic nuclear many-body theory by using the expression

$$S_L(q,\omega) = \frac{1}{\pi A} \Im < \bar{0}|\rho_{L,\mathbf{q}}^\dagger \frac{1}{H - E_0 - \omega - i\eta} \rho_{L,\mathbf{q}}|\bar{0} > . \qquad (2.4)$$

The quantity $(H - E_0)^{-1}$ is expanded in series of $(H_I - \Delta E_0)$, where we have written

$$H - E_0 = (H_0 - E_0^v) + (H_I - \Delta E_0), \qquad (2.5)$$

and H_0 (H_I) is the (non)diagonal part of the hamiltonian in the correlated basis of eq.(1.2), $E_0^v = (0|H|0)$ and $\Delta E_0 = E_0 - E_0^v$. $S_L(q,\omega)$ is then formally obtained by inserting $\sum_N |N> < N| = 1$ between each pair of contiguous operators. The correlated orthogonal states $|N>$ are obtained by a *Schmidt- Lowdin* orthogonalization of the non ortoghonal correlated states of eq.(1. 2) [10]. Divergencies arising from the factor ΔE_0 (which is of the order A) and from unlinked cluster terms of $< 0|\rho_{L,\mathbf{q}}^\dagger|N >$ cancel out and $S_L(q,\omega)$ results to be a sum of *unlinked generalized Goldstone diagrams*.The results presented here have been obtained by using 1p-1h and 2p-2h intermediate orthogonal correlated states.

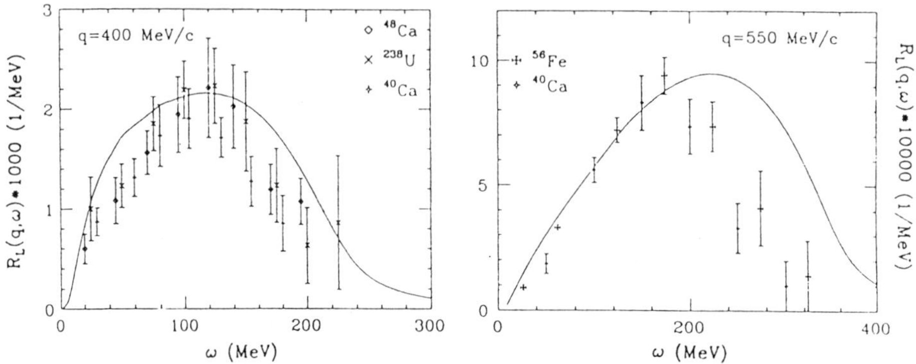

Fig. 1-*From ref.[8]. The longitudinal response of nuclear matter compared with experimental data on complex nuclei*

In fig.(1)) the longitudinal responses of nuclear matter, calculated [8] by disregarding the small neutron term in eq.(2.3), are compared to the available experimental data on complex nuclei. The dipole fit of the proton form factor, including the *Darwin–Foldy* correction [11] has been used.

There is good agreement between the nuclear matter response and the data at low ω, up to the quasifree peak for both the momentum transfers considered. This is an indication that possible modifications of the nucleon form factor in the nuclear medium do not play a significant role. At high q and ω the calculated responses are larger than the experimental data, expecially for ^{40}Ca. These differences are responsible for the disagreement in the integrated responses between theory and experiments, which still stands as an open problem.

At higher momentum transfers an estimate of FSI effects is obtained by comparing the response of eq.(2.2) with that obtained in PWIA and given by

$$S_P(q,\omega) = \frac{4}{(2\pi)^3\rho} \int d\mathbf{k} \int dE \; P_h(k,E)\delta(\omega - E - e(|\mathbf{k}+\mathbf{q}|))\theta(|\mathbf{k}+\mathbf{q}| - k_F), \quad (2.6)$$

where k_F is the Fermi momentum and $P_h(k,E)$ is the hole spectral function, defined as

$$P_h(k,E) = \sum_{\bar{N}} \left| < \bar{0}|a_\mathbf{k}^\dagger|\bar{N}(A-1) > \right|^2 \delta(E - E_N(A-1) + E_0) \quad (2.7)$$

and giving the probability of leaving the final $(A-1)$-nucleon system with an excitation energy E when removing from the nuclear target a nucleon of momentum $\mathbf{k}$ [9]. In eq.(2.7), $|\bar{N}(A-1) >$ are intermediate $(A-1)$-particle states with $E_N(A-1)$ energy and $e(k)$ is the CBF single particle energy. As clearly appears from eq.(2.6), the use of the spectral function amounts to decouple the motion of the struck nucleon, considered as moving freely after being hit, from the remaining nuclear system. In this respect, the CBF S_P contains information about the initial state interactions of the nucleon with the system but FSI are not included.

In orthogonal CBF theory $P_h(k,E)$ is calculated using the expression

$$P_h(k,E) = \frac{1}{\pi}\Im < 0|a_\mathbf{k}^\dagger \frac{1}{H - E_0 - E - i\eta} a_\mathbf{k}|0 >, \quad (2.8)$$

expanding $(H - E_0)^{-1}$ in series of $(H_I - \Delta E_0)$ and introducing closure, as in the case of the dynamical structure function previously discussed.

In fig.(2), $S_L(q,\omega)$ and $S_P(q,\omega)$ are compared at two momentum transfers. At $q = 3.99 fm^{-1}$, S_P largely overestimates the full structure function in the region of the quasielastic peak, whereas, at $q = 7.78 fm^{-1}$, FSI are important only in the low energy region ($\omega \leq 700\ MeV$) not included in the figure.

3. Inclusive scattering at high momentum transfers

The cross section for the inclusive electron nucleus scattering process

$$e + A \rightarrow e' + anything, \tag{3.1}$$

may be written, in Born approximation, as [12]

$$\frac{d^2\sigma}{d\Omega d\epsilon'} = \frac{\alpha^2}{q^4}\frac{\epsilon'}{\epsilon} L^{\mu\nu} W^A_{\mu\nu}(q), \tag{3.2}$$

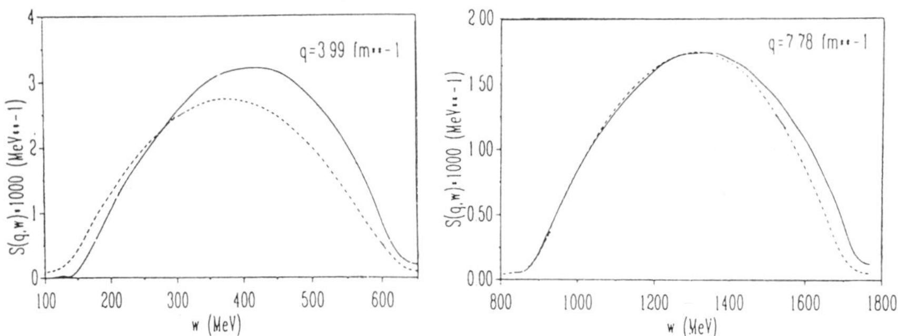

Fig. 2-From ref.[9]. Dynamical structure function of nuclear matter at two momentum transfers. $S_P(q,\omega)$ is given by the solid lines whereas the dashed ones give $S_L(q,\omega)$.

where $\alpha = 1/137$ is the fine structure constant, ϵ and ϵ' are the energies of the incident and scattered electron, respectively, and q is the four momentum transferred by the virtual photon: $q \equiv k - k'$, with $k \equiv \left(\epsilon, \vec{k}\right)$ and $k' \equiv \left(\epsilon', \vec{k'}\right)$. For ultrarelativistic electrons the lepton tensor reduces to

$$L^{\mu\nu} = 2\left(k_\mu k'_\nu + k_\nu k'_\mu - g_{\mu\nu}(k, k')\right), \tag{3.3}$$

where $(k, k') = \epsilon\epsilon' - \vec{k}\vec{k'}$. All the information on the structure of the nuclear target is contained in the tensor $W^A_{\mu\nu}$, whose expression reads as follows:

$$W^A_{\mu\nu}(q) = \sum_N \int d\mathbf{p}_N < 0|J^A_\mu|N ><N|J^A_\nu|0> \delta^{(4)}\left(p_0 + q - p_N\right), \tag{3.4}$$

p_0 and p_N being the four momenta of the initial and final states of the target nucleus $|0>$ and $|N>$.

The explicit calculation of the nuclear tensor in principle requires the consistent relativistic description of the nuclear states $|O>$ and $|N>$, as well as of the nuclear current operator J^A. In practice, various approximation have to be introduced in order to evaluate $W^A_{\mu\nu}$ in different kinematical regimes. The off diagonal matrix elements involved in eq.(3.4) can be evaluated employing nonrelativistic many body wave functions, obtained within nuclear many body theory, to describe both the initial and the final state of the target. This procedure is generally followed to study scattering processes at not too high momentum transfer ($|\vec{q}| \leq 550 MeV/c$. At larger values of $|\vec{q}|$, however, the motion of the struck particle in the final state cannot be described within nonrelativistic approaches. Furthermore, in this kinematical region the nucleon off-shellness and the processes in which excited states of the nucleon are produced are expected to be relevant, and must be carefully taken into account.

A reasonable approximate picture of electron nucleus scattering at high momentum transfer is based on the assumption that, since the electron probes a region of spacial dimensions $\sim 1/|\vec{q}|$ of the nuclear target, the scattering process involves only one nucleon, the residual $(A-1)$-particle system acting as a spectator. Moreover, the final state of the target is written as a product of a one-particle state, describing the free propagation of the struck nucleon, and a $(A-1)$-particle state describing the spectator system. As a consequence, the dynamics of the nuclear target is decoupled from the electromagnetic vertex, and the relativistic description of the motion of the struck hadron reduces to a purely kinematical problem, which can therefore be treated exactly.

The PWIA expression of the nuclear tensor is given by [13]

$$W^A_{\mu\nu,IA}(q) = \int d^3pdE \quad P_h\left(|\vec{p}|,E\right)\left(Z\tilde{W}^p_{\mu\nu}\left(p,E,q\right) + N\tilde{W}^n_{\mu\nu}\left(p,E,q\right)\right), \quad (3.5)$$

where P_h is the nonrelativistic hole spectral function introduced in the previous section and $\tilde{W}^{p(n)}_{\mu\nu}$ denotes the electromagnetic tensor of an off shell proton (neutron). A number of ambiguities are implied in the definition of the off shell nucleon tensors in terms of the free nucleon structure functions $W^{p(n)}_1$ and $W^{p(n)}_2$, which are measured in elastic and inelastic electron proton and electron deuteron scattering experiments.

Since the electron scatters off a bound nucleon, a fraction of the energy transferred by the virtual photon goes into excitation energy of the residual $(A-1)$-particle system, the four-momentum transferred to the struck nucleon being [13] $\tilde{q} \equiv (\vec{q},\tilde{\omega})$. Hence, the nucleon off-shellness is measured by the quantity $\omega - \tilde{\omega}$, where $\tilde{\omega}$ is a function of $\vec{p}$ and E. The most natural procedure to construct $\tilde{W}^{p(n)}_{\mu\nu}$, originally proposed by *De Forest* [13] and employed by *Benhar et al.* [14], is based on the assumption that one can use free nucleon spinors and current operators, but replacing the momentum transfer q with $\tilde{q}$. The gauge invariance $q_\mu\tilde{W}^{\mu\nu} = \tilde{W}_{\mu\nu}q^\nu = 0$ is then used to eliminate the dependence upon the longitudinal current.

Fig.(3) compares the PWIA cross-sections for incident electron energy $E = 3.595 GeV$ and scattering angle $\theta = 30^0$ with the empirical *nuclear matter data* of ref.[2]. Theory and data are in close agreement at $\omega > 1GeV$, whereas sizeable discrepancies occur, both in magnitude and shape, at lower energy loss. It appears that the theoretical curve lies a factor of 3 to 4 below the data at $\omega = .7$ to $.8GeV$, and exhibits a pronounced kink at $\omega \sim .9GeV$, reflecting the threshold for the three-body break-up processes. The origin of this kink can be traced back to the discontinuity of the nuclear matter momentum distribution at $p = k_F$ and FSI seem to be the major source of the observed disagreement between theory and data.

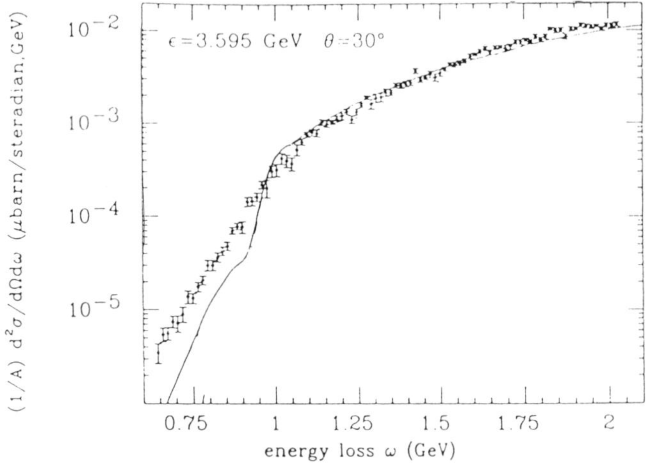

Fig. 3-*From ref.[14]. PWIA inclusive cross–section for incident electron energy $E = 3.595GeV$ and scattering angle $\theta = 30°$, compared with the empirical data.*

FSI effects can be represented by means of an *optical potential $V - iW$* . The simplest approximation for V and W is to use the real and imaginary parts of the optical potential of the struck hadron in nuclear matter. The effects of FSI are very important only at small values of ω where the response is dominated by elastic $e - N$ processes. Dirac optical models fits to the scattering of 0.1 to $1GeV$ protons by nuclei, suggest that $V \sim 25MeV$ in the region of interest, and it has a rather small effect on the response. Thus the main effect of FSI is due to the damping of the motion of the struck nucleon described by the imaginary potential W.

The crudest estimate for W is obtained by taking

$$W\left(p'\right) = \frac{\hbar}{2}\rho v\left(p'\right)\sigma_{NN}\left(p'\right),\tag{3.6}$$

where ρ is the density $(0.16fm^{-3})$ of nuclear matter, $v(p')$ and $\sigma_{NN}(p')$ are respectively the velocity and the scattering cross-section of the struck nucleon. In the region of interest here, v is close to c and σ_{NN} does not depend strongly on p', therefore it is reasonable to approximate $v(p')$ with $v(q)$ and $\sigma_{NN}(p')$ with $\sigma_{NN}(q)$, q being the momentum transfer.

In ref.[14] it has been shown that multiple scattering Glauber theory provides a framework for obtaining approximate expressions for W which take into account the ground state correlations. Correlated Glauber theory (CGT), in first order approximation, provides the following expression of the optical potential

$$V - iW = -\frac{2\pi\rho v(q)}{q}\frac{1}{z}\int_0^z d\xi \int \frac{d\mathbf{k}}{(2\pi)^3} d\mathbf{r}'\, e^{i\mathbf{k}\cdot(\mathbf{r}'-\hat{\mathbf{q}}\xi)} g(r')\, f_q(\mathbf{k}), \qquad (3.7)$$

where $f_q(\mathbf{k})$ is the *free* N–N amplitude and $g(r)$ is the pair distribution function.

The treatment of FSI interaction, although very satisfactory at higher energies, need to be improved to fully include the correlation effects of the struck nucleon with the nuclear matter, as done in the non–relativistic calculation of the response using CBF theory.

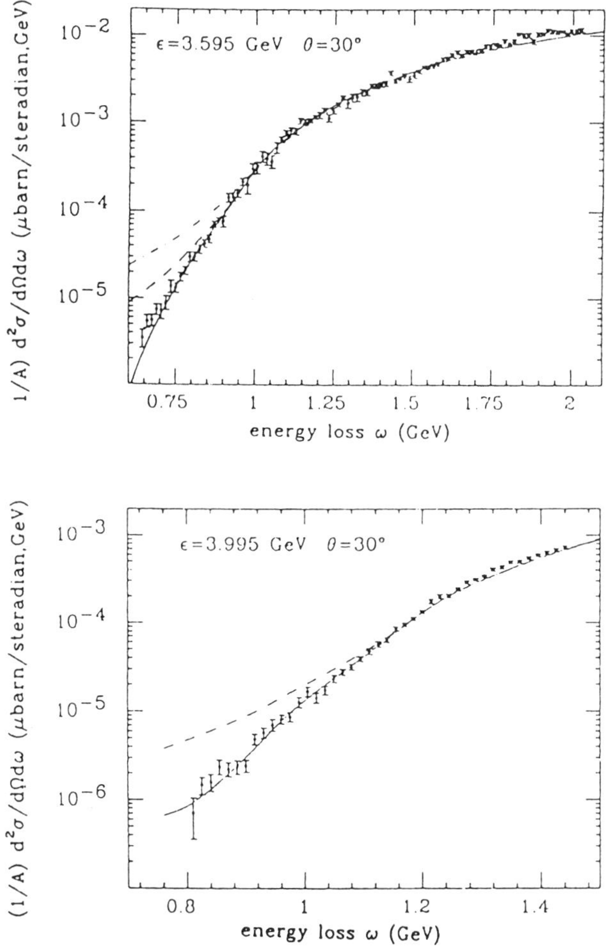

Fig. 4-*From ref.[14]. Inclusive cross–sections of nuclear matter. CGT(dashes) and CGT+colour transparency (solid lines). Dot–dashed from eq.(3.6) to estimate FSI.*

Corrections to σ_{NN} due to the quark structure of the nucleons may be relevant at high energies. The optical potential is sensitive to the interactions of the struck nucleon taking place within a fraction of $1\,fm$ of its interaction with the electron. It is then possible that these interactions differ significantly from the free N–N interactions. Such an effect is also discussed in the context of the *colour transparency*. This effect leads to a time dependence of the cross-section σ_{NN} that we have taken from ref.[15].

Fig.(4) gives the results for the inclusive cross–section of nuclear matter for two kinematics and with different prescriptions for FSI. The effect on the FSI due to correlations is much larger than that coming from *colour transparency*. This feature should be kept in mind when proposing experiments, in which the nucleus is planned to play the role of a detector of non–perturbative QCD phenomena.

Acknowledgements

The authors want to thank V.R.Pandharipande and I.Sick in collaboration with whom some of the results presented here have been obtained.

References

1- R.N.Silver, Phys.Rev. **B38**(1989)2283; A.S.Rinat, Phys.Rev. **B42**(1990)9944; C.Carraro and S.E.Koonin, Phys.Rev.Lett. **65**(1990)2792;

2- D.B.Day *et al.*, Phys.Rev. **C40**(1989)1011;

3- E.Feenberg, *Theory of Quantum Fluids*, Academic Press, N.Y., 1969;

4- J.Carlson, V.R.Pandharipande and R.Schiavilla *Modern Topics in Electron-Scattering*, World Scientific, Singapore, 1991; O.Benhar, A.Fabrocini and S.Fantoni, *ibidem*;

5- R.B.Wiringa, V.Ficks and A.Fabrocini, Phys.Rev. **C38**(1988)1010;

6- S.Fantoni, B.L.Friman and V.R.Pandharipande, Nucl.Phys. **A386**(1982)1;

7- S.Fantoni and V.R.Pandharipande, Nucl.Phys. **A427**(1984)473;

8- A.Fabrocini and S.Fantoni, Nucl.Phys. **A503**(1989)375;

9- O.Benhar, A.Fabrocini and S.Fantoni, Nucl.Phys. **A505**(1989)267;

10- S.Fantoni and V.R.Pandharipande, Phys.Rev. **C37**(1988)1697;

11- C.Ciofi dgegli Atti, Prog.in Part.and Nucl.Phys. **3**(1980)183;

12- C.Itzykson and J.B.Zuber, *Quantum Field Theory*, Mc Graw-Hill, New York, 1980;

13- T.deForest,Jr., Nucl.Phys. **A392**(1983)232;

14- O.Benhar, A.Fabrocini, S.Fantoni, G.A.Miller, V.R.Pandharipande and I.Sick, Phys.Rev C(1991), in press;

15- G.R.Farrar, H.Liu, L.L.Frankfurt and M.E.Strickman, Phys.Rev.Lett. **61**(1988)686.

VARIATIONAL MONTE CARLO CALCULATIONS OF FINITE NUCLEI

R. B. Wiringa

Physics Division
Argonne National Laboratory
Argonne, Illinois 60439 USA

INTRODUCTION

A major goal in nuclear physics is to understand how nuclear structure comes about from the underlying interactions between nucleons. This requires modelling nuclei as collections of strongly interacting particles. Using realistic nucleon-nucleon potentials, supplemented with consistent three-nucleon potentials and two-body electroweak current operators, we calculate nuclear ground-state properties, such as the binding energy, electromagnetic form factors, and momentum distributions. Other properties such as excited states and low-energy reactions are also calculable in this framework.

HAMILTONIAN

One problem in developing a microscopic picture of nuclear structure is that we do not have a working theory for the fundamental forces between nucleons. Although quantum chromodynamics gives some hope of eventually understanding the strong force, it has not yet been developed to the point where it can make quantitative predictions for nucleon-nucleon (NN) interactions. For studying many-body systems we generally rely on potential representations, where the choice of potential forms is guided by meson-exchange theory. To fit NN scattering data and deuteron properties requires a complicated operator structure for the NN potential. Further, meson-exchange theory and the existence of low-energy (≈ 300 MeV) nucleon resonances suggest that there should be significant many-body forces, which require three-nucleon (NNN) or higher-order potentials in a nucleons-only representation. In practice, the NN potential gives a much larger contribution to the energy and induces the major correlations in the nuclear wave function, but the NNN potential contributes a significant fraction of the total binding and is crucial for obtaining quantitative agreement with nuclear masses.

A realistic nuclear Hamiltonian can be written in the form:

$$H = \frac{-\hbar^2}{2m} \sum_i \nabla_i^2 + \sum_{i<j} V_{ij} + \sum_{i<j<k} V_{ijk} \, , \tag{1}$$

where V_{ij} is an NN potential fit to elastic scattering data and deuteron properties, and V_{ijk} is an NNN potential fit to many-body ground state energies. Most realistic NN potentials can be written in an operator form:

$$V_{ij} = \sum_{p=1,n} v_p(r_{ij}) O_{ij}^p \, , \tag{2}$$

where the first fourteen operators are:

$$O_{ij}^p = 1, \; \tau_i\cdot\tau_j, \; \sigma_i\cdot\sigma_j, \; (\sigma_i\cdot\sigma_j)(\tau_i\cdot\tau_j), \; S_{ij}, \; S_{ij}(\tau_i\cdot\tau_j), \; L\cdot S, \; L\cdot S(\tau_i\cdot\tau_j),$$
$$L^2, \; L^2(\tau_i\cdot\tau_j), \; L^2(\sigma_i\cdot\sigma_j), \; L^2(\sigma_i\cdot\sigma_j)(\tau_i\cdot\tau_j), \; (L\cdot S)^2, \; (L\cdot S)^2(\tau_i\cdot\tau_j) \,. \tag{3}$$

Examples are the Reid v_8 potential[1], which uses the first eight operators above, and the Argonne v_{14} potential[2], which uses all fourteen operators shown. In these models the expectation value for the first six operator components of V_{ij} is much larger than for the remaining parts, i.e.:

$$\langle \sum_{p=1,6} v_p(r_{ij})O_{ij}^p \rangle \gg \langle \sum_{p=7,14} v_p(r_{ij})O_{ij}^p \rangle \,. \tag{4}$$

Consequently the first six v_p terms should induce the main correlations in the wave function. Other realistic potentials, such as the Nijmegen[3] and Paris[4] models, use p^2 operators instead of, or in addition to, L^2 operators; however, the contribution of the latter v_p terms is significantly larger in these other models.

The meson-theoretic basis for the NN potential is illustrated in Fig. 1. The dominant long-range interaction is one-pion exchange, which has the operator structure:

$$V_{ij}^\pi = X_{ij}^\pi(\tau_i\cdot\tau_j) = [Y(r_{ij})\sigma_i\cdot\sigma_j + T(r_{ij})S_{ij}]\tau_i\cdot\tau_j \,. \tag{5}$$

At intermediate ranges, two-pion exchange, with the possible excitation of intermediate Δ resonances, is the dominant reaction. Analysis with interaction models that include explicit Δ degrees of freedom[2] show that these processes provide significant attraction, which is predominantly central in character, but includes important components for the first six operators of eq.(3). At shorter ranges the interaction may be dominated by the exchange of heavier mesons, such as the ρ and ω, or the quark substructure of nucleons may start to play some more complicated role; these features contribute to the spin-orbit and higher-order terms in the potential.

Realistic models for the NNN potential also have a non-trivial operator dependence, starting with a long-range two-pion-exchange part of the Fujita-Miyazawa form[5]:

$$V_{ijk}^{2\pi} = A \sum_{cyc} \left(\{X_{ij}^\pi,X_{ik}^\pi\}\{\tau_i\cdot\tau_j,\tau_i\cdot\tau_k\} + \frac{1}{4}[X_{ij}^\pi,X_{ik}^\pi][\tau_i\cdot\tau_j,\tau_i\cdot\tau_k] \right), \tag{6}$$

which is built up from the one-pion-exchange operators. In the Urbana NNN potentials[6] a phenomenological intermediate-range repulsive term is added:

$$V_{ijk}^R = U \sum_{cyc} T^2(r_{ij})T^2(r_{jk}) \tag{7}$$

which can be viewed as an interference of the intermediate-range attraction between two NN pairs sharing one nucleon. Meson-exchange diagrams for the NNN potential are shown in Fig. 2. The overall strengths of the attractive and repulsive parts, A and U, are adjusted to fit nuclear binding energies in many-body calculations. Below we report results for two models: Urbana VII, which was fit to earlier variational calculations of light nuclei and nuclear matter[7] and is used

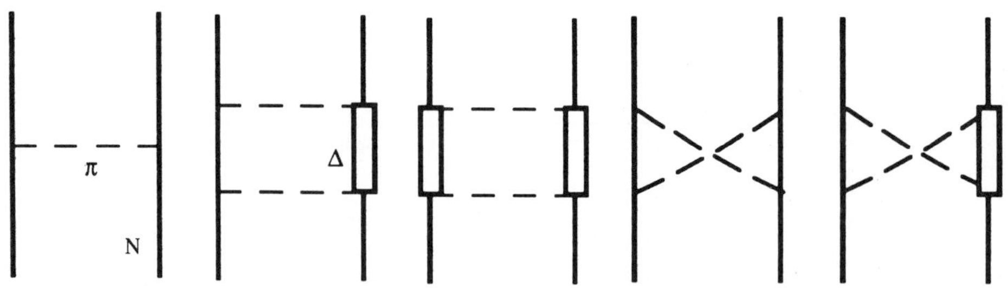

Fig. 1. Diagrams contributing to the NN potential.

here for closed-shell nuclei, and Urbana VIII, which was fit to more recent exact Faddeev and Green's function Monte Carlo (GFMC) calculations and is used here for few-body nuclei[8]. Other NNN potentials are available, such as the Tucson-Melbourne model[9], which has a more complete two-pion-exchange term, but no short-range repulsive part.

VARIATIONAL MANY-BODY THEORY

The variational method can be used to obtain approximate solutions of the many-body Schrödinger equation for Hamiltonians of the kind given above and for a wide range of nuclear systems: few-body nuclei[6-8] such as ^{3}H and ^{4}He, light nuclei[10] such as ^{16}O and ^{40}Ca, nuclear matter[11,12] and neutron stars[13]. A suitably parametrized trial function Ψ_v is used to calculate an upper bound to the energy using the Rayleigh-Ritz variational principle:

$$E_v = \frac{\langle \Psi_v | H | \Psi_v \rangle}{\langle \Psi_v | \Psi_v \rangle} \geq E_0. \tag{8}$$

The parameters in Ψ_v are varied to minimize E_v, and the lowest value is taken as the approximate ground-state energy. The corresponding Ψ_v can then be used to calculate other properties of interest. A better energy may be obtained by using Ψ_v as a starting point for a perturbation or GFMC calculation[14]. The key steps are always to 1) find a good ansatz for Ψ_v, and 2) accurately evaluate the expectation value $\langle H \rangle$.

The strong state-dependence of the interactions induces corresponding correlations in the wave function. A good trial function can be constructed from a product of correlation operators:

$$|\Psi_v\rangle = [\, 1 + \sum_{i<j} U_{ij}^{LS} + \sum_{i<j<k} U_{ijk}^{TNI}\,][S \prod_{i<j}(1+U_{ij})]|\Psi_J\rangle\,, \tag{9}$$

where Ψ_J is a Jastrow wave function, and U_{ij}, U_{ij}^{LS}, and U_{ijk}^{TNI} are two- and three-body correlation operators:

$$|\Psi_J\rangle = \prod_{i<j} f_c(r_{ij})\, |\Phi_A(JMTT_3)\rangle\,, \tag{10}$$

$$U_{ij} = \sum_{p=2,6} [\, \prod_{k \neq i,j} f_3(r_{ij}, r_{jk}, r_{ki})\,]\, u_p(r_{ij}) O_{ij}^p\,, \tag{11}$$

$$U_{ij}^{LS} = \sum_{p=7,8} [\, \prod_{k \neq i,j} f_3(r_{ij}, r_{jk}, r_{ki})\,]\, u_p(r_{ij}) O_{ij}^p\,, \tag{12}$$

$$U_{ijk}^{TNI} = \varepsilon\, V_{ijk}(\tilde{r}_{ij}, \tilde{r}_{jk}, \tilde{r}_{ki})\,. \tag{13}$$

The operators O_{ij}^p are the same as in eq.(3); since they are non-commuting, the symmetrization operator S is required in the product of eq.(9). The first six operators have a nice closed algebra and are conveniently treated together in eq.(11). The spin-orbit (LS) correlations of eq.(12) are

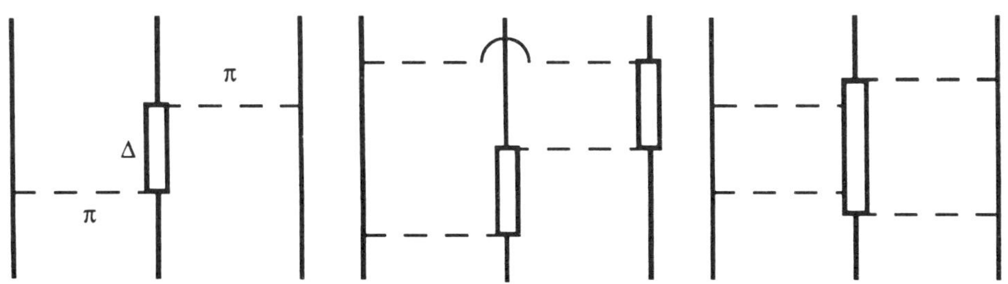

Fig. 2 Diagrams contributing to the NNN potential.

more expensive to compute because of their gradient operators and only a linear term is used in eq.(9), while no correlations have been introduced for L^2 or higher terms in the interaction; this is reasonable for potentials with the property of eq.(4). The $f_3(r_{ij},r_{jk},r_{ki})$ in eqs.(11)-(12) is a supplemental three-body correlation that reduces the noncentral correlation when a third particle comes between the correlated pair. The three-nucleon interaction (TNI) correlation of eq.(13) includes all the operator dependence of eq.(6); ε is a small negative number and $\tilde{r}$ a scaled radial variable.

The Φ is an antisymmetric single-particle state, with appropriate properties for the system of interest, e.g., the quantum numbers $JMTT_3$ for a particular nucleus. For few-body nuclei we have used a single-particle Φ with spin-isospin indices and no spatial dependence, e.g., for ^{4}He:

$$|\Phi_4(JMTT_3)\rangle = |\Phi_\alpha(0000)\rangle = A\, |p\!\uparrow p\!\downarrow n\!\uparrow n\!\downarrow\rangle\,. \tag{14}$$

For closed-shell nuclei like ^{16}O and ^{40}Ca we use a product of four determinants, one for spin-up protons, one for spin-down protons, etc:

$$|\Phi_{16}(JMTT_3)\rangle = |\Phi_O(0000)\rangle = |\ |\phi_{lm}|p\!\uparrow\ |\phi_{lm}|p\!\downarrow\ |\phi_{lm}|n\!\uparrow\ |\phi_{lm}|n\!\downarrow\ \rangle\,. \tag{15}$$

For ^{16}O each determinant is constructed from one 1s (ϕ_{00}) and three 1p (ϕ_{1m}) radial functions, while ^{40}Ca has additional 2s and 1d orbitals. For open-shell nuclei like ^{12}C a fully antisymmetrized Φ_A is much more complicated, being a sum of many products of determinants.

The central $f_c(r)$ and noncentral $u_p(r)$ pair correlation functions reflect the influence of the two-body potential at short distances, while satisfying asymptotic boundary conditions of cluster separability. Reasonable functions are generated by minimizing the two-body cluster energy of a somewhat modified interaction ($\overline{V}-\lambda$), which contains a number of variational parameters. This leads to a set of eight coupled differential equations for the first eight operators[8]. The f_c and $u_6=u_{t\tau}$ are shown in Fig. 3 for several nuclei. Here $f_c(r)$ is small at short distances, to reduce the contribution of the repulsive core of the NN potential, and peaks at an intermediate distance corresponding to the maximum attraction of the NN potential. For light nuclei f_c falls off at larger distances to keep the system confined. For closed shell nuclei confinement is provided by the one-body correlations and f_c goes to a constant value at large distances. The noncentral $u_p(r)$ are all relatively small: the most important is the long-range tensor-isospin part $u_{t\tau}$, which is induced mainly by the one-pion-exchange part of the potential.

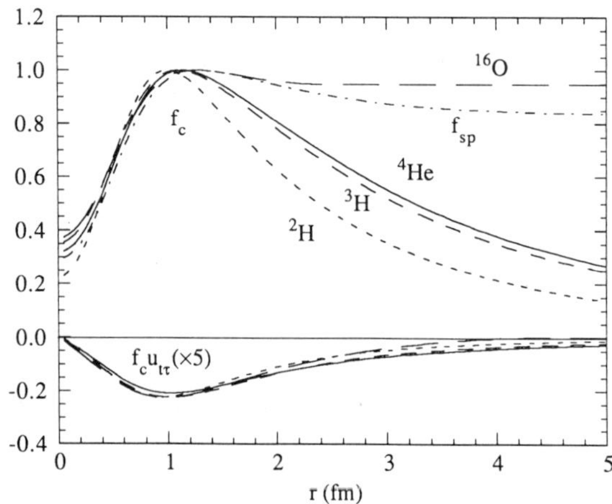

Fig. 3 Correlation functions.

In calculations of five- and six-body nuclei the first four nucleons fill the 1s shell and have no single-particle radial function, while the fifth and sixth nucleons go into the 1p shell. We allow for the possibility of different central correlations f_{ss}, f_{pp}, and f_{sp} for pairs in the s-shell, pairs in the p-shell, and mixed pairs, respectively. For five nucleons,

$$|\Psi_J\rangle = A \left\{ \prod_{1\leq i<j\leq 4} f_{ss}(r_{ij}) \prod_{1\leq k\leq 4} f_{sp}(r_{k5}) |\Phi_\alpha(0000)\times\Phi_p(JMTT_3)\rangle \right\}, \tag{16}$$

$$|\Phi_p(JMTT_3)\rangle = \phi_p(R_{5\alpha}) [Y_1^{m_l}(\Omega_{5\alpha})\times\chi_5(\tfrac{1}{2}m_s)]_{JM}\, \nu_5(\tfrac{1}{2}T_3). \tag{17}$$

There is an explicit antisymmetrization over the five particles, and ϕ_p is a single-particle correlation with $R_{i\alpha} = r_i - R_\alpha^{cm}$. The five-body nuclei ^{5}He and ^{5}Li are not stable against breakup into an α and a nucleon, but we can study the scattering in both $J=3/2$ and $J=1/2$ states, using an R-matrix approach to minimize the energy inside a region with boundary conditions that correspond to an asymptotic scattering state[15]. The difference in energy between these states is a measure of the spin-orbit splitting in nuclei, which is the key to the nuclear shell model.

The six-body nuclei of interest are ^{6}He and ^{6}Li with $JMTT_3 = (001\text{-}1)$ and (1100). They are weakly bound and easily broken into an α and a two-body cluster; ^{6}Li (^{6}He) is only 1.5 MeV (1 MeV) more bound than a separated α and deuteron ($^1S_0$ nn pair). However breakup into a five- plus one-body system costs ~6 MeV because the five-body nuclei are unstable. This dual character can probably best be modelled by a linear combination of shell-model and α-cluster correlations:

$$|\Psi_J\rangle = A \left\{ \prod_{1\leq i<j\leq 4} f_{ss}(r_{ij}) \prod_{1\leq k\leq 4} f_{sp}(r_{k5})f_{sp}(r_{k6})\, f_{pp}(r_{56})\, |a\Phi_{sm}+ b\Phi_{\alpha c}\rangle \right\}, \tag{18}$$

where the shell model wave function is:

$$|\Phi_{sm}\rangle = \Phi_\alpha(0000) \times \phi_p(R_{5\alpha})\phi_p(R_{6\alpha}) \times \tag{19}$$

$$\left\{ [Y_1^{m_l}(\Omega_{5\alpha})\times Y_1^{m_l}(\Omega_{6\alpha})]_{LM_L} \times [\chi_5(\tfrac{1}{2}m_s)\times\chi_6(\tfrac{1}{2}m_s')]_{SM_S} \right\}_{JM} [\nu_5(\tfrac{1}{2}t_3)\times\nu_6(\tfrac{1}{2}t_3)]_{TT_3},$$

while the α-cluster correlation has s- and d-wave components induced by a tensor-like operator:

$$|\Phi_{\alpha c}\rangle = \Phi_\alpha(0000) \times \tag{20}$$

$$\left\{ \left(u(R_{\alpha c}) + w(R_{\alpha c}) [3(\sigma_5\cdot\hat{R}_{\alpha c})(\sigma_6\cdot\hat{R}_{\alpha c}) - (\sigma_5\cdot\sigma_6)] \right) \chi_c(SM_s) \right\}_{JM} \nu_c(TT_3).$$

Here $R_{\alpha c}$ is the separation between the α and cluster centers of mass. The $f_{ss}(r) \approx f_c(^4\text{He})$, $f_{pp}(r) \approx f_c(^2\text{H})$, and $f_{sp}(r)$ is shown in Fig. 3.

CALCULATIONS OF FEW-BODY NUCLEI

The method used to evaluate the energy expectation value varies with the size of the system under consideration. For small systems, $A\leq8$, essentially exact integration is possible using Monte Carlo sampling[16]. For larger nuclei, $8\leq A\leq40$, a cluster expansion is required[10], with Monte Carlo techniques used to evaluate the terms in the expansion. For nuclear matter a cluster expansion is also used, but so far integral equation techniques have been used to sum terms[11-13].

We represent $\Psi_v(R)$ as a vector in spin-isospin space, with $2^A \times [A!/Z!(A-Z)!]$ complex numbers, i.e., a coefficient for each state with definite third components of spin and isospin. The Metropolis Monte Carlo algorithm is used to perform the 3A-dimensional integrals. Sampling is done both on the set of positions, $R=(\vec{r}_1,...\vec{r}_A)$, and the order of operators in the symmetrized product of eq.(9). For few-body nuclei a complete sum over all the spin-isospin variables is made. The computational effort required for this complete sum increases rapidly with A. The time required to calculate the binding energy with a reasonable statistical error for one

nucleus	Hamiltonian	method	binding	breakup
^{3}H	nature	experiment	8.48	
	Reid v_8	variational MC	7.31(2)	
		GFMC	7.54(10)	
		34-ch. Faddeev	7.59	
	Argonne v_{14}	variational MC	8.21(2)	
	+ Urbana VIII	34-ch. Faddeev	8.49	
^{4}He	nature	experiment	28.3	
	Reid	coupled cluster	24.	
	Reid v_8	variational MC	23.6(1)	
		GFMC	24.5(1)	
	Argonne v_{14}	variational MC	27.2(1)	
	+ Urbana VIII	GFMC	28.3(2)	
^{5}He($J=\frac{3}{2}$)	nature	experiment	27.2	$-1.1(\alpha+n)$
	Argonne v_{14}	variational MC	25.1(1)	-2.1
	+ Urbana VIII	GFMC	26.8(2)	-1.5
^{5}He($J=\frac{1}{2}$)	nature	experiment	25.8	$-2.5(\alpha+n)$
	Argonne v_{14}	variational MC	24.1(2)	-3.1
	+ Urbana VIII	GFMC	25.6(2)	-2.7
^{6}Li	nature	experiment	32.0	$1.5(\alpha+d)$
	Argonne v_{14}	variational MC	27.5(3)	-1.9
	+ Urbana VIII			

trial function grows from about 3 minutes for ^{3}H to nearly 4 hours for ^{6}Li on a Cray-YMP. Thus direct integration will probably be limited to A≤8 nuclei.

The results of variational calculations[8] for ground-state binding energies of ^{3}H and ^{4}He are shown in Table 1 along with preliminary results for ^{5}He and ^{6}Li. For comparison we also show 34-channel Faddeev results[17] for ^{3}H and GFMC results[14] for ^{3}H, ^{4}He and ^{5}He. There is also an older coupled-cluster result[18] for ^{4}He. The variational trial function gives upper-bound energies that are 3-4% above the exact results for ^{3}H and ^{4}He. These and other calculations show that realistic NN potentials alone do not give enough binding for few-body nuclei, but it is possible to pick a supplemental NNN potential that will give the correct energy in exact calculations.

For ^{5}He the $J=3/2$ and $J=1/2$ scattering states have been calculated with a boundary condition that has a node in the $\alpha+n$ scattering wave function at 12.5 fm. The corresponding "experimental" energies, which are obtained from phase shift analysis, are 1.1 and 2.5 MeV above the α binding energy, for a spin-orbit splitting of 1.4 MeV. The variational results for the Argonne v_{14} + Urbana VIII model are 2.1 and 3.1 MeV above the variational α binding energy giving a splitting of 1 MeV. The GFMC results are 1.5 and 2.7 MeV above the GFMC α, which gives 1.2 MeV or 85% of the "experimental" spin-orbit splitting with this Hamiltonian. The variational results required 25 hours of Cray-YMP time, including the parameter search, while the GFMC calculation required 200 hours, starting from a reasonable variational trial function.

For ^{6}Li only variational results are available at present, and they are unsatisfactory. The results shown here are for a trial function that gives the correct experimental charge radius, and while the energy is not terrible, it is nearly 2 MeV above that of the corresponding separated α and deuteron. If the radius is unconstrained the trial function expands and the energy asymptotically approaches that of an α plus a deuteron with some residual Coulomb repulsion between them.

CALCULATIONS OF LARGER NUCLEI

For larger nuclei we use a cluster expansion for the noncentral correlation operators U_{ij}, etc., to evaluate the energy expectation value. Consider the expansion for the NN potential:

$$\langle \sum_{i<j} V_{ij} \rangle = \frac{\sum_{i<j} n_{ij} + \sum_{i<j<k} n_{ijk} + \ldots + n_{12\ldots A}}{1 + \sum_{i<j} d_{ij} + \sum_{i<j<k} d_{ijk} + \ldots + d_{12\ldots A}} \tag{21}$$

where the numerator and denominator cluster terms are given by:

$$n_{ij} = \langle \Psi_J | (1+U_{ij}^\dagger) V_{ij} (1+U_{ij}) | \Psi_J \rangle \ , \tag{22}$$

$$d_{ij} = \langle \Psi_J | (1+U_{ij}^\dagger)(1+U_{ij}) | \Psi_J \rangle - 1 \ , \tag{23}$$

$$n_{ijk} = \langle \Psi_J | [S \sum_{cyc} (1+U_{ij}^\dagger)](1+U_{ijk}^\dagger) V_{ij} (1+U_{ijk})[S \sum_{cyc} (1+U_{ij})] | \Psi_J \rangle - \sum_{cyc} n_{ij} \ , \tag{24}$$

etc. The brackets denote a full 3A-dimensional integration, with the central correlations and antisymmetry of the Jastrow wave function treated completely. This may be regrouped as a linked cluster expansion:

$$\langle \sum_{i<j} V_{ij} \rangle = \sum_{i<j} c_{ij} + \sum_{i<j<k} c_{ijk} + \ldots + c_{12\ldots A} \ , \tag{25}$$

$$c_{ij} = n_{ij}/(1+d_{ij}) \ , \tag{26}$$

$$c_{ijk} = [n_{ijk} - \sum_{cyc} c_{ij}(d_{jk}+d_{ki}+d_{ijk})]/(1+\sum_{cyc} d_{ij}+d_{ijk}) \ . \tag{27}$$

The expectation values of these individual cluster terms, n_{ij}, d_{ij}, etc., are evaluated using the same Monte Carlo techniques used for few-body nuclei. However, the spin-isospin algebra is only needed for the particles in the cluster that have noncentral correlation operators. The effort required to evaluate the energy is thus much reduced. The final ^{16}O energy evaluation reported below still required some 80 hours of Cray-2 time, after the trial function had been optimized.

The cluster expansion is not useful unless it converges rapidly. The convergence for ^{16}O is reasonably good at the four-body cluster level, as shown in Table 2, where the contributions of the kinetic energy, NN and NNN potentials are shown at the one-, two-, three-, and four-body cluster levels. The next to last column gives the sum of these terms through the four-body cluster

Table 2. Convergence of cluster expansion for ground-state energy of ^{16}O with Argonne v_{14} + Urbana VII interaction. Energies are in MeV/nucleon.

term	1-body	2-body	3-body	4-body	Σ 1-4	Σ 1-16
T	18.8±0.1	17.2±0.1	−1.7±0.1	0.2±0.2	34.5±0.3	34.5
V_{ij}		−46.8±0.2	8.0±0.1	−1.0±0.2	−39.8±0.2	−39.7
V_{ijk}			−4.0±0.1	2.2±0.1	−1.8±0.1	−2.6
H	18.8±0.1	−29.6±0.1	2.3±0.1	1.4±0.1	−7.1±0.1	−7.8

Table 3. Binding and breakup energies (in MeV/nucleon) for light nuclei with different Hamiltonians and calculational methods.

nucleus	Hamiltonian	method	binding	breakup
^{16}O	nature	experiment	8.0	0.9 (4α)
	Reid	coupled cluster	5.0	−1.0
	Reid v_6	variational MC	6.2	
	Argonne v_{14} + Urbana VII	cluster MC	7.8	0.2
^{40}Ca	nature	experiment	8.5	
	Reid	coupled cluster	6.0	
	Argonne v_{14} + Urbana VIII	cluster MC	8.6 (±1.5)	

level, while the last column gives an extrapolated estimate for the total energy through the sixteen-body cluster level. The kinetic energy converges rapidly, as does the NN potential contribution. The NNN potential has not converged at the four-body level, however, and there is a significant extrapolation in this term. An alternate cluster expansion, which may be more useful for open-shell nuclei like ^{12}C, is under development; in ^{16}O it gives a similar extrapolated result for the V_{ijk} expectation value.

The ground-state binding energy per nucleon of ^{16}O and ^{40}Ca is given in Table 3. A previous calculation of ^{16}O using the coupled cluster method and the Reid potential has a problem similar to the ^{6}Li calculation described above; the nucleus is less bound per nucleon than ^{4}He. A calculation sampling all cluster sizes with a variational wave function for the Reid v_6 interaction also gave too little binding[19]. In the present calculation, with a realistic interaction including a three-nucleon potential, ^{16}O is slightly more bound per nucleon than ^{4}He, thus demonstrating stability for the first time. The result is also quite close to the experimental value, as is the preliminary result for ^{40}Ca with the same Hamiltonian. However, a variational search has not been made for the best trial function in ^{40}Ca and the present result has a large statistical uncertainty, as shown in the table.

ELECTROMAGNETIC FORM FACTORS AND MOMENTUM DISTRIBUTIONS

Other properties of interest which we have studied include the electromagnetic form factors and nucleon momentum distributions. The form factors have significant contributions from two-body charge and current operators attributable to meson exchange (MEC). These include both a "model-independent" (MI) part which is fixed by requiring current conservation with the chosen V_{ij}, and a "model-dependent" (MD) part that includes currents associated with $\rho\pi\gamma$, $\omega\pi\gamma$, and Δ-excitation mechanisms.[20] Results for the magnetic moments of ^{3}H and ^{3}He are given in Table 4 and for the charge form factor of ^{4}He in Fig. 4. These include impulse approximation (IA) calculations with and without MEC for both variational and exact calculations. The results show that the few-body variational trial functions are good for more than just the energy, and that the MEC are essential for quantitative agreement with data. The charge form factor for ^{16}O is also

Table 4. Magnetic moments for A=3 nuclei.

	^{3}H		^{3}He	
	variational	Faddeev	variational	Faddeev
IA	2.586	2.571	−1.778	−1.765
IA+MI	2.983	2.955	−2.139	−2.115
IA+MI+MD	3.058	3.029	−2.202	−2.175
experiment		2.979		−2.127

shown in Fig. 4; here the present variational wave function is not as satisfactory and the charge radius is too small.

The nucleon momentum distribution is given by the integral:

$$\rho_N(k) = \int d\mathbf{R} \sum_i \int d\vec{r}_i{}' \; \Psi(\mathbf{R}') \; \exp[-i\vec{k}\cdot(\vec{r}_i-\vec{r}_i{}')] \; \Psi(\mathbf{R}) \;, \qquad (28)$$

where $\mathbf{R}'=(\vec{r}_1,...,\vec{r}_i{}',...\vec{r}_A)$. The operator expectation values are calculated by direct integration in the few-body nuclei, but a cluster expansion is again needed for larger systems. Results for ^{4}He and ^{16}O are shown in Fig. 5 along with the exact deuteron and a correlated basis function calculation of nuclear matter[21]. All these systems show a remarkably similar high-momentum tail, which is indicative of the strong nuclear correlations; this tail may be useful in understanding phenomena such as low-energy pion and antiproton production.

CONCLUSIONS

Good progress has been made in the microscopic study of nuclear structure for few-body nuclei and nuclear matter in recent years. Current work is now concentrating on the ground states of intermediate size systems, $4<A\leq40$, where the number of nucleons is large enough to make calculations difficult, yet small enough that the finite nature of the system is an important aspect of the physics. Progress is also being made in a variety of few-nucleon reactions, such as the ^{3}He$(n,\gamma)^4$He and ^{3}He$(p,e^+\nu_e)^4$He electroweak capture reactions[22].

The difficulty in explaining the stability of light nuclei may be due to inadequacies in the variational ansatz, or it may be due to the crude parametrization of the NNN potential. Obtaining a consistent description of nuclear systems with realistic interactions remains a challenging problem. Continued progress will require advances in the many-body Hamiltonian, the many-body theory, and the available computational resources.

ACKNOWLEDGMENTS

The work described here has been carried out in collaboration with J. Carlson, V. R. Pandharipande, S. C. Pieper, D. O. Riska, and R. Schiavilla. Computations have been made at the National Energy Research Supercomputer Center in Livermore, California, and at the

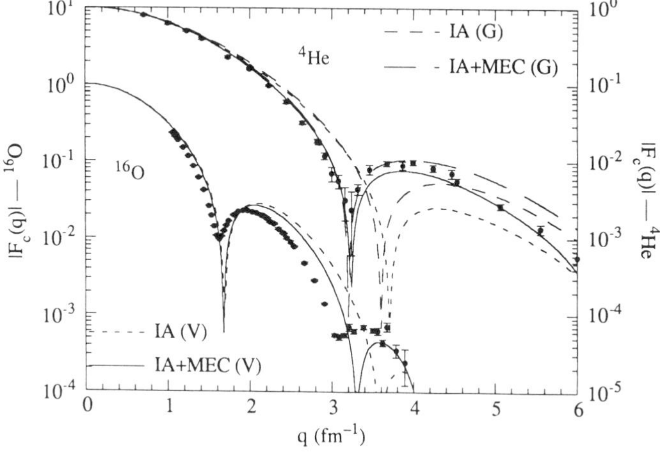

Fig. 4 Charge form factors for ^{4}He and ^{16}O.

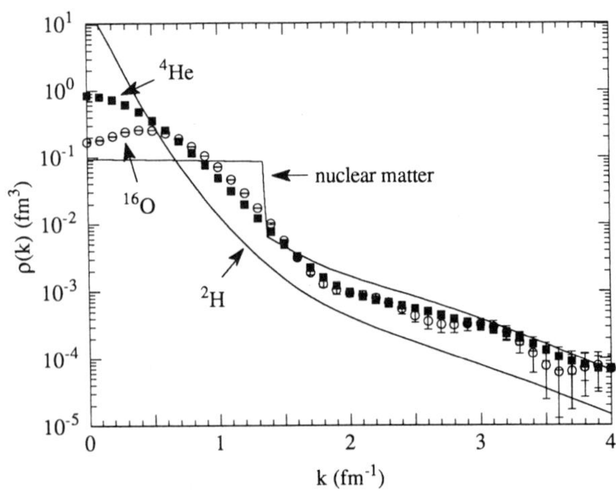

Fig. 5 Nucleon momentum distributions.

National Center for Supercomputer Applications in Urbana, Illinois. This work is supported by the U.S. Department of Energy, Nuclear Physics Division, under contract W-31-109-ENG-38.

REFERENCES

1. R. V. Reid, Ann. Phys. (N.Y.) **50**, 411 (1968).
2. R. B. Wiringa, R. A. Smith, and T. L. Ainsworth, Phys. Rev. C **29**, 1207 (1984).
3. M. M. Nagles, T. A. Rijken, and J. J. de Swart, Phys. Rev. D **17**, 768 (1978).
4. M. Lacombe, B. Loiseau, J. M. Richard, R. Vinh Mau, J. Côte, P. Pirés, and R. de Tourreil, Phys. Rev. C **21**, 861 (1980).
5. J. Fujita and H. Miyazawa, Prog. Theor. Phys. **17**, 360 (1957).
6. J. Carlson, V. R. Pandharipande, and R. B. Wiringa, Nucl. Phys. **A401**, 59 (1983).
7. R. Schiavilla, V. R. Pandharipande, and R. B. Wiringa, Nucl. Phys. **A449**, 219 (1986).
8. R. B. Wiringa, Phys. Rev. **43**, 1585 (1991).
9. S. A. Coon, M. D. Scadron, P. C. McNamee, B. R. Barrett, D. W. E. Blatt, and B. H. J. McKellar, Nucl. Phys. **A317**, 242 (1979).
10. S. C. Pieper, R. B. Wiringa, and V. R. Pandharipande, Phys. Rev. Lett. **64**, 364 (1990).
11. V. R. Pandharipande and R. B. Wiringa, Rev. Mod. Phys. **51**, 821 (1979).
12. I. E. Lagaris and V. R. Pandharipande, Nucl. Phys. **A359**, 349 (1981).
13. R. B. Wiringa, V. Fiks, and A. Fabrocini, Phys. Rev. C **38**, 1010 (1988).
14. J. Carlson, Phys. Rev. C **38**, 1879 (1988); Los Alamos Report No. LA-UR-90-2088 (1990); private communication.
15. J. Carlson, K. E. Schmidt, and M. H. Kalos, Phys. Rev. C **36**, 27 (1987).
16. J. Carlson and R. B. Wiringa, in *Computational Nuclear Physics 1*, edited by K. Langanke, J. A. Maruhn, and S. E. Koonin (Springer Verlag, Berlin, 1991).
17. C. R. Chen, G. L. Payne, J. L. Friar, and B. F. Gibson, Phys. Rev. C **33**, 1740 (1986); private communication.
18. H. Kümmel, K. H. Lührmann, and J. G. Zabolitzky, Phys. Rep. **36C**, 1 (1978).
19. J. Carlson and M. H. Kalos, Phys. Rev. C **32**, 2105 (1985).
20. R. Schiavilla, V. R. Pandharipande, and D. O. Riska, Phys. Rev. C **40**, 2294 (1989); Phys. Rev. C **41**, 309 (1990).
21. S. Fantoni and V. R. Pandharipande, Nucl. Phys. **A427**, 473 (1984).
22. J. Carlson, D. O. Riska, R. Schiavilla, and R. B. Wiringa, Phys. Rev. C **42**, 830 (1990); Phys. Rev. C**44**, 619 (1991).

RELATIVISTIC NUCLEAR MANY-BODY THEORY

Brian D. Serot

Physics Department and Nuclear Theory Center
Indiana University
Bloomington, IN 47405

John Dirk Walecka

Continuous Electron Beam Accelerator Facility
12000 Jefferson Avenue
Newport News, VA 23606

ABSTRACT

Nonrelativistic models of nuclear systems have provided important insight into nuclear physics. In future experiments, nuclear systems will be examined under extreme conditions of density and temperature, and their response will be probed at momentum and energy transfers larger than the nucleon mass. It is therefore essential to develop reliable models that go beyond the traditional nonrelativistic many-body framework. General properties of physics, such as quantum mechanics, Lorentz covariance, and microscopic causality, motivate the use of quantum field theories to describe the interacting, relativistic, nuclear many-body system. Renormalizable models based on hadronic degrees of freedom (*quantum hadrodynamics*) are presented, and the assumptions underlying this framework are discussed. Some applications and successes of quantum hadrodynamics are described, with an emphasis on the new features arising from relativity. Examples include the nuclear equation of state, the shell model, nucleon–nucleus scattering, and the inclusion of zero-point vacuum corrections. Current issues and problems are also considered, such as the construction of improved approximations, the full role of the quantum vacuum, and the relationship between quantum hadrodynamics and quantum chromodynamics. We also speculate on future developments.

INTRODUCTION AND MOTIVATION

The study of atomic nuclei plays an important role in the development of many-body theories. Since early experimental probes of the nucleus were limited to energy scales considerably less than the nucleon mass $M \approx 939\,\mathrm{MeV}/c^2$, the nucleus was initially described as a collection of nonrelativistic nucleons interacting through an instantaneous two-body potential, with the dynamics given by the Schrödinger equation. In this approach, the two-body potential is fit to the empirical properties of the deuteron and low-energy nucleon–nucleon (NN) scattering, and one then attempts to predict the properties of many-nucleon systems. This is a difficult problem, because the NN potential is strong, short-ranged ($R \approx 1\,\text{fm}$), and has a very repulsive central core. Nevertheless, over a period of many years and with the advent of more and more powerful computers, reliable methods have been developed for solving the nonrelativistic nuclear many-body problem. At present, essentially exact solutions exist for both the three-nucleon system and for "nuclear matter," the hypothetical uniform system with equal numbers of neutrons and protons ($N = Z$) obtained by turning off the Coulomb potential and letting the total number of baryons $B = N + Z$ go to infinity. Similar results for up to 8 (and possibly 16) nucleons should be obtained in the near future.

On balance, these microscopic nonrelativistic calculations are reasonably successful. With modern NN potentials, the ^{3}He system is underbound by $\approx 0.8\,\text{MeV}$ (about 10%),[1] and the calculated rms charge radius is $\approx 9\%$ too large. The predicted energy/nucleon of nuclear matter at equilibrium is close to the empirical value

$$(E/B) - M \approx -16\,\text{MeV} , \tag{1}$$

but the saturation density, which is related to the Fermi wavenumber k_{F} through $\rho_{\text{B}} = B/V = 2k_{\text{F}}^3/3\pi^2$, is too high, as the calculated value is $k_{\text{F}} \approx 1.5\,\text{fm}^{-1}$, while the empirical result is $k_{\text{F}} \approx 1.3\,\text{fm}^{-1}$. In addition, some general features of nuclear structure are qualitatively understood, for example, the importance of the Pauli principle in reducing NN correlations, which justifies both the shell model and the single-nucleon optical potential, and the interplay of single-particle and collective degrees of freedom, which determines the shape of the nucleus. By adding a small, phenomenological, density-dependent interaction to the free NN potential, one can quantitatively reproduce the single-particle structure and charge and mass densities of a large number of nuclei.[2] Furthermore, by using nuclear current operators constructed from the properties of free nucleons, an accurate picture of weak and electromagnetic interactions in nuclei has emerged. These successful results have stimulated the development of a diverse array of nuclear models that can be applied to nuclear structure and reactions throughout the periodic table.[3]

In spite of these successes, there are important reasons for developing a relativistic description of nuclear systems. For example, the properties of neutron stars depend on the neutron matter equation of state at densities up to an order of magnitude higher than those observed in ordinary nuclei; nucleon velocities are certainly relativistic at these densities. In addition, although the empirical low-energy NN scattering amplitude has conventionally been parametrized using Galilean invariants, a decomposition using Lorentz invariants reveals surprising new features. In particular, the empirical Lorentz scalar, vector, and pseudoscalar amplitudes are much larger than the amplitudes deduced from the nonrelativistic decomposition. These large amplitudes have important consequences for the spin-, velocity-, and density-dependence of the NN interaction and are at the heart of recent relativistic descriptions of nucleon–nucleus scattering that reproduce spin observables in a very economical fashion.[4–6]

Furthermore, medium-energy accelerators that probe nuclei at distance scales less than 1 fm reveal that the conventional framework is inadequate for a complete understanding of ordinary nuclei. For example, experiments on the electrodisintegration of the deuteron show unambiguously the presence of pion-exchange currents, which arise when the incoming (virtual) photon couples to a pion being exchanged between two nucleons. In addition, medium-energy pion–nucleus and proton–nucleus scattering indicate the importance of baryon resonances, such as the $\Delta(1232)$, in nuclear reactions. Thus both the dynamical nature of the NN interaction and the structure of the nucleon itself become important at an energy scale of several hundred MeV. As supporting evidence for this new dynamics, accurate microscopic meson-exchange models have been constructed to describe the NN interaction. These contain several mesons, the most important of which are the $\pi(0^-,1)$, $\sigma(0^+,0)$, $\omega(1^-,0)$, and $\rho(1^-,1)$, where the indicated quantum numbers denote spin, parity, and isospin, together with both N and Δ degrees of freedom. These meson-exchange models reproduce the large Lorentz-invariant scattering amplitudes mentioned earlier. Nevertheless, if one adheres to a nonrelativistic many-body description of nuclear structure, the new dynamical features can be incorporated into the conventional framework only by reducing them to an approximate form that is appropriate for the Schrödinger equation.

In future experiments, a new generation of accelerators will allow us to study nuclei at higher energies, at shorter distances, and with greater precision than ever before. For example, proton–nucleus collisions will provide complete spin information on both the projectile and the target, electron–nucleus scattering will sample distance scales down to tenths of a Fermi, and ultra-relativistic heavy ion collisions may produce nuclear densities of 10 times equilibrium density at temperatures of 100 to 200 MeV. These experiments will clearly involve physics that goes far beyond the Schrödinger equation, such as relativistic motion of the nucleons, dynamical meson exchanges and baryon resonances, modifications of hadron structure in the nucleus, and the dynamics of the quantum vacuum, which may include the production of a quark-gluon plasma. The challenge to theorists is to develop techniques that describe this new physics, while maintaining the important general properties of quantum mechanics, covariance, electromagnetic gauge invariance, and microscopic causality.

Since quantum chromodynamics (QCD) of quarks and gluons appears to be the fundamental theory of the strong interaction, it is natural to look to QCD as the means to describe this new physics. While this may be desirable in principle, there are many difficulties in practice, primarily because the QCD coupling is strong at distance scales relevant for the vast majority of nuclear phenomena. Although significant progress has been made in the last 10 years in performing strong-coupling lattice calculations, actual QCD predictions at nuclear length scales with the precision of existing (and anticipated) data are not presently available. This situation will probably persist for some time, particularly with regard to many-nucleon systems. Even if it becomes possible to use QCD to describe many-nucleon systems, this description is likely to be awkward, since quarks cluster into hadrons at low energies, and hadrons (not quarks or gluons) are the degrees of freedom actually observed in experiments.

In contrast, a description based on hadronic degrees of freedom is attractive for several reasons. First, these variables are the most efficient at normal densities and low temperatures, and for describing particle absorption and emission. Moreover, hadronic calculations can be calibrated by requiring that they reproduce empirical nuclear properties and scattering amplitudes; we can then extrapolate to the extreme situations mentioned earlier. Finally, although hadronic models must ultimately fail when the quark and gluon dynamics becomes essential, we must understand the limitations of hadronic models to isolate and identify true signatures of subhadronic dynamics.

Our basic goal is therefore to formulate a consistent microscopic treatment of nuclear systems using hadronic (baryon and meson) degrees of freedom. Since the physical phenomena of interest are relativistic and involve particle production and absorption, the only known consistent framework for their description is relativistic quantum field theory based on a local, Lorentz-invariant lagrangian density. We will refer to these hadronic relativistic field theories as *quantum hadrodynamics*, or QHD.[7] QHD is consistent in the same sense as any other relativistic quantum field theory: the assumptions about the relevant degrees of freedom, the form of the lagrangian, and the empirical data that determine the parameters are made at the outset, and one then attempts to extract concrete predictions from the implied formalism. In principle, these assumptions permit the formulation of "conserving approximations"[8,9] that maintain the important general properties mentioned above. Calculations can then be compared to the data to decide where QHD succeeds and where it fails.

We emphasize, however, that QHD still contains strong couplings, as does any treatment of the nuclear many-body problem. It is therefore necessary to develop reliable nonperturbative approximations, so that unambiguous comparisons between theory and experiment can be made. The formulation of practical, reliable techniques for finite-density calculations in strong-coupling relativistic quantum field theories is basically an unsolved problem.[10–12] The development of such tools in a hadronic field theory is not only useful in its own right, but it may also provide insight into similar approaches for QCD. Nuclear many-body theory has had such influence in other areas of physics in the past.

We shall also require that QHD be a renormalizable theory. This means that any QHD model can be characterized by a finite number of coupling constants and masses.[13,14] Thus the model is self-contained, and calculations can be carried out beyond the "tree level" without introducing additional parameters (such as vertex cutoffs) determined solely by short-distance phenomena. This minimizes the sensitivity of calculated results to short-distance input. The dynamical assumption underlying renormalizability is that the quantum vacuum and the internal structure of the hadrons can be described in terms of hadronic degrees of freedom alone. This assumption must ultimately break down, since at very short distances, hadrons are composed of quarks and gluons. For QHD to be useful, nuclear observables of interest must not be dominated by contributions from short distances, where QHD is inappropriate. This conjecture must be tested, and its limitations uncovered, by performing detailed calculations in a consistent relativistic framework.

There are alternative methods for describing hadronic and nuclear systems that go beyond the Schrödinger equation. Several meson-exchange models of the NN interaction have been developed and applied to nuclear systems, both relativistically and nonrelativistically.[15–17] These models focus primarily on the precise reproduction of the two-nucleon data, and they contain many mesonic degrees of freedom and adjustable parameters. These models are not renormalizable, and the number of parameters and the form of the interactions are unconstrained, except by the desire to reproduce the data. Typically, the treatment of the quantum vacuum is either incomplete or neglected entirely,[18] making it impossible to formulate conserving approximations. These features make it difficult to distinguish between hadronic dynamics and dynamics that arises from the underlying QCD degrees of freedom.

For these reasons, we will base our analysis on QHD as defined above, which presumably provides the correct description of many-baryon systems at large distances. Renormalizability restricts the form of the QHD lagrangian and the number of parameters. In a sense, renormalizable QHD is a means of defining a *purely* hadronic theory. By considering only a few hadronic degrees of freedom, we can study the role of meson dynamics and relativity in nuclear systems as simply as possible. Moreover, we have a self-contained model for studying the formal aspects of the relativistic many-body problem with strong couplings. Nevertheless, QHD becomes increasingly complicated at short distances, where a hadronic description must ultimately break down. Hopefully, the general features discussed below will remain valid even if the constraint of renormalizability at the hadronic level turns out to be too restrictive.

Due to limited space, this paper deals with only a fraction of the areas of current interest in QHD. We focus first on the simple model QHD–I,[19] which contains neutrons, protons, and the isoscalar, Lorentz scalar and vector mesons σ and ω. The development is based on the relativistic mean-field and Hartree approximations, and their application to both infinite nuclear matter and the ground states of atomic nuclei. We discuss some successes of this model, including the nuclear equation of state, the shell model, nucleon–nucleus scattering, and the addition of zero-point vacuum corrections. We concentrate on the new features that arise in a relativistic framework and emphasize the important concepts of Lorentz covariance and self-consistency. In the second half of this work, we consider QHD–II,[20,21] the extension of QHD–I to include isovector π and ρ mesons. QHD–II is based on the so-called linear σ model,[22–24] which contains neutrons, protons, pions, and neutral scalar mesons interacting in a chirally invariant fashion. We also discuss extensions beyond the relativistic mean-field and Hartree approximations, as a means for constructing reliable calculational schemes, as well as recent efforts to incorporate the full role of the quantum vacuum and relativistic pion dynamics in a consistent fashion. At the end, we return to consider the relationship between QHD and QCD and to speculate on future developments.

The reader is assumed to have a working knowledge of relativistic quantum field theory, canonical quantization, and the use of path integrals at zero temperature. For general background on relativistic field theory, the reader can turn to any of a number of texts;[25–30] a recent review exists for background on quantum hadrodynamics.[7]

ACCOMPLISHMENTS

A Simple Model

Quantum hadrodynamics as defined above is a general framework for the relativistic nuclear many-body problem. The detailed dynamics must be specified by choosing a particular renormalizable lagrangian density. To illustrate the relativistic formalism as simply as possible, we consider a model called QHD–I, which contains fields for baryons $\left[\psi = \begin{pmatrix} \psi_p \\ \psi_n \end{pmatrix}\right]$ and neutral scalar (ϕ) and vector (V^μ) mesons.[19]

The lagrangian density for this model is given by[7] ($\hbar = c = 1$)

$$\mathcal{L} = \overline{\psi}\left[\gamma_\mu(i\partial^\mu - g_v V^\mu) - (M - g_s\phi)\right]\psi + \tfrac{1}{2}(\partial_\mu\phi\partial^\mu\phi - m_s^2\phi^2)$$
$$- \tfrac{1}{4}F_{\mu\nu}F^{\mu\nu} + \tfrac{1}{2}m_v^2 V_\mu V^\mu + \delta\mathcal{L}, \tag{2}$$

where $F^{\mu\nu} = \partial^\mu V^\nu - \partial^\nu V^\mu$ and $\delta\mathcal{L}$ contains counterterms. The parameters M, g_s, g_v, m_s, and m_v are phenomenological constants that may be determined (in principle) from experimental measurements. This lagrangian resembles massive QED with an additional scalar interaction, so the resulting relativistic quantum field theory is *renormalizable*.[31] The counterterms in $\delta\mathcal{L}$ are used for renormalization.

The motivation for this model has evolved considerably since it was introduced. As discussed in the Introduction, when the empirical nucleon–nucleon (NN) scattering amplitude is described in a Lorentz-covariant fashion, it contains strong isoscalar scalar and four-vector pieces,[4–6] and the simplest way to reproduce these pieces is through the exchange of neutral scalar and vector mesons. The neutral scalar and vector components are the most important for describing bulk nuclear properties, which is our main concern here. Other Lorentz components of the NN interaction, in particular the terms arising from pion exchange, average essentially to zero in spin-saturated nuclear matter and may be incorporated as refinements to the present model. The important point is that even in more refined models, the dynamics generated by scalar and vector mesons will remain; thus it is important to first understand the consequences of these degrees of freedom for relativistic descriptions of nuclear systems.

The field equations for this model follow from the Euler–Lagrange equations and can be written as

$$(\partial_\mu \partial^\mu + m_s^2)\phi = g_s \overline{\psi}\psi \,, \tag{3}$$

$$\partial_\nu F^{\nu\mu} + m_v^2 V^\mu = g_v \overline{\psi}\gamma^\mu \psi \,, \tag{4}$$

$$[\gamma^\mu(i\partial_\mu - g_v V_\mu) - (M - g_s\phi)]\psi = 0 \,. \tag{5}$$

(The counterterms have been suppressed.) Equation (3) is simply the Klein–Gordon equation with a scalar source. Equation (4) looks like massive QED with the conserved baryon current

$$B^\mu \equiv (\rho_\mathrm{B}, \boldsymbol{B}) = \overline{\psi}\gamma^\mu \psi \,, \qquad \partial_\mu B^\mu = 0 \tag{6}$$

rather than the (conserved) electromagnetic current as source. Finally, eq. (5) is the Dirac equation with scalar and vector fields introduced in a minimal fashion. These field equations imply that the canonical energy-momentum tensor $T^{\mu\nu}$ is conserved $(\partial_\mu T^{\mu\nu} = \partial_\nu T^{\mu\nu} = 0)$.

When quantized, eqs. (3)–(5) become *nonlinear quantum field equations*, whose exact solutions are very complicated. In particular, they describe mesons and baryons *that are not point particles*, but rather objects with intrinsic structure due to the implied (virtual) meson and baryon-antibaryon loops. It is here that the dynamical input of renormalizability is apparent, since we are assuming that this intrinsic structure (or at least the long-range part of it) can be described using hadronic degrees of freedom. The validity of this input and its limitations have yet to be tested consistently within the framework of QHD, and we will return to this question later.

We also expect the coupling constants in eqs. (3)–(5) to be large, so perturbative solutions are not useful. Fortunately, there is an approximate nonperturbative solution that should become increasingly valid as the nuclear density increases. Consider a system of B baryons in a large box of volume V at zero temperature. Assume that we are in the rest frame of the matter, so that the baryon flux $\boldsymbol{B} = 0$. As the baryon density B/V increases, so do the source terms on the right-hand sides of eqs. (3) and (4). When the sources are large, the meson field operators can be replaced by their expectation values, which are classical fields:

$$\phi \to \langle\phi\rangle \equiv \phi_0 \,, \qquad V^\mu \to \langle V^\mu \rangle \equiv (V_0, \boldsymbol{0}) \,. \tag{7}$$

For our stationary, uniform system, ϕ_0 and V_0 are *constants* that are independent of space and time, and since the matter is at rest, the classical three-vector field $\mathbf{V} = 0$.

We emphasize the strategy involved in the preceding "mean-field" approximation. First, the resulting mean-field theory (MFT) should give the correct solution to the field equations in the high-density limit. More importantly, however, the MFT serves as a starting point for calculating corrections within the framework of QHD, using Feynman diagrams, path-integral methods, and so forth, as we will discuss later in this work.

The Nuclear Matter Equation of State

When the meson fields in eq. (2) are approximated by the constant classical fields of eq. (7), we arrive at the mean-field lagrangian density

$$\mathcal{L}_{\mathrm{MFT}} = \overline{\psi}[i\gamma^\mu\partial_\mu - g_v V_0\gamma_0 - (M - g_s\phi_0)]\psi - \tfrac{1}{2}m_s^2\phi_0^2 + \tfrac{1}{2}m_v^2 V_0^2 \,. \tag{8}$$

(The counterterms have been suppressed.) The conserved baryon four-current remains as in eq. (6), and the canonical energy-momentum tensor becomes

$$T^{\mu\nu}_{\text{MFT}} = i\overline{\psi}\gamma^\mu\partial^\nu\psi - \tfrac{1}{2}(m_{\text{v}}^2 V_0^2 - m_{\text{s}}^2\phi_0^2)g^{\mu\nu}\,. \tag{9}$$

As discussed by Freedman,[32] there is no need to symmetrize $T^{\mu\nu}$ if we consider only uniform nuclear matter.

Since the meson fields are classical, only the fermion field must be quantized. The Dirac field equation follows from $\mathcal{L}_{\text{MFT}}$:

$$[i\gamma_\mu\partial^\mu - g_{\text{v}}\gamma_0 V_0 - (M - g_{\text{s}}\phi_0)]\psi(t,\mathbf{x}) = 0\,, \tag{10}$$

and since this equation is linear, it can be solved exactly. Note that the scalar field ϕ_0 shifts the baryon mass from M to $M^\star \equiv M - g_{\text{s}}\phi_0$, while the vector field V_0 shifts the energy spectrum. We look for normal-mode solutions with both positive and negative energies, as is natural for the Dirac equation. These solutions can be used to define quantum field operators ψ and $\psi^\dagger$ in the usual fashion, and by imposing the familiar equal-time anticommutation relations, we can construct the baryon number operator $\hat{B} \equiv \int d^3x\,\overline{\psi}\gamma^0\psi$ and the four-momentum operators $\hat{P}^\mu = (\hat{H},\hat{\mathbf{P}}) \equiv \int d^3x\,\hat{T}^{0\mu}$, with the results

$$\hat{H} - \langle 0|\hat{H}|0\rangle = \hat{H}_{\text{MFT}} + \delta H\,, \tag{11}$$

$$\hat{H}_{\text{MFT}} = \sum_{\mathbf{k}\lambda}(\mathbf{k}^2 + M^{\star 2})^{1/2}(A^\dagger_{\mathbf{k}\lambda}A_{\mathbf{k}\lambda} + B^\dagger_{\mathbf{k}\lambda}B_{\mathbf{k}\lambda}) + g_{\text{v}}V_0\hat{B}$$

$$+ \tfrac{1}{2}(m_{\text{s}}^2\phi_0^2 - m_{\text{v}}^2 V_0^2)V\,, \tag{12}$$

$$\delta H = -\sum_{\mathbf{k}\lambda}\left[(\mathbf{k}^2 + M^{\star 2})^{1/2} - (\mathbf{k}^2 + M^2)^{1/2}\right]\,, \tag{13}$$

$$\hat{B} = \sum_{\mathbf{k}\lambda}(A^\dagger_{\mathbf{k}\lambda}A_{\mathbf{k}\lambda} - B^\dagger_{\mathbf{k}\lambda}B_{\mathbf{k}\lambda})\,, \tag{14}$$

$$\hat{\mathbf{P}} = \sum_{\mathbf{k}\lambda}\mathbf{k}\,(A^\dagger_{\mathbf{k}\lambda}A_{\mathbf{k}\lambda} + B^\dagger_{\mathbf{k}\lambda}B_{\mathbf{k}\lambda})\,. \tag{15}$$

Here the index λ denotes both spin and isospin projections. The quantities $A^\dagger_{\mathbf{k}\lambda}$, $B^\dagger_{\mathbf{k}\lambda}$, $A_{\mathbf{k}\lambda}$, and $B_{\mathbf{k}\lambda}$ appearing in these expressions are creation and destruction operators for (quasi)baryons and (quasi)antibaryons with shifted mass and energy, and $\hat{B}$ is the baryon number operator, which clearly counts the number of baryons minus the number of antibaryons. The correction term δH arises from placing the operators in $\hat{H}_{\text{MFT}}$ in "normal order" and represents the contribution to the energy from the filled Dirac sea, where the baryon mass has been shifted by the uniform scalar field ϕ_0.[7] We will return later to discuss this "zero-point energy" correction; for now let us concentrate on the MFT hamiltonian defined by eq. (12).

Since $\hat{H}_{\text{MFT}}$ is diagonal, this model mean-field problem has been solved *exactly* once the meson fields are specified; their determination is discussed below. The solution retains the essential features of QHD: relativistic covariance, explicit meson degrees of freedom, and the incorporation of antiparticles. Furthermore, it yields a simple solution to the field equations that should become increasingly valid as the baryon density increases. Since $\hat{B}$ and $\hat{\mathbf{P}}$ are also diagonal, the baryon number and total momentum are constants of the motion, as are their corresponding densities ρ_{B} and $\mathcal{P}$, since the volume is fixed.

For uniform nuclear matter, the ground state is obtained by filling energy levels with spin-isospin degeneracy γ up to the Fermi momentum k_{F}. (The generalization to finite temperature will be discussed at the end of this section.) The Fermi momentum is related to the baryon density by

$$\rho_{\mathrm{B}} = \frac{\gamma}{(2\pi)^3} \int_0^{k_{\mathrm{F}}} \mathrm{d}^3k = \frac{\gamma}{6\pi^2} k_{\mathrm{F}}^3 \,, \tag{16}$$

where the degeneracy factor γ is 4 for symmetric ($N = Z$) matter and 2 for pure neutron matter ($Z = 0$). The constant vector field V_0 can be expressed in terms of conserved quantities from the expectation value of the vector meson field equation (4):

$$V_0 = \frac{g_{\mathrm{v}}}{m_{\mathrm{v}}^2} \rho_{\mathrm{B}} \,. \tag{17}$$

The expressions for the energy density and pressure now take the simple forms[7]

$$\mathcal{E} = \frac{g_{\mathrm{v}}^2}{2m_{\mathrm{v}}^2}\, \rho_{\mathrm{B}}^2 + \frac{m_{\mathrm{s}}^2}{2g_{\mathrm{s}}^2}\, (M - M^\star)^2 + \frac{\gamma}{(2\pi)^3} \int_0^{k_{\mathrm{F}}} \mathrm{d}^3k \; E^\star(k) \,, \tag{18}$$

$$p = \frac{g_{\mathrm{v}}^2}{2m_{\mathrm{v}}^2}\, \rho_{\mathrm{B}}^2 - \frac{m_{\mathrm{s}}^2}{2g_{\mathrm{s}}^2}\, (M - M^\star)^2 + \frac{1}{3} \frac{\gamma}{(2\pi)^3} \int_0^{k_{\mathrm{F}}} \mathrm{d}^3k \; \frac{k^2}{E^\star(k)} \,, \tag{19}$$

where $E^\star(k) \equiv (\mathbf{k}^2 + M^{\star 2})^{1/2}$. The first two terms in eqs. (18) and (19) arise from the classical meson fields. The final terms in these equations are those of a relativistic gas of baryons of mass $M^\star$. These expressions give the nuclear matter equation of state at zero temperature in parametric form: $\mathcal{E}(\rho_{\mathrm{B}})$ and $p(\rho_{\mathrm{B}})$.

The constant scalar field ϕ_0, or equivalently the effective mass $M^\star$, can be determined thermodynamically at the end of the calculation by minimizing $\mathcal{E}(M^\star)$ with respect to $M^\star$. This produces the *self-consistency condition*

$$M^\star = M - \frac{g_{\mathrm{s}}^2}{m_{\mathrm{s}}^2} \frac{\gamma}{(2\pi)^3} \int_0^{k_{\mathrm{F}}} \mathrm{d}^3k \; \frac{M^\star}{E^\star(k)}$$

$$= M - \frac{g_{\mathrm{s}}^2}{m_{\mathrm{s}}^2} \langle : \overline{\psi}\psi : \rangle \equiv M - \frac{g_{\mathrm{s}}^2}{m_{\mathrm{s}}^2} \rho_{\mathrm{s}} \,, \tag{20}$$

which also defines the scalar density ρ_{s}. Equation (20) is equivalent to the MFT scalar field equation for ϕ_0. Note that the scalar density is smaller than the baryon density [eq. (16)] due to the factor $M^\star/E^\star(k)$, which is an effect of Lorentz contraction. Thus the contribution of rapidly moving baryons to the scalar source is significantly reduced. Most importantly, eq. (20) is a *transcendental self-consistency equation* for $M^\star$ that must be solved at each value of k_{F}. This illustrates the *nonperturbative* nature of the mean-field solution.

An examination of the analytic expression (18) for the energy density shows that the system is unbound ($\mathcal{E}/\rho_{\mathrm{B}} > M$) at either very low or very high densities.[19] At intermediate densities, the attractive scalar interaction will dominate if the coupling constants are chosen properly. The system then *saturates*. Nuclear matter with an equilibrium Fermi wavenumber $k_{\mathrm{F}}^0 = 1.30$ fm^{-1} and an energy/nucleon ($\mathcal{E}/\rho_{\mathrm{B}} - M$) = -15.75 MeV is obtained if the couplings are chosen as[a]

$$C_{\mathrm{s}}^2 \equiv g_{\mathrm{s}}^2\left(\frac{M^2}{m_{\mathrm{s}}^2}\right) = 357.4 \,, \qquad C_{\mathrm{v}}^2 \equiv g_{\mathrm{v}}^2\left(\frac{M^2}{m_{\mathrm{v}}^2}\right) = 273.8 \,. \tag{21}$$

---

[a]The values $C_{\mathrm{s}}^2 = 267.1$ and $C_{\mathrm{v}}^2 = 195.9$ used in ref. 7 yield $k_{\mathrm{F}}^0 = 1.42$ fm^{-1}. This is set L1 in table 1 below.

The nuclear compressibility in this approximation is 545 MeV. Note that only the *ratios* of coupling constants to meson masses enter in eqs. (18), (19), and (20). The resulting saturation curve is shown in fig. 1. In this approximation, the relativistic properties of the scalar and vector fields are responsible for saturation; a Hartree–Fock variational estimate built on the nonrelativistic (Yukawa) potential limit of the interaction shows that such a system is unstable against collapse.

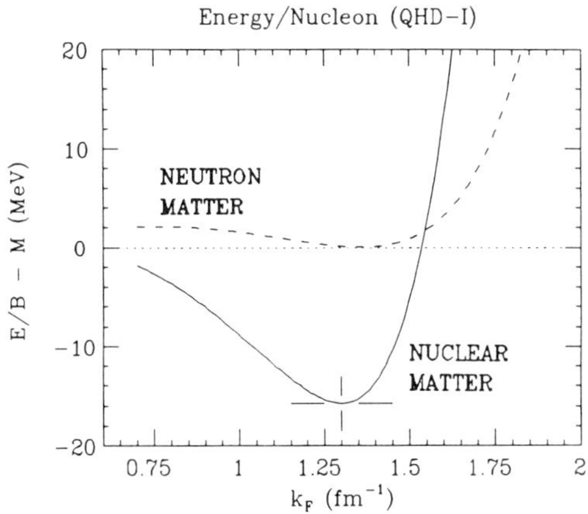

Figure 1.   Saturation curve for nuclear matter. These results are calculated in the relativistic mean-field theory with baryons and neutral scalar and vector mesons (QHD–I). The coupling constants are chosen to fit the value and position of the minimum and are given in row L2 of table 1. The prediction for neutron matter ($\gamma = 2$) is also shown.

The solution of the self-consistency condition (20) for $M^{\star}$ yields an effective mass that is a decreasing function of the density, as illustrated in fig. 2. Evidently, $M^{\star}/M$ becomes small at high density and is significantly less than unity at ordinary nuclear densities. This is a consequence of the large scalar field $g_{\mathrm{s}}\phi_0$, which is approximately 400 MeV and which produces a large attractive contribution to the energy/baryon. There is also a large repulsive energy/baryon from the vector field $g_{\mathrm{v}}V_0 \approx 350$ MeV. Thus *the Lorentz structure of the interaction leads to a new energy scale in the problem*, and the small nuclear binding energy (≈ 16 MeV) arises from the cancellation between the large scalar attraction and vector repulsion. As the nuclear density increases, the scalar source ρ_{s} becomes small relative to the vector source ρ_{B}, and the attractive forces saturate, producing the minimum in the binding curve. Clearly, because of the sensitive cancellation involved near the equilibrium density, corrections to the MFT must be calculated before the importance of this new saturation mechanism can be assessed. Nevertheless, the Lorentz structure of the interaction provides an *additional saturation mechanism that is not present in the nonrelativistic potential limit*, as this limit ignores the distinction between ρ_{s} and ρ_{B}.

The corresponding curves for neutron matter obtained by setting $\gamma = 2$ are also

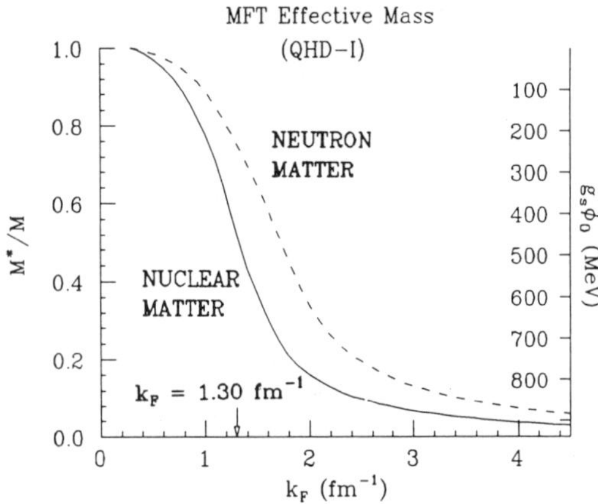

Figure 2.   Effective mass M^*/M as a function of density for nuclear ($\gamma = 4$) and neutron ($\gamma = 2$) matter based on fig. 1.

shown in figs. 1 and 2, and the equation of state (pressure *vs.* energy density) for neutron matter at all densities is given in fig. 3. In this mean-field model, there is a van der Waals (liquid–gas) phase transition, and the properties of the two phases are deduced through a Maxwell construction. At high densities, the system approaches the "causal limit" $p = \mathcal{E}$ representing the stiffest possible equation of state. Thus we have a simple, two-parameter model that is consistent with the equilibrium point of normal nuclear matter and that allows for a covariant, causal extrapolation to any density.

The neutron matter equation of state shown in fig. 3 can be used in the Tolman–Oppenheimer–Volkoff equation for the general-relativistic metric[33] to give the masses of neutron stars as a function of the central density. This MFT gives a maximum neutron star mass of ≈ 2.6 solar masses; this is large enough to accommodate the largest observed neutron stars, which contain roughly 1.5 solar masses. The density at the center of the star is approximately six times larger than the central density in ^{208}Pb, and the asymptotic approach of the equation of state to the causal limit is already relevant in this regime. Moreover, although the low-density behavior of nuclear matter is sensitive to the nearly exact cancellation between attractive scalar and repulsive vector components, the stiff high-density equation of state is determined simply by the Lorentz structure of the interaction; the scalar attraction saturates completely at high densities, producing an essentially massless gas of baryons interacting through a strong vector repulsion, which leads to a stiff equation of state.[34] Note that the onset of the asymptotic regime occurs at densities similar to those in the interiors of neutron stars.

<u>Finite Nuclei</u>

We now generalize the results of the preceding subsection to study atomic nuclei. We continue to work in the mean-field approximation to QHD–I, but since the system

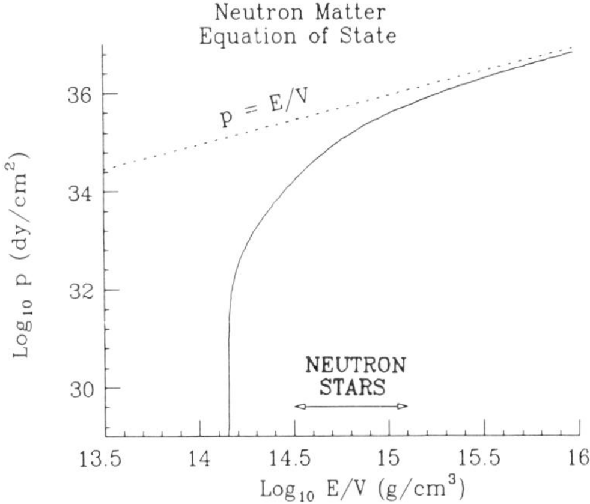

Figure 3. Predicted equation of state for neutron matter at all densities. The dotted line represents the causal limit $p = \mathcal{E}$. The density regime relevant for neutron stars is also shown.

now has finite spatial extent, these fields are spatially dependent. If we initially restrict consideration to spherically symmetric nuclei, the meson fields depend only on the radius, and since the baryon current is conserved, the spatial part of the vector field $\mathbf{V}$ again vanishes.[7] Thus the mean-field QHD–I lagrangian of eq. (8) becomes

$$\mathcal{L}_{\mathrm{MFT}}^{(\mathrm{I})} = \overline{\psi}[i\gamma_\mu\partial^\mu - g_\mathrm{v}\gamma_0 V_0 - (M - g_\mathrm{s}\phi_0)]\psi - \tfrac{1}{2}[(\boldsymbol{\nabla}\phi_0)^2 + m_\mathrm{s}^2\phi_0^2] + \tfrac{1}{2}[(\boldsymbol{\nabla}V_0)^2 + m_\mathrm{v}^2 V_0^2]\,, \quad (22)$$

and the Dirac equation for the baryon field is

$$\left\{i\gamma_\mu\partial^\mu - g_\mathrm{v}\gamma_0 V_0(r) - [M - g_\mathrm{s}\phi_0(r)]\right\}\psi(x) = 0\,. \quad (23)$$

Appropriate values for the scalar and vector couplings (g_s and g_v) and masses (m_s and m_v) will be given below.

Although the baryon field is still an operator, the meson fields are classical; hence eq. (23) is linear, and we may again seek normal-mode solutions of the form $\psi(x) = \psi(\mathbf{x})\,\mathrm{e}^{-iEt}$. This leads to the eigenvalue equation

$$h\psi(\mathbf{x}) \equiv \{-i\boldsymbol{\alpha}\cdot\boldsymbol{\nabla} + g_\mathrm{v}V_0(r) + \beta[M - g_\mathrm{s}\phi_0(r)]\}\psi(\mathbf{x}) = E\psi(\mathbf{x})\,, \quad (24)$$

which defines the single-particle Dirac hamiltonian h. Equation (24) has both positive- and negative-energy solutions $\mathcal{U}(\mathbf{x})$ and $\mathcal{V}(\mathbf{x})$, which allow the field operators to be constructed in the Schrödinger picture. The positive-energy spinors can be written as

$$\mathcal{U}_\alpha(\mathbf{x}) \equiv \mathcal{U}_{n\kappa m t}(\mathbf{x}) = \begin{pmatrix} i\big[G_{n\kappa t}(r)/r\big]\Phi_{\kappa m} \\ -\big[F_{n\kappa t}(r)/r\big]\Phi_{-\kappa m} \end{pmatrix} \zeta_t\,, \quad (25)$$

59

where n is the principal quantum number, $\Phi_{\kappa m}$ is a spin-1/2 spherical harmonic:[35]

$$\Phi_{\kappa m} = \sum_{m_\ell m_s} \langle \ell m_\ell \tfrac{1}{2} m_s | \ell \tfrac{1}{2} j m \rangle Y_{\ell, m_\ell}(\Omega) \chi_{m_s} \,, \tag{26}$$

$$j = |\kappa| - \tfrac{1}{2} \,, \qquad \ell = \begin{cases} \kappa \,, & \kappa > 0 \,, \\ -(\kappa + 1) \,, & \kappa < 0 \,, \end{cases} \tag{27}$$

and ζ_t is a two-component isospinor labeled by the isospin projection t. (We take $t = \tfrac{1}{2}$ for protons and $t = -\tfrac{1}{2}$ for neutrons.) The phase choice in eq. (25) produces real bound-state wave functions F and G for real potentials ϕ_0 and V_0, and the normalization is given by

$$\int_0^\infty dr \left(|G_\alpha(r)|^2 + |F_\alpha(r)|^2 \right) = 1 \,, \tag{28}$$

which ensures unit probability to find each nucleon somewhere in space.

The classical meson field equations follow from eq. (22) and resemble eqs. (3) and (4) restricted to static, spherically symmetric fields. With the general form for the spinors in eq. (25), we can evaluate the nuclear densities, which serve as source terms in the meson field equations. Assume that the nuclear ground state consists of filled shells up to some value of n and κ, which may be different for protons and neutrons; this is appropriate for doubly magic nuclei. In addition, assume that all bilinear products of baryon operators are normal ordered, which removes contributions from the negative-energy spinors $\mathcal{V}_\alpha(\mathbf{x})$. This amounts to neglecting the filled Dirac sea of baryons and defines the mean-field approximation. The zero-point terms arising from $\mathcal{V}_\alpha(\mathbf{x})$ will be included later.

With these assumptions, the local baryon (ρ_B) and scalar (ρ_s) densities become

$$\left.\begin{array}{c} \rho_\mathrm{B}(\mathbf{x}) \\ \rho_\mathrm{s}(\mathbf{x}) \end{array}\right\} = \sum_\alpha^{\mathrm{occ}} \overline{\mathcal{U}}_\alpha(\mathbf{x}) \begin{pmatrix} \gamma^0 \\ 1 \end{pmatrix} \mathcal{U}_\alpha(\mathbf{x}) = \sum_a^{\mathrm{occ}} \left(\frac{2j_a + 1}{4\pi r^2} \right) \left(|G_a(r)|^2 \pm |F_a(r)|^2 \right) , \tag{29}$$

which holds for filled shells, as appropriate for spherically symmetric nuclei. The remaining quantum numbers are denoted by $\{\alpha\} = \{a; m\} \equiv \{n, \kappa, t; m\}$. Notice that since the shells are filled, the sources are spherically symmetric.

The sources produce the meson fields, which satisfy static Klein–Gordon equations:

$$\frac{d^2}{dr^2} \phi_0(r) + \frac{2}{r} \frac{d}{dr} \phi_0(r) - m_\mathrm{s}^2 \phi_0(r) = -g_\mathrm{s} \rho_\mathrm{s}(r) \,, \tag{30}$$

$$\frac{d^2}{dr^2} V_0(r) + \frac{2}{r} \frac{d}{dr} V_0(r) - m_\mathrm{v}^2 V_0(r) = -g_\mathrm{v} \rho_\mathrm{B}(r) \,. \tag{31}$$

The equations for the baryon wave functions follow upon substituting eq. (25) into eq. (24), which produces

$$\frac{d}{dr} G_a(r) + \frac{\kappa}{r} G_a(r) - \left[E_a - g_\mathrm{v} V_0(r) + M - g_\mathrm{s} \phi_0(r) \right] F_a(r) = 0 \,, \tag{32}$$

$$\frac{d}{dr} F_a(r) - \frac{\kappa}{r} F_a(r) + \left[E_a - g_\mathrm{v} V_0(r) - M + g_\mathrm{s} \phi_0(r) \right] G_a(r) = 0 \,. \tag{33}$$

Thus the spherical nuclear ground state is described by coupled, ordinary differential equations that may be solved by an iterative procedure, as discussed in ref. 36. They contain all information about the static ground-state nucleus in this approximation.

The mean-field hamiltonian can be computed just as for infinite matter, and after normal ordering, the ground-state energy is given by

$$E = \int \mathrm{d}^3x \left\{ \frac{1}{2}\left[(\boldsymbol{\nabla}\phi_0)^2 + m_\mathrm{s}^2\phi_0^2\right] - \frac{1}{2}\left[(\boldsymbol{\nabla}V_0)^2 + m_\mathrm{v}^2 V_0^2\right] \right.$$
$$\left. + \sum_\alpha^{\mathrm{occ}} \mathcal{U}_\alpha^\dagger(\mathbf{x})\left[-i\boldsymbol{\alpha}\cdot\boldsymbol{\nabla} + \beta(M - g_\mathrm{s}\phi_0) + g_\mathrm{v}V_0\right]\mathcal{U}_\alpha(\mathbf{x}) \right\}. \qquad (34)$$

Here the meson fields are functions of the radial coordinate, and $\boldsymbol{\alpha}$ and β are the usual Dirac matrices. Notice that if we interpret this expression as an *energy functional* for the Dirac–Hartree ground state, extremization with respect to the meson fields reproduces the field equations (30) and (31), with the densities from eq. (29). Moreover, extremization with respect to the baryon wave functions $\mathcal{U}_\alpha^\dagger(\mathbf{x})$, subject to the constraint

$$\int \mathrm{d}^3x \, \mathcal{U}_\alpha^\dagger(\mathbf{x})\mathcal{U}_\alpha = 1 \qquad (35)$$

for all occupied states (which is enforced by Lagrange multipliers E_α), leads to the Dirac equation (24). This alternative derivation of the Dirac–Hartree equations from an energy functional is useful for extensions of the simple model discussed in this subsection.

Once the solutions to the Dirac–Hartree equations have been found, the ground-state energy (34) can be computed by using the Dirac equation (24) to introduce the eigenvalues E_a and by partially integrating the meson terms to introduce the densities. In the end, the total energy of the system is given by

$$E = \sum_a^{\mathrm{occ}} E_a(2j_a + 1) - \frac{1}{2}\int \mathrm{d}^3x\left[-g_\mathrm{s}\phi_0(r)\rho_\mathrm{s}(r) + g_\mathrm{v}V_0(r)\rho_\mathrm{B}(r)\right]. \qquad (36)$$

Before discussing the Dirac–Hartree solutions, let us generalize the equations to include some additional degrees of freedom and couplings. Although the isoscalar meson fields are the most important for describing general properties of nuclear matter, a quantitative comparison with actual nuclei requires the introduction of some additional dynamics.

For example, it is necessary to include the electromagnetic interaction to account for the Coulomb repulsion between protons. Moreover, since hadronic interactions exhibit an almost exact SU(2) isospin symmetry, the nucleons can couple to isovector mesons in addition to the isoscalar (neutral) mesons of QHD–I. These isovector mesons, for example the π and ρ, come in three charge states $(+, 0, -)$ and couple differently to the proton and neutron. Thus they affect the nuclear symmetry energy, which arises when there are unequal numbers of neutrons and protons.

The construction of a renormalizable lagrangian containing charged, massive vector fields is somewhat complicated and is discussed at length in Abers and Lee;[37] applications to the present model can be found in ref. 7. For our purposes, we require only the classical contributions from these fields, and in this case, the lagrangian simplifies considerably. In particular, since the nuclear ground state has well-defined charge, only the neutral ρ meson field (denoted by b_0) enters, and since the ground state is assumed to have well-defined parity and spherical symmetry, there is no classical π field. Thus the mean-field lagrangian for this extended model, which is called QHD–II, is given by

$$\mathcal{L}_{\mathrm{MFT}}^{(\mathrm{II})} = \overline{\psi}[i\gamma_\mu\partial^\mu - g_\mathrm{v}\gamma_0 V_0 - \tfrac{1}{2}g_\rho\tau_3\gamma_0 b_0 - e\tfrac{1}{2}(1 + \tau_3)\gamma_0 A_0 - (M - g_\mathrm{s}\phi_0)]\psi$$
$$- \tfrac{1}{2}[(\boldsymbol{\nabla}\phi_0)^2 + m_\mathrm{s}^2\phi_0^2] + \tfrac{1}{2}[(\boldsymbol{\nabla}V_0)^2 + m_\mathrm{v}^2 V_0^2]$$
$$+ \tfrac{1}{2}(\boldsymbol{\nabla}A_0)^2 + \tfrac{1}{2}[(\boldsymbol{\nabla}b_0)^2 + m_\rho^2 b_0^2]. \qquad (37)$$

Table 1

Dirac–Hartree Parameter Sets

(All sets use $M = 939$ MeV and $m_v \equiv m_\omega = 783$ MeV.)

Set	g_s^2	g_v^2	g_ρ^2	m_s (MeV)	κ/M	λ
L1	91.64	136.2	36.79	550.	0	0
L2	109.6	190.4	65.23	520.	0	0
NLB	94.01	158.48	73.00	510.	0.852	10
RHA0	53.78	102.58	83.30	456.	0	0

Here A_0 is the Coulomb potential, e is the proton electromagnetic charge, g_ρ is the rho-nucleon coupling constant, and τ_i are the usual isospin Pauli matrices. For now, all of the boson fields are assumed to be functions of the radial coordinate only.

The Dirac–Hartree equations for this extended model can be derived just as before. The Dirac equations for the baryon wave functions now contain b_0 and A_0, and because of the structure of the τ_3 matrix, b_0 couples with opposite sign to protons and neutrons, and A_0 couples only to the protons. In addition to the source terms in eq. (29), which sum over both proton and neutron occupied states, the source term for the ρ meson involves the *difference* between proton and neutron densities, while the Coulomb source involves only protons. These different types of couplings allow for a more accurate reproduction of real nuclei, where the proton and neutron wave functions are not identical. The full set of equations are presented in ref. 7 and are used to compute the results discussed below.

Spherical Nuclei. The solutions of the preceding equations depend on the parameters g_s, g_v, m_s, and g_ρ (when the ρ meson is included). We take the experimental values $M = 939$ MeV, $m_v \equiv m_\omega = 783$ MeV, $m_\rho = 770$ MeV, and $e^2/4\pi = \alpha = 1/137.0$ (which determines the Coulomb potential) as fixed. The free parameters are chosen so that when the Dirac–Hartree equations are solved in the limit of infinite nuclear matter, the empirical equilibrium density ($\rho_B^0 = 0.1484$ fm^{-3}), binding energy (15.75 MeV), and symmetry energy (35 MeV) are reproduced. The empirical equilibrium density ρ_B^0 is determined here from the density in the interior of ^{208}Pb and corresponds to $k_F^0 = 1.3\,\mathrm{fm}^{-1}$. We also fit the empirical rms charge radius of ^{40}Ca ($r_{\mathrm{rms}} = 3.482$ fm), which is determined primarily by m_s. This procedure produces the parameters in the row labeled L2 in table 1, which are taken from ref. 36. This parameter set yields the same values for C_s^2 and C_v^2 as in eq. (21), so that $M^\star/M = 0.541$ and $K \approx 545\,\mathrm{MeV}$ at nuclear matter equilibrium.

Once the parameters have been specified, the properties of all closed-shell nuclei are determined in this approximation. For example, figs. 4, 5, and 6 show the Dirac–Hartree charge densities of ^{16}O, ^{40}Ca, and ^{208}Pb compared with two nonrelativistic calculations[2,38] and the empirical distributions determined from electron scattering.[39–41] The empirical proton charge form factor has been folded with the calculated "point proton" density to determine the charge density. Similar results are obtained for other closed-shell nuclei.[36,42] Evidently, the nonrelativistic and relativistic calculations agree with the data at about the same level of accuracy.

In fig. 7, the predicted energy levels in ^{208}Pb are compared with experimental values derived from neighboring nuclei.[43,44] The relativistic calculations clearly reveal a shell structure. This arises from the spin-orbit interaction that occurs naturally when

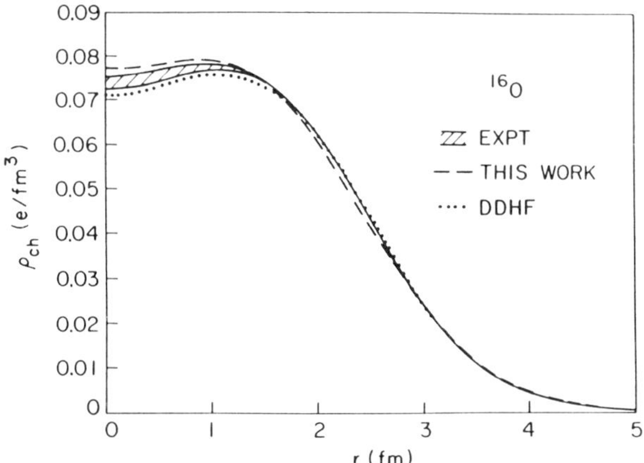

Figure 4. Charge density distribution for ^{16}O. Dirac–Hartree results are indicated by the dashed line. The experimental curve is from ref. 39, and the dotted curve shows the Density-Dependent Hartree–Fock (DDHF) results of Negele.[38]

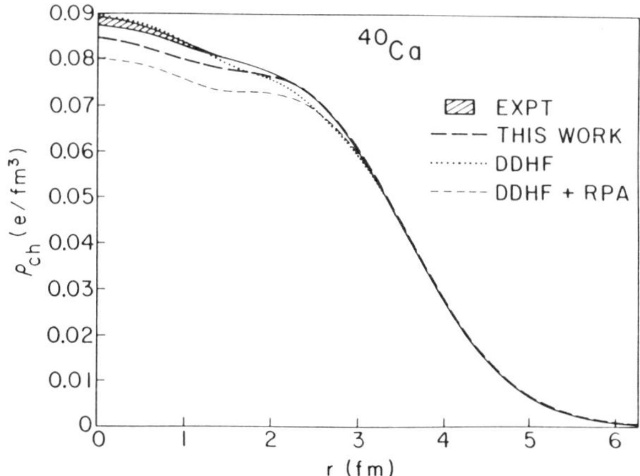

Figure 5. Charge density distribution for ^{40}Ca. The experimental curve is from Sick *et al.*[41] The DDHF results are those of Negele, and the DDHF + RPA calculation is that of Gogny, as indicated in ref. 41. The Dirac–Hartree calculations yield the long-dashed curve.

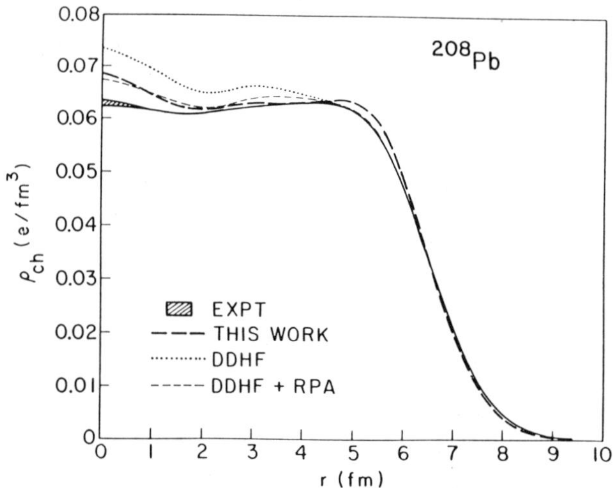

Figure 6. Charge density distribution for ^{208}Pb. The solid curve and shaded region represent the fit to experimental data in ref. 40. Dirac–Hartree results are indicated by the long-dashed line. The DDHF calculations of Gogny[2] are also shown.

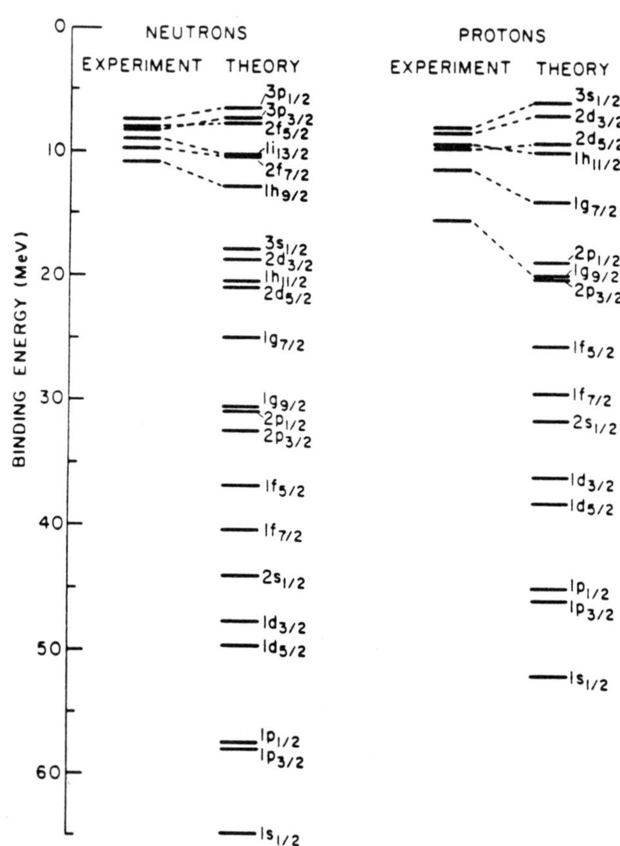

Figure 7. Predicted spectrum for occupied levels in ^{208}Pb. Experimental levels are taken from neighboring nuclei.

a Dirac particle moves in large classical scalar and vector fields.[45–47] Note that whereas $g_s\phi_0$ and $g_v V_0$ tend to cancel in the central potential that determines saturation, they add constructively in the spin-orbit potential. We emphasize that no parameters are adjusted specifically to produce the spin-orbit interaction, as is usually the case in non-relativistic calculations. Thus, with a minimal number of phenomenological parameters determined from bulk nuclear properties, *one derives the existence of the nuclear shell model.*

There are several attractive features in this relativistic model of nuclear structure. First, the calculation of the nuclear ground state is self-consistent. The scalar and vector fields follow directly from the scalar and baryon densities, which are in turn determined by the solutions to the Dirac equation (24) in the classical fields. Second, only four adjustable parameters specify the properties of all closed-shell nuclei in this approximation. Finally, this relativistic shell model is simply one piece of a consistent many-body framework based on QHD. One can therefore systematically investigate corrections to the nuclear ground state, such as those arising from the filled Dirac sea of negative-energy states, as we describe below.

Deformed Nuclei. To study the systematics of this relativistic model of nuclear structure, we extend the preceding equations to deal with deformed, axially symmetric nuclei. This allows us to calculate not only the ground states of nuclei with fully closed shells, but also those for even-even nuclei between closed shells. We will concentrate here on nuclei with $12 \leq B \leq 40$, which includes the $1p$ and $2s$–$1d$ shells.[48] The restriction to azimuthal and reflection-symmetric deformations is reasonable for light, even-even nuclei.

These assumed symmetries of the ground state, together with the assumption of well-defined parity, imply that the nonvanishing meson fields are the same ones that appear in spherical nuclei.[49] Thus the Dirac–Hartree equations are essentially the same as those written earlier, except that all fields now depend on both a radial and an angular coordinate [for example, $\phi_0(r,\theta)$], and all the differential equations become partial differential equations. The source densities are still computed as in eq. (29), but they now depend on r and θ. There are several methods for solving the resulting set of coupled partial differential equations, and the interested reader is directed to the literature for a discussion.[50–52] The equilibrium deformation is obtained by choosing the occupied single-particle states to minimize the energy.

In fig. 8, we show the results of Furnstahl, Price, and Walker for quadrupole moments in the $2s$–$1d$ shell.[52] The parameter sets used are listed as L2 and NLB in table 1; the latter includes scalar-meson self-couplings of the form

$$V(\phi) = \frac{\kappa}{3!}\phi^3 + \frac{\lambda}{4!}\phi^4 \tag{38}$$

in the model lagrangian. (These new couplings allow for some "fine tuning" of the bulk nuclear properties.) Both of these sets produce nuclear saturation at $k_F^0 = 1.30\,\mathrm{fm}^{-1}$ with a binding energy of 15.75 MeV and a symmetry energy of 35 MeV, and they also generate the correct rms radius for ^{40}Ca, just as for the spherical calculations in the preceding subsection. Nevertheless, these two sets produce slightly different values of $M^\star$ in nuclear matter ($M^\star/M = 0.54$ for L2 and $M^\star/M = 0.61$ for NLB), which is sufficient to produce different results for the deformation of nuclei with closed subshells. (The deformation depends sensitively on the level density near the Fermi surface, which is determined essentially by the inverse of $M^\star$.) Observe that set L2 predicts spherical shapes for ^{12}C, ^{28}Si, and ^{32}S, in contradiction to experiment. In contrast, the small modifications afforded by the couplings κ and λ in set NLB lead to

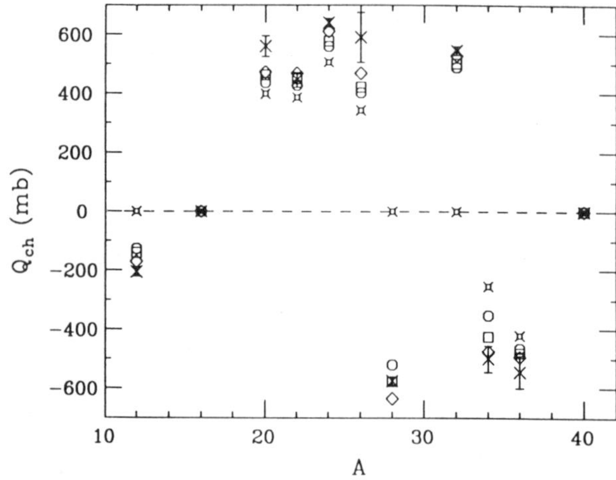

Figure 8.   Intrinsic charge quadrupole moment (mb) in even-even s–d nuclei for parameter sets L2 (stars) and NLB (circles) taken from table 1. The results shown by squares and diamonds follow from similar parameter sets given in ref. 52. Moments derived from experimental measurements of $B(E2)$ values ($\times$) are taken from Leander and Larsson.[53]

excellent agreement with the experimental moments, particularly the systematic trends and the oscillation between oblate and prolate shapes around $B = 32$, while leaving the results for spherical nuclei essentially unchanged. Thus the successful description of spherical nuclei in this relativistic model can be extended to reproduce the observed systematics of light deformed nuclei *with the same parameters*. Finally, the relativistic results are very similar to those obtained from nonrelativistic Skyrme–Hartree–Fock calculations[54] that also use an interaction fitted to nuclear properties.

Nucleon–Nucleus Scattering

The scattering of medium-energy nucleons from nuclei can provide information about both nuclear structure and the nucleon–nucleon (NN) interaction. Since the NN interaction has complex spin, isospin, momentum, and density dependence, nucleon–nucleus scattering exhibits a wide variety of phenomena. As a starting point for describing these phenomena, we use the Dirac–Hartree description of the nucleus, together with the Relativistic Impulse Approximation (RIA), which assumes that the interaction between the projectile and target nucleons has essentially the same form as the interaction between two nucleons in free space. This interaction is used to produce a nucleon–nucleus optical potential that incorporates the leading term in a multiple-scattering series.

Although the simple QHD models discussed above are useful for studying the average properties of the nuclear interaction, they are less useful for describing the detailed quantitative features (such as spin dependence) of the full NN scattering amplitude. These quantitative features are important for any reasonable description of the nucleon–nucleus scattering observables. The RIA allows us to combine the empirical free-space scattering amplitude with a relativistic calculation of the nuclear ground state.

The RIA as originally formulated[4–6] involves two basic procedures. First, the experimental NN scattering amplitude is represented by a particular set of five Lorentz covariant functions[55] that multiply the so-called "Fermi invariant" Dirac matrices. The Lorentz covariant functions are then folded with the Dirac–Hartree target densities to produce a first-order optical potential for use in the Dirac equation for the projectile. We briefly summarize the formalism and then compare observables computed in the RIA with experimental data and with nonrelativistic impulse approximation calculations.

The constraints of Lorentz covariance, parity conservation, isospin invariance, and that free nucleons are on their mass shell imply that the invariant NN scattering operator $\mathcal{F}$ can be written in terms of five complex functions for pp scattering and five for pn scattering. In the original RIA, $\mathcal{F}$ was taken as[7]

$$\mathcal{F} = \mathcal{F}^S + \mathcal{F}^V \gamma^\mu_{(0)} \gamma_{(1)\mu} + \mathcal{F}^{PS} \gamma^5_{(0)} \gamma^5_{(1)} + \mathcal{F}^T \sigma^{\mu\nu}_{(0)} \sigma_{(1)\mu\nu} + \mathcal{F}^A \gamma^5_{(0)} \gamma^\mu_{(0)} \gamma^5_{(1)} \gamma_{(1)\mu} , \tag{39}$$

where the subscripts (0) and (1) refer to the incident and struck nucleons, respectively. Each amplitude $\mathcal{F}^L$ is a complex function of the Lorentz invariants t (four-momentum transfer squared) and s (total four-momentum squared), or equivalently, of the momentum transfer q and incident energy E. It is found empirically that the amplitudes $\mathcal{F}^S$, $\mathcal{F}^V$, and $\mathcal{F}^{PS}$ are much larger than any amplitudes obtained in a nonrelativistic decomposition, which uses Galilean-invariant operators.

The RIA optical potential $U_{\mathrm{opt}}(q, E)$ is defined as

$$U_{\mathrm{opt}}(q, E) = -\frac{4\pi i p}{M} \langle \Psi | \sum_{n=1}^{A} e^{i\mathbf{q}\cdot\mathbf{x}(n)} \mathcal{F}(q, E; n) | \Psi \rangle , \tag{40}$$

where $\mathcal{F}$ is the scattering operator of eq. (39), p is the magnitude of the projectile three-momentum in the nucleon-*nucleus* cm frame (where the scattering observables will be calculated), $|\Psi\rangle$ is the A-particle nuclear ground state, and the sum runs over all nucleons in the target. $\mathcal{F}$ is a function of the momentum transfer q and collision energy E, which we take to be the cm energy for a stationary target nucleon and incident proton at the laboratory energy; this amounts to neglecting nuclear recoil.

With these simplifications, the Dirac optical potential is local, and only diagonal nuclear densities are needed. For a spin-zero nucleus, the only nonzero densities are the baryon and scalar densities of eq. (29), plus a tensor term computed by inserting σ^{0i} between the spinors in eq. (29). Thus the optical potential takes the form

$$U_{\mathrm{opt}} = U^S + \gamma^0 U^V - 2i\boldsymbol{\alpha} \cdot \hat{r} U^T , \tag{41}$$

where $U^L \equiv U^L(r; E)$ for each component. The tensor contribution U^T is small and will be neglected in what follows. The RIA optical potential then has only scalar and vector contributions, and the Dirac equation for the projectile has precisely the same form as eq. (24), with U^V replacing $g_{\mathrm{v}} V_0$ and U^S replacing $(-g_{\mathrm{s}} \phi_0)$:

$$h\,\mathcal{U}_0(\mathbf{x}) = \left\{ -i\boldsymbol{\alpha} \cdot \boldsymbol{\nabla} + U^V(r; E) + \beta \left[M + U^S(r; E) \right] \right\} \mathcal{U}_0(\mathbf{x}) = E\,\mathcal{U}_0(\mathbf{x}) . \tag{42}$$

In practice, one includes in U^V the Coulomb potential computed from the empirical nuclear charge density; other electromagnetic contributions arising from the proton anomalous magnetic moment are of similar size to the tensor term U^T and are neglected here.

For proton scattering from a spin-zero target, three relevant observables may be defined in terms of the proton–nucleus scattering amplitude $T(\theta) = g(\theta) + h(\theta)\,\boldsymbol{\sigma} \cdot \hat{n}$.

These are the differential cross section σ, the polarization (or analyzing power) P, and the spin-rotation function Q:

$$\sigma = |g|^2 + |h|^2 \,, \tag{43}$$

$$P = 2\,\mathrm{Re}(gh^*)/(|g|^2 + |h|^2) \,, \tag{44}$$

$$Q = 2\,\mathrm{Im}(gh^*)/(|g|^2 + |h|^2) \,. \tag{45}$$

In fig. 9, RIA results for proton scattering from ^{40}Ca are shown as solid curves. The target densities are taken from the results for spherical nuclei discussed above, *with no further adjustment of parameters.* The RIA results are compared with non-relativistic impulse approximation calculations (dashed curves) carried out in the conventional Kerman–McManus–Thaler formalism.[56] Evidently, the RIA calculations agree remarkably well with the data. Moreover, although more sophisticated nonrelativistic calculations give better agreement with the data, the relativistic results are superior, particularly for the spin observables, when the impulse approximation is used for both.

The spin dynamics is inherent in the relativistic formalism and arises naturally from the large Lorentz scalar and vector potentials in the Dirac equation (42). This is precisely the same spin dynamics that produces the observed spin-orbit splittings in the bound single-particle levels. *Thus the relativistic Hartree calculations provide a minimal unifying theoretical basis for both the nuclear shell model and medium-energy proton–nucleus scattering—two essential aspects of nuclear physics.*

Nuclear Excited States

We now turn to the calculation of nuclear response functions and the properties of nuclear excited states, as described in the random-phase approximation (RPA) built on the MFT ground state. These excitations arise from the consistent linear response of the ground state, in which the nucleons move coherently in varying classical meson fields that are in turn determined by oscillatory nuclear sources. Many studies of the relativistic nuclear response have been carried out for both infinite matter and finite nuclei, beginning with the pioneering work of Chin,[57] and here we will focus on two issues. The first is the importance of consistency, which implies that the particle-hole interaction in the excited states must be the same (and use the same parameter values) as the interaction in the ground state. This is important for producing a reasonable excitation spectrum when the underlying interaction involves a sensitive cancellation between large attractive and repulsive components. Second, we will see that the relativistic response involves not only the familiar positive-energy particle-hole configurations, but also configurations that mix positive- and negative-energy states. These new configurations are crucial for the conservation of the electromagnetic current and the separation of the "spurious" $J^\pi = 1^-$ state. This emphasizes that the Dirac single-particle basis is complete only when both positive- and negative-energy states are included.

The calculation of the linear response is basically the same as in nonrelativistic many-body theory.[48] The principal idea is to compute the particle-hole (polarization) propagator and to extract the collective excitation energies and transition amplitudes from the poles and residues of this propagator. Several methods have been developed and applied to finite nuclei.[58–61] Existing results are restricted primarily to isoscalar excitations, because fitting the bulk nuclear properties constrains only the isoscalar particle-hole interaction significantly and because the isovector response depends critically on pion dynamics, with the associated complexity that we discuss later. An excellent discussion of the role of consistency is contained in the work of Dawson and Furnstahl,[61] whose results we quote here.

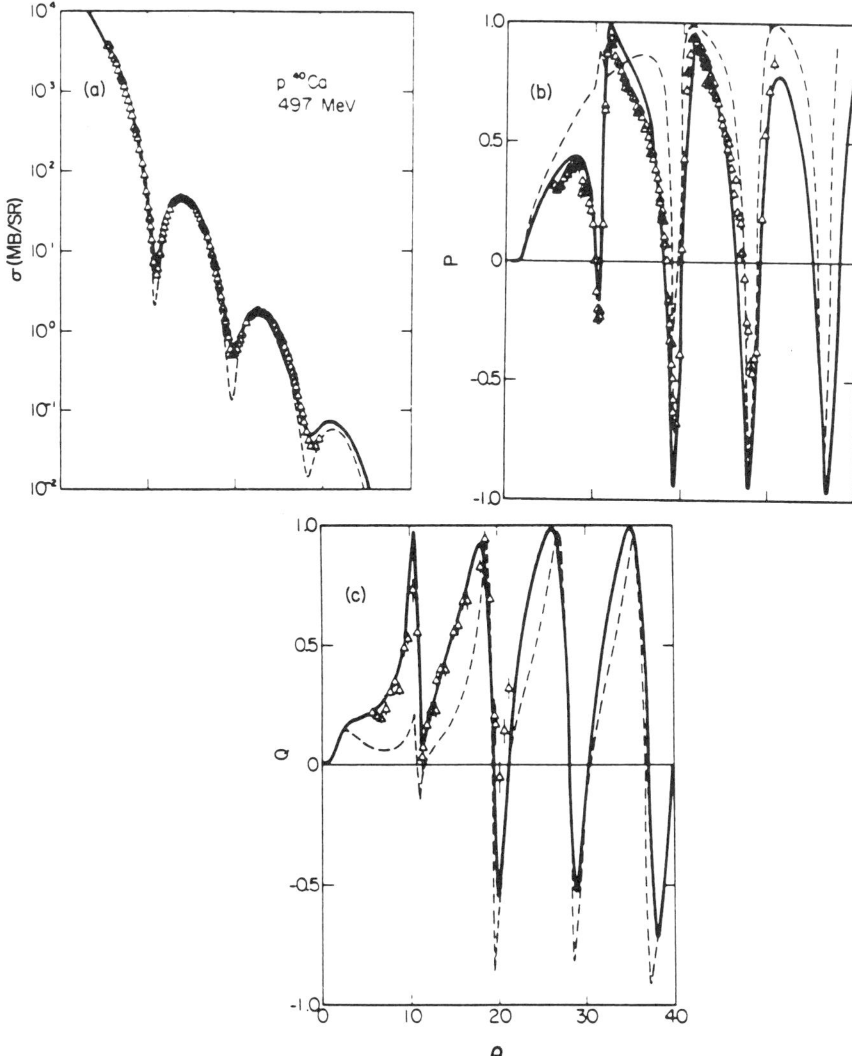

Figure 9.   Calculated cross section (a), analyzing power (b), and spin-rotation function (c) for $p+{}^{40}$Ca at a laboratory kinetic energy of 497 MeV. The RIA results are given by the solid curves, while the dashed curves follow from a nonrelativistic impulse approximation calculation.[7]

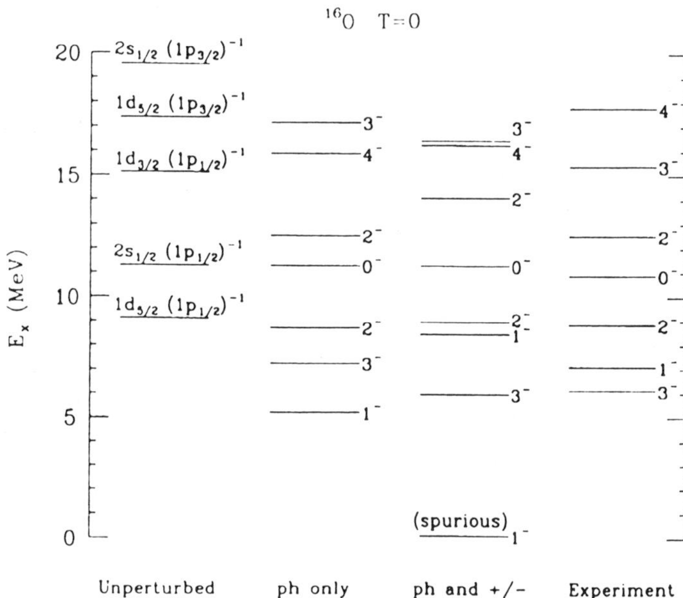

Figure 10.   Energy levels of selected low-lying states in ^{16}O in a spectral RPA calculation. The spurious 1^-0 state does not appear in the second column, since its eigenvalue is imaginary when only particle-hole configurations are included.

In fig. 10 the low-lying, negative-parity, isoscalar states in ^{16}O are compared to several empirical levels that might be reasonably described as particle-hole excitations. (Only the lowest state of each J is shown.) The first column shows the unperturbed excitation energy with the spectrum determined from the ground-state calculation using the parameter set L2 in table 1. The second column contains the RPA spectrum when only positive-energy particle-hole configurations are allowed, and the third column gives the spectrum when negative-energy states are also included. Notice that the full RPA results agree favorably with the empirical values in the fourth column, which is a nontrivial result because of the large cancellations between scalar and vector contributions. Moreover, it is clear that the negative-energy states play an important role in determining the RPA spectrum, particularly for the spurious 1^- state. Lorentz covariance implies that in a consistent RPA calculation, this state should appear at zero excitation energy, which occurs only when the full Dirac basis is maintained.

Figure 11, taken from a calculation by Furnstahl,[62] shows the relativistic RPA result for the ratio of the transition charge density to the longitudinal current density for a particular transition in ^{16}O. The results are plotted as function of the number of configurations, and if the electromagnetic current is conserved, the ratio should be unity. The dotted curve includes only the positive-energy configurations, and the current is evidently not conserved. In contrast, the dashed curve gives the result of the full RPA calculation, including the negative-energy states. The conclusion from this figure is that it is *essential* to include all states in the Dirac basis to maintain the conservation of the current.

The Role of the Quantum Vacuum

We have thus far studied the consequences of the mean-field hamiltonian of eq. (12) and its generalization to finite nuclei. Let us now return to infinite nuclear matter

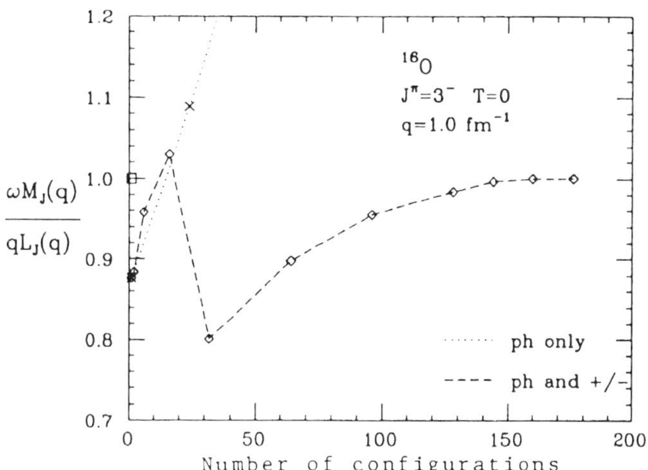

Figure 11.   The ratio of transition charge and longitudinal current densities for the low-lying isoscalar 3^- state in ^{16}O in a spectral RPA calculation, as a function of the configuration space size. The densities are evaluated at $q = 1.0$ fm^{-1}.

and include the contribution from δH in eq. (13), which defines the relativistic Hartree approximation or RHA. (This is also often called the one-loop approximation.) These contributions are an integral part of a fully relativistic description of nuclear structure, and as we have just seen, it is impossible to construct a meaningful nuclear response or consistent nuclear currents without including the negative-energy states. Thus, although the MFT ground state is covariant, causal, and internally consistent by itself, it is hard to justify the omission of the negative-energy contributions.

An inspection of δH reveals that, even with the indicated vacuum subtraction, the sum still diverges. Since the model QHD–I is renormalizable, however, the sum can be rendered finite by including counterterms in the lagrangian (8). These counterterms also appear in the hamiltonian, and they can be grouped with δH, resulting in a correction to the energy density of the form

$$\Delta\mathcal{E}(M^\star) = -\frac{1}{V}\sum_{\mathbf{k}\lambda}\left[(\mathbf{k}^2 + M^{\star 2})^{1/2} - (\mathbf{k}^2 + M^2)^{1/2}\right] - \sum_{n=1}^{4}\frac{\alpha_n}{n!}\,\phi_0^n\,. \tag{46}$$

The counterterms appear as a quartic polynomial in ϕ_0, and the (infinite) coefficients α_n are determined by specifying appropriate renormalization conditions on the energy. Following refs. 57 and 7, we will choose the counterterms to cancel the first four powers of ϕ_0 appearing in the expansion of the infinite sum over energies. Although this procedure is not unique, it has the virtue of minimizing the many-body forces arising from this vacuum correction, and it is easy to verify that only the first four terms in this expansion produce divergent results. The divergences can be defined by converting the sum to an integral and then regularizing dimensionally.[63]

After removing the divergences with the counterterms, the remaining terms are finite, and one finds

$$\Delta\mathcal{E}(M^\star) = -\frac{1}{4\pi^2}\Big\{ M^{\star 4}\ln(M^\star/M) + M^3(M - M^\star) - \frac{7}{2}M^2(M - M^\star)^2$$

$$+ \frac{13}{3}M(M - M^\star)^3 - \frac{25}{12}(M - M^\star)^4 \Big\}. \tag{47}$$

Just as in the MFT, $M^\star \equiv M - g_{\rm s}\phi_0$ is determined at each $\rho_{\rm B}$ by minimization, which produces the one-loop (RHA) self-consistency condition [compare eq. (20)]

$$M^\star = M - \frac{g_{\rm s}^2}{m_{\rm s}^2}\frac{\gamma}{(2\pi)^3}\int_0^{k_{\rm F}}{\rm d}^3k\,\frac{M^\star}{E^\star(k)} + \frac{g_{\rm s}^2}{m_{\rm s}^2}\frac{1}{\pi^2}\Big\{ M^{\star 3}\ln(M^\star/M)$$

$$+ M^2(M^\star - M) - \frac{5}{2}M(M^\star - M)^2 - \frac{11}{6}(M^\star - M)^3 \Big\}. \tag{48}$$

Note that the solution to this equation contains all orders in the coupling $g_{\rm s}$.

Although $\Delta\mathcal{E}$ has historically been called the "vacuum fluctuation correction," this appellation is somewhat unfortunate, since it does not involve any fluctuations. More precisely, $\Delta\mathcal{E}$ is the finite shift in the baryon zero-point energy that occurs at finite density, and is analogous to the "Casimir energy" that arises in quantum electrodynamics. We emphasize that the zero-point (one-loop) vacuum correction is insensitive to the short-distance structure of the baryons, as it arises solely from the change in the baryon mass in the presence of the *uniform* scalar field ϕ_0.[b]

To discuss the size of the one-loop vacuum correction, we can compare predicted quantities using a *fixed* set of parameters determined from the empirical saturation properties of nuclear matter in the MFT. (Alternatively, one could compare results after determining a separate set of RHA parameters that reproduce nuclear matter saturation.) Figure 12 shows the energy/nucleon for the MFT and RHA approximations. Observe that the equilibrium Fermi wavenumber $k_{\rm F}^0$ shifts by roughly 0.25 fm^{-1}, and the binding energy decreases by about 10 MeV when the one-loop vacuum correction is included. Although the latter is small on the scale of the large scalar and vector fields, the modification to the binding energy is significant, reflecting the sensitive cancellation between the attractive and repulsive components in the potential energy. The one-loop vacuum corrections are a direct consequence of the relativistic treatment of the nuclear many-body problem and are absent in a nonrelativistic approach.

In a finite nucleus, the zero-point corrections to the Dirac–Hartree hamiltonian (and the resulting energy functional) arise in just the same way as in eq. (13). The sum over the negative-energy eigenvalues [which augments eq. (36)] must now be computed using the spectrum of eq. (24), and both bound and continuum states are to be included. Since the eigenvalues cannot be determined in closed form due to the spatial dependence of the meson fields, the calculation is considerably more complicated than for nuclear matter. Nevertheless, the corrections have been computed essentially exactly, by starting with a local-density approximation[64] and then by systematically adding corrections from gradients of the meson fields.[65–69]

If the model parameters are adjusted from their previous values to reproduce the desired nuclear matter properties[c] when the zero-point corrections are added to the

---

[b]In addition, this correction cannot be calculated in nonrenormalizable meson–baryon models that are regularized by inserting *ad hoc* form factors at the meson–baryon vertices, since the uniform scalar field involves only zero-momentum-transfer components of the interaction between baryons, and the usual form factors have no effect on these contributions.

[c]We enforce equilibrium at $k_{\rm F} = 1.30\,{\rm fm}^{-1}$ with a binding energy of 15.75 MeV and a symmetry energy of 35 MeV. The scalar mass is again chosen to reproduce the observed rms radius of $^{40}{\rm Ca}$.

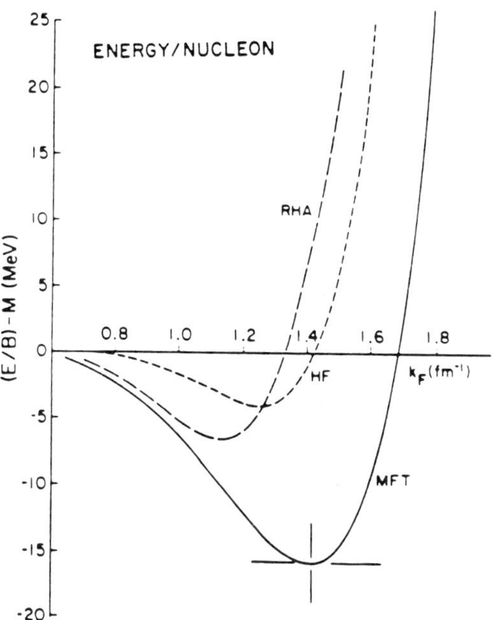

Figure 12.   Energy/nucleon in nuclear matter. The mean-field theory (MFT) results are shown as a solid line. The relativistic Hartree approximation (RHA), which includes the one-loop vacuum correction, produces the long-dashed line, while the short-dashed line is from a relativistic Hartree–Fock calculation (discussed below). All results are computed with parameter set L1 in table 1.

infinite nuclear matter energy density, the resulting parameters[68] are given in the row labeled RHA0 in table 1. The nuclear matter compressibility decreases to $K \approx 452$ MeV, and the baryon effective mass at equilibrium is $M^\star/M \approx 0.73$. How do the additional zero-point corrections change the systematic description of nuclear properties? Since the nuclear compressibility is still rather high, the calculated nuclei remain slightly underbound and the surface thicknesses are too small.[68] The increased $M^\star$ leads to a slightly more compressed spectrum of states near the Fermi surface (in heavy nuclei) and a smaller spin-orbit force. Thus, although predicted nuclear deformations still follow the correct systematics,[52] the spin-orbit splittings are too small by about 50%.[68] The shell oscillations in the nuclear interior are essentially unchanged. The derivative contributions to the local-density approximation are not negligible, but they generate only small changes in nuclear properties once the parameters are re-fitted to the standard input data. We conclude, therefore, that although the zero-point corrections give non-negligible corrections to the simple mean-field results, they do not improve the nuclear systematics.

The Relationship Between QHD and QCD

There is by now overwhelming evidence that hadrons are themselves composed of quarks and gluons. It is also widely believed that the quark-gluon dynamics is described by the nonabelian gauge theory of quantum chromodynamics. A discussion of the QCD lagrangian is beyond the scope of this talk, but this theory is presented in several texts (see, for example, Huang[70] or Rivers[71]), and a brief introduction is given in ref. 7. Quantitative QCD predictions at length scales relevant for hadronic and nuclear phenomena (which would specify precisely the relationship between QCD and QHD) are exceedingly difficult and are currently being pursued at the forefront of relativistic many-body theory. Here we will simply illustrate a simple model calculation of the phase diagram for nuclear matter, where the hadronic phase is described by the MFT of QHD–I, and the quark-gluon phase is described by QCD. This model contains a first-order transition between the hadronic and quark-gluon phases. Although there are some indications from recent QCD lattice simulations that there is no first-order phase transition at zero density for two quark flavors,[72] the nature of the transition at finite density is unknown. Thus a simple two-phase model may still provide a reasonable first approximation to the nuclear equation of state at all temperatures and finite densities.

Our previous discussion of the hadronic equation of state was restricted to zero temperature. The extension to finite temperature is straightforward in the MFT, since the hamiltonian is diagonal and the mean-field thermodynamic potential Ω can be calculated exactly.[d] The results for the scalar density, baryon density, energy density, and pressure are given by[7]

$$\rho_{\mathrm{s}} = \frac{\gamma}{(2\pi)^3} \int \mathrm{d}^3 k \, \frac{M^\star}{E^\star(k)} (n_k + \overline{n}_k) , \tag{49}$$

$$\rho_{\mathrm{B}} = \frac{\gamma}{(2\pi)^3} \int \mathrm{d}^3 k \, (n_k - \overline{n}_k) , \tag{50}$$

$$\mathcal{E} = \frac{g_{\mathrm{v}}^2}{2 m_{\mathrm{v}}^2} \rho_{\mathrm{B}}^2 + \frac{m_{\mathrm{s}}^2}{2 g_{\mathrm{s}}^2} (M - M^\star)^2 + \frac{\gamma}{(2\pi)^3} \int \mathrm{d}^3 k \, E^\star(k)(n_k + \overline{n}_k) , \tag{51}$$

$$p = \frac{g_{\mathrm{v}}^2}{2 m_{\mathrm{v}}^2} \rho_{\mathrm{B}}^2 - \frac{m_{\mathrm{s}}^2}{2 g_{\mathrm{s}}^2} (M - M^\star)^2 + \frac{1}{3} \frac{\gamma}{(2\pi)^3} \int \mathrm{d}^3 k \, \frac{k^2}{E^\star(k)} (n_k + \overline{n}_k) , \tag{52}$$

where the baryon and antibaryon distribution functions are

$$n_k(T, \nu) \equiv \frac{1}{1 + \mathrm{e}^{[E^\star(k) - \nu]/T}} , \quad \overline{n}_k(T, \nu) \equiv \frac{1}{1 + \mathrm{e}^{[E^\star(k) + \nu]/T}} , \tag{53}$$

and the reduced chemical potential is $\nu \equiv \mu - g_{\mathrm{v}} V_0$. (We set Boltzmann's constant $k_{\mathrm{B}} = 1$.) The nuclear matter equation of state at all densities and temperatures for this hadronic MFT model (QHD–I) is shown in fig. 13.

Figure 14 shows the self-consistent nucleon mass obtained my minimizing eq. (51) with respect to $M^\star$. The striking feature is the sudden decrease of the nucleon mass well below $T = M$. Thus, at high temperature (as at high density), the baryons are essentially massless.

For the quark phase, we use a simple model based on QCD and its properties of *asymptotic freedom* and *confinement*. Asymptotic freedom implies that when all of the momenta in a process are large, the renormalized coupling constant for that process becomes small, while confinement reflects the empirical fact that free quarks and gluons are never observed in the laboratory. We limit ourselves to the "nuclear domain," where only u and d quarks are important, and these quarks are assumed to be massless. Since the quark-gluon phase will be relevant only at extremely high temperatures or densities, where the particles have large momenta, we neglect interactions between the quarks except for a constant, positive energy/volume in the vacuum $(E/V)_{\mathrm{vac}} = b$, which models the confinement dynamics. This constant can be interpreted as the energy needed to create a bubble or bag in the vacuum, in which the noninteracting quarks and gluons are confined.

---

[d] We neglect the zero-point corrections from the Dirac sea in this section (see refs. 32 and 73), as well as thermal contributions from the massive isoscalar mesons.

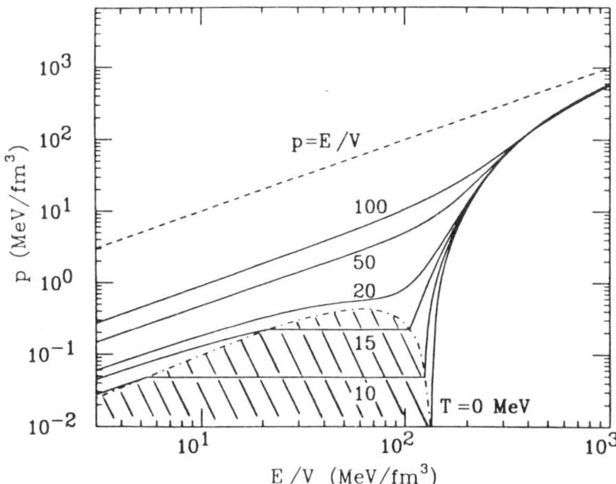

Figure 13.   Nuclear matter equation of state on isotherms. The dashed line represents the causal limit, and the shaded area shows the region of phase separation, as determined by a Maxwell construction. The solid curves are labeled by the temperature, in MeV, and the critical temperature is approximately 18.3 MeV. Here the parameter set L2 is used.

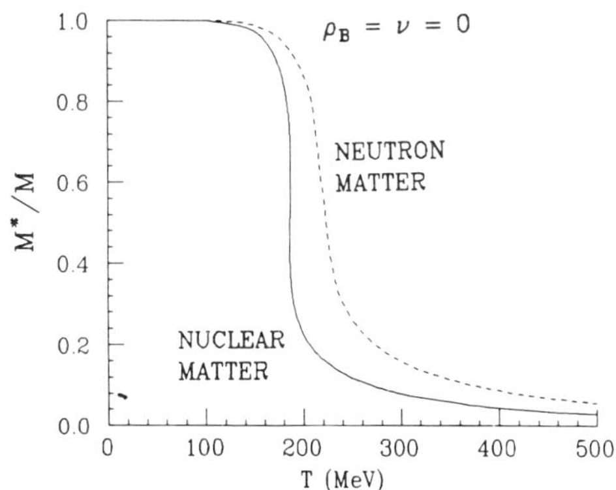

Figure 14.   Self-consistent nucleon mass as a function of temperature at zero baryon density. The value of $M^{\star}/M$ is 0.5 for $T \approx 186$ MeV and $T \approx 222$ MeV in nuclear and neutron matter, respectively. Parameter set L2 is used.

With this model, the QCD equation of state takes the simple form[74]

$$p = \tfrac{1}{3}(\mathcal{E} - 4b) \tag{54}$$

for both nuclear and neutron matter *at all densities and temperatures.* The "bag parameter" b determines the density of the hadron–quark phase transition, and since it represents a bulk property of nuclear systems, it may be different from values determined from the static properties of hadrons. The parameter b is constrained by requiring that at zero temperature and nuclear equilibrium density, the favored phase is the hadronic phase. Here we use $b = 131.2\,\mathrm{MeV/fm}^3$, taken from ref. 7, where it is shown that this value is consistent with the above requirement.

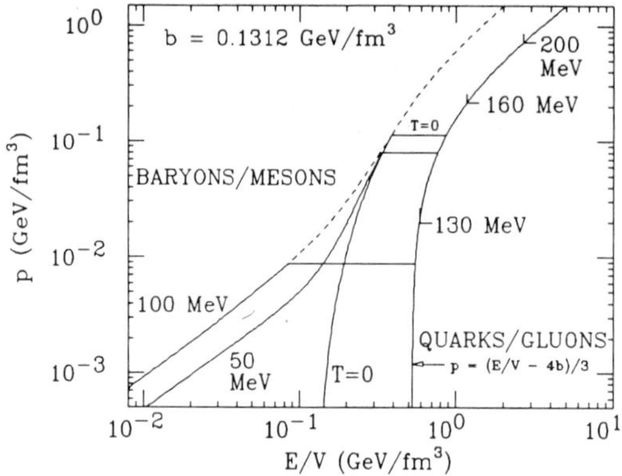

Figure 15. Equation of state isotherms for nuclear matter for the indicated values of the temperature. Phase equilibrium exists along the horizontal segments, which are determined by a Maxwell construction. The left-hand endpoints of the higher-temperature curves correspond to zero baryon density.

Figure 15 shows the resulting curves for this two-phase model of nuclear matter. The hadronic segments are determined from eqs. (51) and (52), together with the couplings and masses from row L2 in table 1. The quark-gluon curve is calculated from eq. (54), and Gibbs' criteria for phase equilibrium ($\mu_1 = \mu_2$, $p_1 = p_2$, $T = $ constant) are used to deduce the region of phase coexistence arising from the first-order transition. The complete phase diagram of nuclear matter is described by a one-parameter model, which allows for a simple correlation of phenomena occurring in very different regimes of the thermodynamic variables. Note that at high enough baryon density or temperature, *one always produces the quark-gluon phase* with an equation of state given by eq. (54).

Although this model is very simple, it has several nontrivial features. First, it is based on a completely relativistic calculation of the nuclear matter phase diagram and phase transition. Second, the statistical mechanics has been done exactly at all temperatures and densities. Third, the QHD model of the hadronic phase successfully describes many bulk properties of nuclear matter and finite nuclei. Fourth, the QCD phase obeys asymptotic freedom.

Finally, the model can be improved by systematically including additional mesonic degrees of freedom (such as thermally excited pions) and by going beyond the mean-field approximation for the hadronic phase. Interactions can also be included in the quark-gluon phase to calculate the corrections to the asymptotically free equation of state. At vanishing density, these corrections can be computed using the techniques of lattice gauge theory, as summarized by Gottlieb.[72] Lattice calculations, however, are not currently practical at finite density, although alternative nonperturbative techniques have been developed for hot QCD.[75,76]

EXTENSIONS AND ISSUES

We have seen that the mean-field approximation to QHD gives a concise and highly successful nuclear phenomenology. In QHD, however, one can in *principle* go beyond the MFT, calculate to arbitrary accuracy, and then compare with experiment. In *practice* this program is extremely difficult, since QHD is a strong-coupling relativistic quantum field theory.[e] Nevertheless, the Feynman rules for the Green's functions are well defined. Thus, just as in nonrelativistic many-body theory, one can use intuition to sum selected infinite sets of diagrams, determine the renormalized coupling constants by refitting nuclear matter properties, and then see whether the MFT results are *stable* under the inclusion of these additional contributions, while investigating new physical phenomena. Many such applications are discussed in ref. 7, where the historical development is presented. This 1986 volume is updated in ref. 77. It is impossible to review all that material and to give an exhaustive list of recent references here. We shall instead present some selected results of extensions beyond the MFT and discuss some of the issues raised by this work.

All of the extensions we discuss involve loop corrections to the MFT of one sort or another. In the MFT, the baryon Green's function can be written as[7]

$$G(k) = (\gamma_\mu k^{*\mu} + M^*)\left\{\frac{1}{k^{*2} - M^{*2} + i\epsilon} + \frac{i\pi}{E^*(k)}\delta[k_0^* - E^*(k)]\theta(k_F - |\mathbf{k}|)\right\}$$

$$\equiv G_F(k) + G_D(k) , \tag{55}$$

where $k^{*\mu} \equiv (k^0 - g_v V_0, \mathbf{k})$ is the kinetic four-momentum[f] and $E^*(k) = \sqrt{\mathbf{k}^2 + M^{*2}}$. The first term $G_F(k)$ is the Feynman propagator for a baryon of mass M^*, and the second term is the contribution arising from baryons already present at finite density; this latter contribution reproduces the MFT results. In discussing the following extensions, we shall frequently distinguish between results obtained with the full baryon propagator $G(k)$ and with just the second, or "density-dependent," contribution $G_D(k)$. Since the three- and four-momenta are constrained in $G_D(k)$, loop integrals over this second term give well-defined, finite results that are direct analogues of the terms arising in nonrelativistic many-body theory.

Relativistic Hartree–Fock

Relativistic Hartree theory is obtained by self-consistently summing the tadpole graphs in the baryon self-energy. Retention of G_D in the tadpoles gives rise to the MFT,

---

[e]Indeed, a theory that is not asymptotically free.

[f]Note that in closed-loop integrals, such as those involved in computing the ground-state energy, a simple shift of integration variables allows one to eliminate the dependence on $g_v V_0$.

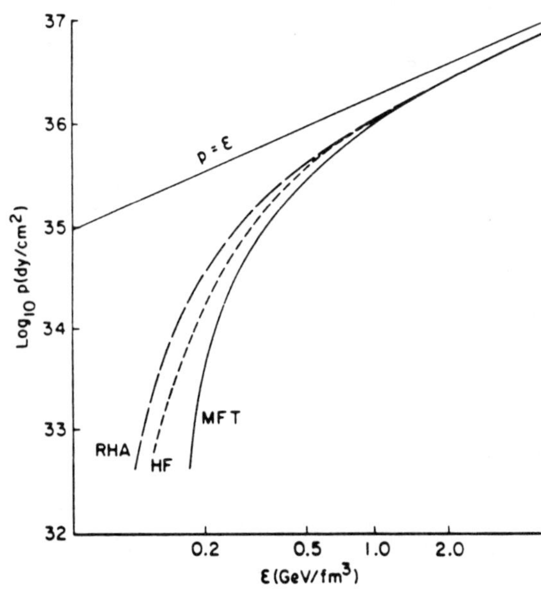

Figure 16.   Nuclear matter equation of state at zero temperature.[7] The curves are labeled as in fig. 12.

while the full G with appropriate counterterms $\delta\mathcal{L}$ produces the RHA, as discussed in the previous section. A characteristic result of QHD–I is that the Lorentz scalar and vector self-energies are very large; these contributions *cancel* in the binding energy but *add* in the spin-orbit interaction. Hartree–Fock (HF) theory is obtained by including the meson emission and reabsorption ("exchange") graphs in the baryon proper self-energy. A calculation of these graphs[78] with G_D for nuclear matter and the coupling constants L1 from table 1 produces the results shown in figs. 12 and 16. Some conclusions from this work are as follows:

1. These HF calculations give exchange terms that are the direct relativistic generalization of those arising when Slater determinants are used to find the best single-particle wave functions in nonrelativistic many-body theory.

2. Once the large scalar and vector self-energies have been established in relativistic Hartree theory, the inclusion of the exchange terms does not qualitatively alter the results.

3. The MFT is thus *stable* under the inclusion of the exchange contributions. In fact, after refitting to the equilibrium nuclear matter properties, the binding energy curves in relativistic Hartree and Hartree–Fock are almost indistinguishable.[7]

4. The HF equation of state approaches that of the MFT at high baryon density.

5. The fully self-consistent Hartree–Fock theory that retains the complete $G(k)$ and meson retardation is complicated;[79] it has not yet been successfully solved.

<u>Relativistic Brueckner–Bethe–Goldstone</u>

In nonrelativistic nuclear many-body theory, the two-nucleon potential V is strong and singular; it has a hard core at short distances. It is therefore necessary to solve for the correlated two-nucleon wave function in the medium. This Bethe–Goldstone wave function vanishes at the core radius and *heals* to the unperturbed wave function at large distances.[g] The product $V\psi$ is thus well defined and leads to a finite energy shift. In QHD–I, however, the interaction is integrable, so that results are sensible even at the mean-field level. The saturation of nuclear matter arises at this level from the saturation of the scalar attraction with increasing density. This scalar attraction is equivalent to an infinite series of velocity-dependent interactions in the nonrelativistic language. It is of interest to see if the MFT results are stable with respect to the inclusion of two-body correlations; in diagrammatic terminology, this amounts to summing the ladder diagrams for the baryon proper self-energy. Several relativistic calculations of this type have been carried out for nuclear matter.[80–82] Some observations, conclusions, and issues are as follows:[80]

1. The box diagram that forms the fundamental unit of the ladder sum involves a loop integration.

2. There are many possible reductions of the relativistic, four-dimensional Bethe–Salpeter equation in the medium to a unitary, three-dimensional integral equation that can be identified as the relativistic extension of the Bethe–Goldstone equation. Numerical results for the binding energy are sensitive to the reduction used.

3. Even after this reduction, calculated results are sensitive to the high-momentum part of the loop integrals, implying important contributions from baryon transitions to states lying well above the Fermi surface. Phenomenological form factors significantly reduce this sensitivity.

4. Since M^* enters in both the positive- and negative-frequency Dirac spinors, the binding energy of nuclear matter is sensitive to the self-consistency condition. At present, it is not known how to construct a self-consistency condition that leads to a conserving approximation when relativistic ladder diagrams are included.

5. Although the shifts in the binding energy of nuclear matter are large, the MFT is again stable with respect to the inclusion of two-baryon correlations on the scale of the large scalar and vector self-energies of the MFT.

6. The equation of state of nuclear matter again becomes the MFT result at high baryon density.

7. For free NN scattering, the corresponding calculation in QHD–I gives results in qualitative agreement with observation, but the charged isovector π and ρ mesons must be included to achieve a quantitative description.[80,82]

8. The fully relativistic Bethe–Salpeter equation in nuclear matter remains to be investigated.

9. In view of the above issues, a quantitative calculation of the binding energy of nuclear matter in QHD is not possible at this stage of the development.

---

[g]The vanishing wave function at short distance implies an insensitivity to what is going on inside the hard-core radius. This insensitivity is responsible for much of the success of nonrelativistic nuclear physics.

Relativistic Random-Phase Approximation (RRPA)

The sum of fermion ring diagrams gives the correct high-density limit for the correlation energy in the electron gas, where the Coulomb interaction is $e^2/\mathbf{q}^2$.[h] One might hope that the corresponding sum of baryon ring diagrams gives the correct high-density correlation energy in QHD–I, when the dimensionless meson masses m^2/k_F^2 become small. The sum of the ring diagrams for nuclear matter in QHD–I was first investigated by Chin.[57] Applications to the spectra of finite nuclei were initiated by Furnstahl,[58,83] and there has been a great deal of recent activity in this area.[84–91,60] We shall refer to the calculation of the rings that keeps only terms with at least one factor of G_D as the RPA; this calculation includes loops at least linear in the density and is the direct relativistic extension of the RPA in nonrelativistic many-body theory. The calculation that also includes the modification of the strong vacuum polarization in the nuclear medium due to the shift $M \to M^*$ will be called RRPA. Some results of this work and issues raised by it are the following:

1. The scalar and vector propagators mix in nuclear matter. Chin showed that at high density, vector meson exchange dominates in QHD–I. The excitation spectrum of nuclear matter in the RPA is that of zero sound, and the sound velocity c_0 approaches the speed of light from below as the baryon density gets large ($c_0 \to 1^-$ as $k_F \to \infty$). This implies that signals in the medium cannot propagate faster than the speed of light, in accord with special relativity. There are other branches in the excitation spectrum corresponding to meson propagation.

2. In nuclear matter in the RRPA, poles appear in the polarization propagator at zero frequency $q_0 = 0$ and finite wavenumber $|\mathbf{q}| \neq 0$; the value of this wavenumber is a few times the nucleon mass in QHD–I.[57,84–86,91] Such poles imply an instability of the system against density fluctuations of the corresponding wavelength. There are several possible interpretations of these results:

    (a) The RRPA for the propagators is inadequate.

    (b) Vertex modifications are important. (Phenomenological form factors affect the numerical results significantly.[84])

    (c) The composite structure of the baryon must be taken into account before one reaches distance scales where these poles develop.

    (d) The instability is real.

3. The polarization propagator governs the linear response of the system. The calculated isoscalar linear response in QHD–I leads to a reduction of the Coulomb sum rule.[87–90,60] There is roughly a 15% reduction in the RPA and an additional 15% reduction in the RRPA.[87] The observed experimental reduction of the Coulomb sum rule is one of the outstanding unsolved problems in traditional nonrelativistic nuclear physics.

4. In finite nuclei, it is essential to admix negative-frequency baryon components into the wave functions to bring the spurious $(1^-, 0)$ state down to zero frequency, to maintain current conservation, and to produce nuclear isoscalar magnetic moments that agree with the Schmidt lines.[61,92]

5. The RRPA calculation involves loop integrations and strong vacuum polarization. Physical effects come from the *modification* of these processes in the nuclear

---

[h] For the present purposes, the sum of rings is equivalent to the random-phase approximation (RPA).

medium. The calculation of hadronic contributions to strong vacuum polarization is a central problem in QHD. More generally, at some distance scale, this vacuum polarization should be calculated in terms of quarks and gluons.

The Quantum Vacuum

One goal in QHD is to systematically calculate the vacuum-loop corrections to the MFT. The one-loop correction from G_F produces the RHA, and we have seen that the MFT is stable against the one-loop vacuum contributions. A path-integral representation of the generating functional shows that the loop expansion is in $\hbar$, with the one-loop contribution the first quantum correction about the classical path of stationary action. One might hope that the two-loop correction would be the next term in a converging expansion. The calculation of the two-loop contribution to the properties of nuclear matter is nontrivial; it is carried out for QHD–I in ref. 12. (See also refs. 77 and 93.) The basic conclusion from this analysis is the following: although formally an expansion in $\hbar$, the parameters characterizing the loop contributions to the properties of nuclear matter are: (i) dimensionless coupling constants ($g_v^2/4\pi\hbar c, g_s^2/4\pi\hbar c^3$), (ii) lengths ($\hbar/m_vc, \hbar/m_sc, \hbar/Mc, 1/k_F$), and (iii) energies ($m_vc^2, m_sc^2, Mc^2, \hbar ck_F$). The loop expansion is essentially an expansion in the dimensionless coupling constants, which are large in QHD. The quantum corrections are correspondingly large, the series is not converging, and the MFT *is not stable against this perturbative loop expansion.* Clearly, an alternative procedure must be found to systematically and reliably calculate vacuum corrections to the MFT results in QHD.

The computation of hadronic contributions to vacuum polarization is a central issue in QHD, and more generally, in all of physics. There *are* indeed hadronic contributions to vacuum polarization; for example, a spectral analysis of the strong-interaction contribution to electromagnetic vacuum polarization shows that the spectral weight function starts at $4m_\pi^2$:[i]

$$\Pi_{\mu\nu}^{\mathrm{str}}(q) = (q_\mu q_\nu - q^2 g_{\mu\nu})\Pi(q^2) \;,$$

$$\Pi(q^2) = \frac{1}{\pi} \int_{4m_\pi^2}^{\infty} \frac{\rho(\sigma^2)}{\sigma^2 - q^2} \, d\sigma^2 \;. \tag{56}$$

In the complex q^2 plane, $\Pi(q^2)$ is an analytic function with a branch cut running along the real axis from $4m_\pi^2$ to ∞. The discontinuity across that cut for $4m_\pi^2 \le q^2 \le 9m_\pi^2$ comes from the electroproduction of two real pions. Thus the low-mass singularities of propagator and vertex functions are most efficiently expressed in terms of hadronic variables.

The contribution of two pions to vacuum polarization at all q^2 can be calculated in QHD–II; it will be well defined and finite. (Vertex modifications will exist in QHD–II and can also be included.) The result will not be meaningful if the dominant contribution to the loop integration comes from very high momenta and short distances, since there a QHD description is clearly wrong.[j] In this instance, one *must* invoke quarks, gluons, and QCD. At short-enough distances, one can use perturbative QCD. On the other hand, at low momenta and long distances, for example, in the low-mass part of the spectral function, QCD is a strong-coupling theory and the effective degrees of freedom are the hadrons, in this case, two pions. Hopefully, once vertex corrections are

---

[i]Pions are included in QHD–II; we use this as the simplest example for the present discussion.

[j]Moreover, for large σ^2 in eq. (56), the two-pion intermediate state by itself is inadequate.

included, QHD has the possibility of describing the low-lying hadronic contributions to
the spectral weight function and vacuum polarization.

The Vertex

As mentioned earlier, vector meson exchange dominates at high baryon density
in QHD–I, where the vector meson is coupled to the conserved baryon current. Milana[94] observes that in a theory with vector coupling, the vertex form factor is a decreasing function of q^2. This implies a decreased sensitivity to the high-momentum
or short-distance contributions to loop integrals and would provide a favorable situation for QHD. It is essential to include vertex corrections in QHD to determine its full
implications.[k] The fully off-shell vertex is complicated in any field theory. Reference
94 is the only vertex study in QHD of which the present authors are aware.

Pions and Chiral Symmetry

The relativistic neutral scalar and vector fields are the most important for determining the bulk properties of nuclear systems. Nevertheless, the lightest and most accessible
meson is the pion, whose interactions with nucleons and nuclei have been extensively
studied at the meson factories. It is therefore impossible to formulate a complete and
quantitative hadronic theory without including pionic degrees of freedom.

Pion interactions are constrained by the observed, nearly exact, SU(2) isospin symmetry. In addition, the soft-pion theorems, the partial conservation of the axial current
(PCAC) in weak interactions, and the theory of QCD indicate that pion dynamics is
also constrained by *chiral* symmetry, which enlarges the symmetry group from SU(2)
to $\mathrm{SU}(2)_L \times \mathrm{SU}(2)_R$. Here "$L$" and "$R$" denote left- and right-handed isospin rotations,
respectively. A thorough discussion of chiral symmetry is beyond the scope of this
talk, but we note that this symmetry has important consequences for the way mesons
interact with themselves, as we will discuss shortly.

Pions can be included in a chiral-invariant manner using the linear σ model.[22–24] This
model contains a pseudoscalar coupling between pions and nucleons and an auxiliary
scalar field σ to implement the chiral symmetry.[l] Weinberg's transformation[95] can
then be used to transform to a pseudovector (derivative) πN coupling multiplied by an
infinite series of nonlinear pion terms. There are two advantages to this transformation:
first, the decoupling of pions as $q_\lambda \to 0$ is now explicit (as are the soft-pion theorems),
and second, the new pseudovector coupling constant is

$$f^2 = g_\pi^2 \left(\frac{m_\pi}{2M} \right)^2 \approx 1.0 \ . \tag{57}$$

This is much smaller than the pseudoscalar coupling constant $g_\pi^2/4\pi \approx 14.4$.

In the limit that the chiral σ mass $m_\sigma \to \infty$, the auxiliary scalar field decouples,
and what remains for the pions and baryons is the nonlinear σ model of Weinberg.[95]
For any finite value of m_σ, however, the theory is renormalizable, and it can be used to
calculate nuclear pion processes.[21,96]

Nevertheless, there is a serious problem here.[7] Suppose the auxiliary scalar field σ in
the chiral-invariant linear σ model is identified with the low-mass scalar field ϕ in QHD–
I. Then the nonlinear meson couplings $(\phi^3, \phi^4, \phi^2\pi^2)$ that remain after spontaneous

---

[k] These vertex corrections reflect the internal hadron structure present in QHD.

[l] The presence of an isoscalar vector field V^μ coupled to the baryon current with minimal coupling
affects none of these arguments.

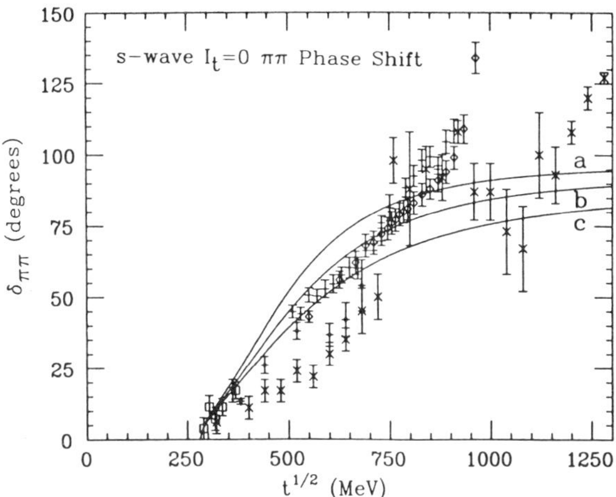

Figure 17.   The s-wave isoscalar $\pi\pi$ phase shift as a function of the total cm energy.[98] The chiral σ masses used here are (a) $m_\sigma = 950$ MeV, (b) 1400 MeV, and (c) 14 GeV.

symmetry breaking are so strong that they preclude a successful MFT of nuclear matter and finite nuclei.[97,21] Equivalently, nonlinear many-body forces are implied that are difficult to reconcile with what is known about nuclear physics.

A possible resolution of this problem is that the chiral σ mass m_σ is very large, reducing the effects of the nonlinear couplings, and that the low-mass scalar field ϕ in QHD–I is generated *dynamically* through the couplings of the chiral σ and the pions. In ref. 98, the process $\pi + \pi \rightarrow \pi + \pi$ is investigated within the framework of the σ model with a high mass σ. The chiral-invariant Born amplitude is unitarized, and the resulting phase shift in the $(0^+, 0)$ channel is shown in fig. 17. One observes a broad, low-mass, near-resonant amplitude in this channel, even though the chiral σ has a large mass.

When this model $\pi\pi$ scattering amplitude is included in the two-pion-exchange part of the NN interaction (see fig. 18), the result is a dynamically generated, broad, low-mass (≈ 600 MeV) peak that resembles the exchange of a light scalar meson. This peak arises even when the *chiral* scalar field has a large mass ($m_\sigma \approx 10$ GeV). It is further demonstrated in ref. 98, within the framework of the linear σ model, that this $(0^+, 0)$ channel leads to the observed intermediate-range attraction in the NN force.

The evident role of the low-mass scalar meson channel in nuclear physics can therefore be understood within the framework of chiral symmetry. The importance of the resulting scalar-isoscalar mean field and optical potential in producing a successful nuclear phenomenology was illustrated in the preceding section. Note, however, that the representation of this effective hadronic degree of freedom through the ϕ field in the local relativistic quantum field theory of QHD–I is a much more sweeping assumption.

For the pion-nucleon interaction, one is left with the nonlinear σ model of Weinberg: derivative couplings to baryons multiplied by $[1 + (f/m_\pi)^2 \boldsymbol{\pi}^2]^{-1}$. In the end, one has a chirally invariant theory including pions that reproduces the soft-pion theorems, that

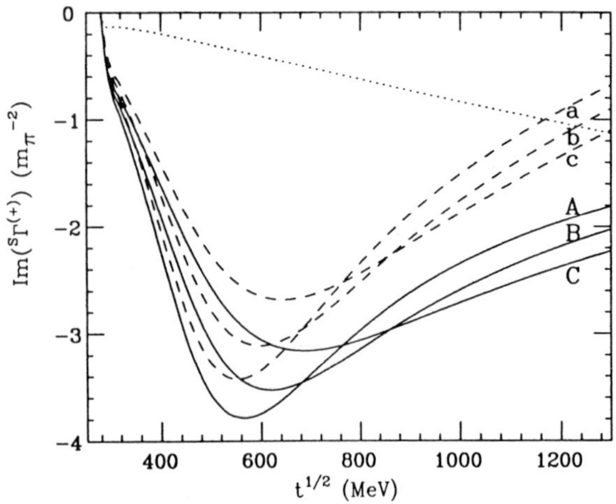

Figure 18.   The spectral weight function for the NN interaction in the scalar-isoscalar channel, from ref. 98. Note that the interaction is attractive. The solid (dashed) curves give the result before (after) the subtraction of the iterated two-pion-exchange box diagram. The chiral mass is $m_\sigma = 950$ MeV, 1400 MeV, and 14 GeV for the curves labeled A (a), B (b), and C (c).

produces a low-mass effective scalar meson dynamically, and that is renormalizable for any large but finite m_σ.

The ρ can be included in this renormalizable theory by starting with a nonabelian Yang–Mills theory based on isospin and then using spontaneous symmetry breaking and the Higgs mechanism to generate the ρ mass, as in the standard model of electroweak interactions. The resulting theory QHD–II, with σ, ω, π, and ρ mesons, is discussed more fully in ref. 7.

Hadronic Degrees of Freedom
<u></u>

The essential phenomenological features of the low-energy πN interaction are that low-momentum pions interact weakly with nucleons (they decouple as $q_\lambda \rightarrow 0$) and that the interaction is dominated by the first pion-nucleon resonance, the $\Delta(1232)$. This resonance represents the first excited state of the baryon, with $(J^\pi, T) = (\frac{3}{2}^+, \frac{3}{2})$. It is essential to have this degree of freedom in the theory, or the results will look nothing like nuclear physics. It is impossible to put a field with these quantum numbers into a simple renormalizable lagrangian. Thus the hope is that this degree of freedom, as with the low-mass scalar field, is again produced *dynamically* within the model. Fortunately, as has been shown in ref. 99, this is indeed the case.

In ref. 99, the sum of πN ladder diagrams with nucleon exchange is investigated within the framework of the chiral πN theory discussed above. Partial-wave dispersion relations are used, the one-baryon-exchange mechanism is input as the driving term, and the resulting integral equations are solved with the N/D method. This is a relativistic extension of nonrelativistic Chew–Low theory. As with Chew–Low, a resonance is found in the $(\frac{3}{2}^+, \frac{3}{2})$ channel.

The box diagram in the ladder sum involves a loop integral, which is finite and well defined in this renormalizable theory. Nevertheless, the loop integration involves significant contributions from high momenta or short distances, and thus the *position* of the resonance is sensitive to the approximations made; the *width* is much less so.[m] It is clear, however, that the first excited state of the nucleon, the $\Delta(1232)$ with $(\frac{3}{2}^+, \frac{3}{2})$, which is the dominant feature of low-energy pion-nucleus interactions, can be generated *dynamically* in QHD.

Other baryon properties arise through different loop integrals in QHD. For example, the vertex diagram consisting of the emission of a pion, its interaction with the virtual electromagnetic field, and its reabsorption by the nucleon contributes to the nucleon's anomalous magnetic moment. The two-pion contribution gives the low-mass, or long-distance, part of the spectral weight function for the anomalous magnetic form factor $F_2(q^2)$:

$$F_2(q^2) = \frac{1}{\pi} \int_{4m_\pi^2}^{\infty} \frac{\rho_2(\sigma^2)}{\sigma^2 - q^2} \, d\sigma^2 \; . \tag{58}$$

Assume that the two-pion contribution arising from this vertex diagram dominates the spectral weight function everywhere. One then gets a semi-quantitative account of the isovector anomalous magnetic moment and its mean square radius.[7][n]

A few comments are relevant here:

1. The dynamical model of ref. 99 allows for the investigation of many interesting questions concerning the behavior of the Δ in the many-body nuclear system, for example, its binding energy in nuclear matter, its optical potential, and the modification of its electroweak properties in the nuclear medium.

2. The production of effective degrees of freedom through dynamical means in QHD, as for the low-mass scalar field with $(0^+, 0)$ and the first excited state of the baryon with $(\frac{3}{2}^+, \frac{3}{2})$, while gratifying, also raises serious issues:

   (a) When does one stop? It is apparently necessary to investigate all possible hadronic channels for resonant behavior; there will be many such resonances.

   (b) Which of these resonances must be included as dynamical input in the generation of others?[o]

   (c) And, a more serious question, which of these hadrons are to be included through local fields in a QHD lagrangian density?

   (d) It is only a hope that one can get a self-contained dynamical description of nuclear physics in the low-energy hadronic sector with a few judiciously chosen hadronic degrees of freedom and a local, relativistic, renormalizable quantum field theory based on these degrees of freedom. It may be an impossible goal.

3. In principle, QCD gives a complete description of the nucleon and all its excited states. In the simple quark model, the nucleon moment arises from the quark

---

[m] Vertex corrections will again modify this sensitivity to the high-momentum behavior.

[n] The *isoscalar* anomalous moment vanishes in this approximation; experimentally it is indeed very small.

[o] This is reminiscent of the *bootstrap* theory of hadronic structure; all hadronic resonances were to be viewed as arising dynamically from the interactions of these same hadrons.

spins, and the first excited state arises from a spin-isospin-flip transition of a quark. Nevertheless, even within QCD, some part of the internal properties of the baryon and its excited states, particularly at large distances, must be equivalent to the strong interaction of hadrons discussed above.

The Relationship Between QHD and QCD

There is now considerable evidence that quantum chromodynamics (QCD), based on quarks and gluons as the underlying degrees of freedom, is the theory of the strong interaction. The colored quarks and gluons are confined to the interior of the hadrons by the strong nonlinear gluon couplings in QCD. At low momenta or large distances, the renormalized coupling constant is large, and QCD is a strong-coupling field theory; in this domain, the observed degrees of freedom are the hadrons. At large momenta, or short distances, QCD is asymptotically free, which implies that the renormalized coupling constant is small; here one can do perturbation theory with free quarks and gluons.

What is the relationship of QHD, the subject of this paper, and the underlying theory of QCD? There are several possibilities:

1. There is an approximate radius R in the hadron outside of which one can use QHD to describe the strong-interaction structure and inside of which one can use quarks and (asymptotically free) QCD. This is the basis of *bag models* of hadrons.[100]

2. One can attempt a similar separation in momentum space. The contribution of the nearby (low-mass) singularities in the spectral representations can be computed from hadronic processes, and the distant contributions can be computed from asymptotically free QCD. The two contributions can then be joined in some manner. This is the basic concept of QCD sum rules.[101]

3. One can assume two models for two different phases of nuclear matter: a baryon/meson phase described by QHD, and a quark/gluon phase described by (asymptotically free) QCD. The two descriptions can be connected with the thermodynamic conditions for phase equilibrium, as discussed earlier in this paper. The investigation of the phase diagram of nuclear matter with the Relativistic Heavy Ion Collider (RHIC) is a top priority for nuclear physics.

More generally, it is probable that at low energies and large distances, QCD can be represented by an *effective field theory* formulated in terms of a few hadronic degrees of freedom. This has indeed been shown to hold in the large-color limit of QCD. In QED, LePage has shown how to construct such a low-energy, large-distance effective field theory.[102] The couping constants in the effective lagrangian are computed from the short-distance behavior of the full, renormalized theory of QED. All possible couplings must be included in the low-energy effective lagrangian, which is then to be used at the "tree level" (that is, without considering loop integrals).

The underlying assumption of QHD is that of a local, relativistic quantum field theory formulated in terms of baryons and the lightest mesons $(\sigma, \omega, \pi, \rho)$. It is assumed that the theory is *renormalizable*, and one then attempts to extract predictions for long-range phenomena by computing both tree-level diagrams *and* renormalized quantum loop corrections. In the end, it may turn out that this assumption is unwarranted, and that the only meaningful interpretation of QHD is as an *effective* theory, to be used at the tree (or one-loop) level. The limitation to renormalizable couplings may then be too restrictive. Nevertheless, the phenomenological success of the MFT of QHD–I in the nuclear domain implies that *whatever the effective field theory for low-energy, large-distance QCD, it must be dominated by linear, isoscalar, scalar and vector interactions.* Recent calculations based on QCD sum rules indicate that this may indeed be true, as they find evidence for large (several hundred MeV) contributions to the scalar and vector parts of the baryon self-energy in nuclear matter.[103]

SUMMARY AND OUTLOOK

This paper is concerned with the theory of relativistic, interacting, nuclear many-body systems (baryon number $B \geq 1$). The only consistent theoretical framework for describing such systems is relativistic quantum field theory based on a local lagrangian density. As in nonrelativistic many-body theory, Feynman rules for the Green's functions allow one to calculate physical observables.

In this work we argue that the most efficient degrees of freedom for extrapolating away from the observed properties of nuclei are the hadrons: baryons and mesons. We require that the theory be renormalizable (*quantum hadrodynamics*); this defines a self-consistent, *purely* hadronic theory and severely constrains the form of the interaction. We focus on simple models: QHD–I, which contains neutrons, protons and the isoscalar, Lorentz scalar and vector (σ, ω) mesons; and QHD–II, the extension to include the isovector π and ρ mesons based on the linear sigma model.

The development starts with the relativistic mean-field (MFT) and Hartree approximations to QHD–I, and their application to both infinite nuclear matter and atomic nuclei. The principal new feature of the relativistic theory is that the baryon self-energy contains large Lorentz scalar and vector pieces, whose effects cancel in the binding energy but add in the spin-orbit interaction. We present some successes of the model, including the nuclear equation of state, the shell model, nucleon–nucleus scattering, and the addition of zero-point vacuum corrections.

We then discuss extensions to include quantum-loop processes, such as the contribution of two-nucleon correlations to the ground-state energy, the relativistic random-phase approximation for nuclear excitations, and two-loop contributions to the correlation energy. We discuss under what situations the MFT is stable against the inclusion of these effects (that is, when the MFT results are qualitatively unchanged) and when it is not. We also examine issues raised by the role of the quantum vacuum in QHD.

Pions are included within the framework of the chiral-invariant sigma model with spontaneously broken chiral symmetry (QHD–II). It is argued that the scalar field of QHD–I is to be associated with the low-mass dynamical enhancement in the $(0^+, 0)$ channel produced by the strong pion couplings, and *not* with the *chiral* scalar field, which may in fact be very massive. It is also shown that the first excited state of the nucleon, the $\Delta(1232)$, which plays such an important role in intermediate-energy nuclear physics, arises dynamically in QHD–II through the summation of nucleon exchange graphs—the relativistic extension of Chew–Low theory.

There is now convincing evidence that *quantum chromodynamics* based on quarks and gluons as the underlying degrees of freedom is the actual theory of the strong interaction; however, the derivation of nuclear structure from the strong-coupling, nonlinear, confining QCD lagrangian is far in the future.[(p)] We discuss the relationship between QHD and QCD. Possibilities include: an approximate separation in coordinate space

---

[(p)]In this regard, contemplate deriving superconductivity or superfluidity directly from the lagrangian of *quantum electrodynamics.*

for hadrons, with an exterior region where one uses QHD and an interior region where QCD is used; a separation in momentum space, where QHD is used for nearby singularities and QCD for those far away (QCD sum rules); and an interpretation in terms of two models for two distinct phases of nuclear matter.

More generally, it is probable that at low energies and large distances, QCD can be represented by an *effective field theory* formulated in terms of a few hadronic degrees of freedom. All possible couplings must be included in the low-energy effective lagrangian, which is then to be used at tree level. The underlying assumption of QHD is that of a local relativistic theory formulated in terms of baryons and the lightest mesons. The theory is assumed to be renormalizable, and one then attempts to extract predictions for long-range phenomena by computing both tree-level diagrams *and renormalized quantum loop corrections.* In the end, it may turn out that this assumption is untenable, and that the only meaningful interpretation of QHD is as an *effective* theory, to be used at the tree or one-loop level. The limitation to renormalizable couplings may then be too restrictive. Nevertheless, the phenomenological success of the MFT of QHD–I in the nuclear domain implies that *whatever the effective field theory for low-energy, large-distance QCD, it must be dominated by linear, isoscalar, scalar and vector interactions.*

What is the outlook? Future work will focus on problems such as:

- The investigation of meson propagation and the behavior of the dynamically induced hadronic resonances in nuclear matter.

- The theoretical search for other dynamically induced resonances.

- The study of the modification of nucleon properties in the nuclear medium. This includes the study of the nucleon–meson vertex functions in QHD.

- The continued attempt to solve QHD–II, including σ, ω, π, and ρ, as a strong-coupling field theory.

- The demonstration that QCD leads to large isoscalar, Lorentz scalar and vector interactions between baryons. (There is already some evidence from QCD sum rules that this is the case.)

- A continued effort to describe the internal structure of hadrons and the phase transition to the quark-gluon plasma through strong-coupling, lattice-gauge-theory simulations of QCD.

- Experimental studies of the behavior of nuclear systems under extreme conditions, which challenge our understanding of the nucleus, through new facilities such as the Continuous Electron Beam Accelerator Facility (CEBAF) and the Relativistic Heavy Ion Collider (RHIC).

REFERENCES

1. B. D. Keister and R. B. Wiringa, Phys. Lett. **173B**, 5 (1986).

2. D. Gogny, in: *Nuclear Physics with Electromagnetic Interactions* (H. Arenhövel and D. Drechsel, eds.), Lecture Notes in Physics, vol. 108, p. 88, Springer, Berlin (1979).

3. J. M. Eisenberg and W. Greiner, *Nuclear Theory*, vols. I–III, North-Holland, Amsterdam (1987).

4. J. A. McNeil, J. R. Shepard, and S. J. Wallace, Phys. Rev. Lett. **50**, 1439 (1983).

5. J. R. Shepard, J. A. McNeil, and S. J. Wallace, Phys. Rev. Lett. **50**, 1443 (1983).

6. B. C. Clark, S. Hama, R. L. Mercer, L. Ray, and B. D. Serot, Phys. Rev. Lett. **50**, 1644 (1983).

7. B. D. Serot and J. D. Walecka, Adv. Nucl. Phys. **16**, 1 (1986).

8. G. Baym and L. P. Kadanoff, Phys. Rev. **124**, 287 (1961).

9. G. Baym, Phys. Rev. **127**, 1391 (1962).

10. E. Dagotto, A. Moreau, and U. Wolff, Phys. Rev. Lett. **57**, 1292 (1986).

11. F. Karsch, Nucl. Phys. **A461**, 305c (1987).

12. R. J. Furnstahl, R. J. Perry, and B. D. Serot, Phys. Rev. C **40**, 321 (1989).

13. W. E. Caswell and A. D. Kennedy, Phys. Rev. D **25**, 392 (1982).

14. J. C. Collins, *Renormalization*, Cambridge University Press, New York (1984).

15. R. Machleidt, K. Holinde, and Ch. Elster, Phys. Rep. **149**, 1 (1987).

16. R. Machleidt, Adv. Nucl. Phys. **19**, 189 (1989).

17. R. Vinh Mau, in: *Mesons in Nuclei*, vol. I (M. Rho and D. H. Wilkinson, eds.), North-Holland, Amsterdam (1979), p. 151.

18. P. G. Reinhard, Rep. Prog. Phys. **52**, 439 (1989).

19. J. D. Walecka, Ann. Phys. (N.Y.) **83**, 491 (1974).

20. B. D. Serot, Phys. Lett. **86B**, 146 (1979); **87B**, 403(E) (1979).

21. T. Matsui and B. D. Serot, Ann. Phys. (N.Y.) **144**, 107 (1982).

22. J. Schwinger, Ann. Phys. (N.Y.) **2**, 407 (1957).

23. M. Gell-Mann and M. Lévy, Nuovo Cim. **16**, 705 (1960).

24. B. Lee, *Chiral Dynamics*, Gordon and Breach, New York (1972).

25. J. D. Bjorken and S. D. Drell, *Relativistic Quantum Mechanics*, McGraw–Hill, New York (1964).

26. J. D. Bjorken and S. D. Drell, *Relativistic Quantum Fields*, McGraw–Hill, New York (1965).

27. C. Nash, *Relativistic Quantum Fields*, Academic, New York (1978).

28. C. Itzykson and J. Zuber, *Quantum Field Theory*, McGraw–Hill, New York (1980).

29. P. Ramond, *Field Theory: A Modern Primer*, Benjamin, Reading, MA (1981).

30. J. I. Kapusta, *Finite-Temperature Field Theory*, Cambridge University Press, New York (1989).

31. D. G. Boulware, Ann. Phys. (N.Y.) **56**, 140 (1970).

32. R. A. Freedman, Ph. D. thesis, Stanford University, 1978.

33. C. W. Misner, K. S. Thorne, and J. A. Wheeler, *Gravitation*, Freeman, San Francisco (1973).

34. Ya. B. Zel'dovich, Sov. Phys. JETP **14**, 1143 (1962).

35. A. R. Edmonds, *Angular Momentum in Quantum Mechanics*, 2nd. ed., Princeton University Press, Princeton, NJ (1957).

36. C. J. Horowitz and B. D. Serot, Nucl. Phys. **A368**, 503 (1981).

37. E. S. Abers and B. W. Lee, Phys. Rep. **9C**, 1 (1973).

38. J. W. Negele, Phys. Rev. C **1**, 1260 (1970); private communication (1982).

39. I. Sick and J. S. McCarthy, Nucl. Phys. **A150**, 631 (1970).

40. B. Frois, J. B. Bellicard, J. M. Cavedon, M. Huet, P. Leconte, P. Ludeau, A. Nakada, Phan Xuan Hô, and I. Sick, Phys. Rev. Lett. **38**, 152 (1977).

41. I. Sick, J. B. Bellicard, J. M. Cavedon, B. Frois, M. Huet, P. Leconte, P. X. Hô, and S. Platchkov, Phys. Lett. **88B**, 245 (1979).

42. J. Boguta, Nucl. Phys. **A372**, 386 (1981).

43. A. Bohr and B. Mottelson, *Nuclear Structure*, vol. I, Benjamin, New York (1969).

44. L. Ray and P. E. Hodgson, Phys. Rev. C **20**, 2403 (1979).

45. W. H. Furry, Phys. Rev. **50**, 784 (1936).

46. L. D. Miller, Phys. Rev. C **9**, 537 (1974).

47. L. D. Miller, Phys. Rev. C **12**, 710 (1975).

48. A. L. Fetter and J. D. Walecka, *Quantum Theory of Many-Particle Systems*, McGraw–Hill, New York (1971).

49. C. E. Price and G. E. Walker, Phys. Rev. C **36**, 354 (1987).

50. S.-J. Lee, J. Fink, A. B. Balantekin, M. R. Strayer, A. S. Umar, P.-G. Reinhard, J. A. Maruhn, and W. Greiner, Phys. Rev. Lett. **57**, 2916 (1986); **59**, 1171(E) (1987); **60**, 163 (1988).

51. W. Pannert, P. Ring, and J. Boguta, Phys. Rev. Lett. **59**, 2420 (1988).

52. R. J. Furnstahl, C. E. Price, and G. E. Walker, Phys. Rev. C **36**, 2590 (1987).

53. G. Leander and S. E. Larsson, Nucl. Phys. **A239**, 93 (1975).

54. D. Vautherin, Phys. Rev. C **7**, 296 (1973).

55. J. A. McNeil, L. Ray, and S. J. Wallace, Phys. Rev. C **27**, 2123 (1983).

56. L. Ray, Phys. Rev. C **19**, 1855 (1979).

57. S. A. Chin, Ann. Phys. (N.Y.) **108**, 301 (1977).

58. R. J. Furnstahl, Ph. D. thesis, Stanford University, 1985; Phys. Lett. **152B**, 313 (1985).

59. P. Blunden and P. McCorquodale, Phys. Rev. C **38**, 1861 (1988).

60. J. R. Shepard, R. Rost, and J. A. McNeil, Phys. Rev. C **40**, 2320 (1989).

61. J. F. Dawson and R. J. Furnstahl, Phys. Rev. C **42**, 2009 (1990).

62. R. J. Furnstahl, in: *Relativistic Nuclear Many-Body Physics* (B. C. Clark, R. J. Perry, and J. P. Vary, eds.), World Scientific, Singapore (1989), p. 337.

63. G. Leibbrandt, Rev. Mod. Phys. **47**, 849 (1975).

64. C. J. Horowitz and B. D. Serot, Phys. Lett. **140B**, 181 (1984).

65. R. J. Perry, Phys. Lett. **182B**, 269 (1986).

66. D. A. Wasson, Phys. Lett. **210B**, 41 (1988).

67. R. J. Furnstahl and C. E. Price, Phys. Rev. C **40**, 1398 (1989).

68. W. R. Fox, Nucl. Phys. **A495**, 463 (1989).

69. P. G. Blunden, Phys. Rev. C **41**, 1851 (1990).

70. K. Huang, *Quarks, Leptons, and Gauge Fields*, World Scientific, Singapore (1982).

71. R. J. Rivers, *Path Integral Methods in Quantum Field Theory*, Cambridge University Press, New York (1987).

72. S. Gottlieb, Nucl. Phys. **B20** (Proc. Suppl.), 247 (1991).

73. R. J. Furnstahl and B. D. Serot, Phys. Rev. C **43**, 105 (1991).

74. G. Baym and S. A. Chin, Phys. Lett. **62B**, 241 (1976).

75. E. Braaten and R. D. Pisarski, Phys. Rev. Lett. **64**, 1338 (1990); Nucl. Phys. **B337**, 569 (1990).

76. R. D. Pisarski, Nucl. Phys. **A525**, 175c (1991).

77. B. D. Serot, in: *From Fundamental Fields to Nuclear Phenomena* (J. A. McNeil and C. E. Price, eds.), World Scientific, Singapore (1991), p. 144.

78. C. J. Horowitz and B. D. Serot, Nucl. Phys. **A399**, 529 (1983).

79. A. F. Bielajew and B. D. Serot, Ann. Phys. (N.Y.) **156**, 215 (1984).

80. C. J. Horowitz and B. D. Serot, Phys. Lett. **B137**, 287 (1984); Nucl. Phys. **A464**, 613 (1987); Nucl. Phys. **A473**, 760(E) (1987).

81. M. R. Anastasio, L. S. Celenza, W. S. Pong, and C. M. Shakin, Phys. Rep. **C100**, 327 (1983).

82. R. Brockmann and R. Machleidt, Phys. Lett. **B149**, 283 (1984); Phys. Rev. C **42**, 1965 (1990).

83. R. J. Furnstahl, Phys. Rev. C **38**, 370 (1988).

84. R. J. Furnstahl and C. J. Horowitz, Nucl. Phys. **A485**, 632 (1988).

85. R. J. Perry, Phys. Lett. **B199**, 489 (1987).

86. T. D. Cohen, M. K. Banerjee, and C.-Y. Ren, Phys. Rev. C **36**, 1653 (1987).

87. C. J. Horowitz and J. Piekarewicz, Phys. Rev. Lett. **62**, 391 (1989); Nucl. Phys. **A511**, 461 (1990).

88. H. Kurasawa and T. Suzuki, Nucl. Phys. **A490**, 571 (1988).

89. K. Wehrberger and F. Beck, Nucl. Phys. **A491**, 587 (1989).

90. X. Ji, Phys. Lett. **B219**, 143 (1989).

91. K. Lim, Ph. D. thesis, Indiana University, 1990.

92. R. J. Furnstahl and B. D. Serot, Nucl. Phys. **A468**, 539 (1987).

93. K. Wehrberger, R. Wittman, and B. D. Serot, Phys. Rev. C **42**, 2680 (1990).

94. J. Milana, Phys. Rev. C **44**, 527 (1991).

95. S. Weinberg, Phys. Rev. Lett. **18**, 188 (1967); Phys. Rev. **166**, 1568 (1968); Physica **A96**, 327 (1979).

96. J. F. Dawson and J. Piekarewicz, Phys. Rev. C **43**, 2631 (1991).

97. A. K. Kerman and L. D. Miller, in: *Second High-Energy Heavy Ion Summer Study*, Lawrence Berkeley Laboratory report LBL–3675 (1974).

98. W. Lin and B. D. Serot, Phys. Lett. **B233**, 23 (1989); Nucl. Phys. **A512**, 637 (1990).

99. W. Lin and B. D. Serot, Nucl. Phys. **A524**, 601 (1991).

100. R. K. Bhaduri, *Models of the Nucleon—From Quarks to Solitons*, Addison–Wesley, Reading, MA (1988).

101. A. Radyushkin, *Lectures on QCD Sum Rules*, CEBAF, Newport News, VA (1991), to be published.

102. P. LePage, in: *From Fundamental Fields to Nuclear Phenomena* (J. A. McNeil and C. E. Price, eds.), World Scientific, Singapore (1991), p. 117.

103. T. D. Cohen, R. J. Furnstahl, and D. K. Griegel, Phys. Rev. Lett. (1991), in press.

CONTINUUM BOUND STATES - NEW PHENOMENA WITHIN QED

James P. Vary, John R. Spence, Charles J. Benesh,

D.K. Ross, Alan J. Sommerer

Department of Physics and Astronomy

Iowa State University, Ames, Iowa 50011, USA

INTRODUCTION

A continuum bound state (CBS) is a normalizable solution of a wave equation which is embedded in the continuum. It appears as a resonance with zero width in the scattering domain. This phenomena was first described by Wigner and von Neumann[1,2] for the Schrödinger equation.

The appearance of a zero width resonance is a mathematical possiblity. However, in the real world, there are perturbations which can induce a finite width when they are added to the calculations. In this work, we present results from relativistic two-body wave equations (RTBWE's) derived from QED which exhibit the CBS phenomena. The neglected higher order effects of QED could make finite contributions to these CBS solutions and further work will be required to estimate those effects.

We organize this presentation along the following path. First, we present calculations of e^+e^- scattering that, without explicit regularization, exhibit resonances. Next, we introduce an explicit regularization to show these resonances are manifestations of the CBS phenomena. We compare these resonance energies with e^+e^- coincidence peaks observed in heavy ion experiments at GSI[3,4]. Up to this point we are summarizing and extending previous results[5]. The electron-proton (e-p) scattering problem is then considered and the CBS phenomena is again obtained[6]. We introduce a model for production of e^+e^- CBS's through collisions of an on-shell positron with an off-shell bound electron[7,8].

ELECTRON-POSITRON SCATTERING - THEORY

We have solved the e^+e^- scattering problem using three different relativistic two-body wave equations, each of which follows directly from QED but with different approximations. We find that each yields $J = 0$, $L = 1$, $S = 1$ CBS's of the e^+e^- system. This work was motivated in part by the calculation[9] of an e^+e^- resonance with a mass of 1.579 MeV and width of 1.5 keV in this same channel using the Kemmer-Fermi-Yang[10] equation. In Ref. 9 the resonance is attributed to a strong attractive (in this channel) short-range dipole-dipole interaction which is motivated by QED.

In the first model we solve an equation (the "TD equation") which is obtained by the Tamm-Dancoff (TD) method.[11] We obtain the second model by dropping a term in the TD equation to yield an equation used in atomic physics and known as the no pair form of the Breit equation.[12] The third model consists of constructing the Blankenbecler-Sugar[13] (Bb-S) form of the second model. These equations are all suitable for describing the bound states of positronium.

Working in the center of momentum (CM) frame, the TD equation represents a complete treatment of QED through the Fock space of physical states with an electron, a positron and with zero or one photon. Since this TD equation is formulated for physical states it is gauge independent. The TD equation is <u>not</u> merely a three-dimensional reduction of a Bethe-Salpeter (B-S) equation.[14]

There are no adjustable parameters in these models. We obtain scattering solutions in a basis of cubic B-splines, and we have verified the results are stable against variations in the B-spline basis. The use of a spline basis can be viewed as solving the scattering problem in an L2 normalizable wave-packet basis. This is desirable to avoid the ultraviolet divergences often encountered in a plane-wave basis for treatment of this type of problem.

In the CM frame, the TD equation for equal mass particles becomes[15]

$$[E - 2E(\vec{q})] \Lambda_a^+(\vec{q})\Lambda_b^+(\vec{q})\phi(\vec{q}) = \Lambda_a^+(\vec{q})\Lambda_b^+(\vec{q}) \int d^3q' V(\vec{q}, \vec{q}')\Lambda_a^+(\vec{q}')\Lambda_b^+(\vec{q}')\phi(\vec{q}') \tag{1}$$

The $\Lambda^+(\vec{q})\,(\Lambda^-(\vec{q}))$ are projection operators that project positive (negative) energy free particle states and are defined by

$$\Lambda_a^{\pm}(\vec{q}) = \frac{1}{2}\left[1 \pm \frac{\vec{\alpha}_a \cdot \vec{q} + \beta_a m}{E(\vec{q})}\right] \tag{2a}$$

$$\Lambda_b^{\pm}(\vec{q}) = \frac{1}{2}\left[1 \pm \frac{-\vec{\alpha}_b \cdot \vec{q} + \beta_b m}{E(\vec{q})}\right] \tag{2b}$$

$$E(\vec{q}) = +(m^2 + \vec{q}^{\,2})^{1/2} \tag{2c}$$

where $\alpha^0 = \beta$, the matrices α_a^μ and α_b^μ are conventional Dirac matrices, and m is the electron mass. The changes needed for the unequal mass case are straightforward[6]. The kernel, $V(\vec{q}, \vec{q}\,')$ is the sum of three terms. Using $\vec{k} = \vec{q} - \vec{q}\,'$ the first term is:

$$V_1(\vec{q}, \vec{q}\,') = \frac{+e^2}{(2\pi)^3} \frac{\vec{\alpha}_a \cdot \vec{\alpha}_b}{k^2} F(\vec{q}, \vec{q}\,'; E) \tag{3}$$

with

$$F(\vec{q}, \vec{q}\,'; E) = 1 - \frac{E(\vec{q}) + E(\vec{q}\,') - E}{k + E(\vec{q}) + E(\vec{q}\,') - E} \tag{4}$$

The second term, $V_2(\vec{q}, \vec{q}\,')$, has the form:

$$V_2(\vec{q}, \vec{q}\,') = \frac{-e^2}{(2\pi)^3} \left(\frac{\vec{\alpha}_a \cdot \vec{k} \otimes \vec{\alpha}_b \cdot \vec{k}}{k^4}\right) F(\vec{q}, \vec{q}\,'; E) \tag{5}$$

Finally, $V_3(\vec{q}, \vec{q}\,')$ is the usual Coulomb term

$$V_3(\vec{q}, \vec{q}\,') = \frac{-e^2}{(2\pi)^3} \frac{1}{k^2} \tag{6}$$

Eqs. (1–6) define the TD equation, our most complete treatment of QED.

In order to elucidate the physics underlying the resonances we consider two approximations to the TD equation, each of which results in a relativistic wave equation which may be more familiar and which can be derived through alternative schemes using the B-S[14] equation as the starting point. The usual derivations starting from the B-S equation involve specific gauge choices. Of course, the phenomena we report here should be investigated with additional relativistic wave equations.

Our second model is well known in atomic physics[16] and we shall refer to it as the "no pair form of the Breit equation."[12] We obtain this model by approximating $F(\vec{q}, \vec{q}\,'; E) = 1$. It has been established that the no pair form of the Breit equation produces a good description of the bound states of positronium.[15]

We can also obtain the second model by making an instantaneous approximation[17] (IA) to the B-S equation in the radiation gauge. We can then obtain the third model by making a corresponding Bb-S reduction.[13] Note that $F(\vec{q}, \vec{q}\,'; E) = 1$ in this third model as well. This results in an equation which we refer to as the "Bb-S equation" and it has the form:[18]

$$\left[E - \frac{\vec{q}^2}{m}\right] \Lambda_a^+(\vec{q})\Lambda_b^+(\vec{q})\bar{\phi}(\vec{q}) = \Lambda_a^+(\vec{q})\Lambda_b^+(\vec{q}) \int d^3q' \bar{V}(\vec{q}, \vec{q}\,')\Lambda_a^+(\vec{q}\,')\Lambda_b^+(\vec{q}\,')\bar{\phi}(\vec{q}\,') \quad (7)$$

The amplitude $\bar{\phi}(\vec{q}\,')$ and the kernel $\bar{V}(\vec{q}, \vec{q}\,')$ are given by

$$\bar{\phi}(\vec{q}\,') = \sqrt{m/E(\vec{q}\,')}\phi(\vec{q}\,') \qquad (8a)$$

$$\bar{V}(\vec{q}, \vec{q}\,') = \sqrt{m^2/E(\vec{q})E(\vec{q}\,')}V(\vec{q}, \vec{q}\,') \qquad (8b)$$

Owing to the presence of the projection operators, all three equations have nonlocal kernels.

For a partial wave with fixed L, S, and J the K-matrix form of the above wave equations may be written as $K = V + VGK$ with

$$K = K(\vec{q}, \vec{q}\,') = < q; LSJM|K|q'; \ LSJM > \qquad (9a)$$

$$V = V(\vec{q}, \vec{q}\,') = < q; LSJM|V|q'; \ LSJM > \qquad (9b)$$

and G is the propagator appropriate to the model. We solved[19] the half-shell K-matrix equations for the $J = 0$, $L = 1$, $S = 1$ channel of positronium by expansion in a basis of 31 cubic B-splines[20] using a Galerkin[21] method developed for the nonrelativistic case.[22] We extended the method of evaluating nonsingular integrals by Hermite interpolation using cubic splines to evaluate singular integrals. In the process, the factor of k^{-4} in V_2 was written as

$$\frac{1}{k^4} = \left(\begin{array}{c} -\lim \\ \mu \to 0 \end{array}\right) \frac{\partial}{\partial\mu^2} \frac{1}{(k^2 + \mu^2)} \qquad (10)$$

and then it was converted to an integrodifferential operator.[23] These analytical and numerical methods were initially developed for the B-S treatment of quarkonium[24] in momentum space where they were found to be very stable. In this way the difficulties due to the singular behavior of the kernel and of the K-matrices at zero momentum transfer and the difficulties due to the relatively large value of the kernels at large momentum transfer could be treated in the spline basis.

We solve for the half-shell K-matrix, $K(q, q')$, since the diagonal elements are proportional to the tangent of the scattering phase shift. For example, in the Bb-S model

$$\tan\delta = -\frac{\pi q}{2}K(q, q) \qquad (11)$$

In order to handle the singular phase shifts involved we evaluate tangent of the full phase shift δ minus the phase shift δ_c due to $V_c = V_1 + V_3$. That is, we evaluate

$$\tan(\delta - \delta_c) = \frac{\tan\delta - \tan\delta_c}{1 + \tan\delta\tan\delta_c} \tag{12}$$

where separate K-matrix solutions, one with the full interaction V and one with V_c are solved and used to evaluate the right-hand side of Eq. (12). We obtain stable solutions to Eq. (12) and $\tan(\delta - \delta_c)$ was nearly zero between resonances.

We define a resonance to occur when $\delta - \delta_c$ changes abruptly by π. Resonances in all three models occur at the same invariant total masses of 1.351, 1.498, 1.659, 1.830, 2.009, and 2.195. We have investigated the mass range below 1.35 MeV and above 2.25 MeV, and it appears there may be additional resonances but further work is needed to assure precision information on those resonances.

The analytical and numerical methods outlined above have been extensively tested and the resonance locations are found to be extremely stable. There is essentially no dependence on the ultraviolet cutoff of this spline basis.

In order to extract the width we now introduce an explicit regularization by giving the photon a small mass, μ. We stress that this explicit regularization was not used to locate the resonances but only to extract the width through the following analytical and numerical procedures.

For this regularized problem we again evaluate Eq. (12) through the resonances and denote the phase shifts by δ_{μ_0} and $\delta_{c\mu_0}$ with the photon mass μ_0 chosen to be small. We adjusted the Born term in the K-matrix for the screened Coulomb interaction to equal the analytic result to within machine accuracy. The same adjustment was then applied to the Born contributions of V_1 and V_2. Working in units of 10 MeV ≈ 20 m, we could use $\mu_0 \approx 10^{-10}$ or an even smaller value in these calculations. It is well known that the Born approximation phase shift $\left(\delta_\mu^B\right)$ for the screened Coulomb potential follows

$$\tan\delta_\mu^B = \frac{\ln\mu}{\ln\mu_0}\tan\delta_{\mu_0}^B \tag{13}$$

Now assume this holds for the full scattering phase shifts from the interaction $V_c = V_1 + V_3$. Then we note that the contributions to the kernel from V_2 have the same dependence on μ at $q = 0$ as V_c except they are multiplied on the right by a second derivative in the momentum.[24] This second derivative does not alter the μ-dependence arising from V_2 in our spline basis treatment. Thus, we argue that Eq.(13) also holds for phase shifts from $V = V_1 + V_2 + V_3$.

Consider the following ratio

$$\frac{\tan(\delta_\mu - \delta_{c\mu})}{\tan(\delta_{\mu_0} - \delta_{c\mu_0})} = \frac{\frac{\ln\mu}{\ln\mu_0}\left(1 + \tan\delta_{\mu_0}\tan\delta_{c\mu_0}\right)}{1 + \frac{\ln^2\mu}{\ln^2\mu_0}\left(\tan\delta_{\mu_0}\tan\delta_{c\mu_0}\right)} \tag{14}$$

Near a resonance, this goes as $\ln\mu_0/\ln\mu$. Next, we note that the tangent of the phase shift goes as $\Gamma/2(E_R - E)$ at a resonance so we arrive at

$$\Gamma_\mu \sim \frac{\ln\mu_0}{\ln\mu}\Gamma_{\mu_0} \tag{15}$$

which implies $\Gamma_\mu \to 0$ as $\mu \to 0$. The addition of any other interaction in the kernel could invalidate this argument and, presumably, give rise to a finite width.

We have verified that Eq. (15) is correct by explicit numerical calculations with $\ln\mu$ varying over many orders of magnitude. Thus, based on these analytical and

numerical results, we argue that all resonances obtained from QED using these kernels
will have zero width and will constitute continuum bound states. This argument
applies to our e^+e^- as well as to our e-p results.

There has been an argument[25] that there can be no resonances in QED in the
energy range of our results which consist solely of an e^+e^- pair. However, the results
of Ref. 25 also show that with even infinitessimal admixtures of higher Fock states,
resonances can indeed exist in QED within our energy range. We readily admit that
such admixtures are expected from improvements to our calculations.

To illustrate the fact that the CBS phenomena, which is non-perturbative, can
occur with weak, long-range potentials, it is worth considering a specific example. It
is easy to show with the Wigner-von Neumann potential[1] that a CBS can be placed at
any desired positive energy and held there as the potential is made arbitrarily weak.
For any finite value of the potential strength, the CBS exists as a normalizable solution
in the continuum. The rms radius of the CBS grows as the inverse strength of the
potential. Only at the point the potential vanishes does the CBS return to a plane
wave.

We note that the claims of Ref. 25 are also in conflict with the theoretical results
of other authors[26,27] who obtain resonances in QED through methods distinctively
different from those we employ.

The approach closest to our work is a set of e^+e^- calculations by Dehnen and
Shahin[26] (which we refer to as "DS") in which they also solved the no-pair equation
as well as an equation different from any we solved for the $J^\pi = 0^+$, and $J^\pi = 0^-$
channels. However they solved them approximately and in coordinate space. DS
neglect the Coulomb interaction and obtain $J^\pi = 0^+$ resonances at total masses of
1.338, 1.410, 1.499, 1.617, 1.776 and 2.010 MeV. DS obtain large widths ~ 10 keV
for these resonances. We note that when we drop the Coulomb term our solutions
develop large widths which are comparable to the widths of DS. Further effort will
be required to see if simulations in momentum space of the other coordinate-space
aproximations invoked by DS can explain the discrepancies in the resonance locations.
For our purposes it is encouraging that calulations by two independent groups with a
total of four relativistic two-body wave equations using vastly different methods each
yield six resonances in the $J^\pi = 0^+$ channel of photonium in approximately the same
energy range.

The authors of Ref. 27 obtain S-wave resonances in a scalar version of QED. The
differences between the approach of Ref. 27 and our own are great so that comparisons
of numerical results are not warranted. However, this same group has developed a
set of arguments[28] to show that RTBWE's derived from QED yield effective local
potentials of the Wigner and von Neumann class which give rise to CBS's. This
analytical connection is instructive and supportive of our results. It is important to
note that we have maintained the full, non-local, forms of our propogators so that our
CBS results have not arisen from a local approximation.

The high momentum behavior of the interaction is dominated by V_2 and when
we drop the V_2 term from any one of the models, the resonances disappear. This
indicates that the V_2 part of the transverse photon physics is essential to the resonance
phenomena. In Fig. 1 we plot, in coordinate space, the square of the amplitude from
the B-S equation for the state at 1.66 MeV. It has a peak near the Compton wavelength
of the electron. Since these calculations are performed with a small photon mass, the
state has a finite width and there is a weak oscillatory tail which is imperceptible in this
figure. This tail is expected to vanish as the photon mass goes to zero. Consequently,
we characterize the size of this CBS as ~ 500 fm. This is considerably larger than
the crude estimate made in Ref. 5 before the coordinate space amplitudes became
available. We have also calculated the amplitudes for the 1.83 MeV state and we find
it is similar in size to the 1.66MeV state. This size scale is two orders of magnitude

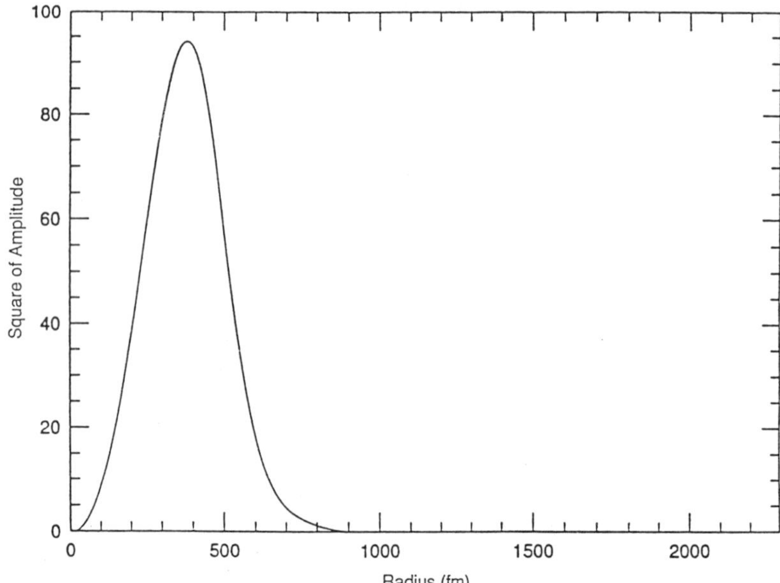

Figure 1 Amplitude squared vs radius in fm of the CBS in photonium at 1.66 MeV. The amplitude is obtained as a solution of the Bb-S equation in momentum space and Fourier transformed to coordinate space before squaring it. See text for discussion.

smaller than the conventional atomic scale and is approximately consistent with the energy spacings. This smaller scale is also indicative of the important role of the transverse photon physics. Since the range of the nuclear force is characterized by the inverse pion mass (1.4 fm), we refer to this new scale as ultra-nuclear or infra-atomic. To distinguish from the well-known positronium system, we refer to this system with continuum bound states as "photonium."

The importance of these high momenta and short distances raises the issue of whether our approximations to QED are adequate. Indeed, we are now preparing to extend the framework to include the effects of other QED processes which could become significant at short distances. This is a challenging enterprise which will involve a major effort. Since experiments are continuing we concentrate on how our current results may be of some utility.

PHOTONIUM - COMPARISON WITH EXPERIMENT

This investigation was originally motivated by the observation of e^+e^- coincidence peaks[3,4] in heavy-ion experiments as well as by numerous[29–34] Bhabha scattering experiments. For heavy ions, the coincidence data have the best resolution and indicate peaks at mass values of 1.642 ± 0.010, 1.772 ± 0.015, and 1.832 ± 0.010 MeV in data from the EPOS group[3] and peaks at 1.562 ± 0.010, 1.662 ± 0.010, 1.738 ± 0.010, 1.831 ± 0.008, and 1.917 ± 0.010 MeV in the data from the ORANGE group.[4] The measured widths range from about 30 to 60 keV and are close to experimental resolution so that they may be taken as upper limits on the natural widths.

Both experiments agree on the strongest peak at $\sim$ 1.83 MeV and on a secondary peak at $\sim$ 1.65 MeV. These two peaks agree with our third and fourth states at 1.83 and 1.66 MeV, respectively. The two experiments disagree with each other on the remaining secondary peaks. None of our other states agree with any of these other secondary peaks. Should some of these other secondary peaks be confirmed, we speculate they may represent e^+e^- resonances in partial waves other than the channel reported here. We present our results in Fig. 2 along with the EPOS data[3] which ranges over a wider kinematic range than the ORANGE data.[4]

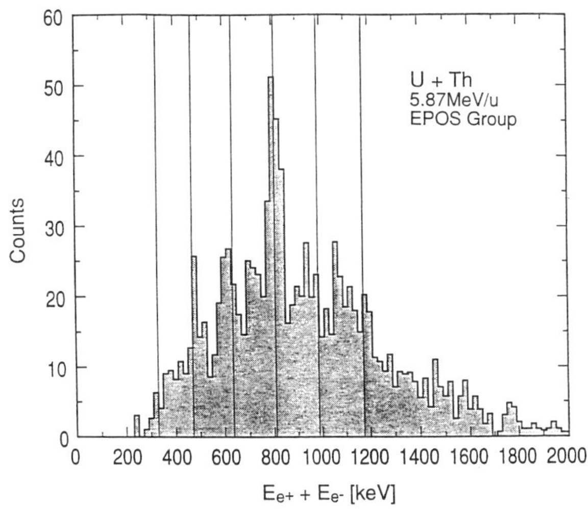

Figure 2 Electron-positron coincidence data (histogram) from Ref. 3 plotted versus the sum of their kinetic energies for U + Th collisions at 5.87 MeV/u. The vertical lines indicate our CBS's after subtracting the electron and positron rest mass.

The two photonium states we predict at the lower masses, and the two at the higher masses, lie either outside the range of masses studied experimentally or in regions where the experimental sensitivities were somewhat weaker due to lower statistics as seen in Fig. 2. Consequently, our prediction of states at 1.351, 1.498, 2.009, and 2.195 MeV constitutes a nontrivial test for our work.

We will discuss below a simple model for describing the interaction of an incident positron with a bound electron to form photonium. This model may be useful to planning and interpreting experiments using positron beams and high Z targets. However, to completely understand the photonium formation rate in heavy ion collisions will require a detailed treatment of the time-dependent effects of the strong Coulomb fields of interacting nuclei which is beyond the scope of the present work. However, it appears that in order to have a connection with our results the strong Coulomb fields of the heavy ions must be capable of significantly enhancing the widths since a large fraction of the produced positrons are concentrated in the observed peak structures under certain experimental conditions.[3,4]

Direct e^+e^- (Bhabha) scattering experiments to search for resonances in narrow mass ranges have been performed with mixed results.[29-34] Resonances have been observed with masses of 1.702[29] 1.684[30] 1.832 MeV[31] 1.662 MeV[32], and at 1.682 MeV[33]. These observed resonances again appear to be consistent with our third and fourth theoretical states. However, resonances are not observed in other experiments[34] which provide stringent upper limits for the free space values of some of these widths.

The apparent inconsistencies of these results presents a challenge. We take note of the difficulty of these experiments and that even the successful experiments report only a weak signal for the resonances. We can, however, develop various scenarios based upon 3-body processes to accomodate the puzzling back-to-back decay in the laboratory frame recently reported.[30,32,33] In one particular scenario the incident positron comes nearly to rest in the nuclear Coulomb field (goes far off shell), forms photonium with an atomic electron which then drifts away and decays. In addition the nuclear Coulomb field enhances the width, as was the case in our interpretation of the heavy-ion results. In another scenario the positron multiple scatters from a correlated electron pair forming photonium with one of the electrons. If one of these scenarios is correct, it may help resolve the apparent conflicts among the Bhabha experiments. For example, some of the experiments with negative results were optimized to detect the decay of a system which had a large recoil velocity in the lab frame as dictated by free two-body kinematics. Our scenarios for stimulating photonium formation via three-body processes imply that experiments using atomic electrons as targets may benefit from designs with enhanced sensitivity for back-to-back decays in the lab system such as the experiment of Refs. 30, 32, and 33. Initial calculations for photonium production in collisions with a correlated electron pair show strong sensitivity to the electron pair correlation distance[35]. In this three-body model, photonium production is strongly enhanced using inner shell electrons of high Z atoms. Below we consider a simple two-body model in which an incoming positron collides with an off-shell electron to produce photonium.

ELECTRON-PROTON RESONANCES

Here we adapt our method for describing photonium within QED to the electron-proton (e-p) system at low energy. Our main goal is to search for e-p resonances analogous to photonium states and to characterize them in order to examine their possible relevance for enhancing fusion rates at temperatures in the eV range.

Treating the proton and electron as pure Dirac particles we employ the unequal mass version of the Bb-S equation described above. We solve this equation for the K-matrix with $0.05 \leq E_{cm} \leq 5.0$ eV in the $J = 0$, $L = 1$, $S = 1$ channel and obtain five very narrow resonances. This equation with $\vec{q}$ as the relative momentum vector and with $\mu = mM/(m + M)$ is written:

$$\left[E - \frac{\vec{q}^{\,2}}{2\mu}\right] \Lambda_e^+(\vec{q})\Lambda_p^+(\vec{q})\overline{\phi}(\vec{q})$$
$$= \Lambda_e^+(\vec{q}\,')\Lambda_p^+(\vec{q}) \int d^3\vec{q}\,' \overline{V}(\vec{q},\vec{q}\,')\Lambda_e^+(\vec{q})\Lambda_p^+(\vec{q})\overline{\phi}(\vec{q}\,'). \tag{16}$$

The $\Lambda^+(\vec{q})(\Lambda^-(\vec{q}))$ are projection operators that project positive (negative) energy-free particle states and are defined by

$$\Lambda_e^{\pm}(\vec{q}) = \frac{1}{2}\left[1 \pm \frac{\vec{\alpha}_e \cdot \vec{q} + \beta_e m}{E_e(\vec{q})}\right] \tag{17a}$$

$$\Lambda_p^{\pm}(\vec{q}) = \frac{1}{2}\left[1 \pm \frac{-\vec{\alpha}_p \cdot \vec{q} + \beta_p M}{E_p(\vec{q})}\right] \tag{17b}$$

$$E_e(\vec{q}) = +(m^2 + \vec{q}^{\,2})^{1/2} \tag{17c}$$

$$E_p(\vec{q}) = +(M^2 + \vec{q}^{\,2})^{1/2} \tag{17d}$$

The amplitude $\overline{\phi}(\vec{q}^{\,\prime})$ and the kernel $\overline{V}(\vec{q}, \vec{q}^{\,\prime})$ are given, in terms of the corresponding quantities from the "no-pair" equation, by

$$\overline{\phi}(\vec{q}^{\,\prime}) = \left[\frac{mM}{E_e(\vec{q}^{\,\prime})E_p(\vec{q}^{\,\prime})} \right]^{1/4} \phi(\vec{q}^{\,\prime}) \tag{18a}$$

$$\overline{V}(\vec{q}, \vec{q}^{\,\prime}) = \left[\frac{m^2 M^2}{E_e(\vec{q}^{\,\prime})E_p(\vec{q}^{\,\prime})E_e(\vec{q})E_p(\vec{q})} \right]^{1/4} V(\vec{q}, \vec{q}^{\,\prime}) \tag{18b}$$

In our work on e^+e^- scattering we solved the TD equation, the no-pair equation *and* the equal mass version of the Bb-S model. The e^+e^- resonance locations agreed through three significant figures among these three treatments.

We expect that our three RTBWE's would yield results that agree equally well in the e-p problem. Although the TD equation represents our most complete treatment of QED, it also involves more than an order of magnitude greater computational effort to solve. Because the relativistic Coulomb scattering problem in momentum space is difficult enough as is and we address the additional challenge of solving it at low energies, we adopted the simplest model from the e^+e^- work. Although the use of any RTBWE is only an approximation, it is the best we can do at present.

We solved the K-matrix equation for the $J = 0$, $L = 1$, $S = 1$ channel of hydrogen since this is the same channel which produced resonances in the e^+e^- problem. We used the same analytical and numerical methods and again checked for numerical stability with respect to variations in the spline basis.

Resonances occur at CM energies equal to 0.748, 1.347, 2.095, 3.032, and 4.707 eV. We have investigated the energy range below 0.70 eV and above 5.0 eV and it appears there may be additional resonances but further work is needed to assure precision information on those resonances. At this stage the calculations contain only an implicit regularization arising from expanding in the spline basis. The calculated widths here are all small compared to the kinetic energy and are consistent with zero using these procedures. In order to extract the width directly we would again need to consider what happens if we introduce an explicit regularization by giving the photon a small mass, μ. However, the analytical and numerical procedures are the same here as in the e^+e^- case. Thus these states of the e-p system are obtained also as zero-width resonances or CBS's. The V_2 term is again essential since, when it is dropped from the equation, the resonances disappear.

Because the resonances in the e-p system are much closer to threshold than those in the e^+e^- system, considerable additional effort is required to extract the amplitude directly through an entirely different calculation than the one we have performed here. We feel that a high momentum scale similar to that of the photonium system is responsible for the non-perturbative effects here. This infers a similar size scale for these resonances. To distinguish these compact continuum bound states of the e-p system from the conventional hydrogen system we will refer to this new system as "protonium."

The importance of these high momenta and short distances again raises the issue of whether our approximations to QED are adequate. Indeed, we are now preparing to extend the framework to include effects such as due to the proton form factors which could become significant at short distances.

In the limit $M \to \infty$ the V_2 term, when sandwiched between projection operators, goes to zero and no resonances would occur. We see, therefore, consistency with the

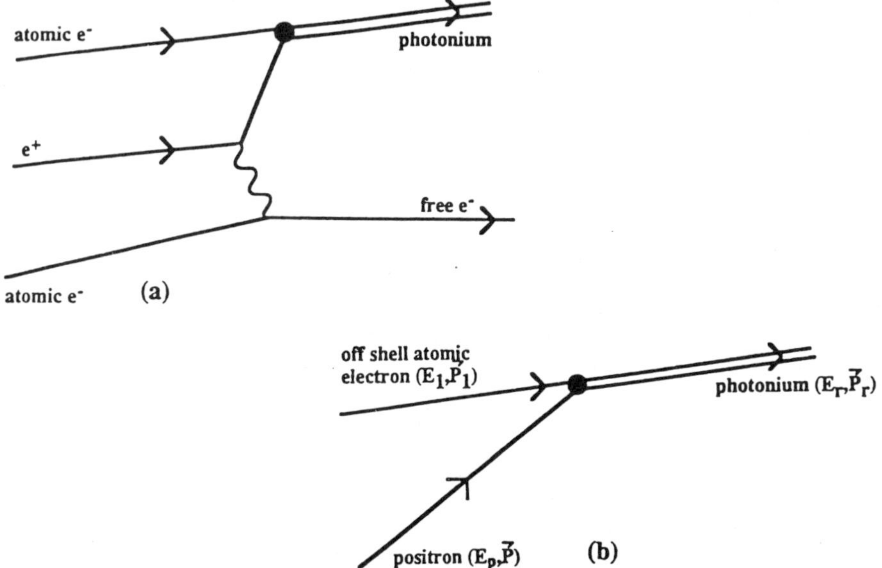

Figure 3 (a) Positron interacting with two correlated atomic electrons to produce photonium and a free electron. (b) Process considered here in which a positron
interacts directly with an off-mass-shell atomic electron to produce photonium.

Dirac equation treatment of e-p scattering where it is well-known that there are no
resonances. It is important to note that we are solving a true two-body equation,
unlike the Dirac equation, and that this seems to be the only way to have a consistent
nonperturbative treatment of two spin-1/2 particles.

RELATION TO EXPERIMENTS ON HYDROGEN

We are not aware of any experimental evidence for, or against, the existence of
these resonances although a photoionization experiment with atomic hydrogen has
recently been completed which scanned the region of these predicted resonances.[36] No
resonances were seen within the 30 meV resolution of the experiment. On the other
hand, there are some unexpected variations in near threshold ionization of hydrogen
through spin-tagged electron-hydrogen scattering.[37] It will require further theoretical
study to explore the possible connection between our theoretical results and these
provocative experiments.

PHOTONIUM FORMATION IN POSITRON SCATTERING

Taking into account the accumulated experimental and theoretical information,
Benesh, Ross, and Vary [35] have given a three-body model for photonium production
in which an incoming positron interacts with a correlated pair of atomic electrons
producing photonium and a free electron (see Fig. 3a). In this model, very strong
sensitivity to the electron pair correlation was obtained. Calculating a cross section
based on the diagram in Fig. 3a is a computationally intensive project involving
the performance of multi-dimensional Monte Carlo integrations. Here we examine a

simple generalization of the original model by considering the scattering of an incident positron from a bound electron[38]. This two-body model maintains many of the desirable features of the original model while remaining computationally simple and allowing rapid comparison with a variety of experimental situations.

The simple two-body model is shown in Fig. 3b in which a positron interacts with an atomic electron and forms photonium directly. The atomic electron is assumed to be in an energy eigenstate with total energy $m_e - \epsilon_1$ where ϵ_1 is the binding energy. Our primary example will be a thorium atom and the main contributions to photonium production are expected from the deeply bound electrons. Hence, ϵ_1 was taken to be 104.7 Kev which is characteristic of the inner electrons of thorium. The atomic electron is also allowed to be off shell to account for all the effects of the interactions with the other electrons and with the nuclear charge in a simplified way. In the illustrative calculations presented herein we invoke a mean-field treatment of the bound electron. The simple model is also an approximation to the more complicated model of Fig. 3a in another sense. In the detailed Monte Carlo integrations associated with Fig. 3a, it was found that a significant contribution was made when the internal positron propogator was close to being on shell[35]. Our simple model discards one of the atomic electrons and simply incorporates an on-shell positron interacting with a single off-shell atomic electron.

Using the notation in Fig. 3b, we can write the differential cross section as

$$d\sigma = \frac{1}{\beta_e^+} \frac{m_e}{E_p} |M|^2 \frac{d^3 P_r n(|\vec{P_1}|)}{2E_r (2\pi)^3} 2\pi \delta(E_1 + E_p - E_r) \tag{19}$$

where over-all momentum conservation was used to integrate out the rest of the atom. The matrix element can be written as

$$M = -\bar{v}(P)u(P_1)g \left[\frac{F^2(P_1^2 - m_e^2)}{((P - P_1)^2 - F^2)m_e^2} \right] \tag{20}$$

where g is a coupling constant and where $n(|P_1|)$ represents the momentum distribution of the electrons in the target atom (the "Compton profile"). For our purposes we will obtain the Compton profile from a sum of absolute squares of Dirac wavefunctions in momentum space.

The expression in square brackets in (20) is our assumed vertex factor. The m_e^2 part of this expression forces the matrix element M to vanish explicitly when the atomic electron is on shell in accord with the calculated zero width for the on-shell resonance. The F^2 part of this expression is the resonance form factor at the vertex describing photonium production. We have $-F^2$ in the denominator since $(P - P_1)^2$ is negative for our off-shell atomic electron.

Averaging $|\bar{v}(P)u(P_1)|^2$ over the spins $\pm S$ and $\pm S_1$ gives $1/4 \, [P_1 \cdot P/m_e^2 - 1]$. The delta function in (19) can be written as

$$\delta\left(m_e - \epsilon_1 + E_p - \sqrt{|\vec{P_r}|^2 + m_r^2} \right)$$
$$= (1 - m_r^2/(E_p + m_e - \epsilon_1)^2)^{-1/2} \times \delta\left(|\vec{P_r}| - \sqrt{(E_p + m_e - \epsilon_1)^2 - m_r^2} \right) \tag{21}$$

and then we can integrate over the magnitude of $|\vec{P_r}|$ to get

$$\frac{d\sigma}{d\Omega} = \frac{1}{16} \frac{m_e g^2}{|\vec{P}|} |\vec{P_r}| \frac{F^4}{[(2E_p - E_r)^2 - (2\vec{P} - \vec{P_r})^2 - F^2]^2}$$
$$\times \left[\frac{E_r E_p - \vec{P_r} \cdot \vec{P} - 2m_e^2}{m_e^2} \right] \times \left[\frac{m_r^2 - 2E_r E_p + 2\vec{P} \cdot \vec{P_r}}{m_e^2} \right]^2 \frac{n(|\vec{P_r} - \vec{P}|)}{(\vec{P_r} - \vec{P})^2} \tag{22}$$

where now $E_r = E_p + (m_e - \epsilon_1)$ and $|\vec{P}_r| = \sqrt{(E_p + m_e - \epsilon_1)^2 - m_r^2}$.

Eq. (22) was integrated numerically over solid angle to give the total cross section per atom to produce photonium when monoenergetic positrons are incident upon an atomic target. For $g^2 = 1$ and $F = 1$ in electron mass units, the results for the cross section as a function of positron kinetic energy for a thorium target are given in Fig. 4. Two different photonium resonance masses are shown. The $m_r = 1.35$ Mev resonance has a cross section which peaks about an order of magnitude higher than the $m_r = 1.66$ Mev resonance and also peaks at lower values of the positron kinetic energy. The thresholds are at .847 electron mass units and 1.453 electron mass units respectively for a binding energy of 104.7 Kev.

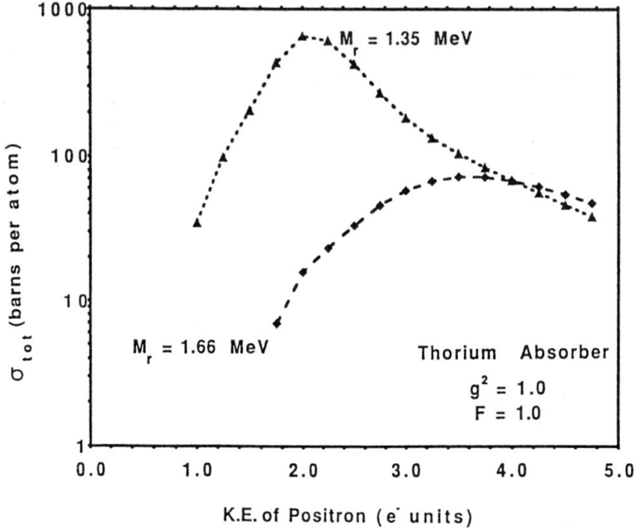

Figure 4 Total cross section per atom for photonium production from a thorium absorber for the 1.35 MeV and 1.66 MeV mass states.

By our results above we might take F to be on the order of 1 to 4 in electron mass units. One can obtain g^2 from an experiment such as the double beta spectrometer experiment by Bargholtz et al[32]. and then use the resulting cross section to predict the results of other experiments. Then the choice of F is not crucial[38] since a different choice leads to a different and largely offsetting value of g^2.

SUMMARY

In summary, we have presented several models in which e^+e^- and e-p continuum bound states are produced by a mechanism which follows directly from QED. There are no free parameters. Certain aspects of transverse photon physics play an essential role. These resonances signal the presence of a new short-distance scale for phenomena in QED. This ultra-nuclear or infra-atomic scale has a characteristic size ~ 500 fm for the e^+e^- system in the mass range we have studied. We have discussed a simple model for describing the interaction of an incident positron with a bound electron to form photonium. This model may be useful to planning and interpreting experiments using positron beams and high Z targets.

Acknowledgement

We thank B.N. Harmon and Z. Zuo for providing the Compton profile of thorium. We are pleased to acknowledge discussions with S. Brodsky, and B. C. Cook. This work was supported in part by the U.S. Department of Energy under Grant No. DE–FG02–87ER40371, Division of High Energy and Nuclear Physics.

References

1. J. von Neumann and E. Wigner, *Z. Phys.* **30**, 465 (1929).
2. For a concise review see F.H. Stillinger and D.R. Herrick, *Phys. Rev. A* **11**, 446 (1975).
3. T. Cowan et al., *Phys. Rev. Lett.* **56**, 444 (1986); T. Cowan and J. S. Greenberg in *Physics of Strong Fields,* edited by W. Greiner (Plenum Press, New York, 1987), p. 111; P. Salabura et al., *Phys. Lett. B* **245,** 153 (1990).
4. W. Koenig, et al., *Phys. Lett. B* **218,** 12 (1989).
5. J.R. Spence and J.P. Vary, *Phys. Lett. B* **254,** 1 (1991).
6. J.R. Spence and J.P. Vary, *Phys. Lett. B* **271,** 27 (1991).
7. C.J. Benesh, D.K. Ross and J.P. Vary, submitted for publication.
8. D.K. Ross and J.P. Vary, in preparation.
9. C.-Y. Wong and R. L. Becker, *Phys. Lett. B* **182,** 251 (1986).
10. N. Kemmer, *Helv. Phys. Acta.* **48,** 10 (1937); E. Fermi and C. N. Yang, *Phys. Rev.* **76,** 1739 (1949).
11. I. Tamm, *J. Phys. U.S.S.R.* **9,** 449 (1945); S. M. Dancoff, *Phys. Rev.* **78,** 382 (1950).
12. G. Breit, *Phys. Rev.* **34,** 553 (1929); **36,** 383 (1930); **39,** 616 (1932).
13. R. Blankenbecler and R. Sugar, *Phys. Rev.* **142,** 1051 (1966).
14. E. E. Salpeter and H. A. Bethe, *Phys. Rev.* **84,** 1232 (1951).
15. H. A. Bethe and E. E. Salpeter, *Quantum Mechanics of One and Two Electron Atoms* (Springer, Berlin, 1957).
16. J. Sucher in *Relativistic, Quantum Electrodynamic, and Weak Interaction Effects in Atoms,* edited by W. Johnson, P. Mohr, and J. Sucher (American Institute of Physics, New York, 1989), p. 28 and references therein.
17. E. E. Salpeter, *Phys. Rev.* **87,** 328 (1952).
18. J. J. Kubis, *Phys. Rev. D* **6,** 547 (1972); K. Erkelenz, *Phys. Reports* **13 C,** 191 (1974); M. J. Zuilhof and J. A. Tjon, *Phys. Rev. C* **22,** 2369 (1980); and R. Machleidt, K. Holinde, and Ch. Elster, *Physics Reports* **149,** 1 (1987).
19. The full details of these treatments of the Coulomb scattering problem in momentum space are of some technical interest and we will present them in a subsequent work.
20. C. DeBoor, *A Practical Guide to Splines* (Springer, Berlin, 1978).
21. K. E. Atkinson, *A Survey of Numerical Methods for the Solution of Fredholm Equations of the Second Kind* (SIAM, Philadelphia, 1976).
22. W. D. Heiss and G. M. Welke, *J. Math. Physics* **27,** 936 (1986).
23. J. R. Spence and J. P. Vary, *Phys. Rev. D* **35,** 2191 (1987).
24. J. R. Spence and J. P. Vary, (submitted for publication); J. R. Spence, Ph.D. thesis, Iowa State University, 1989, available through University Microfilms International, Ann Arbor, MI.
25. M. Grabiak, B. Müller, and W. Greiner, *Annals of Physics* **185,** 284 (1988).
26. H. Dehnen and M. Shahin, *Acta Phys. Pol. B* **21,** 477 (1990); *Int. J. Mod. Phys. A* **6,** 1031 (1991).

27. B. A. Arbuzov, E. E. Boos, V. I. Savrin, and S. A. Shichanin, *Phys. Lett. B* **240**, 477 (1990).

28. B. A. Arbuzov, E. E. Boos, V. I. Savrin, and S. A. Shichanin, *JETP Lett.* **50**, 263 (1989).

29. K. A. Erb, I. Y. Lee, and W. T. Milner, *Phys. Lett. B* **181**, 52 (1986).

30. M. Sakai et al., *Phys. Rev. C* **38**, 1971 (1988).

31. K. Maier et al., *Zeit. Phys.* **A330**, 173 (1988).

32. Chr. Bargholtz et al., *Phys. Rev. C* **40**, 1188 (1989).

33. M. Sakai et al., *Phys. Rev. C* **44**, R944 (1991).

34. R. Peckhaus, Th. W. Elze, Th. Happ, and Th. Dresel, *Phys. Rev. C* **36**, 83 (1987); T. F. Wang, I. Ahmad, S. J. Freedman, R. V. F. Janssens, and J. P. Schiffer, *Phys. Rev. C* **36**, 2136 (1987); E. Lorenz, G. Mageras, U. Steigler, and I. Huszàr, *Phys. Lett. B* **214**, 10 (1988); H. Tsertos et al., *Phys. Lett. B* **207**, 273 (1988); *Zeit. Phys.* **A331**, 103 (1988); *Phys. Rev. D* **40**, 1397 (1989).

35. C.J. Benesh, D.K. Ross and J.P. Vary, to be published.

36. K. T. Norwood, C.-Y. Ng, and J. P. Vary, *J. Chem. Phys.* **93**, 1480 (1990).

37. Q. Guo et al., *Phys. Rev. Lett.* **65**, 1857 (1990).

38. More details of this simple model may be found in: D.K. Ross and J.P. Vary, to be published.

UNCHARGED STATES OF TWO–DIMENSIONAL LATTICE GAUGE HAMILTONIANS

M. L. Ristig and A. Dabringhaus

Institut für Theoretische Physik
Universität zu Köln
D–5000 Köln 41, Germany

ABSTRACT

The method of correlated basis functions may be employed for a semi-analytic study of the ground and excited states of abelian and non-abelian lattice gauge Hamiltonians. We demonstrate at hand of the U(1) and the SU(2) model that ideas and techniques being successful in the variational theory of homogeneous and inhomogeneous boson quantum fluids can be adapted and applied to two-dimensional lattice gauge problems in the uncharged sector. The vacuum ground state is described by a variational trial function of Jastrow type including single-plaquette and plaquette-pair factors. Introducing the one-plaquette density, the plaquette-pair distribution function, and the effective plaquette-plaquette interaction between nearest neighbors we evaluate the ground state energy in terms of these functions. Thereupon we may proceed in close analogy to the variational treatment of quantum fluids such as liquid helium. A set of correlated basis functions may be systematically constructed that represents the elementary excitations in Feynman approximation. We evaluate the associated excitation energies, in particular, the lattice-photon (glueball) masses at zero wave vector. We outline the CBF optimization scheme that leads to a set of Euler-Lagrange equations determining the optimal ground and excited states. Numerical results are presented on the vacuum ground state energy, the excitation energy at zero momentum, the density profile, and the electric field correlation function of the $U(1)_3$ lattice gauge Hamiltonian for a range of coupling parameters running from the strong- to the weak-coupling regime. The results demonstrate that CBF theory is a powerful and efficient tool for studying U(N) and SU(N) lattice gauge models in two spatial dimensions.

The dynamics of lattice models of solid state or quantum field theories[1,2] may be systematically explored and accurately analysed by implementing the method of correlated basis functions (CBF method) developed originally for dealing with quantum fluids such as liquid helium and nuclear matter.[3-5] First attempts in this direction have been described for the one- and two-dimensional Hubbard Hamiltonian[6,7] and the $U(1)_3$ lattice gauge model of quantum electrodynamics[8,9].

In this contribution we employ CBF theory on its variational level for a semianalytic study of the ground and excited states of abelian and non-abelian lattice gauge Hamiltonians

Recent Progress in Many-Body Theories, Vol. 3,
Edited by T.L. Ainsworth et al., Plenum Press, New York, 1992

in the uncharged sector. The U(1) model and its SU(2) version on a square lattice provide simple yet instructive test cases for exploring the dynamics of electromagnetic or color fields with U(N) or SU(N) symmetry.

The behavior of these fields follows from a Hamiltonian of form

$$H = \tfrac{1}{2} \sum_\ell E_\ell^a E_\ell^a + \lambda \sum_q (1 - \cos B_q) \tag{1}$$

for compact QED or QCD on a lattice in the temporal gauge.[10,11] The color electromagnetic field $\{E_\ell^a, B_p\}$ is defined on a square lattice with links ℓ and plaquettes p. The abelian U(1) model is described by a single component ($a = 1$) for the electric field $E_\ell^a = E_\ell$. The non-abelian SU(2) model is characterized by three color electric field operators E_ℓ^a ($a = 1, 2, 3$). The properties of the associated states are depending on the non-negative dimensionless coupling constant λ. The strong-coupling limit is defined by $\lambda \to 0$ and the weak coupling limit by $\lambda \to \infty$.

The vacuum ground and excited states of Hamiltonian (1) are represented by wave functions that depend on the plaquette field variables $\mathbf{B} = \{B_p\} = \{B(\mathbf{n}_p)\}$. The wave functions obey periodic boundary conditions,

$$\Psi(\{B_p\}) = \Psi(\{B_p + 2\pi m_p\}), \tag{2}$$

$$\Psi(\{B(\mathbf{n}_p)\}) = \Psi(\{B(\mathbf{n}_p + L\mathbf{e}_i)\}), \tag{3}$$

where m_p is an integer and the basis vectors $\mathbf{e}_i = \mathbf{e}_1, \mathbf{e}_2$ span an elementary plaquette on a square $L \times L$ lattice. Assuming that the vacuum ground state wave function is real and has no nodes we may represent it by a wave function of Jastrow type,

$$\Psi_0(\mathbf{B}) = \Lambda \exp\left\{\tfrac{1}{2} \sum_q u_q(B_q) + \tfrac{1}{4} \sum_{p \neq q} u_{pq}(B_p, B_q)\right\}. \tag{4}$$

Ansatz (4) contains one-plaquette and pair-plaquette factors and is unit-normalized by an appropriate factor Λ. The sums appearing in eq.(4) extend over $N = L^2$ plaquettes. The wave function incorporates the effects of virtual lattice-photons (glueballs) as well as correlation effects of the field at short range. The Jastrow ansatz is exact in the strong- and weak-coupling limit[9,11] and provides a reasonable description in the region of intermediate coupling strength. In a second improved step we could replace eq.(4) by assuming a more general Feenberg ansatz[12].

To ensure a unique decomposition into genuine one- and two-plaquette contributions we specify that

$$\int_{-\pi}^{\pi} u_q(B_q)\,\sigma(B_q)\,dB_q = 0, \tag{5}$$

$$\int_{-\pi}^{\pi} u_{pq}(B_p, B_q)\,\sigma(B_q)\,dB_q = 0. \tag{6}$$

The measure $dB\sigma(B)$ appearing in eqs.(5) and (6) is determined by the group properties of the assumed model. The two-dimensional U(1) and SU(2) lattice gauge models are, respectively, characterized by the weight functions,

$$\sigma(B) = \frac{1}{2\pi}, \tag{7}$$

$$\sigma(B) = \frac{1}{\pi} \sin^2 B \, . \tag{8}$$

We assume that the ground state trial function does not break the Hamiltonian symmetries (lattice translations and parity). In this case the single-plaquette function $u_p(B_p)$ is independent of the plaquette coordinates,

$$u_p(B_p) = u(B_p) = u\big(B(\mathbf{n}_p)\big) \, , \tag{9}$$

and the pair-plaquette quantity $u_{pq}(B_p, B_q)$ depends on the plaquette positions p, q only through the separation $\mathbf{n} = \mathbf{n}_p - \mathbf{n}_q$ between the plaquettes,

$$u_{pq}(B_p, B_q) = u\big(\mathbf{n}, B(\mathbf{n}_p), B(\mathbf{n}_q)\big) = u\big(-\mathbf{n}, B(\mathbf{n}_p), B(\mathbf{n}_q)\big) \, . \tag{10}$$

In a first step we evaluate the expectation value of Hamiltonian (1) with respect to the wave function (4). CBF theory prescribes the use of the Jackson-Feenberg identity[4] and the Bogoliubov-Born-Green-Kirkwood-Yvon relation for eliminating the single-plaquette quantity $u(B_p)$. This option leads us to the appropriate functional dependence of the ground state energy in terms of the unit-normalized single-plaquette density function

$$\varrho(B_p) = \int_{-\pi}^{\pi} \Psi_0^2(\mathbf{B}) \, \sigma^{(p)}(\mathbf{B}) \, d\mathbf{B}^{(p)} \tag{11}$$

and the pair-plaquette distribution function

$$\varrho(B_p) \, \varrho(B_q) \, g(\mathbf{n}, B_p, B_q) = \int_{-\pi}^{\pi} \Psi_0^2(\mathbf{B}) \, \sigma^{(pq)}(\mathbf{B}) \, d\mathbf{B}^{(pq)} \, . \tag{12}$$

The measures $d\mathbf{B}^{(p)}\sigma^{(p)}(\mathbf{B})$ and $d\mathbf{B}^{(p,q)}\sigma^{(pq)}(\mathbf{B})$ are defined by the products

$$d\mathbf{B}^{(p)}\sigma^{(p)}(\mathbf{B}) = \prod_{r=1 \, (r \neq p)}^{N} dB_r \, \sigma(B_r) \, , \tag{13}$$

$$d\mathbf{B}^{(p,q)}\sigma^{(pq)}(\mathbf{B}) = \prod_{r=1 \, (r \neq p,q)}^{N} dB_r \, \sigma(B_r) \, . \tag{14}$$

The spatial distribution function $g(\mathbf{n}, B_p, B_q)$ aproaches unity for sufficiently large separations $\mathbf{n}$.

In terms of functions (11) and (12) the expectation value of the vacuum ground state energy

$$E = \langle \Psi_0 | H | \Psi_0 \rangle = \langle \Psi_0 | T | \Psi_0 \rangle + \lambda \langle \Psi_0 | V | \Psi_0 \rangle \tag{15}$$

may be written in the form

$$E = N \int_{-\pi}^{\pi} dB \, \sqrt{\varrho(B)} \left\{ -2 \frac{\partial}{\partial B} \, \sigma(B) \frac{\partial}{\partial B} + \lambda \sigma(B)(1 - \cos B) \right\} \sqrt{\varrho(B)} + E_c \, , \tag{16}$$

where

$$E_c = \frac{N}{4} \sum_{\mathbf{n} \neq 0} \int_{-\pi}^{\pi} dB_1 \, dB_2 \, \varrho(B_1)\varrho(B_2) \, v^*(\mathbf{n}, B_1, B_2) g(\mathbf{n}, B_1, B_2). \tag{17}$$

The integral (16) represents the single-plaquette contribution to the vacuum energy containing the 'potential' energy $\lambda\langle\Psi_0|V|\Psi_0\rangle$ and the one-plaquette piece of the 'kinetic' energy. The correlation energy (17) is generated by the effective potential[4,9]

$$v^*(\mathbf{n}, B_1, B_2) = \left\{ \sigma(B_1)D(B_1) + \sigma(B_2)D(B_2) - \tfrac{1}{2}\Delta_\sigma(\mathbf{n}) \frac{\partial^2}{\partial B_1 \partial B_2} \right\} u(\mathbf{n}, B_1, B_2) \tag{18}$$

and is mediated via the pair-plaquette pseudo-potential $u(\mathbf{n}, B_1, B_2)$ of eq.(4). It is driven by an *effective* nearest-neighbor plaquette-plaquette color field interaction being proportional to the strength function $\Delta_\sigma(\mathbf{n})$. The single-plaquette operator $D(B)$ is

$$D(B) = -\varrho^{-1}(B)\frac{\partial}{\partial B}\varrho(B)\sigma(B)\frac{\partial}{\partial B}. \tag{19}$$

The effective color field interaction is caused by the underlying effect of the field operators E_ℓ^a on the field $\mathbf{B}$ of neigboring plaquettes. Its strength $\Delta_\sigma(\mathbf{n})$ is introduced by the relation

$$\Delta_\sigma(\mathbf{n})E_2(\mathbf{n}) = \sum_\ell \left\langle \Psi_0 \Big| (E_\ell^a B_p)(E_\ell^a B_q) \frac{\partial^2}{\partial B_p \partial B_q} u(\mathbf{n}, B_p, B_q) \Big| \Psi_0 \right\rangle. \tag{20}$$

The integral (20) depends, of course, on function $u(\mathbf{n}, B_p, B_q)$ as well as on the properties of the group associated with the algebra of the lattice model in question. We indicate this dependence of the strength function on the group characteristics by a subscript σ, say, the number of group parameters ($\sigma = 1$ for the U(1) and $\sigma = 3$ for the SU(2) model).

To proceed we should evaluate the multi-dimensional integral (20) by appropriate many-body techniques such as the hypernetted-chain scheme[4,5] or quantum Monte Carlo procedures.[11] However, in the present study we do not elaborate on an explicit evaluation of integral (20) but pursue a more convenient but legitimate short cut. Instead of calculating the function $F_\sigma = \sum_\ell \sigma(B_p)\sigma(B_q)(E_\ell^a B_p)(E_\ell^a B_q)$ for subsequently performing the integration (20) we replace this quantity by an appropriate average value that is independent of the link and plaquette variables involved. Adopting such a prescription eq.(20) defines an effective pair-plaquette interaction strength $\Delta_\sigma(\mathbf{n})$ that is characterized by the color field correlation function

$$E_2(\mathbf{n}) = -\tfrac{1}{4}\left\langle \Psi_0 \Big| \sigma^{-1}(B_p)\sigma^{-1}(B_q) \frac{\partial^2}{\partial B_p \partial B_q} u(\mathbf{n}, B_p, B_q) \Big| \Psi_0 \right\rangle. \tag{21}$$

In the following we view the strength $\Delta_\sigma(\mathbf{n})$ as a suitably chosen external parameter. Relation (20) with function $E_2(\mathbf{n})$ defined by eq.(21) is then considered as a consistency condition. It serves as a subcondition that determines the correct value of the strength parameter.

For the U(1) Hamiltonian function F_σ reduces to a single constant $F_1 = (4\pi)^{-2}$ for neighbouring plaquettes being zero otherwise. Consequently, we may calculate the strength factor $\Delta_1(\mathbf{n})$ in a straight-forward manner and obtain $\Delta_1 = -(2\pi)^{-2}$. In this case condition (20) is trivially fulfilled. When dealing with the SU(2) Hamiltonian function F_3 depends on the link variables[11] and a trivial calculation of function $\Delta_1(\mathbf{n})$ — for nearest-neighbor plaquettes — is not possible. However, it is straight-forward to estimate the magnitude of function $\Delta_3(\mathbf{n})$. Exploiting the explicit result on quantity $(E_\ell^a B_q)(E_\ell^a B_p)$ given in Ref. 11 we have $|F_3| \leq (2\pi)^{-2}$ for nearest-neighbor plaquettes (otherwise $F_3 = 0$). Inserting this estimate into eq.(20) and assuming that the derivative $\partial^2 u(\mathbf{n}, B_p, B_q)/\partial B_p \partial B_q$ is negative leads us immediately to the inequality $|\Delta_3(\mathbf{n})| \leq \pi^{-2}$.

The introduction of an effective interaction strength (20) permits us to eliminate the link operators E_ℓ^a that appear in Hamiltonian (1). In terms of plaquette quantities we may equivalently write the Hamiltonian in the form

$$H = H_0 + H_w \tag{22}$$

with the one-plaquette operator

$$H_0 = \sum_p \left\{ -2\sigma^{-1}(B_p)\frac{\partial}{\partial B_p}\,\sigma(B_p)\frac{\partial}{\partial B_p} + \lambda(1 - \cos B_p) \right\} \tag{23}$$

and the nearest-neighbor interaction term

$$H_w = -\frac{1}{2}\sum_{p \neq q} \Delta_\sigma(\mathbf{n})\sigma^{-1}(B_p)\sigma^{-1}(B_q)\frac{\partial}{\partial B_p}\frac{\partial}{\partial B_q}\,. \tag{24}$$

Evaluation of the expectation value of Hamiltonian (22) with respect to the ground state trial function (4) recovers, of course, the results (16)–(19). Expression (24) demonstrates most clearly the role of function $\Delta_\sigma(\mathbf{n})$ as an effective interaction strength.

Starting from the formulation (22)–(24) and working with a suitably fixed effective interaction $\Delta_\sigma(\mathbf{n})$ we may now be guided by the established path of standard CBF theory. In particular, we follow Ref. 9 where the approach has been outlined for the $U(1)_3$ lattice gauge model.

In the following we report briefly the main features of this approach in dealing with the $U(1)$ Hamiltonian. It is entirely straight-forward to generalize the treatment for application to a non-abelian model by taking properly account of the associated weight functions $\sigma(B)$.

A set of excited corrrelated basis functions may be constructed by introducing the fluctuation operator for lattice photons (glueballs)

$$\varrho_\alpha(\mathbf{k}, \mathbf{B}) = \frac{1}{\sqrt{N}} \sum_{\mathbf{n}_p} \exp(-i\mathbf{k}\cdot\mathbf{n}_p)\varrho^{-1/2}\big(B(\mathbf{n}_p)\big)\Phi_\alpha\big(\mathbf{k}, B(\mathbf{n}_p)\big)\,, \tag{25}$$

characterized by the wave vektor $\mathbf{k}$ and a discrete quantum number α for the differing energy branches. The associated wave functions $\Phi_\alpha(\mathbf{k}, B)$ are real and must be orthogonal to the function $\sqrt{\varrho(B)}$ corresponding to the vacuum ground state,

$$\frac{1}{2\pi} \int_{-\pi}^{\pi} dB \sqrt{\varrho(B)}\,\Phi_\alpha(\mathbf{k}, B) = 0\,. \tag{26}$$

For convenience, the function $\Phi_\alpha(\mathbf{k}, B)$ is unit-normalized,

$$\frac{1}{2\pi} \int_{-\pi}^{\pi} dB\,\Phi_\alpha^2(\mathbf{k}, B) = 1\,. \tag{27}$$

In terms of the operators (25) we generate a suitable set of correlated basis functions,

$$\Psi_{\alpha,\mathbf{k}}(\mathbf{B}) = S_\alpha^{-1/2}(\mathbf{k})\varrho_\alpha(\mathbf{k},\mathbf{B})\Psi_0(\mathbf{B}) \tag{28}$$

representing trial excited states of the Bijl-Feynman type. Definition (28) involves the static structure function

$$S_\alpha(\mathbf{k}) = \langle\Psi_0|\varrho_\alpha^\dagger(\mathbf{k},\mathbf{B})\varrho_\alpha(\mathbf{k},\mathbf{B})|\Psi_0\rangle . \tag{29}$$

The elementary excitation energies associated with the states (28) are determined by

$$\omega_\alpha(\mathbf{k}) = \frac{\varepsilon_\alpha(\mathbf{k})}{S_\alpha(\mathbf{k})} , \tag{30}$$

where

$$\varepsilon_\alpha(\mathbf{k}) = \tfrac{1}{2}\langle\Psi_0|\big[\varrho_\alpha^\dagger(\mathbf{k},\mathbf{B}),[H,\varrho_\alpha(\mathbf{k},\mathbf{B})]\big]|\Psi_0\rangle . \tag{31}$$

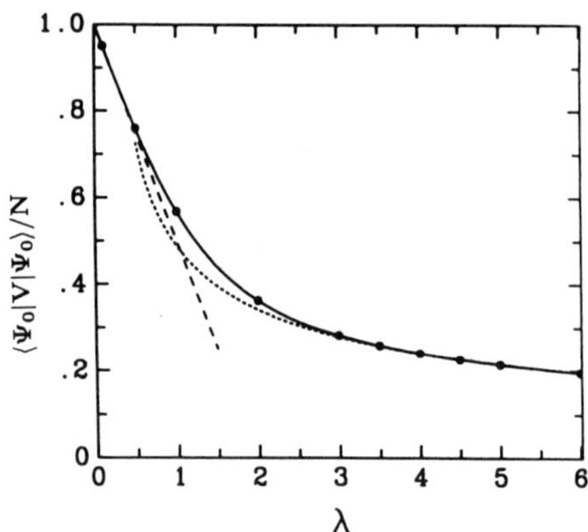

FIG. 1 Single-plaquette energy $\langle\Psi_0|V|\Psi_0\rangle/N$ as a function of the coupling strength λ, calculated with respect to the optimized Jastrow wave function (4), in HNC/0 approximation. The dashed (dotted) line displays the result of the strong (weak)-coupling limit.

The energies $\varepsilon_\alpha(\mathbf{k})$ may be viewed as the analogues of the independent-particle energies $\varepsilon_0(k) = \hbar^2 k^2/2m$ of non-interacting bosons with mass m and momentum $\hbar\mathbf{k}$. Explicit expressions for quantities (29) and (31) in terms of the density $\varrho(B)$ and the spatial distribution function $g(\mathbf{n}, B_1, B_2)$ are given in Ref. 9.

The CBF set of basis states (4) and (28) may be optimized by applying the minimum principle for the ground state energy and the excitation energies. The optimization can be performed by following the standard Euler-Lagrange procedure familiar from the CBF treatment of quantum fluids. Variation of the functional (16) employing the explicit expressions (17)–(19) at fixed strength of the effective interaction with respect to the plaquette density $\varrho(B)$ leads to a renormalized Hartree equation. Similarly, variation of eq.(16) with respect to the pseudo-potential $u(\mathbf{n}, B_p, B_q)$ results in a so-called paired-lattice-photon (paired-glueball) equation. The optimization of the excited basis states (28) may be achieved by a proper variation of the energies (30) with respect to the functions $\Phi_\alpha(\mathbf{k}, B)$. The corresponding Euler-Lagrange equation may be identified as a Feynman eigenvalue equation for functions $\Phi_\alpha(\mathbf{k}, B)$ and the associated energies $\omega_\alpha(\mathbf{k})$. Explicit expressions for the set of these three coupled equations are given in Ref. 9. For practical applications one may follow Feenberg's steps and, preferably, cast the paired-glueball equation into the form of a dressed Bogoliubov equation.[9,13] From this equation one may further extract a useful necessary stability condition on the optimized ground state[14]. We refer the reader to Ref. 9 for an explicit and detailed account of the optimization procedure for the set of corrrelated basis functions describing the ground and excited states of the $U(1)_3$ lattice model in the uncharged sector.

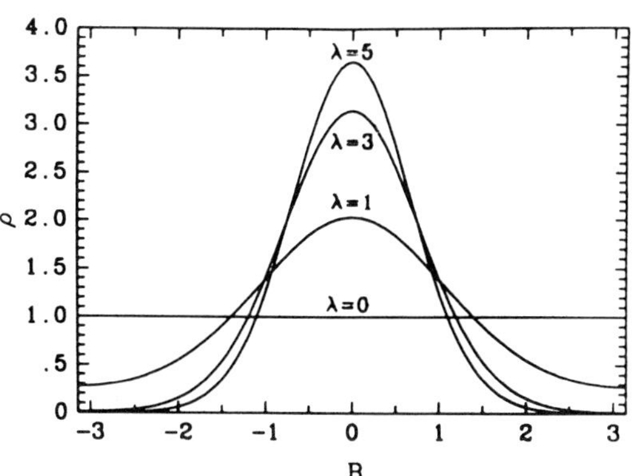

FIG. 2 Single plaquette density $\varrho(B)$ at differing coupling parameters λ, associated with the optimized Jastrow ground-state wave function (4), in HNC/0 approximation.

The optimized correlated basis functions and the corresponding energies may be constructed by solving the coupled set of Euler-Lagrange equations in conjunction with the hypernetted-chain (HNC/0) approximation.[5] Numerical results on the ground state energy, excitation energies, the plaquette density, static structure function, spatial distribution function, and other correlation functions such as the field-field correlation function (21), the stability problem etc. are available for the $U(1)_3$ model on a square lattice. The calculations employ a 20×20 square lattice and are done for coupling parameters $0 \leq \lambda \leq 6$ covering the strong- and intermediate- coupling regime and extending far into the weak-coupling domain. Numerical calculations on the vacuum sector of the $SU(2)$ Hamiltonian are in progress.

Figures 1 and 2 present our results on typical single-plaquette quantities. The plaquette energy (the 'potential' part of the Hamiltonian (22) or the energy (16)) is displayed in Figure 1. We reproduce correctly the exact results in the strong- and the weak-coupling limit. In

the range of intermediate coupling strength our energy results are in excellent agreement with
results of a Green's function Monte Carlo calculation for a 5×5 square lattice[15] and also with
numerical results derived by other authors. Figure 2 depicts results on the plaquette density
$\varrho(B)$ at various coupling strength λ. At $\lambda = 0$ the density is constant in agreement with the
exact strong-coupling result. With increasing values of parameter λ the profile develops a
pronounced peak at zero magnetic field B. Since the plaquette energy and the density $\varrho(B))$
are single-plaquette quantities they are, however, not very sensitive to correlation effects.

Information on the plaquette-plaquette correlations may be drawn more directly from
numerical results on the spatial distribution function $g(\mathbf{n}, B_p, B_q)$ or on the electric field
correlation function (21). Numerical data on quantity $E_2(\mathbf{n})$ as a function of separation
distance $|\mathbf{n}|$ of coupling parameters $\lambda = 3, 4$, and 5 are plotted in Figure 3. The function
$E_2(\mathbf{n})$ vanishes in the strong-coupling limit. In the range of coupling strength $\lambda \geq 3$ our
results are well represented by a Yukawa function,

$$E_2(\mathbf{n}) \sim |\mathbf{n}|^{-1} \exp\left\{-|\mathbf{n}|/r_0(\lambda)\right\}. \tag{32}$$

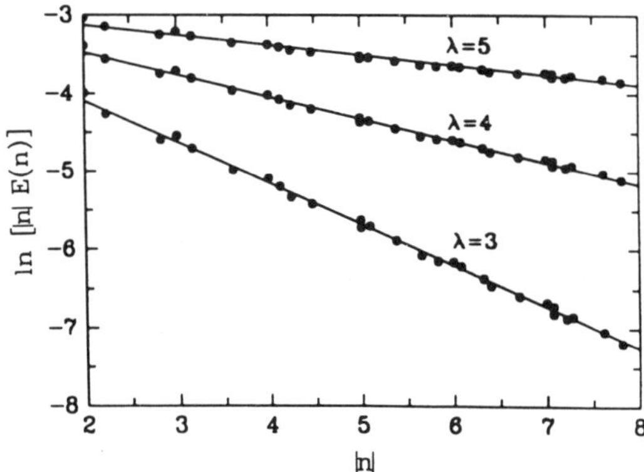

FIG. 3 The electric field correlation function, eq.(21), as a function of plaquette separation
distance $|\mathbf{n}|$, at coupling strength $\lambda = 3, 4, 5$. The numerical results are well represented
by the Yukawa function (32).

The correlation length r_0 is, evidently, a function of the coupling strength λ and increases
with increasing values λ. It, eventually, diverges in the weak-coupling limit. It does not signal
any long-range effects at finite values of the coupling strength. The small fluctuations of our
numerical results (solid dots) are mainly caused by finite size effects in the 20×20 square
lattice (the 20×20 system is not entirely invariant under discrete lattice rotations).

We have calculated the optimal elementary excitation energies (in Feynman approximation) of various branches $\alpha = 1, 2, 3 \cdots$ characterized by even or odd parity and at various
momenta $\hbar\mathbf{k}$. The lowest-lying energy branch $\omega_1(\mathbf{k})$ has a non-zero energy gap above the
vacuum ground state energy. For zero wave vector, $\mathbf{k} = 0$, the gap $\omega_1(0)$ gives the optimal

lattice-photon (glueball) mass. Numerical results on this quantity as a function of the coupling parameter λ are shown in Figure 4. These predictions may be compared with analytic results known in the strong- and weak-coupling limits (indicated in Figure 4 by dashed lines) and with results of Monte Carlo calculations for a 4×4 square lattice[16] (indicated by stars). We observe that the CBF values for the glueball masses lie above the exact values in the strong-coupling regime. This feature is a consequence of the fact that we have carried CBF theory only to the first variational level. In the present approach the excitations are described by Feynman trial functions (28) and the effective interactions between these excitations are entirely ignored. A substantial improvement may be achieved by proceeding to the next stage of the CBF scheme,

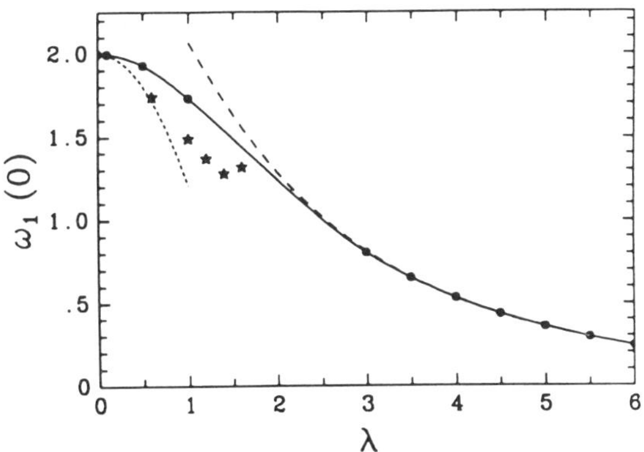

FIG. 4 Numerical results on the lattice-photon (glueball) mass $\omega_1(0)$ as a function of coupling parameter λ. Solid line depicts our CBF results in variational and HNC/0 approximation. The dotted (dashed) curves represent the results of the strong (weak)-coupling limit. Stars give Monte Carlo data of Ref. 16.

where the pertinent effective interactions are systematically included. This may be done by calculating the CBF perturbative corrections to the energies (30) in analogy with refined CBF studies of liquid helium[4]. Another option is to pursue CBF perturbation theory in a basis of functions incorporating Feynman-Cohen correlations.[17, 18]

In future development we may extend the existing CBF theory to treat charged states associated with Hamiltonian (1). Such a study would provide an independent evaluation of the string tension requiring, however more complex and sophisticated formal and numerical manipulations.

REFERENCES

1. P. W. Anderson, in *Frontiers and Borderlines in Many-Particle Physics*, Proceedings of the International School of Physics "Enrico Fermi", Course CIV, Varenna 1987, edited by R. A. Broglia and J. R. Schrieffer (North-Holland, Amsterdam, 1988).
2. J. B. Kogut, in *Elementary Particles*, Proceedings of the International School of Physics "Enrico Fermi", Course XCII, Varenna 1984, edited by N. Cabibbo (North-Holland, Amsterdam, 1987).
3. J. W. Clark and E. Feenberg, Phys. Rev. **113**, 388 (1959).
4. E. Feenberg, *Theory of Quantum Fluids* (Academic Press, New York, 1969).
5. J. W. Clark, in *Progress in Particle and Nuclear Physics*, edited by D. Wilkinson (Pergamon, New York, 1979).
6. X. Q. Wang, S. Fantoni, E. Tosatti, Lu Yu, and M. Viviani, Phys. Rev. **B 41**, 11479 (1990).
7. F. V. Kusmartsev and M. L. Ristig, Phys. Rev. **B 44**, No. 9, September 1991.
8. A. Dabringhaus and M. L. Ristig, in *Condensed Matter Theories*, edited by S. Fantoni and S. Rosati (Plenum), New York, 1991), Vol. 6.
9. A. Dabringhaus, M. L. Ristig, and J. W. Clark, Phys. Rev. **D 43**, 1978 (1991).
10. S. D. Drell, H. R. Quinn, B. Svetitsky, and M. Weinstein, Phys. Rev. **D 19**, 619 (1979).
11. S. A. Chin, O. S. van Roosmalen, E. A. Umland, and S. E. Koonin, Phys. Rev. **D 31**, 3201 (1985).
12. J. W. Clark, Nucl. Phys. **A 328**, 587 (1979).
13. N. N. Bogoliubov, *Lectures on Quantum Statistics* (Gordon and Breach, New York, 1967), Vol.1.
14. L. Szybisz and M. L. Ristig, Phys. Rev. **B 40**, 4391 (1989).
15. D. W. Heys and D. R. Stump, Phys. Rev **D 28**, 2067 (1983).
16. A. C. Irving, J. F. Owens, and C. J. Hamer, Phys. Rev. **D 28**, 2059 (1983).
17. R. P. Feynman and M. Cohen, Phys. Rev. **102**, 1189 (1956).
18. E. Manousakis and V. R. Pandharipande, Phys. Rev. **B 30**, 5062 (1984); **B 33**, 150 (1986).

CORRELATIONS IN QUANTUM SPIN CHAINS AND LATTICES:

A FULLY MICROSCOPIC MANY-BODY APPROACH

R.F. Bishop, J.B. Parkinson and Yang Xian

Department of Mathematics, UMIST
University of Manchester Institute of Science and Technology
P.O. Box 88, Manchester M60 1QD, England

1. INTRODUCTION

Low-dimensional quantum spin systems have long been studied within the framework of magnetism. Two recent developments have greatly added to their interest. These concern the Haldane conjecture[1] for integral-spin one-dimensional (1D) antiferromagnets, and the link between two-dimensional (2D) antiferromagnetism and high-temperature superconductivity. In the latter case, among the various models proposed, the 2D Hubbard model[2] is widely believed to contain the essential correlations of the active electrons in the ceramic oxide materials. Anderson[3] proposed that the high-T_c superconductivity might originate from the non-Fermi liquid behaviour away from half-filling. In the present context, the antiferromagnetic Heisenberg model can itself be derived from the Hubbard model at half filling, and this is believed to describe correctly the electronic properties of the high-T_c materials before doping. This point, first suggested by Anderson[3] soon after their discovery, has now been supported by a large number of experiments.[4]

For the above reasons, it has become important to study these strongly-correlated models at a fundamental *ab initio* level. Although the recent literature has acquired enormous proportions, only a small fraction has been devoted to a fully microscopic approach. In particular, the coupled cluster method (CCM)[5] has proved to be one of the most powerful and universal high-precision techniques in quantum many-body theory. Its advantages include its capacity for systematic improvement via well-defined hierarchies of truncation schemes, and its automatic avoidance of unphysical divergences in the thermodynamic limit for infinite systems. The CCM has been well described elsewhere,[5-13] and it has been applied to a wide range of physical systems. These include problems in nuclear physics, both for finite[9,14,15] and infinite nuclear matter;[16] atomic and molecular systems in quantum chemistry;[7,12,17] and the electron gas;[8,18] as well as such model field-theoretical systems as the quantum anharmonic oscillator treated as a single-mode bosonic field theory,[19] and the relativistic Φ^4 quantum field theory.[20]

In view of this impressive number of successful applications to diverse condensed-matter and field-theoretical systems, it seems timely to apply the CCM to quantum spin-lattice problems. We note that such models in general form very clean systems on which to deploy *any* technique of quantum many-body theory. They contain a nice admixture of exact results and unsolved problems which are both interesting in their own right and relevant to various physical systems. Furthermore, they are among the most quantum-mechanical of all systems, in that their behaviour is often quite different from that of their classical counterparts. Unexpected and counter-intuitive results abound.

Specifically, we mostly study here the spin-$\frac{1}{2}$ anisotropic antiferromagnets described by the XXZ (or Heisenberg-Ising) Hamiltonian, for both 1D chains and the 2D square lattice. These include the Heisenberg model as the special (isotropic) limit. We have already noted the perceived practical importance of the 2D model in connection with high-T_c superconductivity. Although to date the 2D models have not been exactly solved (and are not believed to be exactly integrable), their 1D XXZ-model counterparts *are* exactly soluble by the Bethe ansatz.[21] Hence, we may compare our CCM results with exact results for such important ground-state properties as the energy,[22] the staggered magnetization,[23] and the asymptotic behaviour of the spin-spin correlation functions,[24] as well as the excitation spectrum.[25] We have briefly reviewed some of these exact results elsewhere.[26]

The XXZ model and various preliminary definitions are first described in Sec. 2, after which we discuss the CCM cluster decomposition of the ground ket state in Sec. 3, where we also outline several intuitively appealing and systematic approximation schemes. These are applied to calculate the ground-state (g.s.) energy of the 1D chain and the 2D square lattice respectively in Secs. 4 and 5. One of the key features of the CCM is that the Hilbert space has a biorthogonal structure, rather than the more usual orthogonal basis of alternative theories. Thus, the bra states defined in Sec. 6 are not manifestly Hermitian conjugate to the corresponding ket states at a given level of approximation, and must be calculated separately. With both bra and ket states defined we may then calculate any other ground-state property. As an example, we also calculate in Sec. 6 the staggered magnetization for both the 1D and 2D cases. The corresponding excitation spectra are discussed in Sec. 7. For particular approximation schemes which incorporate long-range correlations we find very strong evidence in both 1D and 2D for quantum phase transitions at critical values of the anisotropy parameter, although with qualitatively different features in the two cases. We briefly discuss extensions of our method to systems of spins with $s > \frac{1}{2}$ in Sec. 8, and then summarize in Sec. 9.

2.   THE MODEL AND PRELIMINARIES

We consider an even number N of quantum spins $\vec{s}_k \equiv \{s_k^a; a = x,y,z\}$ placed on the sites $\{k\}$ of a specified regular lattice in d dimensions whose coordination number is z, with periodic boundary conditions. They obey the usual SU(2) algebra,

$$[s_k^a, s_m^b] = i\delta_{km}\varepsilon_{abc}s_k^c, \tag{1}$$

where ε_{abc} is the usual antisymmetric unit tensor and the summation convention is employed. We shall mostly be interested in the case where the spin quantum number $s = \frac{1}{2}$, with $\vec{s}_k^2 = s(s+1)$, and where $N \to \infty$. Furthermore, we consider only bipartite lattices (i.e., ones in which the lattice can be decomposed into two equivalent sublattices such that all z nearest neighbours of a site on one sublattice belong to the other one), with $\frac{1}{2}N$ spins on each sublattice.

The XXZ-model Hamiltonian,

$$H = \frac{1}{2}\sum (s_\ell^x s_{\ell+\rho}^x + s_\ell^y s_{\ell+\rho}^y + \Delta s_\ell^z s_{\ell+\rho}^z), \tag{2}$$

is a model of nearest-neighbour anisotropic interactions between the localized spins, where the index ℓ runs over all N lattice sites and the index ρ over all z nearest-neighbour lattice vectors. We take both the lattice spacing and the antiferromagnetic coupling constant to be unity. The Ising, Heisenberg, and (XY or) planar models correspond respectively to the three values $\Delta \to \infty$, $\Delta = 1$, and $\Delta = 0$ of the anisotropy parameter Δ. The Hamiltonian of Eq. (2) commutes with the z-component of the total spin operator, $s_z^{total} \equiv \sum_\ell s_\ell^z$, which is hence a good quantum number. We are especially interested in those values of Δ for which the ground state is antiferromagnetic and hence lies in the subspace $s_z^{total} = 0$.

For the spin-$\frac{1}{2}$ models of primary interest we replace the spin operators by the Pauli operators, $\sigma^a \equiv 2s^a$. Furthermore, since we are interested in antiferromagnetic states, the Néel state (in which the spins are perfectly aligned in opposing directions on the sublattices, say along the $\pm$ z-axes) is bound to play a special role. It is thus very convenient to choose from the outset different coordinate axes on the two sublattices, so that all of the spins in the Néel state point in the same named direction (say along the $-z$-axis or in the "down" direction). This is equivalent to performing a purely notional rotation of 180° (about the y-axis, say) of the spins on the "up" sublattice, leading to the transformation $s^x \to -s^x$, $s^y \to s^y$, $s^z \to -s^z$ on this sublattice. We may thus equivalently define raising and lowering operators by,

$$\sigma_i^\pm \equiv \tfrac{1}{2}(\sigma_i^x \pm i\sigma_i^y); \quad \sigma_j^\pm \equiv \tfrac{1}{2}(-\sigma_j^x \pm i\sigma_j^y) , \tag{3}$$

for sites i and j on the "down" and "up" sublattices respectively. This sublattice rotation leaves the commutation relations invariant, namely,

$$[\sigma_k^+, \sigma_m^-] = \sigma_k^z \delta_{km} \; ; \quad [\sigma_k^z, \sigma_m^\pm] = \pm 2\sigma_k^\pm \delta_{km} , \tag{4}$$

for sites k and m on *either* sublattice. In this notation the Néel state $|\Phi\rangle$ has all spins pointing "down", so that $\sigma_k^-|\Phi\rangle = 0$ for all k, and $\sigma_k^+|\Phi\rangle$ is a state with the kth spin reversed with respect to the Néel state. In the Néel basis used henceforth, the Hamiltonian becomes,

$$H = -\tfrac{1}{4} \sum_{\ell,\rho} (\sigma_\ell^+ \sigma_{\ell+\rho}^+ + \sigma_\ell^- \sigma_{\ell+\rho}^- + \tfrac{1}{2}\Delta\sigma_\ell^z \sigma_{\ell+\rho}^z) . \tag{5}$$

3. CCM PARAMETRIZATION OF THE GROUND KET STATE

The Néel state $|\Phi\rangle$ does not represent the g.s. of the XXZ-model Hamiltonian except in the Ising limit $\Delta \to \infty$. For finite values of Δ, quantum fluctuations play a significant role, especially for low-dimensional lattices and for small values of the spin. Equivalently, we may say that the exact ground ket state $|\Psi\rangle$, where

$$H|\Psi\rangle = E_g |\Psi\rangle, \tag{6}$$

has (multi-spin) fluctuations or correlations not present in $|\Phi\rangle$. These are incorporated in the CCM via a correlation operator S,

$$|\Psi\rangle = e^S|\Phi\rangle , \tag{7}$$

which was perhaps first employed for spin-lattice systems by Roger and Hetherington,[27] who were primarily interested in the solid phases of ^{3}He.

As usual in the CCM,[5,8-11] the operator S is constructed from products of creation operators with respect to $|\Phi\rangle$, namely in this case the Néel-basis operators $\{\sigma_k^+\}$ defined in Eq. (3). We thus put

$$S = \sum_{m=1}^{N} S_m , \qquad (8)$$

where S_m creates a linear combination of configurations with m spins flipped with respect to the Néel state. Since the g.s. $|\Psi\rangle$ of the antiferromagnetic models lies in the subspace $s_z^{total} = 0$, the correlation operator S should contain only those partitions S_m with m even, and with $\frac{1}{2}$m creation operators on each sublattice. Henceforth, we use the indices i or $\{i_n\}$ and j or $\{j_n\}$ to indicate sites on the two respective sublattices. We may then decompose the nonzero g.s. cluster correlation operators in the form,

$$S_{2n} = \frac{1}{(n!)^2} \sum_{i_1,\ldots,i_n} \sum_{j_1,\ldots,j_n} S_{i_1\ldots i_n;j_1\ldots j_n} \; \sigma_{i_1}^+ \cdots \sigma_{i_n}^+ \sigma_{j_1}^+ \cdots \sigma_{j_n}^+ , \qquad (9)$$

where any coefficient $S_{i_1\ldots i_n;j_1\ldots j_n}$ with any repeated index may be taken as zero for the present spin-$\frac{1}{2}$ case, in view of the property, $(\sigma^+)^2 = 0$, of the Pauli matrices.

The equations which determine the cluster configuration coefficients $\{S_{i_1\ldots i_n;j_1\ldots j_n}\}$ are now found in the usual CCM way by first rewriting the g.s. Schrödinger equation of Eqs. (6) and (7) as,

$$e^{-S}He^{S}|\Phi\rangle = E_g|\Phi\rangle . \qquad (10)$$

We note that, as usual, the similarity-transformed Hamiltonian, $e^{-S}He^{S}$, can be expanded in terms of the well-known nested commutator series. By making use of the commutation relations of Eq. (4), it is easy to see that this otherwise infinite series now, however, terminates after terms of fourth order in S for the Hamiltonian of Eq. (5). Secondly, the transformed Eq. (10) is then projected in turn with the model Néel state $\langle\Phi|$ and with the Hermitian adjoints of each of the configurations contained in (either the exact or the particular approximation being used for) the cluster operator S. We thereby obtain respectively the *exact* relation,

$$\frac{E_g}{N} = -\frac{z}{8}(\Delta + 2b_1) , \qquad (11)$$

for the g.s. energy, where z is the lattice coordination number and $S_{i;i+\rho} \equiv b_1$ is the nearest-neighbour two-spin-flip coefficient (which is independent of both i and ρ by translational invariance and by the symmetries of the lattice under rotations and reflections, respectively), plus a set of coupled nonlinear algebraic equations for the various cluster configuration coefficients.

Approximations within the CCM now come from truncating the sum over all possible cluster-configuration coefficients in Eqs. (8) and (9) to some (finite or infinite) subset. Several systematic schemes have been investigated by us,[26,28] and we briefly review some of them below. The hierarchy most commonly employed in other CCM applications is the so-called SUBn scheme, in which one keeps only up to n-spin correlations by truncating the sum in Eq. (8) by the replacement $N \to n$,

$$S \to S_{SUBn} = \sum_{\ell=1}^{n/2} S_{2\ell} \ , \tag{12}$$

and in the present case n is an even integer. For example, in SUB4 approximation we now have two coupled sets of equations for the 2-spin-flip and 4-spin-flip coefficients, $\{S_{i;j}\}$ and $\{S_{ii';jj'}\}$ respectively. These are obtained as described above by first replacing $S \to S_{SUB4}$, and then taking the respective overlaps of Eq. (10) with the states $\sigma_i^+ \sigma_j^+ |\Phi\rangle$ and $\sigma_i^+ \sigma_{i'}^+, \sigma_j^+ \sigma_{j'}^+, |\Phi\rangle$. We quote only the final result for the former (2-spin-flip) case,

$$\sum_{\rho} \Big[(1 + 2\Delta S_{i;i+\rho} + 2S_{i;i+\rho}^2)\delta_{i+\rho,j} - 2(\Delta + 2S_{i;i+\rho})S_{i;j}$$

$$+ \sum_{i'} (S_{i;i'+\rho}S_{i';j} + S_{ii';i'+\rho,j}) \Big] = 0 \ . \tag{13}$$

Equation (13) is valid for a spin-$\frac{1}{2}$ bipartite lattice of arbitrary dimensionality. We note also that Eq. (13), like Eq. (11), is actually *exact*, whereas its SUB4 counterpart for the 4-spin-flip coefficients, which we do not quote here, is only approximate since the coupling to the 6-spin-flip terms is neglected at this level.

The lowest (SUB2) approximation in the SUBn scheme is of special interest, as we see below. By using the translational invariance properties, we may write the 2-spin-flip coefficients as $S_{i;i+r} \equiv b_r$, where r is a lattice vector connecting sites on opposite sublattices. The correlation operator S thus has the SUB2 approximant,

$$S \to S_{SUB2} = \sum_i \sum_r b_r \sigma_i^+ \sigma_{i+r}^+ = \frac{1}{2} \sum_\ell \sum_r b_r \sigma_\ell^+ \sigma_{\ell+r}^+ \ , \tag{14}$$

where the indices $\{i\}$ run over one sublattice only, while $\{\ell\}$ run over the entire lattice, as above. The SUB2 equations for the coefficients $\{b_r\}$ can easily be obtained from Eq. (13) by setting the 4-spin-flip coefficients to zero. We obtain

$$2zKb_r - (1 + 2\Delta b_1 + 2b_1^2)\sum_\rho \delta_{r\rho} - \sum_{r'} \sum_\rho b_{r-r'+\rho}b_{r'} = 0 \ ; \tag{15}$$

$$K \equiv \Delta + 2b_1 \ . \tag{16}$$

As explained elsewhere,[26,28] Eq. (15) can be solved exactly in arbitrary dimensionality for any bipartite hyper-cubic lattice, by Fourier transform techniques.

A second quite different set of approximations, called the LSUBn scheme, is based on the localized nature of the interactions. At a given LSUBn level we retain only those configurations in the correlation operator S which contain (any number up to n of) spin-flips, with respect to the Néel state, over a locale or localized region of n contiguous sites on the lattice, and which are compatible with the restriction $s_z^{total} = 0$. Clearly, the lowest member of this sequence, LSUB2, retains only the *single* configuration of two spins flipped on nearest-neighbour sites, and which has the coefficient b_1. By setting to zero all coefficients $\{b_r\}$ in Eq. (15) other than $b_\rho \equiv b_1$, we find the simple quadratic form for the LSUB2 equation,

$$3b_1^2 + 2\Delta b_1 - 1 = 0 \; ; \quad \text{1D chain,} \tag{17a}$$

$$5b_1^2 + 6\Delta b_1 - 1 = 0 \; ; \quad \text{2D square lattice.} \tag{17b}$$

A third approximation scheme, denoted as SUBn+LSUBm (with m > n), is a rather self-evident combination of the previous two schemes.

We have also investigated a fourth (PSUBn) scheme[26] based on "plaquettes" of flipped spins. It was motivated both by the kink structures of spin systems in 1D and the domain-wall structures in higher dimensions. We do not consider it further in the present work.

4. GROUND-STATE ENERGY OF THE SPIN-$\frac{1}{2}$ 1D CHAIN

In 1D, the LSUBn sequence of approximations is defined on "locales" comprising n adjacent sites on the chain. For example, in LSUB4 and LSUB5 approximations we have,

$$S \to S_{\text{LSUB4}} = \sum_{\ell=1}^{N} (b_1 \sigma_\ell^+ \sigma_{\ell+1}^+ + b_3 \sigma_\ell^+ \sigma_{\ell+3}^+ + g_4 \sigma_\ell^+ \sigma_{\ell+1}^+ \sigma_{\ell+2}^+ \sigma_{\ell+3}^+) \; ; \tag{18a}$$

$$S \to S_{\text{LSUB5}} = S_{\text{LSUB4}} + \sum_{\ell=1}^{N} h_5 \sigma_\ell^+ \sigma_{\ell+1}^+ \sigma_{\ell+3}^+ \sigma_{\ell+4}^+ \; ; \tag{18b}$$

while the LSUB6 approximation retains three extra *independent* (under the symmetries) 4-spin-flip configurations, plus a single 6-spin-flip configuration; etc. Results for the g.s. energy, calculated from Eq. (11), using this sequence, are shown in Fig. 1 and Table 1. We note that even the LSUB2 expression from Eq. (17a), $E_g/N = -[\Delta + 2(\Delta^2 + 3)^{\frac{1}{2}}]/12$, gives results of impressive accuracy when compared with the exactly known results[22] for this case. It also gives the correct leading asymptotic form for large Δ,

$$\frac{E_g}{N} \xrightarrow[\Delta \to \infty]{} -\tfrac{1}{4}(\Delta + \Delta^{-1}) \; ; \quad \text{1D chain.} \tag{19}$$

For the Heisenberg model ($\Delta = 1$), the LSUB2 result, $E_g/N = -5/12 \approx -0.417$, may be compared with both the exact result of -0.443 and with the result of -0.428 obtained[29] by actually minimizing the expectation value of the Hamiltonian over all wavefunctions of the LSUB2 type (i.e., with respect to the parameter b_1).

The SUB2 approximation in 1D retains all of the 2-spin-flip configurations, with coefficients $S_{i;i+r} \equiv b_r = b_{-r}$, where r is an odd integer. The ($N \to \infty$) solution of the corresponding SUB2 equations of Eq. (15) in this case is given by,[26,28]

$$b_r = \frac{K}{\pi} \int_0^\pi dq \, \frac{\cos(rq)}{\cos(q)} \left[1 - \sqrt{1 - k^2 \cos^2 q} \right] \; ; \tag{20}$$

$$k^2 \equiv (1 + 2\Delta b_1 + 2b_1^2)/K^2 \; . \tag{21}$$

Putting $r = 1$ in Eq. (20) leads to a self-consistent equation for b_1, which may be solved numerically. In this way we obtain the results for the g.s.

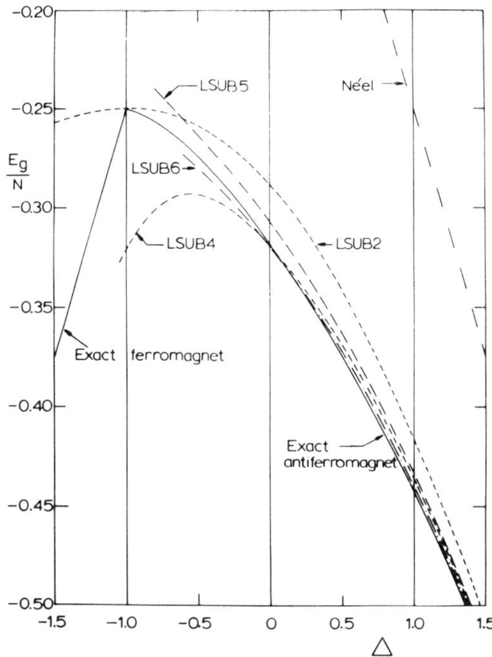

Fig. 1. Ground-state energy per spin for the spin-$\frac{1}{2}$ XXZ model in 1D, as a function of anisotropy parameter Δ. CCM results for various LSUBn approximations are compared with the uncorrelated (classical) Néel result, $E_g/N = -\frac{1}{4}\Delta$, and exact results.[22]

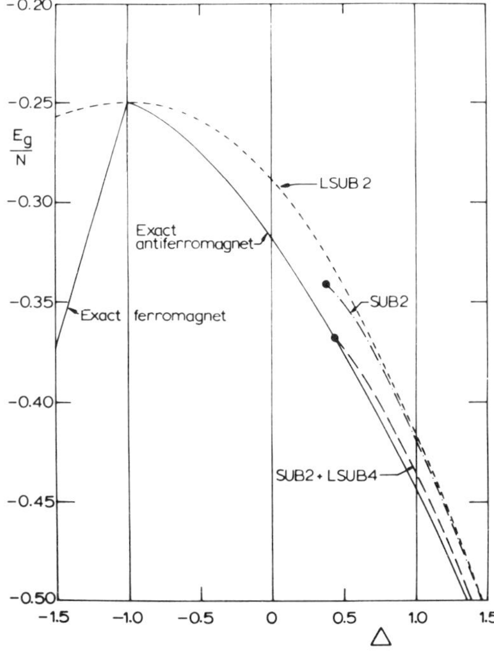

Fig. 2. As in Fig. 1, but showing CCM results in SUB2 and SUB2+LSUB4 approximations, which include long-range correlations. Their corresponding terminating points are clearly indicated.

Table 1. The g.s. energy per spin, E_g/N, for the infinite ($N \to \infty$), spin-$\frac{1}{2}$, 1D XXZ–model chain, for various values of the anisotropy parameter Δ. Results in various CCM approximations are compared with exact values.[22]

Δ	E_g/N						
	LSUB2	LSUB4	LSUB5	LSUB6	SUB2	SUB2+ LSUB4	Exact
-0.5	-0.2588	-0.2932	-0.2638	-0.2797	--	--	-0.2745
0.0	-0.2887	-0.3193	-0.3078	-0.3198	--	--	-0.3183
0.5	-0.3421	-0.3692	-0.3638	-0.3730	-0.3506	-0.3727	-0.3750
1.0	-0.4167	-0.4363	-0.4339	-0.4400	-0.4186	-0.4366	-0.4432
1.5	-0.5069	-0.5195	-0.5185	-0.5218	-0.5075	-0.5196	-0.5234
2.0	-0.6076	-0.6155	-0.6151	-0.6167	-0.6079	-0.6155	-0.6172

energy shown in Table 1 and Fig. 2. We note that the SUB2 energies are surprisingly close to the much simpler LSUB2 estimates for most values of Δ. For example, at $\Delta = 1$ the SUB2 result is $E_g/N \approx -0.419$. Its asymptotic behaviour also agrees with the exact result of Eq. (19).

The most interesting feature of the SUB2 results is that a real solution for b_1 ceases to exist when $k > k_c = 1$, or equivalently when $\Delta < \Delta_c \approx 0.373$. The absence of a solution for $\Delta < \Delta_c$ is a clear signal of a possible phase transition, even though the value of Δ_c is not very close to the exact critical value of 1 at which it is known that the spin-$\frac{1}{2}$ XXZ–model chain undergoes a phase transition. Thus, for $\Delta > 1$ the 1D chain has an ordered, Ising–like, antiferromagnetic phase with exponentially–decaying spin–spin correlation functions at large distances, and a spin–wave excitation spectrum with a finite gap. Conversely, for $|\Delta| < 1$ it has a "critical" phase with correlations which decay algebraically to zero at large distances, and a gapless excitation spectrum.

Clearly, it is desirable to have additional evidence that the breakdown of the SUB2 solution for $\Delta < \Delta_c$ is not merely a mathematical artifice. To this end we calculate the corresponding staggered magnetization and the excitation spectrum below in Secs. 6 and 7. For the moment we note only the highly suggestive changeover at $\Delta = \Delta_c$ in the long–range behaviour of the 2-spin-flip coefficients $\{b_r\}$ from exponential to power–law decay,

$$b_r \xrightarrow[r \to \infty]{} \begin{cases} a/\ell^r, & \Delta > \Delta_c \; ; \\ b/r^2, & \Delta = \Delta_c \, , \end{cases} \qquad (22)$$

where b is a constant, and a and ℓ (> 1) are functions only of Δ.

Finally, we have also considered the SUB2+LSUB4 approximation, since one expects that such a scheme which retains the long–range correlations of the SUB2 approximation, but which also includes the extra short–range 4–spin–flip contribution which seems to be important for increasing the accuracy of the g.s. energy estimation, could continue to show a phase transition. The quantitative results shown in Table 1 are gratifying. Furthermore, the SUB2 critical point is not only not destroyed, but the SUB2+LSUB4 value, $\Delta_c \approx 0.436$ is even slightly closer to the expected value of 1.

Table 2. The g.s. energy per spin, E_g/N, for the spin-$\frac{1}{2}$ Heisenberg model on an infinite ($N \to \infty$) 2D square lattice. Results in various CCM schemes are compared with those from linear spin-wave theory (SWT)[32] and from Green's function Monte Carlo (GFMC) calculations.[30]

LSUB2	SUB2	LSUB4′	SUB2+g_4^a	LSUB4	SWT	GFMC
-0.648	-0.651	-0.653	-0.656	-0.664	-0.658	-0.669

5. GROUND-STATE ENERGY OF THE SPIN-$\frac{1}{2}$ 2D SQUARE LATTICE

We have also performed the counterparts of most of the above calculations on the 1D chain for the 2D square lattice, where very few exact results are known. The LSUBn approximation schemes in particular are now more complicated, however, due to the increased number of spin-flip configurations. For example, we show in Fig. 3 the seven independent configurations that are included in the LSUB4 approximation, together with their respective coefficients in the expansion of the cluster operator S, and two possible measures, w and d of their relative importance. The first is simply the number w of "wrong" bonds, with respect to the Néel state, of a given configuration. Classically, the breaking of each bond in the antiferromagnetic regime costs energy, and configurations with the smallest values of w hence seem likely to be most important. Secondly, we also define a weight parameter, d, as the length of the "domain boundary" of a given configuration of flipped spins. It is defined as the number of lattice bonds crossed by the shortest-path circuit (shown in Fig. 3 by the dashed lines) which encloses all of the flipped spins.

Taken together, the two weights w and d indicate a relative order of decreasing importance of the seven LSUB4 configurations as: b_1; g_4^a; (b_3^a, b_3^b); (g_4^b, g_4^c, g_4^d). With this in mind, we have also performed, apart from full (LSUB2 and) LSUB4 calculations, the subapproximation which keeps only the two most highly weighted coefficients, b_1 and g_4^a. It is denoted as the LSUB4′ scheme. Finally, we have also performed a full SUB2 and a partial SUB2+LSUB4 calculation, namely SUB2+g_4^a. They again yield critical values Δ_c, for values of Δ below which the corresponding (real) solutions do not exist. For the two schemes SUB2 and SUB2+g_4^a, the respective values for the 2D square lattice are $\Delta_c \approx 0.799$ and 0.819, both considerably closer to the classical value of 1 than in 1D.

Other results are shown in Fig. 4, and in Table 2 for the Heisenberg model. Since no exact 2D results are known we make comparison with the best available Green's function Monte Carlo (GFMC) results[30] for $\Delta = 1$, and with other recent Monte Carlo results.[31] We also compare with the corresponding result of Oguchi from Anderson's approximate spin-wave theory (SWT),[32] in which the correction to the classical Néel value is calculated to first order in inverse powers of the spin quantum number s (where s = $\frac{1}{2}$ here). Our results again agree well with the essentially exact GFMC results.

6. CCM PARAMETRIZATION OF THE GROUND BRA STATE; STAGGERED MAGNETIZATION

For reasons that have been well documented elsewhere,[10,11] the CCM

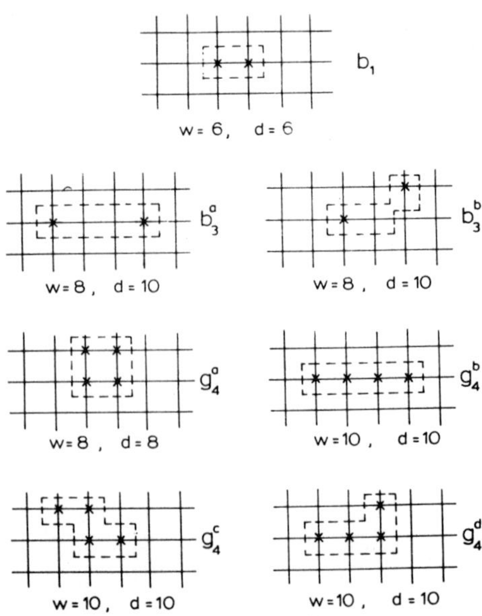

Fig. 3.   The seven independent configurations retained in the LSUB4 scheme for the spin-$\frac{1}{2}$ 2D square lattice, together with their weights w and d, defined in the text.   Crosses indicate the flipped spins, and the dashed lines delineate the corresponding "locale".

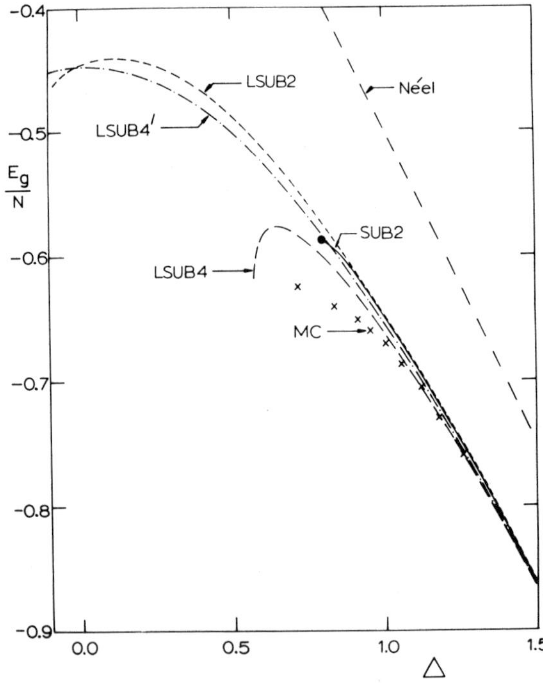

Fig. 4.   Ground-state energy per spin for the spin-$\frac{1}{2}$ XXZ model on the 2D square lattice, as a function of anisotropy parameter Δ, showing CCM results for various approximations described in the text, and Monte Carlo (MC) results of Ref. 31 (indicated by crosses).   The terminating point of the SUB2 scheme is indicated.

parametrizes bra states in a form that is *not* the explicit Hermitian adjoint of their corresponding ket states. One basic reason is to preserve consistency with the important Hellmann–Feynman theorem at any given level of truncation. A disadvantage is that the CCM estimates for the g.s. energy are therefore *not* variational upper bounds (as would be the case in the simpler configuration-interaction method, for example, although this method is not size-extensive and hence not useful for present purposes). However, one also gains the considerable advantage that the expectation value of an arbitrary operator is a finite-order functional of the various configuration amplitudes (by contrast with the case when the bra states are taken as the Hermitian adjoints of the CCM ket states already defined).

In the so-called "normal" version of the CCM which we employ here (to contrast it with the more recent and even more powerful "extended" version[10,11]), the g.s. bra wavefunction corresponding to the ket state $|\Psi\rangle$ is parametrized as,

$$\langle\tilde{\Psi}| = \langle\Phi|\tilde{S}e^{-S} ,\tag{23}$$

where the new *linear* correlation operator $\tilde{S}$ is constructed wholly from products of destruction operators $\{\sigma_\ell^-\}$, apart from a constant term chosen to be unity to preserve the normalization, $\langle\tilde{\Psi}|\Psi\rangle = \langle\Phi|\tilde{S}|\Phi\rangle = 1$. Otherwise, its construction is exactly as in Eqs. (8) and (9), except for the replacement $\{\sigma_\ell^+\} \to \{\sigma_\ell^-\}$. For example, the analogue of Eq. (14) in SUB2 approximation is,

$$\tilde{S} \to \tilde{S}_{SUB2} = 1 + \tfrac{1}{2} \sum_\ell \sum_r \tilde{b}_r \sigma_\ell^- \sigma_{\ell+r}^- ,\tag{24}$$

where the indices $\{\ell\}$ and $\{r\}$ have the same meaning as before. The equations for the bra-state configuration coefficients retained in any particular approximation are now obtained exactly as before by taking the overlap of the Schrödinger equation,

$$\langle\tilde{\Psi}|H = E_g \langle\tilde{\Psi}| ,\tag{25}$$

with each of the ket-state wavefunctions corresponding to the particular configurations retained. For example, in SUB2 approximation, by taking the overlap with the states $\sigma_\ell^+ \sigma_{\ell+r}^+ |\Phi\rangle$, we find,

$$2zK\tilde{b}_r - \left(1 + 2K\tilde{b}_1 - 4\sum_{r'} \tilde{b}_{r'} b_{r'}\right) \sum_\rho \delta_{r\rho} - 2 \sum_r \sum_\rho b_{r-r'+\rho} \tilde{b}_{r'} = 0 ,\tag{26}$$

as the bra-state equivalent of the SUB2 ket-state equation (15). This equation has also been solved by Fourier transform techniques.[26,28] Similarly, the LSUB2 approximation, which now retains only the two coefficients $(b_1, \tilde{b}_1)$, is trivially solved.

We may now calculate the g.s. expectation value of an arbitrary operator in any particular CCM approximation. As an example we quote results for the staggered magnetization defined as,

$$M^z \equiv -\frac{2}{N} \sum_i \langle\sigma_i^z\rangle = -\frac{2}{N} \sum_i \langle\Phi|\tilde{S}e^{-S}\sigma_i^z e^S|\Phi\rangle ,\tag{27}$$

where the summation runs over one sublattice only. If Eq. (27) is evaluated within the SUB2 approximation, we find for a bipartite lattice of arbitrary dimensionality,

$$M^z = 1 - 2 \sum_r \tilde{b}_r b_r \; ; \tag{28a}$$

which reduces to the corresponding LSUB2 expressions,

$$M^z = 1 - 2z\tilde{b}_1 b_1 = \begin{cases} [1 + 2\Delta(\Delta^2+3)^{-\frac{1}{2}}]/3 \, , & \text{1D chain} \; ; \\ [1 + 12\Delta(9\Delta^2+5)^{-\frac{1}{2}}]/5 \, , & \text{2D square lattice} \, . \end{cases} \tag{28b}$$

Both the LSUB2 and full SUB2 results give the correct leading $O(\Delta^{-2})$ correction to the Ising value of 1 for large values of Δ.

The LSUB2 and SUB2 values of M^z as a function of Δ are shown in Fig. 5 for the spin-$\frac{1}{2}$ XXZ model on both the 1D chain and the 2D square lattice. In both cases we compare with results of Anderson's linear (lowest-order) spin-wave theory (SWT).[32] In 1D we also compare with the exact results of Baxter,[23] and in 2D with Monte Carlo results.[30,31] Our main finding is that the behaviour of M^z near Δ_c in our SUB2 calculation now lends strong evidence

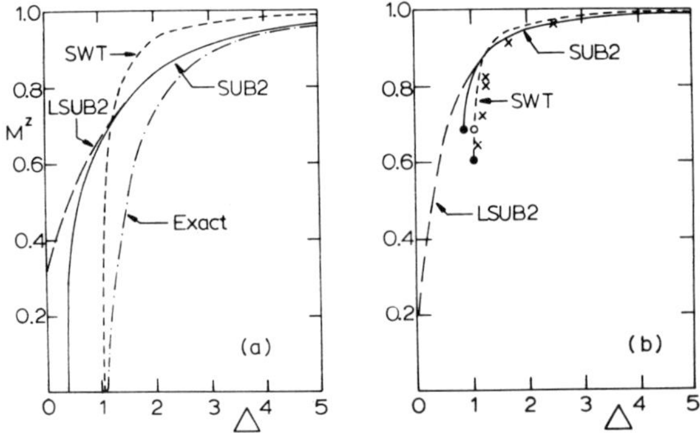

Fig. 5. Staggered magnetization M^z for the spin-$\frac{1}{2}$ XXZ model: (a) in 1D, and (b) on the 2D square lattice, as functions of Δ. Our LSUB2 and SUB2 results are compared with those from linear spin-wave theory (SWT).[32] In 1D, exact results[23] are shown. In 2D, we show the Monte Carlo (MC) results of Ref. 31 (crosses) and of Carlson[30] (circle).

for a phase transition at this point in our approximate calculations. Furthermore, we find that as $\Delta \to \Delta_c$, $M^z \to 0$ in 1D, whereas $M^z \to M_c^z \approx 0.682$ in 2D. This result is important since we believe that the critical points at $\Delta = \Delta_c$ in our calculations approximate those at $\Delta = 1$ for the exact case in 1D and the likely case in 2D. Thus, our results strongly suggest that the Heisenberg antiferromagnet on the 2D square lattice possesses non-vanishing long-range order. This agrees with other calculations for this case. Thus, for example, SWT at $\Delta = 1$ gives[32] $M_c^z \approx 0.606$ (compared to a divergent result, $M^z \to -\infty$ as $\Delta \to 1$, in 1D), while the GFMC results[30] of Carlson and of Trivedi and Ceperley give respectively $M_c^z \approx 0.68 \pm 0.02$, 0.62 ± 0.04 at the same point.

7. CCM PARAMETRIZATION OF EXCITED STATES

Within the CCM excited-state wavefunctions are written in the standard[33] form, $|\Psi_e\rangle = X|\Psi\rangle = Xe^S|\Phi\rangle$, in terms of an excitation operator X which is chosen, like S, to be constructed purely from creation operators. By combining the g.s. Schrödinger equation (6) with its excited-state counterpart, $H|\Psi_e\rangle = E_e|\Psi_e\rangle$, we readily obtain

$$e^{-S}[H,X]e^S|\Phi\rangle = \varepsilon_e X|\Phi\rangle; \quad \varepsilon_e \equiv E_e - E_g \; . \tag{29}$$

The simplest possible form for X which produces excitations with $s_z^{total} = \pm 1$ has the form

$$X \rightarrow X_1 = \sum_i \chi_i \sigma_i^+ \; , \tag{30}$$

where the sum on i runs over one sublattice only, as before. The coefficients $\{\chi_i\}$ are obtained in the by now familiar manner by taking the overlap of Eq. (29) with the bra states $\langle\Phi|\sigma_k^-$. If we combine the approximation of Eq. (30) for X with the SUB2 approximation of Eq. (14) for S, we obtain the set of homogeneous linear equations,

$$\tfrac{1}{2}zK\chi_k - \tfrac{1}{2}\sum_\rho \sum_r b_r \chi_{k-r+\rho} = \varepsilon_e \chi_k \; , \tag{31}$$

which are valid for a bipartite lattice of arbitrary dimensionality, and where indices $\{r\}$ and $\{\rho\}$ have their earlier meanings.

It is trivial to verify that Eq. (31) has lattice plane-wave solutions,

$$\chi_k \rightarrow \chi_{\vec{k}}(\vec{q}) = \exp(-i\vec{k}\cdot\vec{q})\chi(\vec{q}) \; ; \quad \varepsilon_e \rightarrow \varepsilon(\vec{q}) \; . \tag{32}$$

By making further use of the g.s. SUB2 solution of Eq. (20) and its counterpart for higher dimensionality, we obtain the spectrum,

$$\varepsilon(\vec{q}) = \tfrac{1}{2}zK\left[1 - k^2\gamma^2(\vec{q})\right]^{\frac{1}{2}} \; , \tag{33}$$

where the constants K and k^2 are as given in Eqs. (16) and (21), and

$$\gamma(\vec{q}) \equiv \frac{1}{z}\sum_{\vec{\rho}} e^{i\vec{\rho}\cdot\vec{q}} \; . \tag{34}$$

The spectrum of Eq. (33) thus has a finite gap in both 1D and 2D for the case $k^2 < 1$, or equivalently $\Delta > \Delta_c$. The gap disappears precisely at the critical value Δ_c, again exactly as expected at a phase transition. In 1D, we find that as $\Delta \rightarrow \Delta_c$, $\varepsilon(q) \rightarrow \varepsilon_c(q) \approx 1.364|\sin(q)|$, $-\pi < q \le \pi$, which may be compared with the *exact* result[25] at $\Delta = 1$, namely $\varepsilon(q) = \tfrac{1}{2}\pi|\sin(q)|$. Similarly, for the 2D square lattice at $\Delta = \Delta_c$ we have the gapless spectrum,

$$\varepsilon_c(\vec{q}) = 2K_c\left[1 - \tfrac{1}{4}(\cos q_x + \cos q_y)^2\right]^{\frac{1}{2}} \; ; \quad -\pi < q_x, q_y \le \pi \; , \tag{35}$$

with spin-wave velocity $v_s = 2K_c \approx 2.335$. This may be compared with the classical value $v_0 = 2$ at $\Delta = 1$, and with the corresponding results for the Heisenberg square lattice of 2.28 ± 0.1 by Trivedi and Ceperley[30] using GFMC techniques, and 2.316 by Oguchi from Anderson's SWT,[32] evaluating the first-order correction in inverse powers of the spin quantum number.

8. EXTENSION TO SYSTEMS WITH HIGHER SPIN

All of the results presented so far have been for spin-$\frac{1}{2}$ systems. Unlike many other methods, the CCM has no problem in principle or in practice to be extended to systems of higher spin, $s > \frac{1}{2}$. The main changes are that we can no longer use the Pauli matrices, and the raising operator s_k^+ (defined in the Néel basis as before) can now act up to $2s$ times on the same site k. Thus,

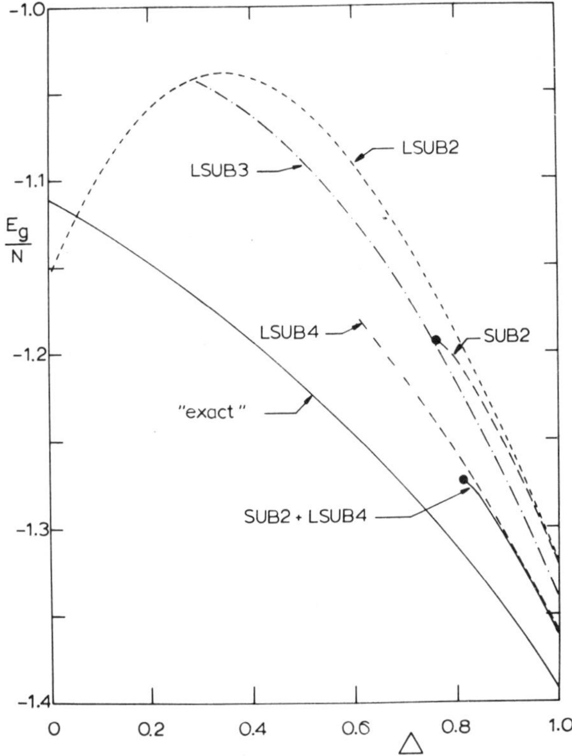

Fig. 6. Ground-state energy per spin for the spin-1 XXZ model in 1D as a function of anisotropy parameter Δ. CCM results in various approximations are compared with "exact" results obtained by us as described in the text.

for example, the cluster configuration coefficients in Eq. (13) can now have a given lattice index occurring up to $2s$ times. Nevertheless, the basic implementation of the CCM remains unchanged, apart from the consequent increase in the number of independent cluster configurations at a given truncation level. For example, on the 1D chain, whereas in the spin-$\frac{1}{2}$ case the LSUB2 and LSUB4 approximations retained 1 and 3 independent configurations respectively, for the spin-1 case these numbers increase to 2 and 7 respectively.

Typical results for the g.s. energy of the spin-1 chain, for example, are shown in Fig. 6 and Table 3 for a number of CCM approximation schemes described previously. Exact results are not known for the spin-1 XXZ model in 1D, but we compare with results extrapolated to the infinite ($N \to \infty$) chain from exact diagonalizations that we have performed on finite chains of length $N \leq 14$. We also compare with results from Anderson's linear spin-wave theory (SWT).[32] As expected, the quantum fluctuations now play a relatively less important role in the Ising-phase region for higher values of the spin quantum number (as also is the case for higher dimensionality lattices). For example, in the SUB2 and SUB2+LSUB4 approximations, the critical values, $\Delta_c \approx 0.759$ and 0.812 respectively, are appreciably closer to the classical value of 1 than their counterparts for the spin-$\frac{1}{2}$ case in Sec. 4.

Table 3. The g.s. energy per spin, E_g/N, for the spin-1 Heisenberg model on an infinite ($N \to \infty$) chain. Results in various CCM schemes are compared with "exact" results extrapolated to $N \to \infty$ from exact diagonalizations on chains of length $N \leq 14$, and with results from Anderson's linear spin-wave theory (SWT).[32]

	E_g/N						
Δ	LSUB2	LSUB3	LSUSB4	SUB2	SUB2+ LSUB4	SWT	"Exact"
0.5	-1.0594	-1.0893	-1.1278	--	--	--	-1.224
1.0	-1.3206	-1.3404	-1.3619	-1.3224	-1.3628	-1.3634	-1.398
1.5	-1.7189	-1.7273	-1.7329	-1.7178	-1.7329	-1.6840	-1.735
2.0	-2.1653	-2.1694	-2.1715	-2.1646	-2.1715	-2.1316	-2.172

We note that the SUB2 approximation takes only independent 2-body correlations into account. Hence, it cannot be expected to distinguish qualitative differences between the $s = \frac{1}{2}$ and $s > \frac{1}{2}$ models. The interesting open question as to whether the $s = 1$ case shows a qualitatively different phase structure in Δ than the $s = \frac{1}{2}$ case, as expected by the Haldane conjecture,[1] could only properly begin to be addressed within the CCM at the SUB4 level. Such full SUB4 calculations have not yet been performed.

9. DISCUSSION

We believe that the results presented in this paper clearly demonstrate the viability of the CCM to deal accurately, and at a fully microscopic level, with the quantum fluctuations of discrete spin-lattice models. The method presupposes no prior knowledge of any phase transitions. Nevertheless, rather low-level approximation schemes seem quite capable of shedding light on such global properties, as well as giving accurate numerical results for such quantities as the g.s. energy and the staggered magnetization. Excitation spectra are also capable of being calculated. We have shown how various hierarchies of approximation schemes can be exploited to yield sets of results of increasing accuracy, by the systematic inclusion of higher-order correlations. We are currently pursuing such schemes as the LSUBn hierarchy to higher values of the truncation index n, using computer-algebraic methods to derive the corresponding coupled sets of equations, in order to examine the convergence properties.

We have shown explicitly how, unlike many other methods, the CCM can be applied to bipartite spin lattices of arbitrary spin and dimensionality. Further work in progress on the spin-1 chain (using the SUB4 approximation, for example) may enable us to discuss the so-called Haldane phase.[1] We are also interested both in extending our calculations on the present systems to include the effects of an external magnetic field, and in enlarging the class of model Hamiltonians. An interesting example for spin-1 systems is one which includes an arbitrary admixture of isotropic Heisenberg (quadratic) and biquadratic exchange terms. Frustrated systems are another example. Finally, we are also interested in extending our methodology to such comparable models as the t-J and Hubbard models, as well as to such lattice gauge field theories as the $U(1)_3$ model of quantum electrodynamics on a square lattice.

ACKNOWLEDGEMENT

One of us (RFB) gratefully acknowledges support for this work in the form of a research grant from the Science and Engineering Research Council (SERC) of Great Britain.

REFERENCES

1. F.D.M. Haldane, Phys. Lett. 93A:464 (1983); Phys. Rev. Lett. 50:1153 (1983); I. Affleck, J. Phys.: Condens. Matter 1:3047 (1989).
2. J. Hubbard, Proc. Roy. Soc. A276:238 (1964); A285:542 (1965); D. Vollhardt, Rev. Mod. Phys. 56:99 (1984).
3. P.W. Anderson, Science 235:1196 (1987).
4. D. Vaknin et al., Phys. Rev. Lett. 58:2802 (1987); G. Shirane et al., ibid. 59:1613 (1987); K.B. Lyons et al., ibid. 60:732 (1988).
5. R.F. Bishop and H.G. Kümmel, Phys. Today 40(3):52 (1987).
6. F. Coester, Nucl. Phys. 7:421 (1958); F. Coester and H. Kümmel, ibid. 17:477 (1960).
7. J. Paldus, J. Čížek and I. Shavitt, Phys. Rev. A 5:50 (1972).
8. R.F. Bishop and K.H. Lührmann, Phys. Rev. B 17:3757 (1978).
9. H. Kümmel, K.H. Lührmann and J.G. Zabolitzky, Phys. Rep. 36C:1 (1978).
10. J.S. Arponen, Ann. Phys. (N.Y.) 151:311 (1983).
11. J. Arponen, R.F. Bishop and E. Pajanne, Phys. Rev. A 36:2519,2539 (1987); in: "Condensed Matter Theories," Vol. 2, P. Vashishta, R.K. Kalia and R.F. Bishop (eds.), Plenum, New York (1987), p. 357.
12. R.J. Bartlett, J. Phys. Chem. 93:1697 (1989).
13. R.F. Bishop, Theor. Chim. Acta (1991) -- to be published.
14. H.G. Kümmel, in: "Nucleon-Nucleon Interaction and Nuclear Many-Body Problems," S.S. Wu and T.T.S. Kuo (eds.), World Scientific, Singapore (1984), p. 46.
15. R.F. Bishop et al., Phys. Rev. C 42:1341 (1990).
16. B. Day, Phys. Rev. Lett. 47:226 (1981); B. Day and J.G. Zabolitzky, Nucl. Phys. A366:221 (1981).
17. R.J. Bartlett, Ann. Rev. Phys. Chem. 32:359 (1981); K. Szalewicz et al., J. Chem. Phys. 81:2723 (1984); A.C. Scheiner et al., ibid. 87:5361 (1987); E.A. Salter, G.W. Trucks and R.J. Bartlett, ibid. 90:1752 (1989).
18. R.F. Bishop and K.H. Lührmann, Phys. Rev. B 26:5523 (1982); K. Emrich and J.G. Zabolitzky, ibid. 30:2049 (1984); J. Arponen and E. Pajane, J. Phys. C 15:2665,2683 (1982).
19. U. Kaulfuss and M. Altenbokum, Phys. Rev. D 33:3658 (1986); R.F. Bishop, M.C. Boscá and M.F. Flynn, Phys. Rev. A 40:3484 (1989); J.S. Arponen and R.F. Bishop, Phys. Rev. Lett. 64:111 (1990).

20. M. Altenbokum and H. Kümmel, Phys. Rev. D $\underline{32}$:2014 (1985); M. Funke, U. Kaulfuss and H. Kümmel, *ibid.* $\underline{35}$:621 (1987).

21. H.A. Bethe, Z. Phys. $\underline{71}$:205 (1931).

22. L. Hulthén, Arkiv Mat. Astron. Fysik A $\underline{26}$:No. 11 (1938); R. Orbach, Phys. Rev. $\underline{112}$:309 (1958).

23. R.J. Baxter, J. Phys. C $\underline{6}$:L94 (1973); J. Stat. Phys. $\underline{9}$:145 (1973).

24. N.M. Bogoliubov, A.G. Izergin and V.E. Korepin, Nucl. Phys. $\underline{B275}$:687 (1986).

25. J. des Cloiseaux and J.J. Pearson, Phys. Rev. $\underline{128}$:2131 (1962); C.N. Yang and C.P. Yang, *ibid.* $\underline{150}$:321 (1966); J.D. Johnson, S. Krinsky and B.M. McCoy, Phys. Rev. A $\underline{8}$:2526 (1973); L.D. Faddeev and L.A. Takhtajan, Phys. Lett. $\underline{85A}$:375 (1981).

26. R.F. Bishop, J.B. Parkinson and Yang Xian, in: "Condensed Matter Theories," Vol. 6, S. Fantoni and S. Rosati (eds.), Plenum, New York (1991), p. 37; Phys. Rev. B. (1991) -- to be published.

27. M. Roger and J.H. Hetherington, Phys. Rev. B $\underline{41}$:200 (1990).

28. R.F. Bishop, J.B. Parkinson and Yang Xian, Phys. Rev. B $\underline{43}$:13782 (1991).

29. S. Sachdev, Phys. Rev. B $\underline{39}$:12232 (1989).

30. J. Carlson, Phys. Rev. B $\underline{40}$:846 (1989); N. Trivedi and D.M. Ceperley, *ibid.* $\underline{41}$:4552 (1990).

31. T. Barnes, D. Kotchan and E.S. Swanson, Phys. Rev. B $\underline{39}$:4357 (1989).

32. P.W. Anderson, Phys. Rev. $\underline{86}$:694 (1952); T. Oguchi, Phys. Rev. $\underline{117}$:117 (1960).

33. K. Emrich, Nucl. Phys. $\underline{A351}$:379,397 (1981).

CLUSTERS - A LINK BETWEEN CONDENSED

MATTER PHYSICS AND NUCLEAR PHYSICS

P. Jena, C. Yannouleas, S.N. Khanna and B.K. Rao

Physics Department
Virginia Commonwealth University
Richmond, VA 23284-2000

ABSTRACT

Clusters consisting of atoms ranging from two to a few thousand exhibit several properties that are strikingly similar to those of nuclei. These include stability, shell structure, spin states, and giant dipole resonances. The physical reasons underlying these similarities will be addressed by reviewing recent experimental evidence and theoretical calculations based on electronic structure. The effect of symmetry and shape on the magnetic properties of atomic clusters is studied and compared with nucleonic clusters. Ways in which atomic clusters can further our understanding of interactions in nuclei is discussed.

INTRODUCTION

The study of atomic clusters has been a topic of great current interest in the last decade. This is due to the fact that the size and composition of clusters can be easily controlled in the laboratory and mass selected clusters can be studied in gas phase, as well as on substrates and in matrices. Furthermore, the shape and dimensionality of the clusters have considerable effect on their electronic properties. The studies on well characterized clusters as a function of size can illustrate how solids form as atoms come together – bridging the gap in our understanding between atomic and molecular physics and solid-state physics. Clusters are also being looked at as new source of novel materials with unique electronic and magnetic properties that depend strongly on their size and composition.

The stability of the clusters of metallic elements, particularly the alkali-metal series, exhibits characteristics that are strikingly similar to those observed in nuclear physics. For example, it was observed [1] that clusters of Na containing 2, 8, 20, 40, ... atoms are unusually stable. These numbers are analogous to the magic numbers associated with stable nuclei. It was shown [1] that the enhanced stability of these clusters is due to electronic shell structure, just as the enhanced stability of magic nuclei is due to nuclear shell structure. Three interesting questions can be posed:

1. Why do metal clusters governed by electrostatic forces show similar features as nuclei which are governed by strong short-range forces?

2. Is it possible that clusters can be used to complement studies in nuclear physics? This will be particularly useful, since the mass numbers and composition of nuclei are limited. Clusters, on the other hand, provide a system where the composition of atoms and their number can in principle be varied without limit.

3. Can nuclear methods be used to study problems in cluster physics?

In this paper, we review briefly some of the experimental and theoretical investigations that address these issues and point out further studies that can strengthen the collaboration between nuclear and condensed-matter physicists.

MASS SPECTRA, SHELL STRUCTURE AND MAGIC NUMBERS

In Fig. 1 we present the mass spectra of Na_N clusters observed by Knight et al. [1] The pronounced peaks or edges at sizes containing 2, 8, 20, 40, ... atoms led Knight et al to propose that these clusters must be unusually stable and that this stability could be due to an electronic shell structure that is similar to the nuclear shell structure. They assumed that a metal cluster can be modelled by a sphere of uniformly distributed positive charges. The density n_0 of these charges is given by $n_0 = Z/\Omega_0$, where Z is the valence and Ω_0 is the atomic volume. In the jellium model, it is conventional to express n_0 in terms of electron-electron distance r_s, namely $1/n_0 = 4\pi r_s^3/3$. For monovalent metals, r_s is equal to the Wigner-Seitz radius. The radius, R, of the jellium cluster is related to r_s through the relation $R = r_s\, N^{1/3}$, where N is the number of atoms. The valence electrons of the atoms, considered completely delocalized, then respond to the background jellium sphere and arrange themselves in electronic shells characterized by their orbital angular quantum number. When one of these shells is filled, the cluster exhibits unusual stability. The closing of one shell and the opening of another is reflected in the pronounced peak or edge in the mass abundances.

The above analysis parallels the stability of nuclei and nuclear magic numbers as derived from nuclear shell structure. The size of the nucleus is given by $R = r_0\, A^{1/3}$, where A is the mass number and $r_0 \approx 1.2$ fm reflects the range of the nuclear force. In contrast, for alkali clusters, r_s is of the order of 2 Å. The success of the electronic shell structure in explaining cluster stability is due to the fact that the long-range electrostatic forces are screened by the conduction electrons and that the range of the screened Coulomb interaction is comparable to the cluster size.

The important difference between clusters and nuclei, however, is that nature has placed an upper limit on the mass number A, but no such limit exists for N, the number of atoms in the cluster. Thus clusters could be used to probe properties of "large nuclei" that don' t exist. For example, in the last decade, a major experimental effort has been to produce artificially superheavy elements so that shell structure in nuclei with $l > 6$ can be observed. While the nuclear experiments were unsuccessful, experiments on alkali clusters consisting of thousands of atoms have verified the existence of shell structure for much higher angular momenta [2] (see Fig. 2).

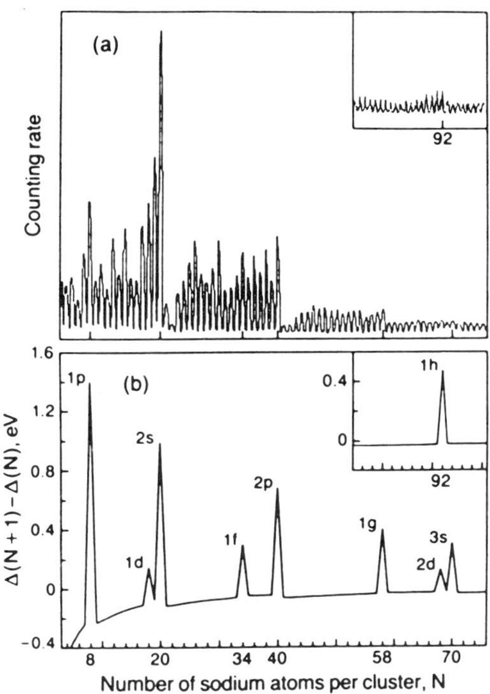

Fig. 1. Na cluster abundance spectrum: (a) experimental, (b) theoretical, using Woods-Saxon-type potential. [After Knight et al.[1]]

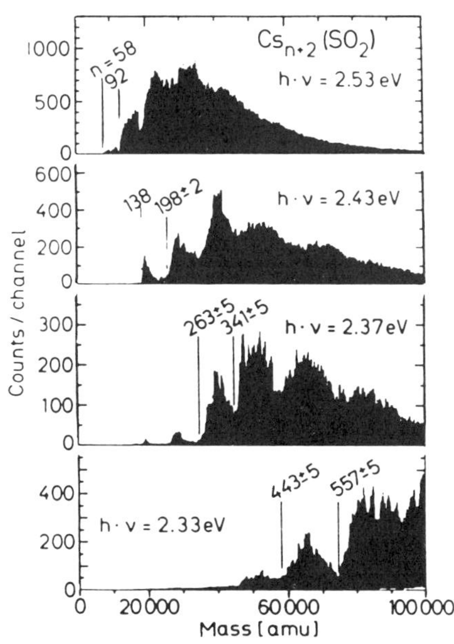

Fig. 2. Mass spectra of $Cs_{n+2}(SO_2)$ clusters. [After Göhlich et al.[2]]

GEOMETRY AND SPIN STRUCTURE

A closer examination of the results in Fig. 1 reveals that although the spherical jellium model reproduces the pronounced features of the mass spectra, it fails to account for other secondary structures in between the magic numbers, specifically the odd-even alternation in the cluster abundances: for open shells, clusters with an even number of atoms being more prevalent than those with an odd number of atoms. It is worthwhile to recall that the jellium results in Fig. 1 were based upon a spin-*unpolarized* calculation. Later Rao et al [3] carried out a self-consistent spin-polarized calculation for the jellium clusters and found that the total energies of the spin-polarized model are indeed lower than the unpolarized model. An analysis of their results on energetics shown in Fig. 3 revealed that not only alkali clusters with closed shells (2, 8, 20, 40, ...) exhibit pronounced stability, but clusters with half-filled shells (5, 13, 27, ...) should also exhibit enhanced stability. The later arises due to the fact that electrons in half filled states maximize their spins due to Hund' s rule and correspondingly the energy is lowered due to the exchange interaction. Although variationally the magnetic state is the ground state, clusters are found experimentally to have lower spin configurations than what is permissible by Hund' s rule. In addition, the mass spectra in Fig. 1 do not indicate that clusters with half-filled states are as stable as neighboring clusters containing an even number of atoms.

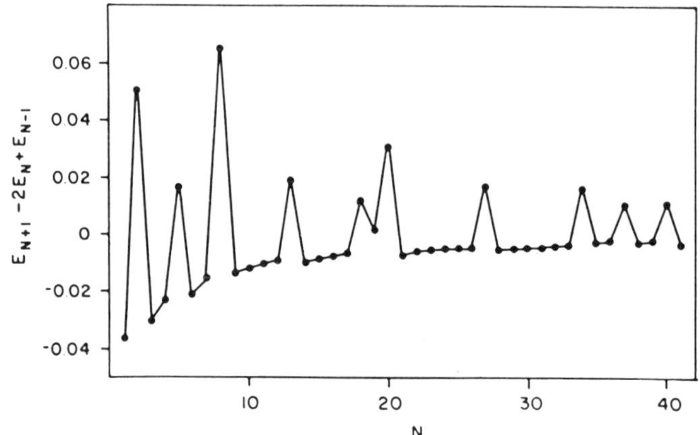

Fig. 3. Second derivative of E(N) vs N for Lithium clusters. [After Rao et al [3]]

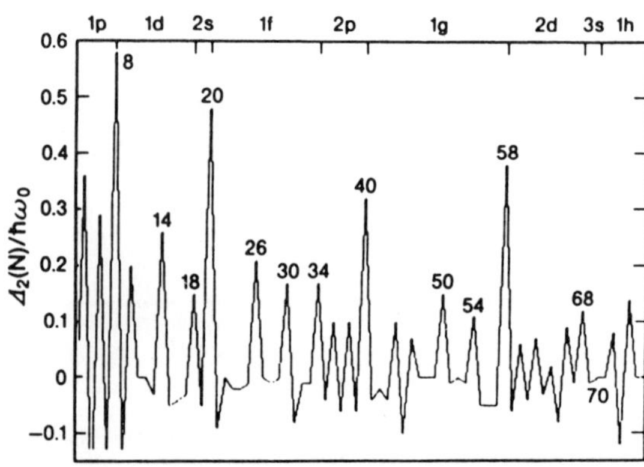

Fig. 4. Spheroidal distortions and Na mass spectrum. [After Clemenger [5]]

Nuclei behave in a similar way. The net spin of a nucleus is not governed by Hund'
s rule and is lower than the maximum permissible value in an open shell. The fact
that clusters have lower than maximum permissible spins implies that the degeneracy
of the energy levels must somehow be lifted. This can be done through the so-called
Jahn-Teller distortion that reduces the energy by distorting the geometry of the cluster.
Such a distortion would split the otherwise degenerate orbitals and the electrons may
not find Hund' s rule to be the best way for minimizing energy. Thus clusters can
depart from sphericity and assume low spin configurations. The Nilsson model [4] for
the nucleus contained the ingredients for such a scheme. For clusters, Clemenger has
shown [5] that the ellipsoidal prolate-oblate distortion away from the spherical jellium
clusters leads indeed to a lowering in energy and the computed variations in energy
account for the mass spectra better than the spherical jellium cluster (see Fig. 4).

The above discussion points out that the ground-state geometry and spin configuration are due to a competition between Jahn-Teller distortion that minimizes the
energy by splitting the degeneracy of the electronic orbitals and the Hund' s rule that
minimizes energy by retaining the orbital degeneracy. Thus an interplay between cluster geometry and spin would be an interesting point for study. For example, consider a
cluster of four Li atoms. When arranged in a tetrahedral configuration, the cluster is in
a spin triplet state. However, when the cluster is allowed to distort to a rhombus form,
the degeneracy of the p-orbital is removed, and the cluster' s ground state assumes a
spin singlet state. Thus as a cluster is transformed from a rhombus to a tetrahedron
by lifting one of the atoms from the initial common plane, a point must come when the
cluster transforms from a non-magnetic to a magnetic state. This is indeed the case as
was found by Rao et al [6] from total energy calculations. Their results are shown in Fig.
5. Note that when the cluster has a dihedral angle of 110°, the magnetic transition
takes place. The magnetic cluster, always higher in energy than the non-magnetic one,
is in a local minimum and should be observable.

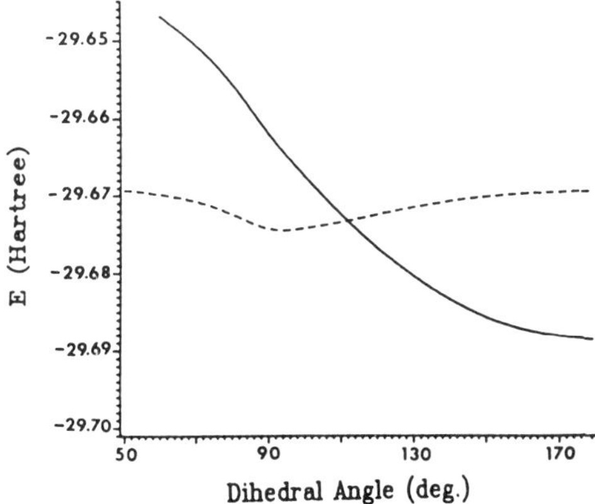

Fig. 5. Energy of Li$_4$ cluster in singlet (solid line) and triplet (dashed line)
states. [After Rao et al [6]]

The importance of cluster geometry on the stability and electronic properties brings
into focus the validity of the jellium model. One would think that as the clusters get
larger, the covalency will decrease and the electrons will be more delocalized. Consequently, the jellium model will become a better approximation for large clusters. This
is, however, not the case. It has been observed experimentally [7] that for very large
clusters (consisiting of a few thousand atoms) the stability is governed by icosahedral

packing rather than by shell closure. The question then remains: what is the limitation on the validity of the jellium model? Recall that the pronounced peaks in the mass spectra are directly linked to shell closings. Thus as long as geometrical distortion does not alter the relative position of the shells, it will have little effect on the magic numbers of clusters. Recently, Khanna and Jena [8] have carried out self-consistent density functional calculations to study the effect of shell splitting due to geometrical distortions in clusters of 3, 4, and 13 atoms. They find that in small clusters the spacings between the successive electronic shells are large. Although the geometrical distortion splits the orbitals, this does not alter the relative ordering. Thus small clusters, even though they depart strongly from sphericity, follow the jellium description of energy-level structure. As clusters become large, the spacings between the shells diminish and their ordering can be easily altered if structural distortion splits the degenerate orbitals. Thus in very large clusters, the energy-level shell structure is more sensitive to the geometry than in small clusters. Consequently, contrary to one' s expectation, the relative stability of clusters derived from the jellium model are appropriate for the small clusters, but not for the very large clusters. [7,9] It should be noted, however, that the average ground-state properties of large clusters are expected to be sufficiently well described by semiclassical modifications of the jellium model. [10-11]

DIPOLE RESONANCES - OPTICAL PROPERTIES - HETEROCLUSTERS

The optical response of metal microclusters also exhibits substantial analogies with corresponding nuclear processes. [12-13] In particular, the photoabsorption proceeds via the excitation of a dipole plasma mode where the valence electrons move collectively against the jellium positive background. This process is quite analogous to the well known giant dipole resonance in nuclei, [13] where the protons move against the neutrons. Fig. 6 provides an overall summary of this analogy between clusters and nuclei, given the different energy scales for the two systems (optical region for clusters, γ-rays for nuclei). There is an interesting difference, however, between the giant dipole resonance in nuclei and in clusters. In nuclei, due to the short-range character of the nuclear forces, the energy of the resonance decreases with increasing size as $100/R$ MeV. On the contrary, in metal clusters, the movement of the electrons against the positive background reduces the screening, and the energy of the resonance is mainly supplied by the long-range part of the Coulomb force. As a result, the plasma resonance slightly increases with size and approaches the Mie limit for a classical metal sphere according to $(\hbar^2 e^2/m_e r_s^3)^{1/2}(1 - 1.5t/R)$, where t is the distance (spill-out) that the electronic cloud extends beyond the radius R of the positive background.

Detailed studies of the photoabsorption profiles have been carried out using techniques directly taken from the nuclear many-body problem [14-16] or closely related to it (time-dependent local-density approximation, TDLDA [17-18]). In particular, the nuclear matrix-RPA (random phase approximation) was adapted to the case of spherical, closed-shell metal clusters. [15] In this approach, the collective motion arises as a linear superposition over many single-particle transitions from the occupied to the unoccupied orbitals of the effective average field. For specific mass numbers and species, like neutral Na_{20} and Na_{40}, this method predicted multipeak photoabsorption profiles more complex than the simple one-peak profile expected from the Mie theory for the charge oscillations of a classical metal sphere. Such profiles have been recently observed experimentally. [12,19] They represent quantum size effects due to the discreteness of the single-particle levels.

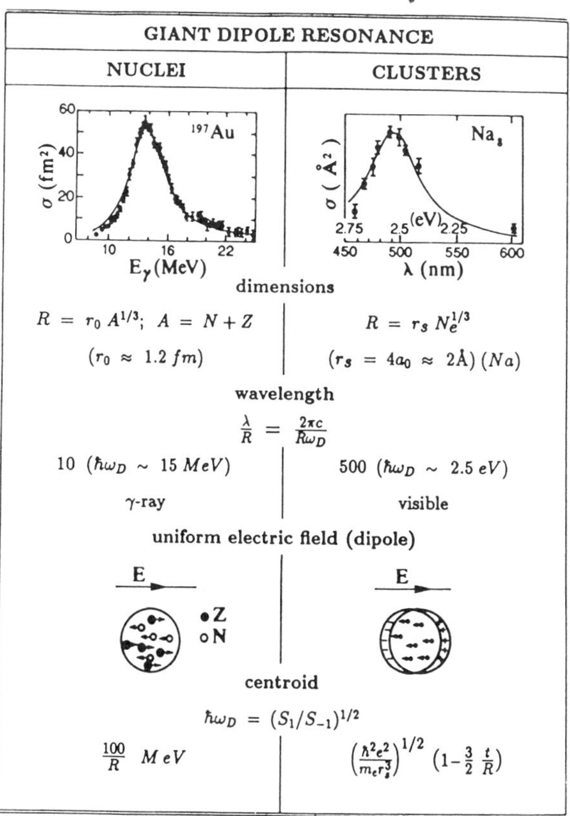

Fig. 6. Analogies between giant dipole resonances in nuclei and metal clusters. In nuclei, the external electromagnetic field sets in motion the protons against the neutrons. In metal clusters, the delocalized valence electrons are set in motion against the positive ionic background. [After Broglia et al [13]]

In particular, it has been found [16] that the split-plasmon that characterizes the photoabsorption profiles of Na_{20} and K_{20} can be treated as a two-level model due to a degeneracy between the collective plasmon and a particular $2s \rightarrow 3p$ particle-hole transition. It is interesting to inquire about perturbing factors that will modify the effective potential of the neutral Na_{20} (or K_{20}) and will lift this degeneracy allowing again for a single dominant photoabsorption peak, as is the case with Na_8. Such a perturbing factor may be provided by the presence of a foreign atom at the center of an otherwise homonuclear cluster. Indeed, the heteroatom at the center of such a compound metallic cluster results in unequal shifts in the single-particle levels with respect to the case of pure clusters. Because the s orbitals have a maximum radial density at the center of the cluster, they are influenced much more than the orbitals with higher angular momenta. If an atom with a smaller Wigner-Seitz radius is at the center, like in NaK_{19}, the 2s level will be pushed downwards, while the same level will be pushed upwards if the central atom has a larger Wigner-Seitz radius, like in CsK_{19}; at the same time the 3p level remains practically unaffected. We expect this movement to control the degree of overlap between the plasmon and the $2s \rightarrow 3p$ transition. Preliminary calculations [20] indicate that this is indeed the case. In fact the RPA oscillator strengths for NaK_{19}, K_{20}, and CsK_{19} are displayed in Fig. 7. In the case of NaK_{19}, there is one dominant line with 80% of the Thomas-Reiche-Kuhn sum rule; the same line is, however, split into two components in the case of K_{20} and CsK_{19}. Variations in the distribution of strength between the two components is apparent as one goes from K_{20} to CsK_{19}. Thus different combinations of heteroatoms are expected to provide substantial leverage to the experimental ability for controlling the appearance (or disappearance) of this doublet which seems to be generic to the optical response of a system with 20 delocalized electrons.

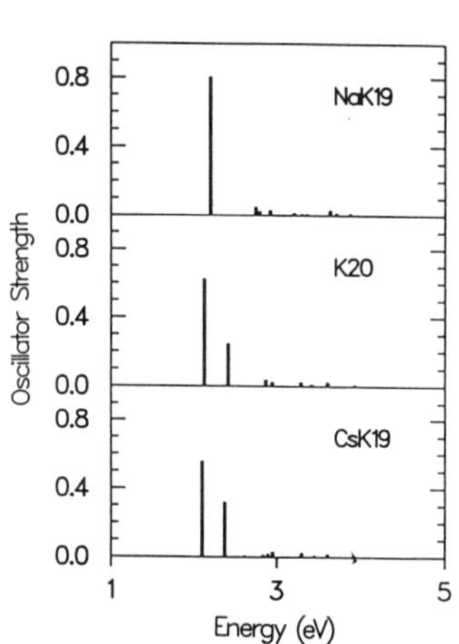

Fig. 7. RPA Oscillator Strengths
[After Yannouleas et al [20]]

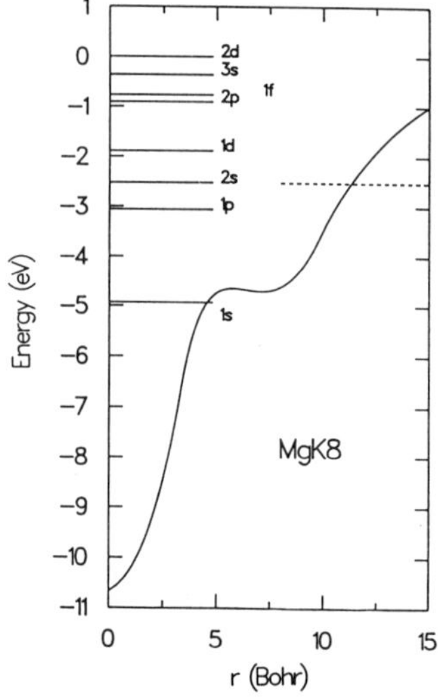

Fig. 8. MgK_8 [From Ref. 20]

It should be mentioned that, if the unequal shifts in the single-particle levels are large enough, a new sequence of magic numbers may arise. An example is offered by MgK_8, which is a magic aggregate [21] in spite of the fact that it corresponds to 10 delocalized valence electrons. The effective potential for MgK_8 is displayed in Fig. 8.

FRAGMENTATION OF CHARGED CLUSTERS - FISSION

Another close analogy between metal clusters and nuclei exists for the process of fragmentation of charged metal clusters. [22-23] This process shares common features with the process of nuclear fission. [22-24] The key in understanding this analogy is again the short-range character of the screened Coulomb force that binds the valence electrons. Because of the short-range binding force, the total binding energy of both systems on the average (after neglecting the variations due to shell effects) can be written as the sum of a volume, surface, and curvature terms (liquid-drop model). Namely, one can write $E = a_v N + a_s N^{2/3} + a_{cr} N^{1/3}$. For nuclei, the coefficients a are well established from experimental data. [25] For the case of clusters, the corresponding coefficients are determined by a fit to the binding energies resulting from jellium calculations. [10-11] When a cluster carries a net total charge Z (like in the case of K_{10}^{2+}), the ensuing electrostatic energy is repulsive and $\sim a_{Cl} Z^2 N^{-1/3}$. As the drop distorts away from the spherical minimum, it is mainly the surface term that competes against the Coulomb repulsion. For a spherical droplet, the fissility parameter, f, defined as the ratio of the Coulomb energy over twice the surface energy determines the stability against symmetric fragmentation. [24] When $f < 0.351$, the droplet is stable, namely the total energy increases after the fragmentation. When $0.351 < f < 1$, the droplet is metastable: after fragmentation the total energy is lower, but there is a barrier to be overcome. When $f > 1$, the droplet fragments spontaneously. Since $f = (de^2/4\pi r_s^3 \sigma)(Z^2/N)$, where d is a constant between 3/5 and 1/2 depending on the assumed distribution of the net total charge and σ is the surface tension, the stability conditions for Na clusters become: [24] $0.39 < Z^2/N$ (unstable), $0.14 < Z^2/N < 0.39$ (metastable), and $Z^2/N < 0.14$ (stable).

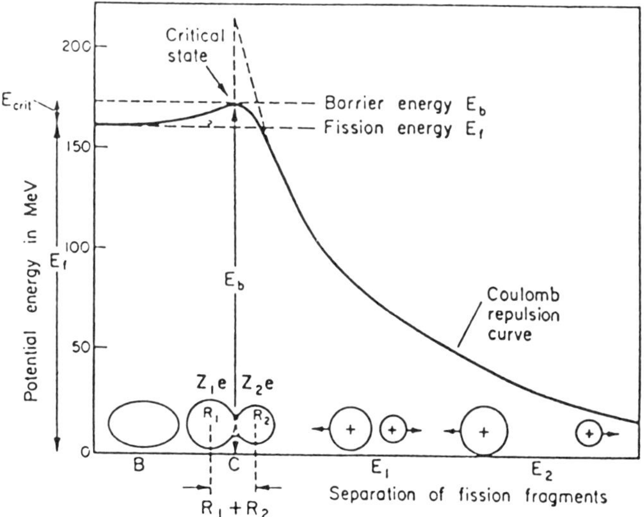

Fig. 9. Energetics of nuclear fission. [From Ref. 27]

In the case of metastable nuclei, the liquid drop model predicts a potential surface displayed in Fig. 9 as a function of the successive shapes from the initial sphere to the two final disconnected fragments. An analogous curve has been calclulated for the metastable case of the doubly ionized gold dimer. [26]

REFERENCES

1. W.A. de Heer, W.D. Knight, M.Y. Chou, and M.L. Cohen, Solid State Physics 40, 93 (1987), and references therein
2. H. Göhlich, T. Lange, T. Bergmann, and T.P. Martin, Phys. Rev. Lett. 65, 748 (1990)
3. B.K. Rao, P. Jena, and M. Manninen, Phys. Rev. B32, 477 (1985); B.K. Rao and P. Jena, Phys. Rev. B32, 2058 (1985)
4. S.G. Nilsson, Mat.-Fys. Medd. K. Dan. Vidensk. Selsk. 29, No. 16 (1955)
5. K. Clemenger, Phys. Rev. B32, 1359 (1985)
6. B.K. Rao, S.N. Khanna, and P. Jena, Chem. Phys. Lett. 121, 202 (1985)
7. T.P. Martin, T. Bergmann, H. Göhlich, and T. Lange, Chem. Phys. Lett. 172, 209 (1990)
8. S.N. Khanna and P. Jena, Chem. Phys. Lett. (1991) in press
9. G.S. Anagnostatos, Phys. Lett. 124A, 85 (1987)
10. E. Engel and J.P. Perdew, Phys. Rev. B43, 1331 (1991)
11. M. Brack, Phys. Rev. B39, 3533 (1989)
12. K. Selby, V. Kresin, J. Masui, M. Vollmer, W.A. de Heer, A. Scheidemann, and W.D. Knight, Phys. Rev. B43, 4565 (1991); K. Selby, M. Vollmer, J. Masui, V. Kresin, W.A. de Heer, and W.D. Knight, Phys. Rev. B40, 5417 (1989)
13. *Optical response of small metal clusters: Atomic analog of nuclear giant resonances*, R.A. Broglia, J.M. Pacheco and C. Yannouleas, Phys. Rev. B, (Sept. 15, 1991) in press
14. G.F. Bertsch, Comput. Phys. Commun. 60, 247 (1990)
15. C. Yannouleas, R.A. Broglia, M. Brack and P-F. Bortignon, Phys. Rev. Lett. 63, 255 (1989)
16. *Collective and single-particle aspects in the optical response of metal microclusters*, C. Yannouleas and R.A. Broglia, Phys. Rev. A, (1991) in press
17. W. Ekardt, Phys. Rev. B31, 6360 (1985)
18. D.E. Beck, Phys. Rev. B43, 7301 (1991)
19. S. Pollack, C.R.C. Wang, and M.M. Kappes, J. Chem. Phys. 94, 2496 (1991)
20. C. Yannouleas, P. Jena, and S.N. Khanna, to be published
21. M.M. Kappes, P. Radi, M. Schär, and E. Schumacher, Chem. Phys. Lett. 119, 11 (1985)
22. B.K. Rao, P. Jena, M. Manninen, and R.M. Nieminen, Phys. Rev. Lett. 58, 1188 (1987); F. Liu, M.R. Press, S.N. Khanna, and P. Jena, Phys. Rev. Lett. 59, 2562 (1987)
23. W.A. Saunders, Phys. Rev. Lett. 64, 3046 (1990)
24. M. Nakamura, Y. Ishii, A. Tamura, and S. Sugano, Phys. Rev. A42, 2267 (1990)
25. Å. Bohr and B.R. Mottelson, "Nuclear Structure", Vol. II, (Benjamin, Reading, Massachusetts, 1975)
26. L. Yi, S.N. Khanna, and P. Jena, Phys. Rev. Lett. 64, 1188 (1990)
27. T.A. Littlefield and N. Thorley, "Atomic and Nuclear Physics", (Van Nostrand Reinhold, Berkshire, England 1979)

QUANTUM INTERFERENCE PHENOMENA IN STRONG LOCALIZATION

Mehran Kardar

Department of Physics
Massachusetts Institute of Technology
Cambridge, Massachusetts 02139, USA

Ernesto Medina

Intevep SA
Apdo 76343
Caracas 1070A, Venezuela

ABSTRACT

The role of quantum interference phenomena is examined for *strongly localized*, non–interacting electrons. We compute, both numerically and analytically, the probability distribution for tunneling between sites separated a distance t, by summing all *forward scattering paths*. We find a probability distribution that is approximately log–normal; its mean proportional to t, and its variance growing as $t^{2\omega}$, with ω depending on the dimension d. Since the mean and variance are independent, *two parameters* are necessary to describe the tunneling probability. We also study the response of the system to a magnetic field B, with and without spin–orbit (SO) scattering. Without SO a magnetic field leads to an increase in the localization length scaling as $B^{1/2}$. With SO, there is still a positive magnetoconductance (initially scaling as $B^2 t^3$), but no change in the localization length. The *universal* characteristics of the probability distribution can be probed by examining its moments. These moments describe the world lines of n attracting bosons in $d - 1$ dimensions– a well known many body problem! Various results for this simple problem are then used to provide analytical information on the distribution for tunneling in strong localization.

INTRODUCTION

The influence of quantum interference phenomena on magnetoconductance (MC) and conductivity fluctuations has been extensively studied for *weakly localized* electrons[1]. In the absence of spin-orbit scattering (SO), a magnetic field causes an increase in the localization length (a positive MC), and a factor of 2 decrease in the conductance fluctuations. These results are attributed to suppression of *backscattering* loops by a magnetic field. With SO, the magnetic field has the opposite effect of decreasing the localization length (a negative MC), but still reduces the conductance fluctuations[2]. Symmetries of the underlying Hamiltonian, and their modification by a magnetic field, can also be invoked to support these conclusions[3].

Recent Progress in Many-Body Theories, Vol. 3,
Edited by T.L. Ainsworth et al., Plenum Press, New York, 1992

By contrast, the behavior of conductivity and its fluctuations for *strongly localized* electrons is more controversial and less well understood. The main mechanism for conductivity in this regime is by thermally activated, variable range, electron tunneling[4]. Nguyen, Spivak, and Shklovskii (NSS)[5] have emphasized that one must account for the quantum interference of *forward scattering* paths to the tunneling probability. Treatments that ignore the correlations between such paths conclude a positive magnetoconductance, but no change in the localization length, whether in the absence[5,6] or presence[7] of SO. On the other hand, a random–matrix approach[8] predicts that a magnetic field leads to a doubling of the localization length ξ (big, positive MC) without SO, but a halving of ξ (big, negative MC) in the presence of SO. Over the past three years we undertook extensive numerical studies of the NSS model. We used a transfer matrix procedure that allows exact summation of forward scattering paths for very large systems ($t \approx 10^3 - 10^4$), and obtained statistics by averaging over many realizations ($\approx 10^3$) of randomness. These results, combined with some analytical insights, provide a coherent description of the behavior of strongly localized electrons[9,10]. A brief description of our approach and its conclusions is presented here, and is contrasted to results obtained by other approaches.

MODEL

The starting point is the Anderson type Hamiltonian

$$\mathcal{H} = \sum_{i,\sigma} \epsilon_i a_{i,\sigma}^\dagger a_{i,\sigma} + \sum_{<ij>,\sigma\sigma'} V_{ij,\sigma\sigma'} a_{i,\sigma}^\dagger a_{j,\sigma'} \quad , \tag{1}$$

where ϵ_i are the random site energies. The *nearest-neighbor* only hopping elements are set to $V_{ij} = V U_{ij} e^{i\mathcal{A}_{ij}}$, where V is a constant, U_{ij} is a randomly chosen $SU(2)$ matrix describing the spin rotation due to a strong SO scatterer[3,7], and $\mathcal{A}_{ij}$ is the magnetic vector potential. The tunneling probability between two sites is related to their overlap, which using a "locator" expansion[11] can be written as

$$< i\sigma|G(E)|f\sigma' > = \sum_\Gamma \prod_{i_\Gamma} \frac{V e^{i\mathcal{A}} U}{E - \epsilon_{i_\Gamma}}. \tag{2}$$

The above expression is a Feynman sum over all possible trajectories Γ between the initial (i) and final (f) sites. Each bond along the path contributes a random spin rotation U, and a phase factor from the magnetic vector potential $\mathcal{A}$.

To simplify the energy denominators, we use the NSS[5] model in which the energy of the initial and final sites is set to zero, while the intermediate ϵ_i take on values of $+W$ or $-W$ with equal probability. A path of length ℓ now contributes an amplitude $W(V/W)^\ell$ to the sum, as well as an overall sign and rotation matrix. For $V/W \ll 1$, corresponding to strongly localized electrons, the sum is convergent[11], and eq.(2) is dominated by the shortest paths connecting the end points. Following NSS, we maximize the interference between such *forward scattering* paths by choosing i and f to lie along the diagonal of a hypercubic lattice, as in depicted for two dimensions in Fig.(1). All shortest paths Γ' now have the same length t, and the tunneling amplitude simplifies to

$$A = < i\sigma|G(0)|f\sigma' > = W(V/W)^t J(t); \qquad J(t) = \sum_{\Gamma'} \prod_{i_{\Gamma'}} \text{sign}(\epsilon_{i_{\Gamma'}}) e^{i\mathcal{A}} U. \tag{3}$$

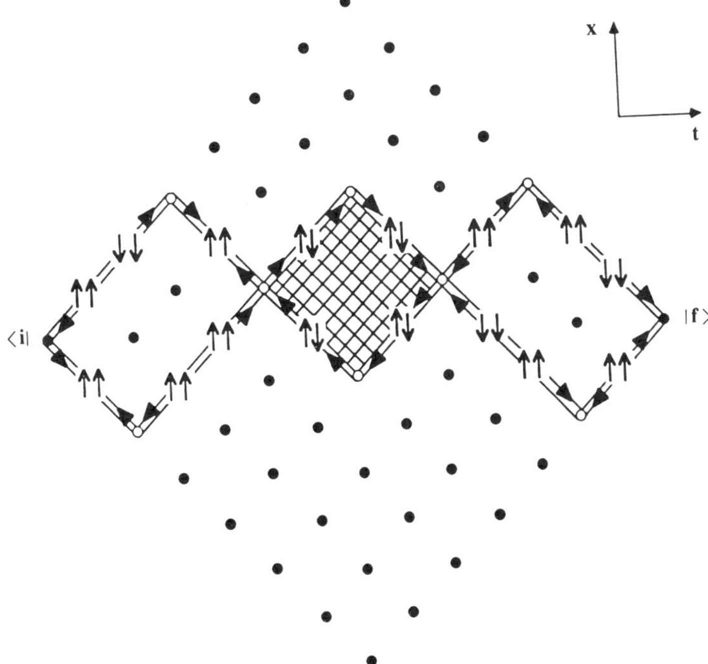

Figure 1 Directed paths contributing to the tunneling between diagonally separated endpoints. The averaging over SO impurities pairs forward and time–reversed paths, and their spins, as indicated.

After averaging over the initial spin, and summing over the final spin, the tunneling probability is

$$T = \frac{1}{2}\mathrm{Tr}(A^\dagger A) = W^2(V/W)^{2t}I(t); \qquad I(t) = \frac{1}{2}\mathrm{Tr}(J^\dagger J). \tag{4}$$

All interference information is now contained in $I(t)$.

THE TUNNELING PROBABILITY DISTRIBUTION

For each realization of randomness, the contribution of forward scattering paths to $J(t)$ in eq.(3) can be computed exactly using a *transfer matrix method*[9,10]. We numerically studied the statistical properties of $I(t)$ for t of up to 2000, and for over 2000 realizations of the random Hamiltonian. In all cases, with and without a magnetic field, in the presence or absence of SO, we find a probability distribution for $I(t)$ that is broad (almost log–normal). The quantity $\ln I(t)$ is similar to a quenched average free energy, and we confirmed that, as any extensive quantity, it grows linearly with t. Since the Green's function in eq.(3) is expected to typically decay as $\exp(-t/\xi)$, the localization length ξ has a contribution from $\ln I(t)$. In fact, we define *local* and *global* contributions to ξ by setting

$$\frac{\langle \ln | < i|G|f > | \rangle}{t} \equiv \xi^{-1} \equiv \xi_0^{-1} + \xi_g^{-1}, \tag{5}$$

147

with

$$\xi_0^{-1} \equiv \ln\left(\frac{W}{\sqrt{2}V}\right) \quad , \text{ and } \quad \xi_g^{-1} \equiv \ln\sqrt{2} - \frac{\langle \ln I(t)\rangle}{2t}. \tag{6}$$

Fluctuations of $\ln I(t)$ also increase with t, but with a slower power law– numerical results show the variance of $\ln I(t)$ scaling as $t^{2\omega}$ with $\omega = 0.33 \pm 0.05$.

Log–normal distributions arise quite naturally for the conductance of one dimensional systems[12]. Indeed, in a one–dimensional description of forward scattering, $I(t)$ is simply the product of the random contributions of t consecutive bonds, and hence manifestly log–normal. However, for such cases the variance of $\ln I(t)$ is proportional to t, i.e. $\omega = 1/2$. We shall argue later on that the exponent ω is a universal characteristic of the distribution which depends on dimensionality ($\omega = 1/3$ in $d = 2$). Another feature of the sum over forward scattering paths in eq.(4) is the appearance of *two independent parameters*. The local factors V and W in the starting Hamiltonian only contribute to ξ_0 in eq.(6). They do not appear in ξ_g or $I(t)$, and hence the fluctuations are completely independent of the Anderson parameter[11] (V/W). This is in agreement with previous results by Cohen et al[13] who argue that a one parameter scaling arises only in the limit of weak disorder. They also argue for two parameter scaling in strong localization.

MAGNETIC FIELD RESPONSE

The response of the system to a magnetic field B (measured in units of quantum flux per plaquette), with and without SO, is plotted in Fig.(2). To emphasize the magnetoconductance (MC) we have subtracted the zero field averages (without SO on the bottom, and with SO on the top portion). The asymptotic slopes on the bottom portion indicate the increase in the global contribution ξ_g^{-1} to the (inverse) localization length. We first note that the introduction of SO (indicated by $+$) is accompanied by a significant increase in tunneling, i.e an increase in ξ. Secondly, the addition of a magnetic field leads to an increase in $\langle \ln I(t)\rangle$, but in qualitatively different manners in the absence or presence of SO. Without SO, there is a change in slope, i.e. the most important effect of the field is to increase the localization length. (The increase in SO seems to saturate to a limit corresponding to addition of random phases (indicated by $*$) on bonds, mimicking random magnetic impurities.) This an enhancement of tunneling that grows exponentially in t. By contrast, with SO, we observe that the slopes in Fig.(2) remain unchanged. Thus the magnetic field enhances the tunneling rate by a t–independent constant, i.e. there is no change in the localization length.

The appropriate scaling laws in a small magnetic field can be obtained by collapsing the data in Fig.(2) for different values of B and t. Such a collapse in the absence of SO is obtained as a function of Bt^2 as indicated in Fig.(3). The linear dependence is consistent with an increase in the global localization length scaling as $B^{1/2}$, and can be fitted to

$$\xi_g^{-1} = (0.053 \pm 0.02) - (0.15 \pm 0.03)\,(\phi/\phi_0)^{1/2}\,. \tag{7}$$

By contrast, the appropriate scaling combination with SO is $Bt^{3/2}$, as demonstrated by collapsing the data from the top of Fig.(3) in Fig.(4). The scaling function now takes the form

$$\langle \ln I(t, B)_{SO} \rangle - \langle \ln I(t, 0)_{SO} \rangle = \begin{cases} cB^2 t^3 & \text{if } B^2 t^3 < 1 \\ C & \text{if } B^2 t^3 > 1 \end{cases}. \tag{8}$$

MOMENT ANALYSIS

We can gain some analytic understanding of the distribution function for $I(t, B)$ by examining the moments $\langle I(t)^n \rangle$. From eqs.(3) and (4) we see that each $I(t)$ represents a forward path from i to f, and a time reversed path from f to i. For $\langle I(t)^n \rangle$, we have to average over the contributions of n such pairs of paths. Averaging

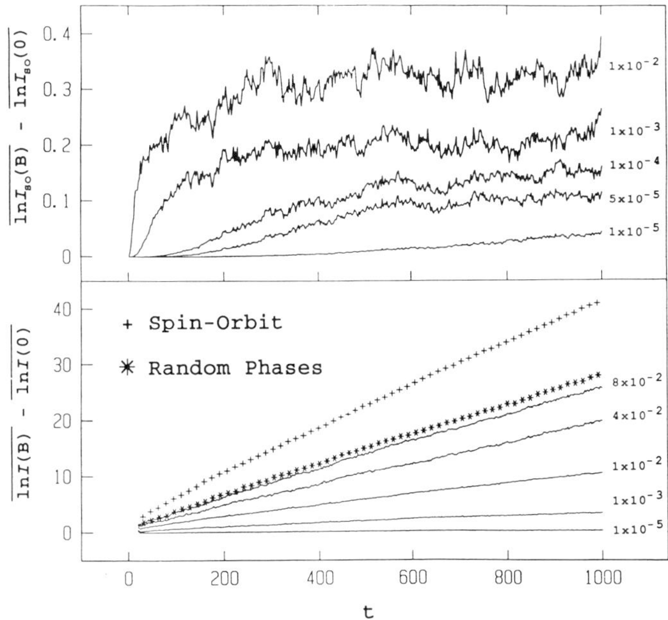

Figure 2 Bottom: Increase in tunneling (MC) without SO (solid curves) from the zero field value I_0. A large MC is also obtained by introducing random phases (*), or SO (+), on the bonds. Top: The analogous MC with SO is much smaller and plotted at enlarged scale.

over the random signs of the site energies forces a *pairing* of the $2n$ paths (since any site crossed by an odd number of paths leads to a zero contribution)[9]. To understand the MC it is useful to distinguish two classes of pairings: (**1**) *Neutral paths* in which one member is selected from J and the other from $J^\dagger$. Such pairs do not feel the field since the phase factors of e^{iA} picked up by one member on each bond are canceled by the conjugate factors e^{-iA} collected by its partner. (**2**) *Charged paths* in which both elements are taken from J or from $J^\dagger$. Such pairs couple to the magnetic field

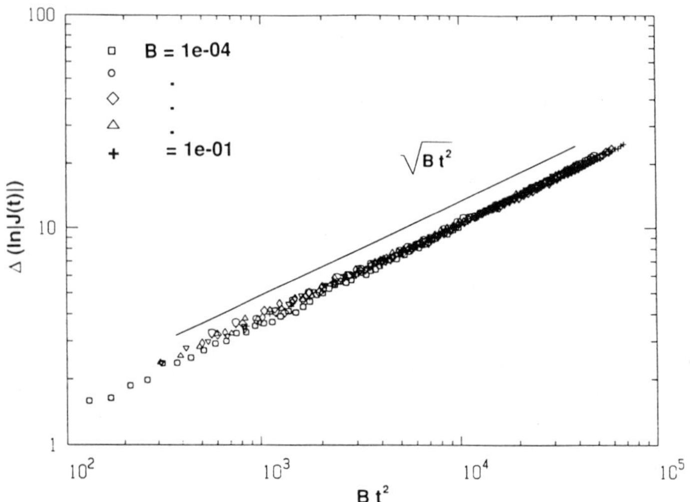

Figure 3 Data collapse of MC for different fields and sizes indicating a scaling of the form $I(B,t) = f(Bt^2)$.

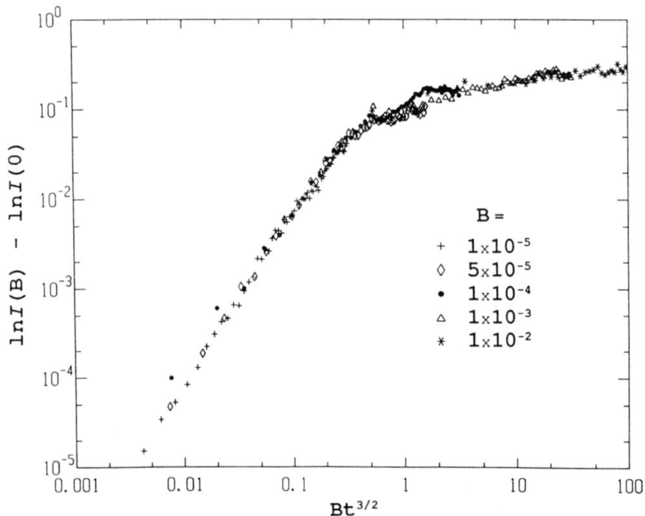

Figure 4 Scaling of MC in the presence of SO.

like particles of charge $\pm 2e$. In the presence of SO, averaging over the random $SU(2)$ matrices further forces neutral paths to carry parallel spins, while the spins on the two partners of charged paths must be antiparallel. These constraints are indicated in Fig.(1) which shows a possible contribution to $\langle I^2 \rangle$ with SO.

Since at each site any path has a choice of two directions, we may naively expect $\langle I(t)^n \rangle$ to asymptotically scale as 2^{nt}. This ignores the correlations between paths which manifest themselves when two paired paths intersect[14]. Since at each intersection the pairs may exchange partners there is an additional statistical attraction for such crossings. The attraction factor depends crucially on the symmetries of the Hamiltonian in eq.(1): For $B = 0$ and without SO, the Hamiltonian has *orthogonal* symmetry. All pairings are allowed and the attraction factor is 3 since an incoming $(12)(34)$ can go out as $(12)(34)$, $(13)(24)$, or $(14)(23)$. Note that even if both incoming paths are neutral, one of the exchanged configurations is charged (see Fig.(1)). A magnetic field breaks time reversal symmetry, discourages charged configurations, and reduces the exchange attraction. The limiting case of a 'large' magnetic field is mimicked by replacing the gauge factors with random phases. In this extreme, the Hamiltonian has *unitary* symmetry, and only neutral paths are allowed. The exchange factor is now reduced to 2; from $(11^*)(22^*) \rightarrow (11^*)(22^*)$, or $(12^*)(21^*)$. With SO, we must also take into account the allowed spin exchanges, and we find that the intersection of two paired paths results[10] in an exchange attraction of $3/2$ (*symplectic* symmetry).

A MANY BODY PROBLEM

Calculating $\langle I(t)^n \rangle$ is now reduced to finding the sum over n paths with the above exchange attractions. Its functional form is most easily calculated in a continuum limit by regarding the paths as world–lines of n attracting particles in $d' = d - 1$ dimensions, subject to a Hamiltonian[15]

$$-\mathcal{H}_n = n \ln 2 + \gamma \sum_{\alpha=1}^{n} \nabla_\alpha^2 + \sigma^2 \sum_{\beta > \alpha} \delta^{d'}(x_\alpha - x_\beta). \tag{9}$$

The first term describes the choice of two directions per step; the second term is a kinetic energy term from particle motion, γ is an effective line tension (in the NSS model it is entropicaly generated from the light cone constraints). The last term describes the contact attraction between replicas due to partner exchange. The n-body partition function is then computed as

$$\langle I^n \rangle = \text{Tr}[\exp(-\mathcal{H}_n t)] \stackrel{t \to \text{large}}{\simeq} \exp(-\epsilon_n^0 t), \tag{10}$$

where ϵ_n^0 is the n body ground state energy (the largest transfer matrix eigenvalue), which dominates the long-t statistics. The solution of this n–body problem is well known for $d' = 1$, as the ground state wave function is given by the Bethe Ansatz

$$\Psi_0 = \exp\left(-\frac{\kappa}{2} \sum_{\beta > \alpha} |x_\alpha - x_\beta| \right). \tag{11}$$

Matching the discontinuities in Ψ_0 when two particles cross with the strength of the attractive interaction gives $4\gamma\kappa = \sigma^2$. The ground state energy is $-\epsilon_n^0 = n \ln 2 + \frac{\gamma\kappa^2}{6} n(n^2 - 1)$. Hence from eq.(10) we obtain the moments

$$\langle I(t)^n \rangle = A(n) 2^{nt} \exp[2\rho n(n^2 - 1)t], \tag{12}$$

where $\rho \equiv \gamma\kappa^2/12$ is an increasing function of the strength of the attraction between paths. We have also included an overall amplitude $A(n)$.

Cumulants C_i of the characteristic function for $\ln I(t)$ are defined by the powers of n in the expression

$$\langle I(t)^n \rangle = \langle \exp(n \ln I(t)) \rangle \equiv \exp\left(\sum_i \frac{n^i}{i!} C_i[\ln I(t)] \right). \tag{13}$$

C_1 is the average, $\langle \ln I(t) \rangle$, while the absence of an n^2 term in eq.(12) indicates no second cumulant at order of t. The n^3 term reflects a third cumulant scaling as t, i.e.

$$\begin{aligned}
\langle \ln I(t) \rangle &= [\ln 2 - 2\rho]\, t, \\
C_2(\ln I(t)) &= 0 \quad (\text{to order } t), \\
C_3(\ln I(t)) &= 12\rho t.
\end{aligned} \tag{14}$$

Comparing the above with eqs.(5) and (6), we note that the global contribution to the localization length is $\xi_g = 1/\rho$, i.e. directly related to the strength of the bound state. Also it follows that $\ln I(t)$ is approximately normal, with fluctuations given by

$$\delta \ln I(t) \sim |\frac{t}{\xi_g}|^{1/3}. \tag{15}$$

We can now appreciate the trends in Fig.(2), as the slopes are indicative of the statistical attraction factors. Without SO, the magnetic field gradually reduces the attraction factor from 3 to 2 leading to the increase in slope. Addition of SO to the Hamiltonian has the similar effect of suddenly decreasing the attraction to $3/2$. Why does the addition of the magnetic field lead to no further change in ρ in the presence of SO? Without SO, the origin of the continuous change in the attraction factor is the type of exchange indicated in Fig.(1) whereby a charged bubble appears from the intersection of two neutral paths[9]. In the presence of SO, it can be shown[10] that the contribution of such configurations is zero. Thus the neutral paths traverse the system without being effected by the magnetic field; their attraction factor, and hence ρ and ξ_g remain unchanged. The smaller positive MC observed in the simulations is due to the quenching of the charged paths by a magnetic field[10]. The resulting change is thus only in the amplitude $A(n)$ of eq.(12).

The exchange attraction between neutral paths can also be computed for $SU(n)$ impurities, and equals $1 + 1/n$, which reproduces 2 for $U(1)$ or random phases, and $3/2$ for $SU(2)$ or SO scattering. The attraction vanishes in the $n \to \infty$ limit, where the paths become independent. The statistical exchange factors are thus universal numbers, simply related to the symmetries of the underlying Hamiltonian. The attractions in turn are responsible for the formation of bound states in replica space, and the universal scaling of the moments in eq.(12). In fact since the single parameter ρ completely characterizes the distribution, the variations in the mean and variance of $\ln I(t)$ should be perfectly correlated. This can be tested numerically by examining respectively coefficients of the mean and the variance from eq.(14) for the different cases studied. The results plotted in Fig.(5) do indeed fall on a single line, parameterized by ρ. The largest value corresponds to the NSS model for $B = 0$

and no SO (orthogonal symmetry, exchange attraction 3). Introduction of a field gradually reduces ρ until saturated at the limit of random phases (unitary symmetry, exchange attraction 2). SO scattering reduces ρ further (symplectic symmetry, exchange attraction 3/2). The final point corresponds to independent paths with $\rho = 0$.

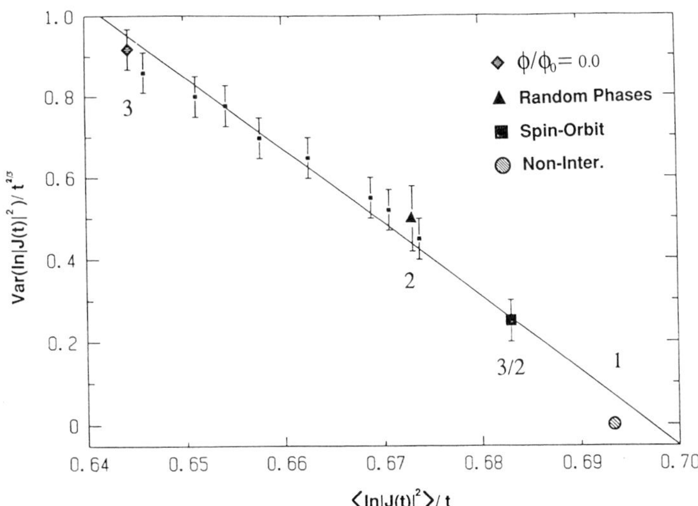

Figure 5 Complete hierarchy of the exchange attractions between paired paths, reflecting the various Hamiltonian symmetries; Diamond: orthogonal; triangle: unitary; square: symplectic; circle: the non-interacting limit corresponding to independent paths, and also to $SU(n)$ with $n \to \infty$. Small squares correspond to finite magnetic fields but no SO.

HIGHER DIMENSIONS

Unfortunately the exact ground state of the Hamiltonian in eq.(9) is not known in higher dimensions. However, it is possible to draw a number of important conclusions for higher d'. It is known that in $d' \leq 2$ any attractive potential will always lead to formation of a bound state. This implies that in three dimensions the localization length (ξ defined through eq.(5)) has both a local and a global contribution. Again the introduction of a magnetic field should lead to a nonuniversal increase in ξ in the absence of SO. With SO, there will be a smaller positive MC, but no change in ξ. Presumably, the scaling forms in eqs.(7) and (8) are still applicable to $d = 3$. For $d' > 2$ a critical amount of attraction is necessary to form a bound state. The phase with no bound state will be self–averaging with small fluctuations, while the formation of the bound state implies anomalous fluctuations of $\ln I(t)$. To specify the distribution function for $\ln I(t)$ completely, we also need the dependence of its fluctuations on t. In the absence of the correct energy, we can make no exact statements about this dependence. However, the same problem arises in the context of directed polymers in random media[16], and a related interface growth problem, where a large body of numerical results is available. Current numerical estimates are consistent with fluctuations in $\ln I$ growing as $t^{1/d+1}$ in the strongly coupled phase in d dimensions[17]. This is consistent with generalizing the small n moment behaviors in eq.(12) to

$$\langle |I(t)|^n \rangle = A(n)\langle |I(t)| \rangle^n \exp[\rho n(n^d - 1)t]. \tag{16}$$

Note that this also reproduces the log–normal distribution in one dimensions[12,13], with $\omega = 1/2$.

COMPARISON TO OTHER RESULTS

(1). In the independent path approximation (IPA)[5,6,7], correlations leading to the nontrivial scaling of moments in eq.(12) are ignored. Hence $\langle I(t)^n \rangle = A(n)2^{nt}$, and the MC reflects the changes in the amplitude $A(n)$. Without SO, all pairings of $2n$ paths contribute equally at $B = 0$ ($A(n, B = 0) = (2n - 1)!!$), while only charged paths survive at finite field ($A(n, B \neq 0) = n!$). The difference between these two numbers at $n = 0$ gives the increase by $\ln 2$ of $\langle \ln I(t) \rangle$ predicted by IPA[6]. Recent experiments at low temperatures[18] observe increases in relative conductance by factors much larger than unity, in disagreement with IPA, but consistent with Fig.(2). Although, IPA clearly fails without SO, it is close to the mark in the presence of SO as the observed initial increase of $B^2 t^3$ (eq.(8)) precisely follows from the quenching of charged paths.

(2). The random matrix approach (RMA) to the problem[3,8] takes as its input only the symmetries of the Hamiltonian, and the assumption of a single parameter scaling. It predicts[8] that the magnetic field results in doubling of the localization length ξ in the absence of SO, and a halving of ξ with SO, in clear contradiction to our results. Despite its simplicity and generality, RMA has a number of shortcomings. First, it is assumed at the outset that a single parameter is sufficient for characterizing the problem. As we demonstrated, description of the localization length (using the tunneling probability) requires both a global and a local factor; hence variations in ξ cannot be universal. Second, by considering the most general random matrices the approach loses all information on spatial connectivity and dimensionality. The correct approach is to consider the ensemble of sparse random matrices in which only elements close to the diagonal are nonzero[19]. Finally, symmetry arguments imply that the addition of a magnetic field, by destroying time reversal symmetry, immediately changes systems with and without SO to one completely described by a unitary ensemble. In fact, from Fig.(2), we do see that the magnetic field in the absence of SO derives the localization length towards its unitary limit (obtained by placing random phases on each bond). However, the change in ξ is gradual, and not immediate. Furthermore, there is no change in ξ with SO. The latter is due to the form of the Hamiltonian. To reach the unitary limit with SO, one has to also include Zeeman splitting terms. Such terms are certainly allowed by symmetry, but involve a much higher energy cost. RMA cannot account for such subtelties.

(3). Recent experiments on Au-doped In_2O_{3-x} films[20] observe a transition from negative MC at small disorder to positive MC at large disorder. This is attributed[20] to an interplay between weak localization effects (causing negative MC with SO) at scales less than ξ, and strong localization effects (positive MC) at scales larger than ξ.

To account for the insensitivity to addition of Au, it is also suggested[20] that addition of SO has little effect on the strongly localized regime. This is in conflict with results of Fig.(2) at high concentrations of SO. However, a more dilute SO concentration may not be in contradiction with the experimental results. We also note that the sign of MC at small fields cannot be changed by what goes on at short distances. The local contribution to $\langle \ln T \rangle$ in eq.(4) must be analytic, and scale as $B^2 t$, while the global change in $\langle \ln I(t) \rangle$ scales as $B^2 t^3$ (eq.(8)). Hence there is always a positive MC at low fields at sufficiently large t (it may crossover to a negative MC at higher fields). Numerical simulations[7] at the scale of ξ are in agreement with such a picture. If a positive MC is a signature of the localized regime, it can only be reconciled with a negative MC predicted by the weak localization theories with strong SO, through a possible phase transition[21]. Could the observed change of sign in the experiments be a finite temperature manifestation of such a zero temperature phase transition? Certainly more experimental and theoretical studies are in order.

ACKNOWLEDGEMENTS

We have benefited from discussions with B. Altshuler, Y. Meir, Y. Shapir, N. Wingreen, and X.R. Wang. This research was supported by the NSF through grant number DMR-90-01519, the PYI program (MK).

REFERENCES

1. For recent reviews see, P.A. Lee and B.L. Altshuler, Physics Today **41**, 36 (1988); P.A. Lee and T.B. Ramakrishnan, Rev. Mod. Phys. **57**, 287 (1985); G. Bergmann, Phys. Rep. **107**, 1 (1984).

2. B.L. Altshuler and B.I. Shklovskii, Zh. Eksp. Teor. Fiz. **91**, 220 (1986) [Sov. Phys. JETP **64**, 127 (1986)]; P.A. Lee, A.D. Stone, and H. Fukuyama, Phys. Rev. B **35**, 1039 (1987); F.J. Wegner, Z. Phys. B **35**, 207 (1979).

3. N. Zannon and J.L. Pichard, J. Phys. (Paris) **49**, 907 (1988).

4. O. Faran and Z. Ovadyahu, Phys. Rev. **B38**, 5457 (1988).

5. V.L. Nguyen, B.Z. Spivak, and B.I. Shklovskii, Pis'ma Zh. Eksp. Teor. Fiz. **41**, 35 (1985) [JETP Lett. **41**, 42 (1985)]; Zh. Eksp. Teor. Fiz. **89**, 11 (1985) [JETP Sov. Phys. **62**, 1021 (1985)].

6. U. Sivan, O. Entin-Wohlman, and Y. Imry, Phys. Rev. Lett. **60**, 1566 (1988); O. Entin-Wohlman, Y. Imry, and U. Sivan, Phys. Rev. B **40**, 8342 (1988).

7. Y. Meir, N.S. Wingreen, O. Entin-Wohlman, and B.L. Altshuler, Phys. Rev. Lett. **66**, 1517 (1991).

8. J.L. Pichard, M. Sanquer, K. Slevin, and P. Debray, Phys. Rev. Lett. **65**, 1812 (1990).

9. E. Medina, M. Kardar, Y. Shapir, and X. R. Wang, Phys. Rev. Lett. **62**, 941 (1989); *ibid*, **64**, 1816 (1990).

10. E. Medina, and M. Kardar, Phys. Rev. Lett. **66**, 3187 (1991); E. Medina, and M. Kardar, MIT preprint (1991).

11. P. W. Anderson, Phys. Rev. **109**, 1492 (1958).

12. P.W. Anderson, D.J. Thouless, E. Abrahams, and D.S. Fisher, Phys. Rev. B **22**, 3519 (1980).

13. A. Cohen, Y. Roth, and B. Shapiro, Phys. Rev. B **38**, 12125 (1988).

14. Y. Shapir and X.R. Wang, Europhys. Lett. **4**, 1165 (1987).

15. M. Kardar, Nucl. Phys. **B290 [FS20]**, 582 (1987).

16. For a general introduction to the literature on the directed polymer and interface growth problems, see M. Kardar, in *New Trends in Magnetism*, M.D. Coutinho-Filho and S.M. Rezende, eds. (World Scientific, Singapore, 1990) p. 277; M. Kardar, in *Disorder and Fracture*, J.C. Charmet, S. Roux, and E. Guyon, eds. (Plenum Press, New York, 1990) p. 3.

17. J.M. Kim and J.M. Kosterlitz, Phys. Rev. Lett. **62**, 2289 (1989).

18. F.P. Milliken and Z. Ovadyahu, Phys. Rev. Lett. **65**, 911 (1990).

19. J. Cook and B. Derrida, Saclay preprint (1990).

20. Y. Shapir and Z. Ovadyahu, Phys. Rev. B **40**, 12441 (1989).

21. N. Evangelou and T. Ziman, J. Phys. C **20**, L235 (1987).

MANY–BODY PROBLEMS

IN HIGH TEMPERATURE SUPERCONDUCTIVITY

Yu Lu

International Centre for Theoretical Physics
Trieste, Italy
and
Institute of Theoretical Physics, Chinese Academy of Sciences
Beijing, People's Republic of China

ABSTRACT

In this brief review the basic experimental facts about high T_c superconductors are outlined. The superconducting properties of these superconductors are not very different from those of the ordinary superconductors. However, their normal state properties cannot be described by the standard Fermi liquid (FL) theory. Our current understanding of the strongly correlated models is summarized. In one dimension these systems behave like a "Luttinger liquid", very much distinct from the FL. In spite of the enormous efforts made in two–dimensional studies, the question of FL vs non–FL behaviour is still open. The numerical results as well as various approximation schemes are discussed. Both the single hole problem in a quantum antiferromagnet and finite doping regime are considered.

1. INTRODUCTION

The revolutionizing discovery of Bednorz and Müller of high temperature superconductivity in cuprates [1] has attracted great attention of physicists and material scientists, resulting in about 20,000 publications in the last five years. Now the frenetic period is over and we face a difficult task to understand the physical mechanism behind this striking phenomenon. In spite of the gigantic efforts made jointly by scientists all over the world, there is still no consensus on the theoretical understanding among scientists. Recently, there have appeared several comprehensive reviews on this outstanding issue [2–4]. The aim of this brief review is to bring some open problems to the attention of traditional "many–body" theorists.

The schematic structures of oxide superconductors are shown in Fig. 1, where the active CuO_2 layers are separated by charge reservoir building blocks. These layers are responsible for the superconducting as well as the anomalous normal state properties, but the charge carrier concentration is controlled by modifying the reservoirs through substitution or addition (removal) of oxygen. For a given class of superconductors the critical temperature is maximized around the doping concentration of (15–20)% .

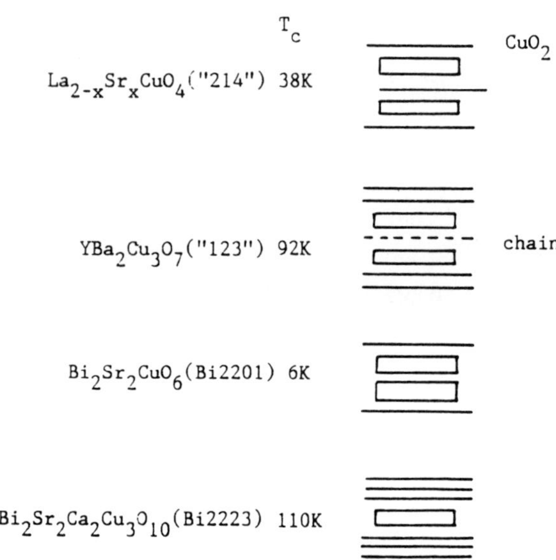

Fig. 1. Schematic structures of oxide superconductors.

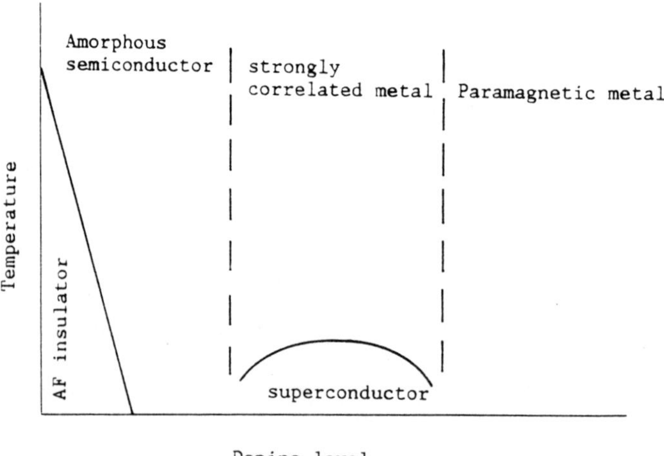

Fig. 2. Generic phase diagrams of oxide superconductors.

The generic phase diagram of oxide superconductors is shown in Fig.2. The reference compound is usually an antiferromagnet. Upon doping the antiferromagnetism vanishes rather rapidly. In region 1 the system shows anomalous semiconducting behaviour, whereas in regime 2 it behaves like a strongly correlated metal, very much different from the ordinary metal like Al. With the lowering of temperature the system becomes superconducting. Upon further doping, in regime 3, more standard metallic behaviour shows up. To understand this rich phase diagram is the fundamental task of the theory. At least in the low–doping region the interplay between magnetism and superconductivity as well as the nature of electronic states in a doped Mott insulator is the key issue.

Since the Cu and O levels are rather close to each other, the covalency effect is playing an important role. The Cu Eg state is split into $d_{x^2-y^2}$ and $d_{3z^2-r^2}$ states due to the Jahn–Teller effect. These orbits, in turn, form bonding and antibonding states with O_{2p} orbits. Within the single particle picture the top most level is only half–filled, hence the reference compound should be a metal. However, in reality it is an insulator, so the single particle picture fails.

To set the stage we will emphasize the following three points as the main characteristic features of the standard FL theory:

(1) There are well defined quasiparticles for which the real part of the excitation energy ($\sim \omega, T$) is much greater than the imaginary part ($\sim \omega^2, T^2$), where ω is the frequency and T the temperature.

(2) The spin $\frac{1}{2}$ and charge e of the quasiparticle are bound together.

(3) The Luttinger theorem (the volume enclosed by the Fermi surface does not change upon switching on the interaction) is valid and there is a jump of the momentum distribution function at the Fermi surface, equal to Z_k, the spectral weight of the quasiparticle.

This last point turned out to be rather subtle and it is difficult to distinguish, at least numerically, between a small jump and a smooth, but steep change.

2. BASIC EXPERIMENTAL FACTS

2.1 Superconducting Properties

First of all, these oxide superconductors share all basic properties of ordinary superconductors like Meissner effect, zero resistance and pairing of quasiparticles with charge $2e$ [5]. This last property is evident from the ac–Josephson effect, flux jump experiment and a direct measurement of flux quantum φ_0. Moreover, the Andreev scattering experiment reveals pairing of particles with opposite momentum. The energy gap $\Delta(T)$ has been measured from tunneling, infrared absorption and photoemission experiments, and the ratio $2\Delta(0)/kT_c$ is in the range $4 - 8$. The gap is anisotropic, but no zeroes have been observed which, probably, excludes $p-$ or $d-$wave pairing.

The phenomenological Ginzburg–Landau theory can well describe various superconducting properties, including critical magnetic field, correlation length ξ and penetration depth λ. The main characteristic features are the anisotropy and the shortness of the correlation length. For $YBa_2Cu_3O_7$, the least anisotropic among high T_c superconductors, one finds $\xi_{ab} = 14 \pm 2$ Å, $\xi_c = 1.5 - 3$ Å, $\lambda_{ab} = 1400$ Å, and $\lambda_c \sim 7000$ Å. Therefore, all oxide superconductors are extreme type–II superconductors.

There are some more subtle coherence effects due to electron pairing, e.g., the "destructive" effect leading to the drastic drop of acoustic absorption coefficient below T_c and the "constructive" effect giving rise to the Hebel–Slichter resonance in the nuclear relaxation rate. The latter has not been observed in oxide superconductors. Some authors attribute this to the damping effect or abrupt opening of the gap [6]. However, recently a peak below T_c has been found in frequency–dependent conductivity [7], although the origin of this peak is still in dispute.

As for the superconducting mechanism, there have been numerous different proposals. Roughly speaking, they can be grouped into two categories: One is more conventional, pairing in the quasi–two–dimensional systems due to attraction mediated by various electric or magnetic excitations. The other one is more exotic, either due to the gauge force between particles obeying fractional statistics – anyons [8], or due to interlayer tunneling [9]. We will not elaborate on this point further.

2.2 Normal State Properties

The normal state properties of oxide superconductors are as perplexing as the high transition temperature itself. On the one hand, there is a well defined Fermi surface from the photo–emission [10] and positron annihilation [11] experiments, and the effective mass determined from the magnetic susceptibility [12] and the Drude component of optical conductivity [13] are consistent with each other. On the other hand, the transport properties, like resistivity and Hall effect, can be interpreted only in terms of a doped Mott insulator, rather than a standard FL [5]. How to reconcile these two apparently contradicting aspects is the main challenge to theory. We will not enumerate all anomalous properties in the normal state, but rather concentrate on this basic dilemma.

The temperature dependence of the in–plane resistivity of several families of oxide superconductors is shown in Fig. 3. Several features are worth mentioning: The resistivities are all close in absolute values; they increase linearly with temperature over a wide range; they are quite small, while extrapolated to absolute zero temperature [14]. One may argue that phonon scattering can give rise to linear temperature dependence of resistivity at $T > \theta/4$, where θ is the Debye temperature. However, the extention of this linear dependence for "$Bi\,2201$" down to 10K, rules out this interpretation.

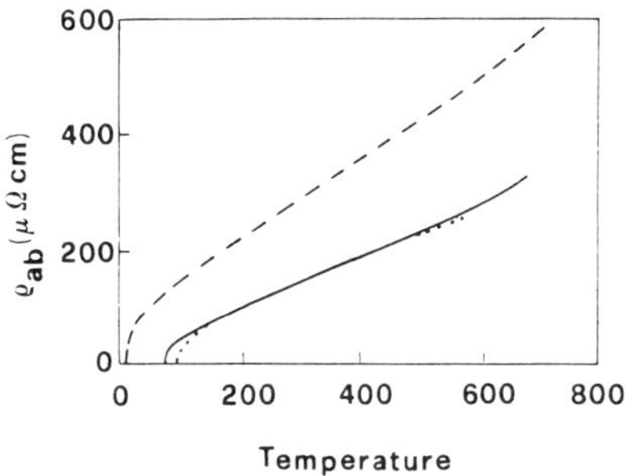

Fig. 3. The temperature dependence of resistivity ρ_{ab} for "Bi 2201" (dashed line), "123" 90K (solid line) and "Bi 2212" (dotted line).

The behaviour of the Hall constant is even more peculiar. Fig.4, reproduced from Ref.15 shows clearly how R_H changes with composition in "electron" and "hole" doped superconductors. The sign of R_H is consistent with the Mott–Hubbard picture and the magnitude deviates from $\frac{1}{X}$ dependence at higher dopings. The most interesting point is the change of R_H sign at certain critical concentration where the superconductivity disappears, while the "normal" metal behaviour recovers. This region of "anomalous" Hall constant coincides with

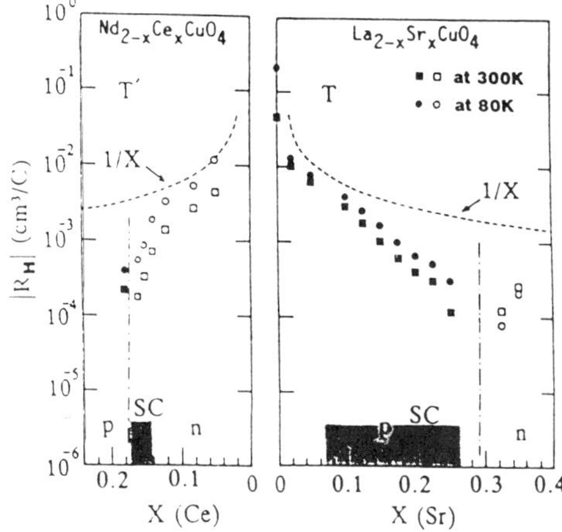

Fig. 4. The absolute value of the Hall coefficient for electron and hole doped superconductors. The shaded regions correspond to superconducting compositions [15].

the regime of linear temperature resistivity, as well as with the linear temperature dependence of R_H^{-1} [16]. Moreover, the slope of R_H^{-1} vs temperature is roughly proportional to the critical temperature itself. The temperature dependence of R_H has also been observed for ordinary metals, but only at low temperatures, for $T < \theta/4$ when anisotropy of phonon scattering plays a role. This comparison is given in Fig.5 following Ong [16].

The angle–resolved photoemission experiment is a very useful tool to study the energy spectrum of electrons. From the energy and angular dependence of emitted electrons one can recover in principle the spectral function $A(\mathbf{k}, \omega)$. First in "$Bi\ 2212$" [10], then in "123" compounds [17] a well–defined Fermi surface has been found in good agreement with the local density functional calculations. This Fermi surface is "large" as requested by the Luttinger theorem. The more subtle question is the precise shape of the spectral function which can check the FL theory in a quantitative way. Unfortunately, the present precision does not allow one to make unambiguous conclusions. The main puzzle comes from the comparison with the transport measurements. As said above, they are consistent with the Mott–Hubbard picture, i.e., the charge carrier concentration is δ instead of $1 - \delta$ (where δ is the doping concentration). Put another way, the Fermi surface, even if it exists, should be a "small" one corresponding to these "pockets". To our knowledge, there is still no good solution of this dilemma. Very recently, Liu and collaborators [18] have studied angle–resolved photoemission spectra for $YBa_2Cu_3O_X$ with $6.35 \leq x \leq 6.9$. These authors found that the spectral weight near E_F, as well as the shape and the dimension of the Fermi surface are relatively constant for $6.4 \leq x \leq 6.9$, but they change abruptly at $x = 6.35$. The spectral weight diminishes and the Fermi surface becomes a small one. It seems that a Mott transition is taking place at this concentration. As we will see later, these results are consistent with our current understanding of the Mott transition. However, the puzzle with highly doped superconductors still remains.

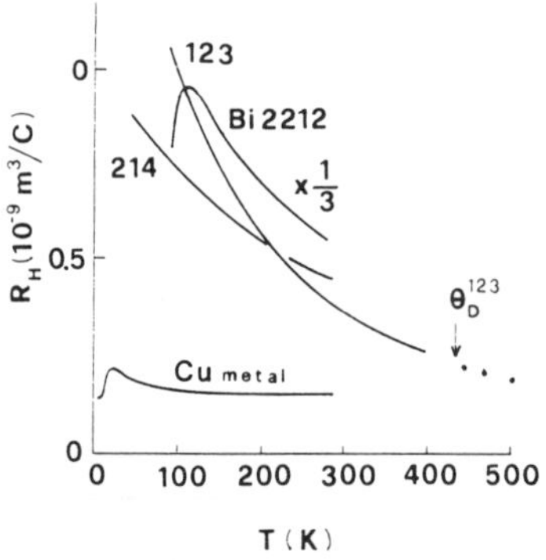

Fig. 5. The schematic plot of temperature dependence of R_H in cuprates compared with that of conventional metal following Ong [16].

Other anomalies in the normal state include [5]: strong background scattering on electronic excitations, observable in frequency–dependent conductivity and Raman scattering; different NMR relaxation behaviour for Cu and O; possible existence of pseudo gap in the spin–excitation; etc. Those who are interested in further details are referred to more complete reviews [2–4].

To summarize the experimental part, it is very interesting to observe that the BCS theory, or the pairing theory in a more general sense, based on the FL theory of Landau seems to be quite adequate to interpret the superconducting properties of oxide superconductors. Nevertheless, the normal state properties of these superconductors resist any simple description within the framework of the same FL theory. The outstanding question is: Do we need to completely write new chapters for the many–body theory to accommodate the strong correlation effects. The answer is probably yes. But it is not completely conclusive yet.

3. STRONGLY CORRELATED ELECTRON MODELS

Very soon after the discovery of high T_c superconductors, P.W. Anderson suggested [19] that one should start from a Mott insulator, because the superconductivity occurs near the metal–insulator transition. Moreover, he proposed that the one–band Hubbard model given by

$$H = -t \sum_{i,j,\sigma} c_{i\sigma}^{+} c_{j\sigma} + U \sum_{i} n_{i\uparrow} n_{i\downarrow} \qquad (1)$$

where

$$n_{i\sigma} = c_{i\sigma}^{+} c_{i\sigma} \,,$$

t is the nearest neighbour hopping and U is the on–site Coulomb repulsion, can bring about the essential physics. Also, this would allow us to describe superconductivity and magnetism in a unified fashion, so far the same electrons are responsible for both phenomena.

In the strong–coupling limit $U \gg t$, the important point is to project out the double occupied sites. This can be achieved via a canonical transformation

$$H \rightarrow e^{iS} H e^{-iS} = H + [iS, H] + \frac{1}{2} [iS, [iS, H]] + \ldots \qquad (2)$$

which has been discussed by many authors [20]. If we neglect the three–site terms, the effective Hamiltonian is the well–known $t - J$ model given by

$$H_{t-J} = -t \sum_{i,j,\sigma} \tilde{c}_{i\sigma}^{+} \tilde{c}_{j\sigma} + J \sum_{i,j} \left(\mathbf{S}_i \cdot \mathbf{S}_j - \frac{1}{4} \right) \qquad (3)$$

where

$$\mathbf{S}_i = c_i^{+} \frac{\sigma}{2} c_i \qquad J = 4 t^2 / U \,, \qquad (4)$$

$$\tilde{c}_{i\sigma} = c_{i\sigma} (1 - c_{i\bar{\sigma}}^{+} c_{i\bar{\sigma}}) \qquad (5)$$

with σ as Pauli matrices. The Hubbard and the $t - J$ model (3) are equivalent to each other only for $U \gg t$, or $t \gg J$ case, but $t - J$ model itself can be studied in the opposite limit $t \ll J$. Sometimes people also study a similar model with next–nearest hopping term t', which is called $t - t' - J$ model, or $t' - J$ model, depending on whether the nearest neighbour hopping is allowed. At half–filling, $t - J$ model reduces to spin $\frac{1}{2}$ Heisenberg model.

One may argue that the actual CuO_2 plane contains one Cu and two O in each unit cell and the more realistic three–band model [21] should be more appropriate.

This model can be written as:

$$
\begin{aligned}
H_{p-d} = {} & \frac{\Delta}{2} \sum_{\ell,\sigma} n_{p\ell\sigma} - \frac{\Delta}{2} \sum_{i,\sigma} n_{di\sigma} \\
& + t_{pd} \sum_{i,\ell,\sigma} (p^+_{\ell\sigma} d_{i\sigma} + h.c.) + t_{pp} \sum_{\ell,\ell',\sigma} p^+_{\ell\sigma} p_{\ell'\sigma} \\
& + U_{dd} \sum_i n_{di\uparrow} n_{di\downarrow} + U_{pp} \sum_\ell n_{p\ell\uparrow} n_{p\ell\downarrow} \\
& + V_{pd} \sum_{i,\ell} n_{di} n_{p\ell} ,
\end{aligned}
\tag{6}
$$

where $n = n_\uparrow + n_\downarrow$, $\Delta = \varepsilon_p - \varepsilon_d$, $n_{p\ell\sigma} = p^+_{\ell\sigma} p_{\ell\sigma}$, $n_{di\sigma} = d^+_{i\sigma} d_{i\sigma}$. The index i runs over Cu sites, while ℓ runs over O sites. The meaning of other parameters is obvious. These parameters can be estimated from cluster calculations and their values are roughly [22]:

$$
\begin{aligned}
\Delta = 3 \text{ eV}, \quad U_{dd} = 10 \text{ eV}, \quad U_{pp} = 5 \text{ eV}, \quad V_{pd} = 1 \text{ eV}, \\
t_{pd} = 1.5 \text{ eV}, \quad t_{pp} = 0.5 \text{ eV} .
\end{aligned}
\tag{7}
$$

If the leading terms in each category (hopping, Coulomb repulsion and site energy) are kept, we end up with a simpler model, containing Δ, t_{pd} and U_{dd} three parameters. Zhang and Rice [23] have argued that if one takes into account the symmetry of the CuO_2 square, the singlet state formed by a symmetrized O orbit with Cu orbit has much lower energy than the triplet and the nonbonding state. Therefore, as far as the low energy physics is concerned, the three–band model is equivalent to the $t - J$ model (5). Experimentally, the Knight shifts for different Cu and O nuclei in "123" compounds can be scaled with the spin susceptibility [24]. This seems to be in favour of a single spin variable, i.e., the effective one–band model. However, not everybody agrees with this viewpoint [25]. Also, if the charge transfer between O and Cu is considered explicitly, one should involve the three–band model [26].

4. ONE–DIMENSIONAL MODELS AND LUTTINGER LIQUIDS

One–dimensional models have attracted a lot of attention both because of their relevance to quasi–one–dimensional systems like organic conductors [27] or conducting polymers [28] and the availability of more rigorous mathematical treatments. Historically, the progress on one–dimensional interacting fermions has been made along two separate lines.

On the one hand, the renormalization group and the bosonization technique have been applied to solve the continuous, Luttinger [29]–type models. The exponents of power–law decay correlation functions and related physical observable quantities, like response functions, relaxation rates, etc., have been extensively studied. Lieb and Mattis have solved exactly the Luttinger model [30]. Mattis, Luther, Emery and Peschel have developed further the technique [31]. This part has been summarized in two review articles [32]. Furthermore, this technique has been polished by Heidenreich *et al.* [33] and by Haldane [34]. On the other hand, the Bethe ansatz method has been used to solve exactly several lattice models the most important of which were the fermion gas model with δ–interaction solved by Yang [35] and the Hubbard model solved by Lieb and Wu [36]. Later on, several authors further studied the ground state properties and the energy spectrum of excitations [37]. However, it is very difficult to obtain correlation functions within this approach.

The deep relations between the above two approaches were first realized by Haldane [38]. He emphasized the existence of a universality class of one–dimensional systems whose low–energy physics is the same as for the Luttinger model. Following Landau who invented the term "Fermi liquid" for interacting systems which can be mapped onto Fermi gas, Haldane coined the term "Luttinger liquid" for this universality class. In particular, several lattice models solvable exactly by Bethe ansatz also belong to this class, and the correlation exponents can be extracted from the energy spectrum. Somehow this important idea did not receive deserved attention 10 years ago.

After the discovery of high temperature superconductors Anderson emphasized this important concept and its relevance for high T_c problem [39]. As we will see later, many properties of high T_c superconductors are consistent with predictions of the Luttinger liquid theory. Following Anderson's proposal, the correlation exponents were first extracted from numerical simulations [40]. Later on, these exponents have been calculated from the Bethe ansatz solution [41–46]. Using the spin–charge decoupling concept, these exponents can be calculated even by standard many–body techniques [47].

In the rest of this section we will sketch some basic ideas. Those interested in details should consult the original literature [32,38,42,45].

4.1 Luttinger and Related Models

In one–dimensional fermion systems, the important physics takes place near the two Fermi points: k_F and $-k_F$. It is natural to linearize the dispersion relation near these two points $\varepsilon_k = v_F(k \pm k_F)$, corresponding to a, b two branches. For noninteracting system, the Hamiltonian is given by

$$H_0 = v_F \sum_{k,s} \{(k - k_F)a_{ks}^+ a_{ks} + (-k - k_F)b_{ks}^+ b_{ks}\} \ . \tag{8}$$

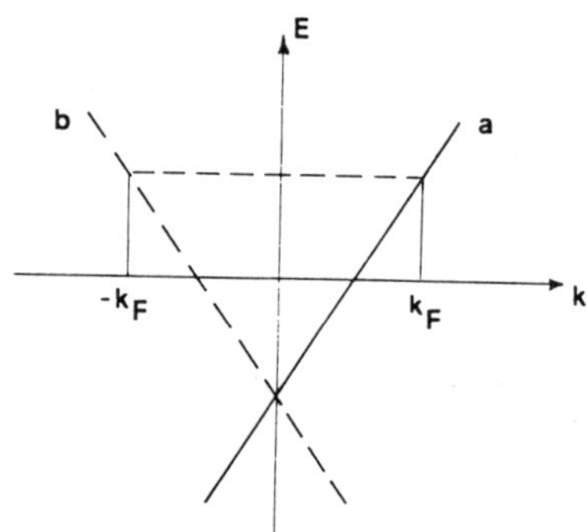

Fig. 6. The Luttinger model.

One can introduce bandwidth cutoff of the order $2\,k_0\,v_F$ around each Fermi point, or extend the band to $-\infty$ to include unphysical states. The latter version has some interesting mathematical properties, and is called Luttinger model (Fig.6).

Since we are mainly interested in the scattering process near these two Fermi points, it is convenient to introduce some coupling constants. For the spinless case, their definitions are given in Fig.7. The study of the phase diagram as a function of coupling constants using various techniques like renormalization group or parquet diagram resummation, is called g–ology [32].

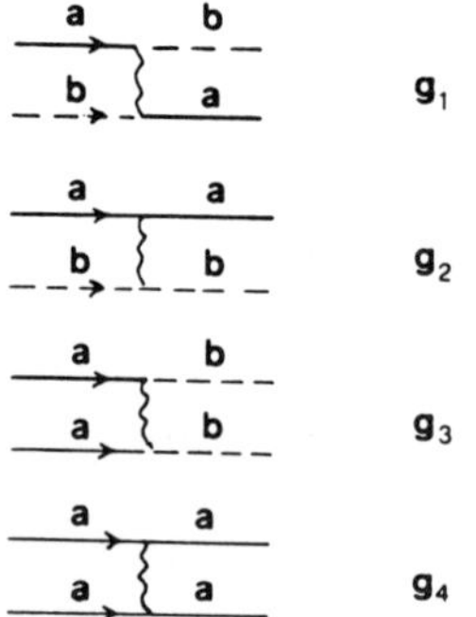

Fig. 7. Definition of coupling constants for spinless fermions. Solid lines correspond to branch a, while the dashed lines represent branch b.

The scattering from one branch to the other is called backscattering characterized by g_1, while the scattering without changing branches is forward scattering, denoted by g_2. The Umklapp process g_3 is important only for commensurate cases, say half–filling. The g_4 process usually only leads to minor renormalization effects. The Luttinger model does not contain backscattering process, i.e., $g_1 = 0$.

At the mean field level, the phase diagram is shown schematically in Fig.8, where CDW means charge–density–wave instability, SDW–spin density wave, SS–singlet superconductivity, TS–triplet superconductivity. However, the mean field result may be wrong in 1D, because there is a logarithmic divergence in the perturbation expansion which should be removed by renormalization group analysis or parquet diagram resummation [32]. To the second order of the coupling constant, the renormalization group equations in terms of energy E are given by

$$
\begin{aligned}
E \frac{dg_1(E)}{dE} &= \frac{1}{\pi v_F} g_1^2(E) , \\
E \frac{dg_2(E)}{dE} &= \frac{1}{2\,\pi v_F} g_1^2(E)
\end{aligned}
\tag{9}
$$

which means $g_1 - 2\,g_2$ is a constant and no renormalization is taking place if $g_1 = 0$. For $g_1 > 0$, $g_1(E) \to 0$, as $E \to 0$, i.e., we have infrared asymptotic freedom. For $g_1 < 0$, $g_1(E)$ at some finite, negative $E \to -\infty$, which signals an instability of the original state.

4.2 Bosonization and Critical Exponents

The basic boson excitations in the Luttinger–type models are density fluctuations,

$$
\rho_a(p) = \sum_k a^+_{k+p} a_k, \quad \rho_b(p) = \sum_k b^+_{k+p} b_k .
\tag{10}
$$

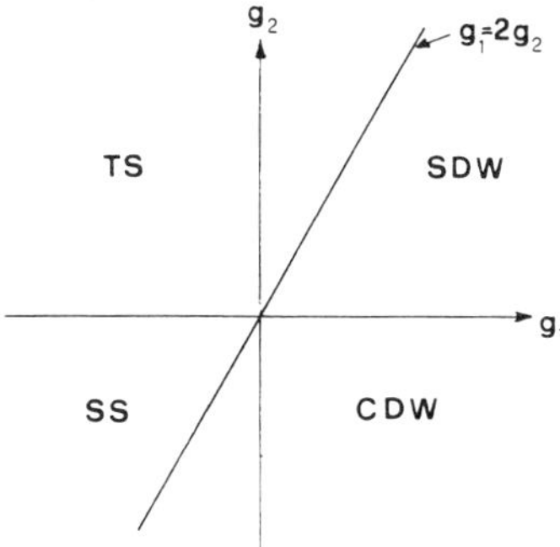

Fig. 8. 1D phase diagram at the mean field level.

These boson variables satisfy nontrivial commutation relations (L is the chain length)

$$[\rho_a(p), \rho_a(-q)] = -\delta_{pq} \frac{pL}{2\pi} ,$$

$$[\rho_b(p), \rho_b(-q)] = \delta_{pq} \frac{pL}{2\pi} , \qquad (11)$$

$$[\rho_a(p), \rho_b(q)] = 0 ,$$

and their low energy excitations correspond to harmonic oscillators with linear dispersion $\omega = uk$. This is for the spinless case. For fermions with spin there are two degrees of freedom: charge and spin. The effective boson Hamiltonian can be written as [42]

$$H = H_\rho + H_\sigma + \frac{2g_1}{(2\pi\alpha)^2} \int dx \cos(\sqrt{8}\phi_\sigma), \qquad (12)$$

$$H_\nu = \int dx \left[\frac{\pi u_\nu k_\nu}{2} \Pi_\nu^2 + \frac{u_\nu}{2\pi k_\nu} (\partial_x \phi_\nu)^2 \right], \quad \nu = \rho, \sigma, \qquad (13)$$

where ϕ is the phase field, defined by

$$\frac{\partial \phi}{\partial x} = \pi \rho(x) \qquad (14)$$

and

$$\phi_{\rho,\sigma} = \frac{1}{\sqrt{2}} (\phi_\uparrow \pm \phi_\downarrow). \qquad (15)$$

Π is the conjugate momentum satisfying the commutation relation

$$[\phi_\nu(x), \Pi_\mu(y)] = i\delta_{\nu\mu} \delta(x - y) , \qquad (16)$$

u_ν is the velocity, while k_ν is the correlation exponent.

For systems without backscattering $g_1 = 0$, the charge and spin are decoupled. For $g_1 > 0$, using the renormalization group one finds the fixed point $g_1^* = 0$. It turns out that the spin dynamics at the fixed point is also trivial, $k_\sigma^* = 1$. The correlation functions for particle and spin are given by

$$< n(x)\, n(0) > = \frac{k_\rho}{(\pi x)^2} + A_1 \cos(2k_F x)\, x^{-1-k_\rho}\, \ell n^{-3/2}(x)$$

$$+ A_2 \cos(4k_F x)\, x^{-1-4k_\rho} + \dots \qquad (17)$$

$$< S(x) \cdot S(0) > = \frac{1}{(\pi x)^2} + B_1 \cos(2k_F x)\, x^{-1-k_\rho}\, \ell n^{1/2}(x) + \dots ,$$

respectively. It is interesting to note that the charge fluctuations control the exponents, while the spin fluctuations only lead to logarithmic corrections [48]. The correlation exponent in the weak coupling case is given by $k_\rho = 1 - U/\pi v_F$. We should mention that there are several subtle points in the bosonization procedure which have been clarified earlier [33,34,38].

As pointed out first by Haldane [38], one can extract the critical exponents from the Bethe ansatz solution, at least for the case of one scaling variable. There is some complication for the case with two scaling variables, but in the absence of external magnetic field, one can still follow the same procedure [44]. Under the assumption that the weak and the strong coupling regimes are determined by the same fixed point $g_1^* = 0$, the velocities u_ρ, u_σ can be determined directly from the Bethe ansatz solution of Lieb and Wu [36], while k_ρ can be determined from the compressibility via the relation [42]

$$\frac{1}{L} \frac{\partial^2 E_0(n)}{\partial n^2} = \frac{\pi}{2} \frac{u_\rho}{k_\rho} , \tag{18}$$

where $E_0(n)$ is the ground state energy, n is the particle number. We will not quote the detailed results here, but only mention that for $U \to 0$, $k_\rho \to 1$, but $k_\rho \to \frac{1}{2}$ for $U \to \infty$ and $n = 0, 1$ for any positive U. For $n \to 0$, this is due to the divergence of the density of states. For $n \to 1$, one should consider holes which behave like low–density spinless fermions.

4.3 Some Properties of Luttinger Liquid

The 1D Luttinger liquid shows some very interesting properties. On the one hand, it looks like Fermi liquid in certain aspects [42]. For example, the specific heat is linear $C(T) = \gamma T$, where the constant γ is related to γ_0 of noninteracting system as

$$\gamma/\gamma_0 = \frac{1}{2} v_F \left(\frac{1}{u_\rho} + \frac{1}{u_\sigma} \right) . \tag{19}$$

Similarly, for the magnetic susceptibility

$$\chi/\chi_0 = v_F/u_\sigma , \tag{20}$$

hence the Wilson ratio

$$R_w = \frac{2 u_\rho}{u_\rho + u_\sigma} . \tag{21}$$

Moreover, the Fermi surface (points in this case) is well defined, and the momentum distribution shows singular behaviour at k_F and a weak singularity at $3 k_F$ [40]

$$n_k = \frac{1}{2} - \beta \, sign(k - k_F)(k - k_F)^\alpha + \beta' \, sign(k - 3 k_F)(k - 3 k_F)^{1+\alpha} + \ldots \tag{22}$$

where

$$\alpha = (k_\rho + 1/k_\rho - 2)/4 . \tag{23}$$

For $U \to \infty$, $n = 0, 1$, $\alpha \to 1/8$. As a consequence, the density of states at the Fermi level vanishes like ω^α. As we mentioned earlier, it is rather difficult to distinguish this singularity at k_F with a jump required by FL theory, at least in numerical simulations.

On the other hand, there is a striking difference from FL: the spin–charge decoupling. It can be made explicit in the large U limit, where the Bethe ansatz wave function can be written as a product of a determinant wave function for spinless fermions (holons) and a wave function for "squeezed" Heisenberg chain [40]. The excitations of the spin chain are spin

$\frac{1}{2}$ neutral objects – spinons. These holons and spinons behave very much the same way as suggested earlier by Anderson and collaborators to account for anomalous properties of high T_c superconductors [39,49]. Moreover, one can calculate the phase diagram for 1D $t-J$ model, which shows even superconductivity [50].

However, the crucial question is whether the 2D Hubbard model belongs to this universality class. Anderson argues strongly in favour of this, but there is no consensus among condensed matter theorists yet.

5. STUDIES OF 2D HUBBARD AND $t-J$ MODELS

In spite of the extreme importance of the issue and the enormous efforts made so far towards its study, we still know very little about it. One result is the ground state at half–filling. Although there is no rigorous proof, there seems to be long range antiferromagnetic order for spin $\frac{1}{2}$ system at least for the square lattice. The evidence comes from the quantum Monte–Carlo simulations [51]. Of course, the ground state is not the classical Néel state, and the staggered magnetization is reduced by about 40% due to quantum fluctuations. The other result is the Nagaoka theorem [52] stating, that in the limit of $U \to \infty$, $J \to 0$, the ground state of a single hole in an otherwise half–filled system should be ferromagnetic.

Apart from these two, other results are still under discussion. In the rest of this section we will briefly review the various approaches used to attack this problem.

5.1 Weak Coupling Approach

In real systems the Hubbard U is close to the bandwidth, but in this approach one assumes $U \ll t$. The phase diagram is the Hartree–Fock approximation is shown in Fig.9 [53]. Near half–filling the Fermi surface of noninteracting system is completely nested. The other characteristic feature of 2D systems is the Van Hove singularity. It is generic in nature, and for the nearest–neighbour hopping model it is located at $\omega = 0$, giving rise to a logarithmic contribution to the density of states. A general analysis, taking into account these two properties, was carried out in the early days of high T_c heat wave [54]. They have found an instability towards a d–wave superconductivity. Later on, Schrieffer and collaborators proposed a spin–bag model based on the weak–coupling approach by considering the SDW ground state [55]. These authors have shown that the spin wave spectrum obtained is identical to what follows from the strong coupling spin wave theory. So these two approaches are consistent at least for the spin fluctuations. However, the single occupancy constraint is not built in, so there is no localization effect existing in the $t-J$ model. The weak coupling approach has been pursued by many authors [56].

Another important point is that in 2D both $e-e$ and $e-h$ channels have similar ℓn^2 singularities. Therefore, the parquet resummation is a suitable scheme to use. However, the full scale parquet equations are too complicated for the available computing facilities.

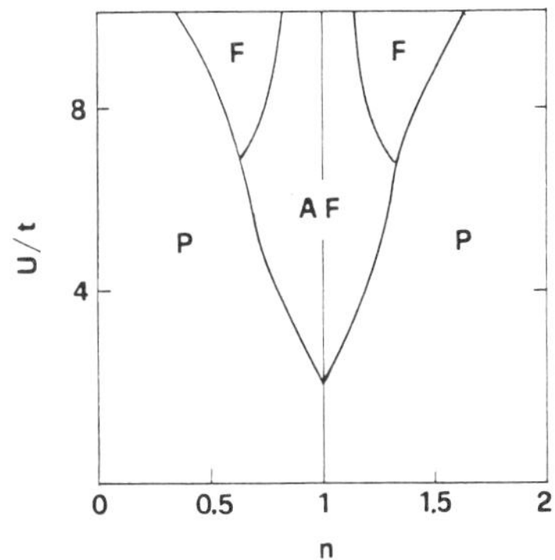

Fig. 9. Hartree–Fock phase diagram for 2D Hubbard model.

Nevertheless, several approximation schemes like "conserving approximation", "pseudo–potential" method, etc., have been tried to attack this problem [57]. The phase diagram obtained is improved compared with the Hartree–Fock case.

The low–density limit can be treated using the Galitzkii ladder diagram approximation. Although the 2D case has been discussed earlier [58], there has been a revival of interest in connection with high T_c problems. Anderson argued [59], that the 2D case is similar to the 1D case where the "orthogonality catastrophe" leads to non–FL behaviour. However, there is no consensus on this issue. Some other authors still find FL behaviour within the standard many–body scheme [60]. In some sense, this is not surprising, because it is difficult to imagine a way to find the breakdown of a scheme within itself. We should mention that the few-body problem (any number of particles) can be solved exactly. The thermodynamic limit of this problem corresponds to zero density. However, if one keeps the formal definition of density, the low–density results are fully recovered [61]. This is also quite surprising.

5.2 Single Hole in a Quantum Antiferromagnet

Since the reference compound is an antiferromagnet, to study the single hole motion on this background is an important issue. However, this simple looking problem turned out to be nontrivial at all. A similar problem for the Ising model was studied a long time back [62]. In the limit $J \rightarrow 0$, the hole does not have coherent propagation, because it leaves behind a string of reversed antiferromagnetic (AF) bonds. The spectral function of the hole does not depend on momentum, i.e., the motion is completely incoherent. We will call this case Brinkman–Rice limit.

Now we are interested in the Heisenberg antiferromagnet where the "wrongly" aligned spins can be cured by spin–flip process. This problem has been studied by many authors using various techniques, including exact diagonalization of small clusters [63], method of restricted Hilbert space [64], variational approach [65], self–consistent Green's functions [66], quantum Bogoliubov–de–Gennes equation [67], and functional integral [68].

There are three effects to be included in the treatment: (1) The distortion of the spin background, or spin polaron effect; (2) the quantum fluctuations showing up as annihilation and creation of spin excitation; (3) the renormalization due to emitting and re–absorbing spin waves. Not all of them have been included in the above mentioned studies. For example, the quasi–classical variational approach [65] does not consider the renormalization process, while the self–consistent Green's function method [66] does not include the spin distortion. The Bogoliubov–de–Gennes formalism includes all these three effects. Due to the self–trapping effects, one can find spin–polaron bound states. Recently, we have found that these bound states still survive even if there is only short range AF order [69]. Some authors assumed long–range dipolar distortion of the spin system around the hole. It seems that the size of spin polaron is rather localized [67,68], and the distortion is rather local.

The crucial question is whether there is a coherent propagation of hole. The answer so far obtained from both cluster and analytic calculations was positive. The bandwidth has been renormalized from the order of t to the order of J. However, the single occupancy constraint was not properly imposed in the analytical calculations, which may change the conclusion. In the cluster calculations finite size effects are important, and one should carry out a careful finite size scaling. The situation is still controversial. The Penn State group [70] claims that Z (spectral weight of coherent motion) remains constant, while the Trieste group [71] finds from quantum Monte–Carlo simulations that Z scales to zero for infinite size system. In one–dimension this problem can be solved exactly in the limit $U \to \infty$ using Bethe ansatz solution [72]. The hole is localized, but in a power fashion, so it can still propagate.

5.3 Strong Coupling Approach. Finite Dopings

The AF long range order disappears upon doping giving rise to superconductivity. Experimentally, there is a critical concentration, below which the AF order survives, but theoretically this question is still under debate. As for the existence of superconductivity in the doped system, the problem is also open. We briefly outline here some relevant results.

5.3.1. Numerical studies

Apart from variational Monte–Carlo [73] and correlated basis function (CBF) – Fermi Hypernetted Chain (FHNC) [74] techniques, more extensively used are the cluster exact diagonalization [63,75] and various realizations of quantum Monte–Carlo [76] method. The limitation for the cluster calculation is the finite size effect, while the difficulty for the quantum Monte–Carlo is the famous "fermion sign" problem. Nevertheless, there are some interesting results, although they are not the final word yet.

A large Fermi surface consistent with the Luttinger theorem has been found in the cluster calculation [77] for the $t - J$ model. The Green's function shows quasiparticle behaviour with a dispersion compatible with band structure calculations. In a certain range of parameters an attractive interaction between holes was found [78], but it has been argued that this corresponds to phase separation, rather to superconductivity [79]. However, the phase diagram obtained by high temperature expansion technique [80] looks somewhat different from Ref.79. The phase separation takes place for large J, but not for small J, as suggested in Ref.79.

As for the Hubbard model, the situation is somewhat different. There is no strong attraction between holes [81] and no phase separation [82]. The d–wave and extended s–wave pairing correlations are enhanced but they do not increase as a function of lattice size [83]. This, probably, means the absence of superconductivity in the Hubbard model per se. One technical question is how to enhance the "sensitivity of the testing antenna" in numerical simulations. This problem has been addressed recently [84].

5.3.2. Slave boson vs slave fermion

The main issue of strong coupling approach is to impose the single occupancy constraint. One possible way is to introduce "slave" particles for electrons $c_{i\sigma} = a_i^+ b_{i\sigma}$, where a_i^+ is the "slave" boson and $b_{i\sigma}$ is the fermion, or vice versa. In this way the non–holonomic constraint

$$\sum_{\sigma} c_{i,\sigma}^+ c_{i,\sigma} \leq 1$$

is converted into a holonomic one

$$a_i^+ a_i + \sum_{\sigma} b_{i,\sigma}^+ b_{i,\sigma} = 1 \ .$$

The slave boson formalism (when a_i is a boson) has been extensively used to study the pristine and doped systems. Within this formalism the ground state of a doped system turned out to be the uniform Resonanting Valence Bond (RVB) state suggested by Baskaran, Zou and Anderson [49] which shows FL like behaviour below the Bose condensation temperature. This is true in the large N expansion, where N is the number of fermion components ($N = 2$ for real electron). Various symmetry–broken phases, including the dimerized phase, flux phase, staggered flux phase, chiral phase with breaking time–reversal symmetry, etc. were considered. Interested readers are referred to the review articles [85]. The serious difficulty of the slave boson approach is the absence of AF correlations. In fact, the ground state in the half–filled case does not satisfy the Marshall sign rule [86].

Alternatively, the Schwinger boson–slave fermion approach (where the boson carries the spin, while the fermion carries the charge) is very useful in describing AF correlations. In the high T_c context, it was first used by Arovas and Auerbach [87], and later on by many others [88–91]. First of all, at half–filling the AF long range order is recovered if the bose condensation is taken into account. The ground state energy obtained is much better than the slave boson case. This state obeys the Marshall sign rule. Secondly, the spin–spin correlations fit rather well the experimental data [87]. Thirdly, the density of states is symmetric with respect to the

Fermi surface [89-91], and the non–FL like behaviour seems to be available [90,91]. However, the Fermi surface in this approach even for finite doping is "small" which is not consistent with experiments.

So far we have discussed the results obtained at the mean field level. It is somehow discouraging that they are quite different for these two approaches which in principle should be equivalent to each other. One possible way out is to go beyond the mean field approximation. Within the slave boson scheme, the next order of $\frac{1}{N}$ expansion has been considered [92]. Although the ground state remains FL–like at the one–loop level, the corrections due to correlations are significant to account for some interesting properties, e.g., the change of sign for Hall constant during doping, etc. Nevertheless, the AF fluctuations are not included explicitly yet. Within the slave fermion scheme, we have considered the loop corrections to the fermion motion due to AF fluctuations [91]. It is strongly renormalized and there is an incoherent peak in the spectral density due to multiple scattering which is consistent with optical data [5].

5.3.3. Gauge field approach

Another way to go beyond the mean field approximation is to consider the gauge field, first considered by Baskaran and Anderson in the high T_c problem [93]. In fact, the slave particle representation is invariant under a local gauge transformation

$$a_j \rightarrow a_j\, e^{i\theta(\mathbf{r}_j,t)}, \qquad b_{j\sigma} \rightarrow b_{j\sigma}\, e^{i\theta(\mathbf{r}_j,t)} .$$

All physical quantities should be invariant with respect to this transformation. This approach has been recently developed further by several authors [94-96]. The effective Hamiltonian (in the continuum fashion) of the coupled holon, spinon and gauge field system can be written as

$$H = \int d^2x \left[-\frac{1}{2m_s}\, f^+ (\nabla - i(\mathbf{a} + \mathbf{A}))^2 f - \frac{1}{2m_b}\, b^+ (\nabla - i\mathbf{a})^2 b \right.$$
$$\left. + \phi(b^+ b + f^+ f - 1) \right], \tag{24}$$

where f is the spinon, b the holon, and m_s, m_b their masses, respectively, while $\mathbf{a}$ is the fictitious gauge field, $\mathbf{A}$ the real electromagnetic field, ϕ the Lagrange multiplier. The single occupancy constraint can be expressed also as a zero total current condition, i.e.,

$$j_s = -j_h = j_{e\ell} .$$

Due to the coupling to fermions the gauge field propagator is overdamped which can explain several important properties in the normal state like the linear temperature dependence of the resistivity, constant thermal conductivity, and so on. Although the single occupancy constraint is imposed only on average, the gauge field description is appropriate for low–energy fluctuations. Whether this approach is successful to explain all main observations, especially the temperature dependence of the Hall constant, remains to be seen. The weakness of this formalism is again the absence of AF fluctuations.

5.3.4. Alternative approaches

Varma and collaborators [97] have put forward a phenomenological theory of marginal Fermi Liquid (MFL). These authors observe that if the imaginary part of the charge and spin polarization function has the following property over a wide range of momentum

$$Im \, \tilde{P}_{\rho,\sigma}(q,\omega) \sim \begin{cases} -N(0) \, \frac{\omega}{T}, & \omega \ll T \\ -N(0), & \omega_c \gg \omega > T \end{cases} \tag{25}$$

a number of experimental results, including resistivity, optical absorption, tunneling density, etc. can be explained in a natural way. Here $N(0)$ is the density of states at the Fermi level and ω_c is the cut–off frequency. An essential assumption is the absence of low energy scale other than the temperature itself.

As a consequence of Eq.(25), the electron self–energy is given by

$$\Sigma(\omega) = \lambda \, \omega \left(\ell n \frac{X}{\omega_c} + i \, \frac{\pi}{2} \, X \, sgn(\omega) \right) \, , \tag{26}$$

where $X = max(|\omega|, T)$ and λ is a coupling constant. Moreover, the spectral weight of the quasi–particle Z becomes

$$Z^{-1} = \left(1 + \lambda \, \ell n \frac{y}{\omega_c} \right) \, , \tag{27}$$

where $y = max[\,|E_k - \mu|, T]$ with E_k being the real part of the quasi particle energy. Therefore, Z vanishes at Fermi surface in a logarithmic fashion, i.e., the FL concept is violated in the weakest way (suggesting the word "marginal").

This approach has been developed further [6,98] to consider a broader class of experiments and to explore further implications. Several authors attempted to check this hypothesis numerically [70,99], the result is still controversial. Recently, an attempt has been made to justify this hypothesis microscopically [100]. All these aspects require further studies.

As an alternative several authors assume that oxide superconductors are very similar to heavy fermion compounds in many ways. According to them [101] the strong correlation effects will lead to some modifications of FL parameters, but one should not give it up completely.

We should also briefly mention here another approach – the infinite dimension expansion. It turns out that many calculation schemes – like perturbation, Gutzwiller variational wave function, etc. become simpler, or even exact in the limit $d \to \infty$ [102]. Therefore, one can construct an expansion in $\frac{1}{d}$. However, the mean field theory in $d = \infty$ turns out to be nontrivial, and several attempts have been made to construct such a theory [103].

Finally, the theory of anyon superconductivity initiated by Laughlin [8] has not been covered in this review. This elegant many–body theory based on Chern–Simons model should certainly have materialization in Nature. If not in high T_c superconductivity, it must be applicable to some equally exciting field. Those interested are referred to recent reviews [104].

6. CONCLUDING REMARKS

We tried in this review to give an overall picture of what is going on in the high T_c studies as a many–body problem. Due to the limitation of my own knowledge and the available space, many aspects have been touched upon only very superficially. However, I would like to emphasize that in spite of the enormous efforts made so far, many questions still remain open. For example: What is the ground state of a doped Mott insulator? Can a hole on an AF background propagate? Is FL description violated for strongly correlated systems? What is the actual mechanism for high T_c superconductivity?

In one–dimension, answers to many questions listed above are clear, mainly because $U = 0$ is a singular point. Any finite U, no matter how small it is, will bring the system into the only available universality class of interacting systems. In two dimensions, the situation is not clear yet.

Historically, one may distinguish "conceptual" and "quantitative" many–body theorists who do not talk to each other very much. Nowadays, this boundary becomes very fuzzy. On the one hand, the problem is so difficult that no one now can present a convincing answer. On the other hand, the sophisticated analytic and numerical tools become more accessible to a larger group of researchers. So everybody is facing the same challenge with an equal opportunity.

ACKNOWLEDGMENTS

I would like to thank my colleagues and friends, especially Y.M. Li, E. Tosatti and Z.B. Su for enlightening discussions on many topics considered here.

REFERENCES

1. J.G. Bednorz and K.A. Müller, Z. Phys. B64: 189 (1988).
2. "Physical Properties of High Temperature Superconductors", Vols.I and II, D.M. Ginsberg ed., World Scientific, Singapore, (1989, 1990).
3. High Temperature Superconductivity, Proc. Los Alamos Symp., 1989, K.S. Bedell, D. Coffey, D.E. Meltzer, D. Pines and J.R. Schrieffer eds., Addison–Wesley, Redwood City, California (1990).
4. June 1991 issue of Physics Today.
5. See, e.g. article by B. Batlogg in Ref.3.
6. Y. Kuroda and C.M. Varma, Phys. Rev. B42: 8619 (1990); P.B. Littlewood and C.M. Varma, J. Appl. Phys. to be published (1991); L. Coffey, Phys. Rev. Lett. 64: 1071 (1990).
7. K. Holczer, L. Forro, L. Mihaly and G. Grüner, Phys. Rev. Lett. 67: 152 (1991); M.C. Nuss, P.M. Mankiewich, M.L. O'Malley, E.H. Westerwick, and P.B. Littlewood, Phys. Rev. Lett. 66: 3305 (1991).
8. R.B. Laughlin, Science 242: 525 (1988); Phys. Rev. Lett. 60: 2677 (1988).

9.   J.M. Wheatley, T.C. Hsu and P.W. Anderson, Phys. Rev. B37: 5897 (1988).

10.   See, e.g., C.G. Olson, R. Liu. D.W. Lynch, R.S. List, A.J. Arko, B.W. Veal, Y.C. Chang,
      P.Z. Jiang, and A.P. Paulikas, Phys. Rev. B42: 381 (1990).

11.   H. Haghighi, J.H. Kaiser, S, Rayner, R.N. West, J.Z. Liu, R. Shelton, S. Solal, and
      M.J. Fluss, Phys. Rev. Lett. 67: 382 (1991).

12.   D.C. Johnston, Phys. Rev. Lett. 62: 957 (1989).

13.   G. Grüner, Physica C 162–164: 8 (1989).

14.   S. Martin, A.T. Fiory, R.M. Fleming, L.F. Schneemeyer and J.V. Waszcsak, Phys. Rev.
      Lett. 60: 2194 (1988).

15.   S. Uchida, H. Takagi, and Y. Tokura, Physica C 162–164: 1677 (1989).

16.   T.R. Chien, D.A. Brawner, Z.Z. Wang, and N.P. Ong, Phys. Rev. B43: 6242 (1991);
      N.P. Ong, Kathmandu Lecture at BCSPIN Summer School, June, 1991.

17.   J.C. Campuzano, G. Jennings, M. Faiz, L. Beaulaigue, B.W. Veal, J.Z. Liu,
      A.P. Paulikas, K. Vandervoort, H. Claus, R.S. List, A.J. Arko, and R.J. Bartlett,
      Phys. Rev. Lett. 64: 2308 (1990).

18.   R. Liu, B.W. Veal, A.P. Paulikas, J.W. Downey, H. Shi, C.G. Olson, C. Gu, A.J. Arko,
      and J.J. Joyce, preprint; R. Liu, private communication.

19.   P.W. Anderson, Science 235: 1196 (1987); in "Frontiers and Borderlines in Many–Body
      Physics", J.R. Schrieffer and R.A. Broglia, eds., North–Holland, Amsterdam (1987).

20.   See, e.g., K.A. Chao et al., J. Phys. C10: L271 (1977);
      J.E. Hirsch, Phys. Rev. Lett. 54: 1371 (1985); A.H. MacDonald, S.M. Girvin and
      D. Yoshioka, Phys. Rev. B37: 9753 (1988).

21.   C.M. Varma, S. Schmitt–Rink, and E. Abrahams, Solid State Commun. 62: 681 (1987);
      V.J. Emery, Phys. Rev. Lett. 58: 2794 (1987);
      J.E. Hirsch, Phys. Rev. Lett. 59: 228 (1987).

22.   M.S. Hybertsen, M. Schlüter, and N.F. Christensen, Phys. Rev. B39: 9028 (1989);
      H. Eskes and G. Sawatzky, in Proceedings of IWEPS '90, Kirchberg, Austria, to be
      published by Springer–Verlag.

23.   F.C. Zhang and T.M. Rice, Phys. Rev. B37: 3759 (1988).

24.   M. Takigawa, A.P. Reyes, P.C. Hammel, J.D. Thompson, R.H. Heffner, Z. Fisk, and
      K.C. Ott, Phys. Rev. B43: 247 (1991).

25.   See, e.g., V.J. Emery and G. Reiter, Phys. Rev. B38: 11938 (1988).

26.   See, e.g., J. Lorenzana and L. Yu, Phys. Rev. B43: 11474 (1991).

27.   D. Jerome and H.J. Schulz, Adv. Phys. 31: 299 (1982).

28.   Yu Lu, "Solitons and Polarons in Conducting Polymers", World Scientific, Singapore
      (1988).

29.   J.M. Luttinger, J. Math. Phys. 4: 1154 (1963).

30.   D.C. Mattis and E.H. Lieb, J. Math. Phys. 6: 304 (1965).

31.   D.C. Mattis, J. Math. Phys. 15: 609 (1974);
      A. Luther and V.J. Emery, Phys. Rev. Lett. 33: 589 (1974);
      A. Luther and I. Peschel, Phys. Rev. B9: 2911 (1974).

32.   J. Solyom, Adv. Phys. 28: 201 (1979);

V.J. Emery, in "Highly Conducting One–Dimensional Solids", J.T. Devreese, ed., Plenum, New York (1979) p.247.

33. R. Heidenreich, B. Schroer, R. Seiler, and D. Uhlenbrock, Phys. Lett. 54A: 119 (1975).

34. F.D.M. Haldane, J. Phys. C: Solid State Phys. 12: 4791 (1979).

35. C.N. Yang, Phys. Rev. Lett. 19: 1312 (1967).

36. E.H. Lieb and F.Y. Wu, Phys. Rev. Lett. 20: 1445 (1968).

37. A.A. Ovchinnikov, Sov. Phys. JETP 30: 1160 (1970);
H. Shiba, Phys. Rev. B6: 930 (1972);
C.F. Coll, III, Phys. Rev. B9: 2150 (1974);
F. Woynarovich, J. Phys. C15: 85 (1982).

38. F.D.M. Haldane, Phys. Rev. Lett. 45: 1358 (1980); Phys. Lett. 81A: 153 (1981);
J. Phys. C14: 2585 (1981).

39. P.W. Anderson, Phys. Rep. 184: 195 (1989), in "Current Trends in Condensed Matter, Particle Physics and Cosmology", J. Pati, Q. Shafi, S. Wadia and Yu Lu, eds., World Scientific, Singapore (1990);
P.W. Anderson and Y. Ren, in Reference 3.

40. M. Ogata and H. Shiba, Phys. Rev. B41: 2326 (1990);
S. Sorella, A. Parola, M. Parrinello, and E. Tosatti, Europhys. Lett. 12: 721 (1990).

41. A. Parola and S. Sorella, Phys. Rev. Lett. 64: 1831 (1990).

42. H.J. Schulz, Phys. Rev. Lett. 64: 2831 (1990); Int. J. of Mod. Phys. B5, 57 (1991).

43. N. Kawakami and S.K. Yang, Phys. Lett. 148A: 359 (1990); Phys. Rev. Lett. 65: 2309 (1990).

44. H. Frahm and V.E. Korepin, Phys. Rev. B42: 10533 (1990); Phys. Rev. B43: 5633 (1991).

45. Y. Ren and P.W. Anderson, Princeton preprint;
Y. Ren, Princeton preprint.

46. F.D.M. Haldane and Y. Tu, La Jolla preprint.

47. Z.Y. Weng, D.N. Sheng, C.S. Ting, and Z.B. Su, preprint.

48. T. Giamarchi and H.J. Schulz, Phys. Rev. B39: 4620 (1989).

49. G. Baskaran, Z. Zou, and P.W. Anderson, Solid State Commun. 63: 973 (1987);
P.W. Anderson and Z. Zou, Phys. Rev. Lett. 60: 132 (1988).

50. M. Ogata, M. Luchini, S. Sorella, and F.F. Assaad, Phys. Rev. Lett. 66: 2388 (1991).

51. J.D. Reger and A.P. Young, Phys. Rev. B37: 5978 (1988);
N. Trivedi and D. Ceperley, Phys. Rev. B40: 2737 (1989);
S.D. Liang, Phys. Rev. B42: 6555 (1990).

52. Y. Nagaoka, Phys. Rev. 147: 392 (1966).

53. J.E. Hirsch, Phys. Rev. B31: 4403 (1985).

54. H.J. Schulz, Europhys. Lett. 4: 609 (1987);
I.E. Dzyaloshinskii and V.M. Yakovenko, Sov. Phys. JETP 67: 844 (1988).

55. J.R. Schrieffer, X.G. Wen, and S.C. Zhang, Phys. Rev. Lett. 60: 944 (1988);
Phys. Rev. B39: 11663 (1989).

56. A. Singh and Z. Tešanović, Phys. Rev. B41: 614, 11604 (E) (1990);

Z.Y. Weng, C.S. Ting, and T.K. Lee, Phys. Rev. B41: 1990 (1990);

W.P. Su and X.Y. Chen, Phys. Rev. B38: 8879 (1988);

G. Vignale and M.R. Hedayati, Phys. Rev. B42: 786 (1990);

D.M. Frenkel and W. Hanke, Phys. Rev. B42: 6711 (1990);

A. Kampf and J.R. Schrieffer, Phys. Rev. B41: 6399 (1990); B42: 7967 (1990).

57. N.E. Bickers and D.J. Scalapino, Ann. Phys. (New York) 193: 206 (1980);

N.E. Bickers, D.J. Scalapino, and S.R. White, Phys. Rev. Lett. 62: 961 (1989);

N.E. Bickers and S.R.,White, Phys. Rev. B43: 8044 (1991);

D.W. Hess and J.W. Serene, preprint;

A.–M.S. Tremblay et al., private communication.

58. P. Bloom, Phys. Rev. B12: 125 (1975).

59. P.W. Anderson, Phys. Rev. Lett. 64: 1839 (1990); 65: 2306 (1990).

60. J.R. Engelbrecht and M. Randeria, Phys. Rev. Lett. 65: 1032 (1990);

H. Fukuyama, Y. Hasegawa, and O. Narikiyo, J. Phys. Soc. Jpn. 60: 2013 (1991).

61. M. Fabrizio, A. Parola, and E. Tosatti, Phys. Rev. B43: 1033 (1991).

62. W.F. Brinkman and T.M. Rice, Phys. Rev. B2: 1324 (1970);

L.N. Bulaevskii, E. Nagaev, and D.L. Khomskii, Sov. Phys. JETP 27: 836 (1968).

63. See, e.g., J. Bonča, P. Prelovsek, and I. Sega, Phys. Rev. B39: 7074 (1989);

E. Dagotto, R. Joynt, A. Moreo, S. Bacci, and E. Gagliano, Phys. Rev. B41: 9049
(1990);

C.–X. Chen and H.B. Schütler, Phys. Rev. B40: 239 (1989);

K.V. Szczepanski, R. Horsch, W.H. Stephan, and M. Ziegler, Phys. Rev. 41: 2017
(1990);

T. Xiang, to be published;

N. Furukawa and M. Imada, J. Phys. Soc. Jpn. 59: 1771 (1990).

64. S.A. Trugman, Phys. Rev. B37: 1597 (1988); B41: 892 (1990); Phys. Rev. Lett. 65:
500 (1990).

65. B.I. Shraiman and E.D. Siggia, Phys. Rev. Lett. 61: 467 (1988); 62: 1564 (1989);

S. Sachdev, Phys. Rev. B39: 12232 (1989).

66. C.L. Kane, P.A. Lee, and N. Read, Phys. Rev. B39: 688 (1989);

S. Schmitt–Rink, C.M. Varma, and A.E. Ruckenstein, Phys. Rev. Lett. 60: 2793 (1988);

G. Martinez and P. Horsch, Phys. Rev. B44: 317 (1991).

67. Z.B. Su, Y.M. Li, W.Y. Lai, and L. Yu, Phys. Rev. Lett. 63: 1318 (1989); Int. J. Mod.
Phys. B3: 1913 (1989).

68. A. Auerbach and B.E. Larson, Phys. Rev. Lett. 66: 2262 (1991).

69. C.Q. Wu, Z.B. Su, and L. Yu, to be published.

70. J. Song and J.F. Annett, Penn State preprint.

71. S. Sorella, A. Parola, and E. Tosatti, SISSA preprint.

72. S. Sorella and A. Parola, to be published in Phys. Rev B.

73. See, e.g. C. Gros, Phys. Rev. B38: 931 (1988); Ann. Phys. (New York) 189: 53 (1989).

74. X.Q. Wang, S. Fantoni, E. Tosatti, L. Yu, and M. Viviani, Phys. Rev. B41: 11479 (1990).

75. See, e.g. the review, E. Dagotto, Int. J. Mod. Phys. B5: 77 (1991).

76. See, e.g., D.J. Scalapino, in Ref.3; S. Sorella, A. Parola, M. Parrinello, and E. Tosatti, Int. J. Mod. Phys. B3: 1875 (1989);
    S.R. White, D.J. Scalapino, R.L. Sugar, E.Y. Loh, J.E. Gubernatis, and R.T. Scalettar, Phys. Rev. B40: 506 (1989);
    M. Imada and Y. Hatsugay, J. Phys. Soc. Jpn. 58: 3752 (1989);
    J.E. Hirsch and S. Tang, Phys. Rev. Lett. 62: 591 (1989).

77. W. Stephan and P. Horsch, Phys. Rev. Lett. 66: 2258 (1991).

78. See, first paper in Ref.63;
    E. Dagotto, J. Riera, and A.P. Young, Phys. Rev. B42: 2347 (1990).

79. V.J. Emery, S.A. Kivelson, and H.Q. Lin, Phys. Rev. Lett. 64: 475 (1990).

80. M. Luchini, M. Ogata, W. Putikka, and T.M. Rice, ETH Zurich, preprint.

81. S. Sorella, A. Parola, and E. Tosatti, Int. J. Mod. Phys. B5: 143 (1991);
    G. Fano, F. Ortolani, and A. Parola, Phys. Rev. B42: 6877 (1990) and preprint.

82. A. Moreo, D.J. Scalapino, and E. Dagotto, Phys. Rev. B43: 11442 (1991).

83. A. Moreo and D.J. Scalapino, Phys. Rev. B43: 8211 (1991) and references therein.

84. E. Dagotto and J.R. Schrieffer, Phys. Rev. B43: 8705 (1991).

85. P.A. Lee in Ref.3; T.M. Rice, Kathmandu and Les Houches lectures, to be published; see also Ref.80.

86. W. Marshall, Proc. Roy. Soc. London, Ser.A 232: 48 (1955);
    S.D. Liang, B. Doucot, and P.W. Anderson, Phys. Rev. Lett. 61: 365 (1988).

87. D. Arovas and A. Auerbach, Phys. Rev. B38: 316 (1988);
    A. Auerbach and D. Arovas, Phys. Rev. Lett. 61: 617 (1988).

88. D. Yoshioka, J. Phys. Soc. Jpn. 58: 32, 1516, 3733 (1989).

89. K. Flensberg, P. Hedegard, and M.B. Pederson, Phys. Rev. B40: 850 (1989);
    P. Hedegard and M.B. Pederson, Phys. Rev. B42: 10035 (1990); 43: 11504 (1991).

90. S. Sarker, C. Jayaprakash, H.R. Krishnamurthy, and M. Ma, Phys. Rev. B40: 5028 (1989);
    C. Jayaprakash, H.R. Krishnamurthy, and S. Sarker, Phys. Rev. B40: 2610 (1990);
    S. Sarker, C. Jayaprakash, and H.R. Krishnamurthy, preprint;
    S. Sarker, preprint.

91. Y.M. Li, D.N. Sheng, Z.B. Su and L. Yu, Mod. Phys. Lett. B, to appear, and preprint.

92. See, e.g., M. Grilli and G. Kotliar, Phys. Rev. Lett. 64: 1170 (1990);
    Z. Wang, Y. Bang, and G. Kotliar, preprint;
    C.A.R. Sà de Melo, Z. Wang, and S. Doniach, preprint.

93. G. Baskaran and P.W. Anderson, Phys. Rev. B37: 580 (1988).

94. L.B. Ioffe and A.I. Larkin, Phys. Rev. B39: 8988 (1989).

95. N. Nagaosa and P.A. Lee, Phys. Rev. Lett. 64: 2450 (1990).

96. L.B. Ioffe and P.B. Wiegmann, Phys. Rev. Lett. 65: 653 (1990);
    L.B. Ioffe and G. Kotliar, Phys. Rev. B42: 10348 (1990);
    L.B. Ioffe, V.A. Kalmeyer, and P.B. Wiegmann, Phys. Rev. B43: 1219 (1991);
    L.B. Ioffe and V.A. Kalmeyer, Phys. Rev. B, to be published.

97. C.M. Varma, P.B. Littlewood, S. Schmitt–Rink, E. Abrahams, and A.E. Ruckenstein, Phys. Rev. Lett. 63: 1996 (1989);
    C.M. Varma, Int. J. Mod. Phys. B3: 2083 (1989).

98. G. Kotliar et al., Europhys. Lett. (1991);

P.B. Littlewood, Les Houches lectures (1991).

99. D.W. Hess and J.W. Serene, preprint.

100. A.E. Ruckenstein and C.M. Varma, M^2S Proceedings, Kanazawa, July, 1991.

101. See, e.g., article by P.A. Lee in Ref.3 and K. Levin, Ju.H. Kim, J.P. Lu, and Q. Si, to appear in Physica C;

B.H. Brandow, Solid State Commun. 69: 915 (1989).

102. See, e.g., the review article, D. Vollhardt, Int. J. Mod. Phys. B3; 2189 (1989).

103. V. Janiš and D. Vollhardt, Aachen preprint;

A. Georges and G. Kotliar, preprint.

104. A. Zee in Ref.3;

F. Wilczek, "Fractional Statistics and Anyon Superconductivity", World Scientific, Singapore (1991).

POINT DEFECT DISORDER IN MODELS OF HIGH TEMPERATURE SUPERCONDUCTIVITY

J. W. Halley*, C. Das Gupta*,†, S. Davis*,# and X.-F. Wang *

*School of Physics and Astronomy
University of Minnesota
Minneapolis, MN 55455

†Indian Institute of Science
Bangalore, India

#Department of Mathematics
Cornell University
Ithaca, NY

Experimental high T_c systems are intrinsically strongly disordered by point defects. Here we consider two possible consequences of this fact: 1) The magnetic phases of prominent models for high T_c are strongly affected. We present numerical calculations of the phase diagram of the Hubbard model as a function of the concentration of point defects and of the band filling in the Hartree Fock approximation. The results may explain the failure to experimentally observe a spin density wave phase predicted by Hartree Fock calculations without defects. 2) The superconductivity of the system may be strongly affected. We consider a model in which the pairing interactions are generated by point defects. We present numerical data on the density of states, T_c and local charge density in the BCS approximation, while taking full account of the disorder. The stability of the BCS state was also checked using a variational Monte Carlo calculation.

INTRODUCTION

A conspicuous feature of the new superconductors [1-2] which is usually assumed to be an inessential complication by theorists is that virtually all of the high temperature materials have a very high degree of spatial disorder leading to mean free paths which are at best a few times the superconducting coherence length. This means at least that one should be developing theoretical methods and models for taking theoretical account of the microscopic disorder. There is also a possibility [4] that some kinds of point defects or twin boundaries might act like "pairing centers" and enhance the pairing interactions leading to

Recent Progress in Many-Body Theories, Vol. 3,
Edited by T.L. Ainsworth et al., Plenum Press, New York, 1992

the superconductivity itself. For both these reasons we have been developing both mean field and correlated calculational methods and models for taking account of the effects of point defects in models of high T_c superconductors.

Here we describe methods for studying strong disorder in both mean field and correlated models of high temperature superconductivity and present results on two examples. In the next section we describe a Hartree Fock calculation of the magnetic phase diagram of the one band Hubbard model. The results provide a possible explanation of the experimental failure to observe a spin density wave predicted for the model without disorder. In the third section we describe results for the mean field BCS theory of a highly dilute tJ model. Finally we briefly explain how to do variational Monte Carlo calculations on the same model and present new results on the same model as an example.

HARTREE FOCK THEORY OF THE ONE BAND HUBBARD MODEL

The effects of point defect disorder on the tendency of the local "Hubbard U" of the one band Hubbard model to localize electrons has been studied by several analytical methods: perturbation theory in the electron-electron interaction strength[5] and mean field approximations to the large U limit of the Hubbard model[6] both suggest that the disorder and Coulomb interactions together enhance the localization effects arising separately from the two effects. The increased localization due to disorder should increase the magnitude of the magnetic moments and thus stabilize magnetic phases. It appears, however, that this has not been much discussed. Without disorder, Schulz[7] showed that, in Hartree Fock approximation, the antiferromagnetic state of the two dimensional Hubbard model becomes an incommensurate spin density wave state as soon as the number of holes increases beyond half filling. Jarrell et al[8] considered the effects of single strontium acceptors in a two band model in which the defects reduced the stability of the antiferromagnetic state as a consequence of the exchange interaction between oxygen holes trapped at acceptors and the copper spins.

Here we report[9] results on the magnetic phase diagram of the two dimensional one band Hubbard model in Hartree Fock approximation in the presence of disorder. Of various possible models for the disorder, we use a point defect model which could represent either acceptor[10] sites associated for example with Sr in high T_c systems, or with donor sites, associated for example with oxygen vacancies[11] . The Hamiltonian is

$$H = -t \sum_{<i,j>,\sigma} c_{i,\sigma}^{\dagger} c_{j,\sigma} + U \sum_i n_{i,\downarrow} n_{i,\uparrow} + V \sum_i' (1 - n_{i,\sigma}) \tag{1}$$

In the usual way, the first sum is on nearest neighbors of a square lattice. The sum in the last term only extends over neighbors of a randomly selected set of sites on the dual lattice with selection probability p per site. This model was selected to roughly describe acceptors such as Sr in $La_{2-x}Sr_xCuO_4$. The results described here are not expected to be qualitatively sensitive to the local geometry of the defects.

In the Hartree Fock approximation we find the eigenstates $|n>$ with energies ϵ_n of the one electron operator

$$H_{HF} = -t \sum_{<i,j>,\sigma} c_{i,\sigma}^{\dagger} c_{j,\sigma} + U \sum_i (n_{i,\downarrow} < n_{i,\uparrow} > + < n_{i,\downarrow} > n_{i,\uparrow} - \tag{2}$$

$$< c_{i,\uparrow}^{\dagger} c_{i,\downarrow} > c_{i,\downarrow}^{\dagger} c_{i,\uparrow} - c_{i,\uparrow}^{\dagger} c_{i,\downarrow} < c_{i,\downarrow}^{\dagger} c_{i,\uparrow} >) + V \sum_i' (1 - n_{i,\sigma})$$

with the self consistency conditions

$$< n_{i,\sigma} >= \sum_{\epsilon_n < \epsilon_F} |<i,\sigma|n>|^2 \tag{3}$$

$$< c_{i,\sigma}^{\dagger} c_{i,-\sigma} >= \sum_{\epsilon_n < \epsilon_F} <i,\sigma|n><n|i,-\sigma> \tag{4}$$

The Fermi level is set by fixing the deviation δ from half filling in the usual notation. We used direct diagonalization for the determination of eigenvalues and eigenvectors and straightforward iteration for self-consistency.

To analyse the results for the magnetic phases, we calculated configuration averaged fourier transforms of the z component of the spin defined as

$$S_z(\vec{k}) = \sum_i e^{i\vec{k}\cdot\vec{r}_i}(< n_{i,\uparrow} > - < n_{i,\downarrow} >) \tag{5}$$

Some examples are shown in Figure 1.

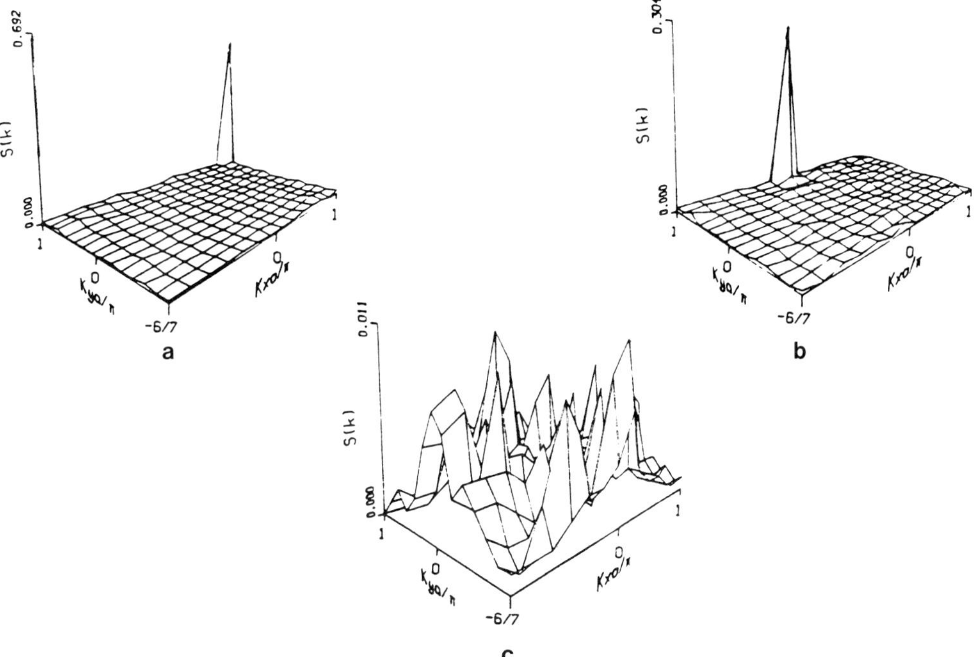

Figure 1. $S_z(\vec{k})$ for some of the phases found. a) Antiferromagnetic phase (AF) b) Spin density wave (SDW) c) "Spin Glass" phase (SG)

The large peak in the corner of the zone (Figure 1a) indicates an antiferromagnetic phase. Elsewhere in the $\delta - p$ plane as discussed below, we find spin configurations as indicated in Figure 1b where the large peak has shifted from the corner of the zone. These are the spin density wave states. Finally we find "spin glass" states in which no peak is found (Figure 1c) and paramagnetic states in which the spins are zero within our numerical accuracy.

For the parameter values $U/t = 8, V/t = -2$ we find the phase diagram shown in Figure 2.

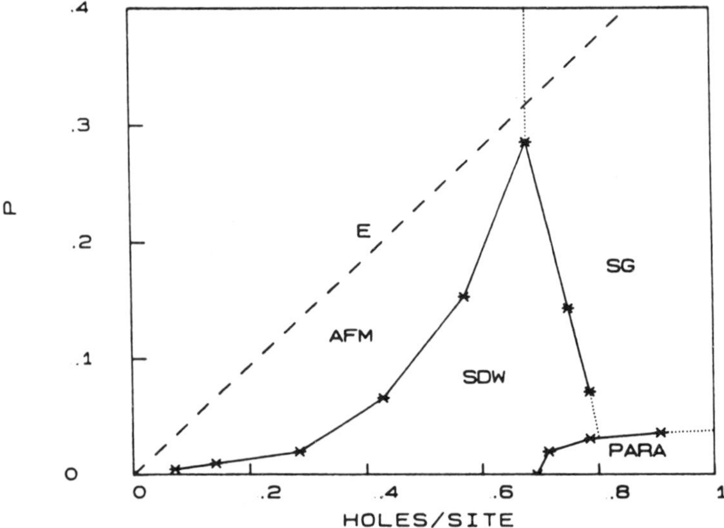

Figure 2. Phase diagram in the p-δ plane.

We note that the antiferromagnetic phase, which only exists at $\delta = 0$ in the ordered, $p = 0$, line in the plane, has been stabilized by the disorder away from $p = 0$. The AF order parameter is observed to decrease with increasing p near small p but then to increase with further increases in p. We presume that the latter effect arises because the disorder has increased the degree of localization and hence the effective magnitude of the exchange coupled local moments. Indication of this effect is seen in the participation ratios(to be reported elsewhere[9]) .

If we regard these as calculations on a toy model for high T_c materials, then it is likely that in real materials, p effectively increases with δ both because the number of acceptor sites increases and because there may be some compensation, leading to donor sites as well, when δ increases. Thus, if a set of experimental samples with various δ lay along a line such as the one marked E in the plane of Figure 2, one might not see the predicted SDW. Indeed, neutron spectroscopy reports a transition directly from the antiferromagnetic to a spin-glass-like phase in high T_c materials.

VERY DILUTE t-J MODEL

In this section we report similar mean field calculations on a point defect disordered model for the superconductivity itself, in which the mean field theory in question is the BCS theory, rather than the Hartree Fock approximation. The model[12] arises if one assumes that the current carrying holes in the copper oxygen planes of high temperature superconductors are attracted to each other through an exchange interaction at the randomly placed sites of oxygen vacancies. Physical arguments that suggest these assumptions are discussed in somewhat more detail elsewhere[12]. Then the model consists of carriers moving with nearest neighbor hopping on a square lattice with an exchange interaction J between a randomly selected fraction p of the bonds. The onsite energy , called ϵ at these special bonds is

186

allowed to be different from that at the other sites. The Hamiltonian is

$$H - \mu N = \sum_{i,\sigma}(\epsilon(i) - \mu)d^{\dagger}_{i,\sigma}d_{i,\sigma} + \sum_{i,j,\sigma}(b(i,j)d^{\dagger}_{i,\sigma}d_{j,\sigma} + h.c.) + . \tag{6}$$

$$- \sum_{i,j}(J(i,j)/2)(d^{\dagger}_{i\uparrow}d^{\dagger}_{j\downarrow} - d^{\dagger}_{i\downarrow}d^{\dagger}_{j\uparrow})(d_{j\downarrow}d_{i\uparrow} - d_{j\uparrow}d_{i\downarrow})$$

We suppose that J(ij) is zero except at a randomly selected fraction p of bonds where its value $J(ij) = J$ is a parameter of the model. b_{ij} is made the same at all nearest neighbor pairs and set equal to t (usually we take t=1 to establish the energy scale). $\epsilon(i)$ is zero at sites away from special bonds but finite and $= \epsilon$ on sites next to special bonds. μ is the chemical potential. We will suppose here, on the basis of the physical arguments mentioned, that μ lies near the top of the band. To specify the mean field theory, we define the retarded functions[13] :

$$G^{ij}_{\sigma}(t) = -i\Theta(t) < \{d_{i,\sigma}(t), d^{\dagger}_{j,\sigma}(0)\} >$$

$$F^{ij}_{\sigma}(t) = -i\Theta(t) < \{d^{\dagger}_{i,\sigma}(t), d^{\dagger}_{j,-\sigma}(0)\} >$$

We obtain equations of motion for the functions $G_{ij} = G^{ij}_{\uparrow} + G^{ij}_{\downarrow}$ and $F_{ij}(t) = F^{ij}_{\uparrow} - F^{ij}_{\downarrow}$:

$$i\hbar dG_{ij}/dt = 2\hbar\delta(t)\delta_{i,j} + \epsilon(i)G_{ij} + \sum_{\delta}b(i+\delta,i)G_{i+\delta j} + \sum_{\delta}(J(i+\delta,i)/2)\Delta_{i,\delta}F_{i+\delta\ j}. \tag{7a}$$

$$i\hbar dF_{ij}/dt = -\epsilon(i)F_{ij} - \sum_{\delta}b(i+\delta,i)F_{i+\delta\ j} + \sum_{\delta}(J(i+\delta,i)/2)\Delta^{*}_{i,\delta}G_{i+\delta\ j}. \tag{7b}$$

with the initial conditions $G_{ij}(0) = -i2\delta_{ij}$, $F_{ij}(0) = 0$. The gap equation is

$$\Delta^{*}_{ij} = -i\int_{-\infty}^{\infty}\frac{F_{ij}(\omega + i\epsilon) - F_{ij}(\omega - i\epsilon)}{e^{\beta\omega} + 1}d\omega. \tag{8}$$

where

$$F_{ij}(\omega + i\epsilon) = 1/2\pi\int_{-\infty}^{\infty}F_{ij}(t)e^{i\omega t}dt$$

In order to reduce the number of equations, we define the sums $F_i = \sum_j c_j F_{ij}$, $G_i = \sum_j c_j G_{ij}$. Then $F_i(0) = 0$, $G_i(0) = -i2c_i$.. To calculate the average Δ [13] at the special sites we choose the c_i as follows. All c_i associated with sites not next to a special site are zero. Associate a random number ϕ_i evenly distributed between 0 and 2π with each site next to a vacancy. Now consider a pair of sites next to a special bond. Label the sites 1 and 2. Set $c_1 = e^{i\phi_2}$ and $c_2 = e^{i\phi_1}$. The phase factors $e^{i\phi}$ are a calculational device only and are not to be confused with the phase factors associated with the gap function itself. Choosing c_i equal to zero away from vacancies is done only because we are choosing to evaluate Δ self-consistently at those sites and does not mean that the gap is zero or neglected elsewhere on the lattice in our calculation.

With choices of the coefficients c_i described above, cancellation of random phases [13] gives the gap equation: We define $F_r(t) = \sum_i \pm e^{-i\phi_i}F_i$ where the plus sign is used if i is adjacent to an x bond and the minus sign is used if i is adjacent to a y bond. Then gap equation takes the form

$$\Delta^* = \frac{-1}{2N_s\beta\hbar}\int_0^{\infty}\frac{F_r(t)}{sinh(t\pi/\beta\hbar)}dt. \tag{9}$$

where N_s is the number of oxygen vacancies. By use of the preceding expressions we obtain the equations of motion for F_i and G_i which are essentially identical to the equation of motion written above.

We solve these equations by simple numerical integration in the time domain. The model favors equal phases of the gap for vacancies on parallel bonds but phases of the gap differing by π for vacancies on perpendicular bonds. After a self-consistent solution for the average gap Δ has been found using the methods outlined above, then the density of states can be found by essentially the same techniques[12].

We first present some results on the nature of the density of states in the normal state. The total density of states is shown in Figure 3 for $\epsilon = 3.5$, $J = 1.5$, $t = 1$ and p= fraction of bonds which are "special"=4% . One sees that the presence of the special sites next to the special bonds results in an extra peak at the top of the band. Looking at the local densities of states reveals the structure underlying this total state density. In Figure 4, we show the average local density of states at the special sites. One sees that there are *two* kinds of states at the special ("copper") sites: The peak near the top of the total density of states is

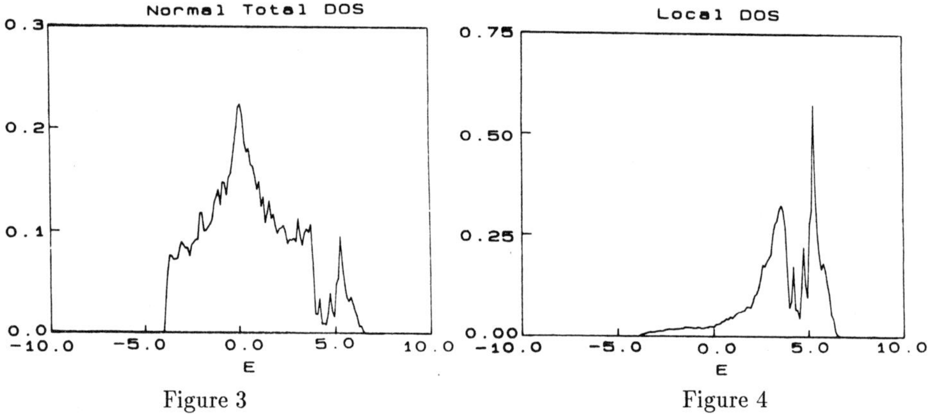

Figure 3Figure 4

associated with a localized state, while there is a second peak in the local density of states for the special sites which is associated with a resonance lying inside the band and which is not localized. Our qualitative picture[12] of the nature of the paired superconducting would suggest that we study cases in which the fermi level lies near, but slightly above the lower peak in the local density of states of the special sites.

Some densities of states for self-consistent paired states are shown in Figure 5. When the Fermi energy is in the metallic region, there is no observable gap in the density of states, even though the BCS equations have a self consistent solution. For larger values of the Fermi energy, when the normal state would be an insulator, we observe a gap like structure. The gap is clearly not sharply defined and there continues to be finite density of states at the fermi level, that is a "gapless" form of superconductivity.

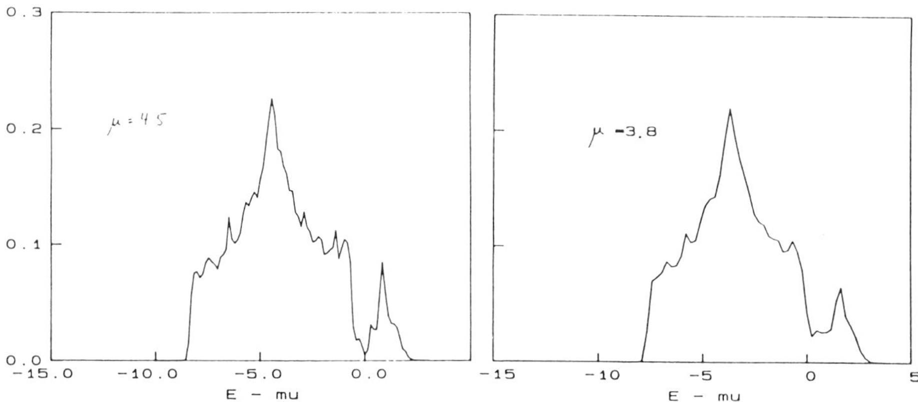

Figure 5. Densities of States of Paired States.

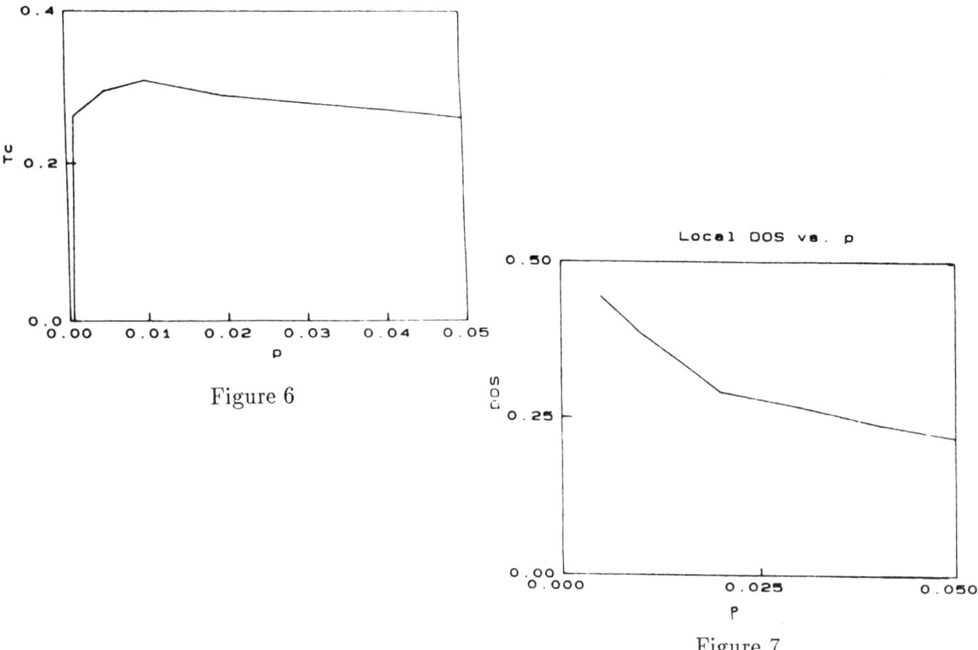

Figure 6

Figure 7

In Figure 6 we show $T_c(p)$. To understand the nonmonotonicity of T_c as a function of p we can look at the density of states at the fermi level at the special as a function of p (Figure 7). The decrease of this local density of states with increasing p appears to be the origin of the drop in $T_c(p)$ at the larger values of p. Calculations reported elsewhere, we found that T_c increases monotonically as a function of hole concentration.

To check whether a pairing state of this sort survives the correlations induced by the large U of the Hubbard model, we have made variational calculations on similarly disordered models by extending the work of Gros[14] to disordered systems. The variational wavefunction $|N>$ is written as

$$|N> \propto P\left(\sum_{j,j'} a(\vec{R}_j, \vec{R}_{j'})c^\dagger_{\vec{R}_j,\uparrow}c^\dagger_{\vec{R}_{j'},\downarrow}\right)^{N/2}|0>$$

Here P is the Gutzwiller projection operator which forbids double occupancy at sites, $c^\dagger_{\vec{R}_{j'},\uparrow}$ is a creation operator for an electron in a tight binding orbital at site j'. We write the one electron eigenfunctions $\phi_n(\vec{r})$ of the disordered system in terms of tight binding functions $\phi_n(\vec{r}) = \sum_j A_{nj}\phi(\vec{r} - \vec{R}_j)$. Then the appropriate form of $|N>$ for a disordered system [15] is obtained by using

$$a(\vec{R}_j, \vec{R}_{j'}) = \sum_n A^*_{n,j_1} A_{n,j_2} v_n/u_n$$

$$v_n/u_n = \Delta_n/[\xi_n + \sqrt{\xi_n^2 + \Delta_n^2}]$$

We have adopted a code used for the homogeneous case to this problem, calculating the eigenstates of the disordered one electron 82 site system by direct diagonalization. We used[15] a variational form of the type

$$\Delta_n = \sum_{i,j} A^*_{n,i}\Delta^{MFT}_{i,j} A_{n,j} \tag{10}$$

where we choose the same functional form for the gap function which we found in the mean field calculation but let the amplitude of the function be a variational parameter.

Results for an 82 site lattice are shown in Figure 8. There were 8 holes (74 electrons) on the lattice. In these calculations we took $J = 1$ at defect bonds. There were 10 randomly distributed defect bonds on the 82 site lattice, corresponding to about 6% of the 164 bonds on the lattice. In the left hand side of the Figure we give the result when J=0.25 at the bonds which are not special (as in reference 15) while in the right-hand side of the Figure we show results when J=0.00 at bond which are not "special". In the calculations in the right hand Figure, we have made 3000 Monte Carlo moves per site as in reference 14 and have averaged over 10 realizations of the disorder at each value of δ. Though our estimates of the variance in the energy indicate that we have not proved the existence of a minimum, these data certainly suggest that the model has a BCS like instability, even in the presence of large Hubbard correlations.

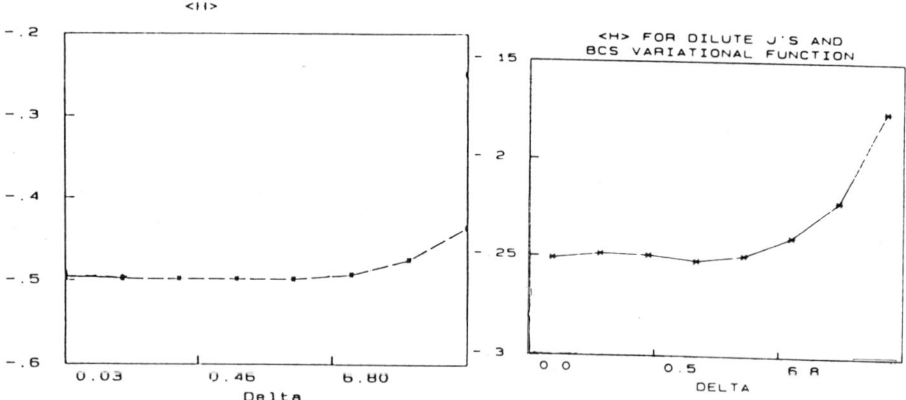

Figure 8. Variational results for the energy in the Variational Monte Carlo calculation.

DISCUSSION AND CONCLUSIONS

These calculations all indicate the dramatic effects which occur when one includes disorder explicitly in models of high temperature superconductivity. The model of high T_c discussed in the last section is highly speculative. Even within its own frame of reference, it will be necessary to take account of a model of the interaction of the oxygen holes with the copper spins in order to obtain a complete picture. Nevertheless, the techniques we have developed are useful in asking more conventional questions, such as the nature of the magnetic phases as discussed in the first section.

We should note some differences between the models used in the first (mean field) and second (variational) calculations described in the last section. In the mean field calculation we assumed that the entire band (physically regarded as an oxygen like band) was nearly full. This is plausible if one recalls that the U for the oxygen sites is not thought to be large. In the variational calculation, we use the Gutzwiller limit in which U is infinite. Then the Hubbard gap is infinite and we can only have a half-filled band at most. Thus the models are really different and we can only draw rather limited inferences about the first from the calculations on the second.

ACKNOWLEDGEMENTS

We wish to thank the Minnesota Supercomputer Institute and the Electric Power Research Institute for support.

REFERENCES

1. J. G. Bednorz and K. Muller, Z. Phys. B64, 189 (1986)

2. M. K. Wu et al, Phys. Rev. Lett. 58, 908 (1987)

3. J. W. Halley and H. B. Shore, Phys. Rev. B37, 525 (1988)

4. J. C. Phillips, Phys. Rev. Lett. 58, 1028 (1987)

5. A. M. Finkelshtein, JETP 57, 97 (1983);

   C. Castellani et al Phys. Rev. B 30, 527 (1984); Phys. Rev. Lett. 59, 323 (1987)

6. G. T. Zimanyi and E. Abrahams, Phys. Rev. Lett. 64, 2719 (1990)

7. H. J. Schulz, Phys. Rev. Lett. 64, 1445 (1990);

   see also D. Poilblanc and T. M. Rice (unpublished)

8. M. Jarrell, D. L. Cox and C. Jayprakash, Phys. Rev. B 40, 8899 (1989)

9. more details appear in C. Das Gupta and J. W. Halley (unpublished)

10. K. J. Von Szczepanski, T. M. Rice and F. C. Zhang, Europhysics Letters 8, 797 (1989)

11. see for example R. Kasowski. W. Y. Hsu and F. Herman, Phys. Rev. B 36, 7248 (1987)

12. J. W. Halley, S. Davis and X-F Wang, (unpublished)

13. J. W. Halley and H. B. Shore, Phys. Rev. B37, 525 (1987)

14. C. Gros, Phys. Rev. B. 38, 931 (1988))

15. J. W. Halley, S. Davis, P. Samsel and R. Joynt, Bull. Mat Sci. (in press) (1991)

A FEW REMARKS ON QUANTUM FLUIDS

Eugene P. Bashkin

*Institut für Theoretische Physyk, Universität zu Köln

D–5000 Köln 41, Germany

INTRODUCTION

Physics of quantum fluids has existed for about half a century. The history of
this branch of physics has known its own ups and downs,revolutions, civil wars and
relatively quiet periods of development. The methods of the theory of quantum fluids are widely used to describe properties of such different systems as liquid helium,
electrons in metals, neutron stars, heavy nuclei, superconductors etc. It would be
very hard, practically impossible, even to briefly outline the most important achievements and the most promising prospects in the whole field of quantum fluids. That
is why one had better restrict oneself to considering, say, some traditional areas in
the field in question like the physics of liquid ^{3}He and ^{4}He, ^{3}He–^{4}He dilute mixtures,
spin–polarized H↓ and D↓ etc. But even these particular topics have their own "fine
structure" and a great variety of different theoretical approaches and treatments are
pertinent to them. In order to make the problem of writing a short paper solvable one
should choose from the very beginning some particular subjects which are going to be
discussed in more detail.

There are two different fundamental approaches to theoretically study properties
of quantum fluids, which are the most successful. The first of them was basically
founded by L. Landau. It is based on the concept of elementary excitations in a
macroscopic system and any excited state of such a system can be described in terms
of quasiparticles. The quasiparticle approach is extremely fruitful in order to calculate
thermal properties , transport coefficients etc. at finite temperatures but it leaves a
problem of the ground state at $T = 0$ unsolved in the cases.

On the other hand there exists a large group of theoretical methods including
variational approaches, Monte–Carlo computations etc, whose starting point is an
ensemble of real particles (not excitations!) forming the system. This way of
thinking about quantum fluids was in very many respects founded by E. Feenberg.
These methods are very successful in calculating ground state properties though there
are no fundamental obstacles prohibiting one to follow the methods at $T \neq 0$ as well.
The recent progress in this field is, indeed, strongly connected with a wide application
of power computers.

Unfortunately the two approaches mentioned above have often been developed
separately from each other.Physicists belonging to one of these schools sometimes are
not aware of the recent achievements of the other one. There is no doubt that such

Recent Progress in Many-Body Theories, Vol. 3,
Edited by T.L. Ainsworth et al., Plenum Press, New York, 1992

a situation is not beneficial for either school is harmful for physics of quantum fluids as a whole. On the contrary a cooperation between them might be extremely fruitful and both the approaches could successfully complement each other.

In this paper the author intends to briefly outline the present "state of the art" and give some new challenges in the field of quantum fluids. The particular prospects which may require common effort from physicists working in different theoretical schools and therefore could help to erect some necessary bridges between them, are going to be considered here. The author should also confess that he himself used to belong to "quasiparticle" kind of mentality. That is why he would like a reader to consider the article only as a look from this theoretical school on the possibility at a "joint venture" with another stream in theoretical physics.

STATE OF THE "ART"

One has to admit that a number of publications on liquid ^{4}He is getting smaller and smaller.This fact, of course, does not exclude the possibility that good works in the field can appear.And fortunately it really happens. It is impossible even to mention all works which deserve to be mentioned. That is why we will try to emphasize only the most general trends which have exhibited themselves for the two last years.

One of such trends is demonstrated by a significant growth in the number of papers on superfluids in restricted geometry (clusters, films, porous media, free surfaces etc). An interesting fact is that this increase in publications occurs simultaneously in both theoretical "streams" discussed above. For instance Chin and Krotscheck carried out the computer variational calculations [1] for ^{4}He droplets consisting of $N = 40, 70, 112$ atoms. The Feynman trial wave function was used. The authors managed to obtain the spectrum of collective excitations and dynamic structure function of this finite ^{4}He system at $T = 0$. On the other hand Pitaevskii and Stringari analytically considered some properties of bigger clusters with $N = 10^3 - 10^5$ at finite temperatures $T = 0.3 - 0.8K$ using the Landau-like approach[2]. They calculated the normal fraction of the cluster as a function of T and N.It was shown that all statistical properties of a helium cluster are determined by surface but not bulk modes.The main features of the obtained results agreed with the path–integral Monte Carlo calculation in small clusters[3].Both kinds of calculation can perhaps be applied to the case of atomic clusters[4] in order to describe shell effects and the phenomenon of magic numbers.

A theory of inhomogeneous quantum fluids has also been developed considerably for the last two years (see, for instance, Ref. 5 and references therein). A remarkable statement concerning an existence of a Bose condensate even in a quasi–2D system in a random potential was made by Lee and Gunn[6]. However the Kosterlitz-Thouless phase transition has totally been ignored in all the papers on numerical calculations of ground state properties of thin ^{4}He films. Of course, traditional variational calculations certainly make sense for thick enough films when the critical temperature of the Kosterlitz–Thouless transition is expected to be small and some temperature range of the applicability of a common theory may exist. But all published results were related to thin (a few angstroms) films. On the other hand there is pretty good agreement between theoretical calculations and available experimental data even for very small thicknesses. Such an agreement probably is not just an accidental coincidence and might be connected with a rather small probability of the creation of virtual vortex pairs at $T = 0$. Nevertheless, one should say that the Kosterlitz-Thouless phase transition is actively being studied in experiments with ^{4}He films and a relevant microscopic description of the phenomenon in question is a very serious challenge to theorists. In fact no progress in this way has been achieved by now. A certain hope appears in conjunction with the shadow wave function method[7–10] which is going to be a rather effective way to microscopically describe vortices in bulk.

When discussing properties of quantum fluids in restricted geometry one should also recall the works on free surfaces of liquid helium. For instance the calculations[11] within a density functional approach with a finite range interaction provide pretty good agreement with experimental data on the surface tension of liquid ^{4}He. The spectrum of excitations in a vapour– liquid ^{4}He interface within the temperature range $T = 0 - 2K$ was calculated in Ref. 12.

We also have to specify some recent papers containing new theoretical ideas which caused a great deal of discussions. Glyde and Griffin[13,14] gave an alternative interpretation of the well–known phonon–maxon–roton quasiparticle energy spectrum. Their conclusions were based upon the analysis of recent experimental data on neutron scattering in superfluid ^{4}He. They argue that the phonon peak in neutron scattering data should be ascribed to sort of a collective zero sound mode (which was first pointed out by D. Pines) whereas the sharp maxon–roton peak is pertinent to a single–particle excitation. Such an interpretation actually doubts the validity of the Landau description of superfluid ^{4}He which has been reigning in condensed matter physics for 50 years and denies the existence of single–particle excitations in this part of the energy spectrum. (Of course, in general, single–particle excitations may exist, for example excited ^{4}He atoms can form so called neutral excitations but the energy of such quasiparticles contains a huge gap of the order of 10 eV).

To an extent an alternative point of view to microscopically describe liquid ^{3}He was suggested by Bouchaud and Lhuillier[15]. They propose to consider liquid ^{3}He as an ensemble of resonating pairs in the strong coupling regime. A striking consequence of this assumption is the conclusion that the Migdal jump in the particle momentum distribution at $k = k_F$ has to disappear. A consideration of strongly correlated pairs only is perhaps disputable (the experimental data in solid phase, say, the "up–up–down–down" magnetic structure, favors four–particle exchange interaction) but it reveals a way of theoretically treating the anomalies in a behavior of liquid ^{3}He.

A very exciting statement on the "shape of fluid" was made by Campbell and Clements[16]. They showed a way to calculate bond orientational order parameter fluctuations in fluids. What they did is actually related to classical liquids but the method can be applied to supercooled fluids where quantum mechanical phenomena may come into effect. The inelastic light scattering from liquid ^{4}He was considered by Halley and Korth[17]. They proved that quasiparticle anharmonicities were more important than the nonlinearity of the density–quasiparticle relation. It may provide us with a new view of the well–known experiments on light scattering in liquid ^{4}He carried out by Greytak et al. The authors of the Ref. 18 used the experimental data on the chemical potential of ^{4}He and obtained the upper– and lower–bounds for the potential and kinetic energy (per particle) of liquid ^{4}He. The intriguing thing is that it was shown that the use of the Lennard–Jones potential to describe the interaction between ^{4}He atoms violated the upper–bound condition. Should one think that the Lennard–Jones potential which has traditionally and so frequently been used in calculations, has nothing to do with the real interaction between two helium atoms?

TRANSPORT PROPERTIES

We have already seen that there exists an extremely wide variety of different variational approaches to calculate ground state properties of liquid helium. But nobody has ever tried even to formulate the problem of how to calculate transport coefficients such as viscosity, thermal conductivity, spin diffusion etc in the framework of numerical microscopic methods. On the other hand, transport properties are one of the most traditional and typical problems in physics of quasiparticles. The way to resolve such problems within "quasiparticle" approach is very well established and understood. In order to do this one has to linearize the Boltzmann transport equation (with some

collision term) for elementary excitations with respect to small gradients of thermodynamic variables and then to calculate the corresponding macroscopic currents. But
in the case of a dense fluid the same procedure for real particles obviously cannot be
done.

Of course there exists a direct way of numerically calculating transport coefficients
by means of corresponding dynamic correlation functions. For example, in order to
obtain thermal conductivity one should find the correlator $< \delta \vec{Q} \delta \vec{Q} >_{\vec{q},\omega}$, where $\vec{Q}$ is
a heat flux. As soon as this correlator has been obtained one can easily reconstruct
thermal conductivity as a function of the frequency ω in general and in the static
limit $\omega = 0$ in particular. Nobody has followed this way up to the present. First,
calculations of dynamic structure functions are not simple at all and normally do not
provide enough accuracy, particularly for the small momentum transfer. But it is the
form–factor for small momentum transfer that substantially contributes to transport
coefficients. Second, calculating transport properties may require computations at
finite temperatures (except probably thermal conductivity which seemingly can be
obtained from the dynamic correlation function for heat flux fluctuations at $T = 0$). It
certainly does not make life easier. Nevertheless the situation does not look completely
hopeless. A great number of calculations of the dynamic structure function $S(\vec{q},\omega)$
for density fluctuations at zero temperature has been done[19−25] and no fundamental
obstacles are seen to prevent extending these methods to the case of $T > 0$ and
carrying out similar computations for the relevant correlations functions.

Calculating static properties of quantum fluids is much simpler and "numerical"
physicists are much more accustomed with this kind of work. But can one infer transport coefficients from static properties of quantum fluids? At first sight the answer to
this question is negative. However it would be a wrong conclusion. Of course at zero
temperature, when the mean free path of elementary excitations in quantum fluids
goes to infinity, one cannot get any information concerning the interaction between
quasiparticles and hence no information about transport properties can be obtained.
But the situation drastically changes when increasing the temperature. At finite temperatures the mean free path of quasiparticles is also finite and static properties of
a fluid exhibit in some way the interaction between excitations. This means that in
principle one can extract the knowledge on transport coefficients from static characteristics of a quantum fluid, say, from the static structure function $S(\vec{q})$ at finite
temperatures. To calculate $S(\vec{q})$ even at $T > 0$ is relatively easy compared to the
dynamic structure function. In order to illustrate how transport coefficients can be
inferred from the static structure function at $T > 0$ let us consider first the case of a
strongly degenerate Fermi liquid.

The static structure function depends strongly on which regime, hydrodynamic
or collisionless, occurs in a fluid. In order to make a judgement about the regime
one has to compare the transferred momentum q with the reciprocal mean free path
of elementary excitations l^{-1}. In the hydrodynamic limit, $ql << 1$, all correlations
functions are substantially local. It yields a constant value for $S(\vec{q})$ when $q \to 0$ which
is determined by compressibility, and gives rise to the well–known expression:

$$S(0) = \frac{NT}{m} \frac{1}{c_T^2} \tag{1}$$

where N is the density, m the mass of a particle and c_T the isothermic velocity of
sound.

For the high momentum transfer, $ql >> 1$, the collisionless regime takes place.
In this case all correlation properties of the system can be found from the collisionless
(no collision term) Boltzmann transport equation of the Fermi liquid theory. Using
the collisionless transport equation one can easily convince oneself that the dynamic

structure factor $S(\vec{q}, \omega)$ is an uniform function of the ratio ω/q:

$$S(\vec{q}, \omega) = f\left(\frac{\omega}{q}\right) \tag{2}$$

and hence the static structure function has the form:

$$S(\vec{q}) = A k_F^2 q \tag{3}$$

Here $k_F = (3\pi^2)^{1/3} N^{1/3}$ is the Fermi momentum and $A = A(F_n)$ is some constant which depends on Fermi liquid harmonics. Let the quantity q_0 be defined as l^{-1}. One regime is obviously replaced by another at $q = q_0$. If the Fermi liquid coefficients do not possess any strong anomalies or singularities one can easily verify that

$$\frac{S(q_0)}{S(0)} \approx const\left(\frac{T}{\epsilon_F}\right) N^{2/3} \sigma << 1 \tag{4}$$

where ϵ_F is the Fermi degeneracy temperature and σ is the total cross–section of the order of the atomic size. It provides us with a very important conclusion that at finite temperatures there has to be a minimum on the curve $S(q)$ at $q \approx q_0$. The appearance of this minimum is caused by the finite mean free path and the position and shape of the hollow on the curve are given by the value of l. Therefore we get an opportunity to obtain information about transport properties when studying the minimum of the static structure function at $T \neq 0$.

What might be the recipe in order to find transport coefficients from the $S(q)$–curve? One could extract, say, viscosity, from the low –momentum part of $S(q)$ in the vicinity of $S(0)$. The exact calculations would imply taking into account the deviation of spatial correlation functions from the $\delta(\vec{r}_1 - \vec{r}_2)$–dependence (i.e. explicitly considering the exponential correlators in a fluid). Of course some extra parameters like the size of a particle or cross–section will enter the final formula when following this way. On the other hand an approximate but much easier procedure may significantly facilitate calculating viscosity from $S(q)$ at finite temperatures. The idea is to use some approximate collision integral in the Boltzmann transport equation which is supposed to meet all the conservation laws and to reproduce all the hydrodynamic and zero sound results when being extrapolated to the low– and high–momentum transfer limits. One of the simplest collision terms of such kind may be chosen in the form:

$$I(\delta n) = \frac{2N\epsilon_F}{5\eta}(\delta n - \overline{\delta n} - 3\overline{\delta n cos\theta}cos\theta) \tag{5}$$

where n is the distribution function, η is the coefficient of viscosity and the overline means averaging over the angles. A similar collision term was used by Abrikosov and Khalatnikov[26] to calculate the sound absorption in liquid ^{3}He and led to a quite satisfactory agreement. Once a collision term in the form of Eq. (5) is available one can calculate a shape of the hollow on the curve $S(q)$. Then, just comparing the numerical data with analytical expressions one can infer the coefficient of viscosity. The easiest way however is not to calculate the shape of the hollow but to directly obtain a position of the minimum from the criterion $\partial S/\partial q = 0$. Unfortunately no numerical calculations of the structure function in ^{3}He fluids at $T \neq 0$ have been done by now.

We have just discussed the possibility to use numerical data on the structure function in order to get transport characteristics of Fermi fluids. But indeed all the conclusions are also valid in the case of Bose quantum fluids. For example liquid ^{4}He at $T < 0.6K$ where all thermodynamic and transport properties are determined by the ensemble of phonons only, undoubtedly should possess an analogous minimum on the curve $S(q)$. In fact in the low–momentum limit $ql << 1$ the quantity $S(0)$ is given by the same hydrodynamical formula (1) as in the case of a Fermi fluid. In the opposite limiting case of the high–momentum transfer $ql >> 1$ where the collisionless regime occurs, the static structure factor takes the form:

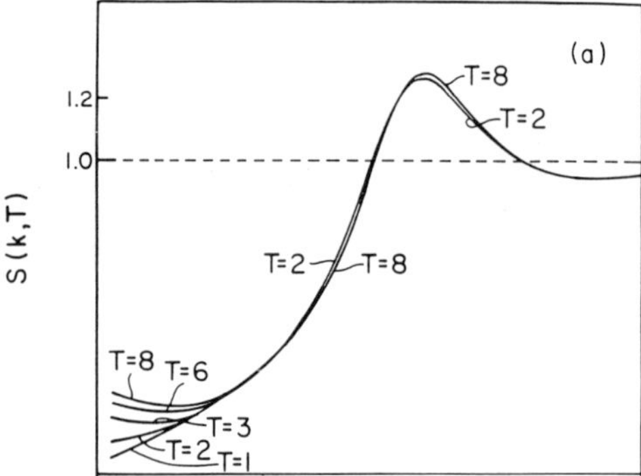

Fig. 1 Calculated $S(k,T)$ versus k in Å^{-1} for several temperatures.[28]

$$S(q) = N\frac{\hbar q}{2mc_0}\coth\frac{\hbar q c_0}{2T} \approx N\frac{\hbar}{2mc_0}q \tag{6}$$

The interaction between phonons in superfluid ^{4}He has been understood quite well[26]. Therefore it is not difficult to figure out that

$$\frac{S(q_0)}{S(0)} = \frac{\hbar c_0}{T}N_{ph}\sigma_{ph-ph} << 1 \tag{7}$$

where N_{ph} is the number of phonons and σ_{ph-ph} is the cross–section for phonon–phonon scattering. So, again we have arrived at the conclusion that there should be a local minimum in the temperature –dependent static structure function at $q \approx l^{-1}$. Of course, strictly speaking this statement has been proved only within the phonon temperature range. But there are some arguments that such a minimum remains even at much higher temperatures. There have been quite a few papers published on the static form factor of liquid ^{4}He at finite temperatures[27,28]. Despite the fact that the accuracy of calculations in Ref. 28 was not very high the authors managed to obtain

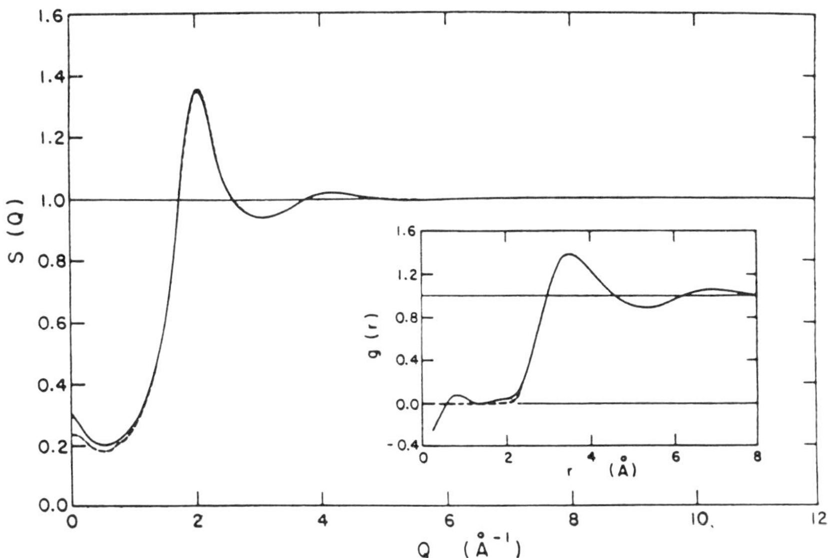

Fig. 2 Experimental $S(Q)$ versus Q at $T = 3.6\,K$.[29]

a rather shallow minimum on the $S(q)$–curve even at $T = 3 - 6K$ (see Fig. 1). It must be also emphasized that the minimum was detected in experiments on neutron scattering in liquid ^{4}He at $T = 3.6K$ as well[29] (see Fig. 2). However the question of how to analytically infer transport coefficients of liquid ^{4}He from the data on $S(q)$ at high enough temperatures is still unclear and not at all simple.

DIMERIZATION AND NEW PHASES IN ^{3}He–^{4}He DILUTE MIXTURES

^{3}He–^{4}He dilute solutions are usually described in terms of a Fermi fluid of impurity ^{3}He excitations dissolved in the superfluid background of ^{4}He. But under the certain conditions ^{3}He quasiparticles can also form a Bose system and the typical features inherent to Bose quantum fluids will be exhibited. It is a well established fact that two ^{3}He quasiparticles in superfluid ^{4}He experience an effective attraction in the s–wave scattering channel. But the magnitude of such an attraction is too small to create a bound state of two ^{3}He atoms in the bulk. Of course even a weak attraction should result in the Cooper pairing in the ^{3}He–subsystem at sufficiently low temperatures and give rise to the superfluid phase transition. In some sense the creation of such a superfluid phase of ^{3}He can be interpreted as a result of Bose–Einstein condensation in the ensemble of the Cooper pairs (though the excitations in superfluid Fermi fluid obey the Fermi–Dirac statistics). The problem of ^{3}He superfluidity in quantum solutions of ^{3}He–^{4}He has actively been studied for a very long time. There is a very large scatter in theoretical estimates for the transition temperature[30] but all experimental attempts have been unsuccessful. We will not consider this topic in any detail in this paper.

Thus the interaction between two isolated ^{3}He atoms dissolved in superfluid ^{4}He cannot lead to a bound state in the bulk. However the situation totally changes if one

199

puts ^{3}He particles on the surface of superfluid ^{4}He (the Andreev states[31]), places them between two parallel walls within liquid ^{4}He or considers properties of thin enough ^{3}He–^{4}He films. In all these or similar cases it provides us with sort of a 2D systems of ^{3}He impurity excitations[32]. We know that in the 2D case any attraction between particles leads to a bound state. In the case of two identical fermions with spin 1/2 as ^{3}He quasiparticles are, it is easy to convince ourselves that an effective attraction causes the appearance of spinless dimers $(^3$He$)_2$ (a bound state of two particles with the opposite spins)[33]. If the temperature of a solution is lower than the binding energy the $(^3$He$)_2$ dimers will significantly influence all thermodynamic properties of a system.

The magnitude of the binding energy Δ can be calculated as:

$$\Delta = \frac{2\hbar^2}{Mr_0^2} exp\left[-\frac{2\hbar^2}{M}\left|\int_0^\infty U_{eff}(\rho)\rho d\rho\right|^{-1}\right], \tag{8}$$

where the effective potential U_{eff} may be expressed in terms of the averaged interaction potential $V(\vec{r})$ for two ^{3}He quasiparticles:

$$U_{eff} = \int V(|\vec{r_1} - \vec{r_2}|)|\psi_1(z_1)|^2|\psi_2(z_2)|^2 dz_1 dz_2 \tag{9}$$

Here M is the effective mass of a bare ^{3}He excitation, r_0 the interaction range,and $\psi_{1,2}(z_{1,2})$ are the wave functions of two ^{3}He quasiparticles, which correspond to the motion in the direction perpendicular to the surface. ^{3}He impurity along z–axis is bigger than the interaction range r_0, the expression (8) obviously reduces to

$$\Delta \approx \frac{2\hbar^2}{Ma^2} exp\left(-\frac{L}{|a|}\right) ,$$
$$L^{-1} = \int |\psi_1(z)|^2 |\psi_2(z)|^2 dz , \tag{10}$$

where a is the s–wave scattering length. For example in the case of dilute ^{3}He–^{4}He mixture which is contained between two parallel walls with the separation d and at temperatures $T << \hbar^2/Md^2$ takes the form:

$$\Delta \approx \frac{2\hbar^2}{Ma^2}e^{-2d/3|a|} , \tag{11}$$

A very rough estimate of Δ in the case of surface ^{3}He states yields $\Delta \sim 10^{-3} - 10^{-2}K$. Of course the precision of this estimate is very low because of the exponential dependence of the binding energy on L and a. However it provides some reasons to be optimistic.

When lowering the temperature $T << \Delta$ we obtain a strongly dimerized 2D fluid of ^{3}He. And it gives us a great variety of new features and phenomena pertinent to a Bose quantum fluid. If the phase diagram of dilute $(^3$He$)_2$–^{4}He mixture looks similar to that of the case of ^{3}He–^{4}He solution in the bulk, i.e.there exists a finite solubility of $(^3$He$)_2$ in superfluid ^{4}He at $T = 0$, one can get a rarefied 2D Bose gas of $(^3$He$)_2$–dimers. Of course the concentration of dimers should be small since as the spatial size of a dimer is exponentially large. When increasing the concentration of dimers one can encounter a gas–liquid phase transition in the system of $(^3$He$)_2$. The corresponding critical temperature can be expressed through the chemical potential of a liquid phase[33],

$$T_{G-L} = \frac{\pi\hbar^2}{2M\lambda}N_0 \; ,$$
$$e^{-\lambda x} + e^{-\lambda} = 1 \; , \qquad\qquad (12)$$
$$x = \frac{2(\mu_L - 2\epsilon_0 + \Delta)M}{\pi\hbar^2 N_0} \; ,$$

where N_0 is the initial areal density of ^{3}He atoms on the surface (we are talking now about the impurity states on the free surface of liquid ^{4}He) and ϵ_0 is the binding energy (at zero momentum) of a ^{3}He-quasiparticle in the bulk. It may of course happen that there is no finite solubility in $(^3\text{He})_2$–^{4}He mixtures at zero temperature. In this case a dense dimerized liquid might exist even at $T = 0$. A 2D Bose fluid cannot experience the Bose–Einstein condensation, but the Kosterlitz–Thouless phase transition to a superfluid state may occur at the temperature:

$$T_C = \frac{\pi\hbar^2}{8M}N_s \; , \qquad\qquad (13)$$

where N_s is the superfluid areal density at $T = T_C$. The value of N_s is not calculated in the framework of the theory. In order to get an estimate let us put N_s to be of the common order of magnitude, $N_s \approx 10^{13}cm^{-2}$. Then we obtain $T_C \approx 1 - 5mK$. Thus one of the natural consequences of the theory, which might be a good subject for an experimental study, is the prediction that quasi–2D systems of ^{3}He–^{4}He dilute mixtures (in restricted geometry) could exhibit two Kosterlitz–Thouless phase transitions (just because now two 2D Bose fluids, ^{4}He and $(^3\text{He})_2$ might be available).

In the experiments[34,35] two superfluid Kosterlitz– Thouless phase transitions in thin ^{3}He–^{4}He films were observed. It sounds very attractive to identify the second transition as the one pertaining to the $(^3\text{He})_2$ system. However the temperature in the experiments seems to be too high. On the other hand Eqs. (10)–(11) for Δ being extrapolated to a thickness d (or L) of the order of 1–2 atomic layers would yield the right order of magnitude. The recent NMR–experiments on the magnetization in thin ^{3}He–^{4}He mixture films[36] could be also interpreted in terms of possible dimerization or BCS–like superfluidity (s–wave or p–wave pairing) in the 2D system of ^{3}He.

The question of what the dimerized ^{3}He fluid looks like depends strongly on the microscopic structure of the liquid. It may be that the dimerized fluid is actually a gas of dimers or a liquid composed of $(^3\text{He})_2$ dimers so that the distance between two neighboring dimers is larger than (or at least comparable to) the radius of a single dimer. In this case one can expect that the coupling between ^{3}He–quasiparticles within a dimer is much stronger than dimer–dimer correlations in the liquid. In other words we deal with a 2D Bose quantum fluid which can reveal the Kosterlitz–Thouless superfluid phase at the temperature given by Eq. (13).

If the density of a dimerized fluid happens to be high enough and the average distance between ^{3}He particles is much less than the radius of a single dimer, we will probably get a 2D Fermi fluid with strong pair correlations at large distances and a pairing of the Cooper type. Indeed the BCS–theory cannot directly be applied to 2D Fermi systems because of critical fluctuations. However it is well known that switching on even a weak interaction with a 3D reservoir (say, an interaction between different 2D planes) makes a 2D Fermi system more stable. In the case of ^{3}He impurities on the surface of ^{4}He the role of a 3D "reservoir" might be played by ^{3}He quasiparticles in the bulk. Thus one could expect that the BCS–like formula for the s–wave pairing,

$$T_{C1} \approx \frac{\pi\hbar^2}{M}N_0 \, exp\!\left(-\frac{L}{4|a|}\right) \; , \qquad\qquad (14)$$

gives the right order of magnitude of a transition temperature in a 2D dimerized (with the pairing) Fermi fluid. It should be noticed that T_{C1} from Eq. (14) is much higher than the binding energy Δ from Eq. (10). It means that if one decreases the temperature of a solution with high concentration a superfluid quasi–2D Fermi liquid of the BCS–like type rather than the a Bose fluid of dimers $(^4He)_2$ will be obtained.

So, one can easily see that there is a vast and interesting field for for both theoretical and experimental research. The analytic theory in terms of 2D ^{3}He excitations can be relatively easily constructed in the case of sufficiently thick films, $L >> r_0$. But from the experimental point of view thin films (a few atomic layers) seem to be the most interesting case. The theory suggested above apparently does not hold in this case and has been just extrapolated to such thin films. That is why it looks to be very important to carry out microscopic numerical calculations in order to study the phenomenon of dimerization in the case of small thicknesses. The main question which should be answered first is: What is the binding energy of a dimer and what is the phase diagram of a quasi–2D ^{3}He–^{4}He system at low enough temperatures (in particular at zero temperature) where the dimerization is expected to play the most significant role? The very detailed calculations of the ground state properties and the equation of state for thin ^{3}He–^{4}He films have recently been done in Refs. 37, 38. These results certainly hold at $T >> \Delta$ for small concentrations and at $T >> T_{C1}$ for higher atomic densities of ^{3}He. It is very tempting to use these methods and to generalize them in order to numerically find out the new phases connected with the phenomena of dimerization.

The dimerization of ^{3}He can also happen in quasi–1D systems such as ^{3}He impurities localized on linear vortices in superfluid ^{4}He or ^{3}He–^{4}He dilute solutions in narrow capillaries (porous media)[33]. An interaction between vortex lines or different channels (say, via phonons) can give rise to a superfluid transition in such systems.

SPIN–POLARIZED QUANTUM FLUIDS

Studying spin–polarized systems is currently one of the most fashionable branches of quantum fluids. Such fast progress follows the recent both theoretical and experimental achievements[39].The most fascinating results obtained for the last decade are connected with the study of transverse spin fluctuations in spin–polarized atomic H↓, gaseous ^{3}He↑, liquid ^{3}He↑ and ^{3}He↑–^{4}He dilute mixtures. A remarkable fact is that under the certain conditions fluctuations of transverse magnetization can propagate through a spin–polarized fluid as weakly dampened collective spin waves. Such oscillations can exist in any paramagnetic fluid independently whether it is a Bose or Fermi system and propagate even in a rarefied gas at high temperatures when particles of a gas obey the classical Boltzmann–Maxwell statistics. The basic theoretical statements concerning spin waves in spin–polarized quantum fluids were formulated in Refs. 40 – 44 and the experimental evidence was obtained in Refs. 45 – 51. In this paper we will restrict ourselves to considering transverse spin modes in a dense spin–polarized Fermi liquid which seems to be the most interesting subject for variational microscopic calculations.

The Boltzmann transport equation of the Landau theory of Fermi liquids allows us to obtain the energy spectrum of magnons[40−42] which takes the form:

$$\omega_q = \Omega_H + bq^2 \ , \ b = \frac{\hbar}{2m^*\alpha}B(Z) \ ,$$

$$B(Z) = \frac{(1 + Z_0)(1 + \frac{Z_1}{3})}{Z_0 - \frac{Z_1}{3}} \tag{15}$$

Here $\Omega_H = 2\beta H/\hbar$ is the Larmor precession frequency, β is the magnetic moment of a ^{3}He atom, m^* is the effective mass of a Fermi excitation, $\alpha = (N_\uparrow - N_\downarrow)/N$

is the degree of polarization, and Z_0 and Z_1 are just the two first harmonics of the spin–dependent part of the Landau interaction function. The spectrum (15) actually corresponds to the Goldstone mode in a system with broken symmetry (the symmetry of ferromagnetic type). Collective modes of this kind is a good example of spin waves in disordered media. The dispersion law in the form (15) holds if the wave number is not too big, $|\Omega_{int}| >> qv_F >> |\Omega_H - \omega_q|$, where v_F is the Fermi velocity and Ω_{int} is defined by the following relationship:

$$\Omega_{int} = \frac{4}{3}\epsilon_F Z_0 \frac{\alpha}{\hbar}. \tag{16}$$

At finite temperatures weakly dampened spin oscillations can propagate provided the degree of polarization is not too small:

$$\frac{\alpha D_0 m^*}{\hbar B(Z)} >> 1\,, \tag{17}$$

where D_0 is the spin diffusion coefficient which is proportional to T^{-2}.

It must be pointed out that all of the above statements of Fermi liquid theory regarding the magnon spectrum are actually not valid in the case of a highly polarized system, $\alpha \sim 1$. There are at least two reasons why Fermi liquid theory fails to describe spin waves. First, all quantities which the Fermi liquid theory is based on, are well defined only on the Fermi surface. In the case of spin– polarized fluid there exist two separated Fermi surfaces for particles with spin up and down. All thermodynamic and transport properties can be expressed in terms of the interaction function $f_{\sigma,\sigma'}(\vec{p}, \vec{p}')$ defined on both Fermi surfaces, i.e. $p, p' = p_{F\uparrow}, p_{F\downarrow}$. But in order to calculate the dispersion law of spin oscillations with the help of the Boltzmann transport equation one would need to make an integration over an entire Fermi sea, i.e. in the depths of the bigger Fermi sphere in obvious contradiction to the basic postulates of the theory. At low degrees of polarization $\alpha << 1$ the separation between two Fermi spheres is negligible and the calculation of spectrum reduces to the integration over the single Fermi surface.

Another reason to expect a failure of the Fermi liquid description of spin waves is that the theory implies a local interaction between quasiparticles (the Landau interaction function is "local" and does not depend on a distance). On the other hand there always exists the "nonlocal" term $D(\nabla \vec{M})^2$ in the total free energy which yields some q^2–corrections to the magnon spectrum (15) too. Similar terms play a very important role in spin dynamics of solid ^{3}He. Inasmuch as the difference between densities of the solid and liquid phases is not so large, one can expect the "nonlocal" terms to exhibit themselves in the case of liquid ^{3}He$\uparrow$ as well. However, these corrections to the spectrum turn out to be proportional to α and for small polarization, $\alpha << 1$, can be ignored.

Up to the present all the results concerning transverse spin excitations in polarized Fermi fluids have been obtained in the framework of the Landau theory of Fermi liquids (either the phenomenological or microscopic description). No numerical many-body calculations in the field have been done. Variational calculations of the excitation spectra in a Fermi liquid is not a simple job because no explicit relationship between the static structure function and dispersion law, like the Feynman–Bijl formula in the case of a quantum Bose fluid, exists. However, when dealing with a spin–polarized Fermi liquid an unique opportunity to obtain the energy spectrum of transverse magnetic excitations from the static correlation properties appears.

A conventional many-body approach may be formulated in terms of the polarization density matrix $n_{\alpha\beta}(\vec{p})$. Let the quantity $n_{\alpha\beta}$ be defined as

$$n_{\alpha\beta} = <a_\beta^+ a_\alpha> \tag{18}$$

where a_β^+ and a_α are the creation and annihilation operators respectively, and the Greek indices numerate spin states. In equilibrium the polarization density matrix is indeed diagonal:

$$n_{\alpha\beta}^{(0)} = \frac{n^+ + n^-}{2}\delta_{\alpha\beta} + \frac{n^+ - n^-}{2}\vec{\sigma}_{\alpha\beta}\frac{\vec{M}}{M} \tag{19}$$

where n^+ and n^- are the occupation numbers for particles with spin up and down respectively, and $\vec{M}$ is the macroscopic magnetization. Considering fluctuations of the magnetic moment means taking into account fluctuations of the spin–dependent part of the density matrix:

$$n_{\alpha\beta}(\vec{p}) = n_{\alpha\beta}^{(0)}(\vec{p}) + \vec{\lambda}_{\vec{p}}(\vec{r},t)\vec{\sigma}_{\alpha\beta} \tag{20}$$

Fluctuations of the transverse magnetization are determined by the off–diagonal components of $n_{\alpha\beta}$, i.e. by the $x-$ and $y-$components of the vector λ. The fluctuating macroscopic magnetic moment can be easily calculated by means of the density matrix:

$$\vec{M} = \int \beta\vec{\sigma}_{\alpha\beta}n_{\beta\alpha}\frac{d^3p}{(2\pi\hbar)^3} \tag{21}$$

Let us define the correlation function for fluctuations of magnetization in the usual manner:

$$< \delta M_i(\vec{r}_1,t_1 \delta M_k(\vec{r}_2,t_2) >= S_{ik}(\vec{r},t) \ , \vec{r} = \vec{r}_1 - \vec{r}_2 \ , t = t_1 - t_2 \tag{22}$$

The classic fluctuation theory being applied to calculate the static form factor for transverse spin fluctuations, yields[52]:

$$S_{ik}(\vec{q}) = \delta_{ik}\beta^2 N\alpha \coth\frac{\hbar\omega_q}{2T} \ , i \ , \ k = x \ , \ y \tag{23}$$

Thus we have obtained a remarkable equation relating the static structure function for transverse magnetization to the excitation energy spectrum. If the temperature is not too small, i.e. $T >> \hbar\omega_q$ for all permitted values of q, which actually reduces to the criterion

$$\epsilon_F >> T >> \beta H \ , \tag{24}$$

the expression (23) becomes simpler

$$\hbar\omega_q = 2\beta^2 N\alpha\frac{T}{S_{xx}(\vec{q})} \tag{25}$$

and one can infer the magnon energy with much better accuracy. If one considers a dynamically polarized Fermi liquid in the absence of an external magnetic field, there will be no Larmor gap in the spectrum (15) and the criterion (24) should be replaced by the condition:

$$1 >> \frac{T}{\epsilon_F} >> \alpha \tag{26}$$

An important point is that in contrast to Eq. (15) for the magnon energy $\hbar\omega_q$ which in fact holds only in the limit $\alpha << 1$, the expressions (23) and (25) relating $\hbar\omega_q$ to $S_{xx}(\vec{q})$ are valid for any degree of polarization.

Thus we can see that numerically calculating the static correlation function for transverse magnetization $S_{xx}(\vec{q}) = S_{yy}(\vec{q})$ would provide us with an unique opportunity to obtain the spectrum of spin wave excitations even in the case $\alpha \sim 1$ where the Fermi liquid theory fails. At low degrees of polarization such a calculation would give an excellent tool to verify the Fermi liquid approach and to find the Fermi liquid parameters Z_1 and Z_2 (taking also into account the data on the static magnetic susceptibility).

ACKNOWLEDGEMENTS

I wish to thank C. Campbell, J. Clark and M. Ristig for stimulating and useful discussions when preparing this manuscript. It was my pleasure to stay for 3 weeks at Texas A & M University where the paper was actually completed. I am grateful to E. Krotscheck, T. Ainsworth and M. Saarela for a very nice scientific cooperation and the warm hospitality extended to me in College Station.

The work was supported in part by the Deutsche Forschungsgemeinschaft under Grant No. Ri 267/14-1.

* The permanent address: Kapitza Institute for Physical Problems, 117334 Moscow, U.S.S.R.

REFERENCES

1. S. A. Chin and E. Krotscheck, Physica B, **165–166**, 531 (1990).
2. L. Pitaevskii and S. Stringari, Physica B, **165–166**, 489 (1990).
3. Ph. Sindzingre, M. L. Klein and D. M. Ceperley, Phys. Rev. Lett. **63**, 1601 (1989).
4. S. Anagnostatos, Condensed Matter Theories, ed. by A. Proto et al, Plenum Press, **6**, (1991), to be published.
5. E. Krotscheck, Recent Progress in Many-Body Theories, ed. by Y. Avishai, Plenum Press, **2**, 183, (1989).
6. D. K. K. Lee and J. M. F. Gunn, Physica B, **165–166**, 509 (1990).
7. M. H. Kalos, Recent Progress in Many-Body Theories, this volume.
8. S. Vitiello, Ibidem.
9. L. Reatto, Ibidem.
10. M. Bernasconi, A. Ferrante, X. Q. G. Wang and S. Fantoni, Ibidem.
11. J. Dupont–Roc, M. Himbert, N. Pavloff and J. Treiner, Physica B, **165–166**, 515 (1990).
12. K. A. Gernoth and M. L. Ristig, Recent Progress in Many-Body Theories, this volume.
13. H. R. Glyde, Excitations in Two–Dimensional and Three–Dimensional Quantum Fluids. Ed. by A. F. G. Wyatt and H. J. Lauter, Plenum Press, 1, (1991).
14. A. Griffin, Ibidem, 15 (1991).
15. J. P. Bouchaud, Recent Progress in Many-Body Theories, ed. Y. Avishai, Plenum Press, **2**, 331 (1990).
16. C. E. Campbell and B. E. Clements, Condensed Matter Theories, ed. by S. Fantoni and S. Rosati, Plenum Press, **6**, 79 (1991).
17. M. S. Korth and J. W. Halley, Recent Progress in Many-Body Theories, this volume.
18. J. Boronat, A. Fabrocini and A. Polls, Phys. Rev. B, **39**, 2700 (1989).
19. Momentum Distribution, Ed. by R. N. Silver and P. E. Sokol, Plenum Press (1989).
20. E. Manousakis and V. R. Pandharipande, Phys. Rev. B, **30**, 5062 (1984).
21. J. W. Clark and R. N. Silver, Proceedings of the Vth International Conference on Nuclear Reaction Mechanisms, Varenna, Italy (1988).
22. P. Whitlock and R. M. Panoff, Can. J.Phys. **B65**, 1409 (1987).
23. O. Benhar, A. Fabrocini and S. Fantoni, In "Momentum Distribution", eds. R. N. Silver and P. E. Sokol, Plenum Press (1989).
24. B. E. Clements, E. Krotscheck, C. J. Tymczak and C. E. Campbell, Recent Progress in Many-Body Theories, this volume.
25. A. D. Jackson, A. Lande and R. A. Smith, Phys. Rept. **86**, 55 (1982).

26. I. M. Khalatnikov, An Introduction to the Theory of Superfluidity, Benjamin, New York (1965).
27. L. Reatto, Phys. Lett. **66A**, 484 (1978).
28. G. Senger, M. L. Ristig, K. E. Kü rten and C. E. Campbell, Phys. Rev. B, **33**, 7562 (1986).
29. E. C. Svensson,V. F. Sears, A. D. B. Woods and P. Martel, Phys. Rev. B, **21**, 3638 (1980).
30. E. Østgaard and E. P. Bashkin, Physica B, (1991), to be published.
31. A. F. Andreev, Sov. Phys. JETP, **23**, 939 (1966).
32. D. O. Edwards and W.F. Saam, Prog. Low Temp. Phys. **7A**, 285 (1978).
33. E. P. Bashkin, Sov. Phys. JETP, **51(1)**, 181 (1980).
34. D. J. Bishop and J. D. Reppy, Phys. Rev. **B22**, 5171 (1980).
35. X. Wang and F. M. Gasparini, Phys. Rev., **B34**, 4916 (1986); **38**, 11245 (1988).
36. R. H. Higley, D. T. Sprague and R. B. Hallock, Phys. Rev. Lett., **63**, 2570 (1989).
37. E. Krotscheck, M. Saarela and J. L. Epstein, Phys. Rev. Lett., **61**, 1728 (1988); **64**, 427 (1990).
38. R. H. Anderson and M. D. Miller, Recent Progress in Many-Body Theories, this volume.
39. Spin–Polarized Quantum Systems, eds. S. Stringari and I. S. I., World Scientific, (1988).
40. V. P. Silin, Sov. Phys. JETP, **8**, 870 (1959).
41. A. J. Leggett, J. Phys. C, **3**, 448 (1970).
42. E. P. Bashkin, Sov. Phys. JETP Lett. **33(1)**, 8 (1981); Phys. Lett., **101A**, 164 (1984); Sov. Phys. USPEKHI, **29(3)**, 238 (1986).
43. C. Lhuillier and F. Laloe, J. Phys. (Paris), **43**, 197, 225, 833 (1982).
44. L. P. Levy and A. R. Ruckenstein, Phys. Rev. Lett., **52**, 1512 (1984); **53**, 302 (1984).
45. B. R. Johnson, J. B. Denker, N. Bigelow, L. P. Levy, J. H. Freed and D. M. Lee, Phys. Rev. Lett., **52**, 1508 (1984); **53**, 302 (1984).
46. P. J. Nacher, G. Tastevin, M. Leduc, S. B. Crampton and F. Laloe., J. Phys. Lett. (Paris), **45**, L–441 (1984).
47. N. Masuhara, D. Candela, D. O. Edwards, R. F. Hoyt, H. N. Scholz, D. S. Sherrill and R. Combescot, Phys. Rev. Lett. **53**, 1168, (1984).
48. J. R. Owers–Bradley, H. Chocholacs, R. M. Mü ller, M. Kubota and F. Pobell, Phys. Rev. Lett. **51**, 2120 (1983).
49. H. Ishimoto, H. Fukuyama, N. Nishida, Y. Miura, Y. Takano, T. Fukuda, T. Tazaki and S. Ogawa, Phys. Rev. Lett. **59**, 904 (1987).
50. W. J. Gully and W. J. Mullin, Phys. Rev. Lett., **52**, 1810 (1984).
51. D. Candela, L–J. Wei, D. R. McAllaster and W. J. Mullin, Phys. Rev. Lett. **67**, 330 (1991).
52. E. P. Bashkin, Sov. Phys. JETP., **60(6)**, 1122 (1984).

VARIATIONAL THEORY OF ^{3}He-^{4}He MIXTURES: EQUATION OF STATE AND STABILITY

M. Saarela

Department for Theoretical Physics

University of Oulu, SF-90570 Oulu, Finland

E. Krotscheck

Center for Theoretical Physics and Department of Physics

Texas A&M University, College Station, TX 77843, USA

INTRODUCTION

The finite solubility of ^{3}He in ^{4}He near absolute zero has provided an interesting experimental and theoretical problem for many decades. The challenge is to understand the behavior of a strongly correlated quantum fluid mixture where the fermion concentration can be varied over a reasonable wide range. Excellent review articles have been published by Ebner and Edwards [1], Baym and Pethick[2] and most recently by Ouboter and Yang [3] where quantities such as equation of state, heat of mixing, sound velocities etc. have been analyzed as a function of concentration and pressure.

The first theoretical model for treating dilute mixtures of helium particles proposed by Landau and Pomeranchuck[4] over fifty years ago was based on the free quasiparticle concept. Later Bardeen, Baym and Pines[5] (BBP) pointed out that it in necessary to assume an effective interaction between quasiparticles in order to explain transport properties. The BBP model has been further refined by including corrections due to the momentum dependence of the interaction between the quasiparticles[1,5] and improvements beyond the Hartree-Fock approximation [6]. A very successful semiphenomenological theory of ^{3}He-^{4}He mixtures was recently formulated by Hsu et al.[7] in the framework of a generalized Aldrich-Pines pseudopotential model. The theory is based on an effective interaction between the helium isotopes which is derived by combining theoretical considerations like short range repulsion, Pauli principle corrections and exact sum rules with experimental information like the density and concentration dependence of the speed of sound in the mixture.

A model independent analysis of the existing experimental data was performed by Ouboter and Yang[3]. These authors came to the conclusion that the expansion of the energy density as a function of ^{3}He concentration is improved when fractional powers inherited from the Landau-Pomeranchuck model are used.

In this work we present a microscopic variational calculation using the Jastrow-Feenberg type correlated wave function. This method has developed in a long sequence of publications started by Massey, Woo and Tan[8] who studied the binary boson mixtures. These calculations were later improved by using hypernetted chain summation (HNC) summation techniques for the many-body diagrams[9] and including the three-body correlations[10]. The fermion character of at least one of the mixed components ignored in these works is crucial for the stability in the case where the interparticle potential is the same between all the particles as is the case in the helium fluids[11] and in Coulomb systems[12,13].

In order to make quantitative predictions for the strongly correlated helium fluids it is essential to include 3-body correlations in the wave function and to add elementary diagrams to the HNC equations[14,15,16]. This is even more true in the case of dilute mixture where quantities which depend on energy derivatives like the pressure and the chemical potentials gain important corrections from these terms. On the other hand, the fact that the ^{3}He - component of the mixture is very dilute, permits simplifying approximations in the Fermi-HNC equations.

We start by deriving the set of variational Euler-Lagrange equations for the dilute fermion-boson mixture which give as a solution the structure functions in the three channels —^{4}He-^{4}He, ^{4}He-^{3}He and ^{3}He-^{3}He—. The driving term in these equations is the particle-hole potential matrix. In the limit where the correlation functions are assumed to have only a weak density dependence these potentials can be shown to correspond to the pseudopotentials derived by Pines *et al.*. We also derive the optimizing matrix equation for the three-body correlation functions generalizing the method used by Chang and Campbell to the binary mixture. From the set of elementary diagrams we calculate the four- and five-body diagrams and give a simple estimate of the higher order terms trough a multiplicative factor. This approach is shown to give a good quantitative agreement with experimental equations of state in the case of pure ^{4}He[17] and for one ^{3}He impurity in ^{4}He [18].

Our main results are the equation of state, from which we can calculate other thermodynamic properties such as chemical potentials, and the particle-hole potentials which determine the sound velocities. By increasing the ^{3}He concentration —keeping the total density fixed — we approach the *spinodal point* where the mixture becomes unstable against infinitesimal fluctuations of the ^{3}He-concentration. We also study the actual phase separation which occurs when the chemical potential of the ^{3}He component in the mixture becomes larger than the chemical potential of pure ^{3}He.

THE VARIATIONAL THEORY

We assume that the particle motion is determined by a nonrelativistic Hamiltonian

$$H = -\sum_{\alpha} \sum_{i=1}^{N_\alpha} \frac{\hbar^2}{2m_\alpha} \nabla_i^2 + \frac{1}{2} \sum_{\alpha\beta} {\sum_{i,j}}^{\prime\, N_\alpha,N_\beta} V^{\alpha\beta}(|\mathbf{r}_i - \mathbf{r}_j|). \tag{1}$$

As a general convention, we use Greek subscripts $\alpha, \beta, \ldots \in \{3, 4\}$ to refer to the *species* (a ^{3}He or a ^{4}He particle), and Latin subscripts $i, j, \ldots$ as in the $\mathbf{r}_i$ to enumerate the individual particles. The quantities N_α are the numbers of particles of each species, and $N = N_3 + N_4$. The ^{3}He concentration will be denoted by x:

$$N_3 = xN, \qquad N_4 = (1 - x)N. \tag{2}$$

The prime on the summation symbol indicates that no two pairs (i, α), (j, β) can be the same. In our case, the interaction $V^{\alpha\beta}(|\mathbf{r}_i - \mathbf{r}_j|)$ is independent of the particle type, and depends only on the distances between particles. We will use the Aziz[19] interaction throughout this paper.

For the microscopic description of strongly interacting quantum systems, specifically the helium liquids, the Jastrow-Feenberg variational method[14,20,21] is today the method of choice. The starting point of the theory is an *ansatz* for the ground state wave function of the form

$$\Psi_0(\{r_i\}) = e^{\frac{1}{2}U(\{r_i\})}\Phi_0(\{r_i\})$$

$$U(\{r_i\}) = \frac{1}{2!}\sum_{\alpha\beta}\sum_{i,j}{}' u_2^{\alpha\beta}(\mathbf{r}_i,\mathbf{r}_j) + \frac{1}{3!}\sum_{\alpha\beta\gamma}^{N_\alpha,N_\beta,N_\gamma}\sum_{i,j,k}{}' u_3^{\alpha\beta\gamma}(\mathbf{r}_i,\mathbf{r}_j,\mathbf{r}_k). \qquad (3)$$

Here $\Phi_0(\{\mathbf{r}_i\})$ is a Slater determinant of plane waves ensuring the antisymmetry of the Fermion-component of the wave function. The pair correlation functions $u_2^{\alpha\beta}(\mathbf{r}_i,\mathbf{r}_j)$ and the triplet correlation functions $u_3^{\alpha\beta\gamma}(\mathbf{r}_i,\mathbf{r}_j,\mathbf{r}_k)$ are most effectively[15,17] determined by the variational principles,

$$\frac{\delta E}{\delta u_2^{\alpha\beta}(\mathbf{r}_i,\mathbf{r}_j)} = \frac{\delta E}{\delta u_3^{\alpha\beta\gamma}(\mathbf{r}_i,\mathbf{r}_j,\mathbf{r}_k)} = 0, \qquad (4)$$

where

$$E = \frac{\langle\Psi_0|H|\Psi_0\rangle}{\langle\Psi_0|\Psi_0\rangle} \qquad (5)$$

is the variational energy expectation value which can be written as

$$\frac{E}{N} = \frac{T_F}{N} + \sum_{\alpha\beta}\frac{\rho_\alpha\rho_\beta}{2\rho}\int d^3 r\, g^{\alpha\beta}(r)\left[V^{\alpha\beta}(r) - \left(\frac{\hbar^2}{8m_\alpha} + \frac{\hbar^2}{8m_\beta}\right)\nabla^2 u_2^{\alpha\beta}(r)\right]$$

$$+ \frac{T_{(3)}}{N} + \frac{T_{JF}}{N}. \qquad (6)$$

The first term, T_F, is the kinetic energy of the free Fermi gas, and $T_{(3)}$ is the kinetic energy due to three-body correlations,

$$T_{(3)} = -\frac{1}{3!}\sum_{\alpha\beta\gamma}\rho_\alpha\rho_\beta\rho_\gamma\int d^3 r_1 d^3 r_2 d^3 r_3\, g_3^{\alpha\beta\gamma}(\mathbf{r}_1,\mathbf{r}_2,\mathbf{r}_3)$$

$$\times\left[\frac{\hbar^2}{8m_\alpha}\nabla_1^2 + \frac{\hbar^2}{8m_\beta}\nabla_2^2 + \frac{\hbar^2}{8m_\gamma}\nabla_3^2\right]u_3^{\alpha\beta\gamma}(\mathbf{r}_1,\mathbf{r}_2,\mathbf{r}_3). \qquad (7)$$

The last term, T_{JF}, is a pure Fermi contribution, originating from the Jackson-Feenberg identity. The single exchange loop approximation in the Fermion channels leads to the form

$$\frac{T_{JF}}{N} = \frac{\hbar^2\rho_3^2}{8m_3\rho}\int d^3 r\, \Gamma^{33}(r)\nabla^2 g_F(r), \qquad (8)$$

where $g_F(r)$ is the pair distribution function of the non-interacting Fermi gas, and $\Gamma^{33}(r)$ is a direct dressed correlation function defined shortly. The notation ρ refers to the total density and ρ_α with a greek subscript to the density of one of the components.

Starting from the expression (6) for the energy, we see that the Euler equation for the two- and three-body correlations consist of two parts: One part originates from the *explicit* appearance of the correlation functions in the kinetic energy terms, the other one from the *implicit* dependence of the two- and three-body distribution functions on the correlation functions. The two-body Euler equation has the general form

$$\left[\frac{\hbar^2}{8m_\alpha} + \frac{\hbar^2}{8m_\beta}\right]\nabla^2 g^{\alpha\beta}(r) = g'^{\alpha\beta}(r) \qquad (9)$$

where

$$g'^{\alpha\beta}(r) = \frac{\rho}{\rho_\alpha\rho_\beta} \sum_{\gamma\delta} \rho_\gamma\rho_\delta \int d^3r' \frac{\delta g^{\gamma\delta}(r')}{\delta u_2^{\alpha\beta}(r)} \left[V^{\gamma\delta}(r') - \left(\frac{\hbar^2}{8m_\gamma} + \frac{\hbar^2}{8m_\delta} \right) \nabla^2 u_2^{\gamma\delta}(r') \right]$$
$$+ \frac{2\rho}{\rho_\alpha\rho_\beta} \frac{\delta}{\delta u_2^{\alpha\beta}(r)} \left[\frac{T_{JF}}{N} + \frac{T_{(3)}}{N} \right]. \tag{10}$$

The (F)HNC equation provide the connection between the pair and triplet correlation functions $u_2^{\alpha\beta}(\mathbf{r}_i, \mathbf{r}_j)$ and $u_3^{\alpha\beta\gamma}(\mathbf{r}_i, \mathbf{r}_j, \mathbf{r}_k)$ and the pair distribution functions $g^{\alpha\beta}(r)$. They describe the self-consistent summation of two types of diagrams: Chain connections and parallel connections. The coordinate space equations define a "direct" dressed correlation function $\Gamma^{\alpha\beta}(r)$ through

$$\Gamma^{\alpha\beta}(r) = e^{\left[u_2^{\alpha\beta}(r)+E^{\alpha\beta}(r)+N^{\alpha\beta}(r)\right]} - 1 \tag{11}$$

where $E^{\alpha\beta}(r)$ are sets of "elementary diagrams" containing pair and/or triplet correlations, and the $N^{\alpha\beta}(r)$ are the sets of "nodal" diagrams. The functions $\Gamma^{\alpha\beta}(r)$ take care of the short-ranged correlations, and are related to the pair distribution functions in the respective channels by

$$g^{\alpha\beta}(r) = \left[1 + \Gamma^{\alpha\beta}(r)\right] C^{\alpha\beta}(r) \tag{12}$$

in which the coefficient functions $C^{\alpha\beta}(r)$ are corrections due to Fermi statistics. The simplest approximation, $C^{44}(r) = C^{34}(r) = 1$ and $C^{33}(r) = g_F(r)$, turns out to be accurate for this work.

The equations describing the "chaining" process are most conveniently written in a matrix form in momentum space. We define the 2×2 matrices of *non-nodal* ($X^{\alpha\beta}(k)$) and *nodal* functions ($N^{\alpha\beta}(k)$)

$$\tilde{\mathbf{X}} \equiv \left(\sqrt{\rho_\alpha} X^{\alpha\beta}(k) \sqrt{\rho_\beta} \right)$$
$$\tilde{\mathbf{N}} \equiv \left(\sqrt{\rho_\alpha} N^{\alpha\beta}(k) \sqrt{\rho_\beta} \right) \tag{13}$$

and

$$\tilde{\boldsymbol{\Gamma}} \equiv \tilde{\mathbf{X}} + \tilde{\mathbf{N}} = \left(\sqrt{\rho_\alpha} \Gamma^{\alpha\beta}(k) \sqrt{\rho_\beta} \right). \tag{14}$$

Similarly, we write for the static structure function

$$\mathbf{S} \equiv \left(S^{\alpha\beta}(k) \right) \tag{15}$$

The structure function of the *non-interacting* Fermi gas in the mixture has only a non-trivial (3,3)-component:

$$\mathbf{S_F} \equiv \begin{pmatrix} S_F(k) & 0 \\ 0 & 1 \end{pmatrix} \tag{16}$$

With these definitions, the FHNC//0 equation describing the chaining of diagrams has the simple form[21]

$$\tilde{\mathbf{N}} = \tilde{\mathbf{X}} * \mathbf{S_F} * \tilde{\boldsymbol{\Gamma}} \tag{17}$$

The structure function in this approximation is

$$\mathbf{S} = \mathbf{S_F} + \mathbf{S_F} * \tilde{\boldsymbol{\Gamma}} * \mathbf{S_F}. \tag{18}$$

Combining these relations leads to the following solution for the direct correlation function

$$\tilde{\mathbf{X}} = \mathbf{S_F}^{-1} - \mathbf{S}^{-1}. \tag{19}$$

The next problem is to manipulate the Euler equation, Eq. (9), into a form that can be solved numerically. We will outline only a few steps. It is convenient to work in momentum space. We define

$$\mathbf{H_1} \equiv \begin{pmatrix} \frac{\hbar^2 k^2}{2m_3} & 0 \\ 0 & \frac{\hbar^2 k^2}{2m_4} \end{pmatrix} \tag{20}$$

and write the Euler equation in matrix form

$$\mathbf{S}' = -\frac{1}{4}\left[\mathbf{H_1} * (\mathbf{S} - \mathbf{1}) + (\mathbf{S} - \mathbf{1}) * \mathbf{H_1}\right] \tag{21}$$

where $\mathbf{S}'$ is the (matrix of) structure functions obtained from $g'^{\alpha\beta}$. The effect of differentiated exchange lines is most conveniently included by introducing the quantity

$$\mathbf{S_F}' \equiv -\frac{1}{4}\left[\mathbf{H_1} * (\mathbf{S_F} - \mathbf{1}) + (\mathbf{S_F} - \mathbf{1}) * \mathbf{H_1}\right]. \tag{22}$$

We also define the "primed" analog of the HNC-equation[15], Eq. (10),

$$X'^{\alpha\beta}(r) = \left[1 + \Gamma^{\alpha\beta}(r)\right]\left\{V^{\alpha\beta}(r) + E'^{\alpha\beta}(r) - \left[\frac{\hbar^2}{8m_\alpha} + \frac{\hbar^2}{8m_\beta}\right]\nabla^2 u_2^{\alpha\beta}(r)\right\}$$
$$+ \Gamma^{\alpha\beta}(r)N'^{\alpha\beta}(r). \tag{23}$$

It is very useful to introduce two new quantities. The *particle-hole potential* is related to the non-nodal sum of diagrams,

$$V_{p-h}^{\alpha\beta}(r) = \left[\frac{\hbar^2}{8m_\alpha} + \frac{\hbar^2}{8m_\beta}\right]\nabla^2 X^{\alpha\beta}(r) + X'^{\alpha\beta}(r), \tag{24}$$

and the *induced potential* is similarly related to the nodal sum of diagrams,

$$w_I^{\alpha\beta}(r) = \left[\frac{\hbar^2}{8m_\alpha} + \frac{\hbar^2}{8m_\beta}\right]\nabla^2 N^{\alpha\beta}(r) + N'^{\alpha\beta}(r). \tag{25}$$

Inserting these definitions into Eq. (23) gives the coordinate space expression of the particle-hole potential,

$$V_{p-h}^{\alpha\beta}(r) = \left[1 + \Gamma^{\alpha\beta}(r)\right]\left[V^{\alpha\beta}(r) + \Delta V^{\alpha\beta}(r)\right] + \left[\frac{\hbar^2}{2m_\alpha} + \frac{\hbar^2}{2m_\beta}\right]\left|\nabla\sqrt{1 + \Gamma^{\alpha\beta}(r)}\right|^2$$
$$+ \Gamma^{\alpha\beta}(r)w_I^{\alpha\beta}(r). \tag{26}$$

We have abbreviated with

$$\Delta V^{\alpha\beta}(r) = E'^{\alpha\beta}(r) + \left[\frac{\hbar^2}{8m_\alpha} + \frac{\hbar^2}{8m_\beta}\right]\nabla^2 E^{\alpha\beta}(r) \tag{27}$$

the contribution from triplet correlations and elementary diagrams.

Combining the primed equation together with the definitions (15), (19), (24), and (25) gives us the final form of the Euler equation in the FHNC//0 approximation,

$$\mathbf{S}^{-1}\mathbf{H_1}\mathbf{S}^{-1} - \mathbf{S_F}^{-1}\mathbf{H_1}\mathbf{S_F}^{-1} = 2\tilde{\mathbf{V}}_{p-h}, \tag{28}$$

and the induced potential,

$$\tilde{\mathbf{w}}_{\mathbf{I}} = -\tilde{\mathbf{V}}_{p-h} - \frac{1}{2}\left[\mathbf{S_F}^{-1}\mathbf{H_1}\tilde{\mathbf{\Gamma}} + \tilde{\mathbf{\Gamma}}\mathbf{H_1}\mathbf{S_F}^{-1}\right]. \tag{29}$$

For any given choice of elementary diagrams and three-body correlations, the equations (26), (27), (28), and (29) form a closed set of equations that can be solved by iteration until convergence is reached.

The derivation of the Euler equation for the triplet correlations follows the same pattern established above and in the previous works on optimized correlation functions[15,17,18]. We limit here the discussion to the convolution approximation of the triplet distribution function[15] and use the "Lado" approximation, $S_F(\mathbf{k_1}, \mathbf{k_2}, \mathbf{k_3}) = S_F(k_1)S_F(k_2)S_F(k_3)$, for the triplet structure function of the noninteracting Fermi system. A lengthy algebra with the three-body primed equations gives a simple final result

$$\sum_{\lambda\mu\nu} D^{\alpha\beta\gamma,\lambda\mu\nu}(\mathbf{k_1}, \mathbf{k_2}, \mathbf{k_3})\tilde{u}_3^{\lambda\mu\nu}(\mathbf{k_1}, \mathbf{k_2}, \mathbf{k_3}) = \tilde{n}_3^{\alpha\beta\gamma}(\mathbf{k_1}, \mathbf{k_2}, \mathbf{k_3}), \tag{30}$$

where

$$D^{\alpha\beta\gamma,\lambda\mu\nu}(\mathbf{k_1},\mathbf{k_2}, \mathbf{k_3}) = \frac{\hbar^2 k_1^2}{4m_\alpha}\delta_{\alpha\lambda}S^{\beta\mu}(k_2)S^{\gamma\nu}(k_3)$$

$$+\frac{\hbar^2 k_2^2}{4m_\beta}\delta_{\beta\mu}S^{\alpha\lambda}(k_1)S^{\gamma\nu}(k_3) + \frac{\hbar^2 k_3^2}{4m_\gamma}\delta_{\gamma\nu}S^{\alpha\lambda}(k_1)S^{\beta\mu}(k_2), \tag{31}$$

and

$$\tilde{n}^{\alpha\beta\gamma}(\mathbf{k_1}, \mathbf{k_2},\mathbf{k_3}) = \sum_\lambda \frac{\hbar^2}{4m_\lambda\sqrt{\rho_\lambda}}\left\{ S^{\alpha\lambda}(k_1)S^{\beta\lambda}(k_2)S^{\gamma\lambda}(k_3)\tau_{123}^{(\lambda)}\right.$$

$$-[k_1^2\delta^{\alpha\lambda}S^{\beta\lambda}(k_2)S^{\gamma\lambda}(k_3) + \text{cycl.}]\xi_{123}^{(\lambda)}$$

$$\left.-[\mathbf{k_1}\cdot\mathbf{k_2}\delta^{\alpha\lambda}\delta^{\beta\lambda}S^{\gamma\lambda}(k_3) + \text{cycl.}]\right\} \tag{32}$$

Cyclic permutation here means permutation of the coordinate index and the particle type index together. For a compact presentation of the equations we have introduced the notations,

$$\xi_{123}^{(4)} = 1,$$

$$\xi_{123}^{(3)} = \frac{S_F(\mathbf{k_1}, \mathbf{k_2}, \mathbf{k_3})}{S_F(k_1)S_F(k_2)S_F(k_3)}$$

$$\tau_{123}^{(4)} = \frac{1}{2}(k_1^2 + k_2^2 + k_3^2),$$

$$\tau_{123}^{(3)} = \frac{k_1^2}{S_F(k_1)}\xi_{123}^{(3)} + \frac{\mathbf{k_1}\cdot\mathbf{k_2}}{S_F(k_1)S_F(k_2)} + (\text{cycl.}). \tag{33}$$

In the calculation of the three-body energy correction it is important to note that one has to include besides the kinetic energy term $T_{(3)}$ defined in Eq. (7) also contributions from elementary diagrams which contain one or two triplet correlation functions. These additional mixed diagrams which are topologically "elementary" and contain triplet correlations, have been spelled out explicitly, for the one-component system, in Ref. 17. The expression for the full three-body energy is also simplified when the solution of the Euler equation (30) is used and can be written in the form,

$$\frac{E_{(3)}}{N} = -\frac{1}{2 \cdot 3!} \int \frac{d^3 k_1 d^3 k_2 d^3 k_3}{(2\pi)^6 \rho} \delta(\mathbf{k}_1 + \mathbf{k}_2 + \mathbf{k}_3) \times$$

$$\times \sum_{\alpha\beta\gamma} \tilde{u}_3^{\alpha\beta\gamma}(\mathbf{k}_1, \mathbf{k}_2, \mathbf{k}_3) D^{\alpha\beta\gamma,\alpha'\beta'\gamma'}(\mathbf{k}_1, \mathbf{k}_2, \mathbf{k}_3) \tilde{u}_3^{\alpha'\beta'\gamma'}(\mathbf{k}_1, \mathbf{k}_2, \mathbf{k}_3). \tag{34}$$

Finally we give the three-body contribution to the particle-hole potential defined in Eq. (27).

$$\Delta \tilde{V}_3^{\alpha\beta}(k) = \int \frac{d^3 k_2 d^3 k_3}{(2\pi)^6 \rho} \delta(\mathbf{k}_1 + \mathbf{k}_2 + \mathbf{k}_3) \times$$

$$\left\{ \sum_{\lambda\mu\mu'} \tilde{u}_3^{\alpha\lambda\mu}(\mathbf{k}_1, \mathbf{k}_2, \mathbf{k}_3) \frac{\hbar^2 k_2^2}{4m_\lambda} S^{\mu\mu'}(k_3) \tilde{u}_3^{\beta\lambda\mu'}(\mathbf{k}_1, \mathbf{k}_2, \mathbf{k}_3) \right.$$

$$+ \left[\frac{\hbar^2}{8m_\beta \sqrt{\rho_\beta}} \sum_{\mu\nu} u_3^{\alpha\mu\nu}((\mathbf{k}_1, \mathbf{k}_2, \mathbf{k}_3) \right.$$

$$\times \left[-S^{\mu\beta}(k_2) S^{\nu\beta}(k_3) \tau_{123}^{(\beta)} + 2k_2^2 \delta_{\mu\beta} S^{\nu\beta}(k_3) \xi_{123}^{(\beta)} + \mathbf{k}_2 \cdot \mathbf{k}_3 \delta_{\mu\beta} \delta_{\nu\beta} \right]$$

$$\left. \left. + (\alpha \leftrightarrow \beta) \right] \right\}. \tag{35}$$

THE EQUATION OF STATE

Our main result is the energy/particle as a function of the total density ρ and the concentration x. That form of the solution is somewhat incovenient because the derivatives needed for the calculation of the chemical potentials are taken at constant pressure. That is why we first perform the transformation from the density to the pressure variable. We expanded the energy/particle in cubic powers of the density for each concentration,

$$\frac{E}{N} = \sum_{i=0}^{3} a_i(x) \rho^i. \tag{36}$$

The pressure is calculated by differentiating the energy/particle with respect to total density.

$$P = \rho^2 \frac{\partial(E/N)}{\partial\rho} \tag{37}$$

giving

$$P = P(\rho, x) = \sum_{i=1}^{3} i a_i(x) \rho^{i+1} \tag{38}$$

The solution of this equation gives $\rho = \rho(P, x)$. Inserting the result back to Eq. (36) gives the desired energy/particle as a function of P and x which is needed for the calculation of chemical potentials.

The chemical potentials have the usual definitions:

$$\mu_4 = H - x\frac{\partial H}{\partial x}\Big|_P$$

$$\mu_3 = H + (1 - x)\frac{\partial H}{\partial x}\Big|_P. \tag{39}$$

The enthalpy H is related to the energy/particle and the pressure,

$$H = \frac{E}{N} + \frac{P}{\rho}. \tag{40}$$

In Fig. 1 we give our results for the chemical potentials. One has to remember that we have one parameter which is used to estimate the higher order (≥ 6) elementary diagrams. This parameter is chosen a value 1.4 such as the ^{4}He chemical potential at zero pressure fits the experiment. A high accuracy of the fit is seen also at higher pressures. A sensitive test of accuracy is the calculation of the ^{3}He chemical potential.

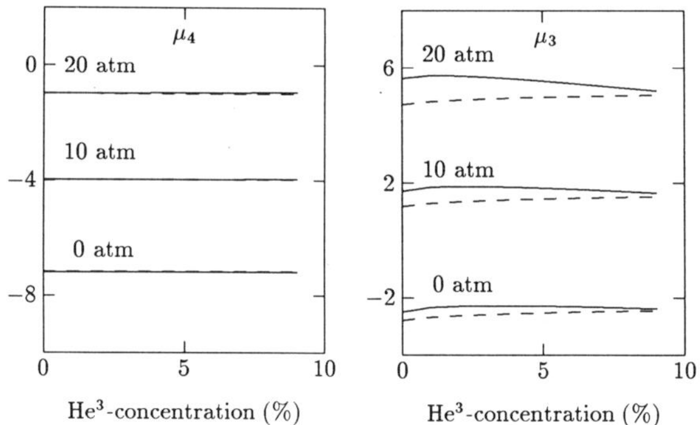

Fig. 1. Chemical potential μ_4 and μ_3 of ^{4}He and ^{3}He, respectively, as a function of ^{3}He concentration at three different pressures, P = 0, 10, and 20 atm. The solid lines are the results of this work and the dashed lines are calculated from an analytic expression fitted into experiments[3].

There the three-body and elementary diagram corrections are very important since the plane HNC result is +9K at P = 0 atm and x=0. Also in this case our results are in reasonable agreement with experiments.

It is also interesting to compare our calculated densities as a function of concentration with experiments. The volume increase of the mixture when the concentration increases is known to be linear. For the total density this gives the following expression

$$\rho(P, x) = \frac{\rho_4^0(P)}{1 + x(\beta(P) - 1)} \tag{41}$$

where $\rho_4^0(P)$ is the ^{4}He density at x=0 and $\beta = V_3\rho_4^0$ is the volume excess factor. A typical value for the β-factor is < 1.3 which means that at small concentrations the density is a linear function of concentration, too. In Fig. 2. we show our solutions for the density from Eq. (38). One notices that the calculated results deviate slightly from the linear behavior which may be due to the numerics. The least square fit to these curves yields an estimate of the volume excess factor shown also in Fig. 2.

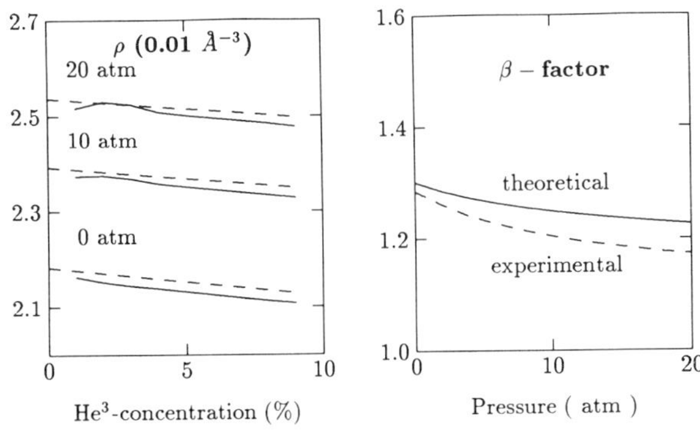

Fig. 2. Total density ρ as a function concentration at indicated pressures. The solid line is the calculated result and the dashed one is taken from experiments[3]. In the figure on the right we give the experimental β-factor and the result of the linear fit to the calculates densities.

PARTICLE-HOLE POTENTIALS AND STABILITY

One of the guiding principles in the development of modern microscopic many-body theories[23] is that these theories do not allow solutions for systems that are physically unstable. The quantum liquid mixture of ^{3}He and ^{4}He is, depending on the external pressure, stable up to only six to seven percent concentration[1] of ^{3}He. For example, the mixture would spontaneously phase separate if Fermi statistics were turned off. Thus, when attempting a microscopic description of the ^{3}He-^{4}He mixture, one must *a priori* anticipate that one is working close to a point where the *physical system* becomes unstable against phase separation. Consequently, we not only expect, but even require, that our *theoretical description* will exhibit some kind of an instability. This instability of the theory close to a physical instability is quite desirable, but it requires, of course, a careful analysis of the underlying equations and of the numerical methods used to solve the equations.

When studying the stability of the quantum liquid mixture, we have to distinguish between *global* and *local* instabilities. A *global* instability means that another phase of the system of a lower energy exists, which the system can reach by a *macroscopic* perturbation of its configuration. In our case, when the chemical potential of a ^{3}He atom in the mixture becomes larger than the chemical potential of the atom in bulk ^{3}He, the mixture will phase separate. When the concentration of the ^{3}He component is increased, this the instability will occur first. More specifically, the experimental result[3] for the chemical potential in pure ^{3}He is -2.473K at P=0. Our result for the ^{3}He chemical potential in the mixture is -2.48K meaning that ^{3}He and ^{4}He would barely mix at P=0 and would phase separate at higher pressure. The corresponding experimental number is -2.778K which allows mixing up to 6.6%. The question is of tiny energy differences which are beyond the accuracy of our numerics.

The second, *local* instability occurs when the system becomes unstable against *infinitesimal* fluctuations about its equilibrium configuration. Such an instability is normally indicated by the softening of a collective excitation and should be reflected by an instability of the theory. This second type of instability is theoretically the more interesting one since it provides a test for the consistency of the mathematical description.

For the mixture to be *locally* stable with respect to changes of the density and the concentration, the following second derivative matrix of the energy must be positive definite.

$$\left(\frac{\sqrt{\rho_\alpha \rho_\beta}}{\Omega} \frac{\partial^2 E}{\partial \rho^\alpha \partial \rho^\beta} \right)_{\alpha\beta} = \left(\begin{matrix} \hat{V}^{33}(0+) + mc_F^2 & \hat{V}^{34}(0+) \\ \hat{V}^{34}(0+) & \hat{V}^{44}(0+) \end{matrix} \right). \tag{42}$$

where we have separated the total energy (cf. Eq. (6)) into two parts,

$$E = T_F + E_c[\rho_3, \rho_4], \tag{43}$$

The second derivative of the free Fermi energy, T_F, with respect to ^{3}He density is $m_3 c_F^2 = \hbar^2 k_F^2 / 3m_3$ where $c_F = \hbar k_F / \sqrt{3}m_3$ is the addiabatic velocity of sound in the free Fermi gas. The differentiation of the total correlation energy, E_c, gives the potentials $\hat{V}^{\alpha\beta}(0+)$,

$$\hat{V}^{\alpha\beta}(0+) = \frac{\sqrt{\rho_\alpha \rho_\beta}}{\Omega} \frac{\partial^2 E_c}{\partial \rho^\alpha \partial \rho^\beta}. \tag{44}$$

The similarity of the notations for the $\hat{V}^{\alpha\beta}$ and the $\tilde{V}^{\alpha\beta}_{p-h}$ introduced in Eqs. (24) and (26) is, of course, intentional. One can show[24] that in the limit of keeping the pair correlation functions fixed the potential $\hat{V}^{\alpha\beta}(0+)$ and $\tilde{V}^{\alpha\beta}_{p-h}(0+)$ should be identical.

The eigenvalues of the matrix (42) can be related to the zero and second sound velocities, c_0 and c_2, respectively.

$$m_4 c_0^2 \approx \hat{V}^{44}(0+)$$

$$m_3 c_2^2 \approx m_3 c_F^2 - \left(\hat{V}^{33}(0+) - \frac{\left[\hat{V}^{34}(0+) \right]^2}{m_4 c_0^2} \right). \tag{45}$$

The stability of the bulk liquid requires that $m_4 c_0^2$ should always be positive. In the mixture the stability against the concentration fluctuations requires that $m_3 c_2^2 > 0$.

In the actual calculations it is the Euler equation (28) which determines the instability. In order to have a solution for that equation , we must guarantee that the 2×2 matrix

$$\tilde{V}_{p-h} + \frac{1}{2} \mathbf{S_F}^{-1} \mathbf{H_1} \mathbf{S_F}^{-1} \tag{46}$$

is positive definite. We calculate the matrix in the zero-momentum limit where the instability will first show up, and obtain

$$\lim_{k \to 0+} \left[\tilde{V}_{p-h} + \frac{1}{2} \mathbf{S_F}^{-1} \mathbf{H_1} \mathbf{S_F}^{-1} \right] = \left(\begin{matrix} \tilde{V}^{33}(0+) + \frac{4}{3}m_3 c_F^2 & \tilde{V}^{34}(0+) \\ \tilde{V}^{34}(0+) & \tilde{V}^{44}(0+) \end{matrix} \right). \tag{47}$$

The discrepancy in the (3,3) matrix element between Eqs. (42) and (47) is obvious. The origin of this is in the mean field approximation which should be replaced by the correct treatment of the ring diagrams.

Insisting on the argument that the pair correlation functions depend only weakly on the density we can relate our particle-hole potentials to the pseodupotentials used by Hsu and Pines[7]. In their notations $\rho f^s_{k,4} \to \tilde{V}^{44}_{p-h}(k)$, $\rho U_k \to \tilde{V}^{34}_{p-h}(k)$, and $\rho V^{\uparrow\downarrow}_k \to \tilde{V}^{33}_{p-h}(k)$. In Fig. 3. we show the comparison of these qualities. The limiting values of the (3,4) and (3,3) components of the particle hole potential at k=0 are related to the volume excess factor[5].

$$\tilde{V}^{34}_{p-h}(0) \approx \beta m_4 c_0^2$$
$$\tilde{V}^{33}_{p-h}(0) \approx (2\beta - 1)m_4 c_0^2. \tag{48}$$

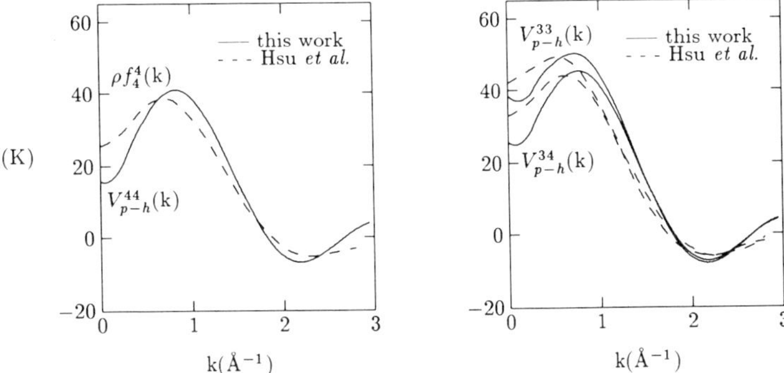

Fig. 3. Particle-hole potentials defined in Eq. (26) (solid lines) are compared with the corresponding quantities of the pseudopotential theory by Hsu and Pines[7].

In the work by Hsu and Pines[7] the values of the pseudopotentials at k=0 are fitted to the experimental zero sound velocity and the β-factor whereas our values are the results from the calculations. The 20% discrepancy in the zero sound velocity is magnified here.

The crucial quantity for the transport properties is the effective ^{3}He-^{3}He interaction. That is calculated by subtracting the "phonon" induced interaction $\tilde{w}^{33}_I(k)$ from the "direct" interaction, $\tilde{V}^{33}_{p-h}(k)$. We are mainly interested in the small concentration and low momentum limit where the induced potential can be written in the following form,

$$\tilde{w}^{33}_I(k) = \frac{\left[\tilde{V}^{34}_{p-h}(k)\right]^2}{\tilde{V}^{44}_{p-h}(k) + \frac{\hbar^2 k^2}{4m_4}}. \tag{49}$$

The result,

$$V_{eff} = V_{p-h}^{33} - \frac{\left[\tilde{V}_{p-h}^{34}(k)\right]^2}{\tilde{V}_{p-h}^{44}(k) + \frac{\hbar^2 k^2}{4m_4}}, \tag{50}$$

is plotted in Fig. 4.(a) together with the most recent experimental determination of the parameters in the quadratic form[25] $V_{eff} = V_0(1 - (k/k_0)^2)$ with $V_0 = 2.15K$ and $k_0 = 0.403\mathring{A}^{-1}$

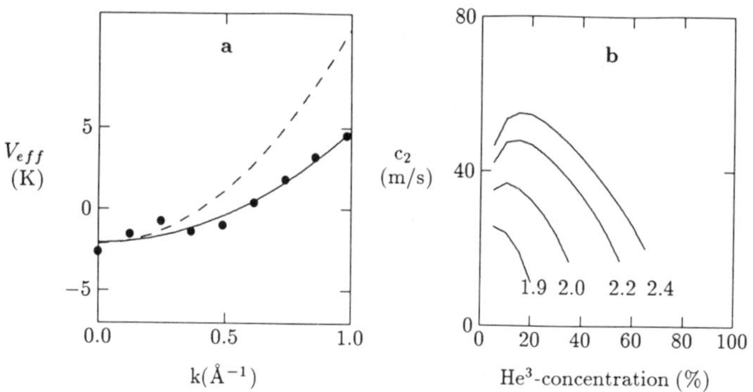

Fig. 4. In figure (a) we show the effective ^{3}He-^{3}He interaction as a function of momentum. The black dots are the calculated values and the solid curve is the least square quadratic fit to those values. The dashed curve is the experimental result explained in the text. In figure (b) we present the second sound velocities as a function of concentration at different densities indicated by the numbers in units $(10^{-2}\mathring{A}^{-3})$.

Finally, in Fig. 4.(b) we show the results for the second sound – calculated from the eigenvalues of the matrix (46) – as a function of concentration at various densities. At all given densities there is a clear signature of the instability. As discussed earlier the increasing concentration of ^{3}He decreases the density for a given pressure and we find that the mixture becomes locally unstable when the concentration becomes larger than 35% at P=0.

ACKNOWLEDGMENTS

This work was supported in part by a research grant from the Finnish Academy of Science (to M. S.), and the National Science Foundation under Contract PHY-8806265 and the Texas Advanced Research Program under Grant No. 010366-01 (to E.K.). We thank A. Kallio and P. Pietiläinen for many valuable discussions.

REFERENCES

1. C. Ebner and D. O. Edwards, Phys. Rep. **2**, 77 (1971).
2. G. Baym and C. Pethick, in *The Properties of Liquid and Solid Helium, Volume 2* , edited by K. H. Bennemann and J. B. Ketterson, p. 123 (Wiley, New York, 1978).
3. R. de Bruyn Ouboter and C. N. Yang, Physica **144B**, 127, (1986).
4. L. D. Landau and I. Pomeranchuk, Dokl. Akad. Nauk. SSSR **59**, 669 (1948).
5. J. Bardeen, G. Baym, and D. Pines, Phys. Rev. **156**, 207 (1967).
6. H. H. Fu and C. J. Pethick, Phys. Rev. **B14**, 3837 (1976).
7. W. Hsu and D. Pines, J. Stat. Phys. **38**, 273 (1985), and W. Hsu, D. Pines, and C. H. Aldrich, Phys. Rev. **B32**, 7179 (1985).
8. W. E. Massey, C.W. Woo, and M. T. Tan, Phys. Rev. **A1**, 519 (1970).
9. K. E. Kürten and C. E. Campbell, Phys. Rev. **B26**, 124 (1982).
10. A. Fabrocini and A. Polls, Phys. Rev.**B30**, 1200 (1984).
11. A. Fabrocini and A. Polls, Phys. Rev.**B25**, 4533 (1982).
12. T. Chakraborty, A. Kallio, L.J. Lantto, and P.Pietiläinen, Phys. Rev. **B27**, 3061 (1983), and L. Lantto, Phys. Rev. **B36**, 5160 (1987).
13. C. E. Campbell and J. G. Zabolitzky, Phys. Rev.**B29**, 123 (1984).
14. E. Feenberg, *Theory of Quantum Liquids* (Academic, New York, 1969).
15. C. C. Chang and C. E. Campbell, Phys. Rev. **B15**, 4238 (1977).
16. R. A. Smith, A. Kallio, M. Puoskari, P. Toropainen, Nucl. Phys. **A328** 186 (1979).
17. E. Krotscheck, Phys. Rev. **B33**, 3158 (1986).
18. M. Saarela, in *Recent Progress in Many Body Theories Volume 2*, edited by Y. Avishai, p. 337 (Plenum Press, New York, 1990).
19. R. A. Aziz, V. P. S. Nain, J. C. Carley, W. J. Taylor, and G. T. McConville, J.Chem. Phys. **70**, 4330 (1979).
20. C. E. Campbell, in *Progress in Liquid Physics* , edited by C. A. Croxton (Wiley, London 1977) Chapter 6.
21. J. W. Clark, in *Progress in Particle and Nuclear Physics*, edited by D. H. Wilkinson (Pergamon, Oxford 1979), Vol. 2, p. 89.
22. E. Krotscheck, J. Low Temp. Phys. **27**, 199 (1977).
23. A. D. Jackson, *Ann. Review Nucl. Part. Sci.* **33**, 105 (1983).
24. E. Krotscheck, Annals of Physics **155**, 1 (1984).
25. D. Candela, L-J. Wei, D.R. McAllaster, and W. J. Mullin, Phys. Rev. Letters **67**, 330 (1991).

ON SHADOW WAVE FUNCTIONS FOR CONDENSED PHASES OF HELIUM

L. Reatto

Dipartimento di Fisica, Università degli Studi di Milano
20133 Milano, Italy

INTRODUCTION

In the variational approach to quantum systems like liquid or solid ^{4}He it is customary to write the trial ground state function in the form

$$\psi_o(\vec{r}_1..\vec{r}_N) = F(\vec{r}_1..\vec{r}_N)\phi(\vec{r}_1..\vec{r}_N) \tag{1}$$

where ϕ is the ground state of a suitable idealized system and has the proper behavior under exchange of particles, i.e. it is symmetric for bosons and antisymmetric for fermions. In the case of bosons in a bulk fluid state ϕ is just a constant and all the correlations are induced by the factor F which can be written in the Feenberg form

$$F(\vec{r}_1..\vec{r}_N)=\exp\left\{-\tfrac{1}{2}\sum_{i<j}u(r_{ij}) - \tfrac{1}{2}\sum_{i<j<\ell} u^{(3)}(\vec{r}_i,\vec{r}_j,\vec{r}_\ell)+..\right\}. \tag{2}$$

If only the pair term is kept this gives the well known Jastrow function and large part of the work on the ground state of ^{4}He, for instance, is based on such ansatz. It is clear that this approach is useful only if the higher order terms in ln (F) are rapidly decreasing in importance because terms beyond the triplet ones are impractical.

Recently a new family of wave functions has been introduced for ^{4}He, the shadow function [1,2]. Here particles are correlated not only directly via a term of the Jastrow form but also indirectly via a coupling to subsidiary variables, the shadows. This function, in principle, is a particular case of (2) in which part of the two body pseudopotential u(r) and pseudopotentials of higher order are implicitly induced via integration over the subsidiary variables. The reason why this should be a useful representation of the higher order terms in (2) is that the shadow variables give a simple way to take into account the effect of the quantum hole due to delocalization and hard core interaction of the particles. Given this picture it is natural to assume that this quantum hole should be present also in the excited states of low energy and, in fact, suitable wave functions have been written for roton[3] and for vortex excitations[4]. Here we discuss some aspects of the structure of the shadow function for the ground state and for roton excited states. We show how to construct from these the low temperature density matrix of ^{4}He. Finally we discuss the structure of shadow functions for fermions.

Ground state

Vitiello, Runge and Kalos[1] wrote the unnormalized shadow function for ^{4}He in the form

$$\Psi_o(R) = \psi_r(R) \int dS \, \psi_s(S) K(R,S), \qquad R=(\vec{r}_1,\ldots\vec{r}_N), \; S=(\vec{s}_1,\ldots\vec{s}_N) \tag{3}$$

where the correlating factors for particles and shadows are taken of the Jastrow form

$$\psi_r(R) = \Pi_{i<j} \, e^{-\frac{1}{2} u(r_{ij})}, \quad \psi_s(S) = \Pi_{i<j} \, e^{-u_s(s_{ij})} \tag{4}$$

and the pseudopotentials have an inverse power form

$$u(r) = (b/r)^5, \quad u_s(s) = (b_s/s)^m. \tag{5}$$

The particle-shadow correlating factor was taken of the gaussian form

$$K(R,S) = \Pi_i \, e^{-C|\vec{r}_i-\vec{s}_i|^2} \tag{6}$$

so that each shadow variable is strictly localized in the neighborhood of an associated particle. Therefore when we square $\Psi_o(R)$ in order to take expectation values each particle has two shadow variables in its neighborhood. b, b_s, m and C are variational parameters.

A possible justification[1,5] for such a wave function derives from consideration of the discrete path integral representation of the density matrix. If the system is at a finite temperature $T=1/K_B\beta$ we can write

$$<R|e^{-\beta H}|R> = \sum_P \int dR^{(1)} dR^{(\nu-1)} <R|e^{-\beta H/\nu}|R^{(1)}> \ldots <R^{(\nu-1)}|e^{-\beta H/\nu}|R>,$$

$$R^{(\alpha)} = (\vec{r}_1^{\,(\alpha)}, \ldots \vec{r}_N^{\,(\alpha)}) \tag{7}$$

where ν is an arbitrary integer ≥ 2. Each factor in the r.h.s. of (7) is just the density matrix of the same system but at the higher temperature νT so that if ν is large enough one can use the quasi classical approximation

$$<R^{(\alpha)}|e^{-\beta H/\nu}|R^{(\alpha+1)}> \simeq e^{-\beta V(R^{(\alpha)})/2\nu} \Pi_i e^{-K_\nu T|\vec{r}_i^{\,(\alpha)} - \vec{r}_i^{\,(\alpha+1)}|^2}$$
$$e^{-\beta V(R^{(\alpha+1)})/2\nu}, \quad K_\nu T = \frac{mk_B \nu T}{\hbar^2}, \tag{8}$$

where $V(R)$ is the potential energy and m the mass of the atoms. Therefore to each position $\vec{r}_i$ of an atom we find associated $\nu-1$ variables $(\vec{r}_i^{\,(1)}, \ldots \vec{r}_i^{\,(\nu-1)})$ which are connected together by a harmonic spring in a chain. We recognize then that this quantum problem is isomorphous to a problem in classical statistical mechanics of interacting and flexible chain polymers. To recover the quantum problem all the monomer degrees of freedom have to be traced out but for the monomers of one species, that corresponding to the set $R=(\vec{r}_1 \ldots \vec{r}_N)$. In the case of Bose statistics in (7) one has also to sum over all permutations of particle index in each $R_\alpha=(\vec{r}_1^{\,(\alpha)} \ldots \vec{r}_N^{\,(\alpha)})$ so that the chains can be cross linked.

The steric hindrance due to the hard core repulsion between atoms strongly depresses the probability of cross linking when the density is large as in liquid or solid ^{4}He and the chains have a rather compact shape in the cage of the neighbors. The idea then is to represent as an approximation each chain in terms of its center of mass and of a probability distribution of the position of a monomer around this center of mass. By extending this picture to the ground state we can interpret the shadow variables as the centers of mass and the gaussians in (6) as the probability distribution of the particles (the "monomers") around their center of mass: a particle has to carry along a quantum hole which is the effect of quantum delocalization and of the hard core repulsion and this hole is represented by the shadow variable.

If the previous picture is correct we should expect that the shadow variables are more strongly localized and correlated than the particles and this is indeed found[1,2] to be the case when the expectation value of the energy is minimized with respect to the variational parameters in the shadow function. This is so much so that at large enough density the shadow induce crystalization[1,2] in the system while the particles still enjoy a rather large motion around their equilibrium positions.

If we take the path integral representation of the density matrix as underlying the shadow function form of the ground state, this suggests some other aspects on the structure of the ground state. First is the relation between the interatomic interaction $v(r)$ and the pseudopotentials $u(r)$ and $u_s(s)$. $u(r)$ should reflect the monomer-monomer interaction in which the hard core part of the interaction is effective but the attractive part of $v(r)$ is much less so. Therefore the choice (5) for u is appropriate even if a shorter range form[6] like that given be the solution of the two body Schroedinger equation should be preferable. On the other hand we believe that u_s should reflect not only the repulsive part of $v(r)$, what makes impenetrable the quantum holes, but also its attractive part when $v(r)$ has such part as in the case of He. In fact the correlation between two centers of mass in the polymer representation reflects the interaction between ν monomers so that one recovers the full strength of $v(r)$ from the reduced one $v(r)/\nu$ between a couple of monomers. The spread of the chains, though, will somewhat weaken the depth of $v(r)$ and its minimum will be displaced to larger distance. We conjecture then that u_s should be attractive, i.e. $e^{-u_s(s)}$ should overshoot unity, at intermediate distances. A simple ansatz for u_s is the form

$$u_s(s) = v(ds)/A \tag{9}$$

where A and d are variational parameters. d is expected to be slightly smaller than unity and A has the role of an effective temperature and is expected to be of the order of the kinetic energy per particle.

Vitiello and Mac Farlaine have performed a variational computation with the pseudopotential (9). The preliminary results at equilibrium density give a substantial reduction of E_o as shown in the table and the

Table1. Ground state energy of ^{4}He at equilibrium density from GFMC with the HFDHE2 potential[7], for an optimal Jastrow[15] and Jastrow plus triplet[15], for the repulsive shadow pseudopotential[2] eq.(5) and for the attractive one eq.(9) (parameters C=5.5, b=1.12, d=0.94, A=10.5 K). All energies are in Kelvins.

GFMC	J	J+T	Shadow eq.(5)	Shadow eq.(9)
−7.12	−5.94	−6.86	−6.19	−6.62

optimal parameters A and d are in nice agreement with the expectation. In addition there is a substantial improvement in the radial distribution function g(r) which now is in good agreement with the exact g(r) as given by Green Function Monte Carlo[7] (GFMC). On the contrary the original shadow function with the repulsive u_s gives a rather poor g(r).

The presence of attractive correlations between shadows has an important consequence on the property of self boundness of the system. With standard wave functions of the Jastrow or of the Jastrow plus triplet terms one finds that $E_o(\rho)$ as function of density has a negative minimum for a certain density ρ_o and this represents the equilibrium density within the approximation. However these wave functions do not describe a self bound state in the sense that the density has to be imposed by the boundary conditions, for instance by periodic b.c. in a simulation. In fact if the pseudopotential in a Jastrow function is purely repulsive, i.e. $u(r) > 0$, then N particles will fill uniformly any volume which is available. In the case of droplets or of films one has to impose the finite density by introducing single particle factors which constrain the volume. Also the shadow function with the choice (5) of the pseudopotentials does not describe a self bound state. The situation can be different with the pseudopotential (9) because the system can condense to a well defined density if the attraction between shadows is strong enough. The mechanism is the same as in condensation of classical particles. It is not yet known if this shadow function (9) with the parameters minimizing the energy gives condensation or not but it seems likely so. In fact if we take $r_{ij} = s_{ij}$ for the purpose of estimation of the overall strength of the attractive correlations, the total pseudopotential $2u_s(r) + u(r)$ for Ψ^2 has a minimum -1.4 at $r/\sigma = 1.3$ ($\sigma = 2.556$ Å). The equivalent classical system (the one with pair interaction $v^*(r)$ and at temperature T^* such that $v^*(r)/k_B T^* = 2u_s(r) + u(r)$) is in a state well below its critical temperature and therefore in a condensed state.

We return now to the path integral representation and consider the cross linking of the chains. This is specific to Bose statistics and it is quite essential[8] in order to have Bose Einstein condensation (BEC). The possibility of cross linking means that a monomer representing a ^{4}He atom is not always in the neighborhood of the same center of mass. Actually one could say that there is no center of mass at all. However if the path starting from a particle is made of blobs connected by stretched regions where two particles exchange each other then by centers of mass we mean now the positions of these blobs. Since a particle is not always in the

neighborhood of the same center of mass, this implies for the ground state
that the particle-shadow term should not be confining as in (6) but a more
appropriate form for the particle-shadow factor $K(R,S)$ in (3) is

$$K(R,S) = \prod_{i,j} \lambda(|\vec{r}_i - \vec{s}_j|).$$ (10)

λ should favor configurations in which particles are close to shadows and
a convenient form is

$$\lambda(z) = \exp\left\{B\, e^{-Cz^2/B}\right\},$$ (11)

If $B \gg 1$ the resulting shadow function is equivalent to the previous one
because the attraction is so large that essentially each particle is
confined in the neighborhood of one shadow. For a smaller value of B there
is a finite probability that a particle migrates from the neighborhood of
one shadow to the neighborhood of another one. As Fantoni discussed at
this meeting with this form of shadow function it is easier to extend the
diagrammatic method and HNC formalism because the problem is isomorphous
to that of a three components mixture of classical particles in which one
component (the "particles" of the quantum problem) has the tendency to
associate with the two other components (the two sets of shadows in Ψ_o^2).

What we are considering now is a more general shadow function and we
might expect that it gives a lower energy. This gain might be not too
large because by itself localization of a particle around one shadow or
around another one does not lower the energy since the shadow label is a
dummy index. What we predict is that the shadow function with the non
confining form (10) can have a substantially larger BEC than with the
confining form (6). In our previous discussion we have put in relation the
shadow function with the confining form (6) to the path integral without
cross linking between paths. But absence of cross linking implies absence
of BEC and one might wonder what happens with the shadow function in this
respect. The point is that the previous connection was made between the
diagonal part of the density matrix and Ψ_o^2. For the momentum distribution
we need the off diagonal part

$$\Psi_o(\vec{r}'_1, \vec{r}_2 .. \vec{r}_N)\Psi_o(\vec{r}_1, \vec{r}_2 .. \vec{r}_N)$$ (12)

and the two shadows of particle 1 need not to be close to each other since
there is no direct coupling between them in (12). These two shadows have
only to be close, respectively, to $\vec{r}_1$ and to $\vec{r}'_1$ and one can show[5] that
the one particle reduced density matrix has a non zero limit when $|\vec{r}_1 - \vec{r}_2| \to \infty$

so that BEC is present. However both $\vec{r}_1$ and $\vec{r}'_1$ carry a shadow which has to fit in the interstices left by the other shadows so that we might expect a rather low value of BEC. In fact with the function (3-6) the condensate at equilibrium density is $4.5\%^2$, a factor of 2 smaller than the value given by GFMC. The situation is different with the non confining form (10,11) because when we displace particle 1 from $\vec{r}_1$ to $\vec{r}'_1$ we do not have to carry along at the same time also the shadow variable and we expect a substantially larger value of the condensate if the parameter B in (11) is not too large.

<u>Excited states and the density matrix</u>

If the shadow variables represent the quantum holes due to the delocalization of the particles we should expect that suitable shadow functions should be useful in the representation of low energy excited states of the system. For instance if we want to introduce a density fluctuation in the system it appears preferable to introduce this in term of the shadow variables in the form[3]

$$\Psi_{\vec{q}}(R) = \psi_r(R) \int dS \, \psi_s(S) K(R,S) \, \sigma_{\vec{q}}, \tag{13}$$

$$\sigma_{\vec{q}} = \sum_j e^{i\vec{q}\cdot\vec{s}_j} , \tag{14}$$

where the factors ψ_r, ψ_s and $K(R,S)$ are the same as in the ground state. The wave function (13) is an eigenstate of linear momentum and has a resemblance with the Feynman form[9] but the density fluctuation is in terms of the shadow variables and not of the particles. The excitation energy with this wave function has been computed[3] and the roton energy has been found in quite good agreement with experiment not only at the equilibrium density but also at higher density. One important aspect with the wave function (13) is that even if it contains a simple density fluctuation in the shadow variables it implicitly contains backflow with respect to the particle variables. In fact analysis[3] of this wave function shows that it contains terms of all orders in the density fluctuation $\rho_{\vec{q}} = \sum_j \exp(i\vec{q}\cdot\vec{r}_j)$ of the <u>particle</u> variables.

Starting from the Feynman form of the excited states it is possible to construct the low temperature density matrix ρ_T[11,12] and this leads to what is known as the Penrose-Reatto-Chester (PRC) form. unfortunately the Feynman form is a poor approximation for rotons so that the PRC form of ρ_T cannot be expected to be very accurate. If the backflow is introduced by starting with the Feynman-Cohen form[12] it is still possible[13] to construct ρ_T which now contains also three and four body correlations so that it is rather difficult to use. Therefore it is interesting to enquire if it is possible to construct ρ_T starting from the shadow excited states which give a substantially better representation[3] of a roton than the Feynman-Cohen form.

We start by writing the normalization of the excited state (13) in the form

$$N_q = N_o \langle \sigma'_{-\vec{q}} \sigma_{\vec{q}} \rangle_{RSS'} \tag{15}$$

where N_o is the normalization constant of the ground state and $\langle ... \rangle_{RSS'}$ represents the normalized average in the extended configuration space (R,S,S') with respect to the weight

$$\psi_r^2(R)\psi_s(S)\psi_s(S')K(R,S)K(R,S'). \tag{16}$$

$\sigma'_{\vec{q}}$ is the density fluctuation (14) for the $\{\vec{s}'_j\}$ variables. The states corresponding to multiple excitations with population numbers $\{n_{\vec{q}}\}$ can be written in the form

$$\psi_{\{n_q\}}(R) = \psi_r(R)\int dS\, \Pi_{\vec{q}}(\sigma_{\vec{q}})^{n_{\vec{q}}} \psi_s(S)K(R,S)\Big/ N_{\{n_q\}}^{1/2} \tag{17}$$

and the normalization constant reads

$$N_{\{n_q\}} = N_o \langle \Pi_{\vec{q}} \left[\sigma'_{-\vec{q}}\, \sigma_{\vec{q}} \right]^{n_{\vec{q}}} \rangle_{RSS'} . \tag{18}$$

Averages of this kind have been already analyzed[14,13], here we have only to consider these averages of powers of density fluctuations in the wider configuration space and the result is

$$N_{\{n_q\}} = N_o\, \Pi_{\vec{q}}\, n_{\vec{q}}!\, (n_{\vec{q}}/N_o)^{n_{\vec{q}}}. \tag{19}$$

At this point we are ready to write the density matrix in the form

$$\langle R'|\rho_T|R\rangle = Z^{-1}\psi_r(R')\psi_r(R)\int dS'dS\,\psi_s(S')\psi_s(S)K(R,S')K(R,S) \sum_{\{n_{\vec{q}}\}}$$

$$e^{-\beta \Sigma_{\vec{q}} n_{\vec{q}} \varepsilon_q} \; \frac{\Pi_{\vec{q}}(\sigma_{-\vec{q}}\sigma_{\vec{q}})^{n_{\vec{q}}}}{N_0 \Pi_{\vec{q}} n_{\vec{q}} (n_{\vec{q}}/N_0)^{n_{\vec{q}}}} \tag{20}$$

where ε_q is the excitation energy of state $\Psi_{\vec{q}}$ and Z is the partition
function. The sum over excitation numbers in (20) is formally the same of
the PRC case but the quantity $\rho_{\vec{q}}/\sqrt{NS(q)}$ is replaced here by
$\sigma_{\vec{q}}/\sqrt{\langle \sigma'_{-\vec{q}}\sigma_{\vec{q}}\rangle_{RSS'}}$ and $\rho_{-\vec{q}}/\sqrt{NS(q)}$ by $\sigma'_{-\vec{q}}/\sqrt{\langle \sigma'_{-\vec{q}}\sigma_{\vec{q}}\rangle_{RSS'}}$. Therefore the same
summation technique can be used and after some algebra we get the form

$$\langle R'|\rho_T|R\rangle = \psi_r(R')\,\psi_r(R)\int dS'\,dS\,\psi_s(S')\psi_s(S)K(R,S')K(R,S) \; \times$$

$$\Pi_{i<j} e^{-\chi_1^T(s'_{ij})-\chi_1^T(s_{ij})} \; \Pi_{i,j} e^{-\chi_2^T(|\vec{s}'_i-\vec{s}_j|)}/Q_T \tag{21}$$

where Q_T is the normalization constant which follows from

$$\int dR \; \langle R|\rho_T|R\rangle = 1. \tag{22}$$

ψ_r, ψ_s and $K(R,S)$ define the shadow ground state as given, for instance, by
(4) and (6) and the temperature dependent correlating factors read

$$\chi_1^T(s) = \int \frac{d^3q}{\rho(2\pi)^3} \; \frac{\gamma_q^2}{1-\gamma_q^2} \; \frac{N}{\langle \sigma'_{-\vec{q}}\sigma_{\vec{q}}\rangle_{RSS'}} \; e^{i\vec{q}\cdot\vec{s}} \; , \tag{23}$$

$$\chi_2^T(s) = \int \frac{d^3q}{\rho(2\pi)^3} \; \frac{-\gamma_q}{1-\gamma_q^2} \; \frac{N}{\langle \sigma'_{-\vec{q}}\sigma_{\vec{q}}\rangle_{RSS'}} \; e^{i\vec{q}\cdot\vec{s}} \; , \tag{24}$$

$$\gamma_q = \exp(-\beta\varepsilon_q). \tag{25}$$

From (21) it is clear that thermal excitations modify the intershadow
pseudopotential and, more important, they introduce a coupling between the
two kinds of shadows, the S and the S', via the pseudopotential χ_2^T. This
function is negative at short distance and has an absolute minimum at zero
distance so that the factors $\exp\left[-\chi_2^T(|\vec{s}'_i-\vec{s}_i|)\right]$ in (21) favor the overlap
of the two shadows of the same particle. From this it follows that the
range of ρ_T in the off diagonal direction becomes smaller for increasing T
so that also the BEC fraction is a decreasing function of T. It should be
noticed that the density matrix (21) is appropriate only at low temperature where the assumption of independent excitations is a reasonable approximation. Computation of the temperature dependence of the structure
factor S(q) with this shadow density matrix is in progress.

Fermions

The standard variational approach to a fermion fluid like ^{3}He takes for the factor ϕ in the wave function (1) the ideal gas form, i.e. a Slater determinant of plane waves filling the Fermi sphere in the case of the spin polarized state or the product of two Slater determinants in the case of an unpolarized state, one determinant for the spin up atoms and one for the spins down. The simplest choice for the correlating factor F in (1) is a Jastrow form. We address the question if a shadow function can be expected to be useful also for fermions and which form should it have. One choice is to introduce the shadow variables in the symmetric part F. This leaves unchanged the nodal structure of the free gas system and computationally it is the simplest choice. However from the path integral analogy we get an indication that we should proceed in a different way. In fact the path integral representation (7) is valid also for fermions if we introduce a sign factor $(-1)^P$ under the summation over the permutations P. Therefore each cross link of the chains introduces a minus sign. However the paths are still made of blobs so that one can define local centers of mass but now each time there is a stretched part corresponding to a cross link there is a change of sign. Therefore when the monomer representing a ^{3}He atom goes from the neighborhood of one shadow to that of another one there should be a change of sign. This suggests that the Slater determinant should be in terms of the shadow variables. An additional reason for this choice is that the plane waves in shadow variables implicitly contain backflow in the particle variables. As discussed[3] in the case of roton excitation in order to have a variable amount of backflow one should introduce the variable $\vec{s}_i+\alpha\,(\vec{s}_i-\vec{r}_i)$ where α is an additional variational parameter. Then for the spin polarized state the unnormalized trial wave function in configuration space reads

$$\Psi_F(\vec{r}_1..\vec{r}_N) = \psi_r(R)\int dS\ \psi_s(S)K(R,S)\det\{e^{i\vec{k}_\nu\cdot\left[\vec{s}_j+\alpha\,(\vec{s}_j-\vec{r}_j)\right]}\} \tag{26}$$

where the factors ψ_r, ψ_s and $K(R,S)$ can have the same functional form of the Bose case. It will be interesting to perform a computation of the energy with this new wave function whose nodal structure can be rather different from that of the ideal gas when $\alpha\neq-1$.

Acknowledgements: I would like to acknowledge many useful discussions with M.H. Kalos and S.A. Vitiello who have also generously shared their results. This work has been partially supported by Consorzio INFM, by MURST and by CNR under the CNR-NSF cooperative science program.

References

1. S.A. Vitiello, K. Runge and M.H. Kalos, Phys. Rev. Lett. $\underline{60}$, 1970 (1988).

2. S.A. Vitiello, K.J. Runge, G.V. Chester and M.H. Kalos, Phys. Rev.B $\underline{42}$, 228 (1990).

3. W. Wu, S.A. Vitiello and L. Reatto in "Monte Carlo Methods in Theoretical Physics", S. Caracciolo and A. Fabrocini eds. Giardini Editore (in press); W. Wu, S.A. Vitiello, L. Reatto and K.H. Kalos, Phys. Rev. Lett. (in press).

4. S.A. Vitiello, M.H. Kalos and L. Reatto in "Condensed Matter Theories" vol. V, V.C. Anguilera-Navarro ed., Plenum 1990, p.141; S.A. Vitiello, in this volume.

5. L. Reatto and G.L. Masserini, Phys. Rev. B $\underline{38}$, 4516 (1988).

6. M.H. Kalos, D. Levesque and L. Verlet, Phys. Rev. A $\underline{9}$, 2178 (1974).

7. M.H. Kalos, M.A. Lee, P.A. Whitlock and G.V. Chester, Phys. Rev. B $\underline{24}$, 115 (1981).

8. E.L. Pollock and D.M. Ceperley, Phys. Rev. B $\underline{36}$, 8343 (1987).

9. R.P. Feynman, Phys. Rev. $\underline{94}$, 262 (1954).

10. O. Penrose in Proc. Int. Conf. in Low Temperature Physics, J.R. Dillinger ed., Wisconsin University Press, 1958, p.117.

11. L. Reatto and G.V. Chester, Phys. Rev. $\underline{155}$, 88 (1967).

12. R.P. Feynman and M. Cohen, Phys. Rev. $\underline{102}$, 1189 (1956).

13. S. Battaini and L. Reatto, Phys. Rev. B $\underline{28}$, 1263 (1983).

14. F.Y. Wu, J. Math. Phys. $\underline{12}$, 1923 (1971).

15. S.A. Vitiello and D.E. Schmidt (to be published).

HNC THEORY FOR SHADOW WAVE FUNCTIONS

A. Ferrante[1], M. Bernasconi[1], X.Q.G. Wang[1],
S. Fantoni[1,3] and E. Tosatti[1,2]

[1]International School for Advanced Studies, Trieste, Italy
[2]International Centre for Theoretical Physics, Trieste, Italy
[3]Istituto Nazionale di Fisica Nucleare, Sezione di Trieste

Abstract

The hyper-netted-chain theory is applied to the study of strongly interacting Bose and Fermi systems described by the shadow-correlated wave functions. A new class of shadow wave functions is also proposed which, in spite of their greater generality, requires a much simpler HNC scheme to compute distribution functions, density matrices and the energy expectation value. Moreover, HNC-theory allows for a full optimization of the correlations functions, particularly in the case of this new class of extended shadow wavefunctions. Possible applications of the shadow wave functions to the study of the Fractional Quantum Hall effect are also discussed.

1. Introduction

The description of strong interparticle correlations in both Bose or Fermi systems with continuous degrees of freedom is a long-standing problem of much current interest.

In the last few years extensive studies have been carried out on the properties of dense quantum many-body systems, such as liquid and solid helium at zero temperature or nuclear and neutron matter. The strong correlations present in these systems can hardly be treated by means of conventional perturbation theories and have been proved to be reasonably well described with a Jastrow correlation operator by variational calculations. Improvements upon Jastrow-type wave functions have been made either including triplet, backflow and spin correlations in variational calculations[1,2,3] or employing the full machinery of Green Function Monte Carlo method[4,5].

More recently the shadow wave function (SWF) has been proposed[6,7] as a new variational ansatz to compute the properties of solid and liquid ^4He at zero temperature. The form of SWF has been suggested by the structure assumed by the ground state wave function after having iterated once the Green Function integral equation, with an initial wave function of the Jastrow type, and it is given by

$$\Psi_B(\mathbf{r}_1, ..., \mathbf{r}_N) = F(\mathbf{r}_1, ..., \mathbf{r}_N)F_S(\mathbf{r}_1, ..., \mathbf{r}_N), \qquad (1.1)$$

Recent Progress in Many-Body Theories, Vol. 3,
Edited by T.L. Ainsworth et al., Plenum Press, New York, 1992

where

$$F(\mathbf{r_1}, ..., \mathbf{r_N}) = \prod_{i<j}^{N} f_{pp}(r_{ij}), \qquad (1.2)$$

is the Jastrow correlation operator, and

$$F_S(\mathbf{r_1}, ..., \mathbf{r_N}) = \int \prod_{i=1}^{N} \Theta(|\mathbf{r_i} - \mathbf{s_i}|) \prod_{i<j}^{N} f_{ss}(s_{ij}) \, d\mathbf{S}, \qquad (1.3)$$

where $\mathbf{s_i}$ are the so called "shadow" variables; $f_{ss}(s_{ij})$ has the same structure as $f_{pp}(r_{ij})$, namely it heals out to 1 at large intershadow distances, whereas the correlation $\Theta(x)$ between a particle and its associated shadow is taken of the Gaussian form, normalized to 1, namely $\Theta(x) = (c/\pi)^{3/2} \exp(-cx^2)$, where c is a variational parameter.

The presence of the factors F and F_S in Ψ_B allows for a good description of the short-range repulsion, as well as of the more than two-body correlations necessary for solidification. In fact the SWF have been proved to be very efficient in describing both the liquid _and_ solid phase of ^4He, without the need of an ad-hoc Nosanow one-particle factor[6,8]. This feature makes SWF particularly appealing as a tool to study for instance liquid-solid phase transition phenomena in the ground state, metal-insulator transitions, etc. .

Physically, the shadow variables $\mathbf{s_i}$ can be thought of as mimicking the quantum correlation "holes" which the particles carry around themselves in the dense system[9]. This physical interpretation of the shadow variables as well as the request of more variational freedom and of a full symmetry under the exchange of particles and of holes, suggests further extended forms for the shadow wave function (ESWF), in which the factor F_S is now replaced by

$$F_{ES}(\mathbf{r_1}, ..., \mathbf{r_N}) = \int \prod_{i,j}^{N,M} f_{ps}(|\mathbf{r_i} - \mathbf{s_j}|) \prod_{i<j}^{M} f_{ss}(s_{ij}) \, d\mathbf{S}, \qquad (1.4)$$

where M is the number of shadow particles.

The extension which (1.4) represents over the standard SWF of (1.3) concerns two aspects. First, in F_{ES} of (1.4) all the shadow particles are correlated with all real particles, rather than being in a one to one correspondence as in (1.3). This allows the possibility for the hole _number_ and _locations_ to become different from those of the real particles. In the crystallization problem, for example, it may be much more convenient to work with hole shadows which occupy the sites of the dual lattice, rather than the particle lattice. The second aspect, which is however related to the first, is to assume a more general correlation function f_{ps} than the original Gaussian form of eq. (1.3). If the shadow must occupy the dual lattice, f_{ps} must contain a hard core part, very much like f_{pp}.

In Eq (1.4) all the three correlation functions $f_{pp}(x)$, $f_{ps}(x)$ and $f_{ss}(x)$ are taken to heal out to 1 at large values of x. The particle-shadow correlation $f_{ps}(x)$ may then, as discussed above, have a maximum at some value d, to describe a proper binding between holes and particles.

In this contribution we present a first attempt to apply the HNC theory to SWF and ESWF for both Bose and Fermi systems. The wave function for the Fermi case is assumed of the form

$$\Psi_F(\mathbf{r}_1,...,\mathbf{r}_N) = \Psi_B(\mathbf{r}_1,...,\mathbf{r}_N)\Phi_{FG}(\mathbf{r}_1,...,\mathbf{r}_N), \qquad (1.5)$$

where Φ_{FG} is the uncorrelated Fermi Gas (Slater Determinant) wave function, and Ψ_B is given in eq. (1.1).

The existing work on SWF applied to ^4He has been based so far on the Variational Monte Carlo (VMC) method[10,8]. Of course, this is the method of choice for very accurate samplings if the system has to be simulated by a finite number of particles. However, there is clearly a scope in trying to devise alternative approaches, even if more approximate, which could help in taking full advantage of the great additional flexibility of this class of correlated wave functions.

The potential advantage of (F)HNC methods, as proved by previous studies, is to permit an approximate evaluation of averages which is simple and therefore very suitable for a variational approach, where the correlation functions need to be efficiently adjusted, while working strictly at the thermodynamic limit[2,3,11] .

Apart from these considerations, a straight comparison between the recent SWF energy estimates for liquid ($E = -6.24$ at $\rho\sigma^3 = 0.365$) and solid ^4He ($E = -3.563$ at $\rho\sigma^3 = 0.55$)[10], the best Jastrow + Triplets ($E = -6.96$ at $\rho\sigma^3 = 0.365$)[2] and Nosanow + Jastrow + Triplets ($E = -3.786$ at $\rho\sigma^3 = 0.55$)[10] and the GFMC results ($E = -7.12$ at $\rho\sigma^3 = 0.365$)[5] indicates that more work needs to be done if one wants to exploit the full power of the SWF approach, even for the simple Bose case. It is worth noticing that the optimal forms of f_{pp}, f_{ps} and f_{ss} may be easily found within HNC theory, by solving appropriate Euler-Lagrange equations[12], whereas any such optimization is all but handy in VMC.

The plan of this paper is the following. The (F)HNC for the SWF is discussed in Sect. 2, and Sect. 3 presents the (F)HNC scheme for ESWF. It is shown that ESWF are actually much better suited for a simple (F)HNC treatment than SWF. In fact the distribution functions are those of a three-component mixture in the Bose case and of a two-boson and one-fermion component mixture in the Fermi case. This is not true for the simple SWF form of (1.3), which leads to a much more involved HNC scheme. A discussion of possible applications of the HNC schemes for Shadow Wave Function to the Fractional Quantum Hall effect is given in Sect. 4. Particularly, it is shown that the hierarchical states [13.14] are of the ESWF type and that the proposed HNC theory can be used to calculate the energy of such states. Sect. 5 is left to the discussion and conclusions.

2. HNC Treatment for SWF

The cluster expansions of the distribution functions for the SWF are based on the reference state Ψ_{MF} obtained from Ψ_B of eq. (1.1) by setting the correlations $f_{pp}(x)$ and $f_{ss}(x)$ equal to 1, with the result that $\Psi_{MF} = 1$ for Bose systems and $\Psi_{MF} = \Phi_{FG}$ for Fermi systems.

The cluster expansion of a particular distribution function is obtained by expanding it in power of $h_{pp}(x) = f_{pp}^2(x) - 1$ and $h_{ss}(x) = f_{ss}(x) - 1$. Each term of the cluster expansion is called cluster term and is more conveniently represented by a diagram made up of bonds and points. The limitation due to the one to one correspondence between shadow and real particles in eq. (1.3) leads to a variety of topologically different points in the cluster diagrams and, consequently, to a quite involved HNC scheme which, however, can be handled numerically.

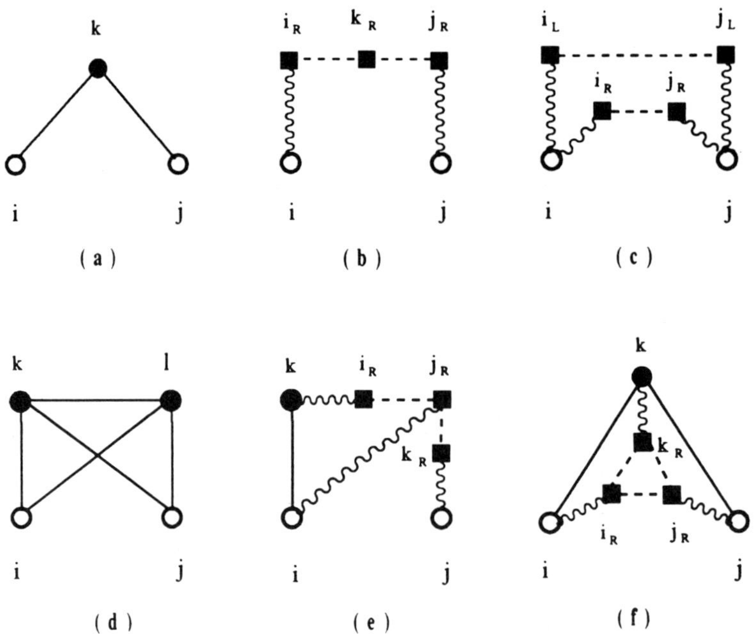

Fig.1-*Cluster diagrams contributing to $g_{pp}(r_{ij})$ for SWF. The solid, dashed and wavy lines represent h_{pp}, h_{ss} and Θ bonds respectively. The particle variables are represented by circles and the shadow by squares.*

2a) Bose case

In the Bose case one must distinguish five types of bonds in correspondence to the five functions $h_{pp}(x)$, $h_{ss}^R(x)$, $h_{ss}^L(x)$, $\Theta^R(x)$ and $\Theta^L(x)$, where h_{ss}^R and Θ^R come from Ψ_B while h_{ss}^L and $\Theta^L(x)$ from Ψ_B^*. Although $h_{ss}^R = h_{ss}^L$ and $\Theta^R = \Theta^L$, diagrammatically they need to be distinguished. Correspondingly, there are eight different types of points:

(i) four types of particle-points, denoted as p_i ($i = 0, 3$); p_0 is reached by h_{pp}-bonds only (p-points of diagrams (1a) and (1d)), p_1 is also reached by Θ^R (p-points of diagrams (1b), (1e) and (1f)), p_2 by Θ^L and p_3 by both Θ^R and Θ^L (p-points of diagram (1c));

(ii) two types of right (left) shadow-points, denoted as $s^{R(L)}$ ($i = 0, 1$); $s_0^{R,L}$ is reached by $h_{ss}^{R(L)}$ only (s-point of diagram (1b)), and $s_1^{R(L)}$ is also reached by $\Theta^{R(L)}$ (s-points of diagrams (1c), (1e) and (1f)).

The following diagrammatical rules result from the properties of the cluster expansion:

(a) a point p can be reached by an unlimited number of h_{pp}-bonds;

(b) a point $s^{R(L)}$ can be reached by an unlimited number of $h_{ss}^{R(L)}$-bonds. An h_{ss}-bond can never join s^R with s^R points;

(c) h_{pp}-bonds are never connected to s-points as well as h_{ss}-bonds are never connected to p-points;

(d) $\Theta^{R(L)}$ connects a p-point with its associate $s^{R(L)}$-point; there may be at most one $\Theta^{R(L)}$ reaching $s^{R(L)}$ and at most one Θ^R and one Θ^L connected to a point p.

Few examples of cluster diagrams of the nodal (diagrams (1a), (1b) and (1e)), composite (diagram (1c)) and bridge (diagrams (1d) and (1f)) type are displayed in Fig. 1. One can see from diagrams (1e) and (1f) that triplet correlations are induced by pair correlations with shadow particles.

The diagrammatical rules listed above lead to the following HNC equations:

$$N_{\alpha_i \beta_j}(x_{12}) = \sum_{\gamma_k \delta_l} A_{\gamma_k \delta_l} \int \left(g_{\delta_c \beta_j}(x_{32}) - N_{\delta_c \beta_j}(x_{32}) - \delta_{l0}\delta_{j0} \right)$$
$$\left(g_{\alpha_i \gamma_k}(x_{13}) - \delta_{i0}\delta_{k0} \right) \, dx_3, \tag{2.1}$$

where the indices $\alpha_i, \beta j, \ldots$ denote the particular type of external point of the nodal (N) or distribution (g) functions. The greek letters $\alpha, \beta, \ldots$ are here used to indicate p, s^R or s^L, whereas the latin letters are used for the subindices and run from 0 to 3 for p and from 0 to 1 for both s^R and s^L. The matrix A embodies the diagrammatical rules. It is given by

$$A_{p_i p_j} \equiv T_{ij} \equiv \begin{pmatrix} \rho & \rho & \rho & \rho \\ \rho & o & o & o \\ \rho & o & \rho & o \\ \rho & \rho & o & o \end{pmatrix}, \tag{2.2}$$

$$A_{s_i^L s_j^L} = A_{s_i^R s_j^R} \equiv S_{ij} \equiv \begin{pmatrix} \rho & 1 \\ 1 & 0 \end{pmatrix}, \tag{2.3}$$

and it vanishes elsewhere. The 64 distribution functions $g_{\alpha_i \beta_i}(x)$ may be written in the following highly compact form:

$$g_{\alpha_0 \beta_0}(x) = (1 + \delta_{\alpha\beta} h_{\alpha\alpha}(x)) \, e^{S_{\alpha_0 \beta_0}(x)}, \tag{2.4}$$

$$g_{\alpha_i \beta_j}(x) = g_{\alpha_0 \beta_0}(x) \sum_{<ij>} \prod_n S_{\alpha_{i(n)} \beta_{j(n)}}(x), \tag{2.5}$$

with the functions $S_{\alpha_i \beta_j}(x)$ given by

$$S_{\alpha_i \beta_j}(x) = \left(\delta_{\alpha_i p_1} \delta_{\beta_j s_1} + \delta_{\alpha_i s_1} \delta_{\beta_j p_1} \right) \Theta(x) + N_{\alpha_i \beta_j}(x) + E_{\alpha_i \beta_j}(x), \tag{2.6}$$

where the functions $E_{\alpha_i \beta_j}(x)$ denotes the bridge (or elementary) diagrams. In eq.(2.5) the summation $\sum_{<ij>}$ is extended over all the possible products of S-functions, with the exclusion of $S_{\alpha_0 \beta_0}(x)$, under the conditions $\sum i(n) = i$ and $\sum j(n) = j$, where each set $\{i(n)\}$, $\{i(n)\}$ may contain the values 1 and 2 at most once. For instance, $i(n) \equiv \{0, 0, 3\}$ or $\{0, 1, 2\}$ are allowed combinations, whereas $\{0, 1, 1\}$ is not. Such an algebraic rule ensures that the superposition of subdiagrams $S_{\alpha_i \beta_j}$ satisfies the diagrammatical rules $(a) - (b)$.

Considering the symmetry $N_{\alpha\beta} = N_{\beta\alpha}$ and that $h_{ss}^R = h_{ss}^L$ and $\Theta^R = \Theta^L$, one ends up with only 21 different distribution functions. Therefore, the HNC scheme requires the solution of 21 coupled integral equations, which is a completely feasible numerical problem. In fact the main difficulty with this method is on the approximation of the bridge diagrams, which are expected to be not negligible. However, it should be noted that, already at the level of HNC/0 approximation ($E_{\alpha\beta} = 0$), the effect of the induced triplet and higher order correlations is taken into account (see diagram (1e)).

Following ref. [2] a reasonable approximation for the bridge functions may consist on taking $E_{p_0 p_0} \approx s E_{p_0 p_0}^{(4)}$, $E_{s_0 s_0} \approx t E_{s_0 s_0}^{(4)}$ and $E_{\alpha_i \beta_j} \approx E_{\alpha_i \beta_j}^{(4)}$ for $(\alpha_i \beta_j) \neq (p_0 p_0), (s_0 s_0)$ where the elementary diagrams $E_{\alpha\beta}^{(4)}$ are calculated using the $g - 1$ bonds. The scaling factors s and t are calculated by equating the Jackson-Feenberg (JF), the Pandharipande-Bethe (PB) and the Clark-Westhaus (CW) kinetic energy expectation values among themselves.

The full particle-particle distribution function $g_{pp}(x)$ is obtained by summing up all the distribution functions of the pp-type:

$$g_{pp}(x) = \sum_{i,j=0,3} g_{p_i p_j}(x). \tag{2.7}$$

2b) Fermi case

In addition to the bosonic links, previously discussed, one has to include also the fermionic exchange link $\frac{1}{2} l(k_F x)$, where k_F is the Fermi momentum and $l(x) = \frac{3}{x} j_1(x)$ is the Slater function. This link connects only the points p_i among themselves. As in standard FHNC theory[15,16] the points p_i reached by exchange lines (at most two) need to be distinguished from those connected to bosonic bonds only.

We denote such points with e_i or c_i whether the related exchange loop is closed or not. Obviously, the exchange loops in any cluster diagrams are always closed. However, as it is well known, FHNC theory introduces subdiagrams which may have open exchange loops and that is where the point c_i come in. The subindex i in both e_i and c_i runs from 0 to 3 as in p_i. The extra-diagrammatical rules due to the exchange bonds trivially follows from FHNC theory.

The nodal equations have the same structure as in eq. (2.1). The indices $\alpha_i, \beta_j, \ldots$ run here over a larger class of points, namely p_i $(i = 0, 3)$; e_i $(i = 0, 3)$; c_i $(i = 0, 3)$, s_i^R $(i = 0, 1)$, s_i^L $(i = 0, 1)$ so that A is a 16x16 matrix, given by

$$A_{\alpha\beta} \equiv \begin{pmatrix} T & T & 0 & 0 & 0 \\ T & 0 & 0 & 0 & 0 \\ 0 & 0 & T & 0 & 0 \\ 0 & 0 & 0 & S & 0 \\ 0 & 0 & 0 & 0 & S \end{pmatrix}, \tag{2.8}$$

where the 4x4 matrix T and the 2x2 matrix S are given in eqns. (2.2) and (2.3) respectively.

The distribution functions $g_{\alpha_i \beta_j}(x)$ have the same structure as in eqns. (2.4) and (2.5). In addition to the 21 distribution functions g_{pp}, g_{ss}, g_{ps} of the Bose case, there are here 38 new functions having e or c as indices, namely g_{pc}, g_{ee}, g_{ce} and g_{cs}. The explicit expressions of these distribution functions can be trivially obtained by generalizing eq. (2.4) to the Fermi case by using standard FHNC theory[16] and will not be reported here for the sake of brevity.

The above FHNC scheme for SWF, although feasible, is quite involved numerically, and one would like to switch to the much simpler case of ESWF, discussed in the next section.

3. Extended Shadow Wave Functions

The ESWF are characterized by a Bose factor Ψ_B, given by eq.(1.4), in which all the shadow particles are correlated to all the real particles and, consequently, all the three correlations $f_{pp}(x_{ij})$, $f_{ss}(x_{ij})$ and $f_{ps}(x_{ij})$ heal out to 1 at large values of x_{ij}.

As in the case of SWF, discussed in the previous section, the reference state Ψ_{MF} is 1 in the Bose case and Φ_{FG} in the Fermi case, and the links are constituted by the five bosonic functions $h_{pp}(x_{ij}) = f_{pp}^2(x_{ij}) - 1$, $h_{ss}^R(x_{ij}) = f_{ss}(x_{ij}) - 1$, $h_{ps}^R(x_{ij}) = f_{ps}(x_{ij}) - 1$, $h_{ps}^L(x_{ij}) = h_{ps}^R(x_{ij})$, $h_{ss}^L(x_{ij}) = h_{ss}^R(x_{ij})$ and the exchange function $\frac{1}{2} l(k_F x_{ij})$.

The cluster diagrams are characterized by only three different types of points p, s^R and s^L (which coincide with p_0, s_0^R, s_0^L of the SWF case). Due to the fact that Ψ_{ES} is symmetric under the exchange of the shadow variables $\mathbf{s_i}$, irrespective of $\mathbf{r_i}$, there are only four independent HNC quantities, i.e., N_{pp}, $N_{ps^R} = N_{ps^L}$, $N_{s^R s^R} = N_{s^L s^L}$, $N_{s^R s^L}$ for the Bose case plus four extra quantities N_{pc}, N_{sc}, N_{ee}, N_{cc} for the Fermi case. This implies the solution of four (Bose) or eight (Fermi) coupled integral equations, to compute pair correlation functions and the energy expectation value. In fact the normalization in the Bose case is

$$\langle \Psi_{ES} \mid \Psi_{ES} \rangle = \tag{3.1}$$

$$= \int \prod_{i<j}^{N} f_{pp}^2(r_{ij}) \prod_{i<j}^{M} f_{ss}(s_{ij}^R) \, f_{ss}(s_{ij}^L) \prod_{i,j}^{N,M} f_{ps}(|\mathbf{r_i} - \mathbf{s_j^R}|) \, f_{ps}(|\mathbf{r_i} - \mathbf{s_j^L}|) \, d\mathbf{R} \, d\mathbf{S^R} \, d\mathbf{S^L},$$

and coincides with the partition function of a classical three-component (p, s^R, s^L) system interacting via the (pseudo)potentials $u_{pp} = -lnf_{pp}^2$, $u_{s^R s^R} = u_{s^L s^L} = -lnf_{ss}$, $u_{p s^R} = u_{p s^L} = -lnf_{ps}$, $u_{s^R s^L} = 0,$. The integral equations to calculate the distribution functions are well known[17]. The nodal and closure equations have a much simpler structure than the corresponding eqns. (2.1-5) and are given by

$$N_{\alpha\beta}(x_{12}) = \sum_{\gamma} \rho_{\gamma} \int (g_{\alpha\gamma}(x_{13}) - 1)(g_{\gamma\beta}(x_{32}) - N_{\gamma\beta}(x_{32}) - 1) \, dx_3 \tag{3.2}$$

where $\rho_{\gamma} = (\rho_p, \rho_s, \rho_s)$ and ρ_s, ρ_p are respectively the densities of shadows and particles, and

$$g_{\alpha\beta}(x_{12}) = e^{-u_{\alpha\beta}(x_{12}) + N_{\alpha\beta}(x_{12}) + E_{\alpha\beta}(x_{12})}. \tag{3.3}$$

Similarly, in the Fermi case, one has a mixture of two Bose fluids (s^R, s^L) and one Fermi fluid (p), whose FHNC treatment can be easily derived from that proposed by Fabrocini and Polls[18] for the case of the Boson-Fermion mixture.

The energy per particle

$$\frac{\langle E \rangle}{N} = \frac{\langle T \rangle}{N} + \frac{\langle V \rangle}{N}, \tag{3.4}$$

can be written in terms of the two and three-body distribution functions only. This differs from the case of the Jastrow+Triplet trial wave function, which requires the evaluation of up to the five-body distribution functions if the PB form of kinetic energy is employed. We give in the following the expressions of $<T>^{JF}$, $<T>^{PB}$ and $<V>$ for the Bose case. The potential energy expectation value is given by

$$\frac{\langle V \rangle}{N} = \frac{\rho_p}{2} \int g_{pp}(r)v(r) \, dr, \tag{3.5}$$

where $v(r)$ is the interatomic potential. The kinetic energy expectation value can be split into a two-body plus a three body term

$$\frac{\langle T \rangle}{N} = T_{2B} + T_{3B}, \tag{3.6}$$

where

$$T_{2B} = \int [\rho_p \, t_{pp}(r)g_{pp}(r) + \rho_s \, t_{ps}(r)g_{ps}(r)] \, dr, \tag{3.7}$$

and the two-body kinetic energy operators $t_{pp}(r)$ and $t_{ps}(r)$ are given by

$$t_{pp}^{JF}(r) = \frac{\hbar^2}{8m} \left\{ u''_{pp}(r) + \frac{2}{r}u'_{pp}(r) \right\}, \tag{3.8}$$

$$t_{ps}^{JF}(r) = \frac{\hbar^2}{4m} \left\{ u''_{ps}(r) + \frac{2}{r}u'_{ps}(r) - [u'_{ps}(r)]^2 \right\}, \tag{3.9}$$

in the case of the JF kinetic energy expression, and by

$$t_{pp}^{PB}(r) = \frac{\hbar^2}{4m}\left\{ u''_{pp}(r) + \frac{2}{r}u'_{pp}(r) - \frac{1}{2}\left[u'_{pp}(r)\right]^2 \right\}, \qquad (3.10)$$

$$t_{ps}^{PB}(r) = 2\,t_{ps}^{JF}(r), \qquad (3.11)$$

for the PB kinetic energy expression. The three-body terms are given by:

$$T_{3B}^{JF} = -\frac{\hbar^2\rho_s^2}{4m}\int \left[g_{p_sL\,s_R}^{(3)}(\mathbf{r_{12}},\mathbf{r_{13}}) - g_{p_sR\,s_R}^{(3)}(\mathbf{r_{12}},\mathbf{r_{13}}) \right] u'_{ps}(r_{12})$$
$$u'_{ps}(r_{13})\,\hat{\mathbf{r}}_{12}\cdot\hat{\mathbf{r}}_{13}\,\mathbf{dr_{12}}\,\mathbf{dr_{13}}, \qquad (3.12)$$

$$T_{3B}^{PB} = -\frac{\hbar^2}{8m}\int \left[\rho_p^2\,g_{ppp}^{(3)}(\mathbf{r_{12}},\mathbf{r_{13}})\,u'_{pp}(r_{12})u'_{pp}(r_{13}) \right.$$
$$- 4\rho_s^2\,g_{p_sR\,s_R}^{(3)}(\mathbf{r_{12}},\mathbf{r_{13}})\,u'_{ps}(r_{12})u'_{ps}(r_{13})$$
$$\left. - 4\rho_s\rho_p\,g_{pps}^{(3)}(\mathbf{r_{12}},\mathbf{r_{13}})\,u'_{pp}(r_{12})u'_{ps}(r_{13}) \right]\hat{\mathbf{r}}_{12}\cdot\hat{\mathbf{r}}_{13}\,\mathbf{dr_{12}}\,\mathbf{dr_{13}}. \qquad (3.13)$$

Similar expressions are found for the Fermi case.

The three-body terms of eqns. (3.12),(3.13) are most conveniently evaluated by first calculating the quantities $\int \hat{\mathbf{r}}_{13}\,u'(r_{13})g_{\alpha\beta\gamma}^{(3)}(\mathbf{r_{12}},\mathbf{r_{13}})\,\mathbf{dr_{13}}$, and then performing the last integration on $\mathbf{dr_{12}}$. Such "effective" distribution functions can be obtained by using HNC theory, as shown in ref. [19] or more recently by Lado[20].

Reliable approximations to the elementary diagrams are needed in order to obtain a true variational estimate of the energy; three approaches are the most successfully used for classical liquid-mixtures, that is (i) resorting to some kind of interpolating closure[21], or (ii) parametrizing $E_{\alpha\beta}$ from a suitable reference hard-spheres system[22,23], or, finally, (iii) evaluating directly $E_{\alpha\beta}^{(4)}$ with $g-1$ links and then scaling or correcting them at short range in a proper way [24,25]. The free parameters present in these approaches can be fixed by imposing convenient consistency relations, satisfied by the exact distribution functions, like for instance $<T>^{PB} = <T>^{JF}$. The (F)HNC scheme discussed above can be easily handled numerically, and work for both ^3He and ^4He is in progress[26].

Moreover, the Euler-equations $\delta <H>/\delta f_{\alpha\beta} = 0$, have a simple structure if $<H>$ is calculated in the HNC/0 approximation. We hope to learn, by solving these equations in the correlation functions f_{pp}, f_{ps} and f_{ss} the main features of their analytical behaviour at various densities ρ_p and different phases.

4. Application to the Fractional Quantum Hall effect

A further interesting application of the ESWF is provided by the Fractional Quantum Hall effect. The hierarchical states [27,28,29] are described microscopically by electron wave function which can be expressed as an ESWF. Different formulations for the FQH hierarchical states have beeen proposed; one possible choice for the second level of the hierachy has been provided by MacDonald et al.[13,14] in terms of the following wave function

$$\Psi_\nu = \prod_{j<k}^{N}(z_j - z_k)^{p+1}\,e^{-\sum_{j=1}^{N}|z_j|^2/4l_0^2}\int \prod_{j,k}^{N,M}(z_j - s_k)\prod_{j<k}^{M}(s_j^* - s_k^*)^m(s_j - s_k)$$
$$e^{-\sum_{j=1}^{M}|s_j|^2/2l_0^2}\,\mathbf{dS},$$

where $l_0 = \sqrt{\hbar c/eB}$ is the magnetic length, $N + M = mN$, and both the coordinates z_k and s_k are in the complex notation $x_k - iy_k$. The wave function Ψ_ν is given by the product of the particle-hole conjugate of $\Psi_{\nu_0=1/m} = \prod (z_j - z_k)^m$ with the polynomial $\prod (z_j - z_k)^p$ and therefore[13] its filling ν is given by $\nu_{-1} = (1-\nu_0)^{-1} + p$. For instance $m = 3$ and $p = 2$, $\nu = 2/7$.

The exponential factors in (4.1) are relevant only in the trivial long wavelength limit, therefore Ψ_ν is an ESWF with $f_{pp}(jk) = (z_j - z_k)^{p+1}$, $f_{ps}(jk) = (z_j - s_k)$ and $f_{ss}(jk) = (s_j^* - s_k^*)^m (s_j - s_k)$.

As in the HNC treatment of the $\nu = 1/m$ states[30], $f_{\alpha\beta}(jk)$ are written in the exponential form $exp(-u_{\alpha\beta}(jk))$ and then each $u_{\alpha\beta}(jk)$ is separated into a short-range integrable term $u_{\alpha\beta}^s(jk)$ and a long-range term $u_{\alpha\beta}^l(jk)$ which can be analytically handled in the HNC theory. In fact, this important feature, well known in the study of the classical (two-dimensional) one-component plasma (see, for instance ref. [31]), takes care of the overall charge neutrality of the system, in spite of the fact that the $f_{\alpha\beta} - 1$ have a long-range behavior.

A major difference between the well known HNC treatment for the $\nu = 1/m$ states and the present for the ESWF of eq. (4.1), is that the various HNC quantities are complex functions and depend on both x_{jk} and y_{jk} components of the interparticle distance t_{jk}. Both the correlations $f_{ps}(t_{jk})$ and $f_{ss}(t_{jk})$ depend on the polar angle θ_{jk}, namely $f_{ps}(jk) = |t_{jk}| \exp(-i\theta_{jk})$ and $f_{ss}(jk) = |t_{jk}|^4 \exp(2i\theta_{jk})$. As a consequence, the long range parts $u_{ps}^l(jk)$ and $u_{ss}^l(jk)$ contain respectively the terms $-i\theta_{jk}$ and $2i\theta_{jk}$, which brings the angular dependence into all the HNC quantities.

It turns out that the nodal functions $N_{\alpha\beta}(jk)$ and the composite functions $c_{\alpha\beta}(jk) = g_{\alpha\beta}(jk) - 1 - N_{\alpha\beta}(jk)$ have a long range behavior exactly given by $-u_{\alpha\beta}^l(jk)$ and $+u_{\alpha\beta}^l(jk)$, so that all the distribution functions $g_{\alpha\beta}(jk)$ are short ranged. The HNC equations can be easily solved numerically, exploiting the usual device of doing analytically the Fourier transform of the long range part $N_{\alpha\beta}^l(jk)$, needed in the algorithm of solution[31]. Work in this direction is in progress[32].

The HNC procedure briefly discussed above is not limited to the particular form of the wave function as given in eq. (4.1). The wave functions obtained through the Fractional Statistic Transformation[33] is also of the ESWF form, only the detailed structure of f_{ss} changes. The HNC procedure can provide much insight into the different formulations through an explicit calculation of the properties of the ground states wave functions. Moreover, one can consider a completely general $f_{ss}(jk)$ and then determine its shape by solving a proper Euler equation. This may turn out to be extremely important to improve upon the hierarchical ansatz, or to study the state at $\nu = 1/2$, whose nature is of great interest[34,35].

5. Discussion and conclusions

In this contribution we have applied the HNC theory to strongly interacting Bose and Fermi systems described with SWF. The solution of the (F)HNC integral equations turns out to be numerically feasible and therefore this method is alternative to VMC and it is expected to be very powerful to exploit the great flexibility of SWF.

The interpretation of shadow variables as correlation holes which particles carry around themselves, suggests a more general form of shadow function, the ESWF, in which all the shadow particles are correlated to all the real particles. In addition this wave function, having a full symmetry under the exchanges of particles and of shadows, requires a much simpler (F)HNC scheme than the original SWF. The calculation of the distribution functions is similar to that for a three-component mixture. In this case the solution of the Euler equation to optimize the correlation functions is a much more manageable numerical problem.

The FHNC scheme presented in this contribution can be easily extended to the case of an inhomogeneous system to study atomic droplets or liquid-solid interfaces. The integral equations in the case of a one-body density $\rho(\mathbf{r})$ which is not a constant, can be derived from the FHNC procedure proposed for the case of the simple Jastrow ansatz[36] and applied to study nuclei[37,38] or surface properties[39].

Finally, we have given a preliminary discussion of a possible direct application of ESWF to the Fractional Quantum Hall effect.

REFERENCES

[1] K. Schmidt, M.H. Kalos, M.A. Lee, G.V. Chester: *Phys. Rev. Lett.* **45**, 573 (1980); *Phys. Rev. Lett.* **47**, 807 (1981).

[2] Q.N.Usmani, S. Fantoni, V.R. Pandharipande: *Phys. Rev.* **B 26**, 6123 (1982).

[3] E.Manousakis, S.Fantoni, V.R.Pandharipande, Q.N.Usmani: *Phys. Rev.* **B 28**, 3770 (1983);
M. Viviani, E. Buendia, S. Fantoni, S. Rosati: *Phys. Rev.* **B 38**, 4523 (1988).

[4] M.H. Kalos, M.A. Lee, P.A. Whitlock, G.V. Chester: *Phys. Rev.* **B 24**, 115 (1981).

[5] M.A Lee, K.E Schmidt, M.H. Kalos: *Phys. Rev. Lett.* **46**, 728 (1981).

[6] S. Vitiello, K. Runge, M.H. Kalos: *Phys. Rev. Lett.* **60**, 1970 (1988).

[7] L. Reatto, G.L. Masserini: *Phys. Rev.* **B 38**, 4516 (1988).

[8] L. Reatto, contribution to this volume.

[9] O. Gunnarson, B.I. Lundqvist: *Phys. Rev.* **B 13**, 4274 (1976).

[10] S.A Vitiello, K.J. Runge, G.V.Chester, M.H. Kalos: *Phys. Rev.* **B 42**, 228 (1990).

[11] X. Q. Wang, S. Fantoni, E. Tosatti, L. Yu and M. Viviani: *Phys. Rev.* **B 41**, 11479 (1990).

[12] L.J. Lantto, P.J. Siemens: *Phys. Lett.* **B 68**, 308 (1977).

[13] S.M. Girvin: *Phys. Rev.* **B 29**, 6012 (1984).

[14] A.H. Mac Donald, G.C. Aers, M.W.G. Dharma-wardana: *Phys. Rev.* **B 31**, 5529 (1985); A.H. Mac Donald, D.B. Murray: *Phys. Rev.* **B 32**, 2707 (1985).

[15] J. W. Clark: in *Progress in Particle and Nuclear Physics*, edited by D. H. Wilkinson (Pergamon, Oxford, 1979), Vol 2.

[16] A. Fabrocini and S. Fantoni: in *First International Course on Condensed Matter*, edited by D. Prosperi, S. Rosati and G. Violini (World Scientific, Singapore, 1986), ACIF series, Vol. 8.

[17] J. P. Hansen, I. R. McDonald: *Theory of Simple Liquids* (2nd Ed., Academic Press, London, 1986).

[18] A. Fabrocini, A. Polls: *Phys. Rev.* **B 25**, 4533 (1982); *Phys. Rev.* **B 30**, 1200 (1984).

[19] S. Fantoni, S. Rosati: *Phys. Lett.* **B 84**, 23 (1979).

[20] F. Lado: *Mol. Phys.* **72**, 1387 (1991).

[21] J.P. Hansen, G. Zerah: *J. Chem. Phys.* **84**, 2336 (1986).

[22] Y. Rosenfeld, N.W. Ashcroft: *Phys. Rev.* **A 20**, 1208 (1979).

[23] F. Lado: *Phys. Rev.* **A 8**, 2548 (1973).

[24] S.Ichimaru, H.Iyetomi: *Phys. Rev.* **A 27**, 3241 (1983).

[25] P. Ballone, G. Pastore, M. P. Tosi: *J. Chem. Phys.* **81**, 3174 (1984).

[26] A. Ferrante, private communication.

[27] B.I. Halperin: *Phys. Rev. Lett.* **52**, 1583, 2390 (1984).

[28] R.B. Laughlin: *Surf. Sci.* **141**, 11 (1984).

[29] F.M.D. Haldane: *Phys. Rev. Lett.* **51**, 605 (1983).

[30] R.B. Laughlin: *Phys. Rev. Lett.* **50**, 1395 (1983).

[31] F. Lado: *Phys. Rev.* **B 17**, 2827 (1978).

[32] M. Bernasconi, private communication.

[33] B. Blok, X.G. Wen: *Phys. Rev.* **B 43**, 8337 (1991).

[34] C.D. Chen, E. Tosatti: *Physica* **A 172**, 336 (1991); G. Fano, F. Ortolani, E. Tosatti: *Nuovo Cimento* **9**, 1337 (1987).

[35] M. Greiter, X.G. Wen, F. Wilczek: *Phys. Rev. Lett.* **66**, 3205 (1991).

[36] S. Fantoni, S. Rosati: *Nucl. Phys.* **A 328**, 478 (1979).

[37]   E. Krotscheck: *Nucl. Phys.* **A 465**, 461 (1987).

[38] G.P. Co' et al., SISSA preprint (1991).

[39]   E. Krotscheck, G.X.Qian, W. Kohn: *Phys. Rev.* **B 31**, 4267 (1985); *Phys. Rev.* **B 32**, 5693 (1985).

MANY-BODY PROBLEMS IN ATOMIC PHYSICS

Ingvar Lindgren

Department of Physics
Chalmers University of Technology/University of Gothenburg
S-412 96 Göteborg, Sweden

INTRODUCTION

During the last decade a new and exciting field has opened up in atomic physics, namely the field of heavy-ion spectroscopy. Highly ionized atoms have been studied for a long time, for instance, using astrophysical light sources. More recently, however, new experimental tools have become available, such as laser-produced plasmas and plasma discharges in Tokamaks and similar devices, which make it possible to study long iso-electronic sequences, such as copper-like ions up to uranium (for a review, see, for instance, Martinsson 1989). In heavy-ion accelerators, like UNILAC at GSI in Darmstadt, GANIL in Caen and the BEVALAC at Berkeley, highly stripped ions can be produced, up to hydrogen-like uranium. Quite recently, for instance, very accurate experiments have been reported from Berkeley on Li-like uranium (Schweppe et al 1991). Here, the Lamb shift in the 2s-2p transition is about 40 eV, to be compared with the experimental uncertainty of 0.1 eV. Experimental results of this kind demonstrate the importance of developing a many-body procedure, which incorporates quantum-electrodynamic (QED) effects into the many-body formalism in a consistent way.

The problem of treating interacting many-electron systems quantum-mechanically has been a challenge since the advent of quantum mechanics, and many different procedures have been developed during the years. Two schemes, based on the variational principle, were introduced very early, namely the Hylleraas procedure with the interelectronic distance explicitly in the wave function and the self-consistent-field (SCF) technique, introduced by Hartree and further developed by Slater and Fock. These techniques are still of great importance in atomic physics. The traditional configuration-interaction (CI) technique, where the mixing coefficients - but not the orbitals as in MCHF or MCDF - are varied, is still frequently used in molecular applications, but less so in atomic problems. The Hylleraas procedure, which in practice is limited to few-electron systems - essentially He-like ones - was used with great success by Pekeris in the 1950's and has now been developed to extreme precision particularly by Drake (1982, 1988). A related procedure is the hyperspherical method (Fano 1983), where new coordinates are introduced, particularly suited for representing systems with two excited electrons. The SCF technique in the multi-configurational form is a powerful many-body procedure, which has been developed particularly by Froese-Fischer (1977) in the non-relativistic case (MCHF) and by Desclaux (1975) and Grant et al (1980) in the relativistic case (MCDF).

The second class of many-body procedures is non-variational, and these can be either perturbative (of order-by-order type) or non-perturbative (all-order type). The perturbative scheme most frequently used is the Brueckner-Goldstone or Linked-Diagram Expansion (LDE), based on graphical technique (Brueckner 1955, Goldstone 1957, Lindgren-Morrison 1986). This scheme is commonly referred to as Many-Body Perturbation Theory (MBPT). All perturbative schemes, however, are more or less prohibited beyond third-order wave function or fourth-order energy, while in many atomic problems it is necessary to go to beyond that order to match the experimental accuracy. This has led to the development of various nonperturbative schemes, where certain effects are evaluated to all orders in the perturbation and other effects are taken to low order or left out completely. The coupled-cluster approach, first developed in nuclear physics by Coester and Kümmel (1960) and introduced into quantum chemistry by Čižek (1966), is the most powerful and most widely used non-perturbative scheme. Several groups have contributed significantly to the development of this technique and carried out extensive calculations. (For references, see, for instance, Bartlett 1991).

The SCF and MBPT schemes are nowadays also extensively used in the relativistic framework. No explicit relativistic many-body hamiltonian exists, but approximate procedures can be derived with arbitrary accuracy from quantum-electrodynamics (QED). The simplest relativistic many-body procedures is based on the Dirac single-electron equation and the nonrelativistic Coulomb interaction,

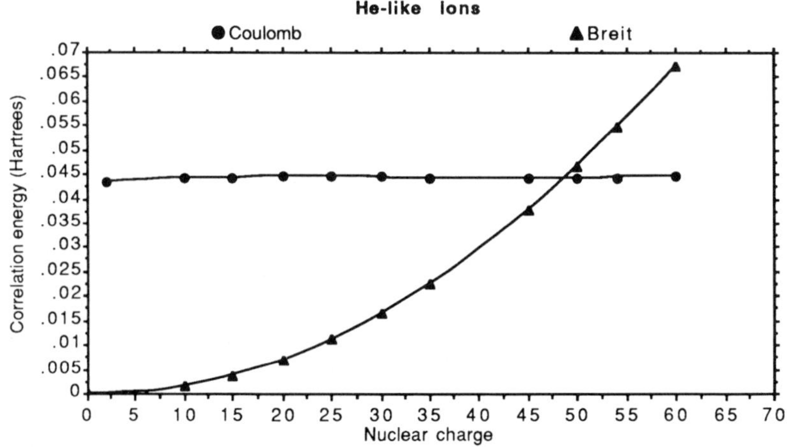

Figure 1. Contributions to the correlation energy of the ground-state of He-like ions due to Coulomb and Breit interactions (from Gorceix et al 1987).

and this has been the basis for many multi-configuration Dirac-Fock (MCDF) calculations as well as some early relativistic MBPT calculations. Already for medium-heavy elements, however, it is important to take the relativistic (magnetic and retardation) effect on the electron-electron interaction into account, usually represented by the Breit interaction. In order to illustrate this effect, we consider the contributions to the ground-state energy of He-like ions, shown in Fig 1, taken from Gorceix et al (1987). This shows that the relativistic effect on the electron-electron interaction becomes quite appreciable already for $Z \approx 20$, and becomes comparable to the non-relativistic second-order energy for $Z \approx 50$. Nowadays, the Breit interaction is often included in a self-consistent way in such calculations, although the effect of multiple Breit interactions is quite small. This is done in the no(-virtual)-pair approximation, which implies that the effect of negative energy states (electron-positron-pair production) is omitted. Radiative effects, such as the

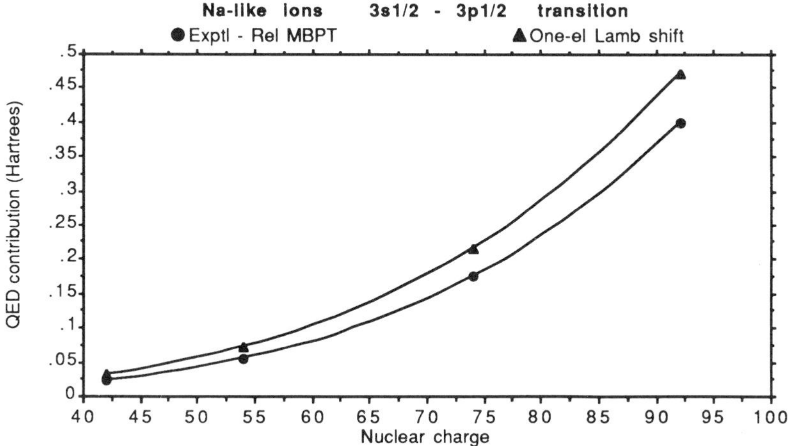

Figure 2. The difference between experimental and relativistic MBPT (no Lamb shift) results for the binding energy of the 3s electron in Na-like ions, compared with the calculated one-electron (unscreened) Lamb shift. (From Johnson et al 1988b and Seely and Wagner 1990).

Lamb shift, and also left out - or included only on the one-electron level. Recent experimental information concerning highly charged ions, however, show significant deviations from the predictions of the relativistic many-body theory, even when one-electron Lamb-shift contributions are included. This is illustrated by the results for Na-like systems in Fig. 2, taken from the many-body calculations of Johnson et al (1988b) and the analysis of recent experimental data by Seely and Wagner (1990). This indicates that also *many-body* radiative effects, such as *screening* of the Lamb shift, are quite significant. The problem of including QED effects beyond the first-order level into the many-body procedure in a consistent way is now one of the hottest issues in this field of research and of vital importance for future progress in the field.

In this paper, we shall first briefly review the non-relativistic MBPT and summarize the recent progress in that field. In the second part of the paper, we shall describe various relativistic MBPT procedures, starting from the QED formalism. In particular, we shall indicate how effective electron-electron interactions to be used in MBPT procedures can be derived in a systematic way. Indications will also be given how QED effects *beyond* standard MBPT, such as effects of Lamb shift and virtual electron-positron pairs, might be incorporated in such a procedure.

NON-RELATIVISTIC MANY-BODY THEORY

Basic concepts

The aim of atomic many-body theory is to solve the Schrödinger equation

$$H\,\Psi = E\,\Psi \tag{I.1}$$

for an interacting many-electron system with sufficient accuracy. Non-relativistically the hamiltonian for an N-electron atom is of the form

$$H = \sum_{i=1}^{N} h_S(i) + \sum_{i<j=1}^{N} 1/r_{ij} \tag{I.2}$$

where h_S is the Schrödinger hamiltonian for a single electron in the field of the nucleus

$$h_S = -\tfrac{1}{2}\nabla^2 - \frac{Z}{r} \tag{I.3}$$

(Hartree atomic units are used throughout this paper with $e=m=\hbar=4\pi\varepsilon_0=1$).

In perturbation theory the hamiltonian is partitioned into an unperturbed hamiltonian, H_0, and a perturbation, V,

$$H = H_0 + V \tag{I.4}$$

Normally, H_0 is a sum of single-particle hamiltonians

$$H_0 = \sum_{i=1}^{N} h_0(i) \quad ; \quad h_0 = -\tfrac{1}{2}\nabla^2 - \frac{Z}{r} + u(r) \tag{I.5}$$

where u(r) is a central (local or non-local) potential representing some average interaction of the other electrons (such as Hartree-Fock).

Taking a closed-shell system as an illustration, the exact wave function can be expanded as

$$\Psi = \Phi + \sum_{a,r} c_a^r \Phi_a^r + \sum_{\substack{a<b \\ r<s}} c_{ab}^{rs} \Phi_{ab}^{rs} + \dots \tag{I.6}$$

where Φ is the unperturbed wave function (single determinantal product of single-particle functions) and an eigenfunction of H_o

$$H_o \, \Phi = E_o \, \Phi \tag{I.7}$$

Φ_a^r, Φ_{ab}^{rs} represent single, double ... substitutions

$$\Phi_a^r = a_r^\dagger a_a \Phi; \quad \Phi_{ab}^{rs} = a_r^\dagger a_s^\dagger a_b a_a \Phi \quad \text{etc} \tag{I.8}$$

using the notation of second-quantization. The single-particle functions (orbitals) are eigenfunctions of h_o, and we use a, b, .. (r, s, ..) to represent spin-orbitals (un)occupied in Φ. This gives

$$\Psi = (1 + \sum_{a,r} c_a^r a_r^\dagger a_a + \sum_{\substack{a<b \\ r<s}} c_{ab}^{rs} a_r^\dagger a_s^\dagger a_b a_a + ...) \Phi$$

$$= (1 + \sum_{a,r} x_a^r a_r^\dagger a_a + \frac{1}{2} \sum_{\substack{a,b \\ r,s}} x_{ab}^{rs} a_r^\dagger a_s^\dagger a_b a_a + ...) \Phi \tag{I.9}$$

In the configuration-interaction technique (CI) the expectation value of the hamiltonian is minimized by varying the mixing amplitudes c_a^r, c_{ab}^{rs}, .. In perturbation theory the corresponding amplitudes $[c_a^r = x_a^r$, $c_{ab}^{rs} = x_{ab}^{rs} - x_{ab}^{sr}]$ are instead determined by successive approximations.

The expression in the parentheses of (I.9) represents the wave operator, which is the key operator in many-body theory. This operator transforms the unperturbed wave function into the exact one,

$$\Psi = \Omega \, \Phi \tag{I.10}$$

and in the present case it can be expressed

$$\Omega = 1 + \sum_{a,r} x_a^r a_r^\dagger a_a + \frac{1}{2} \sum_{\substack{a,b \\ r,s}} x_{ab}^{rs} a_r^\dagger a_s^\dagger a_b a_a + ... \tag{I.11}$$

This operator is normal-ordered with respect to the true vacuum (with the creation operators, $a^\dagger$, to the left of the annihilation operators, a).

In the closed-shell case we have only two kinds of orbitals - those occupied and those unoccupied in Φ. In the open-shell case we also have *valence* orbitals, partly filled and partly empty in the model space. This leads to a degeneracy problem, which implies that we have to solve the Schrödinger equation for a number of states simultaneously

$$H \, \Psi^{(a)} = E^{(a)} \, \Psi^{(a)} \qquad (a=1,2,.....d) \tag{I.12}$$

The zeroth-order wave functions (ZOWF), $\Psi_o^{(a)}$, are confined to a model space (P), which must contain all eigenfunctions degenerate with the state(s) under consideration. On the other hand, the model space need not be completely degenerate. It can contain several energy levels or configurations and is then referred to as a multi-configuration or "quasi-degenerate" model space (Lindgren and Morrison 1986). In the standard procedure we consider here, the model space is assumed to be complete in the sense that it contains all possible occupancies of the valence orbitals. (Sometimes this leads to an inconveniently large model space, and different ways of getting around this problem will be discussed at the end of this section).

The wave operator transforms *all* zeroth-order functions of the model space to the corresponding exact ones

$$\Psi^{(a)} = \Omega \, \Psi_0^{(a)} \qquad (a=1,2,.....d) \tag{I.13}$$

The functions $\Psi^{(a)}$ form a corresponding target space, and we assume a one-to-one correspondence between the states of the two subspaces.

In the general case, the ZOWF are not known in advance but can be obtained by means of the effective hamiltonian, H_{eff}, operating entirely within the model space, with the property

$$H_{eff}\Psi_0^{(a)} = E^{(a)}\Psi_0^{(a)} \quad (a=1,2,.....d) \tag{I.14}$$

Its eigenvectors, $\Psi_0^{(a)}$, are the ZOWF, and its eigenvalues, $E^{(a)}$, are the corresponding *exact* energies.

The wave operator satisfies the Bloch Equation (Bloch 1958a,b; Lindgren 1974; Kvasnička 1974, 1977; Lindgren-Morrison 1986)

$$\left[\Omega,H_0\right]P = \left(V\Omega - \Omega V_{eff}\right)P \tag{I.15}$$

where P is the projection operator for the model space and V_{eff} is the effective interaction

$$V_{eff}P = \left(H_{eff} - H_0\right)P \tag{I.16}$$

Equivalent expressions can also be obtained without partitioning the hamiltonian (I.4). The Bloch eqn (I.15) and the effective hamiltonian then become

$$H\Omega P = \Omega H_{eff}P \; ; \quad H_{eff}P = \Omega^{-1}H\Omega P \tag{I.17}$$

In intermediate normalization (IN) the ZOWF are the projections of the full wave functions onto the model space, which leads to

$$\Psi_0^{(a)} = P\Psi^{(a)} \; ; \quad P\Omega P = P \quad \text{and} \quad V_{eff}P = PV\Omega P \tag{I.18}$$

Most calculations performed so far have employed this normalization in combination with a complete model space. Below we shall discuss schemes where both these restrictions are relaxed. The formulas given in this section, however, (apart from I.18) are generally valid.

Perturbation theory

In perturbation theory the wave operator is expanded order by order

$$\Omega = \Omega^{(0)} + \Omega^{(1)} + \Omega^{(2)} + \dots \tag{I.19}$$

[where $\Omega^{(0)}=1$ in IN]. Inserting this into the Bloch eqn

$$\left[\Omega^{(n)}, H_o\right]P = \left(V\Omega - \Omega V_{eff}\right)^{(n)}P \tag{I.20}$$

generates the Rayleigh-Schrödinger (RS) expansion.

The RS expansion contains unphysical "unlinked" terms, which cancel to a large extent, as first shown by Brueckner (1955) and Goldstone (1957). This leads to the so-called linked-diagram expansion (LDE), which was generalized to the general open-shell case by Brandow (1967). Using the Bloch eqn above this theorem can be formulated

$$\left[\Omega^{(n)}, H_o\right]P = \left(V\Omega - \Omega V_{eff}\right)^{(n)}_{linked}P \tag{I.21}$$

which formally can be regarded as an expansion of

$$\left[\Omega, H_o\right]P = \left(V\Omega - \Omega V_{eff}\right)_{linked}P \tag{I.22}$$

The exact meaning of the terms "linked" and "unlinked" will not be given here, but the reader is referred to standard texts in the field (Lindgren-Morrison 1986).

Non-Perturbative or All-Order Procedures

The order-by order expansion of the wave operator often leads to slow convergence. An alternative - and usually much more effective - way is to expand the wave operator in terms of zero-, one-, two-, .. body terms, defined by means of second-quantization,

$$\Omega = \Omega_o + \Omega_1 + \Omega_2 + \dots \tag{I.23}$$

and to treat these (with some truncation) to all orders of perturbation theory.

In the general open-shell case it is convenient to express the second-quantized wave operator in the following form (Lindgren-Morrison 1986)

$$\Omega = \Omega_o + \sum_{i,j} x^i_j \left\{a^\dagger_i a_j\right\} + \frac{1}{2}\sum_{ijkl} x^{ij}_{kl} \left\{a^\dagger_i a^\dagger_j a_1 a_k\right\} + \dots \tag{I.24}$$

where the curly brackets represent *normal-ordering with respect to a closed-shell "vacuum"*. In this form the *particle-hole (p-h)* creation operators appear to the left of the p-h annihilation operators, defined with respect to a closed-shell system. The second term in (I.24) represents the one-body term, the third term the two-body term and so on. The Bloch eqn (I.22) can then be separated into a set of coupled n-body equations:

$$\left[\Omega_n, H_o\right]P = \left(V\Omega - \Omega V_{eff}\right)_{n,\,linked} P \tag{I.25}$$

which (up to some n_{max}), can be solved iteratively. This will then be equivalent to evaluating the corresponding effects to *all orders* of perturbation theory.

The electron correlation is heavily dominated by the effective two-electron interaction. Therefore, a very powerful and efficient all-order procedure is the pair correlation approach with truncation after the two-body term

$$\Omega = \Omega^{(0)} + \Omega^{(1)} + \Omega^{(2)} \tag{I.26}$$

For closed-shell systems, the one-body effect is rather small, if Hartree-Fock orbitals are used in zeroth order, due to the Brillouin theorem. In open-shell cases, on the other hand - or if a non-Hartree-Fock potential is used - the one-body term is quite important. It has the effect of changing the input orbitals to approximate Brueckner orbitals (Löwdin 1962; Lindgren et al 1976; Lindgren and Morrison 1986, Sects 12.2 and 14.5).

The Coupled-Cluster Approach

It was observed by Sinanŏglu early in the 1950's that the most important effects left out in a CI calculation with single and double excitations are *quadruple* excitations of a particular kind. Denoting the effect of the double excitations by T_2, he found that the quadruple excitations could largely be regarded as products of double excitations (disconnected diagrams) and represented by the operator $\frac{1}{2}T^2$. This was generalized by Hubbard (1957) and particularly by Coester and Kümmel (1960) in a very elegant way to the exponential Ansatz

$$\Omega = \exp S = 1 + S + \frac{1}{2!}S^2 + \frac{1}{3!}S^3 \ldots \tag{I.27}$$

and first applied to nuclear problems. The operator S (or T), which is then completely connected, is often referred to as the cluster operator. This formalism was introduced into quantum mechanics by Čižek and further developed and applied by Paldus and Čižek (1975), Bartlett and Purvis (1978, 1982) and others.

In the open-shell case there is one complication, since the creation and annihilation operators do not commute for the valence orbital(s). As a consequence, S^2 does not have the same form as the corresponding four-body term, which is normal-ordered (I.24) with all p-h creation operators to the left. A factorization of the type (I.27) then leads to spurious diagrams with connections between the valence lines (Offermann et al 1976). The diagrams do, however, factorize into normal-ordered products, and the spurious diagrams do not appear, if normal (rather than ordinary) products are used in the cluster products. This leads to the normal-ordered exponential Ansatz (Ey 1978; Lindgren 1978)

$$\Omega = \left\{\exp S\right\} = 1 + S + \frac{1}{2!}\left\{S^2\right\} + \frac{1}{3!}\left\{S^3\right\} \ldots \tag{I.28}$$

The cluster operator satisfies an analogous Bloch eqn

$$\left[S,H_o\right]P = Q\left(V\Omega - \Omega V_{eff}\right)_{connected} P \tag{I.29}$$

which can be separated into n-cluster equations in analogy with the wave-operator equation

$$\left[S_n,H_o\right]P = \left(V\Omega - \Omega V_{eff}\right)_{n,connected} P \tag{I.30}$$

Truncating after the two-body term - coupled-cluster pair approximation -

$$S = S_1 + S_2 \tag{I.31}$$

leads to the coupled one- and two-electron equations

$$\left[S_1,H_o\right] = \left\{V + VS + \tfrac{1}{2}VS_1^2 + VS_1S_2 + \ldots - S_1V_{eff,1}\right\}_{1,connected} \tag{I.32a}$$

$$\left[S_2,H_o\right] = \left\{V + VS + \tfrac{1}{2}VS_1^2 + VS_1S_2 + \tfrac{1}{2}VS_2^2 + \ldots - S_2V_{eff,2} - \ldots\right\}_{2,connected} \tag{I.32b}$$

Terms with more than two clusters are here indicated by the dots. These are represented graphically in Fig. 3. This approximation has been applied in a number of non-relativistic calculations (see, for instance, Bartlett and Purvis 1982, Lindgren 1985, Salomonson and Öster 1989a). Relativistic applications will be discussed in the next section.

Recent developments and applications

The many-body procedure described above, utilizing a complete model space and intermediate normalization, can be regarded as the standard model of MBPT (and CCA), and by far most applications have been employing this scheme. The connectivity criteria are fulfilled in this scheme, which implies that the effective hamiltonian and the cluster operator (but not the wave operator) are completely connected. This scheme also fulfills the important condition of size-consistency (Primas 1965; Pople et al 1976) and size-extensivity (Bartlett 1981). Size-extensivity implies that the results scale *linearly* with the size of the system, while size-consistency has a somewhat wider meaning (Mukherjee and Pal 1989). These properties are particularly important in molecular applications - less so in atomic problems. A technique that is not size extensive might yield, for instance, considerably erroneous results for the molecular dissociation energy.

Although the standard model of MBPT works quite well in most cases, it has certain disadvantages,

    1. intruder states are likely to appear, causing convergence problems

    2. the effective hamiltonian is non-hermitian.

In order to get a good representation of the ZOWF - and then fast convergence - it is desirable to make the model space large. This, however, may lead to intruders, i.e. states which do not belong to the manifold considered (target space) but nevertheless fall in the same energy span. Such states cause the perturbation

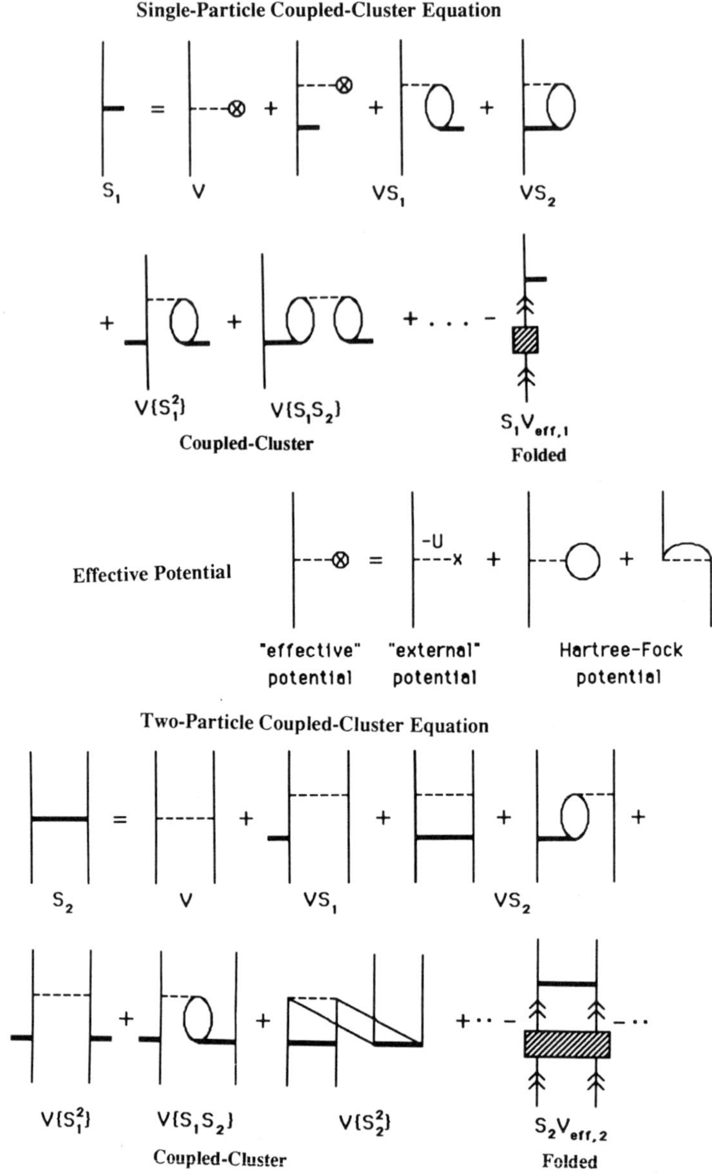

Figure 3. One- and two-particle clusters equations in the pair approximation (I.31). Only examples of diagrams of each category are shown. The lines at the bottom represent incoming core or valence orbitals and the lines at the top outgoing virtual or valence orbitals. (At least one outgoing line must be virtual.) The incoming lines of the folded diagrams must be valence lines.

expansion to diverge, and also non-perturbative approaches encounter serious difficulties (Schucan and Weidenmüller 1972, 1973). Normally, however, one is not interested in *all* target states associated with a large model states, and a way to reduce the probability for intruders - without loosing significant information - is then to reduce the model and target spaces and make the model space *incomplete*. With intermediate normalization (IN), however, *disconnected* diagrams will then appear in the cluster operator and effective hamiltonian, and size-extensivity will be lost.

Various schemes for handling incomplete model spaces have been developed during the last decade. Hose and Kaldor (1979, 1980, 1981) were the first to address the problem seriously. They developed a MBPT procedure by introducing a wave operator for each determinant of the model space, eliminating the concept of valence orbital entirely. Certain disconnected diagrams then appear, but these were claimed not to affect the size extensivity. Monkhorst et al (Jeziorski and Monkhorst 1981; Stolarczyk and Monkhorst 1985) have developed a coupled-cluster formalism based on a similar approach. It was first pointed out by Mukherjee (1986a,b), however, that the source of the disconnectivity is the intermediate normalization (IN). By abandoning that condition, he could show that connectivity could be completely restored for a general incomplete model space with essentially the same formalism as for complete model space. The general criteria for connectivity have been analysed in detail by Lindgren and Mukherjee (1987).

When IN is not used, it is no longer possible to get a simple expression for the effective interaction (I.18). Instead, one has to consider both Q and P projections of the Bloch eqn

$$Q[S,H_o]P = Q\left(V\Omega - \chi V_{eff}\right)_{conn} P \tag{I.33a}$$

$$P[S,H_o]P = P\left(V\Omega - \Omega V_{eff}\right)_{conn} P \tag{I.33b}$$

where $\chi = \Omega - 1$. From the latter eqn, a recursive formula for the effective interaction can be obtained

$$V_{eff} P = P\left(V\Omega - \chi V_{eff} - [S,H_o]\right)_{conn} P \tag{I.34}$$

This represents no major complication in practical applications.

Another very interesting way of handling the intruder problem is to employ the intermediate-hamiltonian (IH) formalism, introduced by Malrieu, Durand and coworkers (Malrieu et al 1985, Durand and Malrieu 1987). Here, the effective hamiltonian is defined in such a way that it reproduces the exact energies only for a *subgroup* of the target states. With this technique one can utilize the larger model space with its good representation of the ZOWF and simultaneously to a large extent avoid the intruder problem. A special version of this technique is to break the one-to-one correspondence between the model and target spaces, as recently described by Evangelisti et al (1991).

Heully and Daudey (1988) have demonstrated the idea of the IH on the well-known problem of the ground-state energy of the Be atom. Here, it is known that a single-configuration model space ($2s^2$) leads to slow convergence due to the quasi-degeneracy of the 2s and 2p levels. An extended model space with the configurations $2s^2$ and $2p^2$ gives good results in second order but diverges thereafter, due to intruders of the type 2sns (Salomonson et al 1980). The IH approach, on the other hand, based on a two-configuration model space, combines good result in second order with convergence of the higher orders. Of course, the IH technique does not help in the case one is interested in the upper ^{1}S state of the Be atom with dominating p^2 character.

Another way to improve the ZOWF is to start from orbitals generated by MCHF - rather than ordinary HF. This has been applied by Morrison and Froese-Fischer (Morrison 1986, 1988; Morrison and Froese-Fischer 1987) to evaluate the second-order correlation energy of the ground state of the Be atom. Numerically their technique is based on the solution of inhomogeneous pair equations. The method has recently been extended by Liu and Kelly (1991) to the relativistic case, using the spline technique. A major problem with this approach is to match the multi-configurational zeroth-order functions to the many-body procedure in a consistent way. Once this problem can be solved, the approach has several attractive features.

As mentioned, intermediate normalization makes the effective hamiltonian *non-hermitian*. Various hermitian coupled-cluster schemes have been developed (Kvasnička 1981; Pal et al 1984 and Haque and Kaldor 1986; Lindgren 1989a, 1991). One way to achieve hermiticity is to apply the Jørgensen condition (1975)

$$P\Omega^\dagger \Omega P = P \qquad\qquad (I.35)$$

rather than the IN condition (I.18), $P\Omega P=P$. However, this is not a necessary condition, and other schemes have been proposed.

The non-hermiticity caused by IN leads to an asymmetry in the representation at a particular level, which can be illustrated by the diagrams in Fig. 4. In the pair-correlation procedure (I.31), where only single and double excitations are considered in the cluster operator, diagram (a) will be included but not its hermitian adjoint (b). The reason for this is that the corresponding wave-operator diagram (d) contains a *triple* excitation, before it is closed. The analogous diagram (c), on the other hand, which after closing gives rise to the effective-hamiltonian diagram (a), contains only single and double excitations and will therefore appear in the pair approximation. This means that the standard pair-correlation approach does *not* contain all energy contributions in third order. The cause of this asymmetry is the intermediate normalization. Physically, of course, diagrams (a) *and* (b) do represent two-body effects and both should be included in a pair-correlation procedure. This can be achieved in a hermitian approach (Lindgren 1989a, 1991), based on the Jørgensen condition (I.35). The effective hamiltonian will then get the manifestly hermitian form

$$H_{eff} P = P\Omega^\dagger H\Omega P \qquad\qquad (I.36)$$

By adding symmetrizing terms to the Bloch eqn

$$Q[S,H_0]P = Q\left(V\Omega - \chi V_{eff} + \chi^\dagger \left(V\Omega - \chi V_{eff}\right)_+\right)_{conn} P \qquad\qquad (I.37)$$

(where the + sign represents effects in the intermediate state outside the approximation employed), the effective hamiltonian will be hermitian also for each truncation. Then all terms indicated in Fig. 4 will be included in the pair correlation. In

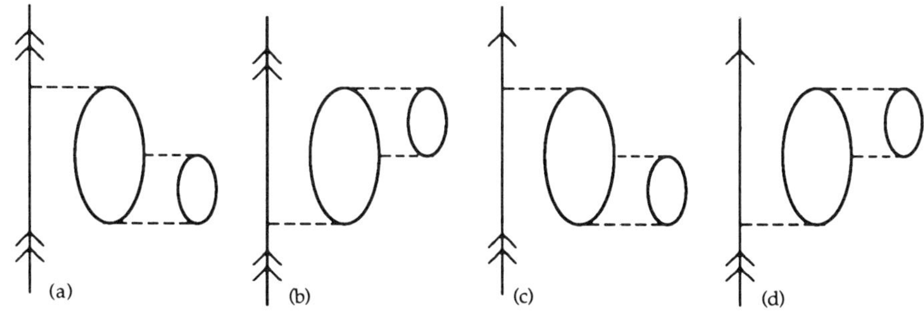

Figure 4. In the pair-correlation approach (I.31) , the effective-operator diagram (a) would be generated in intermediate normalization but not the corresponding hermitian adjoint diagram (b). Similarly, the wave-operator diagram (c) would be included but not the analogous diagram (d), which involves a triple excitation. In the "symmetrized" hermitian formulation (I.35) all these diagrams would be included in the pair approach.

this scheme, *all* third-order diagrams of the effective hamiltonian are obtained within the pair correlation (with singles and doubles). [Expressions similar to (I.37) are obtained by Bartlett et al (1989) with the unitary CC method.] Furthermore, it can be shown that the normalization according to the Jørgensen condition is compatible with connectivity also for an incomplete model space (Lindgren 1991).

The hermitian formulation based on the Jørgensen condition is convenient also for the evaluation of an additional perturbation, h, such as the hyperfine interaction or the isotopic shift. Since the full wave function is normalized in this scheme, the corresponding effective operator - to first order in h - is given by

$$h_{eff} = \Omega^\dagger h \Omega \qquad (I.38)$$

and the expectation value of the perturbation by

$$\left\langle \Psi(a) \middle| h \middle| \Psi(a) \right\rangle = \left\langle \Psi_o(a) \middle| \Omega^\dagger h \Omega \middle| \Psi_o(a) \right\rangle \qquad (I.39)$$

As mentioned, the coupled-cluster pair approach, CCSD, (I.31,32) - treated carefully - usually represents a very good computational scheme, capable of reproducing 95% or more of the correlation effect. Recently, some accurate calculations have been carried *beyond* that approximation.

Bartlett and coworkers (Watts et al 1989; Kucharski and Bartlett 1989) have developed a series of approximations, including three- and four-body clusters (CCSDT, CCSDTQ) and applied them with small basis sets. The results on the Ne atom, for instance, show that the effects of triples is of the order of 3 mH and of quadruples about 0.1 mH, which should be compared with the total correlation energy of about 0.4 H.

Another calculation beyond CCSD has been performed by Salomonson and Ynnerman (1991) on the sodium atom. Here, the Dirac-Fock value for the ground-state ionization energy is 182 033 μH, while the experimental value is 188 858 μH. The CCSD contribution is evaluated to be 6428 μH, which gives a discrepancy of 397 μH or about 6% of the correlation effect. The "hermitian-correction diagrams" in third-order of the kind illustrated in Fig. 4 have been evaluated by Blundell (1990) and found to be 69 μH. Salomonson and Ynnerman have evaluated the most important "genuine" three-body (S_3) effects and found them to amount to 313 μH. With a relativity correction of 30 μH, this gives a total theoretical value of 188 873 μH, which agrees with the experimental value within 1 part in 10^4 or about 0.2% of the correlation. The same technique has also been applied to the potassium atom, where the correction for non-hermiticity seems to be more important.

RELATIVISTIC COMPUTATIONAL SCHEMES

The many-body procedures described above can be used also relativistically - with some modifications - the main difference being the hamiltonians used. As mentioned, no exact relativistic hamiltonian exists, but approximate hamiltonians with various degree of sophistication can be derived from QED.

Until recently, most relativistic many-body calculations were based upon the Dirac-Coulomb hamiltonian

$$H = \sum h_D + \sum \frac{1}{r_{ij}} \qquad (II.1)$$

where the single-electron Schrödinger hamiltonian, h_S, of (I.2) is replaced by the corresponding Dirac hamiltonian

$$h_D = c\boldsymbol{\alpha}\cdot\mathbf{p} + \beta mc^2 - \frac{Z}{r} \qquad (II.2)$$

This has been the basis for many multi-configuration Dirac-Fock (MCDF) calculations (Desclaux 1975, Grant et al 1980) and some early relativistic MBPT calculations. The eigenvalues of this hamiltonian, however, are not bound from below, with the consequence that the eigenstates may dissolve into the negative continuum (Brown-Ravenhall 1951). In the MCDF procedure this is avoided by proper boundary conditions. For a systematic many-body procedure, on the other hand, this hamiltonian does not form a sound basis.

A more rigorous basis for relativistic many-body work is the *projected* Dirac-Coulomb(-Breit) hamiltonian (Sucher 1980)

$$H = \Lambda_+ \left[\sum h_D + \sum \left(\frac{1}{r_{ij}} + B_{ij}\right) \right] \Lambda_+ \qquad (II.3)$$

Here, Λ_+ is the projection operator for positive energy states, which prevents the negative-energy states from entering into the wave function. This approximation is also known as the no-(virtual-)pair approximation (NVPA), since virtual electron-positron pairs are not allowed in intermediate states. Nowadays, it is customary to include in this hamiltonian also the Breit interaction, B_{ij}, which represents the first-order relativistic effects on the electron-electron interaction, i.e., magnetic interactions and retardation of the (instantaneous) Coulomb interaction. This approximation can be derived from QED, and - as will be indicated below - it can be shown to yield all effects of the order α^2 (in Hartree atomic units), when the Coulomb gauge is used. Higher-order QED effects, such as radiative effects (Lamb shifts) and effects of negative energy states, appear first in order α^3 and are omitted in this scheme. The NVPA scheme is now widely used and expected to be quite accurate for most neutral and weakly ionized atoms of naturally abundant elements. It forms the basis for the modern versions of the MCDF procedures (Quiney et al 1987; Gorceix et al 1987; Gorceix and Indelicato 1988), and it has in recent years also been employed in MBPT and coupled-cluster calculations (Johnson et al 1988a,b, 1990; Quiney et al 1989, 1990; Lindroth 1988, 1991; Salomonson and Öster 1989b).

One possible way to include QED effects into the many-body procedure in a more systematic and consistent way is to express the electron-electron interaction in terms of irreducible potential interactions

$$V_{eff} = V_{eff}^{(1)} + V_{eff}^{(2)} + \ldots \qquad (II.4)$$

which, in principle - when used iteratively - would generate the many-body and QED effects to the desired accuracy. The first part, $V_{eff}^{(1)}$, represents the exchange of a *single,* virtual photon. The second part, $V_{eff}^{(2)}$, corresponds to the *"irreducible"* part of the *two-photon* exchange, i.e., the part which is not taken care of by iterating the first part, and so on (Lindgren 1988, 1989b). [In principle, there will also be effective *three-, four-* ... *body* interactions in this scheme, but these effects are found to be extremely small in normal cases (Mittleman 1971; Zygelman and (Mittleman 1986)]. Such a scheme would make it possible to utilize much of the developments that have been made in non-relativistic MBPT and CCA also in the relativistic framework. In the next section we shall indicate how these effective interactions can be derived from QED.

QED EFFECTS

Single-photon exchange

Let us first consider the exchange of a single, virtual photon between two bound-state electrons, which can be represented by the Feynman diagram in Fig. 5, including two different time-orderings. The scattering amplitude is with standard notations given by

$$S^{(2)} = -e^2 \iint d^4x_1 d^4x_2 \ T[(\Psi^\dagger \alpha^\nu A_\nu \Psi)_2 \ (\Psi^\dagger \alpha^\mu A_\mu \Psi)_1]$$

(II.5)

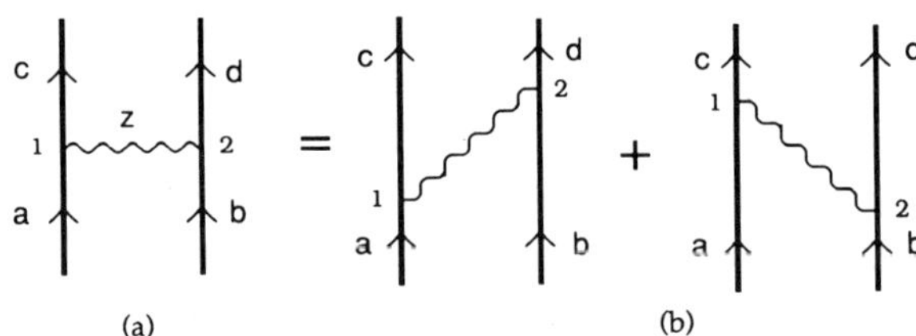

Figure 5. The Feynman representation of the exchange of a single, virtual photon between the electrons (a) involves two different time-orderings (b). The wavy line represents the photon and the heavy, vertical lines the electronic states in the bound-interaction picture.

The field operators are here expressed in the Furry bound-interaction picture (Furry 1951)

$$\Psi = \sum a_i \phi_i(x) \ ; \ \Psi^\dagger = \sum a_i^\dagger \phi_i^*(x)$$

(II.6)

The space part of the orbitals is here generated in an external potential $V(x)$

$$[c\boldsymbol{\alpha}\cdot\mathbf{p}+\beta mc^2 + V(\mathbf{x})]\,\phi(\mathbf{x}) = \varepsilon\,\phi(\mathbf{x}) \tag{II.7}$$

The contraction between the vector-potential operators in (II.5) - denoted by the hook - can be expressed by means of the photon propagator

$$\overline{A_\nu(x_2)\,A_\mu(x_1)} = \left\langle 0\left|T[A_\nu(x_2)A_\mu(x_1)]\right|0\right\rangle = \frac{i}{c}D_{F\nu\mu}(x_2 - x_1) \tag{II.8}$$

Time integrations lead to

$$\left\langle cd\left|S^{(2)}\right|ab\right\rangle = -2\pi i\,\delta(\varepsilon_a + \varepsilon_b - \varepsilon_c - \varepsilon_d)\left\langle cd\left|\alpha_1^\mu\alpha_2^\nu\,e^2c\,D_{F\nu\mu}(\mathbf{x}_2 - \mathbf{x}_1, c\omega_{ac})\right|ab\right\rangle \tag{II.9}$$

where $D_{F\nu\mu}(\mathbf{x}_2 - \mathbf{x}_1, c\omega_{ac})$ is the fourier transform of the photon propagator, defined by

$$D_{F\nu\mu}(x_2 - x_1) = \int\frac{dz}{2\pi}\,e^{-iz(t_2 - t_1)}\,D_{F\nu\mu}(\mathbf{x}_2 - \mathbf{x}_1, z) \tag{II.10}$$

and the omegas represent the orbital-energy differences (in frequency units)
$\omega_{ac} = (\varepsilon_a - \varepsilon_c)/c = \omega_{db} = (\varepsilon_d - \varepsilon_b)/c$.

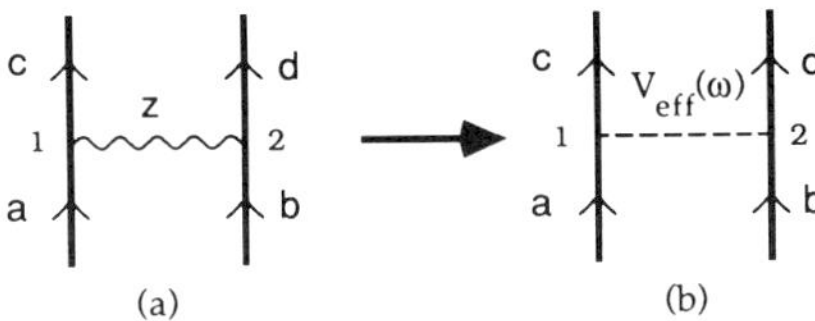

Figure 6. The exchange of a single photon between two electrons is compared with an effective-potential interaction (b).

The result of single-photon exchange (II.9) can be compared with single-potential scattering (Fig. 6)

$$\left\langle cd\left|S^{(1)}\right|ab\right\rangle = -2\pi i\,\delta(\varepsilon_a + \varepsilon_b - \varepsilon_c - \varepsilon_d)\left\langle cd\left|V_{12}\right|ab\right\rangle \tag{II.11}$$

Identification shows that the single-photon exchange is equivalent to the scattering of an (energy-dependent) "effective" potential interaction

$$V_{eff}(\omega) = \alpha_1^\mu\alpha_2^\nu\,e^2c\,D_{F\nu\mu}(\mathbf{x}_2 - \mathbf{x}_1, c\omega) \tag{II.12}$$

The form of the photon propagator depends on the gauge used. Here, we shall consider two gauges, the Feynman gauge and the Coulomb gauge, in which cases the effective interaction becomes

$$V_{eff}^{F}(\omega) = \frac{e^2}{4\pi\varepsilon_o r_{12}}(1-\boldsymbol{\alpha}_1\cdot\boldsymbol{\alpha}_2)\,e^{i|\omega|r_{12}} \tag{II.13a}$$

$$V_{12}^{C}(\omega_{ac}) = \frac{e^2}{4\pi\varepsilon_o}\left\{\frac{1}{r_{12}}-\boldsymbol{\alpha}_1\cdot\boldsymbol{\alpha}_2\frac{e^{i|\omega|r_{12}}}{r_{12}}+\left[\boldsymbol{\alpha}_1\cdot\boldsymbol{\nabla}_{1'}\left[\boldsymbol{\alpha}_2\cdot\boldsymbol{\nabla}_{2'}\frac{e^{i|\omega|r_{12}}-1}{\omega^2 r_{12}}\right]\right]\right\} \tag{II.13b}$$

respectively. The *unretarded* or frequency-independent parts of these interactions, obtained by letting $\omega \to 0$, become

$$V_{eff}^{F}(\omega=0) = \frac{e^2}{4\pi\varepsilon_o r_{12}}\left(1-\boldsymbol{\alpha}_1\cdot\boldsymbol{\alpha}_2\right) \tag{II.14a}$$

$$V_{eff}^{C}(\omega=0) = \frac{e^2}{4\pi\varepsilon_o r_{12}}\left[1-\tfrac{1}{2}\boldsymbol{\alpha}_1\cdot\boldsymbol{\alpha}_2-\frac{(\boldsymbol{\alpha}_1\cdot\mathbf{r}_{12})(\boldsymbol{\alpha}_2\cdot\mathbf{r}_{12})}{2r_{12}^2}\right] \tag{II.14b}$$

These interactions are known as the Coulomb-Gaunt and Coulomb-Breit interactions, respectively.

In principle, the results of QED are *gauge independent* in each order. Nevertheless, it is found that the two interactions above lead to significantly different results, when used in SCF or MBPT calculations (Gorceix and Indelicato 1988; Lindroth and Mårtensson-Pendrill 1989). It should be noted, however, that the single-photon exchange involves energy conservation (II.9), which implies that the results are strictly valid only for evaluating the first-order energy contribution and cannot be used without further analysis in an iterative procedure to evaluate higher-order contributions. This well-known fact, which is usually not appreciated by many-body practitioners, is of vital importance for interpretation of the results, as will be further analysed below.

Two-photon exchange

In order to find the complete form of the effective single-photon interaction that can be used in an iterative process, it is necessary to find also the *non-diagonal* elements, i.e. when energy is *not* conserved. For this purpose we have to investigate the two-photon exchange (Feinberg and Sucher 1988). Here, there are two distinct Feynman diagrams, the ladder and the crossed-photon diagram (Fig. 7).

The scattering amplitude corresponding to the ladder diagram (Fig. 7a) can be represented by means of an effective interaction, V_L, in analogy with the single-photon exchange,

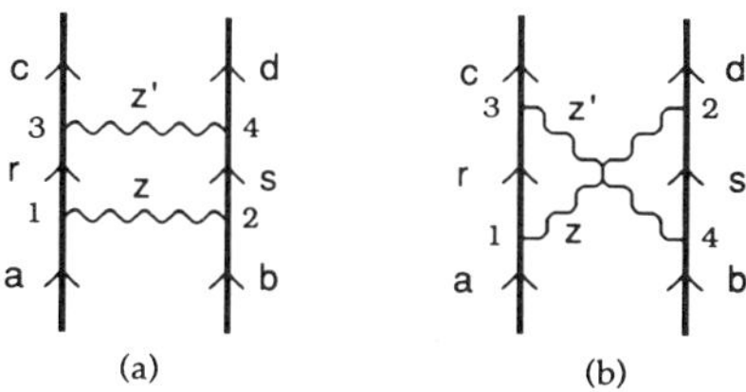

Figure 7. The two-photon exchange between the electrons is represented by two Feynman diagrams, (a) the "ladder" and (b) the "crossed-photon" diagram.

$$\left\langle cd \left| S^{(4)} \right| ab \right\rangle = -2\pi i \, \delta(\varepsilon_a + \varepsilon_b - \varepsilon_c - \varepsilon_d) \left\langle cd \left| V_L \right| ab \right\rangle \qquad (II.15)$$

The interaction V_L is to order α^2 given by (Lindgren 1989b, 1990)

$$\left\langle cd \left| V_L \right| ab \right\rangle = \sum_{rs} \frac{\left\langle cd \left| \frac{1}{r_{12}} \right| rs \right\rangle \left\langle rs \left| V^M \right| ab \right\rangle + \left\langle cd \left| V^M \right| rs \right\rangle \left\langle rs \left| \frac{1}{r_{12}} \right| ab \right\rangle}{\varepsilon_a + \varepsilon_b - \varepsilon_r - \varepsilon_s} \qquad (II.16)$$

where V^M is the Mittleman interaction (Mittleman 1972)

$$\left\langle rs \left| V^M \right| ab \right\rangle = \tfrac{1}{2} \left\langle rs \left| V(\omega_{ar}) + V(\omega_{bs}) \right| ab \right\rangle \qquad (II.17)$$

$V(\omega_{ar})$ and $V(\omega_{bs})$ are the retarded single-photon interactions (II.12,13). In the case of energy conservation this interaction becomes identical to the one-photon interaction (II.12,13).

To order α^2 the interaction (II.16) can also be expressed

$$\left\langle cd \left| V_L \right| ab \right\rangle = \sum_{rs} \frac{\left\langle cd \left| V^M \right| rs \right\rangle \left\langle rs \left| V^M \right| ab \right\rangle}{\varepsilon_a + \varepsilon_b - \varepsilon_r - \varepsilon_s} \qquad (II.18)$$

which is the standard second-order perturbation expression with the interaction V^M. This shows that when this interaction is used in an iterative or perturbative procedure, the ladder diagram will come out correctly to order α^2. This represents the *reducible* part of that diagram. The remaining, irreducible part is of higher order.

However, in order to get all pieces of order α^2, we also have to consider the *crossed* diagram (Fig. 7b). This can be represented by an analogous interaction V_X,

$$\left\langle cd\left|V_X\right|ab\right\rangle = -\sum_{rs} \frac{\left\langle cd\left|\frac{1}{r_{12}}\right|rs\right\rangle\left\langle rs\left|V^{diff}\right|ab\right\rangle + \left\langle cd\left|V^{diff}\right|rs\right\rangle\left\langle rs\left|\frac{1}{r_{12}}\right|ab\right\rangle}{\varepsilon_a + \varepsilon_b - \varepsilon_r - \varepsilon_s} \qquad (II.19)$$

where the interaction V^{diff} is given by

$$\left\langle rs\left|V^{diff}\right|ab\right\rangle = \tfrac{1}{2}\left\langle rs\left|V(\omega_{ar}) - V(\omega_{bs})\right|ab\right\rangle \qquad (II.20)$$

Note that this interaction differs from the Mittleman interaction by the minus sign. This has the consequence that V^{diff} has no frequency-independent part and hence that *only the retardation contributes*. Therefore, this diagram does not have any reducible part.

As mentioned, the retardation contributes to order α^2 in the Feynman gauge but *not* in the Coulomb gauge. Therefore, the crossed-photon diagram is of this order in the Feynman gauge - even with only positive-energy intermediate states - but of higher order in the Coulomb gauge. In order α^2 it is then in the Coulomb gauge sufficient to consider the ladder diagram, while in the Feynman gauge it is necessary to include also the crossed diagram. When a potential interaction is used iteratively - either in a self-consistent procedure or perturbatively - only the *reducible* part of the diagrams will be included. The crossed diagram, which has no reducible part, will then not appear. The fact that this diagram is of order α^2 in the Feynman gauge explains the apparent gauge dependence (Sucher 1988; Lindgren 1990).

The result of this analysis can be summarized in the following way:

The frequency-independent Coulomb-Breit interaction (II.14b) used iteratively will yield all effects to order α^2, provided the Coulomb gauge is used. If the Feynman gauge is used, not even the frequency-dependent interaction (II.13a) will be sufficient to achieve this accuracy. Then also the irreducible two-photon exchange (crossed photons) has to be included.

We have now found the first (one-photon) interaction in the expansion (II.4). To find the complete *two-photon* interaction it is necessary to analyse *three-photon* exchange (as well as the two-photon exchange in more detail) and so on. In principle, any gauge can be used in this procedure, but it seems that the Coulomb gauge will yield the be best result at each level.

Recent developments and applications

The Dirac-Coulomb hamiltonian (II.1) has been the basis for MCDF procedures, developed by Desclaux (1975) and Grant and coworkers (1980) and applied to many systems. The differential equations involved are here solved by the finite-difference method. An MBPT program based on the same hamiltonian and the solution of inhomogeneous one- and two-particle equations with explicit projection operators - also using the the finite-difference method - has been developed by Lindroth (1988; Lindroth et al 1987). This technique, however, is difficult to carry beyond the no-virtual-pair approximation, due to the complicated form of the projection operators.

Grant and coworkers have extended their relativistic self-consistent procedure to include also the frequency-independent Breit interaction in a self-consistent manner - not only as a first-order perturbation (Quiney et al 1987). Treating the Breit interaction on the same level as the Coulomb interaction makes the procedure more covariant (Ishikawa et al 1991). The new version is furthermore based on the algebraic approximation with finite basis set. A similar basis set has also been used by the same group in a relativistic MBPT procedure (Quiney et al 1989, 1990; Wilson 1991). The finite basis set gives a discrete representation of the entire spectrum - without any continuum - and the calculations can be performed entirely algebraicly.

Desclaux and coworkers have also recently included the effect of single photon exchange in their self-consistent procedure and demonstrated the gauge-dependence of this method for He-like ions, as discussed previously (Gorceix et al 1987; Gorceix and Indelicato 1988). The procedure has also been applied to Li-like ions, where comparison is made with available experimental data. The latter calculations are performed in the Coulomb gauge, which - according to the analysis above - must be used in order to yield the correct results to order α^2. When the Lamb-shift contribution - with screening - is included, the agreement with experimental data is very good (see further below).

Relativistic MBPT calculations, utilizing a discrete basis set based on *splines* (piece-wise polynomial fitting, deBoor 1978) has been performed by Johnson et al (1988a,b, 1990) and applied to long sequences of Li-, Na- and Cu-like ions (see Fig 2 above). The calculations are carried out to third-order in the Coulomb interaction and to second order in the Coulomb-Breit (one Coulomb and one Breit interaction). For high nuclear charges (Z>10), this gives sufficient convergence, and the difference from experimental data for the transition energies can be almost entirely explained as QED effects, mainly Lamb shifts, as will be further discussed in the following section.

Another kind of basis set has been introduced by Salomonson and Öster (1989a,b, 1990) by discretizing the radial space. This technique is similar in spirit to the spline method and can be regarded as a limiting case of that method (single-point representation rather than polynomial fitting). This yields one basis function for each grid point. The method, which is very accurate and quite simple to handle, has been applied to some He- and Be-like systems. The results for the He atom are illustrated in Table I, together with corresponding results by Blundell et al (1989). The method of Salomonson and Öster has also been used for evaluation of hyperfine structure, transition energies and isotopic field shifts (Hartley and Mårtensson-Pendrill 1990, 1991) and of parity-non-conservation effects (Hartley et al 1990). The method has recently been extended to include the Breit interaction self-consistently in the generation of Hartree-Fock-Breit orbitals (Lindroth et al 1989) and and in the pair-correlation procedure, combined with Coulomb interactions to all orders (Lindroth 1991). This version has been applied to the ground-state energy of the Be atom (Lindroth et al 1991) and to some transitions in He-like Ar (Lindroth and Salomonson 1990) and Be-like Fe and Mo (Lindroth and Hvarfner 1991). The many-body part is quite accurate, and comparison with available experimental data in the last case shows significant QED effects.

Table I. Ground-state energy of the He atom (in Hartrees).

	Non-relativ.	Relativistic	Rel. corr. (μH)	
		Salomonson-Öster (1989a,b)	Blundell et al (1989)	
First order	2.75	2.750 114 97	114.97	
Correlation	0.153 724 39	0.153 741 54	17.15(5)	17(1)
Total	2.903 724 39	2.903 856 51	132.12(5)	132(1)
Drake (1988)	2.903 724 377	2.903 856 49	132.11	

Table II. Two-photon contribution for He-like ions (in Hartrees)
(from Blundell, Johnson, Mohr, Sapirstein 1990).

	NO VIRTUAL PAIRS			VIRTUAL PAIRS	
	Coulomb-	Coulomb-	Breit-		
Z	Coulomb	Breit	Breit	Single	Double
25	-0.1607	-0.0164	+0.0005	+0.0004	-0.0016
50	-0.1730	-0.0610	-0.0101	+0.0185	-0.0245
75	-0.2030	-0.1349	-0.0101	+0.0185	-0.0245
100	-0.2797	-0.2442	-0.0544	+0.0546	-0.0500

All many-body calculations discussed above are performed in the no-virtual-pair approximation, where the effect of negative-energy states (electron-positron-pair production) is neglected. The effect of such states has been evaluated for He-like ions by Blundell et al (1990), and the result is shown in Table II. This analysis shows that the effect of virtual pairs becomes important only for quite heavy systems. It also shows that the effect of multiple Breit interactions (in the no-virtual-pair approximation) is essentially negligible for Z<50.

The electron self energy

Before meaningful comparison with experiments can be made of many-body calculations for heavy ions, it is necessary to take QED effects - primarily the Lamb shift - into account. The Lamb shift contains two parts, the electron self energy and the vacuum polarization, which in first order are represented by the Feynman diagrams in Fig. 8.

The vacuum polarization can with good accuracy be treated as a potential correction (Uehling potential; Mohr 1982, Soff and Mohr 1988) and will not be further considered here. The self energy, on the other hand, is considerably more complicated to evaluate in the general case, and we shall indicate here how this can be done.

After time integrations, the scattering amplitude corresponding to the first-order self-energy diagram yields the energy shift

$$\Delta E_{SE} = i \iint d^3x_1 \, d^3x_2 \, \phi_a^*(x_2) iec\alpha^\nu \int \frac{dz}{2\pi} iS_F(x_2,x_1,\varepsilon_a - z) iec\alpha^\mu \phi_a(x_1) \frac{i}{c} D_{Fv\mu}(x_2-x_1,z) \qquad \text{(II.21)}$$

where $S_F(x_2,x_1,\varepsilon_a - z)$ is the electron propagator or Green's function

$$S_F(x_2,x_1,\varepsilon_a - z) = \sum_t \frac{\phi_t(x_2)\phi_t^*(x_1)}{\varepsilon_a - z - \varepsilon_t(1-i\eta)} \qquad \text{(II.22)}$$

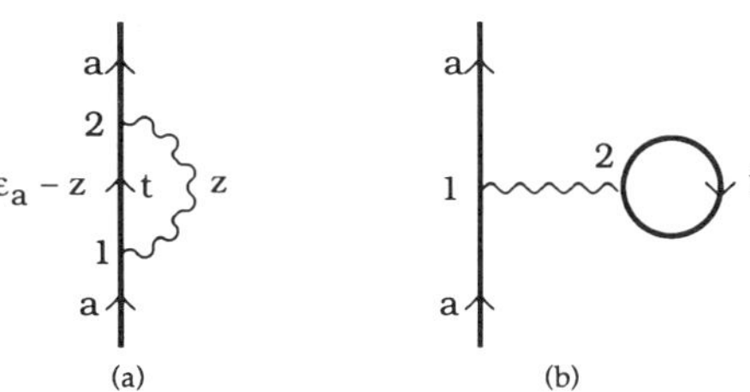

Figure 8. Feynman diagrams for the first-order self energy (a) and vacuum polarization (b) for a bound electron.

Here t runs over all single-particle states - positive as well as negative. $D_{Fv\mu}(x_2-x_1,z)$ is the fourier transform of the photon propagator (II.10)

$$D_{Fv\mu}(x_2-x_1,z) = -\frac{c}{\varepsilon_o} g_{v\mu} \int \frac{d^3k}{(2\pi)^3} \frac{e^{ik\cdot(x_2-x_1)}}{z^2 - c^2 k^2 + i\eta} = g_{v\mu}\frac{e^{i|z|r_{12}/c}}{4\pi\varepsilon_o \, c \, r_{12}} \qquad \text{(II.23)}$$

where $r_{12} = |x_2 - x_1|$. In the Feynman gauge, this leads to with $k = |\mathbf{k}|$.

$$\Delta E_{SE} = -\frac{e^2 c}{2\varepsilon_o} \int \frac{d^3k}{(2\pi)^3} \frac{1}{k} \sum_t \frac{\langle a|\alpha_\mu e^{-ik\cdot x}|t\rangle\langle t|\alpha^\mu e^{ik\cdot x}|a\rangle}{\varepsilon_a - \varepsilon_t - ck\,\mathrm{sgn}\,\varepsilon_t} \qquad \text{(II.24)}$$

The free-electron self energy is contained in the physical electron mass and has to be subtracted from bound-electron self energy, which is the *mass renormalization*. This is illustrated in Fig. 9.

In the non-relativistic limit the bound-electron self energy (II.24) becomes

$$\Delta E_{SE} = -\frac{e^2}{3c\varepsilon_o} \int \frac{d^3k}{(2\pi)^3} \frac{1}{k} \sum_t \frac{\langle a|p\,e^{-ik\cdot x}|t\rangle\cdot\langle t|p\,e^{ik\cdot x}|a\rangle}{\varepsilon_a - \varepsilon_t - ck} \qquad \text{(II.25)}$$

After renormalization, neglecting retardation ($e^{ik\cdot x} \approx 1$), this reduces to the result of Bethe (1947).

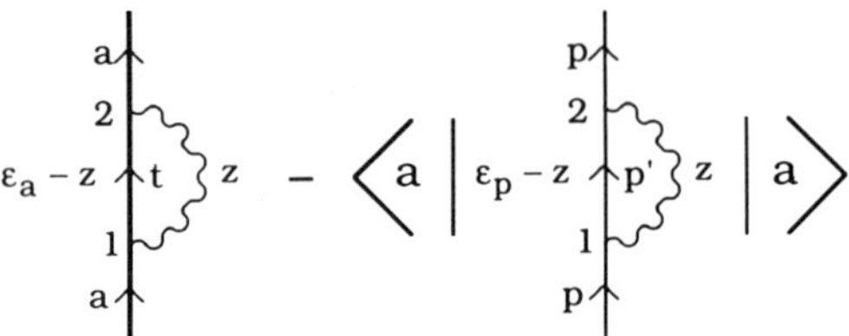

Figure 9. In the mass renormalization the free-electron part of the electron self energy, averaged over the state considered, is eliminated.

In the general case, renormalization is a very delicate problem, due to the strong divergences involved. The standard procedure, introduced by Brown, Langer and Schaefer (1959), is to expand the bound-electron propagator as shown in Fig 10. Here, diagrams (a) and (b) contain infinities, which are cancelled by infinities in the free-electron part, while diagram (c) is finite.

Most Lamb-shift calculations for heavy ions are based on the method of Brown, Langer and Schaefer (BLS), which in practice is limited to $Z \geq 10$. Very accurate calculations have been performed by Mohr (1974a,b, 1982, 1988) for n=1 and 2 on a number of heavy hydrogen-like ions, using a modification of this method. These calculations have recently been extended to n=3-5 by Mohr and Kim (1991).

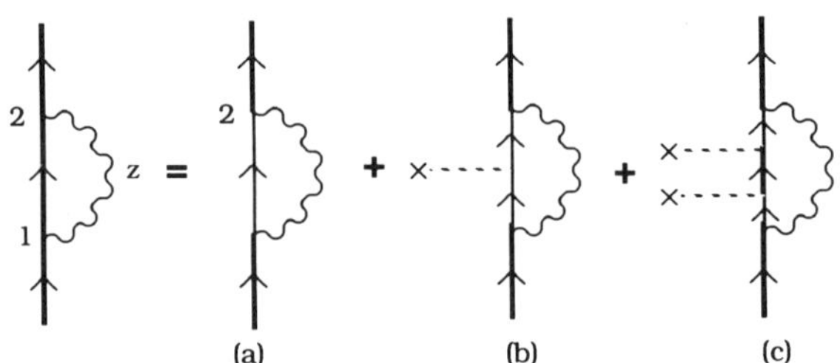

Figure 10. Expansion of the bound-electron self energy according to Brown, Langer and Schaefer (1959).

For systems with more than one electron, the *screening* of the Lamb shift due to the other electrons becomes important. This is a many-body effect, but for alkali-like systems, with one electron outside closed shells, it can in the first approximation be treated as a one-body effect with a modified potential. Calculations of the screening in this approximate manner have recently been performed by several groups. Indelicato and Mohr (1990) treated the spherically averaged Coulomb potential from the core electrons as a small perturbation. Cheng et al (1991) calculated screening by using a local Hartree-Fock-Slater potential, and Indelicato and Desclaux (1990) used a semi-classical approximation for the screening. As an illustration, we consider the 2s-2p$_{1/2}$ transition in Li like uranium, where the Lamb-shift contribution is about 40 eV and the screening about 2 eV, compared to the experimental uncertainty of 0.10 eV (Schweppe et al 1991). By including screening, Indelicato and Mohr (1990) obtained a transition energy of 282.1 eV, Indelicato and Desclaux (1990; Indelicato et al 1987) 281.6(9) eV and Cheng et al (1991) 280.7eV, which all agree well with the experimental result 280.59(10) (Schweppe et al 1991).

For Na-like systems the difference between experiments and relativistic MBPT (without Lamb shift) is compared with the one-electron Lamb-shift in Fig. 2 in the introduction. Cheng et al (1991) have calculated the effect of screening for Na-like platinum to be -5.4 eV, which brings the theoretical value in almost exact agreement with the experimental result.

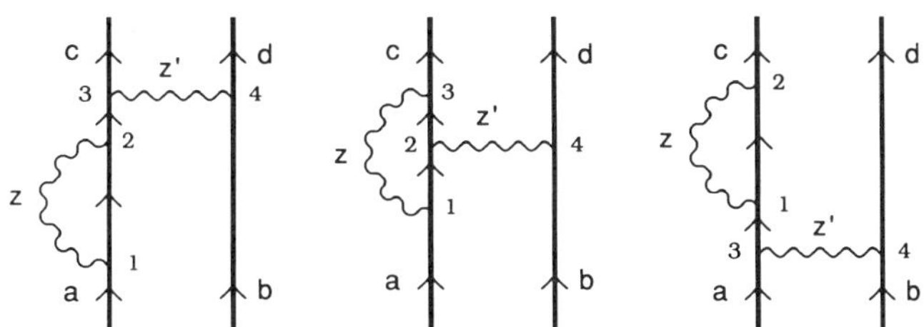

Figure 11   *Higher-order self-energy diagrams, representing   two-body effects (screening) of the Lamb shift.*

For lighter elements (Z<10), it is necessary to treat the bound orbitals more accurately, without relying on the BLS expansion in Fig. 10. Here, the relativistic effects are less dominant, and a non-relativistic approach might be meaningful. It is well-known that, while the relativistic approach converges after renormalization, the non-relativistic method of Bethe leads to *logarithmic divergence*. It has shown by Au and Feinberg (1974), however, that inclusion of retardation makes also the non-relativistic approach convergent. Such an approach has recently been applied by Persson et al (1991), using the above-mentioned discretizing method of Salomonson and Öster (1989a). Here, the renormalization is made numerically in each grid for each photon momentum (k) and each partial wave (l) in the expansion of the photon propagator, before the integrations/summations are performed.

This method has been applied to H and He$^+$, and the results are shown in Table III and compared with the corresponding analytical results of Bethe and Salpeter (1957). By using a local potential ("optimized Hartree-Fock-Slater"), which simulates the HF potential very well, a screening effect of 10% is obtained for the neutral He atom in the ground state. This gives a *one-body* Lamb shift of 190 000 Mhz in the He atom, which should be compared with the value 150 000 MHz obtained by Drake (1982), using accurately correlated wave functions. The difference - about -40,000 MHz - gives an estimate of the size of the genuine *two-particle* effects on the Lamb shift. Such effects are represented by diagrams of the type shown in Fig 11, and for more accurate treatment of the Lamb shift in systems with more than one electron it will be necessary to take these diagrams explicitly into account. Eventually, it might be possible to include such effects in the effective multi-photon interactions discussed previously (II.4).

Table III.   First-order self-energy for hydrogen and helium
(in MHz, from Persson et al 1991)

System	l≤50	extrapol	Bethe-Salpeter
H 1s	7730	7900	7930
H 2s	995	1025	1015
He 1s unscreened	102 100	104 500	102 900
He 1s screened		95 000	
He $1s^2$ (Drake 1982)		150 000	

SUMMARY AND OUTLOOK

There has been a continuous development of atomic many-body theory and its application during the past decades. In the late 50's and early 60's the Brueckner-Goldstone form of MBPT was developed and introduced into atomic physics (Kelly). During the 60's also the closed-shell form of the coupled-cluster approach was introduced (Coester, Kümmel, Čižek) as well as the general open-shell form of MBPT (Brandow).

In the 70's the open-shell form of the coupled-cluster approach was formulated for a complete model space. In this decade also the SCF methods were developed in the multi-configurational form, non-relativistically (Froese-Fischer) as well as relativistically (Desclaux, Grant), and accurate numerical procedures for calculating Lamb shift for heavy, hydrogenic ions were introduced by (Mohr).

In the 80's coupled-cluster approaches for incomplete model space were formulated, and the root of the connectivity problem was localized to the intermediate normalization (Mukherjee). Also other ways of getting around the intruder problem were suggested, such as the intermediate hamiltonian (Malrieu, Durand). Extremely accurate wave functions for two-electron systems were developed in this decade (Drake). The SCF methods were extended to include magnetic interactions and retardation (Breit interactions), and relativistic forms of MBPT and coupled-cluster approach were introduced, in all cases based on the no-virtual-pair approximation. By including one-electron Lamb-shift corrections in an approximate way, it became possible to make meaningful comparisons with accurate experimental data for highly charged ions that had become available.

In the 90's it is expected that the merging of QED and many-body techniques will be of increasing importance. Some preliminary calculations of the screening of the Lamb shift, which in principle is a many-body effect, have already appeared. This technique, however, has to be further developed, and real many-body-QED (multi-photon exchange) effects have to be considered in order to match the increasing accuracy of the experimental data.

Acknowledgments

The author wishes to express his thanks to his collaborators Eva Lindroth, Ann-Marie Mårtensson-Pendrill, Hans Persson, Sten Salomonson and Anders

Ynnerman for stimulating discussions and valuable comments on the manuscript.

REFERENCES

Au C-K and Feinberg G 1974, *Effects of Retardation on Electromagnetic Self-Energy of Atomic States*, Phys Rev A9:1794

Bartlett R J 1981, Ann Rev Phys Chem 32:359

Bartlett R J (ed) 1991, *Proceedings from the Workshop on Coupled Cluster Theory and the Interface of Atomic Physics and Quantum Chemistry*, Harvard 6-11 August 1990, to be published in Theor Chim Acta

Bartlett R J, Kucharski A and Noga J 1989, *Alternative Coupled-Cluster Ansätze II. The Unitary Coupled-Cluster Method*, Chem Phys Lett 155:133

Bartlett R J and Purvis G D 1978, Int J Quant Chem 14:561

1980, *Molecular Applications of Coupled Cluster and Many-Body Perturbation Methods*, Physics Scripta 21:255

1982, *The Full Coupled-Cluster Singles and Doubles Model: The Inclusion of Disconnected Triples*, J Chem Phys 76:1910

Bethe H A 1947, *The Electromagnetic Shift of Energy Levels*, Phys Rev 72:339

Bethe H A and Salpeter E E 1957, *Quantum Mechanics of One- and Two-Electron Atoms*, Springer-Verlag

Bloch C 1958a, *Sur la Théorie des Perturbations des États Liés*, Nucl Phys 6: 329

1958b, *Sur la Détermination de l'État Fondamental d'un Système de Particules*, Nucl Phys 7: 451

Blundell 1990, private communication

Blundell S A, Johnson W R, Liu Z W and Sapirstein J 1989, *Relativistic All-Order Equation for Helium* Phys Rev A39:3768

Blundell S A, Johnson W R, Mohr P and Sapirstein J, 1990, unpublished

Brandow B H 1967, *Linked-Cluster Expansion for the Nuclear Many-Body Problem*, Rev Mod Phys 39: 771

Brown G E, Langer J S and Schaefer D G 1959, *Lamb Shift of a Tightly Bound Electron. I. Method*, Proc Roy Soc London A251:92

Brown G E and Ravenhall D G 1951, *On the Interaction of Two Electrons*, Proc Roy Soc London A208:552

Brueckner K A 1955, *Many-Body Problem for Strongly Interacting Particles. II. Linked Cluster Expansion*, Phys Rev 100:36

Cheng K T, Johnson W R and Sapirstein J 1991, *Screened Lamb Shift Calculations for Lithiumlike Uranium, Sodiumlike Platinum, and Copperlike Gold*, Phys Rev Lett 66:2960

Čižek J 1966, *On the Correlation Problem in Atomic and Molecular Systems. Calculations of Wavefunction Components in Ursell-Type Expansion using Quantum-Field Theoretical Methods*, J Chem Phys 45:4256

Coester F and Kümmel H 1960, *Short-Range Correlations in Nuclear Wave Functions*, Nucl Phys 17:477

deBoor C 1978, *A Practical Guide to Splines*, Applied Math Series 27, Springer N Y

Drake G W F 1982, *Quantum Electrodynamics Effects in Few-Electron Atomic Systems*, Adv Atomic Mol Phys 18:399

1988, *High Precision Variational Calculations for the $1s^2\,^1S$ State of H^- and the $1s^2\,^1S$, $1s2s\,^1S$ and $1s2s\,^3S$ States of Helium*, Nucl Instr Meth B31:7

Desclaux J P 1975, *A Multiconfiguration Relativistic Dirac-Fock Program*, Comput Phys Commun 9:31

Durand Ph and Malrieu J P 1987, *Effective Hamiltonians and Pseudo-Operators as Tools for Rigorous Modelling*, Adv Chem Phys 67:321

Evangelisti S, Durand Ph and Heully J L 1991, *Breaking of the One-to-one Correspondence between Model and Target Spaces in Effective Hamiltonian Theory: Generalized Effective Hamiltonians*, Phys Rev A43:1258

Ey W 1978, *Degenerate Many Fermion Theory in expS Form (III). Linked Valence Expansion,*-Nucl Phys A296:189

Fano U 1983, *Correlations of Two Excited Electrons*, Rep Prog Phys 46:97

Feinberg G and Sucher J 1988, *Two-Photon-Exchange Force between Charged Systems: Spinless Particles*, Phys Rev D38:3763

Froese-Fischer Ch 1977, *The Hartree-Fock Method for Atoms*, Wiley

Furry W H 1951, *On Bound States and Scattering in Positron Theory*, Phys Rev 81:115

Goldstone J 1957, *Derivation of The Brueckner Many-Body Theory*, Proc Roy Soc London A239: 267

Gorceix O and Indelicato P 1988, *Effect of the Complete Breit Interaction on Two-Electron Ion Energy Levels*, Phys Rev A37:1087

Gorceix O, Indelicato P and Desclaux J P 1987, *Multiconfiguration Dirac-Fock Studies of Two-Electron Ions: I. Electron-Electron Interaction*, J Phys B20:639

Grant I P, McKenzie B J, Norrington P H, Mayers D F and Pyper N C 1980, *An Atomic Multiconfigurational Dirac-Fock Package*, Comput Phys Commun 21:207

Haque M A and Kaldor U 1986, *Open-Shell Coupled-Cluster Method: Variational and Nonvariational Calculations of Ionization Potentials*, Int J Quant Chem 29:425

Hartley A C, Lindroth E and Mårtensson-Pendrill A-M 1990, *Parity Non-conservation and Electric Dipole Moments in Caesium and Thallium*, J Phys B23:1990

Hartley A C and Mårtensson-Pendrill A-M 1990, *Relativistic Correlation Effects on the Hyperfine Structure and Electric Dipole Transitions in Cs and Tl*, Z Phys D15:309

1991, *Calculations of Isotope Shifts in Caesium and Thallium using Many-Body Perturbation Theory*, J Phys B24:1193

Heully J-L and Daudey J-P 1988, *Many-Body Perturbation Calculation on Be using a Multiconfiguration Model Space and an Intermediate Hamiltonian*, J Chem Phys 88:1046

Hose and Kaldor 1979, *Diagrammatic Many Body Perturbation Theory for General Model Spaces*, J Phys B12:3827

1980, *A General Model-Space Diagrammatic Perturbation Theory*, Physica Scripta 21:357

1981, *The Shifted Scheme in the General-Model-Space Diagrammatic Perturbation Theory*, Chem Phys 62:469

Hubbard J 1957, *The Description of Collective Motions in Terms of Many-Body Perturbation Theory*, Proc Roy Soc London A240:539

Indelicato P and Desclaux J P 1990, *Multiconfiguration Dirac-Fock Calculations of Transition Energies with QED Corrections in Three-Electron Ions*, Phys Rev A42:5139

Indelicato P, Gorceix O and Desclaux J P 1987, *Multiconfigurational Dirac-Fock Studies of Two-Electron Ions: II. Radiative Corrections and Comparison with Experiments*, J Phys B20:651

Indelicato P and Mohr P J 1990, *Electron Screening Corrections to the Self Energy in High-Z Atoms*, 12th Int Conf on Atomic Physics, Ann Arbor, Mich, and Theor Chim Acta to appear

Ishikawa Y, Quiney H M and Malli G L 1991, *Dirac-Fock-Breit Self-Consistent-Field Method: Gaussian Basis-Set Calculations on Many-Electron Atoms*, Phys Rev A43:3270

Jeziorski B and Monkhorst H J 1981, *Coupled-Cluster Method for Multi-Determinantal Reference States*, Phys Rev A24:1668

Johnson W R, Blundell S A and Sapirstein J 1988a, *Many-Body Perturbation-Theory Calculations of Energy Levels along the Lithium Isoelectronic Sequence*, Phys Rev A37:307,2764

1988b, *Many-Body Perturbation-Theory Calculations of Energy Levels along the Sodium Isoelectronic Sequence*, Phys Rev A38:2699

1990, *Many-Body Perturbation-Theory Calculations of Energy Levels along the Copper Isoelectronic Sequence*, Phys Rev A42:1087

Jørgensen F 1975, *Effective Hamiltonians*, Mol Phys 29:1137

Kinze R J 1991, *Measurements of QED Effects in Z=24 to 34 Lithiumlike Ions*, Phys Rev A43:1637

Kucharski S A and Bartlett R J 1989, *Coupled-Cluster Methods that include Connected Quadruple Excitations*, T_4: *CCSDTQ-1 and Q(CCSDT)*, Chem Phys Lett 158:550

Kvasnička V 1974, *Construction of Model Hamiltonians in Framework of Rayleigh-Schrödinger Perturbation Theory*, Czech J Phys B24: 605

1977, *Application of Diagrammatic Quasidegenerate RSPT in Quantum Molecular Physics*, Adv Chem Phys 36 345

1981, *Coupled-Cluster Approach for Open-Shell Systems*, Chem Phys Lett 79:89

Lindgren I 1974, *The Rayleigh-Schrödinger Perturbation Theory and the Linked-Diagram Theorem for a Multi-Configurational Model Space*, J Phys B 7: 2441

1978, *A Coupled-Cluster Approach to the Many-Body Perturbation Theory for Open-Shell Systems*, Int J Quantum Chem S12:33

1985, *Accurate Many-Body Calculations on the Lowest ^2S and ^2P States of the Lithium Atom*, Phys Rev A31:1273

1988, *Towards a Relativistic Many-Body Procedure*, Nucl Instr Meth B31:102

1989a, *A Relativistic Coupled-Cluster Approach with Radiative Corrections* in "Many-Body Methods in Quantum Chemistry", Lecture Notes in Chemistry 52:293, Springer-Verlag

1989b, *Effective Potentials in Relativistic Many-Body Theory* in Proceedings of "Program on Relativistic, Quantum Electrodynamics and Weak Interaction Effects in Atoms", ITP Santa Barbara, Jan-June 1988, American Institute of Physics Conference Series no 189, p 371

1990 *Gauge Dependence of Interelectronic Potentials*, J Phys B23, 1085

1991 *Hermitian Formulation of the Coupled-Cluster Approach*, J Phys B 24: 1143

Lindgren I, Lindgren J and Mårtensson A-M 1976, *Many-Body Calculations of the Hyperfine Interaction of some Excited States of Alkali Atoms, using Approximate Brueckner or Natural Orbitals*, Z Phys A276:113

Lindgren I and Morrison J 1986, *Atomic Many-Body Theory*, second ed. Springer-Verlag

Lindgren I and Mukherjee D 1987, *On the Connectivity Criteria in the Open-Shell Coupled-Cluster Theory for General Model Spaces*, Physics Reports 151:93

Lindgren I and Salomonson S 1980, *A Numerical Coupled-Cluster Procedure applied to the Closed-Shell Atoms Be and Ne*, Physica Scripta 21:335

Lindroth E 1988 *Numerical Solution of the Relativistic Pair Equation*, Phys Rev A37:316

Lindroth E 1991, to be published

Lindroth E, Heully J-L, Lindgren I and Mårtensson-Pendrill A-M 1987, *A Relativistic Pair Equation Projected onto Positive Energy States*, J Phys B20:1679

Lindroth and Hvarfner 1991, *Relativistic Calculation of the 2^1S_0-$2^{1,3}P_1$ Transitions in Beryllium-like Molybdenum and Beryllium-like Iron*, submitted to Phys Rev A

Lindroth E, Mårtensson-Pendrill A-M 1989, *Further Analysis of the Complete Breit Interaction*, Phys Rev A39:3794

Lindroth E, Mårtensson-Pendrill A-M, Ynnerman A and Öster P 1989, *Self-Consistent Treatment of the Breit Interaction, with Application to the Electric Dipole Moment in Thallium*, J Phys B 22:2447

Lindroth E, Persson H, Salomonson S and Mårtensson-Pendrill A-M 1991, *Corrections to the Beryllium Ground State Energy*, submitted to Phys Rev

Lindroth and Salomonson S 1990, *Relativistic Calculations of the $2\,^3S_1$ - $1\,^1S_0$ Magnetic Dipole Transition Rate and Transition Energy for Heliumlike Argon*, Phys Rev A41:4659

Liu Z W and Kelly H P 1991, *Atomic Many-Body Perturbation Method based on Multiconfiguration Dirac-Fock Wave Functions*, Phys Rev A43:3305

Löwdin P O 1962, *Studies in Perturbation Theory. IV. Solution of the Eigenvalue Problem by Projection Operator Formalism*, J Math Phys 3:969

Malrieu J P, Durand Ph and Daudey J 1985, *Intermediate Hamiltonians as a New Class of Effective Hamiltonians*, J Phys A18:809

Mårtensson-Pendrill A-M 1989, *Numerical Determination of Non-Relativistic and Relativistic Pair Correlation*, in Defranceschi M and Delhalle (eds), *Numerical Determination of the Electronic Structure of Atoms, Diatomic and Polyatomic Molecules*, Kluwer Acad Publ

Martinsson I 1989, *The Spectroscopy of Highly Ionized Atoms*, Rep Prog Phys 52:157

Mittleman M H 1971, *Structure of Heavy Atoms: Three Body Potentials*, Phys Rev A4:893

1972, *Configuration-Space Hamiltonian for Heavy Atoms and Corrections to Breit Interaction*, Phys Rev A5:2395

Mohr P J 1974a, *Self Energy Radiative Corrections in Hydrogen-Like Systems*. Ann Phys (New York) 88:26

1974b, *Numerical Evaluation of the $1\,S_{1/2}$-State Radiative Level Shift*, Ann Phys (New York) 88:52

1982, *Self-Energy of the $n=2$ States in a Strong Coulomb Field*, Phys Rev A26:2338

1988, *Structure of High-Z One- and Two Electron Atoms* Nucl Instr Meth B31:1

Mohr P J and Kim Y-K 1991, to appear

Morrison J 1986, *Calculations Combining the Multi-Configuration Hartree-Fock Method and Many-Body Perturbation Theory*, Physica Scripta 34:423

1988, *Calculations Combining the Multi-Configuration Hartree-Fock Method and Many-Body Perturbation Theory: III. Higher-Order Effects Involving Correlation of the Core*, J Phys B21:2915

Morrison J and Froese-Fischer C 1987, *Multi-Configuration Hartree-Fock Method and Many-Body Perturbation Theory: A Unified Approach*, Phys Rev A35:2429

Mukherjee D 1986a, *The Linked-Cluster Theorem in the Open-Shell Coupled-Cluster Theory for Incomplete Model Spaces*, Chem Phys Lett 125: 207
1986b, *Aspects of Linked Cluster Expansion in General Model Space Many-Body Perturbation Theory and Coupled-Cluster Theory*, Int J Quant Chem S20:409

Mukherjee D and Pal S 1989, *Use of Cluster Expansion Methods in the Open-Shell Correlation Problem*, Adv Chem Phys 20:291

Offermann R, Ey W and Kümmel H 1976, *Degenerate Many Fermion Theory in expS Form (I). General Formalism*, Nucl Phys A273:349

Pal S, Prasad M D and Mukherjee D 1984, *Development of a Size-Consistent Energy Functional for Open Shell States*, Theor Chim Acta 66:311

Paldus J and Čižek J 1975, *Time Independent Diagrammatic Approach to Perturbation Theory of Fermion Systems*, Adv Quant Chem 9:105

Persson H, Lindgren I and Salomonson S 1991, unpublished

Pople J A, Binkley J S and Seeger R 1976, *Theoretical Models Incorporating Electron Correlation*, Int J Quant Chem S10:1

Primas H 1965 in Sinanoglu O (ed), *Modern Quantum Chemistry*, Acad Press

Quiney H M, Grant I P and Wilson S 1987, *The Dirac Equation in the Algebraic Approximation: V. Self-Consistent Field Studies including the Breit Interaction*, J Phys B20:1413
1989, *The Accuracy of Dirac-Hartree-Fock Calculations using Analytic Basis Sets*, J Phys B22:L15
1990, *Relativistic Many-Body Perturbation Theory using Analytic Basis Functions*, J Phys B23:L271

Salomonson S, Lindgren I and Mårtensson A-M 1980, *Numerical Many-Body Perturbation Calculations on Be-like Systems using a Multiconfigurational Model Space*, Physica Scripta 21:351

Salomonson S and Ynnerman A 1991, *Coupled-Cluster Calculations of Matrix Elements and Ionization Energies of Low-lying States in Sodium*, Phys Rev A43:88

Salomonson S and Öster P 1989a, *Solution of the Pair Equation using a Finite Discrete Spectrum*, Phys Rev A40:5559
1989b, *Relativistic All-Order Pair Functions from a Discretized Single-Particle Dirac Hamiltonian*, Phys Rev A40:5548

Salomonson S and Öster P 1990, *Numerical Solution of the Coupled-Cluster Single- and Double-Excitation Equations with Applications to Be and Li$^-$*, Phys Rev A41:4670

Schucan T H and Weidenmüller H A 1972, *The Effective Interaction in Nuclei and its Perturbation Expansion: An Algebraic Approach*, Ann Phys NY 73:108
1973, *Perturbation Theory for the Effective Interaction in Nuclei* Ann Phys NY 76:483

Schweppe J, Belkacem A, Blumenfield L, Claytor N, Feinberg B, Gould H, Kostroun V E, Levy L, Misawa S, Mowat J R and Prior M H 1991, *Measurement of the Lamb Shift in Lithiumlike Uranium (U^{89+})*, Phys Rev Lett 66:1434

Seely J F and Wagner R A 1990, *QED Contributions to the 3s-3p transition in Highly Charged Na-like Ions*, Phys Rev A41:5246

Soff G and Mohr P J 1988, *Vacuum Polarization in a Strong External Field*, Phys Rev A38:5066

Stolarczyk L Z and Monkhorst H J 1985, *Coupled-Cluster Method in Fock Space. I. General Formalism*, Phys Rev A32:725

Sucher J 1980, *Foundations of the Relativistic Theory of Many Electron Atoms*, Phys Rev A22:348

1988, *On the Choice of the Electron-Electron Potential in Relativistic Atomic Physics*, J Phys B21:L585

Watts J D, Trucks G W and Bartlett R J 1989, *The Unitary Coupled-Cluster Approach and Molecular Properties. Applications of the UCC(4) Method*, Chem Phys Lett 157:359

Wilson S 1991, in *The Effects of Relativity in Atoms, Molecules and the Solid State*, by Wilson S, Grant I P and Gyorffy (eds), Plenum Press

Zygelman B and Mittleman M H 1986, *Contributions of Three-Body Potentials to the Binding Energy of Heavy Atoms*, J Phys B19:1891

MANY BODY EFFECTS IN FAST ION-ATOM COLLISIONS

John F. Reading and A.L. Ford

Center for Theoretical Physics
Department of Physics
Texas A&M University, College Station TX, 77843, USA

INTRODUCTION

Traditionally in atomic physics the many body aspect of the problem has focussed on energy levels of atoms, where correlation plays a role at the few per cent level. In this paper we present an overview for non-specialists of the much larger effect that correlation has on cross sections for excitation, ionization and charge transfer in an ion atom collision. New methods for time dependent perturbations of many body systems have been developed which might be of interest to a many body audience. But more importantly, we hope to stimulate work by many body experts in the problems which yet have to be solved.

An ion is much heavier than an electron; it is thus easy to arrange an experimental situation in which it is accurate to describe an ion atom collision as if the ion passes the the atom at some impact parameter B on a classical trajectory, which is unperturbed by electronic excitation of the atom; i.e. the semi-classical approximation is accurate [1]. This classical assumption immediately introduces time into the quantum mechanical problem. The electrons in the atom, located at r_i, are perturbed by the coulombic potential of a bare projectile ion, $Z_p/|r_i - R|$, and R, the relative displacement of the projectile nucleus and target atomic nucleus is a function of time defined by the trajectory. Here Z_p is the projectile charge. If electrons are also initially, or finally, on the projectile they too perturb the atom and are pertubed in their turn by it. A unique feature of an ion-atom many body collision is that not only is the collision simplified by the presence of the heavy nuclei but also the high charge of these particles implies that a mean field approximation is very accurate; thus great progress has been made with a static independent particle model (IPM) in which each electron is assumed to move in some average shielded field, usually chosen as the initial asymptotic mean field of the system. For example in the collision of a fast fully stripped lithium ion with a neon atom producing an inner shell vacancy, the electrons are assumed to move in the hartree fock field of the ground state of neon and the field of the bare projectile; this is in spite of the fact that in reality there is some probability that electrons will jump onto the projectile and change the shielding.

Within the IPM the only correlation that can play a role is Pauli correlation, which seems at first to be simplistic and devoid of interest. But the accuracy of the IPM allows three important concepts to be introduced: that of an inclusive cross section, a partially inclusive cross section, and an exclusive cross section. An exclusive cross section is that in which the final electronic configuration is completely determined experimentally. An inclusive cross section is one in which the final state is characterised by one single electron quantum state, e.g. there is at least one vacancy in the K-shell of Neon; this is measured experimentally by observing the emission of a characteristic K-shell X-ray or Auger electron. In the inclusive process other vacancies may have been produced during the collision, but, because of the accuracy of the IPM, the energy of the X-ray is not materially effected by this and these different cross sections are all lumped together in the observation. An interesting theorem is that Pauli correlation plays a very simple role in an inclusive cross section: the single electron theorem. But, in a partially inclusive cross section, e.g. at least two vacancies produced, Pauli correlation can be very important producing coherent enhancement or suppression of cross sections, known as CESIME processes.

A third unique feature of an ion-atom collision is the experimental ability to vary the coupling constant by using a series of different projectiles. In nuclear physics the differences in scattering a deuteron and an alpha particle from a nucleus are not simply that the deuteron has half the nuclear charge of the alpha particle, but in an atomic collision, because of the long range nature of the coulombic force, this is precisely the situation. This feature is no more dramatically exposed than in the recent use of anti-protons for atomic collisions as compared to equal velocity protons; here we can change the sign of the coupling constant. This flexibility has been used to begin the study of the effect of dynamic correlation.

An interesting process to try to picture is the elastic, large angle collision between a fast energetic projectile and a composite particle, e.g. a proton and a nucleus. It is interesting because we know the fundamental process is the interaction of the proton with the individual nucleons. The question is how do **all** the nucleons in the target nucleus manage to turn through the correct angle and diverge from the collision each in its original pristine place in momentum space except with the addition of the precise collective momentum displacement needed to keep the recoiling nucleus in its ground state? Two pictures emerge. If the interaction between the proton and the target nucleons is weak then we can anticipate that that few nucleons are struck and the dominant processes are intra-nuclear collisions, i.e. the few excited nucleons pull their correlated passive fellows with them onto a new path. If the interaction is strong then many nucleons take part in the collision, each prepared, by its interaction, for its proper final disposition. The first mechanism is called an internally induced multi-fermion process, IIMP, the latter an externally induced multi-fermion process, EIMP.

This is one of many such examples where several of the individual particles that make up the composite have changed their state but only one or two have interacted with the external probe. Measurement of the IIMP component in such collisions can be used to study correlation in the initial or final state. Or, to put it another way, dynamic correlation during the collision must be understood if one wishes to describe accurately processes dominated by IIMP. In nuclear physics EIMP and IIMP can only be separated by turning the EIMP down somewhat by using the one variable available, the energy of the collision. However, because of hard nuclear cores in the nucleon-nucleon force the energy can never be high enough for EIMP to be completely suppressed.

In ion-atom collisions at sufficiently high inter ion-atom velocities, v, the projectile-electron force as measured by $Z_p e^2/\hbar v$ is vanishingly weak; IIMP can be made as dominant as one wishes by varying the velocity or projectile.

A set of crucial experiments have been performed with MeV/amu protons and anti-protons doubly ionizing helium. At 1MeV anti-protons are twice as efficient as protons in this reaction because the EIMP and IIMP interfere destructively and constructively as the coupling constant is changed in sign [2,3]. Above 10MeV the ratio of double ionization to single ionization ionization settles to the IIMP limit of 0.0027, a fundamental constant of nature and a unique measurement of correlation. Experiments are now planned to study three and four electron ionization in a single IIMP event. Whole shells might be released.

To understand these phenomena new theoretical methods must be devised. Correlated continuum states must be constructed, and then transition amplitudes,which must include second and higher order terms in the born series, calculated. Some progress has been made on two electron systems in describing ionization; a so called forced impulse method(FIM) accurately treats the time dependent perturbation of a two electron correlated system [4]. But, for example, no satisfactory treatment has yet been devised for two electron excitation. The basic problems to be addressed include:

How to construct accurate many electron correlated continuum channel wavefunctions.

How to calculate transitions to such channels when first order perturbation theory is inaccurate.

The plan of this short introduction to the area is as follows. In the next section we discuss Pauli correlation, the single electron theorem and CESIME processes. We then describe briefly the forced impulse method as an example of finite hilbert basis state (FHBS)calculations that have been used to try to make a start on these systems.

THE SINGLE ELECTRON THEOREM AND CESIME PROCESSES

The single electron theorem addresses the role that Pauli correlation, plays in the collision of a many electron system; it sharpens the concepts of exclusive and inclusive reactions [5]. It states that:-

In the IPM the *inclusive* cross section for producing a given vacancy in an ion-atom collision is given by calculating the cross section for the single electron, in the vacancy orbital, to be ionised, or excited to any state on the target or projectile not *initially* occupied.

For definiteness consider the problem of measuring and calculating the production of a characterisic target K-shell x-ray, when a lithium ion, with two electrons in its K-shell, is incident on a neon target atom. During the collision many things can happen especially if the ion is moving with a speed comparable to that of the orbiting target K-shell electron. Multiple target vacancies might be produced; the lithium ion may have one electron left after the collision in its K-shell, two in its L-shell and so on. None of these things will necessarily be important to the experimentalist. He will observe an X-ray, or the corresponding Auger electron, for each target K-shell vacancy that is produced. This is the *inclusive* cross section the theorist must calculate.

If the probe is weak, e.g. the velocity is very high, the probablity of EIMP may be negligible. In this case the single electron theorem is trivial. All excitations excepting that of the target K-shell electron can be ignored, because if that electron is not struck then no K-shell hole can be produced. We know exactly where the passive electrons are:- in their initial states. As long as we do not allow these as final states for the ejected K-shell electron, we have correctly included the Pauli correlation.

As we turn up the interaction, EIMP becomes important. We must certainly modify our treatment of the K-shell electron and treat its excitation to all orders in the Born series. However now we have to worry about the role that the other electrons play. Should we not introduce Pauli blocking operators in the Born series? Now we can make a hole in the L-shell, should we not allow this as a possible final state for the K-shell electron?. The answer to these questions is yes, if we wish to calculate *exclusive* cross sections, in perturbation theory. However it is not necessary to use many electron perturbation theory, and we are only interested in an inclusive cross section. We know that each orbital electronic wavefunction, $\psi_i(t)$ developes independently from the others and the correct description of the system is given by forming a simple determinant of these time dependent orbitals. To calculate the required cross section we have to project onto all multi-electron states that have a K-shell vacancy, and sum the resulting probabilities. The simple single electron theorem results.

As an example of the use of this theorem consider an experiment designed to test the theoretical understanding of single electron charge transfer between the K-shell of neon and the K-shell of lithium. Ideally the experiment should be performed with one electron and two bare nuclei. It is this single particle cross section that the theorist can calculate most easily. But the experiment is most easily performed in the environment of a neutral neon atom. It is the effect of this environment that we wish to consider.

An obvious experiment one might design is to send in a fully stripped lithium ion into neon [5]. The signature that a neon K-shell vacancy has been produced will be the detection of a K-shell x-ray. The signature that an electron has been captured is a coincidental change in charge of the lithium ion. Some corrections must be made before experiment can be used to test theory. For example, there is a finite probability that an electron is captured from the L-shell at the same time that the K-shell electron was ionized.

Compare this to the much simpler experiment where the x-ray production cross section is measured with one electron in the lithium K-shell, and again with two electrons in the K-shell. The difference in these cross sections according to the theorem is due to the closing down of K to K shell charge transfer, and gives the desired single electron cross section.

This all may seem rather trivial, until we remember that in this collision the lithium electron has a high probability of being removed, and that this does not matter. It might be suprising to the reader though that the *coincidence* experiment described, even with the corrections contemplated, does not in fact give a measure of single electron charge transfer. The very act of measuring two things in the experiment has destroyed the inclusive nature of the cross section [5]. Pauli correlation immediately reasserts its influence. This occurs at the transition amplitude level and adding or subtracting probabilites *incoherently* does not give the correct answer. For example the inclusive probability, P_1 for creating a hole in single particle state ψ_1 is given by

$$P_1 = 1 - \sum_n |u_{n,1}|^2,$$

where the sum is taken over the initially occupied states, and $u_{n,1}$ is the u-matrix element for the transition to a state ψ_n. The probabilty, $P_{1,2}$, for making two holes inclusively is just the determinant of the two by two matrix the elements $M_{i,j}$ of which are given by

$$M_{i,j} = \delta_{i,j} - \sum_n u^*_{i,n} u_{j,n}.$$

with a similar result for three or more holes. Clearly

$$P_{1,2} < P_1 P_2.$$

For definite occupancy, as opposed to a vacancy there is a corresponding enhancement in the cross section. The procedure for calculating a partially exclusive, partially inclusive many particle-hole cross section is complicated. This general task has however been carried out [6]. Processes where there is *coherent enhancement or supression of independent multi-fermion events* are called CESIME processes [7]. They are predicted to be important when electrons are moved about with a speed characteristic of the surrounding passive electrons. To this date no clear experimental example of a CESIME process has been identified [6–10]. It remains a challenge for future experiments.

Two lessons can be learned. Firstly, the more exclusive the measurement the more likely correlation will play a role; and secondly it is difficult to expose this if the correlation effect is just a correction to an already uncertain calculation.

An experiment which avoids these difficulties is the high energy double ionisation of helium by fast ions. Information about the final states of both electrons is measured, and at asymptotic energies dynamical correlation is the dominant mechanism by which the reaction procedes. Further as the energy of the projectile is lowered interference between IIMP and EIMP can be exposed by the factors of two differences in the cross sections produced by protons and anti-protons. These are not reproduced by calculations which ignore dynamical correlation [4].

The Forced Impulse Method, FIM

How are we to calculate dynamic correlation? In the ion-atom sector of fermion reactions we have yet to get past the few body problem at this level of sophistication. This is remarkable as much work in nuclear physics has focused on a natural generalisation of the single determinant ideas used above namely time dependent hartree-fock, TDHF [11]. Why is the idea of an optimised time varying mean field not useful here. A simple answer is that TDHF is fine if there is one dominant channel, and we are interested in following that channel. But the correlation effects of IIMP are often to be found in small cross sections, not the dominant channel.

For example consider a TDHF calculation to understand the double ionization of helium by a very fast, weakly interacting, proton. Shake-off effects are poorly described, because hartree-fock is not sufficiently accurate to describe even the ground state of helium. Furthermore, as the most likely thing to happen with such a weak probe is nothing at all, the mean field remains that of the ground state of helium throughout the collision. What is needed is a detailed tracking of that particular and unlikely path that lead to IIMP.

The forced impulse method offers an exact quantum mechanical treatment of a correlated collision [4]. So far the only systems it has been practical to apply it to have just two electrons. The basic idea of the FIM is that for high projectile speeds the struck electrons have no time to move very far before the collision is over. Therefore they do not *sense* the binding forces that act on them, be they due to correlation or the mean field. We are familiar with the high energy techniques that exploit this fact:-

the Glauber [1] , Distorted Wave and Impulse approximations [12]. A basic criticism
of all of these methods is that it is not clear how to systematically improve them so
one can be certain that they are reliable as the velocity is lowered. The FIM does
not rely on the speed of the projectile for accuracy. Rather the collision is chopped
into consecutive time steps. For each time step an impulse like approximation is
made. In principle the steps can be made smaller and smaller until the impulse
approximation is *forced* to be accurate. Perhaps the best analogy is that the usual
impulse approximation fails at low velocity in the same way that a photograph of a
moving object is blurred. The FIM method makes a movie picture by tying together
several fast frames, and tracks the detailed developement of the system.

One practical way to implement the calculation of all the coherent amplitudes
needed in the IPM is to expand the wavefunction in a finite hilbert basis set. We di-
agonalise the single electron hamiltonian to form a set of single particle pseudo-states,
χ_n. These states have a discrete spectrum, ϵ_n: positive energy states representing
continuum states.

$$\chi_n(t) = \chi_n e^{-i\epsilon_n t/\hbar}.$$

The many electron states can be formed from suitably antisymmetrised single particle
pseudo-states; we shall call these states χ_N.

The time dependent single particle wavefunction, ψ_n is written as

$$\psi_n(t) = \sum_{n'} u_{n',n}(t, -t_L)\chi'_n(t),$$

with boundary condition that at some large negative time, $-t_L$,

$$u_{n',n}(-t_L, -t_L) = \delta'_n, n.$$

The u-matrix elements, at t_L, are calculated variationally by solving a set of linear,
time dependent coupled differential equations. As the projectile-electron potential is
single particle, we may now form the time dependent many electron state wavefunction, Ψ_N by taking an antisymmetrised product of the ψ_n.

If we wish to introduce dynamic correlation we can in principle diagonalize the
many electron hamiltonian using the states χ_N as a basis, to produce eigenstates of
the projected, correlated many electron hamiltonian χ_M. Thus

$$\chi_M = \sum_N a_M^N \chi_N,$$

where

$$\sum_{N'} [(\epsilon_N - \epsilon_M)\delta_{N,N'} + <\chi_N|V_{RI}|\chi_{N'}>]a_M^{N'} = 0,$$

where we have adopted a transparent notation for the eigenenergies of the many
particle states, and we keep the notation that

$$\chi_M(t) = \chi_M e^{-i\epsilon_M t/\hbar}.$$

Again, in principle, we could integrate out the time dependent schrodinger
equation deriving the amplitudes, $U_{M,M'}$, for transitions between the many particle pseudo-states

$$\Psi_M = \sum_{M'} U_{M',M}(t, -t_L)\chi_{M'}(t).$$

Similar equations, of course, hold for the uncorrelated system described by Ψ_N.

However because of the large number of pseudo-states involved for even two electron systems such a procedure is not practical.

We return to the approximate methods introduced earlier, and make use of ideas due to Reisenfeld and Watson and Glauber. We try to write a solution for Ψ_M in terms of Ψ_N when we apply an impulse to the system with the projectile between times t_1 and t_2. Before the impulse the system is in a stationary state χ_M.

$$\Psi_M = \sum_N b_M^N(t_1,t_1)\chi_N(t)e^{i(\epsilon_N-\epsilon_M)t/\hbar},$$

where

$$b_M^N(t_1,t_1) = a_M^N.$$

After t_1 we write

$$\Psi_M = \sum_N b_M^N(t,t_1)\Psi_N e^{i(\epsilon_N-\epsilon_M)t/\hbar}.$$

Calculating b variationally gives the equation

$$\frac{i\hbar\partial b_M^N(t,t_1)}{\partial t} = \sum_{N'} e^{i(\epsilon_{N'}-\epsilon_N)t/\hbar}[(\epsilon_N - \epsilon_M)\delta_{N,N'}$$

$$+ < \Psi_N V_{RI}\Psi_{N'} >]b_{N'}^M(t,t_1).$$

From this equation we make the observation that the derivitive of b is zero at t_1. This implies that we may replace b_M^N by a_M^N for a short period of time giving

$$\Psi_M = \sum_N a_M^N \Psi_N e^{i(\epsilon_N-\epsilon_M)t/\hbar}.$$

Thus we are able to follow the developement of the wavefunction without having to refer to the correlating forces! As Ψ_N is a simple product of single particle wavefunctions the computing effort need to construct Ψ_N is characteristic of a single electron problem, not a many electron problem.

To find the amplitude $U_{M',M}(t,t_1)$ we substitute our approximation for Ψ_M into the t-matrix integral and derive the algorithm

$$U_{M',M}(t_2,t_1) = \delta_{M',M} + \sum_{M',M} a_{M'}^{N'*} \int_{t_1}^{t_2} dt\, e^{i\Delta t/\hbar} \frac{\partial U_{N',N}(t,t_1)}{\partial t} a_M^N,$$

where

$$\triangle = \epsilon_{M'} - \epsilon_M - \epsilon_{N'} + \epsilon_N.$$

If the collision is truely fast enough then we may consider the interval to be extended to $(t_L,-t_L)$. But unlike the analytical approximation schemes discussed before we can drive this method to convergence by stringing together consecutive short intervals each T_C in length. We force the impulse-like approximation to be true by chosing T_C to be short enough. Hence the name of the procedure the forced impulse method(FIM). It defines a crude time step T_C.

$$\mathbf{U}_{(t_L,-t_L)} = \mathbf{U}(t_L,t_L - T_C)\mathbf{U}(t_L - T_C,t_L - 2T_C) - - - \mathbf{U}(T_C - t_L,-t_L).$$

Of course if T_C needed for convergence is as short as the time interval, T, needed to integrate out the many body coupled equations derived from the conventional FHBS then nothing has been gained. Fortunately rather long time intervals seem to be able to be used [4].

The FIM has succesfully reproduced the fact that anti-protons are much more efficient at doubly ionising helium than protons [2].

Further experiments, such as double excitation and differential studies [13], have been performed, or are being planned, as the realisation grows that the IIMP exposed in this way are very interesting. At asymptotically high projectile energies the ratio of double to single ionization for helium has now been measured and offers a new test of the correlation in the wavefunction of both the initial and final state. We now understand that the cross section is not dominated by dipole interactions so that the ionic studies offer new testing areas not accessible to photon experiments. An obvious generalisation is to determine the ratio of triple, double and single ionization in lithium. We would then be testing three electron correlation. Perhaps it is possible to measure the removal of every electron in a shell produced by the projectile interacting with one electron.

Less ambitious is the task of exploiting the two electron capability that the FIM offers. Instead of reducing a many electron system to a product of single electron orbitals perhaps we can reduce it to a product of geminal two electron orbitals. This would be a step to addressing the many body problem which is of course the goal of many theorists in widely different areas. The beauty of ion-atom collisions is that we may turn the many body complications on in a careful sequential manner and not confront all the difficulties at once.

ACKNOWLEDGEMENTS

This work was supported by U.S. National Science Foundation grant PHY-9009717, and a grant from the Texas A&M Center for Energy and Mineral Resources.

REFERENCES

1. R. Glauber *Lectures in Theoretical Physics* vol. II, eds. W.E. Brittin et al., New York: Interscience (1959).

2. H. Knudsen, L.H. Andersen, P. Hvelplund, G. Astner, H. Danared, L. Liljebi and K.G. Rensfelt, J. Phys. B: At. Mol. Phys. **17**, 3545 (1984); L.H. Andersen, P. Hvelpund, H. Knudsen, S.P. Moller, K.G. Elsener, K.G. Reensfelt, and E. Uggerhoj, Phys. Rev. Lett. **57**, 2147 (1986).

3. J.H. McGuire, Phys. Rev. Lett. **49**, 1153 (1982); J. Phys. B: At. Mol. Phys. **17**, 1779 (1984).

4. J.F. Reading and A.L. Ford, J. Phys. B: At. Mol. Phys. **20**, 3747 (1987); Phys. Rev. Lett. **58**, 543 (1987).

5. J.F. Reading, Phys. Rev. **A8**, 3262 (1973); J.F. Reading and A.L. Ford, *ibid* **A21**, 124 (1980); J. Reinhardt, B. Muller, W. Greiner and G.Soff, Phys. Rev. Lett. **43**, 1307 (1979).

6. R.L. Becker, A.L. Ford and J.F. Reading, Phys. Rev. **A29**, 3111 (1984).

7. J. F. Reading, A. L. Ford and R. L. Becker, Nucl. Inst. Meth. **169**, 273 (1980); R. L. Becker, A. L. Ford and J. F. Reading, J. Phys. B: Atom. Mol. Phys. **13**, 4059 (1980); Phys. Rev. A **23**, 510 (1981).

8. M. Rodbro, E.H. Pedersen, C.L. Cocke and J.R. McDonald, Phys. Rev. **A19**, 1936 (1979).

9. C.L. Cocke and P.K. Gardner, Phys. Rev. **A16**, 2248 (1977).

10. J.R. McDonald, C.L. Cocke and W.W. Eidson, Phys. Rev. Lett. **32**, 648 (1974); M. Rodbro, E.H. Pedersen, C.L. Cocke and J.R. McDonald, Phys. Rev. **A19**, 1936 (1979).

11. J.D. Garcia, Nuclear Instruments and Methods A**240**, 552 (1985).

12. I.M. Cheshire, Proc. Soc. Phys. **84**, 89 (1964); W.B. Riesenfeld and K.M. Watson, Phys. Rev. **102**, 1157 (1956).

13. J.P. Giese and Erik Horsdal, Phys. Rev. Lett. **60**, 2018 (1988); E.Y. Kamber, C.L. Cocke, S. Cheng and S.L. Varghese, Phys. Rev. Lett. **60**, 2026 (1988).

EXTENSION OF COUPLED CLUSTER METHODOLOGY TO OPEN SHELLS: STATE UNIVERSAL APPROACH

J. Paldus,[a,b] **P. Piecuch,**[a,*] **B. Jeziorski**[c] **and L. Pylypow**[b]

[a]*Department of Applied Mathematics, University of Waterloo, Waterloo, Ontario, N2L 3G1, Canada*
[b]*Department of Chemistry and Guelph-Waterloo Center for Graduate Work in Chemistry, Waterloo Campus, University of Waterloo, Waterloo, Ontario, N2L 3G1, Canada*
[c]*Department of Chemistry, University of Warsaw, Pasteura 1, 02-093 Warsaw, Poland*

ABSTRACT

A brief description of basic models and methods that are currently exploited in molecular electronic structure calculations indicates the desirability of extending the single-reference coupled-cluster (CC) approach to the multi-reference (MR) case. The recently developed explicit formalism of orthogonally spin-adapted, state-universal (or Hilbert space) MR-CC theory for the special case of closed-shell type references is applied to a simple four-electron model system formed by two interacting hydrogen molecules. Varying the geometry of the nuclear framework from the square configuration to a linear one this model simulates the breaking of a single chemical bond and enables a continuous transition to be made between the degenerate and non-degenerate regimes. The simplicity of the model enables a comparison with the exact solution, obtained by the full configuration interaction method. In this way the performance of the MR-CC approach at both linear and non-linear levels of approximation, nature and multiplicity of MR-CC solutions and the role played by the intruder states can be examined in considerable detail.

INTRODUCTION

The many-electron correlation problem arising in finite molecular systems, even when only a very small number of electrons and nuclei is involved, represents a truly challenging many-body problem. With present day methodology and computational facilities, a highly reliable and a rather accurate description can be achieved for systems with 4-10 electrons, particularly for those of a closed shell type near their equilibrium

---

* Permanent address: Institute of Chemistry, University of Wroclaw, F. Joliot-Curie 14, 50-383 Wroclaw, Poland.

Recent Progress in Many-Body Theories, Vol. 3,
Edited by T.L. Ainsworth et al., Plenum Press, New York, 1992

geometries. However, an extension of the existing procedures and algorithms to larger or open-shell systems is far from being satisfactory.

Standard quantum chemical methods[1,2] that are being employed for a quantitative description of the molecular electronic structure invariably rely on non-relativistic, finite-dimensional, electronic model Hamiltonians invoking the Born-Oppenheimer (or clamped nuclei) approximation. These electronic Hamiltonians are best described using the second quantization formalism, when they take the following form (see, for example, Ref. 3)

$$\hat{H} = \sum_{\mu,\nu}\langle\mu|\hat{z}|\nu\rangle\sum_{\sigma}X_{\mu\sigma}^{\dagger}X_{\nu\sigma} + \tfrac{1}{2}\sum_{\kappa,\lambda,\mu,\nu}\langle\kappa\lambda|\hat{v}|\mu\nu\rangle\sum_{\sigma,\tau}X_{\kappa\sigma}^{\dagger}X_{\lambda\tau}^{\dagger}X_{\nu\tau}X_{\mu\sigma}\,, \tag{1}$$

where $\hat{z}$ represents the electronic kinetic energy and the potential energy in the electrostatic field of nuclei, while $\hat{v}$ describes the interelectronic Coulomb repulsion and the creation (annihilation) operators $X_{\mu\sigma}^{\dagger}(X_{\mu\sigma})$ are associated with a finite set of (usually non-orthogonal, in which case the standard second quantization formalism must be properly modified) atomic spin-orbitals $\chi_{\mu\sigma}(\mathbf{x}) = \langle\mathbf{x}|\mu\sigma\rangle = \langle\mathbf{r}|\mu\rangle\langle s|\sigma\rangle$, $\sigma = \pm 1/2$, $\mathbf{x} \equiv (\mathbf{r},s)$, represented by the product of atomic orbitals (AO's) $|\mu\rangle$ and spin eigenfunctions $|\sigma\rangle$, $\mathbf{r}$ and s designating space and spin coordinates, respectively.

Thus, in the so-called *ab initio approaches*, the model Hamiltonian $\hat{H}$, Eq. (1), is specified by selecting a suitable finite set of AO's (referred to, depending on its size relative to the electron number N, as minimal, double-zeta, double-zeta plus polarization, etc., basis set), spanning a chosen finite-dimensional subspace of the general one-electron Hilbert space, and the one- and two-electron integrals, $\langle\mu|\hat{z}|\nu\rangle$ and $\langle\kappa\lambda|\hat{v}|\mu\nu\rangle$, respectively, are then computed using this AO basis. Since, generally, up to 4-center two-electron integrals can arise, one often represents the AO's as a linear combination of Gaussians (one requires, roughly, 3-4 Gaussians in lieu of a Slater-type orbital).

For larger systems, the ab initio description becomes unmanageable and one reverts to *semiempirical approaches*. In this case one employs a hypothetical minimum basis set of AO's by treating the one- and two-electron integrals in (1) as semiempirical parameters (after their number has been drastically reduced using various simplifying assumptions, such as the tight-binding approximation for one-electron integrals, zero differential overlap for two-electron integrals, etc.). Moreover, while the ab initio Hamiltonian always represents the total electronic energy of the system (to which the constant electrostatic nuclear energy can be added to obtain the total energy), various semiempirical Hamiltonians usually describe only some not very precisely defined component of the total energy (e.g., the π-electron energy in the case of the Pariser-Parr-Pople (PPP) or Hückel Hamiltonians[4]).

The basic methodologies for finding approximate eigenvalues and corresponding eigenstates of the Hamiltonian (1) can be divided into the molecular orbital (MO) and valence bond (VB) types. While the latter ones presently enjoy a definite rennaisance,[1,5,6] the vast majority of all applications continues to rely on the MO scheme. Here the first step consists in finding an orthonormal set of independent particle model (IPM) one-electron states, referred to as molecular orbitals (MO's) and represented as linear combinations of AO's (LCAO approximation) using the Hartree-Fock-Roothaan selfconsistent field procedure.[7] This step is nowadays routinely carried out even for rather large systems (with 200 electrons or more) and for smaller systems, a very large AO basis set can be employed in order to approach the true Hartree-Fock (HF) limit.

The HF approximation yields, in most cases, very accurate total energies (usually within a few per cent of the exact ones). Unfortunately, the accuracy afforded by the HF (or SCF) approximation is not sufficient for a reliable determination of various observables of chemical interest, such as dissociation energies, dipole or quadrupole moments, polarizabilities or hyperpolarizabilities, etc. This fact was first established in the early sixties[8,9] when it was found that, for example, the F_2 molecule is not stable within the HF approximation (which gives a negative dissociation energy in this case). Since that time, the importance of correlation effects was found to be crucial in most problems of chemical interest, particularly in calculating various activation, bond or dissociation energies, excitation or ionization energies, electron affinities, etc. In fact, correlation effects are even important for an accurate determination of equilibrium geometries, particularly for excited species or reaction intermediates.

The methods employed in going beyond the HF approximation (within the MO formalism) can be generally classified into variational and perturbative categories. The prototype of variational approaches is the configuration interaction (CI) method (referred to as the "shell model" by nuclear physicists). Again, one stays within a finite dimensional N-electron model space represented by an antisymmetric component of the N-th tensor power $\mathcal{V}_1^{\otimes N}$ of the one-electron space $\mathcal{V}_1$ employed, and relies on the simple Ritz variational principle.[10] The matrix representative of a spin-independent Hamiltonian, Eq. (1), in this space will block into subproblems characterized by the total spin quantum numbers S and S_z. While the Slater determinants constructed from MO's are automatically eigenstates of $\hat{S}_z$, proper linear combinations must be chosen to obtain spin-adapted states relative to $\hat{S}^2$. In this respect the representation theory of the symmetric group $\mathcal{S}_N$ or the orbital unitary groups U(n), $2n = \dim \mathcal{V}_1$, plays a very useful role.[11-14] The latter proved to be very beneficial and stimulating not only in constructing the N-electron spin-adapted states, but also in providing efficient algorithms for the construction (or an implicit construction in the so-called direct CI approaches[15]) of the Hamiltonian matrix representatives and their diagonalization,[12-14] since the pertinent Hamiltonian takes the form[12,16]

$$\hat{H} = \sum_{i,j} \langle i|\hat{z}|j \rangle E_j^i + \tfrac{1}{2} \sum_{i,j,k,\ell} \langle ij|\hat{v}|k\ell \rangle E_{k\ell}^{ij}, \tag{2}$$

where

$$E_j^i = \sum_{\sigma} X_{i\sigma}^{\dagger} X_{j\sigma} \tag{3}$$

are U(n) generators and

$$E_{k\ell}^{ij} = E_k^i E_\ell^j - \delta_k^j E_\ell^i \tag{4}$$

are elements of its enveloping algebra, the creation and annihilation operators being now defined in the basis of orthonormal molecular spin-orbitals $|i\rangle|\sigma\rangle$. This formulation found its usefulness not only in CI (particularly large scale CI) approaches, but in perturbation-type approaches as well.

The great advantage of CI-type approaches is their versatility and universality since they can be applied to closed- or open-shell systems in their ground or excited states (characterized by an arbitrarily high spin multiplicity) with the same ease.[10] Unfortunately, the dimensionality of the required N-electron subspaces rapidly increases with increasing n and N, thus requiring a drastic truncation in spite of the capability of modern computers to diagonalize matrices of extremely high orders (routinely $10^4 - 10^6$ for sparse CI matrices). Almost all truncation schemes that are employed in practical applications are based on the excitation order of individual configurations relative to one or a few [for the so-called multi-reference (MR) CI] reference configurations. The consequence of such a truncation is the fact that the resulting energies are not size extensive. For example, for a system consisting of non-interacting two-electron subsystems, the CID energy (CI truncated at the doubly excited level) increases with N as $\sqrt{N}$ when $N \to \infty$. Thus, the correlation energy per particle that is recovered by such a procedure vanishes for extended systems and the resulting wave function is orthogonal to the exact one in this approximation. In spite of these shortcomings, the CI procedure can provide a wealth of useful results, particularly for excited states, that would be difficult to obtain otherwise. Nonetheless, the lack of size-extensivity is sorely felt when the method is applied to phenomena in which the number of particles in subsystems is not conserved, such as various dissociative, associative or reaction processes.

The problem of size extensivity can be automatically avoided when using the second type of correlated methods, namely the perturbative-type approaches. The basis for these methods is represented by the Rayleigh-Schrödinger many-body perturbation theory (MBPT) as developed by Brueckner,[17] Goldstone,[18] Hubbard,[19] Hugenholtz,[20], Brandow[21] and others.[22] Indeed, each individual connected (and thus linked) Goldstone or Hugenholtz energy diagram represents a size-extensive energy contribution, as does any truncation of the MBPT scheme, order by order or diagram by diagram. Unfortunately, an order by order evaluation of the MBPT series[2] is computationally extremely demanding when going beyond the fourth order[23a] and has its limitations,[23,24] especially in quasidegenerate situations, when the MBPT series may poorly converge or not converge at all.[25,26] In these cases the so-called coupled cluster (CC) methods[27–31] that sum certain types of MBPT diagrams to infinite order provide much better results.

The basic concept for CC methodology, at least when applied to non-degenerate closed shell ground states, is based on the MBPT structure of the exact wave function $|\Psi\rangle$[19] that can be expressed in terms of *connected* terms through the exponential ansatz, namely

$$|\Psi\rangle = \exp(\hat{T})|\Phi\rangle , \tag{5}$$

where $|\Phi\rangle$ designates the IPM wave function and T represents the so-called cluster operator that is given by all i-body connected cluster components of the exact wave function,

$$\hat{T} = \sum_{i=1}^{N} \hat{T}_i . \tag{6}$$

An independent source for the exponential ansatz is statistical mechanics.[32] In view of this ansatz, the higher excited components of the wave function consist of both connected and disconnected parts, the latter being most important in case of tetra-excitations (as symbolically expressed by the inequality $\hat{T}_4 \ll \frac{1}{2}\hat{T}_2^2$). Thus, the basic approximation of the single reference CC approach,[27] Eq. (5), is obtained by setting $\hat{T} \cong \hat{T}_2$ and is referred to as the CCD (CC with doubles) approach. Using the exponential ansatz (5) with $\hat{T} \cong \hat{T}_2$ in the time-independent Schrödinger equation, one obtains an energy independent system of non-linear algebraic equations for the pair cluster components $\langle rs | \hat{t}_2 | ab \rangle$,

$$\hat{T}_2 = \tfrac{1}{4} \sum_{r,s,a,b} \langle rs | \hat{t}_2 | ab \rangle E_{ab}^{rs} , \tag{7}$$

where we designated the particle (hole) orbitals relative to the IPM reference $|\Phi\rangle$ by $r, s\,(a, b)$. These equations represent, in fact, recursion formulas for the generation of higher and higher order MBPT diagrams of a certain type. Thus, solving these equations, one not only automatically generates the appropriate MBPT diagrams but also simultaneously evaluates them. Once the pair cluster components are known, the energy is easily calculated from the asymmetric energy formula

$$\begin{aligned} E \;&= \langle \Phi | \hat{H} \hat{T}_2 | \Phi \rangle \\[2mm] &= \tfrac{1}{4} \sum_{a,b,r,s} \langle ab | \hat{v} | rs \rangle [\langle rs | \hat{t}_2 | ab \rangle . \end{aligned} \tag{8}$$

Higher approximations result when the $\hat{T}_1$ (CCSD) and $\hat{T}_3$ (CCSDT) components are also accounted for. The latter method is computationally extremely demanding and can only be carried out for relatively small systems. Various approximate CCSDT approaches are thus often employed.[29-31]

Although the CCD or CCSD approximations usually provide highly reliable and accurate energies for closed shell non-degenerate ground states,[29] and are nowadays extensively employed,[30] an extension of this formalism to general open shell systems proved to be very difficult.[31,33] We shall briefly summarize these developments in the next section and illustrate the performance of one possible generalization on a simple four electron model system.[34-39,25,26]

MULTI-REFERENCE CC APPROACHES

An extension of the cluster ansatz (5) to the multi-reference (MR) case, that is required when open-shell systems or nearly-degenerate states are involved, is by no means unambiguous and easy. Generally, the MO set employed is partitioned into three disjoint subsets consisting of *core*, *valence* or *active*, and *excited* or *virtual* orbitals. The core orbitals are fully occupied and virtual ones unoccupied in all reference configurations $|\Phi_p\rangle (p = 1, \cdots, M)$ that differ in occupancies of valence orbitals and span the model space $\mathcal{M}_0$, $\dim \mathcal{M}_0 = M$. One defines the projection operator $P = \sum_q P_q, \; P_q = |\Phi_q\rangle\langle\Phi_q|$ of the N-electron Hilbert space $\mathcal{H}_N$ (or the whole Fock space $\mathcal{F}$) onto $\mathcal{M}_0$ and its orthogonal complement $Q = 1 - P$, and identifies an M-dimensional subspace $\mathcal{M}$ of $\mathcal{H}_N$ that is spanned by the exact eigenstates $|\Psi_p\rangle$ into which $|\Phi_p\rangle$'s evolve when all the interactions in the Hamiltonian $\hat{H}$ are properly accounted for. Thus, the mapping $\tilde{P} : \mathcal{M} \to \mathcal{M}_0$, where $\tilde{P}$ represents a restriction of P to $\mathcal{M}$, is a bijection, whose inverse we designate by $\tilde{U}$, $\tilde{U} = \tilde{P}^{-1}$. Extending, next, $\tilde{U}$ to the whole of $\mathcal{H}_N$ (or to $\mathcal{F}$) by defining it as the zero operator on $\mathcal{M}_0^{\perp}$, we find that both P and U are idempotent and $PU = P$ while $UP = U$. The operator U thus represents the MR generalization of the wave operator $W = e^T P_0$, $P_0 = |\Phi_0\rangle\langle\Phi_0|$ of the single reference ($|\Phi_0\rangle$) theory, producing some linear combination $|\tilde{\Psi}_p\rangle$ of exact states $|\Psi_p\rangle$ from $\mathcal{M}$,

$$|\tilde{\Psi}_p\rangle = U|\Phi_p\rangle . \tag{9}$$

In view of the property $PU = P$ we are also assured that these states are intermediately normalized, i.e.

$$\langle\tilde{\Psi}_p|\Phi_q\rangle = \delta_{pq} . \tag{10}$$

The fundamental idea that enables one to "transform" the exact problem into the one formulated in $\mathcal{M}_0$ is that of an *effective* or *model Hamiltonian*. One easily finds that the exact energies E_p associated with the eigenstates $|\Psi_p\rangle$, $(p = 1, \cdots, M)$ are given by the eigenvalues of the following effective Hamiltonian (defined on $\mathcal{M}_0$)

$$\hat{H}^{(\mathrm{eff})} = PHU = PHUP , \tag{11}$$

with the wave operator U being determined by the Bloch equation

$$UHU = HU . \tag{12}$$

While all genuine MR-CC approaches are based on the effective Hamiltonian (11) and the Bloch equation (12), ambiguities arise when we try to extend the cluster ansatz (5) to the MR case. It is beyond the scope of this brief overview to outline various, often conflicting, developments that paved the tortuous path through literature (for more detail, see e.g. Refs. 31,33 and references therein). We shall only state that the existing MR-CC approaches that employ a model Hamiltonian formalism are basically of two types: the so-called *Fock space* or *valence universal* methods[40−46,23,33] and *Hilbert space* or *state universal* methods.[47−52,38,39]

The valence universal ansätze employ a single − valence universal − cluster operator, similarly as in the single reference case mentioned earlier. However, appropriate precautions must be made, since there is no "natural" vacuum to be used as a reference.[22] Thus, while in the single reference case the cluster operators involve only creation operators when the hole-particle formalism is employed, both creation and annihilation operators associated with active orbitals will appear in the MR case. There are several possible avenues that are opened to resolve this problem, but all of them essentially rely on the assumption of valence universality if only a single cluster operator is to be employed. This, in fact, implies that together with the studied system, we must simultaneously consider all its ions, up to and including the one in which only core orbitals are occupied. Thus, the entire inter-related sequence of systems of CC equations must be solved. The apparent advantage of simultaneously obtaining information about ionized species is, unfortunately, at least partially illusory, since a satisfactory handling of such systems within the finite-dimensional MO scheme usually requires that quite different orbital bases be employed for the parent molecule or radical, and its various ions. So far this approach found its greatest use in the computation of various "differential" properties, such as excitation or ionization energies, rather than in the determination of potential energy surfaces of a group of interacting states over a wide range of nuclear framework geometries. Recently, the so-called cluster conditions that ascertain the negligibility of higher excited cluster components in this scheme were derived[53] and examined[54] (note that in the MR case, the excitation order is only defined relative to each reference configuration). A study of the performance of the valence universal scheme for the model system examined in this paper is in progress.[55]

The state universal or Hilbert space approach,[47] on the other hand, employs a valence space with a fixed number of electrons and the wave operator is represented as a super-position of exponential operators corresponding to individual reference configurations $|\Phi_q\rangle$,

$$U = \sum_{q=1}^{M} \exp(\hat{T}^{(q)}) P_q \,. \tag{13}$$

Using this ansatz in the Bloch equation (12) then yields a system of equations of general form

$$\sum_r \Lambda_p^{(r)}(G_i) = \sum_r \Xi_p^{(r)}(G_i) \,, \tag{14}$$

where G_i designates a general excitation operator (that involves at least one non-valence orbital) and

$$\Lambda_p^{(0)}(G_i) = \langle G_i \Phi_p | H | \Phi_p \rangle \,,$$

$$\Lambda_p^{(1)}(G_i) = \langle G_i \Phi_p | [H, T^{(p)}] | \Phi_p \rangle \,, \tag{15}$$

$$\Lambda_p^{(2)}(G_i) = \tfrac{1}{2} \langle G_i \Phi_p | [[H, T^{(p)}], T^{(p)}] | \Phi_p \rangle \,,$$

while

$$\Xi_p^{(1)}(G_i) = \sum_{q(\neq p)} \langle G_i \Phi_p | T^{(q)} - T^{(p)} | \Phi_q \rangle H_{qp}^{(\mathrm{eff})} \,,$$

$$\Xi_p^{(2)}(G_i) = \tfrac{1}{2} \sum_{q(\neq p)} \langle G_i \Phi_p | (T^{(q)} - T^{(p)})^2 + [T^{(q)}, T^{(p)}] | \Phi_q \rangle H_{qp}^{(\mathrm{eff})} \,, \tag{16}$$

where we restricted ourselves to at most quadratic terms. The left-hand side terms $\Lambda_p^{(r)}(G_i)$ are recognized to be of the same form as in the single reference case, while the right-hand side terms $\Xi_p^{(r)}(G_i)$ represent coupling terms. The detailed structure of these equations for the case of orthogonally spin-adapted references and excitation operators G_i was given elsewhere.[38,50,52] We shall now apply this formalism to a simple four electron model system.[34−39,25,26]

METHOD AND MODEL DESCRIPTION

We wish to explore the basic characteristics and potential of the Hilbert space MR-CC approaches for the determination of potential energy hypersurfaces of various molecular systems. Since this type of CC approach has seen few applications so far,[38,39,49,56] we direct our attention to a simple model system[34] that can be described by the simplest possible version of MR-CC theory, namely that involving a two-dimensional model space spanned by closed-shell type configurations.[50,38] This situation arises when we have two active orbitals of different spatial symmetry.[50] The required formalism in orthogonally spin-adapted form was first worked out in detail for a linearized version of the theory[50] (although in the effective Hamiltonian all terms were included). It was soon realized that while the linear version works well in the degenerate or at least highly quasidegenerate regimes, it breaks down in the non-degenerate case when intruder states are present. In this way it behaves in an exactly complementary way to single reference CC theory. The linear CC(L-CC) version[50] was thus supplemented by quadratic terms,[38] which were derived in two completely independent ways relying, respectively, on algebraic formulation[50] via spin-adapted replacement operators (forming the enveloping algebra of $U(n)$) and via diagrammatic techniques[52] combining the usual Goldstone-Hugenholtz formalism with graphical methods of spin algebras. The simple two-reference version of the MR-CC theory can be easily extended to an arbitrary number of closed shell references,[52] as well as to a completely general MR case when we forego the orthogonally spin-adapted form. The latter form of the general theory is much more difficult to derive. For the present purposes, however, the two-reference version is perfectly adequate. We thus rely on the formulation given in Refs. 50 and 52 and restrict the cluster operators $T^{(q)}$, $q = 1, 2$, to mono- and bi-excited configurations relative to each reference. Thus, in this approximation, a different set of excited configurations $G_i|\Phi_p\rangle$ is considered for each p, $p = 1, 2$.*)

Our preliminary results[38] included only quadratic pair clusters $\frac{1}{2}T_2^{(q)2}$ in both direct ($\Lambda_p^{(2)}$) and coupling ($\Xi_p^{(2)}$) terms. Here we wish to explore separately the role of pair cluster interaction in the direct terms only as well as the effect of all the linear and quadratic terms involving $\hat{T}_1^{(p)}$ and $\hat{T}_2^{(p)}$ clusters relative to the reference $|\Phi_p\rangle$, $p = 1, 2$. For this purpose we observe that only the following terms will contribute to the coupling terms $\Xi_p^{(r)}$, Eq. (16), assuming that we restrict ourselves to at most biexcited operators G_i, namely

$$\Xi_p^{(1)}(G_i) = \sum_{n=1}^{2} R_n^{(p)}(G_i), \tag{17}$$

where

$$R_n^{(p)}(G_i) = \langle G_i \Phi_p | T_n^{(q)} | \Phi_q \rangle H_{qp}^{(\text{eff})}, \tag{18}$$

and

$$\Xi_p^{(2)}(G_i) = \sum_{\substack{n,n'=1 \\ (n \leq n')}}^{2} B_{nn'}^{(p)}(G_i) + \tilde{B}_{12}(G_i), \tag{19}$$

with

$$B_{nn}^{(p)}(G_i) = \tfrac{1}{2}\langle G_i \Phi_p | (T_n^{(q)})^2 | \Phi_q \rangle H_{qp}^{(\text{eff})},$$

$$B_{12}^{(p)}(G_i) = \langle G_i \Phi_p | T_2^{(q)}(T_1^{(q)} - T_1^{(p)}) | \Phi_q \rangle H_{qp}^{(\text{eff})}, \tag{20}$$

$$\tilde{B}_{12}(G_i) = \langle G_i \Phi_p | [T_2^{(q)}, T_1^{(p)}] | \Phi_q \rangle H_{qp}^{(\text{eff})}.$$

We shall also test the role of various approximations for the effective Hamiltonian $H_{qp}^{(\text{eff})}$. The explicit expressions for $H_{qp}^{(\text{eff})}$ were given in Section VI of Ref. 50. The diagonal terms are clearly identical with the single reference CC expressions for the energy and involve at most quadratic $(T_1^{(p)})^2$ terms. The off-diagonal elements $H_{21}^{(\text{eff})}$ and $H_{12}^{(\text{eff})}$, however, can involve up to quartic terms in $T_1^{(p)}$ clusters. The pertinent expressions are easily obtained from those given in Ref. 50 [Eqs. (84)-(102)] by neglecting all the triply and quadruply excited cluster components.

Let us, finally, describe the four-electron model employed[34] that has been extensively studied using the single reference CC theory[34−35] and perturbation theory[25] as well as

---

*) Note that the linear version of the theory[50] was formulated for a more general case, whose performance it would also be interesting to assess.

MR-CC theory[38,39] and MR-MBPT[26,36] and other approaches.[37] We restrict ourselves in this paper to the so-called H4 model[34] in which two hydrogen molecules are brought together so that the protons form a square configuration with the nearest-neighbor separation (or "bond length") a. In addition to the (nearly) equilibrium bond length $a = 1.6$ a.u. we also consider the stretched ($a = 2.0$ a.u.) and compressed ($a = 1.2$ a.u.) cases. The interacting H_2 molecules are then rotated by an angle $\phi = \alpha\pi$ so that the nuclear framework forms an isosceles trapezoid (see Fig. 2 of Ref. 34). Proceeding from the square conformation ($\alpha = 0$), in which both reference configurations $|\Phi_p\rangle$, $p = 1, 2$, are degenerate, to the linear one ($\alpha = 1/2$), in which no degeneracy is present and which is well described by the single reference theory,[34,35,25] we in fact model a potential energy surface involving a dissociation of a single chemical bond. The other models[34] of this type that we examined and that characterize a simultaneous breaking of two or more single bonds,[57] as well as non-planar geometries,[58] will be given elsewhere. Here we also restrict ourselves to a simple minimum basis set model[34] and employ restricted HF MO's throughout. Since we compare our results with the exact solution, given by the full CI method, the basis employed is not as important as when the comparison with experiment is sought, while all the basic phenomena that we wish to explore (such as the singular behavior due to the intruder states or the role of various terms arising in the theory) are present already in the simplest minimum basis set (cf. Refs. 34 and 35).

RESULTS AND DISCUSSION

As might be expected,[50] the linear version of the state-universal MR-CC formalism provides very good correlation energies in the vicinity of the degenerate limit, as was already pointed out earlier,[38,51] and as may be seen from Figs. 1-3 that correspond to three distinct internuclear separations, $a = 1.2$, 1.6 and 2.0 a.u., respectively. However, as the non-degenerate regime is approached, the second reference state energy becomes degenerate with the next lowest lying state that plays the role of an intruder state. Consequently, the linear MR-CC energy undergoes a singular behavior, just as the single-reference L-CCD or L-CCSD scheme does in the quasidegenerate region (where, in fact, our second reference $|\Phi_2\rangle$ may be regarded as an "intruder state" for $|\Phi_1\rangle$). This singularity appears in the vicinity of the geometry for which the energy of the $|\Phi_2\rangle$ configuration, as given by the diagonal CI matrix element $\langle\Phi_2|\hat{H}|\Phi_2\rangle$, intersects with the corresponding energy of the intruder configuration. We can thus foresee this behavior by simply observing the dependence of the non-interacting configuration energies, as given by the diagonal CI matrix elements, on the geometry of the nuclear framework. It should also be noted that once we pass the region of singular behavior, the second root of the effective Hamiltonian describes the third state, to which $|\Phi_2\rangle$ now primarily contributes. It is remarkable, though, that the ground state is again well described by the lowest root of the effective Hamiltonian matrix. It is also interesting to observe that the "width" of the singularity becomes narrower and narrower as we compress the nuclear framework together. This is precisely the reason why in our earlier work[34,35,25,38] we always considered slightly stretched ($a = 2.0$ a.u.) H_2 molecules.

Just as in the single reference case,[34,35] the singular character of the linear approximation can be removed by accounting for the quadratic (or higher order) terms of the CC theory. This is illustrated graphically in Figs. 1-3 and in greater detail numerically in Table I. This table compares the results obtained with three different non-linear approximations for three different internuclear separations ($a = 1.2, 1.6$ and 2.0 a.u.), and the exact result given by the FCI solution. In the first approximation (MRCCSD-1), in addition to all the linear and absolute terms, the quadratic term involving pair clusters $\frac{1}{2}(T_2^{(p)})^2$ is included only in the direct term [i.e., only the

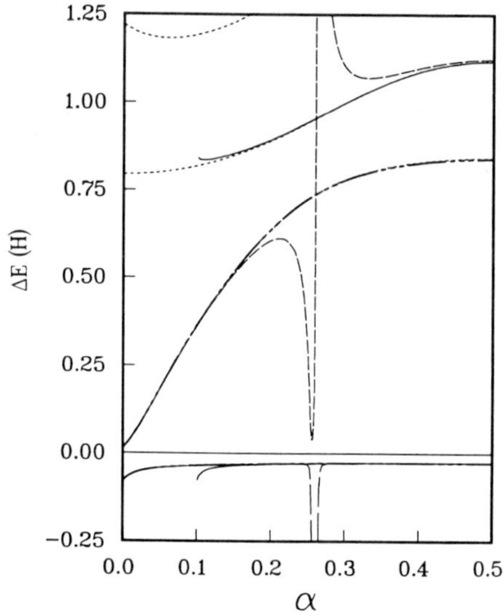

Figure 1. A comparison of the full CI (short dashes) and various MR-CCSD correlation energies ΔE (in Hartrees) relative to the ground state restricted Hartree-Fock (RHF) energy, $\Delta E = E - E_0^{\mathrm{RHF}}$, for the symmetric low energy singlet states of the minimum basis set H4 model considered over the whole range of geometries defined by the angle α (in radians), Fig. 2 of Ref. 34. The internuclear H-H separation is $a = 1.2$ a.u. The L-MRCCSD energies are represented by long dashed lines and display a singular behavior around $\alpha \cong 0.26$. The first and the second pair of the MRCCSD-3 solutions are represented by the thick long dash - short dashes and by the thin solid lines, respectively. See the text for more detail.

term $\Lambda_p^{(2)}$, Eq. (15), with $T^{(p)} = T_2^{(p)}$ is considered, while the quadratic component of the coupling term is neglected, $\Xi_p^{(2)} \equiv 0$]. In the second approximation (MRCCSD-2), the $\frac{1}{2}(T_2^{(p)})^2$ clusters are included in both direct and coupling terms (so that both $\Lambda_p^{(2)}$ with $T^{(p)} = T_2^{(2)}$ and B_{22} terms are included). Finally, the third approximation

(MRCCSD-3) represents the fully quadratic approximation including the monoexcited $T_1^{(p)}$ clusters, in which all the terms listed in Eqs. (14-16) or (14) and (17-20) are taken into account. In all calculations reported here, $|\Phi_1\rangle$ represents the restricted HF solution for the model considered and the corresponding MO's are used throughout.

Table I. Comparison of the full CI and MRCCSD energies relative to the ground state restricted Hartree-Fock (RHF) energy E_0^{RHF}, $\Delta E = E - E_0^{\mathrm{RHF}}$ (in mH) for the first two symmetric singlet states of the minimum basis set H4 model with the H-H internuclear separations $a = 1.2, 1.6$ and 2.0 a.u. Three different variants of the MRCCSD method are defined in the text (MRCCSD-3 representing the fully quadratic approximation).

	α	FCI		MRCCSD-1		MRCCSD-2		MRCCSD-3	
		ΔE_1	ΔE_2	ΔE_1	ΔE_2	ΔE_1	ΔE_2	ΔE_1	ΔE_2
	0.000	-78.424	14.047	-78.384	14.029	-78.375	14.032	-78.375	14.032
	0.010	-62.143	40.203	-62.106	40.191	-62.097	40.194	-62.096	40.194
	0.050	-41.920	184.303	-41.904	184.329	-41.892	184.334	-41.887	184.334
$a=1.2$	0.100	-35.363	363.174	-35.359	363.234	-35.340	363.239	-35.326	363.242
	0.200	-29.815	638.474	-29.864	638.475	-29.819	638.477	-29.798	638.489
	0.300	-27.706	779.422	-27.847	778.421	-27.733	778.416	-27.832	778.433
	0.500	-26.745	842.472	-26.913	839.850	-26.689	839.840	-27.211	839.849
	0.000	-97.028	6.296	-96.964	6.278	-96.930	6.283	-96.930	6.283
	0.010	-80.823	30.725	-80.768	30.718	-80.732	30.724	-80.730	30.724
	0.050	-57.203	158.568	-57.216	158.650	-57.169	158.660	-57.131	158.663
$a=1.6$	0.100	-48.258	301.982	-48.363	302.076	-48.284	302.083	-48.190	302.099
	0.200	-41.293	466.333	-41.830	465.466	-41.627	465.453	-41.410	465.505
	0.300	-38.966	520.443	-39.971	517.801	-39.620	517.774	-39.442	517.835
	0.500	-37.824	536.434	-39.019	532.632	-38.579	532.602	-38.644	532.658
	0.000	-117.621	-7.268	-117.686	-7.263	-117.575	-7.266	-117.575	-7.266
	0.010	-102.307	14.250	-102.419	14.274	-102.299	14.270	-102.287	14.271
	0.050	-76.429	118.950	-76.841	118.989	-76.673	118.987	-76.473	119.000
$a=2.0$	0.100	-65.321	221.428	-66.328	220.710	-66.023	220.678	-65.538	220.742
	0.200	-57.260	310.816	-60.074	306.354	-59.349	306.252	-58.363	306.376
	0.300	-54.775	333.480	-58.539	326.890	-57.579	326.768	-56.522	326.899
	0.500	-53.690	337.643	-57.567	330.525	-56.573	330.409	-55.705	330.550

Remarkably enough, the direct $(T_2^{(p)})^2$ terms are entirely sufficient to remove the singular behavior of the linear approximation. In fact, the ground state energies provided by all three approximations differ at most by ~ 2 mH when $a = 2.0$ a.u. (less than ~ 0.5 mH for $a = 1.2$ and 1.6 a.u.). The difference between the MRCCSD-2 and 3 approximations that is entirely due to monoexcited clusters $T_1^{(p)}$ vanishes in the square configuration[*], since the $T_1^{(p)}$ components vanish due to the symmetry in the $\alpha = 0$

---

[*] Note that this is not the case for the results given in Ref. 39, which would seem to indicate some internal inconsistency.

limit. The effect of monoexcited clusters then increases the ground state energy up
to $\alpha \cong 2$ or 3 a.u., when a rapid decrease sets in (which is particularly apparent for
the compressed $a = 1.2$ a.u. geometry). However, the inclusion of T_1 clusters does not
always improve the agreement with the full CI result, particularly for the compressed
geometry $(a = 1.2$ a.u.). Even smaller differences between these approximations
are found for the second root of the effective Hamiltonian corresponding to the first

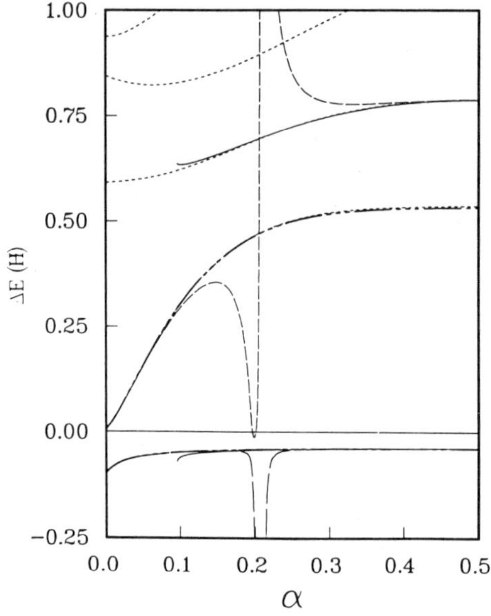

Figure 2. Same as Fig. 1 for $a = 1.6$ a.u.

excited state of the same symmetry as the ground state (maximum difference in all
cases is less than ~ 0.15 mH). When compared with the exact FCI energies, the
agreement is again the best in the highly quasidegenerate region and deteriorates
as the non-degenerate limit $(\alpha = 0.5)$ is approached. An overall agreement also
deteriorates with the increasing internuclear separation a, as expected. Also, in the
degenerate region, we get invariably the better result for the second state rather than
for the ground state while the opposite is true in the non-degenerate limit. It would
be interesting to investigate the results provided by these approximations when using
other MO's than those corresponding to the $|\Phi_1\rangle$ HF ground state.

We must also mention that in order to obtain the MR-CCSD solutions given in Table
I, we cannot employ the linear MRCCSD (L-MRCCSD) solution as a starting iteration once the singularity region is reached. In fact, we have resorted to an "analytic

continuation" procedure (cf., e.g., Refs. 59,60) using the converged MRCCSD amplitudes for the nearby geometry as a starting iteration. Even in the highly degenerate region, where the linear MRCCSD solutions represent a very good approximation, the "analytic continuation" procedure is faster since it requires fewer iterations, as might be expected (assuming that a sufficiently small step in increasing the angle α is used).

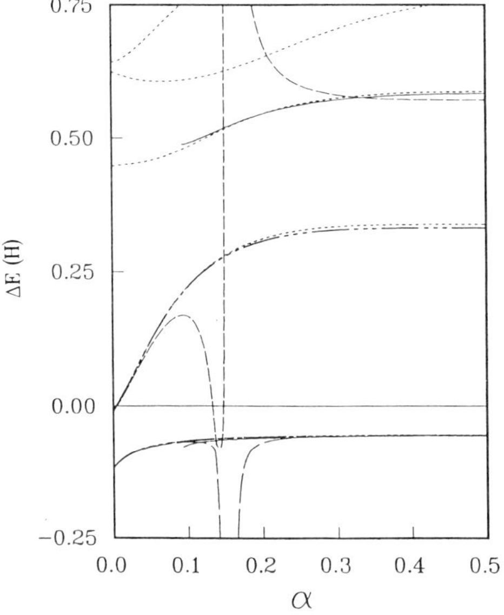

Figure 3. Same as Fig. 1 for $a = 2.0$ a.u.

If we employ the L-MRCCSD solution as a starting iteration in the non-degenerate limit ($\alpha \cong 0.5$), the MR-CCSD method converges to a different pair of solutions, the lowest energy one again describing the ground state but the second one describing the second excited state rather than the first one (as does the L-MRCCSD result). We can then "analytically continue" this solution towards the degenerate limit obtaining a very good description for both states as shown in Figs. 1-3. Only when highly degenerate regime is reached ($\alpha \cong 0.1$), the energies obtained begin to rapidly deviate from the full CI result and soon no converged solution can be obtained. This behavior is very much reminiscent of that found with the standard single reference CCD or CCSD approach in highly quasidegenerate situations, such as represented by the semiempirical cyclic polyene models.[59,60] In this case it can be shown that the CC(S)D equations cease to possess a real solution and the energy bifurcates into two complex solutions (the relevant singularity being the first order branch point).[59,60]

Let us finally examine the role of higher order terms in the expression for the effective Hamiltonian. Here we have again examined three distinct approximations, referred to as MRCCSD-3x, x = a,b,c in Table II. As this notation indicates, we always use the approximation involving all quadratic terms in the MRCCSD equation (referred to above as MRCCSD-3), but truncate the expression for the effective Hamiltonian. The first approximation (MRCCSD-3a) considers at most bilinear terms in cluster amplitudes; in fact we require that the entire expression for the direct and coupling terms (taking into account the dependence of the effective Hamiltonian on cluster amplitudes) contains at most bilinear terms. In the MRCCSD-3b approximation we simply neglect cubic and quartic terms in the effective Hamiltonian matrix elements $H_{12}^{(\mathrm{eff})}$ and $H_{21}^{(\mathrm{eff})}$, so that both MRCCSD equations and the expression for the effective Hamiltonian contain at most bilinear terms. Finally, when all the terms in the effective Hamiltonian are retained, we obtain the MRCCSD-3c approximation, which is thus identical with the MRCCSD-3 approximation considered earlier.

Table II. Comparison of the full CI and MRCCSD-3 energies ΔE (in mH), $\Delta E = E - E_0^{\mathrm{RHF}}$, for the H4 ($a = 2.0$ a.u.) model, obtained with different approximations for the effective Hamiltonian. The 3a approximation considers at most bilinear terms in the cluster amplitudes in both direct and coupling terms, the 3b approximation neglects higher than quadratic terms in the effective Hamiltonian, while the approximation 3c (equivalent with MRCCSD-3 of Table I) considers all the terms in the effective Hamiltonian (see the text for details).

α	FCI		MRCCSD-3a		MRCCSD-3b		MRCCSD-3c	
	ΔE_1	ΔE_2	ΔE_1	ΔE_2	ΔE_1	ΔE_2	ΔE_1	ΔE_2
0.000	-117.621	-7.268	-117.654	-7.105	-117.575	-7.266	-117.575	-7.266
0.010	-102.307	14.250	-102.367	14.448	-102.287	14.271	-102.287	14.271
0.050	-76.429	118.950	-76.511	119.268	-76.474	119.001	-76.473	119.000
0.100	-65.321	221.428	-65.485	221.374	-65.539	220.743	-65.538	220.742
0.200	-57.260	310.816	-57.953	308.413	-58.361	306.376	-58.363	306.376
0.300	-54.775	333.480	-55.882	329.805	-56.519	326.898	-56.522	326.899
0.500	-53.690	337.643	-55.051	333.743	-55.704	330.550	-55.705	330.550

Table II reveals immediately that there is practically no difference between the approximations b and c, the differences in the resulting energies amounting to at most couple of μH's. It is, however, interesting that the simplest approximation MRCCSD-3a yields in fact the best agreement with the exact full CI energies (except for the excited state in the highly degenerate region), although the differences amount to at most a few mH's. This may be related to the fact that in this approximation the same truncation scheme is used at every level of the theory, thus providing a balanced truncation scheme.

We can thus conclude that the Hilbert space MRCC approach, when carried out to the same level of approximation as the CCSD single reference approach, can provide an excellent description of the potential energy surfaces even when the dissociation of chemical bond is involved. The full treatment of this and similar models will be presented elsewhere.[56,57]. It is hoped that this very encouraging result will stimulate further applications and development of the theory, which will eventually enable a reliable description of the potential energy hypersurfaces involving breaking of multiple bonds and multiradical reaction intermediates.

ACKNOWLEDGEMENTS

Continued support by NSERC (J.P.) is much appreciated. Wholehearted thanks are due to Mrs. Helen Warren for her immaculate typesetting of the camera-ready manuscript. Great appreciation and sincere thanks (J.P.) are also due to Drs. C.E. Campbell and E. Krotscheck for creating the most stimulating atmosphere at the MB VII International Conference and for giving us the opportunity to take part in it. We are also most indebted to Dr. Tom Ainsworth for his patience, understanding and help in submitting our contribution.

REFERENCES

1. R. McWeeny, "Methods of Molecular Quantum Mechanics", 2nd ed., Academic Press, New York (1989).

2. S. Wilson, "Electron Correlation in Molecules", Clarendon, Oxford (1984).

3. J. Paldus and J. Čížek, *Adv. Quantum Chem.*, **9**:105 (1975).

4. R.G. Parr, "The Quantum Theory of Molecular Electronic Structure", Benjamin, New York (1963).

5. D.L. Cooper, J. Gerratt, and M. Raimondi, *Nature*, **323**:699 (1986); *Adv. Chem. Phys.*, **69**:319 (1987); *Int. Rev. Phys. Chem.*, **7**:59 (1988); *Topics in Current Chemistry*, **153**:41 (1990); *Chem. Rev.*, **91**:929 (1991).

6. X. Li and J. Paldus, *J. Mol. Struct. (Theochem)*, **229**:249 (1991).

7. C.C.J. Roothaan, *Rev. Mod. Phys.*, **23**:69 (1951).

8. A.C. Wahl, *J. Chem. Phys.*, **41**:2600 (1964).

9. H.F. Schaefer III, "The Electronic Structure of Atoms and Molecules: A Survey of Rigorous Quantum Mechanical Results", Addison-Wesley, Reading, MA (1972).

10. I. Shavitt, *in* "Methods of Electronic Structure Theory", H.F. Schaefer, III, ed., Plenum Press, New York (1977), pp. 189-275.

11. W. Duch and J. Karwowski, *Comp. Phys. Reports*, **2**:93 (1985).

12. J. Paldus, *in*: "Theoretical Chemistry: Advances and Perspectives", Vol. 2, H., Eyring and D. Henderson, eds., Academic Press, New York (1976), pp. 131-290; *in*: "Mathematical Frontiers in Computational Chemical Physics", IMA Series, Vol. 15., D.G. Truhlar, ed., Springer-Verlag, New York (1988), pp. 262-299.

13. I. Shavitt *in*: "New Horizons in Quantum Chemistry", P.-O. Löwdin and B. Pullman, eds., Reidel, Dordrecht, Holland (1983), pp. 279-293; *in* "Mathematical Frontiers in Computational Chemical Physics", IMA Series, Vol. 15., D.G. Truhlar, ed., Springer-Verlag, New York (1988), pp. 300-349.

14. F.A. Matsen and R. Pauncz, "The Unitary Group in Quantum Chemistry", Elsevier, Amsterdam (1986).

15. B.O. Roos and P.E.M. Siegbahn, *in*: "Methods of Electronic Structure Theory", H.F. Schaefer III, ed., Plenum Press, New York (1977), pp. 277-318.

16. J. Paldus and B. Jeziorski, *Theor. Chim. Acta*, **73**:81 (1988).

17. K.A. Brueckner, *Phys. Rev.*, **100**:36 (1955).

18. J. Goldstone, *Proc. Phys. Soc. (London) A*, **239**:267 (1957).

19. J. Hubbard, *Proc. Phys. Soc. (London) A*, **240**:539 (1957); **243**:336 (1958); **244**:199 (1958).

20. N.M. Hugenholtz, *Physica*, **23**:533,481 (1957).

21. B. Brandow, *Rev. Mod. Phys.*, **39**:771 (1967); *Adv. Quantum Chem.*, **10**:187 (1977).

22. I. Lindgren and J. Morrison "Atomic Many-Body Theory", Springer-Verlag, Berlin (1982).

23. (a) S.A. Kucharski and R.J. Bartlett, *Adv. Quantum Chem.*, **18**:281 (1986); (b) P. Čárský and M. Urban "Ab Initio Calculations. Methods and Applications in Chemistry", Lecture Notes in Chemistry, Vol. 16, Springer-Verlag, Berlin (1980).

24. K. Jankowski, *in*: "Methods in Computational Chemistry", Vol. 1, S. Wilson, ed., Plenum Press, New York (1987), pp. 1-116.

25. S. Wilson, K. Jankowski, and J. Paldus, *Int. J. Quantum Chem.*, **23**:1781 (1983); **28**:525 (1985).

26. S. Zarrabian and J. Paldus, *Int. J. Quantum Chem.*, **38**:761 (1990).

27. J. Čížek, *J. Chem. Phys.*, **45**:4256 (1966); *Advan. Chem. Phys.*, **14**:35 (1969); *in*: Ref. 30.

28. J. Paldus, J. Čížek, and I. Shavitt, *Phys. Rev. A*, **5**:50 (1972).

29. R.J. Bartlett, *Annu. Rev. Phys. Chem.*, **32**:359 (1981); *J. Phys. Chem.*, **93**:1697 (1989), and references therein.

30. R.J. Bartlett, ed., Special issue of Theor. Chim. Acta with Proceedings of the Workshop on CC Theory at the Interface of Atomic Physics and Quantum Chemistry, Institute for Theoretical Atomic and Molecular Physics at the Harvard-Smithsonian Center for Astrophysics, Cambridge, MA, in press.

31. J. Paldus, "Coupled Cluster Theory", *in*: "Methods in Computational Molecular Physics", NATO ASI Series, S. Wilson and G.H.F. Diercksen, eds., Plenum Press, New York (1992).

32. J.E. Mayer and M.G. Mayer, "Statistical Mechanics", Chap. 13, J. Wiley and Sons, New York (1940).

33. D. Mukherjee and S. Pal, *Adv. Quantum Chem.*, **20**:292 (1989).

34. K. Jankowski and J. Paldus, *Int. J. Quantum Chem.*, **18**:1243 (1980).

35. J. Paldus, P.E.S. Wormer, and M. Bénard, *Coll. Czech. Chem. Commun.*, **53**:1919 (1988).

36. U. Kaldor, *Int. J. Quantum Chem.*, **28**:103 (1985).

37. N. Iijima and A. Saika, *Int. J. Quantum Chem.*, **27**:481 (1985).

38. J. Paldus, L. Pylypow, and B. Jeziorski, *in*: "Many-Body Methods in Quantum Chemistry", U. Kaldor, ed., Lecture Notes in Chemistry, Vol. 52, Springer-Verlag, Berlin (1989), pp. 151-170.

39. A. Balková, S.A. Kucharski, L. Meissner, and R.J. Bartlett, *in*: Ref. 30.

40. F. Coester, *in*: "Lectures in Theoretical Physics", Vol. 11B, K.T. Mahanthappa and W.E. Brittin, eds., Gordon and Breach, New York, pp. 157-186.

41. R. Offerman, W. Ey, and H. Kümmel, *Nucl. Phys. A*, **273**:349 (1976); R. Offerman, *Nucl. Phys. A*, **273**:368 (1976); W. Ey, *Nucl. Phys. A*, **296**:189 (1978).

42. D. Mukherjee, R.K. Moitra, and A. Mukhopadhyay, *Mol. Phys.*, **33**:955 (1977); A. Haque and D. Mukherjee, *J. Chem. Phys.*, **80**:5058 (1984).

43. I. Lindgren, *Int. J. Quantum Chem.*, Symp. **12**:33 (1978); *J. Phys. B*, **24**:1143 (1991); I. Lindgren and D. Mukherjee *Phys. Rep.*, **151**:93 (1987).

44. W. Kutzelnigg, *J. Chem. Phys.*, **80**:822 (1984); W. Kutzelnigg, D. Mukherjee, and S. Koch, *Chem. Phys.*, **87**:5902 (1987).

45. L. Stolarczyk and H.J. Monkhorst, *Phys. Rev. A*, **32**:725,743 (1985); **37**:1908, 1926 (1988).

46. A. Haque and U. Kaldor, *Chem. Phys. Lett.*, **117**:347 (1985); **120**:261 (1985); *Int. J. Quantum Chem.*, **29**:425 (1986); U. Kaldor, *J. Chem. Phys.*, **87**:467 (1987).

47. B. Jeziorski and H.J. Monkhorst, *Phys. Rev. A*, **24**:1668 (1981).

48. J. Paldus, *in*: "New Horizons of Quantum Chemistry", P.-O. Löwdin and B. Pullman, eds., Reidel, Dordrecht, Holland (1983), pp. 31-60.

49. W.D. Laidig and R.J. Bartlett, *Chem. Phys. Lett.*, **104**:424 (1984); W.D. Laidig, P. Saxe, and R.J. Bartlett, *J. Chem. Phys.*, **86**:887 (1987).

50. B. Jeziorski and J. Paldus, *J. Chem. Phys.*, **88**:5673 (1988).

51. L. Meissner, K. Jankowski, and J. Wasilewski, *Int. J. Quantum Chem.*, **34**:535 (1988); L. Meissner, S.A. Kucharski, and R.J. Bartlett, *J. Chem. Phys.*, **91**:6187 (1989).

52. P. Piecuch and J. Paldus, *Theor. Chim. Acta*, in press.

53. B. Jeziorski and J. Paldus, *J. Chem. Phys.*, **90**:2714 (1989).

54. K. Jankowski, J. Paldus, and J. Wasilewski, *J. Chem. Phys.*, **95**:3549 (1991).

55. K. Jankowski, J. Wasilewski, and J. Paldus, unpublished results.

56. A. Balková, S.A. Kucharski, and R.J. Bartlett, *Chem. Phys. Lett.*, **182**:511 (1991).

57. J. Paldus, P. Piecuch, L. Pylypow, and B. Jeziorski, unpublished results.

58. P. Piecuch and J. Paldus, unpublished results.

59. J. Paldus, M. Takahashi, and R.W.H. Cho, *Phys. Rev. B*, **30**:4267 (1984).

60. P. Piecuch, S. Zarrabian, J. Paldus, and J. Čížek, *Phys. Rev. B*, **42**:3351 (1990).

MONTE CARLO CALCULATIONS OF ATOMS, MOLECULES, AND IONS

K.E. Schmidt and Jiong Xiang

Department of Physics and Astronomy
Arizona State University
Tempe, Arizona 85287 U.S.A.

and

J.W. Moskowitz

Department of Chemistry
New York University
4 Washington Place
New York, New York 10003 U.S.A.

ABSTRACT

By requiring an average backflow correlation, we select a particular Boys and Handy form for a variational wave function. This form explains roughly 75 percent of the correlation energy of the atoms Helium through Argon and some molecules. Good agreement of our calculated ionization potentials and electron affinities shows that this wave function is behaving well in the chemically important valence region. We also explore techniques to visualize where the variational wave functions have high variance.

INTRODUCTION

We will describe some variational calculations on atoms, ions, and molecules. The main goal of this work is to understand the electronic structure of these systems and to develop low variance variational trial functions. A more detailed description of some of the methods is given elsewhere[1].

The questions we wish to address here are: 1) How can we represent electronic correlations with simple variational trial functions; and 2) what are the physical and mathematical motivations for selecting a particular form for the correlations. A successful approach to these problems will also provide low variance trial functions that can make calculations on large electronic systems feasible.

Our work is closely related to that of Umrigar, Wilson, and Wilkins[2,3] our wave functions are a subset of theirs. The major differences are in motivation and selection of terms. Similar results using integral equation techniques have been obtained by Pang, Campbell and Krotscheck[4].

THE VARIATIONAL METHOD

We use the nonrelativistic Hamiltonian for the electrons interacting via coulomb potentials with each other and with fixed nuclei. The variational method is usually based on the Rayleigh-Ritz variational principle which produces an upper bound to the ground-state energy E_0,

$$E_0 \leq \int dR E_L(R) P(R), \tag{1}$$

where R represents the 3N electron coordinates, E_L is the local energy corresponding to position R,

$$E_L(R) = \frac{H\Psi_T(R)}{\Psi_T(R)}, \tag{2}$$

with H the Hamiltonian. $P(R)$ is the probability density for R which is simply

$$P(R) = \frac{\Psi_T^2(R)}{\int dR \Psi_T^2(R)}. \tag{3}$$

We use ideas successfully applied to the study of helium fluids and droplets[5-7] Some of these ideas can be brought over directly to the electronic structure problem. However, some modifications are needed to describe electrons. The variational wave functions consist of an antisymmetric model function, such as a Hartree-Fock wave function, and a set of correlation operators or functions.

The correlations that have been most successful in describing the helium liquids[8] are the two-body correlation that is a function of the interpair distance, three-body correlations which arise from backflow effects[9,10] and in fermion systems, explicit Feynman-Cohen backflow correlations[5,6,11] In the electronic case the two-body correlation is generated primarily by the electron-electron coulomb repulsion. The three-body correlation (that is the correlation between three electrons) is probably weak and we neglect it here. The backflow correlations are derived by looking at the motion of a particle through the fluid. Local current conservation produces a backflow correlation corresponding to the classical motion of the fluid around the moving particle. These correlation should be important in electronic systems. Here we argue that an average effect of backflow is to induce an electron-electron-nucleus correlation.

We will argue in the next section that a Boys and Handy form[12-18] for the correlation function can be used to produce both the two-body correlation and the average backflow correlation.

We will parameterize the correlation and vary the parameters to optimize them[1]. Since we will often have 10 to 20 parameters, we need a good optimization method. As noted by Umrigar, Wilson, and Wilkins[2], and previously by others[19-21] minimizing the variance of the local energy can be used as efficient optimization method. Variance minimization can be thought of as maximizing a lower bound to the ground-state energy by using , for example, the Stevenson[22,23] or Temple[24] bounds. If we apply both the upper and lower bound principles to a good trial function, we find that the lower bound is almost always much farther from the true energy. So minimizing the error from both above and below tends to maximize the lower bound preferentially, and thus minimize the variance. The local variance also has the property that it is a positive quantity which makes it inherently more stable to use with the Monte Carlo method.

The actual method applied to variance minimization seems to be rather unimportant. We minimize the unreweighted variance,

$$V=\sum_i (E_L(R_i)-E_0)^2 \tag{4}$$

using Newton's method or a modified Newton method. Typically, only a few iterations are required to converge the energy as can be seen in Table 1. The R_i in Eq. (4) are the sampled electron positions.

Throughout our work here, we use the Hartree-Fock orbitals to define our antisymmetric model function that the correlation functions multiply. Since we do not vary these orbitals, we do not have the optimal choice. Eventually, we will include backflow effects directly into the orbitals. Our main reason for using the Hartree-Fock orbitals now is that they have a firm physical motivation. We do not want to introduce variational parameters without having some idea of their physical effects.

THE JASTROW-SLATER WAVE FUNCTION

The Slater-Jastrow wave function is a common starting point for variational Monte Carlo calculations. The correlation function contains a product over all pairs of electrons of a two-body correlation function, often called a Jastrow function. The addition of the two-body correlation tends to destroy the correct electronic density; the Hartree-Fock wave function tends to have the correct electronic density except in the tails. To compensate, we add an electron-nuclear one-body term. This is equivalent to multiplying all the orbitals by a common function. The Jastrow-Slater wave function for an atom is,

$$\Psi_T(R)=\prod_{i<j} f_2(|\vec{r}_{ij}|)\prod_i f_1(|\vec{r}_i|)\Phi. \tag{5}$$

with obvious generalizations to molecules. Φ is a Slater determinant or the linear combination needed to obtain the correct quantum numbers. We take f_2 and f_1 to be slight generalizations of commonly used forms,

Table 1. The energy and its statistical error in atomic units as a function of the Newton iteration number for the Be atom using a 9 term correlation function, starting from the Hartree-Fock wave function.

Iteration	Energy
0	-14.5639 ± 0.0093
1	-14.6318 ± 0.0019
2	-14.6303 ± 0.0021
3	-14.6339 ± 0.0019
4	-14.6336 ± 0.0017

$$f_1 = \exp\left(\sum_{m=2}^{L} a_m \bar{r}_i{}^m\right) \tag{6}$$

and,

$$f_2 = \exp\left(\sum_{m=1}^{L} c_m \bar{r}_{ij}{}^m\right), \tag{7}$$

with

$$\bar{r}_{ij} = \frac{b r_{ij}}{1 + b r_{ij}} \tag{8}$$

and,

$$\bar{r}_i = \frac{d r_i}{1 + d r_i}, \tag{9}$$

with b and d mostly taken to be 1.0 a.u. since this is a reasonable size for the atoms we study. They of course can be variational parameters. We take a_m and c_m to be variational parameters.

When two electrons are close together, their coulomb repulsion dominates. There-fore, we take $c_1 = 1/2b$ to satisfy the unlike-spin cusp condition. Similarly, if Φ already satisfies the nuclear cusp conditions, there is no a_1 term.

We have optimized this Jastrow-Slater wave function for L=4, i.e. a 7 term correlation with 6 independent parameters. Our idea was to include sufficient terms to be reasonably certain that the simple Jastrow-Slater wave function was well optimized. Further improvements when adding other correlations can then be assessed.

Results for the percent of the exact correlation energy that our wave function gives for the atoms He through Ne are shown in Table 2. The correlation energy as usual is the difference between the exact energy and the Hartree-Fock energy. Our correlation energies are calculated from our total energies using the Clementi-Veillard[25] values for the experimental ground-state energy corrected for nonrelativistic electrons and fixed nuclei.

The Slater-Jastrow form gives about 50 percent of the correlation energy. The rest of the correlation energy must come from other sources

BACKFLOW CORRELATIONS

A simple way to include backflow correlations is to replace the single particle orbitals by

Table 2. Slater-Jastrow correlation energies calculated with a 7-term correlation function for the atoms through Neon.

H 100%							He 90%
Li 92%	Be 56%	B 52%	C 53%	N 54%	O 53%	F 51%	Ne 57%

$$\phi_n(\vec{r_i}) \rightarrow \phi_n(\vec{r_i} + \sum_{j \neq i} \vec{r_{ij}} \eta(r_{ij})). \tag{10}$$

In liquid helium, η is approximately

$$\eta \ \alpha \ \hat{r}_{ij} \cdot \frac{\vec{\nabla}_i f_{ij}}{f_{ij}}. \tag{11}$$

The actual η is related to the difference between the angular momentum zero and one solutions of the two-body equation. This difference is approximately the gradient above[5]. If we expand Eq. (10) in η, the linear term gives a correction proportionate to

$$\frac{\vec{\nabla}_i \phi_n}{\phi_n} \cdot \frac{\vec{\nabla}_i f_{ij}}{f_{ij}}. \tag{12}$$

This is the same kind of term we get in the local energy when ∇^2 operates on the Jastrow-Slater Ψ_T. Backflow tends to cancel these terms and decrease the variance of the local energy. Backflow gives different correlations depending on the single-particle states of the electrons i and j, and requires a particular form for the correlation.

Rather than use a full state-dependent backflow correlation, we first try an average backflow correlation. We can approximately average the $\vec{\nabla}\phi_n^2$ terms to obtain $\vec{\nabla}\rho$ where ρ is the local electronic density. The average backflow correlation is approximately

$$\frac{\vec{\nabla}_i \rho(\vec{r_i})}{\rho(\vec{r_i})} \cdot \frac{\vec{\nabla}_i f_{ij}}{f_{ij}}. \tag{13}$$

This term is zero for helium liquids since ρ is constant there. For helium droplets, it would contribute, but ρ is nearly constant except at the surface so it tends to be unimportant. In atoms and molecules, the electronic density changes enormously, so we expect that a major effect of backflow will involve these density gradient correlations. The basic structure of these terms is

$$\vec{r_i} \cdot \vec{r_{ij}} g_1(r_i) g_2(r_{ij}) \tag{14}$$

where g_1 and g_2 are arbitrary functions. Rewriting the dot product as

$$2\vec{r_i} \cdot \vec{r_{ij}} = r_j^2 - r_i{}^2 - r_{ij}{}^2, \tag{15}$$

power series expanding g_1 and g_2, and replacing all r terms by the corresponding $\bar{r}$ of Eqs. (8) and (9), gives a simple parameterization of the average backflow correlations. The terms in the average backflow correlation are then like

$$(\bar{r}_j{}^2 - \bar{r}_i{}^2 - \bar{r}_{ij}{}^2)\bar{r}_i{}^s \bar{r}_j{}^t, \tag{16}$$

with s and t integers.

To satisfy the cusp conditions, we take no terms with exponents of 1. Terms with s=t=0 are simply one- and two-body correlations already included in the Jastrow-Slater form. All of the terms in our correlation can now be written in the Boys and Handy form[12-18]

$$\Psi_T = \prod_{i<j} \exp[\sum_l c_l \bar{r}_i{}^{m(l)} \bar{r}_j{}^{n(l)} \bar{r}_{ij}{}^{o(l)}]\Phi. \tag{17}$$

We try using just the lowest order average backflow terms with m,n,o = 2,2,0 and m,n,o = 2,0,2 and 0,2,2. This is our 9-term correlation. To test convergence, we also add 8 additional terms to get a 17 term correlation by expanding g_1 and g_2 to higher order[1]. If our average backflow arguments are correct, we expect our 9-term wave function to give a substantial improvement over the 7-term Jastrow-Slater form.

AVERAGE BACKFLOW RESULTS

Results for the percent of the correlation energy for the atoms through Argon with our best wave functions are shown in Table 3. These use the 17-term wave function for the atoms through neon, and the 9-term wave function for the atoms Sodium through Argon. The 9-term wave function for the atoms through neon gives only a few percent less correlation energy than the 17-term wave function indicating that the dominant physical effect is the average backflow correlation. The results of Table 3 show that our wave function is improved for closed shell systems so that Mg and Al give better results than Be and B. Also clear from Table 3 is that our wave function slowly gets worse as Z increases with Ar giving almost 10 percent less of the correlation energy than Ne.

Comparing Tables 2 and 3, we see that the average backflow correlation gives about 25 percent additional correlation energy. Coefficients of the 9-term correlation operator are given for the neon atom in table 4. For this closed shell case, the average backflow correlation gives an additional 28 percent of the correlation energy from the two lowest order terms.

An important question is whether these wave functions adequately describe the valence electrons which are of primary importance in understanding molecules. It would be possible to have a wave function that lowered the energy by improving the wave function only for the core electrons. To test whether our correlations improve the wave function in the valence region, we have calculated positive and negative ions. We show the ionization potentials for the atoms lithium through neon in Table 5, and the electron affinities of boron through fluorine in Table 6. Both are calculated by simply subtracting the corresponding neutral atom variational energy from the ion variational energy. Good agreement is found for the ionization potentials with the 9-term correlation indicating that we are correlating the outer electrons properly. The electron affinities, however, are improved using the 17-term correlation. This may be due to the greater extent of the outer orbital for these negative ions, or it may indicate the need for full state-dependent backflow correlations.

We have applied this wave function to a few molecules all with the obvious generalization of the 9-term wave function. Molecules that are nearly spherical give results that are similar to the atom results above. For example, for HF we get 86 percent of the correlation energy. This is slightly better than the 82 percent of the correlation energy we

Table 3. Our best correlation energies for the atoms through Argon using the Boys and Handy form for the average backflow correlation.

H 100%							He 100%
Li 97%	Be 68%	B 69%	C 72%	N 77%	O 80%	F 82%	Ne 85%
Na 83%	Mg 81%	Al 74%	Si 72%	P 77%	S 74%	Cl 72%	Ar 76%

Table 4. The coefficients of the 9-term correlation function for the neon atom. The source of the correlation and the approximate amount of correlation energy for the terms is also shown.

Coefficient	Term	Source
0.500	$\bar{r}_{ij}$	Coulomb repulsion 57% of correlation
-0.91877	$\bar{r}_{ij}^2$	
1.00830	$\bar{r}_{ij}^3$	
-1.52616	$\bar{r}_{ij}^4$	
0.05298	$\bar{r}_i^2$	Rescale Density
0.43018	$\bar{r}_i^3$	
-1.34785	$\bar{r}_i^4$	
-2.98409	$\bar{r}_i^2\bar{r}_j^2$	Average
2.07252	$\bar{r}_i^2\bar{r}_{ij}^2$	Backflow 28% of correlation

Table 5. The ionization energy of the Li through Ne computed by using the variational Monte Carlo method with a 9-term correlation function. The energy is given in eV. Experimental results are from Moore's compilation[26].

Atom	9 term	Experiment
Li	5.27±0.02	5.392
Be	8.55±0.02	9.322
B	8.36±0.03	8.298
C	11.29±0.03	11.260
N	14.47±0.02	14.534
O	13.33±0.02	13.618
F	17.40±0.03	17.422
Ne	21.71±0.02	21.564

Table 6. The electron affinity of B through F computed using the variational Monte Carlo method with 9-term and 17-term correlation functions. The energy is given in eV. Experimental results are from Hotop and Lineberger[27].

Atom	9 term	17 term	Experiment
B	0.045±0.01	0.237±0.02	0.277
C	1.14±0.03	1.27±0.02	1.263
O	1.01±0.02	1.30±0.02	1.461
F	3.14±0.02	3.46±0.04	3.399

got for the atom. This is in agreement with the standard idea that closing the shell makes these wave functions more accurate since there is less possibility of configuration mixing.

However, molecules that are less spherical give worse results. Calculations on LiH, Li_2, and N_2 give 84, 70, and 55 percent of the correlation energy, respectively. This is in contrast to the atoms where we obtained 97 and 77 percent of the correlation energy for Li and N, respectively. We have tried adding an electron-nucleus-nucleus term to compensate for density changes for the nonspherical density caused by our correlations. This term does not seem to change the results significantly. Clearly, much more work needs to be done to understand the electronic structure of molecules.

LOOKING FOR THE VARIANCE

Finally, we have been investigating methods to try to find methods to predict new correlations. One method we are applying is to look at surfaces of energy variance for a single electron with the others fixed. Specifically, we calculate

$$p(\vec{r})=(E_L(\vec{r},\vec{r_2},\cdots,\vec{r_N})-E_0)^2\Psi_T^2(\vec{r},\vec{r_2},\cdots,\vec{r_N}), \tag{18}$$

with $\vec{r_2}$ through $\vec{r_N}$ fixed at sampled positions. We calculate the values of $p(\vec{r})$ on a grid of positions. These values are then used to construct surfaces of constant $p(\vec{r})$ which can then be displayed, along with the electron positions. We use a Silicon Graphics personal Iris to view, rotate, clip, and switch between surfaces to try to understand the structure of the variance.

Unfortunately, the transparencies shown during the workshop cannot be reproduced here. They show clearly the defects in the Hartree-Fock wave function with no correlations. They also indicate that regions of high variance are concentrated around regions of the nodes in the trial function of the Boys and Handy form. The requirement for full state dependent backflow correlations would have exactly this kind of signature. These methods may allow us to visualize and then improve the wave function of complex molecules.

CONCLUSIONS

The simple Jastrow-Slater wave function gives about 50 percent of the correlation energy for the atoms through Ne if the Hartree-Fock orbitals are used. The addition of average backflow correlations represented by the Boys and Handy form of the wave function add about another 25 percent of the correlation energy. Our wave function generalizes easily to ions and molecules and gives similar results for spherical and nearly spherical systems. For molecules with charge distributions that are far from spherical, we get significantly worse results (although still more than half of the correlation energy). Methods of scientific visualization can also be helpful in determining what kind of terms are needed to improve trial wave functions.

ACKNOWLEDGEMENTS

This work was supported by the National Science Foundation through grants CHE-9015337 and DMR-9012143.

REFERENCES

1. K.E. Schmidt and J.W. Moskowitz, *J. Chem. Phys.* **93**, 4172 (1990).

2. C.J. Umrigar, K.G. Wilson, and J.W. Wilkins, *Phys. Rev. Lett.* **60**, 1719 (1988).

3. C.J. Umrigar, K.G. Wilson, and J.W. Wilkins, in *Computer Simulation Studies in Condensed Matter Physics: Recent Developments*, ed. D. P. Landau, K. K. Mon and H. B. Schuttler,Springer Verlag, New York (1988).

4. T. Pang, C.E. Campbell, and E. Krotscheck, *Chem. Phys. Lett.* **163**, 537 (1989).

5. K.E. Schmidt and V.R. Pandharipande, *Phys. Rev.* **B19**, 2504 (1979).

6. K.E. Schmidt, M.A. Lee, M.H. Kalos, and G.V. Chester, *Phys. Rev. Lett.* **47**, 807 (1981).

7. V.R. Pandharipande, S.C. Pieper, and R.B. Wiringa, *Phys. Rev.* **B34**, 4571 (1986).

8. See, e.g. the references in, K.E. Schmidt, and D.M. Ceperley, in *The Monte Carlo Method in Condensed Matter Physics*, ed. K. Binder,Springer-Verlag, Berlin (To be published, 1991).

9. V.R. Pandharipande, *Phys. Rev.* **B18**, 218 (1978).

10. K.E. Schmidt, M.H. Kalos, M.A. Lee, and G.V. Chester, *Phys. Rev. Lett.* **45**, 573 (1980).

11. R.P. Feynman and M. Cohen, *Phys. Rev.* **102**, 1189 (1956).

12. S. F. Boys and N. C. Handy, *Proc. Roy. Soc.* **A309**, 209 (1969).

13. S. F. Boys and N. C. Handy, *Proc. Roy. Soc.* **A310**, 43 (1969).

14. S. F. Boys and N. C. Handy, *Proc. Roy. Soc.* **A310**, 63 (1969).

15. S. F. Boys and N. C. Handy, *Proc. Roy. Soc.* **A311**, 309 (1969).

16. N. C. Handy, *J. Chem. Phys.* **51**, 3205 (1969).

17. N. C. Handy, *J. Chem. Phys.* **58**, 279 (1973).

18. N. C. Handy, in *International Review of Science, Physical Chemistry Series Two, Theoretical Chemistry*, ed. A. D. Buckingham and C. A. Coulson,Butterworths, London (1975).

19. H. Conroy, *J. Chem. Phys.* **41**, 1331 (1964).

20. H. Conroy, *J. Chem. Phys.* **41**, 1336 (1964).

21. R. L. Coldwell and R. E. Lowther, *Int. J. Quant. Chem. Symp.* **12**, 329 (1978).

22. A.F. Stevenson, *Phys. Rev.* **53**, 199 (1938).

23. A.F. Stevenson and M.F. Crawford, *Phys. Rev.* **54**, 375 (1938).

24. G. Temple, *Proc. Roy. Soc. London* **119**, 276 (1928).

25. A. Veillard and E. Clementi, *J. Chem. Phys.* **49**, 2415 (1968).

26. C.E. Moore, *Ionization Potentials and Ionization Limits Derived from the Analyses of Optical Spectra,* National Bureau of Standards, Report NSRDS-NBS34 (1970).

27. H. Hotop and W.C. Lineberger, *J. Phys. Chem. Ref. Data* **14**, 731 (1985).

EXACT QUESTIONS TO SOME INTERESTING

ANSWERS IN MANY BODY PHYSICS

D. P. Arovas

Department of Physics
University of California at San Diego
La Jolla, CA 92093-0319

S. M. Girvin

Department of Physics
Indiana University
Bloomington, IN 47405

I. INTRODUCTION

In many body physics, we are usually faced with the intractable problem of finding the ground state wavefunction of a thermodynamically large system of interacting particles. Typically this problem is solved perturbatively. That is, we start with some simple noninteracting Hamiltonian and its ground state (*e.g.* the free Fermi gas) and perturb in the interactions (*e.g.* the Coulomb potential):

$$\text{Hamiltonian } H \longrightarrow \text{ perturbation theory } \longrightarrow \text{ ground state } | \Psi \rangle .$$

Below, we shall do just the opposite — start with a ground state and solve for the Hamiltonian, matching questions to answers[1] as in the game show *Jeopardy*:

$$\text{ground state } | \Psi \rangle \longrightarrow \text{ magic } \longrightarrow \text{ Hamiltonian } H .$$

The magic consists of a clever choice of wavefunction $| \Psi \rangle$, one that guarantees that it will be the ground state of some nontrivial interacting Hamiltonian. The attentive reader might point out that given *any* wavefunction $| \Phi \rangle$, it is always possible to choose a Hamiltonian $\mathcal{H} = -| \Phi \rangle \langle \Phi |$ which renders $| \Phi \rangle$ its exact nondegenerate ground state. Such a Hamiltonian, however, will generally involve interactions among arbitrarily large numbers of particles and over arbitrarily long distances, and will not be the sort of model that captivates the interest of one's colleagues.

The central idea is this: suppose one finds a state $| \Psi \rangle$ which is annihilated by a set of local Hermitian operators $\{Q_\Gamma\}$, *i.e.* $Q_\Gamma | \Psi \rangle = 0$ for any Γ, where Γ denotes a group of particles, such as a pair of neighboring spins, two helium atoms, *etc.* Let us further suppose that the operator Q_Γ is positive semidefinite, meaning that is has no negative eigenvalues. One can now construct a Hamiltonian

$$H = \sum_\Gamma Q_\Gamma \tag{1.1}$$

which renders $| \Psi \rangle$ a zero-energy eigenstate. It is a trivial matter to prove that H can have no negative eigenvalue: let $| \psi_i^\Gamma \rangle$ and λ_i^Γ $(i = 1, 2, \ldots)$ denote the eigenstates and eigenvalues of Q_Γ. Then for any $| \phi \rangle$

$$\langle \phi | H | \phi \rangle = \sum_\Gamma \sum_i \left| \langle \phi | \psi_i^\Gamma \rangle \right|^2 \lambda_i^\Gamma \geq 0 \tag{1.2}$$

Recent Progress in Many-Body Theories, Vol. 3,
Edited by T.L. Ainsworth et al., Plenum Press, New York, 1992

since each λ_i^Γ is non-negative. Therefore $|\,\Psi\,\rangle$ must be a ground state of H, although it need not necessarily be unique.

Armed with a Hamiltonian and a ground state, one can construct trial excited states of the form $|\,k\,\rangle \equiv \rho_k |\,\Psi\,\rangle$, where ρ_k carries wavevector k, in analogy to the Bijl-Feynman theory of phonons and rotons in liquid ^{4}He.[2,3] ρ_k is usually chosen so as to create some physical elementary excitation, such as a charge or spin density wave. The trial excitation energy is

$$
\begin{aligned}
\omega_k &= \langle\,k\,|\,H\,|\,k\,\rangle \Big/ \langle\,k\,|\,k\,\rangle \\
&= f(k)/s(k) \ ,
\end{aligned}
\tag{1.3}
$$

where the oscillator strength $f(k)$ and static structure factor $s(k)$ are given by

$$
\begin{aligned}
f(k) &= \tfrac{1}{2} \langle\,\Psi\,|\, \left[[\rho_k^\dagger, H]\,, \rho_k \right] \,|\,\Psi\,\rangle \\
s(k) &= \langle\,\Psi\,|\,\rho_k^\dagger \rho_k\,|\,\Psi\,\rangle \ .
\end{aligned}
\tag{1.4}
$$

Since ω_k is the ratio of the lowest two moments of the dynamic structure factor $S(k,\omega)$,

$$
\begin{aligned}
\omega_k &= \int_0^\infty d\omega\, \omega\, S(k,\omega) \Big/ \int_0^\infty d\omega\, S(k,\omega) \\
S(k,\omega) &= \sum_j \left| \langle\,j\,|\,\rho_k\,|\,\Psi\,\rangle \right|^2 \delta(\omega - \omega_k) \ ,
\end{aligned}
\tag{1.5}
$$

ω_k is an *exact* upper bound to the first excitation energy at wavevector k. If only one mode $|\,j\,\rangle$ were to contribute from the sum over states in eq(1.5), the approximation would be exact, and we would have $S(k,\omega) = s(k)\delta(\omega - \omega_k)$. For this reason, this treatment is known as the *single mode approximation*, or SMA. It has been used to great effect in the theory of liquid ^{4}He,[2] quantum magnets,[4] and the fractional quantum Hall effect.[5]

Some of the models we will discuss exhibit the phenomenon of *off-diagonal long-range order* (ODLRO), indicating the presence of a Bose condensate. ODLRO is reflected in the long-distance behavior of off-diagonal density matrix,

$$
\begin{aligned}
g(|r - r'|) &\equiv \langle\,\Psi^\dagger(r)\,\Psi(r')\,\rangle \\
\lim_{r \to \infty} g(r) &\neq 0 \ ,
\end{aligned}
\tag{1.6}
$$

where the operator $\Psi(r')$ removes an object at r' and $\Psi^\dagger(r)$ reinserts it at a different position r. Ψ must carry Bose statistics if eq(1.6) is to make sense. In superfluid ^{4}He, $\Psi^\dagger$ creates a helium atom, which is a boson, while in a superconductor or Fermi superfluid, $\Psi^\dagger$ creates a *pair* of fermions (which acts as a boson at low energies). In the fractional quantum Hall effect, $\Psi^\dagger$ creates a composite object carrying both electric charge and magnetic flux, while in some of the magnetic systems we shall discuss, the nature of the condensate is not clear.

Each of the examples discussed below furnished a paradigm for a distinctly nonclassical type of order due to strong quantum fluctuations which are present even at zero temperature. The examples come from diverse systems – magnets, rotors, and electron gases in strong magnetic fields.

II. THE MAJUMDAR-GHOSH MODEL

Consider a one-dimensional chain of spin $S = \tfrac{1}{2}$ particles in which alternate pairs of sites are formed into singlets $\frac{1}{\sqrt{2}}(|\uparrow_i\downarrow_j\rangle - |\downarrow_i\uparrow_j\rangle)$. This state can be pictorially represented as

$$
|\,\Psi\,\rangle = |\cdots \ \bullet\!\!-\!\!\bullet \ \ \bullet\!\!-\!\!\bullet \ \ \bullet\!\!-\!\!\bullet \ \ \cdots\rangle \ ,
\tag{2.1}
$$

where $\bullet\!\!-\!\!\bullet$ represents a singlet pair.[6] The state $|\,\Psi\,\rangle$ has no classical analog and is macroscopically distinct from, say, the ferromagnetic state $|\cdots\uparrow\uparrow\uparrow\uparrow\uparrow\uparrow\cdots\rangle$ or the Néel state $|\cdots\uparrow\downarrow\uparrow\downarrow\uparrow\downarrow\cdots\rangle$.[7]

The answer is $|\Psi\rangle$, but what is the question? To find it, note that any consecutive trio of sites in $|\Psi\rangle$ must contain a singlet pair, and hence the *total* spin of any consecutive three sites must be one half. Now the eight states available from three spin-$\frac{1}{2}$ objects can be arranged into two $S=\frac{1}{2}$ doublets and one $S=\frac{3}{2}$ quadruplet, *viz.*

$$\tfrac{1}{2} \otimes \tfrac{1}{2} \otimes \tfrac{1}{2} = \tfrac{1}{2} \oplus \tfrac{1}{2} \oplus \tfrac{3}{2} \, , \tag{2.2}$$

so in general we should expect some $S=\frac{3}{2}$ component from any group of three sites. What is special about $|\Psi\rangle$ is that no such component exists for *any* consecutive trio $(n-1, n, n+1)$, meaning that $|\Psi\rangle$ is annihilated by a projection operator,

$$\mathrm{P}_{3/2}(n-1,n,n+1)\,|\Psi\rangle = 0 \, , \tag{2.3}$$

where $\mathrm{P}_{3/2}$ is the projector onto the total spin-$\frac{3}{2}$ state.

What is $\mathrm{P}_{3/2}(n-1,n,n+1)$? Well, the total spin operator for the three sites under consideration is $\boldsymbol{J} = \boldsymbol{S}_{n-1} + \boldsymbol{S}_n + \boldsymbol{S}_{n+1}$. We know that $\boldsymbol{J}^2 = J(J+1)$, where J is either $\frac{1}{2}$ or $\frac{3}{2}$. Therefore, the operator

$$\begin{aligned}
\mathrm{P}_{3/2}(n-1,n,n+1) &\equiv \tfrac{1}{3}(\boldsymbol{J}^2 - \tfrac{3}{4}) \\
&= \tfrac{1}{2} + \tfrac{2}{3}\boldsymbol{S}_{n-1} \cdot \boldsymbol{S}_n + \tfrac{2}{3}\boldsymbol{S}_n \cdot \boldsymbol{S}_{n+1} + \tfrac{2}{3}\boldsymbol{S}_{n-1} \cdot \boldsymbol{S}_{n+1} \, ,
\end{aligned} \tag{2.4}$$

normalized so that $(\mathrm{P}_{3/2})^2 = \mathrm{P}_{3/2}$, annihilates any state with $J=\frac{1}{2}$ and preserves any state with $J=\frac{3}{2}$. The question is now obvious:

$$\begin{aligned}
H &= \sum_n \mathrm{P}_{3/2}(n-1,n,n+1) \\
&= \tfrac{4}{3}\sum_n \left[\boldsymbol{S}_n \cdot \boldsymbol{S}_{n+1} + \tfrac{1}{2}\boldsymbol{S}_n \cdot \boldsymbol{S}_{n+2} \right] + \tfrac{1}{2}N \, .
\end{aligned} \tag{2.5}$$

This model was first discussed by Majumdar and Ghosh.[8]

We have just proved that the $S=\frac{1}{2}$ chain with antiferromagnetic first– and second–neighbor interactions with $J_2 = \frac{1}{2}J_1$ has a simple ground state composed of consecutive singlets as depicted in eq(2.1). One might not have guessed this *a priori*! Of course, the state $|\Psi\rangle$ breaks translational invariance, and there are in reality two degenerate ground states $|\Psi_1\rangle$ and $|\Psi_2\rangle$, with $t\,|\Psi_{1,2}\rangle = |\Psi_{2,1}\rangle$, where t is the (one unit) lattice translation operator. Although we have not proven so, it turns out that $|\Psi_{1,2}\rangle$ are the only zero energy eigenstates of H.[9] These ground states exhibit what is known as "spin-Peierls" order, for which $\langle \boldsymbol{S}_n \cdot \boldsymbol{S}_{n+1} - \boldsymbol{S}_{n-1} \cdot \boldsymbol{S}_n \rangle \sim (-1)^n\, g$, where for $|\Psi_{1,2}\rangle$ the order parameter is $g = \pm\frac{3}{4}$, the sign depending on which ground state is chosen. By taking linear combinations of the form $|\pm\rangle = \frac{1}{\sqrt{2}}(|\Psi_1\rangle \pm |\Psi_2\rangle)$, one can form eigenstates of the lattice translation operator with crystal momenta 0 and π. The normalized spin-spin correlation functions are then translationally invariant and easily calculable:

$$\frac{\langle \pm |\, \boldsymbol{S}_l \cdot \boldsymbol{S}_{l'} \,| \pm \rangle}{\langle \pm | \pm \rangle} = \tfrac{3}{4}\delta_{l,l'} - \tfrac{3}{8}\delta_{|l-l'|,1} \, . \tag{2.6}$$

The correlation function vanishes over distances greater than one lattice spacing, indicating that the correlation length ξ is zero.

Discussion

First, let us recall some well-known results. The nearest neighbor model ($J_2 = 0$), solvable by Bethe's *Ansatz*,[10] exhibits algebraically decaying correlations with $\langle \boldsymbol{S}_0 \cdot \boldsymbol{S}_n \rangle \sim (-1)^n/|n|$ (up to logarithmic corrections). The absence of long-range order is due to the existence of quantum fluctuations in the Néel state, which are present even at $T=0$, due to the transverse part of the Heisenberg interaction, $S_i^+ S_j^- + S_i^- S_j^+$. The quantum Heisenberg model in one dimension is related to the classical nonlinear sigma model in two dimensions at finite temperature $T \sim 1/S$,[11] and the destruction of long-range order due to quantum fluctuations is precisely analogous to the destruction of classical order by thermal fluctuations. In two dimensions, there can be no spontaneous breaking of a continuous symmetry at any finite temperature,[12] and so the *ground state* of the one-dimensional quantum model should be *disordered*. (Of course the ferromagnetic ground state is always ordered, but the transverse Heisenberg exchange annihilates the ferromagnetic state, so there are no quantum fluctuations present in the ferromagnetic ground state.) Now the decay of the correlations in Bethe's $J_2 = 0$ ground state is rather slow — this is sometimes referred to as 'quasi- long-range order'. In fact, the addition of an infinitesimal amount of easy-axis anisotropy is enough to induce Néel order in that model.[13] An additional antiferromagnetic second-neighbor interaction frustrates the system's attempts to order. One expects the quantum fluctuations to be enhanced and the correlation functions to decay more rapidly (with a higher power of the distance) as J_2 is increased. Eventually, a critical point is reached and the system acquires spin-Peierls (dimer) order. The dimer phase breaks a *discrete* symmetry and is *not* precluded by the Mermin-Wagner theorem. The phase transition has been described by Haldane,[13] who analyzed an XXZ version of the J_1–J_2 model using bosonization techniques and found the critical point to lie at $J_2 \sim \frac{1}{6}J_1$. The Majumdar-Ghosh model, with $J_2 = \frac{1}{2}J_1$, lies on the other side of the critical point from the Bethe *Ansatz* soluble model, and furnishes us with a simple paradigm for spin-Peierls order in a frustrated $S = \frac{1}{2}$ antiferromagnetic chain.

Excitations

The two-fold degeneracy of the ground state suggests the existence of soliton-like defects which interpolate between the two degenerate ground states. The dispersion of these defects was first considered by Shastry and Sutherland,[14] who used simple trial states of the form

$$| \psi \rangle = | \cdots \;\; \bullet\!\!-\!\!\bullet \;\; \bullet\!\!-\!\!\bullet \;\; \uparrow \;\; \bullet\!\!-\!\!\bullet \;\; \bullet\!\!-\!\!\bullet \;\; \cdots \rangle \tag{2.7}$$

to describe the solitons. If we fix the boundary conditions so that

$$\begin{aligned} | \psi(n \to -\infty) \rangle &= | \Psi_1 \rangle \\ | \psi(n \to +\infty) \rangle &= | \Psi_2 \rangle \,, \end{aligned} \tag{2.8}$$

the solitons are forced to live on even numbered sites, and it proves convenient to work in a doubled unit cell scheme where the state $| 2n, + \rangle$ describes an $S^z = +\frac{1}{2}$ soliton living on site $2n$. It is easy to work out the overlap matrix in both real space

$$\langle 2m, \sigma | 2n, \sigma' \rangle = 2^{-|m-n|} \delta_{\sigma,\sigma'} \tag{2.9}$$

and momentum space

$$| K, \sigma \rangle \equiv \sqrt{\frac{2}{N}} \sum_n e^{2iKn} | 2n, \sigma \rangle$$

$$\langle K, \sigma | K', \sigma' \rangle = \frac{3}{5 - 4 \cos 2K} \delta_{K,K'} \delta_{\sigma,\sigma'} \,, \tag{2.10}$$

where $K \in [-\frac{1}{2}\pi, \frac{1}{2}\pi)$ lives in the 'little zone'. To evaluate the trial soliton energy, we only need the matrix element

$$\langle \;\; \bullet\!\!-\!\!\bullet \;\; \uparrow \;\; \bullet\!\!-\!\!\bullet \;\; | P_{3/2}(-1,0,1) | \;\; \bullet\!\!-\!\!\bullet \;\; \uparrow \;\; \bullet\!\!-\!\!\bullet \;\; \rangle = \tfrac{1}{2} \tag{2.11}$$

(where the soliton lies on site 0), and one finds the simple result

$$\varepsilon_K = \frac{\langle K, \sigma | H | K, \sigma \rangle}{\langle K, \sigma | K, \sigma \rangle} = \tfrac{1}{6}(5 - 4 \cos 2K) \,. \tag{2.12}$$

Sadly, the states $| K, \sigma \rangle$ are not exact excited states. (Corrections to these simple trial solitons have been considered by Caspers, Emmett, and Magnus.[9]) This feature distinguishes the extent to which this model is soluble — only the ground state is known exactly[15] — from the complete solubility of the integrable models,[16] in which all the excited states are known.

318

On topological grounds, we expect that soliton excitations should exist. If we describe each spin as a fermion, with $|\sigma\rangle = c_\sigma^\dagger |0\rangle$ and associate a unit 'background charge' to each site which is neutralized when one fermion is present, then the solitons described above represent neutral $S = \frac{1}{2}$ defects. This reversed charge-spin relation is characteristic of such topological excitations; another example occurs in the Su–Schrieffer-Heeger model of polyacetylene (CH_x).[17]

The single mode approximation provides us with another way of making trial excited states. We define

$$S_k^\alpha = \frac{1}{\sqrt{N}} \sum_n e^{ikn} S_n^\alpha$$

$$|k, \alpha, +\rangle \equiv S_k^\alpha |+\rangle \ .$$

(2.13)

Here $\alpha = x$, y, or z is an SU(2) index, and $k \in [-\pi, \pi)$ ranges over the full Brillouin zone. The state $|k, \alpha\rangle$ has total spin $J = 1$.[18] S_k^α creates a magnon, rather than a soliton, and does not affect the topological order on either end of the chain. We choose to start with the ground state $|+\rangle$, although this is arbitrary. A linearly independent set of excitations with wavevector $k + \pi$ are described by $S_k^\alpha |-\rangle$.

Application of the SMA formulae $(1.3 - 1.5)$ yields an structure factor $s(k) = \frac{1}{4}(1 - \cos k)$ and an oscillator strength $f(k) = \frac{1}{3}(1 - \cos k)$. The trial dispersion is therefore $\omega_k = \frac{4}{3}$, independent of the wavevector k. This is just the energy it takes to promote one of the singlet bonds to a triplet. It is interesting to compare this energy to the bottom of the two soliton continuum. One can form two-soliton states (both singlets and triplets) at wavevector k at an energy

$$\Omega_{k,q} = \varepsilon_{\frac{1}{2}k+q} + \varepsilon_{\frac{1}{2}k-q}$$

$$= \frac{5}{3} - \frac{4}{3} \cos\tfrac{1}{2}k \, \cos q \ ,$$

(2.14)

and so the bottom of the two-soliton band lies at an energy $\Omega_k^{\min} = \frac{5}{3} - \frac{4}{3} |\cos \frac{1}{2}k|$. The magnon, which will in general mix with the $S = 1$ soliton pair states, lies above the bottom of the two-soliton band throughout most of the Brillouin zone, with the exception of a small region about $k = \pi$, where the magnon lies lower $(\omega_\pi / \Omega_\pi^{\min} = \frac{4}{5})$. This analysis suggests that the magnon exists as a true bound state only in the vicinity of $k = \pi$, and that elsewhere it is only a resonance.

Suppose one wishes to perturb about the Majumdar-Ghosh point and examine a model with $J_2/J_1 = \frac{1}{2} + \epsilon$. If one wishes to use the perturbative machinery of quantum field theory, one is immediately faced with the problem of how to construct a Fock space of excited states. In principle, this can be done. Consider the states formed by multiple application of the operator S_k^α, $i.e.$

$$|k_1, \alpha_1; k_2, \alpha_2; \ldots, k_r, \alpha_r; \pm\rangle = S_{k_1}^{\alpha_1} S_{k_2}^{\alpha_2} \cdots S_{k_r}^{\alpha_r} |\pm\rangle \ .$$

(2.15)

Let us assume that, on an even-membered ring ($i.e.$ one without an odd number of solitons), this set of states is complete. We construct a bosonic (symmetric) Fock space in the following manner: Starting from $r = 2$, we symmetrize the r-magnon states of eq(2.15) and then orthogonalize them to all states of smaller or equal r. $I.e.$ the $J = 0$ and $J = 1$ components of the two-magnon spectrum can mix with the 0- and 1-magnon states, and therefore one should orthogonalize the 2-magnon states to the 0-magnon and 1-magnon states. The orthogonalization at level r requires knowledge of the $2r$-point ground state correlation functions. In practice, this severely limits the extent to which this procedure can be carried out. Once one has generated a bosonic Fock space, one can take matrix elements of the Hamiltonian in this basis and generate a set of interaction vertices, at which point the usual field-theoretic perturbation treatment can be applied.

Haldane[19] has devised a scheme by which one may approximately perturb about the soluble point. His idea is to approximate the SMA states as free bosonic excitations, which is tantamount to taking

$$S_k^\alpha \approx \sqrt{s(k)} \left(b_{k,\alpha}^\dagger + b_{-k,\alpha} \right) \ ,$$

(2.16)

where $s(k) = \frac{1}{4}(1 - \cos k)$ is the static structure factor at the Majumdar-Ghosh point (see above). Eq(2.16) is an approximation inasmuch as it violates the known commutation relations

$[S^\alpha_k, S^\beta_{k'}] = i\,\varepsilon_{\alpha\beta\gamma}\, S^\gamma_{k+k'}/\sqrt{N}$. This goes beyond the lowest order 'backflow' corrections of the type considered in the theory of the phonon-roton spectrum of liquid ^{4}He, and is similar (but not precisely equivalent) to truncating the above field-theoretic construction by restricting the class of interaction vertices considered. Close to the Majumdar-Ghosh point, we write

$$
\begin{aligned}
H &= \tfrac{4}{3} \sum_n \left\{ (1 + \epsilon_1)\, \boldsymbol{S}_n \cdot \boldsymbol{S}_{n+1} + (\tfrac{1}{2} + \epsilon_2)\, \boldsymbol{S}_n \cdot \boldsymbol{S}_{n+2} \right\} \\
&= H_0 + \tfrac{4}{3} \sum_k (\epsilon_1 \cos k + \epsilon_2 \cos 2k)\, S^\alpha_k\, S^\alpha_{-k} \;,
\end{aligned}
\tag{2.17}
$$

where H_0 is the Hamiltonian at the Majumdar-Ghosh point, approximated by the free boson model $H_{\text{free}} = \sum_{k,\alpha} \omega^\dagger_k\, b^\dagger_{k,\alpha} b_{k,\alpha}$. The perturbed Hamiltonian is thus approximated as

$$
H \approx \sum_{k,\alpha} \omega_k\, b^\dagger_{k,\alpha} b_{k,\alpha} + \tfrac{4}{3} \sum_{k,\alpha} s(k)\,(\epsilon_1 \cos k + \epsilon_2 \cos 2k)\,(b^\dagger_{k,\alpha} + b_{-k,\alpha})(b_{k,\alpha} + b^\dagger_{-k,\alpha})
\tag{2.18a}
$$

$$
= \tfrac{1}{2} \sum_{k,\alpha} \left[(\omega_k + \Delta_k)\,(b^\dagger_{k,\alpha} b_{k,\alpha} + b^\dagger_{-k,\alpha} b_{-k,\alpha}) + \Delta_k\,(b^\dagger_{k,\alpha} b^\dagger_{-k,\alpha} + b_{k,\alpha} b_{-k,\alpha}) \right] + E_0
\tag{2.18b}
$$

with $\omega_k = \tfrac{4}{3}$ as above and

$$
\begin{aligned}
\Delta_k &= \tfrac{8}{3}\, s(k)\,(\epsilon_1 \cos k + \epsilon_2 \cos 2k) \\
&= \tfrac{2}{3}\,(1 - \cos k)\,(\epsilon_1 \cos k + \epsilon_2 \cos 2k) \;.
\end{aligned}
\tag{2.19}
$$

This is easily diagonalized by a Bogoliubov transformation, yielding

$$
E_k = \tfrac{4}{3} \sqrt{1 + (1 - \cos k)\,(\epsilon_1 \cos k + \epsilon_2 \cos 2k)} \;.
\tag{2.20}
$$

The structure factor becomes $\tilde{s}(k) = s(k) \cdot (\omega_k/E_k)$. To simplify matters, consider the case when $\epsilon_2 = 0$. The dispersion is then $E_k = \tfrac{4}{3}\sqrt{1 + \epsilon_1 \cos k\,(1 - \cos k)}$, and the $k = \pi$ gap $E_\pi = \sqrt{1 - 2\epsilon_1}$ collapses at $\epsilon_1 = \tfrac{1}{2}$, corresponding to a ratio $J_2/J_1 = \tfrac{1}{2}/(1 + \epsilon_1) = \tfrac{1}{3}$, which is at least in the same ballpark as the value predicted by bosonization,[13] $J_2/J_1 = \tfrac{1}{6}$. The Hamiltonian of eq(2.17) is really only dependent on a single parameter, $(\tfrac{1}{2} + \epsilon_2)/(1 + \epsilon_1)$, which is not true for the model of eqs(2.18 – 2.20).

III. VALENCE BOND SOLID ANTIFERROMAGNETS

The answer is

$$
|G\rangle = |\cdots \!-\!\!\bullet\!\!-\!\!\bullet\;\;\bullet\;\;\bullet\;\;\bullet\!\!-\!\!\cdots\rangle \;.
\tag{3.1}
$$

This diagram depicts a many-body state of a quantum $S = 1$ chain. One can consider $S = 1$ as two symmetrized $S = \tfrac{1}{2}$ objects,

$$
|(\alpha, \beta)\rangle \equiv \frac{1}{\sqrt{2\delta_{\alpha,\beta}}}\,(|\alpha\rangle \otimes |\beta\rangle + |\beta\rangle \otimes |\alpha\rangle)
\tag{3.2}
$$

and then form the state

$$
|G\rangle = \cdots \epsilon_{\beta\gamma}\, \epsilon_{\delta\epsilon}\, \epsilon_{\zeta\eta} \cdots |(\alpha,\beta)\rangle_0 \otimes |(\gamma,\delta)\rangle_1 \otimes |(\epsilon,\zeta)\rangle_2 \cdots
\tag{3.3}
$$

by antisymmetrizing each of the two labels with a corresponding label from each of the two neighboring sites. What this means is that the $J = 2$, $J^z = 2$ configuration $|\uparrow\uparrow\rangle_n \otimes |\uparrow\uparrow\rangle_{n+1}$ never appears on any two consecutive sites. And since the state $|\Psi\rangle$ is manifestly SU(2)-symmetric, we conclude that there can be no $J = 2$ component to any bond. Like the Majumdar-Ghosh state, this is special, because the tensor product of two spin-1 objects,

$$
1 \otimes 1 = 0 \oplus 1 \oplus 2
\tag{3.4}
$$

in general results in states of total spin $J = 0, 1$, and 2.

It is simpler to describe the state $|\Psi\rangle$ in terms of the Schwinger representation of SU(2),

$$S_n^+ = a_n^\dagger b_n \qquad S_n^z = \tfrac{1}{2}(a_n^\dagger a_n - b_n^\dagger b_n)$$
$$S_n^- = a_n b_n^\dagger \qquad \hat{S}_n = \tfrac{1}{2}(a_n^\dagger a_n + b_n^\dagger b_n) \,, \tag{3.5}$$

in which each spin is defined by two bosons, together with the constraint that the total Bose occupation $n_a + n_b$ is constrained to be $2S$. If $|\psi\rangle = f(a^\dagger, b^\dagger)|0\rangle$, where f is homogeneous and of degree $2S$, then ψ represents a spin-S object. In this language, $|\Psi\rangle$ can be written as

$$|\Psi\rangle = \prod_n (a_n^\dagger b_{n+1}^\dagger - b_n^\dagger a_{n+1}^\dagger)|0\rangle \,. \tag{3.6}$$

The state $|\Psi\rangle$ is identical (aside from an overall normalization) to that defined in eq(3.3). It describes a $S = 1$ chain with total spin 0 (a global singlet) with the peculiar feature that no bond contains any $J = 2$ component to its spin, *i.e.* $\text{P}_2(n, n+1)|\Psi\rangle = 0$, where P_2 is the projector onto bond spin $J = 2$.

To obtain an expression for P_2, consider the general case of the tensor product of two spin-S objects:

$$S \otimes S = 0 \oplus 1 \oplus 2 \oplus \cdots \oplus 2S \,. \tag{3.7}$$

There are thus $(2S+1)$ different possible values for the combined spin J of the pair. The projection operator P_J is then

$$\text{P}_J(ij) = \prod_{\substack{k=0 \\ (k \neq J)}}^{2S} \frac{2\boldsymbol{S}_i \cdot \boldsymbol{S}_j + 2S(S+1) - k(k+1)}{J(J+1) - k(k+1)} \tag{3.8}$$

since $\boldsymbol{J}^2 = 2S(S+1) + 2\boldsymbol{S}_i \cdot \boldsymbol{S}_j$. For $S = 1$ and $J = 2$, we have

$$\text{P}_2(n, n+1) = \tfrac{1}{3} + \tfrac{1}{2}\boldsymbol{S}_n \cdot \boldsymbol{S}_{n+1} + \tfrac{1}{6}(\boldsymbol{S}_n \cdot \boldsymbol{S}_{n+1})^2 \tag{3.9}$$

and so the exact question to our answer is

$$\begin{aligned} H &= \sum_n \text{P}_2(n, n+1) \\ &= \tfrac{1}{2}\sum_n \left[\boldsymbol{S}_n \cdot \boldsymbol{S}_{n+1} + \tfrac{1}{3}(\boldsymbol{S}_n \cdot \boldsymbol{S}_{n+1})^2\right] + \tfrac{1}{3}N \,. \end{aligned} \tag{3.10}$$

This model and its soon-to-be-mentioned generalizations were discovered by Affleck, Kennedy, Lieb, and Tasaki.[20] The interactions extend only over a single bond length, although a biquadratic term is necessary to make $|\Psi\rangle$ an exact ground state. $|\Psi\rangle$ breaks no discrete symmetries, and has been proven to be the exact nondegenerate ground state of eq(3.10).[20] Some exact excited states have been described in ref[21].

<u>The General VBS State</u>

The general VBS state is written[20,4]

$$|\Psi(\mathcal{L}, m)\rangle = \prod_{\langle ij\rangle \in \mathcal{L}} (a_i^\dagger b_j^\dagger - b_i^\dagger a_j^\dagger)^m |0\rangle \,, \tag{3.11}$$

where $\mathcal{L}$ represents a lattice (*e.g.* simple cubic) and m is a positive integer. The expression $\prod_{\langle ij\rangle \in \mathcal{L}}(a_i^\dagger b_j^\dagger - b_i^\dagger a_j^\dagger)^m |0\rangle$ is homogeneous of degree zm in the creation operators $(a_i^\dagger, b_i^\dagger)$ at each site, where z is the coordination number (number of nearest neighbors) of the lattice $\mathcal{L}$, meaning that $|\Psi\rangle$ represents a spin $S = \tfrac{1}{2}zm$ state. What is special about it? Well, first of all note that $\mathcal{A}_{ij}^\dagger \equiv (a_i^\dagger b_j^\dagger - b_i^\dagger a_j^\dagger)$ transforms as an SU(2) singlet, indicating that $|\Psi\rangle$ is an overall singlet. Based on our investigation of the $S = 1$ VBS chain, we should ask what is the maximum value of the total spin of any bond $\langle ij\rangle$. Due to the global spin isotropy of $|\Psi\rangle$, we can answer this by finding the maximum value of J^z for any bond. Since the m operators $\mathcal{A}^\dagger_{ij}$ connecting sites i and j carry no net spin (*i.e.* $[\mathcal{A}^\dagger_{ij}, \boldsymbol{S}_i + \boldsymbol{S}_j] = 0$) the only contribution to $J^z(ij)$ comes from the remaining $2(z-1)$ bonds which are connected to either i or j. A moment's thought gives

$$\begin{aligned}
J_{\mathrm{max}} = J^z_{\mathrm{max}} &= \tfrac{1}{2}m \cdot 2(z-1)\\
&= 2S - m \,,
\end{aligned} \tag{3.12}$$

which is significant because J_{max} is smaller that the maximum possible total bond spin (which is $2S$, according to eq(3.7)). So the question is this:

$$H = \sum_{\langle ij \rangle} \sum_{J=2S-m+1}^{2S} V_J \mathrm{P}_J(ij) \tag{3.13}$$

where $S = \frac{1}{2}zm$, and $V_J \geq 0$. H is constructed so that $H\,|\,\Psi\,\rangle = 0$, and so $|\,\Psi\,\rangle$ must be a ground state of H.

The general VBS states lead us to an entire family of soluble models provided that the spin S is an integer multiple of half the lattice coordination number. The interactions only extend across a single bond length, but, as we saw in eqs(3.9,3.10), include higher order powers of the Heisenberg interaction. For example, on the honeycomb lattice ($z=3$), we can describe a spin-$\frac{3}{2}$ model by taking $m=1$ in eq(3.11). There is then no bond which contains any projection onto total spin $J=3$, despite the fact that $\frac{3}{2} \otimes \frac{3}{2} = 0 \oplus 1 \oplus 2 \oplus 3$, and so the interaction we seek is the projection operator $\mathrm{P}_3(ij)$. According to eq(3.8),

$$\begin{aligned}
\mathrm{P}_3(ij) &= \frac{(\boldsymbol{J}^2 - 0)(\boldsymbol{J}^2 - 2)(\boldsymbol{J}^2 - 6)}{(12 - 0)(12 - 2)(12 - 6)}\\[2mm]
&= \tfrac{1}{90}(\boldsymbol{S}_i \cdot \boldsymbol{S}_j)^3 + \tfrac{29}{270}(\boldsymbol{S}_i \cdot \boldsymbol{S}_j)^2 + \tfrac{27}{160}\boldsymbol{S}_i \cdot \boldsymbol{S}_j + \tfrac{11}{128} \,.
\end{aligned} \tag{3.14}$$

By discouraging states of large J on each bond, the interactions are antiferromagnetic.

The correlations within the VBS states are effectively analyzed using a coherent state approach. A review of the properties of SU(2) coherent states is included in Appendix A. One finds that the coherent state wavefunction is

$$\begin{aligned}
\Psi(\hat{\boldsymbol{\Omega}}_1, \ldots, \hat{\boldsymbol{\Omega}}_N) = \langle\,\{\hat{\boldsymbol{\Omega}}_i\}\,|\,\Psi\,\rangle &= \prod_{\langle ij \rangle} (u_i v_j - v_i u_j)^m\\[2mm]
\left|\Psi(\hat{\boldsymbol{\Omega}}_1, \ldots, \hat{\boldsymbol{\Omega}}_N)\right|^2 &= \prod_{\langle ij \rangle} \left(\frac{1 - \hat{\boldsymbol{\Omega}}_i \cdot \hat{\boldsymbol{\Omega}}_j}{2}\right)^m \,,
\end{aligned} \tag{3.15}$$

where each $\hat{\boldsymbol{\Omega}}_i$ lies on the unit sphere. The probability density $|\Psi|^2$ is maximized when $\hat{\boldsymbol{\Omega}}_i \cdot \hat{\boldsymbol{\Omega}}_j = -1$ for every bond, *i.e.* the system wants to be antiferromagnetic.

For the one-dimensional chain, a VBS model exists for any integer S. The spin-spin correlation functions of the general VBS chain are easily computable:[4]

$$\langle\,\Psi\,|\,\boldsymbol{S}_0 \cdot \boldsymbol{S}_n\,|\,\Psi\,\rangle = (-1)^n (S+1)^2 \left(\frac{S}{S+2}\right)^{|n|} \,, \tag{3.16}$$

where we have taken the thermodynamic limit with n finite and nonzero. The correlations decrease exponentially with a correlation length $\xi = 1/\ln(1+2/S)$, in agreement with Haldane's prediction of massive behavior in integer Heisenberg spin chains.[22] The VBS chains provide us with a paradigm for the type of order Haldane envisioned for the integer spin Heisenberg chains. Although Haldane's analysis is based on the Heisenberg model with purely bilinear exchange,[11] the VBS states which are ground states of Hamiltonians which include biquadratic and higher order exchange interactions, are useful for understanding the Heisenberg model because they exhibit the same type of magnetic order, *i.e.* they lie in the same region of some generalized phase diagram and can be adiabatically connected to one another without crossing any phase boundaries. Consider again the $S = 1$ chain, this time with Hamiltonian

$$H = \tfrac{1}{2} \sum_n \left[\boldsymbol{S}_n \cdot \boldsymbol{S}_{n+1} + \lambda \left(\boldsymbol{S}_n \cdot \boldsymbol{S}_{n+1} \right)^2 \right] \tag{3.17}$$

where λ is arbitrary. For $\lambda = \tfrac{1}{3}$, the model has exponentially decaying correlations and is rigorously known to have an excitation gap.[20] As one decreases λ towards the Heisenberg point $\lambda = 0$, the gap will vary; the issue is whether or not it collapses before $\lambda = 0$ is reached. Numerical work[23,24] suggests that the ground state of eq(3.17) remains massive for all $\lambda \in (-1, +1)$. We shall return to this point later on. The static structure factor of the spin-S VBS chain is

$$
\begin{aligned}
s_{\alpha\beta}(k) &= \frac{1}{N} \sum_{n,n'} e^{ik(n-n')} \langle \, \Psi \, | \, S_n^\alpha \, S_{n'}^\beta \, | \, \Psi \, \rangle \\
&= \tfrac{1}{3}(S + 1) \frac{1 - \cos k}{1 + \cos k + \frac{2}{S(S+2)}} \, \delta_{\alpha\beta} \equiv \tfrac{1}{3} \, s(k) \, \delta_{\alpha\beta} \; ,
\end{aligned}
\tag{3.18}
$$

which is analytic in the neighborhood of the real axis, and which vanishes for $k = 0$, reflecting the fact that $| \, \Psi \, \rangle$ is a global singlet.

It is in general impossible to compute the correlation functions in the higher-dimensional VBS states. However, we can gain much insight into the magnetic order present if we interpret the square of the wavefunction $|\Psi|^2$ as a Boltzmann weight for some classical model at finite temperature. The technique of *approximating* many body wavefunctions by pair products like eq(3.15) first arose in studies of liquid $^4\text{He}^3$ and nuclear matter,[25] and was more recently used with brilliant success in the theory of the fractional quantum Hall effect.[26] In the case of the VBS states, the pair product wavefunction is exact. Writing

$$
\begin{aligned}
\left| \Psi(\hat{\boldsymbol{\Omega}}_1, \ldots, \hat{\boldsymbol{\Omega}}_N) \right|^2 &= \prod_{\langle ij \rangle} \left(\frac{1 - \hat{\boldsymbol{\Omega}}_i \cdot \hat{\boldsymbol{\Omega}}_j}{2} \right)^m \equiv \exp\left(-\beta \Phi(\hat{\boldsymbol{\Omega}}_1, \ldots, \hat{\boldsymbol{\Omega}}_N) \right) \\
\Phi(\hat{\boldsymbol{\Omega}}_1, \ldots, \hat{\boldsymbol{\Omega}}_N) &= - \sum_{\langle ij \rangle} 2 \ln \sin^2 \tfrac{1}{2} \vartheta_{ij} \; ,
\end{aligned}
\tag{3.19}
$$

we find that the *ground state correlations* in the VBS models can be described by the *thermal averages* of a related classical model *in the same number of dimensions*. The effective temperature is $T = \beta^{-1} = 2/m$ and the classical model describes a set of unit length spins interacting via a potential $v(\vartheta) = -2 \ln \sin^2 \tfrac{1}{2} \vartheta$, where $\vartheta_{ij} = \cos^{-1} \hat{\boldsymbol{\Omega}}_i \cdot \hat{\boldsymbol{\Omega}}_j$ is the angle between $\hat{\boldsymbol{\Omega}}_i$ and $\hat{\boldsymbol{\Omega}}_j$. The potential $v(\vartheta)$ diverges at $\vartheta = 0$ (when neighboring spins are aligned) and thereby strongly encourages antiferromagnetic ordering. $| \, \Psi \, \rangle$ will exhibit long-range antiferromagnetic order if the fictitious temperature $T = 2/m$ is lower than the Néel temperature for the associated classical model. A crude mean field theory, obtained by approximating $v(\vartheta) \approx \left(1 + \hat{\boldsymbol{\Omega}}_i \cdot \hat{\boldsymbol{\Omega}}_j \right)$ and taking $\langle \hat{\boldsymbol{\Omega}} \rangle = \pm \boldsymbol{\eta}$, results in the mean field equation[4]

$$\eta = \mathrm{ctnh}\,(\eta S) - (\eta S)^{-1} \; , \tag{3.20}$$

leading to a critical value $S_\mathrm{c} = 3$; VBS states with $S < S_\mathrm{c}$ are disordered while $S > S_\mathrm{c}$ implies Néel order (within this mean field treatment). At or below two dimensions, however, the Mermin-Wagner theorem[12] precludes ordering of the classical model at *any* finite temperature, and so one concludes that all the VBS states with $d \le 2$ are disordered, which almost certainly implies an excitation gap. This is noteworthy, because it has been rigorously proven then the ground state of the Heisenberg antiferromagnet $H = + \sum_{\langle ij \rangle} \boldsymbol{S}_i \cdot \boldsymbol{S}_j$ possesses Néel order for all $S \ge \tfrac{3}{2}$ on the honeycomb lattice and for all $S \ge 1$ on the square lattice.[20,27] This implies that, starting from the VBS model and perturbing in $\left(H_{\mathrm{Heis}} - H_{\mathrm{VBS}} \right)$, a phase boundary is encountered at which point long-range order sets in. This phase transition is usually accompanied by a collapse of the gap.

Excitations

For an isotropic Hamiltonian $H = \sum_{\langle ij \rangle} Q(\boldsymbol{S}_i \cdot \boldsymbol{S}_j)$, the magnon spectrum will be grouped into SU(2) multiplets. The elementary magnon carries spin-1, and is approximated within the SMA as

$$| \boldsymbol{k}, \alpha \rangle \equiv S_{\boldsymbol{k}}^{\alpha} | \Psi \rangle \ . \tag{3.21}$$

The energy is again $\omega_{\boldsymbol{k}} = f(\boldsymbol{k})/s(\boldsymbol{k})$, where

$$\begin{aligned}
f(\boldsymbol{k}) &= \tfrac{1}{6} \langle \Psi | \left[S_{-\boldsymbol{k}}^{\alpha}, [H, S_{\boldsymbol{k}}^{\alpha}] \right] | \Psi \rangle = \tfrac{1}{6} z D \left(1 - \gamma_{\boldsymbol{k}} \right) \\
D &= \tfrac{1}{4} \langle \Psi | \left[S_i^{\alpha} - S_j^{\alpha}, [Q(\boldsymbol{S}_i \cdot \boldsymbol{S}_j), S_i^{\alpha} - S_j^{\alpha}] \right] | \Psi \rangle
\end{aligned} \tag{3.22}$$

where i and j are nearest neighbors. $\gamma_{\boldsymbol{k}}$ is given by $\gamma_{\boldsymbol{k}} = z^{-1} \sum_{\delta} e^{i \boldsymbol{k} \cdot \boldsymbol{\delta}}$, where the sum is over nearest neighbor displacement vectors. The energy $\omega_{\boldsymbol{k}}$ provides an exact upper bound to the lowest excitation energy at wavevector $\boldsymbol{k}$. This proves gaplessness at the zone center if $s(\boldsymbol{k})$ vanishes more slowly than k^2 (assuming lattice inversion symmetry), and at any point in the Brillouin zone where $s(\boldsymbol{k})$ diverges. For the $S = 1$ VBS chain with $Q(\boldsymbol{S}_i \cdot \boldsymbol{S}_j) = P_2(ij)$, one obtains a trial magnon dispersion of $\omega_{\boldsymbol{k}} = \frac{5}{27}(5 + 3 \cos k)$ which is quite accurate throughout the region $\frac{1}{2}\pi \leq k \leq \pi$. The SMA value of the gap, $\omega_{\pi} = \frac{10}{27} = 0.370 \ldots$, is remarkably close to the numerical value of 0.350 deduced from exact diagonalization of finite sized (≤ 12 sites) chains.[19]

Haldane has examined the neighborhood of the VBS point using SMA perturbation theory.[19] Taking

$$H_{\epsilon} = H_{\text{VBS}} + \epsilon \sum_{\langle ij \rangle} \boldsymbol{S}_i \cdot \boldsymbol{S}_j$$

$$H_{\text{VBS}} \approx \sum_{\boldsymbol{k}, \alpha} \omega_{\boldsymbol{k}} \, b_{\boldsymbol{k}, \alpha}^{\dagger} b_{\boldsymbol{k}, \alpha} \tag{3.23}$$

$$S_{\boldsymbol{k}}^{\alpha} \approx \sqrt{s(\boldsymbol{k})} \left(b_{\boldsymbol{k}, \alpha}^{\dagger} + b_{-\boldsymbol{k}, \alpha} \right) ,$$

one arrives at the standard form of eq(2.18b) with $\Delta_{\boldsymbol{k}} = 2 \epsilon s(\boldsymbol{k}) \cos k$. The Bogoliubov spectrum is given by $E_{\boldsymbol{k}} = \sqrt{\omega_{\boldsymbol{k}}^2 + 2 \omega_{\boldsymbol{k}} \Delta_{\boldsymbol{k}}}$. The gap is predicted to collapse when $\omega_{\boldsymbol{k}} + 2\Delta_{\boldsymbol{k}} = 0$, and, assuming this occurs at the zone corner $\boldsymbol{k} = \boldsymbol{\pi}$, this condition gives $\chi^{\text{SMA}}(\boldsymbol{\pi}) = \epsilon^{-1}$, where χ is the static spin susceptibility,

$$\chi(\boldsymbol{k}) = 2 \int_0^{\infty} \frac{d\omega}{\omega} S(\boldsymbol{k}, \omega)$$

$$\chi^{\text{SMA}}(\boldsymbol{k}) = 2 \, s(\boldsymbol{k})^2 / f(\boldsymbol{k}) \ . \tag{3.24}$$

In an antiferromagnet, the static susceptibility is peaked at the zone corner. The effect of an additional positive antiferromagnetic Heisenberg coupling should be to increase the tendency to order, and so the gap collapse condition makes good sense. As applied to the $S = 1$ VBS chain, this approximation predicts a gap collapse at $k = \pi$ at $\epsilon_+ = 5/54$ (towards the Heisenberg point), and a gap collapse at an incommensurate wavevector $k = \pm 0.374\pi$ for $\epsilon_- = -200/27$. Although the critical values of ϵ are way off, some of the features of the general bilinear–biquadratic exchange model are nicely identified by this simple approach. One such feature is the appearance of incommensurate correlations for negative ϵ; such behavior is observed in numerical studies.[24] Another such feature is a dimensional crossover in the long-distance behavior of the correlation functions. One usually expects a d-dimensional quantum model to be related to a $(d+1)$-dimensional classical model, yet the VBS state correlation functions are given by those of a classical model in the *same* number of dimensions. The classical Ornstein-Zernike expression for the correlation function in the disordered phase is $C(\boldsymbol{r}) \sim r^{(1-d)/2} \exp(-r/\xi)$. Eq(3.16) shows that the one-dimensional VBS states exhibit purely exponential correlations with no power law correction, which is expected based on the aforementioned correspondence with the associated one-dimensional classical model. SMA perturbation theory correctly reproduces the $|n|^{-1/2}$ power law correction as soon as one moves off the VBS point.

IV. THE FRACTIONAL QUANTUM HALL EFFECT

The fractional quantum Hall effect (FQHE) is a phenomenon observed in two-dimensional electron gases in high magnetic fields.[28] Typically, the electron gas resides in an inversion layer of a semiconductor heterojunction. The Drude expression for the Hall conductivity, $\sigma_{xy}^{\text{Drude}} = nec/B$ is linear in the electron density n. In the quantum limit, where the temperature and density are both very small, it is convenient to express the density in terms of a dimensionless quantity $\nu = 2\pi \ell^2 n$, where $\ell = \sqrt{\hbar c/eB}$ is called the *magnetic length*. One then has $\sigma_{xy}^{\text{Drude}} = \nu\, e^2/h$, where $e^2/h = (25813\,\Omega)^{-1}$ is the quantum of conductance. One can vary ν by changing the field strength B at fixed electron density n, and from the Drude model one expects a linear dependence of σ_{xy} on ν. What one observes, however, is that in the vicinity of certain simple rational values of ν ($\nu = p/q$), σ_{xy} is *quantized* at $(p/q) \cdot e^2/h$ over a range of ν values. Along with the plateau in σ_{xy}, one observes that the longitudinal conductivity σ_{xx} becomes thermally activated, behaving as $\sigma_{xx} \sim e^{-\Delta/k_B T}$ at low temperatures.

For integer p/q (*i.e.* $q = 1$), it is natural to conjecture that this behavior is connected somehow with the filling of a Landau level. The fractional effect is much more difficult to understand, because there is nothing within a single electron picture that would sanctify a certain set of rational filling fractions. To understand this, one must invoke collective many–body effects.

If the cyclotron energy $\hbar\omega_c$ is large compared with the characteristic electron-electron interactions, a situation which is approximately true in FQHE experiments, it is reasonable to neglect inter-Landau level excitations. Laughlin[26] proposed the following trial wavefunction to describe the nature of the electronic state at fillings $\nu = 1/q$:

$$\Psi_q(\boldsymbol{r}_1,\ldots,\boldsymbol{r}_N) = \prod_{i<j}(z_i - z_j)^q \, \exp\left(-\frac{1}{4\ell^2}\sum_k |z_k|^2\right) , \qquad (4.1)$$

where q is an odd integer, in accordance with Fermi statistics. This many body state is comprised solely of lowest Landau level single particle wavefunctions, yet is highly correlated in such a way as to prevent the electrons from getting too close to one another. For $q = 1$, this state is a Vandermonde determinant describing a completely filled Landau level. For $q = 3, 5, \ldots$, the Landau level is only partially filled. Interpreting $|\Psi|^2 \equiv e^{-\beta\Phi}$ as a Boltzmann weight, we find that for $\beta = 1/q$,

$$\Phi(\boldsymbol{r}_1,\ldots,\boldsymbol{r}_N) = -2q^2 \sum_{i<j} \ln|z_i - z_j| + \frac{q}{2\ell^2}\sum_i |z_i|^2 , \qquad (4.2)$$

which describes a two-dimensional one-component plasma (2DOCP)[29] of charge $\tilde{e} = \sqrt{2}\,q$ objects interacting via a repulsive 2-d Coulomb potential $v(\boldsymbol{r}) = -\ln r$. The one-body term in eq(4.2) represents a uniform charged background; $\nabla^2(r^2/2\sqrt{2}\,\ell^2) = -2\pi\varrho_{\text{back}}$ gives the background charge density $\varrho_{\text{back}} = -1/\sqrt{2}\,\pi\ell^2$. The 2DOCP obeys several sum rules; first and foremost among them is the charge neutrality sum rule, which states that in equilibrium the total charge is zero. This fixes the mean density n by the relation $\sqrt{2}\,qn = -\varrho_{\text{back}}$. Thus we arrive at $n = 1/2\pi\ell^2 q$ or $\nu = 1/q$, as promised.

The 2DOCP will crystallize when the long-range interactions win out over the effects of thermal fluctuations. This competition is expressed as the ratio $\Gamma = \tilde{e}^2/k_B T$, known as the *plasma parameter*. For our 2DOCP, we have $\tilde{e}^2 = 2q^2$ and $k_B T = q$, giving $\Gamma = 2q$. The critical value of Γ above which crystalline order sets in is $\Gamma \simeq 138$,[29] meaning that the Laughlin wavefunction describes a fluid state for $\nu \gtrsim 1/70$.

Haldane[30,31] first pointed out that the Laughlin state is special because as a function of the relative coordinate $(z_i - z_j)$ it vanishes as $(z_i - z_j)^q$. For $q > 1$ this means that no pair of particles has any projection onto states or relative angular momentum $l < q$. In the lowest Landau level, any interaction $v(\boldsymbol{r}_i - \boldsymbol{r}_j)$ can be expanded in terms of relative angular momentum projection operators. The coefficients V_s in the expansion

$$\Pi_0\, v(\boldsymbol{r}_i - \boldsymbol{r}_j)\,\Pi_0 = \sum_{s=0}^{\infty} V_l\, P_l(ij) , \qquad (4.3)$$

where $P_l(ij)$ is the projector onto relative angular momentum l for the pair (ij) and Π_0 is the projector onto the lowest Landau level, can be manipulated to as to render the Laughlin state Ψ_q an exact zero-energy ground state of the lowest-Landau level restricted Hamiltonian $H = \Pi_0 \sum_{i<j} v(\mathbf{r}_i - \mathbf{r}_j) \Pi_0$ — one simply takes $V_l = 0$ for all $l \geq q$. (For fermions, only the odd l terms will contribute.) To determine the 'pseudopotential coefficients'[30], we use the results of Appendix B to project $v(\mathbf{r}_i - \mathbf{r}_j)$ onto the $n = 0$ Landau level,

$$v(\mathbf{r}_i - \mathbf{r}_j) = \int \frac{d^2k}{(2\pi)^2}\, v(\mathbf{k})\, e^{i\mathbf{k}\cdot(\mathbf{r}_i - \mathbf{r}_j)}$$

$$\Pi_0\, v(\mathbf{r}_i - \mathbf{r}_j)\, \Pi_0 = \int \frac{d^2k}{(2\pi)^2}\, v(\mathbf{k})\, \exp\left[\frac{i\ell}{\sqrt{2}}\,(k_x + ik_y)\left(b_i - b_j\right)\right] \exp\left[\frac{i\ell}{\sqrt{2}}\,(k_x - ik_y)\left(b_i^\dagger - b_j^\dagger\right)\right]$$

$$= \int \frac{d^2k}{(2\pi)^2}\, v(\mathbf{k}) \sum_{n=0}^{\infty} \frac{1}{n!^2}\, (-k^2\ell^2)^n\, (J_{ij}^-)^n\, (J_{ij}^+)^n\ ,$$

$$(4.4)$$

where $J_{ij}^\pm$ are the relative angular momentum L_{ij} raising and lowering operators:

$$J_{ij}^+ = \frac{1}{\sqrt{2}}\left(b_i^\dagger - b_j^\dagger\right)$$

$$J_{ij}^- = \frac{1}{\sqrt{2}}\left(b_i - b_j\right) \qquad\qquad (4.5)$$

$$L_{ij} = J_{ij}^+ J_{ij}^- = \Pi_0\left(\mathbf{r}_i - \mathbf{r}_j\right) \times \tfrac{1}{2}(\mathbf{p}_i - \mathbf{p}_j)\cdot \hat{\mathbf{z}}\,\Pi_0\ .$$

If we place the identity operator $1 = \sum_l P_l(ij)$ to the right of the last of eq(4.4), we obtain

$$\Pi_0\, v(\mathbf{r}_i - \mathbf{r}_j)\, \Pi_0 = \sum_{l=0}^{\infty} \frac{P_l}{l!} \int \frac{d^2k}{(2\pi)^2}\, v(\mathbf{k}) \sum_{n=0}^{\infty} \frac{(n+l)!}{n!^2}(-k^2\ell^2)^n\ , \qquad (4.6)$$

and the sum can be performed, leading to the result

$$V_l = \int \frac{d^2k}{(2\pi)^2}\, e^{-k^2\ell^2}\, L_l(k^2\ell^2)\, v(\mathbf{k}) \qquad\qquad (4.6)$$

where $L_s(x)$ is a Laguerre polynomial. For the Coulomb potential $v(\mathbf{r}) = e^2/r$, the pseudopotential components are

$$V_s^{\text{Coul}} = \frac{\sqrt{\pi}e^2}{2\ell}\, \frac{(2l-1)!!}{2^l\, l!}\ , \qquad\qquad (4.7)$$

which decreases as $1/\sqrt{l}$ when l gets large. Thus, while the exact question to Laughlin's answer has $V_1 > 0$ and $V_3 = V_5 = \cdots = 0$, for the physical Coulomb interaction one finds $V_3^{\text{Coul}}/V_1^{\text{Coul}} = 5/8$, $V_5^{\text{Coul}}/V_1^{\text{Coul}} = 63/128$, $etc.$ Numerical work, however, shows that the Laughlin state is an excellent model for the physics of the FQHE.[30]

Excitations

There are two types of excitations to discuss. The first type is the quasiparticle, a charged excitation resembling a 'bubble' in the Laughlin ground state created by adiabatic insertion (quasihole) or removal (quasielectron) of a quantum hc/e of magnetic flux. One of the most striking features of the FQHE is that the quasielectrons and quasiholes possess fractional charge $\mp e/q$. For a detailed description, see ref[26].

In addition to the quasiparticles, there are also neutral collective excitations analogous to the phonons and rotons in liquid ^{4}He. To describe these excitations, Girvin, MacDonald, and Platzman[5] proposed to modify the usual SMA formula $\rho_k\,|\,\Psi\,\rangle$ by projecting the resultant density wave into the lowest Landau level. The reason is that the usual density operator $\rho_k = \sum_i e^{i\mathbf{k}\cdot\mathbf{r}_i}$ can create inter-Landau level excitations. Indeed Kohn's theorem guarantees that as $\mathbf{k} \to 0$ all the oscillator strength is saturated by the cyclotron mode.[32] Naïve application of the SMA formulae then leads to

$$f(\mathbf{k})/s(\mathbf{k}) = \hbar\omega_c \left\{ 1 - \frac{q-2}{4}k^2\ell^2 + \mathcal{O}(k^4\ell^4) \right\} \qquad\qquad (4.8)$$

which is purely *inter*-Landau level in the long wavelength limit.

The authors of ref[5] succeeded in creating a purely *intra*-Landau level excitation by taking $|\,k\,\rangle = N^{-1/2}\bar{\rho}_k\,|\,\Psi\,\rangle$. Here, $\bar{\rho}_k = \Pi_0\,\rho_k\,\Pi_0$ is the projected density operator:

$$
\begin{aligned}
\rho_k &= \sum_i \exp\left[\frac{i\ell}{\sqrt{2}}\,(k_x + ik_y)\left(a_i^\dagger + b_i\right)\right] \exp\left[\frac{i\ell}{\sqrt{2}}\,(k_x - ik_y)\left(a_i + b_i^\dagger\right)\right] \\
\bar{\rho}_k &= \sum_i \exp\left[\frac{i\ell}{\sqrt{2}}\,(k_x + ik_y)\,b_i\right] \exp\left[\frac{i\ell}{\sqrt{2}}\,(k_x - ik_y)\,b_i^\dagger\right] .
\end{aligned}
\tag{4.9}
$$

The cyclotron ladder operators a_i and $a_i^\dagger$, which promote inter-Landau level transitions, have been projected away in $\bar{\rho}_k$. Working entirely within the lowest Landau level, then, the kinetic energy is completely quenched, and the Hamiltonian becomes

$$
\begin{aligned}
H &= \Pi_0 \sum_{i<j} v(\boldsymbol{r}_i - \boldsymbol{r}_j)\,\Pi_0 \\
&= \tfrac{1}{2}\int \frac{d^2k}{(2\pi)^2}\,v(\boldsymbol{k})\left(\bar{\rho}_k\,\bar{\rho}_{-k} - N e^{-\frac{1}{2}k^2\ell^2}\right) .
\end{aligned}
\tag{4.10}
$$

The projected oscillator strength and structure factor are then given by

$$
\begin{aligned}
\bar{f}(k) &= N^{-1}\,\langle\,\Psi\,|\,\left[\bar{\rho}_k^\dagger,[H,\bar{\rho}_k]\right]\,|\,\Psi\,\rangle \\
\bar{s}(k) &= N^{-1}\,\langle\,\Psi\,|\,\bar{\rho}_k^\dagger\,\bar{\rho}_k\,|\,\Psi\,\rangle
\end{aligned}
\tag{4.11}
$$

and can be evaluated from the relations[5]

$$
\begin{aligned}
\bar{f}(k) &= \int \frac{d^2p}{(2\pi)^2}\,v(\boldsymbol{p})\left\{1 - \cos(\ell^2\,\boldsymbol{k}\times\boldsymbol{p}\cdot\hat{\boldsymbol{z}})\right\}\left(\bar{s}(\boldsymbol{k}+\boldsymbol{p})\,e^{\ell^2\,\boldsymbol{k}\cdot\boldsymbol{p}} - \bar{s}(\boldsymbol{p})\,e^{-\frac{1}{2}k^2\ell^2}\right) \\
\bar{s}(\boldsymbol{k}) &= s(\boldsymbol{k}) - 1 + e^{-\frac{1}{2}k^2\ell^2}
\end{aligned}
\tag{4.12}
$$

where $s(\boldsymbol{k})$ is the static structure factor of the 2DOCP,

$$
s(\boldsymbol{k}) = \tfrac{1}{2}k^2\ell^2 + \frac{q-2}{8}\,k^4\ell^4 + \mathcal{O}(k^6\ell^6) ,
\tag{4.13}
$$

whose long-wavelength behavior is constrained by the charge neutrality, perfect screening, and compressibility sum rules.[29] Since both $\bar{f}(k)$ and $\bar{s}(k)$ vanish as k^4 at long wavelengths, the SMA predicts a gap in the collective mode spectrum $\omega_k = \bar{f}(k)/\bar{s}(k)$. We stress that the f-sum rule $f(k) = \hbar^2 k^2/2m$ applies only to the *unprojected* oscillator strength and not to $\bar{f}(k)$. One finds that the dispersion ω_k exhibits a minimum at the inverse interparticle spacing $k^* \sim \sqrt{n}$, reminiscent of the roton in ^{4}He.

Away from the solvable point, we can apply SMA perturbation theory. Inverting eq(4.6), we have

$$
v(\boldsymbol{k}) = \sum_m 4\pi\ell^2\,V_m\,L_m(k^2\ell^2) .
\tag{4.14}
$$

Let us write $v(\boldsymbol{k}) = v_0(\boldsymbol{k}) + \delta v(\boldsymbol{k})$, where $v_0(\boldsymbol{k})$ represents the truncated pseudopotential for which the Laughlin state is the exact ground state. Approximating the magnetophonon/magnetoroton excitations as independent bosons,

$$
\bar{\rho}_k \approx \sqrt{N\,\bar{s}(k)}\,\left(b_k^\dagger + b_{-k}\right) ,
\tag{4.15}
$$

one arrives at the Hamiltonian

$$
H = \frac{N}{2n}\int \frac{d^2k}{(2\pi)^2}\,\omega_k\,(b_k^\dagger b_k + b_{-k}^\dagger b_{-k}) + \frac{N}{2}\int \frac{d^2k}{(2\pi)^2}\,\bar{s}(k)\,\delta v(k)\,(b_k^\dagger + b_{-k})\,(b_{-k}^\dagger + b_k)
\tag{4.16}
$$

which yields a dispersion $E_{\boldsymbol{k}} = \sqrt{\omega_{\boldsymbol{k}}^2 + 2\,n\,\omega_{\boldsymbol{k}}\,\bar{s}(\boldsymbol{k})}$. The gap collapses when $\chi^{\text{SMA}}(\boldsymbol{k})\cdot n\,\delta v(\boldsymbol{k}) = 1$, where $\chi^{\text{SMA}}(\boldsymbol{k}) = -2\bar{s}(\boldsymbol{k})^2/\bar{f}(\boldsymbol{k})$ is the SMA static susceptibility.[5] This makes good physical sense — increasing the strength of the potential in the vicinity of the magnetoroton minimum (where χ^{SMA} is largest) causes the magnetoroton to go soft. We have examined the gap collapse within a model in which only V_1 and V_3 are nonzero, and find a critical value of $(V_3/V_1)_{\text{c}} \simeq 2.0$. Rezayi has numerically diagonalized this model for six particles on the surface of a sphere, and finds that the SMA overestimates $(V_3/V_1)_{\text{c}}$ by about a factor of four.[33]

V. EXTENDED VALENCE BOND SOLID MAGNETS

Recently there has been an explosion of interest in quantum magnetism, prompted by the discovery of high temperature superconductivity in doped antiferromagnetic Mott insulators. Many theoretical attempts to understand quantum magnetism[34,35,36] have resorted to 'large-N' techniques,[37] where one extends one's horizons from the physical case of SU(2) spins to the fantasyworld of SU(N). The reason for doing this is that in many instances the large-N limit can be solved exactly, from which one can gain valuable intuition about the SU(2) problem.

Extensive studies of the large-N SU(N) quantum Heisenberg model were performed by Read and Sachdev.[36] These authors identified two possible types of magnetic order — (1) Néel order with a concomitant gapless spin wave, and (2) a massive phase — and investigated the phase diagram as a function of the SU(N) representation and the lattice structure. They concluded that the massive phase of the SU(N) Heisenberg antiferromagnet should exhibit spin-Peierls order with a ground state degeneracy which is periodic in n_{c} mod z, where n_{c} is the number of columns in the Young tableau describing the SU(N) representation, and z is the lattice coordination number. For $n_{\text{c}} = 0$ mod z, the massive phase was predicted to have a nondegenerate ground state breaking no discrete symmetries. The SU(2) version of this degeneracy argument (where $n_{\text{c}} = 2S$) was predicted by Haldane,[11] who argued that the non-Néel phase of a square lattice SU(2) model should exhibit a ground state degeneracy of $(1,4,2,4)$ for $2S$ mod $4 = (0,1,2,3)$. Note that the condition $2S$ mod $z = 0$ is precisely the requirement for the existence of a VBS model, as we discussed in section 3.

Recently, Affleck $et\ al.$ discovered a series of solvable SU($2n$) models.[38] The interactions are again local and of the projection operator form, but rather than breaking a discrete lattice symmetry, as for spin-Peierls systems, the ground states of these 'extended valence bond solids' (XVBS) break the internal SU($2n$) symmetry of charge conjugation. XVBS models can be defined on any lattice provided that n is an integer multiple of the lattice coordination number ($n = mz$), and SU($2n$) in a totally antisymmetric self-conjugate representation described by a Young tableau with one column ($n_{\text{c}} = 1$) consisting of n boxes. We remind the reader that for SU(2) all representations are self-conjugate, and that the maximum number of boxes in a given column is N for SU(N). Just as in eq(3.7), multiplying two self-conjugate representations generates many more representations, among which lies the singlet.

SU(N)

The group SU(N) has $N^2 - 1$ traceless generators S_{β}^{α} which satisfy

$$[S_{\beta}^{\alpha}, S_{\nu}^{\mu}] = \delta_{\beta}^{\mu}\, S_{\nu}^{\alpha} - \delta_{\nu}^{\alpha}\, S_{\beta}^{\mu}\ . \tag{5.1}$$

In describing totally antisymmetric representations it is convenient to employ N species of fermions and define $S_{\beta}^{\alpha} = c^{\dagger\alpha} c_{\beta} - \frac{1}{2}\delta_{\beta}^{\alpha}$. Under the charge conjugation operation $C\colon c_{\alpha} \to c^{\dagger\alpha}$, the spin operators transform as

$$S_{\beta}^{\alpha} \longrightarrow c_{\alpha}c^{\dagger\beta} - \tfrac{1}{2}\delta_{\alpha}^{\beta} = -S_{\alpha}^{\beta} \tag{5.2}$$

and the vacuum as $|\,0\,\rangle \to (2n!)^{-1}\epsilon_{\alpha_1\alpha_2\cdots\alpha_{2n}}\, c^{\dagger\alpha_1}c^{\dagger\alpha_2}\cdots c^{\dagger\alpha_{2n}}\,|\,0\,\rangle$. In matrix notation, we have simply $S \to -S^{\text{t}}$, where S^{t}, where S^{t} denotes the transpose of S. This transformation property may be contrasted to what happens under the time-reversal (T) operation which simply maps S into itself and complex conjugates c-numbers. (If an operator $\mathcal{O}$ goes to $\mathcal{O}^T$ under time-reversal, then broken time-reversal symmetry is marked by $\langle\mathcal{O}^T\rangle \neq \langle\mathcal{O}\rangle^*$. The

(*) operation in this equation is essential; otherwise, if one were to redefine the T-breaking order parameter $\mathcal{O}_T \to i\mathcal{O}_T$, one would conclude that T never breaks.) Note that charge-conjugation in the SU(2) case actually corresponds to a global SU(2) rotation by an angle π about the y-axis, since $\epsilon_{\alpha\beta} = i\,\sigma^y_{\alpha\beta}$ where σ^y is the Pauli spin matrix. Thus C-breaking in this case would correspond simply to a breaking of the continuous SU(2) symmetry. SU($2n$) C-breaking ground states that preserve the SU($2n$) symmetry therefore are realized only when $n > 1$.[38]

The simplest example of an XVBS model is an SU(4) chain, where each site is in the completely antisymmetric 6-dimensional $\begin{smallmatrix}\square\\\square\end{smallmatrix}$ representation. Each site is occupied by two fermions of four possible 'flavors',

$$
\begin{aligned}
\big|\, \underset{\alpha\ \beta}{\times} \,\big\rangle &\equiv \frac{1}{2\sqrt{3}}\,\epsilon_{\alpha\beta\mu\nu}\, c^{\dagger\mu} c^{\dagger\nu}\,|\,0\,\rangle \\
\big|\, \overset{\alpha\ \beta}{\bullet} \,\big\rangle &\equiv \frac{1}{\sqrt{3}}\, c^{\dagger\alpha} c^{\dagger\beta}\,|\,0\,\rangle
\end{aligned}
\tag{5.3}
$$

and the two C-breaking ground states are given by

$$
\begin{aligned}
|\,\Psi_1\,\rangle &= \big|\cdots \overset{\alpha\ \beta}{\bullet}\ \underset{\beta\ \gamma}{\times}\ \overset{\gamma\ \delta}{\bullet} \cdots\big\rangle \\
|\,\Psi_2\,\rangle &= \big|\cdots \underset{\mu\ \nu}{\times}\ \overset{\nu\ \rho}{\bullet}\ \underset{\rho\ \sigma}{\times} \cdots\big\rangle\ .
\end{aligned}
\tag{5.4}
$$

The charge conjugation operator C for SU(k) satisfies $C^\dagger = C^{-1} = (-1)^{\frac{1}{2}k(k-1)}C$. It is easy to show that

$$
\begin{aligned}
C\,\big|\, \overset{\alpha\ \beta}{\bullet} \,\big\rangle &= \big|\, \underset{\alpha\ \beta}{\times} \,\big\rangle \\
C\,\big|\, \underset{\alpha\ \beta}{\times} \,\big\rangle &= \big|\, \overset{\alpha\ \beta}{\bullet} \,\big\rangle
\end{aligned}
\tag{5.5}
$$

and $C\,|\,\Psi_{1,2}\,\rangle = |\,\Psi_{2,1}\,\rangle$. To find the question to which $|\,\Psi_{1,2}\,\rangle$ is the answer, consider the product of $\begin{smallmatrix}\square\\\square\end{smallmatrix}$ representations on two neighboring sites,

$$
\underset{6}{\begin{smallmatrix}\square\\\square\end{smallmatrix}} \otimes \underset{6}{\begin{smallmatrix}\square\\\square\end{smallmatrix}} = \underset{1}{\bullet} \oplus \underset{15}{\begin{smallmatrix}\square&\square\\\square&\\\square&\end{smallmatrix}} \oplus \underset{20}{\begin{smallmatrix}\square&\square\\\square&\square\end{smallmatrix}}\ .
\tag{5.6}
$$

Of the four fermion labels associated with any two consecutive sites in $|\,\Psi_{1,2}\,\rangle$, three are contracted with an ϵ-symbol, meaning that the $\boxplus$ representation is absent on every bond. The question is then

$$
\begin{aligned}
H &= \sum_n \mathrm{P}_{\boxplus}(n, n+1) \\
&= \sum_n \left\{ \tfrac{5}{12} + \tfrac{1}{2} S^\alpha_\beta(n)\, S^\beta_\alpha(n+1) + \tfrac{1}{12}\Big[S^\alpha_\beta(n)\, S^\beta_\alpha(n+1)\Big]^2 \right\}\ .
\end{aligned}
\tag{5.7}
$$

Again, as in the case with the VBS models, the Hamiltonian features more than just bilinear spin-spin interactions. Excitations and SMA perturbation theory in the SU(4) chain, as well as generalized XVBS states, are discussed in ref[38].

The XVBS states furnish a paradigm for a more exotic (*e.g.* neither Néel nor spin-Peierls) phase of quantum antiferromagnetism. The C-breaking phases do not fit directly into the Read-Sachdev picture because the XVBS Hamiltonians contain more than just the bilinear interaction term. Of course, introducing additional higher-order terms with small coefficients should not immediately destroy the excitation gap — the Read-Sachdev picture should remain valid within a neighborhood of the Heisenberg point. However, when the coefficients $V_{k>1}$ of $[S^\alpha_\beta(i)\, S^\beta_\alpha(j)]^k$ get sufficiently large, a phase boundary may be reached. An analogy may be drawn to the case of SU(2) chains, where a hierarchy of gapless integrable models[16] exists for all S. Haldane's field theoretic treatment,[11] which leads to the prediction of a gap for integer S chains, is applicable only in a neighborhood of the (purely bilinear) Heisenberg point. Sufficiently large biquadratic amplitude in *e.g.* the $S = 1$ chain collapses the gap and leads to a different magnetic phase beyond the integrable point.

VI. 'SUPERSYMMETRIC' QUANTUM XY MODEL

The quantum XY rotor model used to describe superconductors and Josephson junction arrays is

$$H_0 = -U \sum_j \frac{\partial^2}{\partial \theta_j^2} - J \sum_{\langle ij \rangle} \cos(\theta_i - \theta_j) \, , \tag{6.1}$$

where $\langle ij \rangle$ refers to near-neighbor pairs on the lattice of sites. This model is essentially equivalent to the boson Hubbard model. The kinetic energy of the rotors represents the potential energy of the bosons (the Hubbard U term) and the potential energy of the rotors represents the kinetic energy of hopping of the bosons. The connection is seen by noting that the integer units of angular momentum of the rotors represents the integer number of bosons (Cooper pairs) on a site (relative to some large integer mean number). The analog of the boson creation operator is the angular momentum raising operator $b_j^\dagger = e^{i\theta_j}$ (see Appendix C). The ideas presented below are quite general, but we restrict our discussion here to the particular case of $d = 2$ space dimensions.

A trial variational wave function for the ground state of this model is the following:

$$\Psi = \exp \left\{ \lambda \sum_{\langle ij \rangle} \cos(\theta_i - \theta_j) \right\} \, , \tag{6.2}$$

where λ is a variational parameter whose optimal value is $\lambda \approx J/2U$. The choice of this *Ansatz* is motivated by the fact that for a simple harmonic oscillator, the trial state involving the exponential of the potential energy,

$$\psi(x) = e^{-\lambda V(x)} \, , \tag{6.3}$$

is *exact*. For large J/U, where the ground state is highly spin (phase) ordered, we can make the 'spin-wave' approximation and expand the Hamiltonian to second order in the notionally small quantity $(\theta_i - \theta_j)$. This gives a coupled harmonic oscillator problem whose solution is the wave function presented above (again with the exponent expanded to second order to yield a Gaussian). The advantage of this variational form however is that, unlike the simple spin-wave approximation ground state, it respects the symmetry of the problem under $\theta_j \to \theta_j + 2\pi$. This symmetry is closely associated with the existence of topologically non-trivial vortex excitations which will be more extensively discussed below.

The wavefunction of eq(6.2) is of the pair product (Jastrow) form. The problem of finding a Hamiltonian for such a wavefunction has been considered previously by Sutherland[39] and more recently by Kane, Kivelson, Lee, and Zhang.[40] Given any Jastrow state

$$\Psi = \exp \left\{ -\sum_{\langle ij \rangle} g(x_i - x_j) \right\} \tag{6.4}$$

one can construct a Hamiltonian which renders Ψ a ground state. Differentiating Ψ with respect to x_i gives

$$\frac{\partial \Psi}{\partial x_i} = -\left\{ {\sum_j}' g'(x_i - x_j) \right\} \Psi \, , \tag{6.5}$$

where the prime restricts the sum to the nearest neighbors of i. Defining

$$\begin{aligned} Q_i &= +\frac{\partial}{\partial x_i} + {\sum_j}' g'(x_i - x_j) \\ Q_i^\dagger &= -\frac{\partial}{\partial x_i} + {\sum_j}' g'(x_i - x_j) \, , \end{aligned} \tag{6.6}$$

one can design a Hamiltonian[40]

$$H = \sum_i Q_i^\dagger Q_i$$

$$= -\sum_i \frac{\partial^2}{\partial x_i^2} - 2 \sum_{\langle ij \rangle} g''(x_i - x_j) + 2 \sum_{\langle ij \rangle} \left\{ g'(x_i - x_j) \right\}^2 + \tag{6.7}$$

$$+ \sum_i \sum_{j \neq k}{}' g'(x_i - x_j)\, g'(x_i - x_k)$$

which renders Ψ a zero energy ground state. Notice that H involves a three-body term. Kane *et al.* have recently argued that this three-body term is irrelevant to the long-wavelength properties of Ψ.[40] For our rotor/XY model, we should take

$$g(\theta_i = \theta_j) = -\lambda \sum_{\langle ij \rangle} \cos(\theta_i - \theta_j) \;, \tag{6.8}$$

in which case Ψ in eq(6.2) is the ground state of $H = H_0 + V$, where H_0 is the XY model Hamiltonian of eq(6.1) and V is a perturbation

$$V = 2\lambda^2 \sum_{\langle ij \rangle} \sin^2(\theta_i - \theta_j) + \lambda^2 \sum_i \sum_{j \neq k}{}' \sin(\theta_i - \theta_j)\sin(\theta_i - \theta_k) \;. \tag{6.9}$$

In a classical XY model at finite temperature, we know that the probability of finding the system with some particular configuration of the phases is simply proportional to the Boltzmann factor

$$P(\theta_1, \ldots, \theta_N) = \exp\left\{ -\beta J \sum_{\langle ij \rangle} \cos(\theta_i - \theta_j) \right\} \;. \tag{6.10}$$

This system undergoes the well-known Kosterlitz-Thouless phase transition in which vortex defects in the spin-ordering unbind above the critical temperature. In the quantum XY model at zero temperature, the probability of finding the system in a particular configuration is given by the square of the quantum ground state wave function,

$$P(\theta_1, \ldots, \theta_N) = |\Psi|^2 = \exp\left\{ 2\lambda \sum_{\langle ij \rangle} \cos(\theta_i - \theta_j) \right\} \;. \tag{6.11}$$

For this particular ground state we see that the quantum probability distribution is identical to that of a classical XY model with the parameter U/J playing the role of temperature.

This gives us a very nice illustration of how quantum fluctuations induce vortices. For small U/J the spins are highly aligned and (in two dimensions) the spin-spin correlation function (which is the analog of the off-diagonal density matrix for the bosons — see Appendix C)

$$G(ij) = \langle e^{i\theta_i} e^{-i\theta_j} \rangle \sim \langle b_i^\dagger b_j \rangle \;, \tag{6.12}$$

shows algebraic order

$$G(ij) \sim |\vec{r}_i - \vec{r}_j|^{-\eta} \tag{6.13}$$

with the exponent η increasing continuously with the coupling U/J. Thus we have a super-conducting ground state for small U/J.

A classical system at zero temperature would have true ODLRO with $\eta = 0$ and then η continuously increasing with temperature, again showing that the quantum fluctuations in the present model act like thermal fluctuations in a corresponding classical model. As the coupling U/J is raised the system undergoes the Kosterlitz-Thouless phase transition to a state with exponentially decaying spin correlations

$$G(ij) \sim e^{-|\vec{r}_i - \vec{r}_j|/\xi} \;, \tag{6.14}$$

where the correlation length ξ diverges at the critical temperature. In the extreme limit $U \to \infty$ the wave function becomes very simple:

$$\Psi(\theta_1, \ldots, \theta_N) = 1 \ . \tag{6.15}$$

This simplest of all states can be understood in two different ways. The first is to say that it means that every rotor has zero angular momentum. In the boson particle language this means that every site has no deviation in charge from the background value. We have a charge insulator with a large excitation gap. If we move a particle from one site to another, it costs energy U on each site. The upper and lower 'Hubbard bands' are split by $2U$ less some small bandwidth associated with the hopping term. A second interpretation of this wave function is that all spin configurations are nearly equally likely, so a typical configuration has randomly oriented spins. Such a configuration would be described in a classical model as being full of vortices as if the temperature were very high. We may use a duality language[41] to say that the disordered state of the spins is an ordering (condensation) of the vortices. The vortices condense in order to increase the phase uncertainty and thus decrease the particle number uncertainty (thereby lowering the charging energy U).

One of the interesting features of the connection between classical statistical mechanics and quantum wave functions (and path integrals) is that one expects on general grounds that the universality class of a zero temperature (quantum phase transition) in a d-dimensional system to correspond to that of a $(d+1)$-dimensional classical system. This follows most easily from the path integral representation of the partition function which, since the various terms in the Hamiltonian do not commute, necessarily involves a time-ordered integral over 'imaginary' or 'Euclidean' time from $\tau = 0$ to $\tau = \hbar\beta$, where β is the inverse temperature. For zero temperature, the duration of this time interval diverges and time effectively plays the role of an extra dimension. Based on this argument one expects the 2-dimensional boson Hubbard model with Hamiltonian H_0 considered above to be in the universality class of the 3-dimensional classical XY model.[42] The soluble model we have found here differs from H_0 by what appears to be a 'benign' (to be defined precisely shortly) perturbation V, and yet we have just shown that its transition is in the universality class of the 2- not the 3-dimensional classical XY model. The same can be said for the freezing transition of the Laughlin plasma wave function[26] and for the plasma in the AKLT spin chain consider earlier.

Understanding the paradox requires consideration of several aspects of the problem. First, it seems obvious that square of *any* ground state wave function of any kind gives a particle distribution which has nothing to do with any integrations over Euclidean time and hence must just correspond to some classical Boltzmann factor for a problem in only d-dimensions:

$$|\Psi|^2 = \exp\left(-\beta H_{\rm cl}\right) \ , \tag{6.16}$$

with $\beta = 2$ and

$$H_{\rm cl} \equiv -\log|\Psi| \ . \tag{6.17}$$

Of course in general, $H_{\rm cl}$ will *not* be recognizable as a simply-defined statistical mechanics problem because it will contain various (possibly long-range) multi-particle interactions or related unorthodox features. What we really mean when we say that the problem is equivalent to a classical problem in $(d+1)$ dimensions is the following: If we view the problem as $(d+1)$-dimensional (through the path integral for example), we obtain a 'reasonable looking' statistical mechanics problem. 'Reasonable' may be in the eye of the beholder, but for example, we typically obtain a statistical mechanics problem with short-range two-body interactions.

This point is well-illustrated by considering variational wave functions for the boson Hubbard model, superfluid ^{4}He, or Heisenberg antiferromagnets[43] which capture the long-range correlations in the ground state better than the wave function considered here. The essential physics in these systems at long-wavelength is the existence of a linearly dispersing collective mode. In the Bijl-Feynman theory of the collective mode, the variational excited state for the density-wave mode is[2,5] $\Psi_k = \rho_k\Psi$, where Ψ is the exact ground state wave function and ρ_k is the Fourier transform of the density

$$\rho_k = \sum_j e^{ik\cdot r_j} \ . \tag{6.18}$$

The variational expression for the excitation energy is found using the oscillator strength sum rule to be

$$\Delta(\boldsymbol{k}) = \frac{\hbar^2 k^2}{2m\, s(\boldsymbol{k})} \tag{6.19}$$

where the static structure factor expresses the ground state density-density correlations:

$$s(\boldsymbol{k}) = \frac{1}{N} \langle \Psi \mid \rho_{-k}\, \rho_k \mid \Psi \rangle \ . \tag{6.20}$$

As discussed in the Introduction, this is a variational *Ansatz*, based essentially on the assumption that a single collective mode absorbs all of the oscillator strength. However, general arguments suggest that it should be exact at wavelengths long compared to the interparticle spacing. In order to obtain the expected linearly dispersing collective mode, it is clear that the static structure factor must have the special property of vanishing linearly at small wave vectors. For an ordinary classical fluid we expect the structure factor to go to a constant (given by the compressibility) rather than vanishing linearly. For a fluid with long-range (Coulomb) repulsion in which the collective mode has a gap, we expect the structure factor to vanish quadratically. A simple Jastrow wave function for the ground state of the superfluid would correspond to a statistical mechanics problem with only two-body interactions:

$$\Psi = \exp\left\{ -\sum_{i<j} f(\boldsymbol{r}_i - \boldsymbol{r}_j) \right\} \ . \tag{6.21}$$

It is clear that in order to have special correlations in the density at long wavelengths the effective interaction potential f must (generically) have a long-range tail.[44] For example, in three dimensions $f(\boldsymbol{r}) \sim \alpha/r^2$, where α is a constant related to the compressibility (or speed of sound). This long-range tail appears even though the physical quantum Hamiltonian contains only short-range potentials. In contrast to this d-dimensional case, the $(d+1)$-dimensional statistical mechanics model derived from the path integral only involves the short-range interactions of physical Hamiltonian.

We now turn back to our soluble model which violates the generic rules described above since it has as its exact ground state, a two-body Jastrow form with only short-range interactions. We presume that the difference is due to the existence of the supposedly 'benign' three-body term V in the quantum Hamiltonian. Let us first address the question of why we expect that generically, this perturbation is in fact 'benign.' Kane *et al.*[40] have shown that if one expresses the density as a mean value plus fluctuations, the three-body term can be expanded into terms involving 1, 2, and 3 powers of the density fluctuations. Lumping the first two in with the regular two-body parts of the Jastrow factor, they show that the remaining ('connected') three-body piece is irrelevant in the renormalization group sense. That is, the three-body potential has no essential effect on the nature of the long-range correlations in the state. Thus we are lead to a paradox. The three-body correlations seem irrelevant and yet they change the universality class of the critical point.

The paradox is resolved by noting that Kane *et al.* have shown that the three-body term is only *perturbatively* irrelevant. That is, it is irrelevant at the level of approximation where one neglects the vortices which enforce the discreteness of the particle density in the effective action which they use in their path-integral description of the ground state.[45] We have seen already that the Kosterlitz-Thouless transition in the ground state wave function comes about precisely because of vortex effects. Hence it is not surprising that the perturbative argument breaks down near the critical point.

We are now in a position to turn to the question of the nature of the collective mode for our soluble model. We again want to make the single-mode approximation. It is inconvenient however to evaluate the required structure factor. We therefore consider a different excited-state wave function which appears naively to describe a single-particle excitation $\Psi_k = b_k^\dagger \Psi$ where we have defined the 'creation operator'

$$b_k^\dagger \equiv \frac{1}{\sqrt{N}} \sum_j e^{i\boldsymbol{k}\cdot\boldsymbol{r}_j}\, e^{i\theta_j} \ . \tag{6.22}$$

We say that this is only naively a single-particle excitation because in an ordinary, weakly repulsively interacting bose fluid, the single-particle excitations are mixed with the collective mode by scattering from the condensate and do not exist independently (at long wavelengths).[46] Thus we might expect that this wave function has some reasonable physical connection with the true collective mode wave function.

The analog of the oscillator strength is

$$f(\mathbf{k}) = \langle\, \Psi \,|\, [b_{\mathbf{k}}, H]\, b_{\mathbf{k}}^{\dagger} \,|\, \Psi \,\rangle \; . \tag{6.23}$$

It is convenient to express this as a double commutator. To do so, consider the symmetry operation $(\theta, \mathbf{k}) \to (-\theta, -\mathbf{k})$ which results in

$$\begin{aligned} \Psi &\longrightarrow \Psi \\ b_{\mathbf{k}} &\longleftrightarrow b_{\mathbf{k}}^{\dagger} \\ H &\longrightarrow H \end{aligned} \tag{6.24}$$

and

$$f(-\mathbf{k}) = \langle\, \Psi \,|\, b_{\mathbf{k}}\, [b_{\mathbf{k}}^{\dagger}, H] \,|\, \Psi \,\rangle \; . \tag{6.25}$$

Under the assumption of parity symmetry $f(\mathbf{k}) = f(-\mathbf{k})$, and using the fact that the ground state energy is precisely zero, we obtain the desired double commutator

$$f(\mathbf{k}) = \tfrac{1}{2} \langle\, \Psi \,|\, \left[b_{\mathbf{k}}, [H, b_{\mathbf{k}}^{\dagger}] \right] \,|\, \Psi \,\rangle \tag{6.26}$$

which is readily evaluated and found to be $f(\mathbf{k}) = 1$ independent of the coupling constant λ.

With this result, the single-mode approximation gives $\Delta(\mathbf{k}) = 1/s(\mathbf{k})$. The analog of the structure factor is the spin-spin correlation function

$$s(\mathbf{k}) = \frac{1}{N} \sum_{i,j} e^{i\mathbf{k}\cdot(\mathbf{r}_i - \mathbf{r}_j)} \langle e^{i\theta_i}\, e^{-i\theta_j} \rangle \; . \tag{6.27}$$

Above the Kosterlitz-Thouless point $(\lambda < \lambda_c)$ we expect

$$\begin{aligned} s(\mathbf{k}) &\sim \int d^2 r\, e^{i\mathbf{k}\cdot\mathbf{r}}\, e^{-r/\xi(\lambda)} \\ &\sim \xi^2 + a_1\, k^2 + a_2\, k^4 + \dots \; , \end{aligned} \tag{6.28}$$

and an excitation gap $\Delta(0) \sim \xi^{-2}$. Kosterlitz-Thouless theory gives $\xi \sim \exp\left(b/\sqrt{|\lambda - \lambda_c|} \right)$, where b is a positive constant,[47] and so we have the gap scaling to zero at the critical point as

$$\Delta(0) \sim \exp\left(-\frac{2\,b}{\sqrt{|\lambda - \lambda_c|}} \right) \; . \tag{6.29}$$

On the superfluid side of the transition $(\lambda > \lambda_c)$ we have algebraic decay of the spin-spin correlation function, with $s(\mathbf{k}) \sim k^{-2+\eta}$, *i.e.*

$$\Delta(\mathbf{k}) \sim k^{2-\eta} \tag{6.30}$$

at long wavelengths. Since η rises continuously from zero to its final value of $\tfrac{1}{4}$ at the critical point, we see that the dispersion is never linear anywhere in the superfluid phase. The dispersion is intermediate between the linear result obtained for the case of a true condensate and the quadratic result obtained for free bosons. The Goldstone mode is attempting to 'Bogoliubov itself into existence,' but never quite makes it before the transition occurs. This suggests that the argument that the single-particle excitations are not independent of the collective excitations may not apply here, since true condensation does not occur.

In summary, the 'supersymmetric' quantum XY model has a simple ground state wave function. Using the connection between this wave function and the known statistical mechanics of the classical XY model, we are able to make definite statements about the collective excitations. These excitations are much softer than in the ordinary XY model which would have a wave function whose corresponding statistical mechanics problem contains long-range forces. The corresponding $(d+1)$-dimensional statistical mechanics problem associated with the quantum path integral would however, contain only short-range forces.

VII. OFF-DIAGONAL LONG-RANGE ORDER

The exotic quantum fluids we are discussing are unusual not only in terms of their physical properties, but also in terms of the mathematical nature of their order parameters. Recent theoretical progress[48,49,50,51,41] has elucidated the unusual nature of the order parameters and given a fairly complete understanding of the origin of the physical properties.

As a 'warm-up' exercise, let us consider the ordering of the $d = 2$ XY model which has been understood since the work of Kosterlitz and Thouless[52] and Berezinskii[53] to be of an unusual 'topological' type. For dimension $d > 2$, the XY spins magnetize below the critical temperature so that the spin-spin correlation function goes to a constant

$$G(\boldsymbol{r}_i - \boldsymbol{r}_j) \equiv \langle \boldsymbol{S}_i \cdot \boldsymbol{S}_j \rangle = \langle e^{i\theta_i} e^{-i\theta_j} \rangle \sim M^2 \tag{7.1}$$

at large separations. We say that the system spontaneously breaks rotational symmetry

$$\langle \cos\theta_i\,\hat{\mathbf{x}} + \sin\theta_i\,\hat{\mathbf{y}} \rangle = \boldsymbol{M} \;, \tag{7.2}$$

and the magnetization $\boldsymbol{M}$ is identified as a 'local order parameter.' For $d = 2$ this simple picture is spoiled by infra-red divergences associated with $d = 2$ being the lower critical dimension. The Mermin-Wagner theorem[12] guarantees that there can be no continuous broken symmetry in two dimensions and hence the system can never magnetize. Nevertheless there is still a phase transition at the Kosterlitz-Thouless temperature T_{KT} where the decay of the spin correlations becomes quasi-long-range:

$$\begin{aligned} G(\boldsymbol{r}_i - \boldsymbol{r}_j) &\sim \exp\left\{-|\boldsymbol{r}_i - \boldsymbol{r}_j|/\xi(T)\right\} \qquad (T > T_{\mathrm{KT}}) \\ &\sim |\boldsymbol{r}_i - \boldsymbol{r}_j|^{-\eta(T)} \qquad\qquad\quad (T < T_{\mathrm{KT}}) \end{aligned} \tag{7.3}$$

The decay exponent η decreases continuously from $\eta(T_{\mathrm{KT}}) = \frac{1}{4}$ to $\eta(0) = 0$. This continuous line of critical points is easily understood within the context of the spin-wave approximation.[54] At this level however, there is no phase transition at all to a disordered phase, no matter how high the temperature is raised! The spin-wave approximation misses the symmetry of the system under $\theta_i \to \theta_i + 2\pi$ and therefore misses the existence of vortices. A vortex is a topological defect in which the phase $\theta(\boldsymbol{r})$ winds by 2π as one moves around a path enclosing the defect. It is the fact that spins with orientation $\theta + 2\pi$ can be smoothly joined onto spins with orientation θ that allows vortices to exist. The thermal activation of vortices at high temperatures destroys the slow algebraic decay of magnetic correlations and gives an exponential decay with the correlation length being essentially the mean distance between vortices. Vortices also exist below T_{KT}, but they are confined (by their logarithmic attraction) into neutral (topological charge zero) pairs which serve only to renormalize the temperature dependence of the algebraic decay exponent $\eta(T)$.

Since the ordering of the XY model is not visible in the magnetization itself ($\langle \boldsymbol{S}_i \rangle = 0$) but is only manifested through confinement of topological defects, the model is said to exhibit 'topological order.' This topological order is not visible in the form of a local order parameter, but only in terms of correlations over large length scales.

Clearly the soluble $d = 2$ *quantum* XY model we have discussed previously has a ground state which orders (as U/J is varied) in a manner identical (by construction) to that of the classical XY model. We emphasize that, as mentioned earlier, the ordinary quantum XY model is in a different universality class: that of the $d+1 = 3$ dimensional classical XY model and hence exhibits ordinary magnetization in the 'low temperature' ($U \ll J$) phase.

With this introduction let us now consider the somewhat analogous, but still totally unprecedented topological ordering which occurs in the FQHE. In the original paper of Byers and Yang[55] defining the concept of off-diagonal long-range order (ODLRO), it is made clear that only bosons (or pairs of fermions) can condense. Single fermions can not. Knowing that magnetic fields induce vortices, and having seen above that in the XY model vortices induce disorder, we anticipate that there should be no (ordinary) ODLRO for the FQHE. This is indeed the case. We have interacting fermions in a magnetic field and, provided the ground state density is uniform, one can readily show that the one-body density matrix decays as a Gaussian:[56]

$$\rho(\boldsymbol{r}, \boldsymbol{r}') \equiv N \int d^2 r_2 \int d^2 r_3 \cdots \int d^2 z_N \, \overline{\Psi(\boldsymbol{r}', \boldsymbol{r}_2, \ldots, \boldsymbol{r}_N)} \, \Psi(\boldsymbol{r}, \boldsymbol{r}_2, \ldots, \boldsymbol{r}_N)$$

$$= \frac{\nu}{2\pi\ell^2} \, e^{-|r-r'|^2/4\ell^2} \, e^{(\bar{z}z' - \bar{z}'z)/4\ell^2} \, , \tag{7.4}$$

where ν is the filling fraction and $\ell = \sqrt{\hbar c/eB}$ is the magnetic length, as discussed in section IV. For the particular case of Laughlin's variational wave function at filling factor $\nu = 1/m$,

$$\Psi_m = \prod_{i<j} (z_i - z_j)^m \, \exp\left(-\frac{1}{4\ell^2} \sum_k |z_k|^2\right) \, , \tag{7.5}$$

these properties are likewise fulfilled. One readily sees that (as Halperin[57] noted very early on), the high magnetic field induces a large density of vortices which disorder the electron phases. In fact, the field is so high that there are $\nu^{-1} = m$ times as many vortices as particles! This hints however at the essential physics: For certain densities and B fields the *commensuration* between particles and vortices allows the formation of an insulating state ($\sigma_{xx} = 0$) which Anderson[58] has described as analogous to the Mott-Hubbard insulating phase which can occur for a commensurate ratio of electrons to lattice sites in a Hubbard model.

To understand the essence of the Laughlin state, imagine freezing the positions of particles $2, 3, \ldots, N$ and moving particle 1 around. As Halperin first noted,[57] this (and every) particle sees m vortices (analytic zeros of the wave function) located at the positions of the *other* particles. It is this 'binding' of zeros to particles which is the 'topological order' within the FQHE.[49] In order to relate this to the XY model topological order which we discussed previously and to bose condensation, we need to somehow map the problem onto bosons in zero magnetic field. As it stands, we have fermions and we have a finite density of vortices, all of the same sign.

The desired mapping can be achieved by taking advantage of the ambiguity of statistics in two dimensions.[59,60,61] We attach an odd integer $\nu^{-1} = m$ number of flux quanta to each particle which changes their statistics into that of bosons without any effect whatsoever on the physics. The new Hamiltonian after this 'singular gauge transformation' is

$$H = \frac{1}{2m} \sum_{j=1}^{N} (\boldsymbol{p}_j + \tfrac{e}{c} \boldsymbol{A}_j + \tfrac{e}{c} \boldsymbol{a}_j)^2 + V \, , \tag{7.6}$$

where the extra vector potential seen by the jth particle due to the flux tubes on the other particles obeys

$$\hat{\boldsymbol{z}} \cdot \nabla_j \times \boldsymbol{a}_j = -m\phi_0 \sum_{\substack{k \\ (k \neq j)}} \delta^2(\boldsymbol{r}_j - \boldsymbol{r}_k) \tag{7.7}$$

and the vector potential due to the uniform external field $\boldsymbol{B} = b\hat{\boldsymbol{z}}$ satisfies

$$\hat{\boldsymbol{z}} \cdot \nabla_j \times \boldsymbol{A}_j = B = \phi_0/2\pi\ell^2 \tag{7.8}$$

where $\phi_0 = hc/e$ is the Dirac flux quantum. The wave function for the composite bosons (electron plus flux tubes) is now permutation symmetric,

$$\Psi_{\text{boson}} = \prod_{i<j} \frac{(\bar{z}_i - \bar{z}_j)^m}{|z_i - z_j|^m} \, \Psi_{\text{fermion}} \, , \tag{7.9}$$

but physical observables are unaffected because the statistical phase has been replaced by a dynamical (Aharanov-Bohm) phase due to the interaction between the particles and the flux tubes.[61]

Within the mean field approximation for the statistical interaction,[62,49,63] we replace the density of flux tubes by its mean value

$$\langle \hat{\mathbf{z}} \cdot \nabla_j \times \mathbf{a}_j \rangle = -m\phi_0 \cdot n \tag{7.10}$$

and see that this average value precisely cancels the external B-field. Hence we have successfully mapped the problem onto bosons in zero (mean) field and anticipate that ODLRO will manifest itself in this representation. One way to see that this is so, at least for the specific case of the Laughlin variational wave function, is to note that the singular-gauge phase factor in eq(7.9) exactly cancels the phase part of the Laughlin wave function so Ψ_{boson} is purely real. The exact ground state is often close to, but not precisely equal to, the Laughlin state and Ψ_{boson} is then no longer purely real as will be discussed in more detail further below.

A different perspective can be obtained by examining the singular-gauge-transformed density matrix which expresses the density matrix for the composite bosons in terms of the original electron wave functions, *viz.*

$$\tilde{\rho}(\mathbf{r}, \mathbf{r}') \equiv N \int \prod_{k=2}^{N} d^2 r_k \, \overline{\Psi(\mathbf{r}', \mathbf{r}_2, \ldots, \mathbf{r}_N)} \, \exp\left\{ \frac{2\pi i}{\phi_0} \int_r^{r'} d\mathbf{u} \cdot \mathbf{a}(\mathbf{u}) \right\} \Psi(\mathbf{r}, \mathbf{r}_2, \ldots, \mathbf{r}_N) \tag{7.11}$$

Normally we think of the density matrix as having some uninteresting gauge dependence. However singular gauge transformations are quite peculiar. Since $\mathbf{a}_1$ depends on the positions of all the other particles (*cf.* Eq(7.7)), $\tilde{\rho}$ is a highly non-trivial N-body correlation function rather than a one-body object. Furthermore, since $\mathbf{a}_1$ has non-zero curl, the line integral in eq(7.11) is path-dependent. However the ambiguities are always multiples of 2π and so the exponential is well-defined.

If we view the m zeros seen by particle 1 bound to the positions of the other particles as vortices, then the vector potential term in $\tilde{\rho}$ represents m antivortices attached to each particle. In general, the exact ground state wave function has (in the FQHE state) the zeros bound near, but not precisely on, the positions of the particles. Then $\tilde{\rho}$ measures the ODLRO in a system of vortices and antivortices bound together in a manner analogous to that in the $2d$ XY model for $T < T_{\text{KT}}$. From the Laughlin plasma analogy[26,64] we know that when the zeros are bound to the particles, the particles 'see' each other as 2d Coulomb charges and hence form an incompressible fluid with a finite excitation gap.[5] Because of the incompressibility of charged bosons, even at zero temperature, the system exhibits only algebraic ODLRO.[49] Read[50] has developed a different formulation of the problem in which the order parameter exhibits true ODLRO.

For short-range repulsive interactions, the Laughlin state is the exact ground state,[31,30] but as the range of the interaction increases, the zeros are bound less and less tightly to the particles. Numerical calculations by Haldane and Rezayi[65] show that there is a critical point beyond which the vortices and antivortices unbind (ODLRO in $\tilde{\rho}$ disappears), the excitation gap collapses, and the overlap between the exact ground state and the Laughlin state also simultaneously collapses, thus confirming this picture of the topological ordering in the FQHE.

In summary, the singular gauge density matrix $\tilde{\rho}$ is an infinite-body correlation function which measures the hidden topological order represented by the binding of particles and zeros in the FQHE ground state. When $\tilde{\rho}$ exhibits power-law ODLRO, the system has the correlations of an incompressible Coulomb plasma and therefore exhibits the excitation gap which is the *sine qua non* of the FQHE. Related ideas and further developments may be found in refs[50,51, 66,41].

We turn now to an examination of the order which appears in the ground state of the $S = 1$ VBS spin chain considered earlier. Just as in the FQHE, the ground state of this system is a disordered liquid with a finite excitation gap and only short-range correlations

$$G_{ij} = \langle S_i^z S_j^z \rangle \sim \left(-\tfrac{1}{3}\right)^{|i-j|} . \tag{7.12}$$

We see from the sign alternation that there are antiferromagnetic correlations but the decay length is less than one lattice spacing. Just as in the FQHE however, there is a hidden multi-particle correlation function which is long-range. Den Nijs and Rommelse[67] discovered that if one examines any spin configuration which has non-zero weight in the ground state, and if one removes all the sites with $S^z = 0$, then the remaining sites have perfect antiferromagnetic order. That is every $S^z = +1$ site is followed by some number of $S^z = 0$ sites and then necessarily a $S^z = -1$ site, never another $S^z = +1$ site. A typical configuration is shown below:

$$\uparrow 0 \downarrow 0 0 \uparrow 0 0 0 \downarrow \uparrow \downarrow \uparrow 0 0 \downarrow 0 0 0 \uparrow 0 \downarrow 0 \uparrow 0 0 0 0 \downarrow$$

This order is invisible in the ordinary spin-spin correlation function because the 'random' insertion of $S^z = 0$ sites destroys the ordinary Néel order. However the order is visible in the following analog of the FQHE singular-gauge density matrix

$$\tilde{G}_{ij} \equiv - \langle\, S_i^z \prod_{k=i+1}^{j-1} e^{i\pi S_k^z}\, S_j^z \,\rangle\ . \tag{7.13}$$

The extra phase factor plays the role of the line integral of the vector potential and for $i \neq j$ we have the remarkable result $\tilde{G}_{ij} = \frac{4}{9}$, *independent* of the distance $|i - j|$. This N-body 'string correlation function' exhibits perfect long-range order in the VBS ground state.

Recall that in the FQHE we found perfect binding of zeros and particles in the Laughlin state. As the Hamiltonian was varied away from the exactly soluble point, the order became less perfect and eventually disappeared indicating the collapse of the excitation gap and the destruction of the FQHE. The same holds here. Consider the general $S = 1$ chain of eq(3.17). At the soluble VBS point we have $\lambda = \frac{1}{3}$ and perfect ODLRO in $\tilde{G}$. As λ is reduced towards zero (the Heisenberg point) the hidden long-range order is reduced but still survives. It has been demonstrated numerically[68,69] that $\tilde{G}_{ij}$ asymptotically reaches a value of about 85% of its VBS value. Since the ODLRO survives the transition from AKLT to Heisenberg, we expect that the Haldane gap[22] in the Heisenberg model has the same physical origins as the VBS gap. Evidently it costs a finite amount of energy to introduce a defect in the hidden topological order of the ground state. This gap is believed[70,23] to exist in the entire region from $-1 < \lambda < +1$. (The end points of this interval represent soluble models which are known to be gapless.[71,16])

The reason that the string correlation function decreases in amplitude as one moves away from the AKLT point is analogous to the vortices and antivortices beginning to fluctuate away from each other in the FQHE and in the XY model. The ground state away from the AKLT point begins to contain small amplitudes for finding pairs of $S^z = \pm 1$ sites in the 'wrong' order. However $\tilde{G}_{ij}$ does not decay exponentially as long as these pairs of defect sites are 'confined.' This is the analog of the algebraic ODLRO in the XY model when vortices and antivortices are confined.

The reason for the confinement of the defects can be understood in the following way.[72] Suppose that we break two of the valence bonds of the ground state leaving 'unpaired' spins on sites i and j. This corresponds to the defect pair in the topological order discussed above. Near the AKLT point, the cheapest way to do this involves keeping as many singlet bonds as possible. This can be done only by having alternating pairs of double and single bonds running in the line between i and j as shown here[73]

The pairs of sites without any singlets connecting them have some amplitude for having maximally parallel spins and hence cost some energy in the VBS Hamiltonian. Thus there is a 'string tension' and the energy cost of separating the defect pair rises linearly with the separation. This finite string tension 'confines' the defect pairs and sustains the hidden ODLRO. As we move away from the VBS point, the ground state contains defect pairs, but the string tension remains finite until the excitation gap collapses at $\lambda = \pm 1$.

To summarize, there are deep analogies between the FQHE and antiferromagnetic integer spin chains. Both have a disordered liquid ground state with a finite excitation gap. Both contain a hidden 'topological order' associated with the confinement of topological defects. The liquid state in both cases is robust in the sense that there is a finite region in parameter space around the exactly soluble point in which the gap and the hidden ODLRO remain finite.

ACKNOWLEDGEMENTS

We have benefited from discussions with many colleagues over the years. We would particularly like to thank F. D. M. Haldane, A. H. MacDonald, S. Kivelson, M. P. A. Fisher, and N. Read. DPA was supported in part by NSF grant DMR-8957993 and by a fellowship from the Alfred P. Sloan Foundation. SMG was supported in part by NSF grant DMR-8802383.

APPENDIX A: COMPENDIUM OF RESULTS FOR SPIN COHERENT STATES

The normalized coherent state $|\hat{\Omega}\rangle$ is defined by

$$|\hat{\Omega}\rangle = (2S!)^{-1/2}(\bar{u}a^{\dagger} + \bar{v}b^{\dagger})^{2S}|0\rangle \,, \tag{A.1}$$

where $u = \cos(\theta/2)\exp(+i\phi/2)$ and $v = \sin(\theta/2)\exp(-i\phi/2)$ are spinor coordinates, and a, b are Schwinger boson operators. The unit vector $\hat{\Omega}$ is defined by the polar angle θ and the azimuthal angle ϕ. The overlap between states is easily found to be

$$\langle\hat{\Omega}|\hat{\Omega}'\rangle = (u\bar{u}' + v\bar{v}')^{2S} \,. \tag{A.2}$$

If $|\Phi\rangle$ is an arbitrary spin-S state, then the coherent state overlap $\langle\hat{\Omega}|\Phi\rangle$ is given by

$$|\Phi\rangle = f(a^{\dagger}, b^{\dagger})|0\rangle \equiv \sum_{k=0}^{2S} f_k (a^{\dagger})^k (b^{\dagger})^{2S-k}|0\rangle \tag{A.3}$$

$$\Phi(u, v) = \langle\hat{\Omega}|f\rangle = \sqrt{2S!}\, f(u, v) \,.$$

Despite their nonorthogonality, there is a compact expression for the resolution of the identity operator,

$$1 = \frac{2S + 1}{4\pi}\int d\Omega\,|\hat{\Omega}\rangle\langle\hat{\Omega}| \,. \tag{A.4}$$

In calculating the matrix element of an operator $\hat{A}$ which commutes with the total spin, it is first convenient to normal order $\hat{A}$ so that it has the form

$$\hat{A} = \sum_{k,l,j} A_{klj}\hat{T}_{klj} \tag{A.5}$$

$$\hat{T}_{klj} = (a)^k (b)^l (a^{\dagger})^{k+j}(b^{\dagger})^{l-j} \,.$$

One then finds

$$\langle\Psi|\hat{A}|\Phi\rangle = \frac{2S + 1}{4\pi}\int d\Omega\,\langle\Psi|\hat{\Omega}\rangle\langle\hat{\Omega}|\hat{A}|\Phi\rangle, \tag{A.6}$$

which can be evaluated by

$$\langle\hat{\Omega}|\hat{T}_{klj}|\Phi\rangle = \left(\frac{\partial}{\partial u}\right)^k \left(\frac{\partial}{\partial v}\right)^l u^{k+j} v^{l-j}\, \Phi(u, v) \,. \tag{A.7}$$

Proceeding from eq(A.7), it can furthermore be shown that

$$\langle\Psi|\hat{T}_{klj}|\Phi\rangle = \left[\frac{(2S + k + l + 1)!}{(2S + 1)!}\right]\frac{2S + 1}{4\pi}\int d\Omega\,\overline{\Psi(u, v)}\,\bar{u}^k\bar{v}^l u^{k+j} v^{l-j}\,\Phi(u, v) \,. \tag{A.8}$$

Under a rotation $R(\xi; \hat{\mathbf{n}}) = \exp(i\xi\hat{\mathbf{n}}\cdot\mathbf{S})$, the coherent states transform as

$$\exp(i\,\xi\,\hat{\mathbf{n}}\cdot\mathbf{S})|\hat{\Omega}\rangle = |\hat{\Omega}'\rangle$$

$$\begin{pmatrix}\bar{u}'\\\bar{v}'\end{pmatrix} = \exp\left(i\tfrac{1}{2}\xi\,\hat{\mathbf{n}}\cdot\boldsymbol{\sigma}\right)\begin{pmatrix}\bar{u}\\\bar{v}\end{pmatrix} \,, \tag{A.9}$$

where $\boldsymbol{\sigma}$ are the Pauli matrices. Note also that $\Omega^{\alpha} = u_i\,\sigma_{ij}^{\alpha}\,\bar{u}_j$, with $u_1 \equiv u$ and $u_2 \equiv v$.

APPENDIX B: CYCLOTRON AND GUIDING CENTER VARIABLES

Consider an electron confined to the (x, y) plane in the presence of a uniform magnetic field $\boldsymbol{B} = -B\,\hat{\boldsymbol{z}}$. The one-body Hamiltonian is $H = (\boldsymbol{p} + \frac{e}{c}\boldsymbol{A})^2/2m$, where $A_\alpha = \frac{1}{2}B\varepsilon_{\alpha\beta}r^\beta$. The cyclotron and guiding center variables are defined by

$$\pi_\alpha = p_\alpha + \frac{e}{c}A_\alpha = p_\alpha + \frac{\hbar}{2\ell^2}\varepsilon_{\alpha\beta}r^\beta$$

$$\kappa_\alpha = p_\alpha - \frac{e}{c}A_\alpha = p_\alpha - \frac{\hbar}{2\ell^2}\varepsilon_{\alpha\beta}r^\beta$$

(B.1)

and satisfy

$$[\pi_\alpha, \pi_\beta] = i\hbar^2\ell^{-2}\varepsilon_{\alpha\beta} \qquad\qquad [\pi_\alpha, \kappa_\beta] = -i\hbar^2\ell^{-2}\varepsilon_{\alpha\beta} \tag{B.2}$$

with $[\pi_\alpha, \kappa_\beta] = 0$. One can now define the ladder operators

$$a = \frac{i\ell}{\sqrt{2}\hbar}\left(\pi_x + i\pi_y\right) = \sqrt{2}\,\ell\left(\bar{\partial} + \frac{z}{4\ell^2}\right)$$

$$b = \frac{i\ell}{\sqrt{2}\hbar}\left(\kappa_x - i\kappa_y\right) = \sqrt{2}\,\ell\left(\partial + \frac{\bar{z}}{4\ell^2}\right) ,$$

(B.3)

for which $[a, a^\dagger] = [b, b^\dagger] = 1$ and in terms of which $H = \hbar\omega_c\left(a^\dagger a + \frac{1}{2}\right)$, where $\omega_c = eB/mc$ is the cyclotron frequency. The operators a and $a^\dagger$ raise and lower the Landau level index; the macroscopic degeneracy of each Landau level is due to the fact that b and $b^\dagger$ do not appear in H.

Note that $a = \sqrt{2}\,\ell\,e^{-z\bar{z}/4\ell^2}\,\bar{\partial}\,e^{+z\bar{z}/4\ell^2}$, so the equation $a\,\psi(\boldsymbol{r}) = 0$ has the solution $\psi(\boldsymbol{r}) = f(z)\,e^{-z\bar{z}/4\ell^2}$, where $f(z)$ is any analytic function. This defines the lowest Landau level. A convenient basis on the plane is the angular momentum eigenstates, $f_m(z) = z^m$.

APPENDIX C: QUANTUM ROTORS AND BOSE OSCILLATORS

The commutation relations for the quantum rotor are

$$\left[\frac{1}{i}\frac{\partial}{\partial\theta}, e^{i\theta}\right] = e^{i\theta} . \tag{C.1}$$

The rotor wavefunctions are periodic in $\theta \to \theta + 2\pi$ and can therefore be written as functions of the variable $e^{i\theta}$: $\psi = \psi(e^{i\theta})$. The commutation relations of eq(C.1) are equivalent to those of the Bose oscillator,

$$[a, a^\dagger] = 1 \qquad [a^\dagger a, a^\dagger] = a^\dagger \tag{C.2}$$

through the correspondence

$$a^\dagger \longleftrightarrow e^{i\theta}$$

$$a \longleftrightarrow e^{-i\theta}\frac{1}{i}\frac{\partial}{\partial\theta} = \frac{\partial}{\partial(e^{i\theta})} . $$

(C.3)

However, the adjoint relationship of a and $a^\dagger$ is not respected by the usual metric for the rotor system,

$$\langle \phi \,|\, \psi \rangle_{\text{rotor}} = \int\limits_{-\pi}^{\pi}\frac{d\theta}{2\pi}\,\overline{\phi(e^{i\theta})}\,\psi(e^{i\theta}) . \tag{C.4}$$

Rather, we must define the inner product by

$$\langle \phi \,|\, \psi \rangle = \int\frac{d^2r}{\pi}\,e^{-r^2}\,\overline{\phi(re^{i\theta})}\,\psi(re^{i\theta})$$

$$= \int\frac{d^2r}{\pi}\,e^{-z\bar{z}}\,\overline{\phi(z)}\,\psi(z) ,$$

(C.5)

where $z \equiv r e^{i\theta}$. The adjointness is now recovered:

$$
\begin{aligned}
\langle \phi \,|\, a^{\dagger}\psi \rangle &= \int \frac{d^2r}{\pi}\, e^{-z\bar{z}}\, \overline{\phi(z)}\, z\, \psi(z) \\
&= \int \frac{d^2r}{\pi}\, e^{-z\bar{z}}\, \overline{\phi(z)} \cdot z\psi(z) \\
&= -\int \frac{d^2r}{\pi}\, \frac{\partial}{\partial \bar{z}}\left(e^{-z\bar{z}} \right) \overline{\phi(z)}\, \psi(z) \\
&= \int \frac{d^2r}{\pi}\, e^{-z\bar{z}}\, \overline{\phi'(z)}\, \psi(z) \\
&= \langle a\phi \,|\, \psi \rangle \ .
\end{aligned}
\tag{C.6}
$$

Of course, eq(C.5) is the usual inner product for Bose coherent states. This correspondence between bosons and rotors has been exploited by Fisher and Lee to elicit a correspondence between two-dimensional bosons and a bulk superconductor in a magnetic field.[41]

Consider now a many rotor wavefunction $\Psi(e^{i\theta_1}, \ldots, e^{i\theta_N})$. The mean Bose occupancy at site i is given by the expectation of $a_i^{\dagger} a_i$, which is to say

$$
\langle n_i \rangle = \langle \Psi \,|\, \frac{1}{i}\frac{\partial \Psi}{\partial \theta_i} \rangle \ .
\tag{C.7}
$$

Now there is absolutely no reason why this should be a positive quantity for a generic rotor wavefunction Ψ. The reason is that a given rotor may be in a state of negative angular momentum, which corresponds to an unphysical negative occupancy state for the boson. Thus, only non-negative angular momentum states are to be allowed the rotors if they are to correspond to a physical many boson state. One way to practically accomplish this is to 'wind up' all the rotors by multiplying Ψ by the product $\prod_j e^{iN_k\theta_k}$, where each N_k is very large. If we start with the wavefunction of the solvable rotor model of section VI, we have

$$
\Psi(e^{i\theta_1}, \ldots, e^{i\theta_1}) = \prod_k e^{iN_k\theta_k}\, \exp\left\{ \lambda \sum_{\langle ij \rangle} \cos(\theta_i - \theta_j) \right\}
\tag{C.8}
$$

which *still* contains components in which some particles have negative angular momentum. However, the amplitude of these unwanted configurations is fantastically small when we take each N_j to be thermodynamically large, which is appropriate in describing superconducting grains containing macroscopic numbers of Cooper pairs. In this case, we can argue that amplitude fluctuations can be neglected when evaluating the inner product of eq(C.5). The reason is that all averages are weighted by a product $\prod_k |z_k|^{2N_k} e^{-|z_k|^2}$, which for large N_k is sharply peaked about $|z_k|^2 = N_k$. The RMS fluctuations in $|z_k|^2$ are proportional to $\sqrt{N_k}$ — the usual story. If for example we were to evaluate the off-diagonal density matrix,

$$
\begin{aligned}
\langle \Psi \,|\, a_i^{\dagger} a_j \,|\, \Psi \rangle &= \langle a_j^{\dagger}\Psi \,|\, a_i^{\dagger}\Psi \rangle \\
&= \int \prod_k \left(\frac{d^2r_k}{\pi}\, e^{-r_k^2} \right) \bar{z}_j\, z_i \left| \Psi(\{r_l\}) \right|^2 ,
\end{aligned}
\tag{C.9}
$$

we would in a first approximation neglect amplitude fluctuations in the $|z_k|$ and evaluate the phase propagator $\langle e^{i\theta_i} e^{-i\theta_j} \rangle$. The interesting issue is then one of long-range *phase coherence* in Ψ; this is discussed in section VI.

REFERENCES

[1] We are indebted to C. L. Kane for suggesting the clever title of this article.

[2] R. P. Feynman, *Statistical Mechanics*, (Benjamin, New York, 1972).

[3] G. D. Mahan, *Many Particle Physics*, second edition (Plenum, New York, 1990) and references therein.

[4] D. P. Arovas, A. Auerbach, and F. D. M. Haldane, *Phys. Rev. Lett.* **60**, 531 (1988).

[5] S. M. Girvin, A. H. MacDonald, and P. M. Platzman, *Phys. Rev. Lett.* **54**, 581 (1985); *Phys. Rev. B* **33**, 2481 (1986); see also S. M. Girvin in Chapter 9 of Ref[28].

[6] To fix the sign of $|\,\bullet\!-\!\bullet\,\rangle$, we take $|\,\bullet\!-\!\bullet\,\rangle = \frac{1}{\sqrt{2}}(\,|\uparrow_{\rm left}\downarrow_{\rm right}\rangle - |\downarrow_{\rm left}\uparrow_{\rm right}\rangle\,)$.

[7] By "macroscopically distinct", we mean that the overlap between the two states decreases exponentially with the size of the system.

[8] C. K. Majumdar and D. K. Ghosh, *J. Math. Phys.* **10**, 1388 and 1399 (1969); *J. Phys. C* **3**, 911 (1970); P. M. van den Broek, *Phys. Lett.* **77A**, 261 (1980).

[9] W. J. Caspers, K. M. Emmett, and W. Magnus, *J. Phys. A* **17**, 2697 (1984).

[10] H. Bethe, *Z. Phys.* **71**, 205 (1931).

[11] F. D. M. Haldane, *Phys. Rev. Lett.* **61**, 1029 (1988).

[12] N.D. Mermin and H. Wagner, *Phys. Rev. Lett.* **17**, 1144 (1966).

[13] F. D. M. Haldane, *Phys. Rev. B* **25**, 4295 (1982).

[14] S. S. Shastry and B. Sutherland, *Phys. Rev. Lett.* **47**, 964 (1981). See also Ref[9].

[15] A handful of exact excited states are known. See C. K. Majumdar, K. Krishnan, and V. Mubayi, *J. Phys. C* **5**, 2896 (1972) and also W. J. Caspers and W. Magnus, *Phys. Lett.* **88A**, 103 (1982).

[16] P. Kulish and E. Sklyanin, in *Lecture Notes in Physics* (Springer-Verlag, Berlin, 1982), Vol. 151, p. 61; P. Kulish, N. Yu. Reshetikhin and E. Sklyanin, *Lett. Math. Phys.* **5**, 393 (1981); L. Takhtajan, *Phys. Lett.* **90A**, 479 (1982); J. Babudjian, *Nucl. Phys.* **B125**, 317 (1983).

[17] W. P. Su, J. R. Schrieffer, and A. J. Heeger, *Phys. Rev. Lett.* **42**, 1698 (1979).

[18] One can say that $S_k^\alpha\,|\pm\,\rangle$ creates a state transforming by the adjoint representation of SU(2).

[19] F. D. M. Haldane, unpublished.

[20] Ian Affleck, T. Kennedy, E. H. Lieb, and H. Tasaki, *Phys. Rev. Lett.* **59**, 799 (1987); *Comm. Math. Phys.* **115**, 477 (1988).

[21] D. P. Arovas, *Phys. Lett.* **137A**, 431 (1989).

[22] F. D. M. Haldane, *Phys. Lett.* **93A**, 464 (1983); *Phys. Rev. Lett.* **50**, 1153 (1983).

[23] J. Sólyom, *Phys. Rev. B* **36**, 8642 (1987).

[24] T. Kennedy, *J. Phys.: Cond. Matter* **2**, 5737 (1990).

[25] R. Jastrow, *Phys. Rev.* **98**, 1479 (1955).

[26] R. B. Laughlin, *Phys. Rev. Lett.* **50**, 1395 (1983) and also in Chapter 7 of Ref[28].

[27] E. J. Neves and J. F. Perez, *Phys. Lett.* **114A**, 331 (1986).

[28] *The Quantum Hall Effect* 2nd ed., edited by R. E. Prange and S. M. Girvin (Springer-Verlag, New York, 1990).

[29] J. M. Caillol, D. Levesque, J. J. Weis, and J. P. Hansen, *J. Stat. Phys.* **28**, 2 (1985); M. Baus and J. P. Hansen, *Phys. Rep.* **59**, 1 (1980); B. Jancovici, *Phys. Rev. Lett.* **46**, 386 (1981).

[30] F. D. M. Haldane, *Phys. Rev. Lett.* **51**, 605 (1983) and also in Chapter 8 of Ref[28].

[31] See also S. A. Trugman and S. Kivelson, *Phys. Rev. B* **31**, 5280 (1985).

[32] W. Kohn, *Phys. Rev.* **123**, 1242 (1961).

[33] E. H. Rezayi, private communications.

[34] I. Affleck, *Phys. Rev. Lett.* **54**, 966 (1985); I. Affleck and J. B. Marston, *Phys. Rev. B* **37**, 3774 (1988).

[35] D. P. Arovas and A. Auerbach, *Phys. Rev. B* **38**, 316 (1988).

[36] N. Read and S. Sachdev, *Nucl. Phys.* **B316**, 609 (1989); *Phys. Rev. Lett.* **62**, 1694 (1989); *Phys. Rev. B* **42**, 4568 (1990).

[37] N. Read and D. M. Newns, *J. Phys. C* **16**, 3273 (1983); N. E. Bickers, p59, 845, 1987.

[38] I. Affleck, D. P. Arovas, J. B. Marston, and D. A. Rabson, "SU($2n$) Quantum Antiferromagnets with Exact C-Breaking Ground States" (accepted for publication in *Nucl. Phys. B*).

[39] B. Sutherland, *J. Math. Phys.* **12**, 246 (1971); *J. Math. Phys.* **12**, 251 (1971); *Phys. Rev. A* **5**, 1372 (1972).

[40] C. L. Kane, S. Kivelson, D. H. Lee, and S. C. Zhang, *Phys. Rev. B* **43**, 3255 (1991).

[41] D. H. Lee and M. P. A. Fisher, *Phys. Rev. Lett.* **63**, 903 (1989); M. P. A. Fisher and D. H. Lee, *Phys. Rev. B* **39**, 2756 (1989); X. G. Wen and A. Zee, *Int. J. Mod. Phys. B* **4**, 437 (1990).

[42] Min-Chul Cha, Matthew P. A. Fisher, S. M. Girvin, Mats Wallin, and A. Peter Young, *Phys. Rev. B* (October 1, 1991).

[43] Nandini Trivedi and D. M. Ceperley, *Phys. Rev. B* **40**, 2737 (1989); *Phys. Rev. B* **41**, 4552 (1990).

[44] L. Reatto and G. V. Chester, *Phys. Rev.* **155**, 88 (1967).

[45] C. L. Kane, private communication.

[46] Phillipe Nozières and David Pines, *The Theory of Quantum Liquids, II* (Addison-Wesley, Reading, MA, 1989).

[47] Petter Minnhagen, *Rev. Mod. Phys.* **59**, 1001 (1987).

[48] S. M. Girvin, in Chapter 10 of Ref[28].

[49] S. M. Girvin and A. H. MacDonald, *Phys. Rev. Lett.* **58**, 1252 (1987).

[50] N. Read, *Phys. Rev. Lett.* **62**, 86 (1989).

[51] S. C. Zhang, T. H. Hansson, and S. Kivelson, *Phys. Rev. Lett.* **62**, 82 (1989).

[52] J. M. Kosterlitz and D. J. Thouless, *J. Phys. C* **5**, L124 (1972).

[53] V. L. Berezinskii, Zh. Eksp. Teor. Fiz. **61**, 1144 (1971) [Sov. Phys. JETP **34**, 610 (1972)].

[54] J. V. José, L. P. Kadanoff, S. Kirkpatrick, and D. R. Nelson, *Phys. Rev. B* **16**, 1217 (1977).

[55] N. Byers and C. N. Yang, *Phys. Rev. Lett.* **7**, 46 (1961); C. N. Yang, *Rev. Mod. Phys.* **34**, 694 (1962).

[56] A. H. MacDonald and S. M. Girvin, *Phys. Rev. B* **38**, 6295 (1988).

[57] B. I. Halperin, Helv. Phys. Acta **56**, 75 (1983).

[58] P. W. Anderson (private communication).

[59] J. M. Leinaas and J. Myerheim, *Il Nuovo Cimento* **37B**, 1 (1977).

[60] F. Wilczek, *Phys. Rev. Lett.* **48**, 11144 (1982); *ibid.* **49**, 957 (1982).

[61] G. S. Canright and S. M. Girvin, *Science* **247**, 1197 (1990).

[62] D. P. Arovas, J. R. Schrieffer, F. Wilczek, and A. Zee, *Nucl. Phys.* **B251**, 117 (1985).

[63] R. B. Laughlin, *Phys. Rev. Lett.* **60**, 2677 (1988); *Science* **242**, 525 (1988).

[64] S. M. Girvin, in Appendix I of Ref[28].

[65] E. H. Rezayi and F. D. M. Haldane, *Phys. Rev. Lett.* **61**, 1985 (1988).

[66] J. K. Jain, *Phys. Rev. B* **41**, 7653 (1990).

[67] K. Rommelse and M. den Nijs, *Phys. Rev. Lett.* **59**, 2578 (1987); M. Den Nijs and K. Rommelse, *Phys. Rev. B* **40**, 4709 (1989).

[68] S. M. Girvin and D. P. Arovas, *Physica Scripta* **T27**, 156 (1989).

[69] Y. Hatsugai and M. Kohmoto, "Numerical Study of the Hidden Antiferromagnetic Order in the Haldane Phase", ISSP preprint, 1991.

[70] I. Affleck, *Nucl. Phys.* **B265[FS15]**, 409 (1986).

[71] B. Sutherland, *Phys. Rev. B* **12**, 3795 (1975).

[72] N. Read (private communication).

[73] Shoucheng Zhang and Daniel P. Arovas, *Phys. Rev. B* **40**, 2708 (1989).

INHOMOGENEOUS PARQUET THEORY

Roger Alan Smith and Hong-Wei He

Center for Theoretical Physics, Department of Physics

Texas A & M University, College Station, Texas 77843, U.S.A.

1. Introduction

The parquet summation of Feynman diagrams provides a powerful approach for calculating the properties of many-particle systems. For inhomogeneous boson systems, we have showed previously[1] that the parquet theory, with suitable approximations, could lead to the inhomogeneous variational theory of Krotscheck, Qian and Kohn[2] (KQK). In this paper, we focus more clearly on the nature of the approximations which were implicitly made in the previous work and discuss in more detail the nature of the single-particle self-energy which is a natural part of the parquet theory which is less clearly manifested in the corresponding variational theory. At the same time, we introduce a new notation which enables rather transparent connections between the diagrammatic quantities and equations of the perturbation theory and the corresponding entities of the variational theory.

We take for granted that the bosons system can be treated as the high-degeneracy limit of a fermion system to order $\frac{1}{N}$, where N is the number of particles, as proposed by Brandow[3]. This simplifies the analysis by allowing the use of the fermion Green's functions.

The Hamiltonian describing the system with spin-independent forces may be written in second-quantized form as

$$
H = H_0 + \lambda H'
$$
$$
H_0 = \int d^3 r_i \psi_\sigma^\dagger(i)(-\frac{1}{2}\nabla_i^2)\psi_\sigma(i) + \int d^3 r_i \psi_\sigma^\dagger(i)U_{\text{ext}}(i)\psi_\sigma(i) \tag{1.1}
$$
$$
H' = \frac{1}{2}\int d^3 r_i \int d^3 r_j \psi_\sigma^\dagger(i)\psi_\tau^\dagger(j)V(i,j)\psi_\tau(j)\psi_\sigma(i).
$$

The two-body interaction H' is to be treated as a perturbation of strength λ, with the physical problem corresponding to $\lambda = 1$.

2. The One-Body Green's Function

In perturbation theory, the one-body Green's function is formally expanded in terms of the non-interacting one-body Green's function,

$$
[G^0(\omega)]^{-1} = \omega - H_0. \tag{2.1}
$$

Recent Progress in Many-Body Theories, Vol. 3,
Edited by T.L. Ainsworth et al., Plenum Press, New York, 1992

The Dyson equation relates the full Green's function to the self-energy with

$$\left[G(\omega)\right]^{-1} = \omega - H_0 - \Sigma(\omega).$$ (2.2)

In the *instantaneous approximation*, Σ is independent of ω and it is convenient to use basis states eigenstates $|\alpha\rangle$ with eigenvalues $\omega_\alpha lpha$ of the operator

$$H_0' = H_0 + \Sigma.$$ (2.3)

In this basis, the single-particle Green's function takes the simple form

$$G_{\alpha,\alpha'}(\omega) = \delta_{\alpha,\alpha'}\left[\frac{p(\alpha)}{\omega - \epsilon_\alpha + i\eta} + \frac{h(\alpha)}{\omega - \epsilon_\alpha - i\eta}\right].$$ (2.4)

An instantaneous self-energy $\Sigma(i,j)$ is Hermitian but not necessarily local in coordinate space. The familiar Hartree-Fock approximation, for example, features a local Hartree term and a non-local Fock term. Of course for systems with strong short-range repulsions, the Hartree-Fock self-energy is totally inadequate, since the short-range repulsion must be screened. For the parquet theory to be useful in discussing such systems, a generalization of the Hartree-Fock approximation is needed. Our parquet theory retains the instantaneous approximation but at the outset further specification of the form of Σ is largely unnecessary. At the end, the specification of Σ can be made in two ways: by requiring the ground-state energy to be a minimum or by requiring the self-energy to be consistent (in some sense) with the self-energy actually constructed from the parquet two-body vertex. The compatibility of these methods will be discussed later.

3. Parquet Summation for the Full Two-Body Vertex

The next fundamental diagrammatic quantity is the two-body Green's function. It consists of the free direct and exchange diagrams plus Γ, the direct part of the two-body vertex. Γ can be expressed as a sum of two-body reducible and irreducible diagrams. A diagram for Γ is reducible in the s-channel, t-channel, or u-channel if cutting two internal Green's functions can separate the top from the bottom, the left from the right, or one diagonal from the other. Otherwise, the diagram is irreducible. The simplest irreducible diagram is the bare interaction. All other irreducible diagrams are at least of fourth order in the bare interaction. The s-channel diagrams are particle-particle ladder diagrams, t-channel diagrams include chain diagrams as well as vertex corrections on either side of a simpler vertex. The u-channel diagrams are a kind of particle-hole ladder. The exact parquet theory gives Γ as a sum of irreducible diagrams given by[4-5]

$$\begin{aligned}
\Gamma &= \Gamma_0 + \Gamma l \Gamma_0 + \Gamma_0 \, r\, \Gamma + \Gamma l \Gamma_0 \, r\, \Gamma \\
\Gamma_0 &= I + L + U + C \\
L &= (\Gamma - L)\, s\, \Gamma \\
U &= (\Gamma - U)\, u\, \Gamma \\
C &= (I + L + U)\, b\, (I + L + U + C).
\end{aligned}$$ (3.1)

I, L, C and U represent sums of irreducible diagrams, ladders, chains, and particle-hole ladders, respectively. The elementary operators s, u, c, l and r are shown in fig. 1, and the composite operator b is defined by $\alpha\, b\, \beta = \alpha\, c\, \beta + \alpha\, c\,(\Gamma l \beta)$.

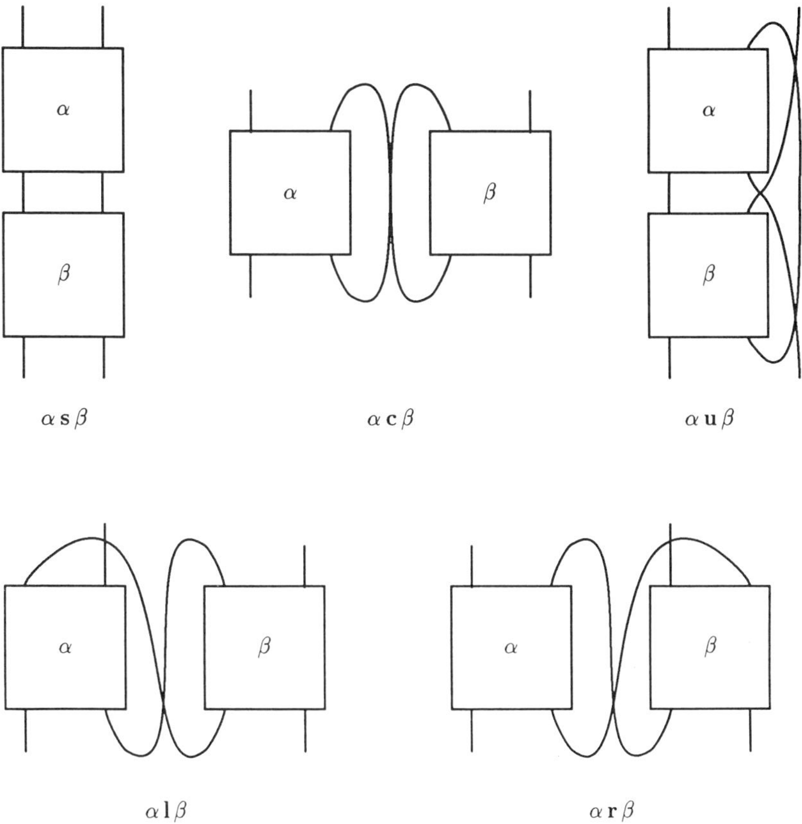

$$\alpha\,\mathbf{s}\,\beta \qquad \alpha\,\mathbf{c}\,\beta \qquad \alpha\,\mathbf{u}\,\beta$$

$$\alpha\,\mathbf{l}\,\beta \qquad \alpha\,\mathbf{r}\,\beta$$

Fig. 1. The different kinds of two-body reducibility.

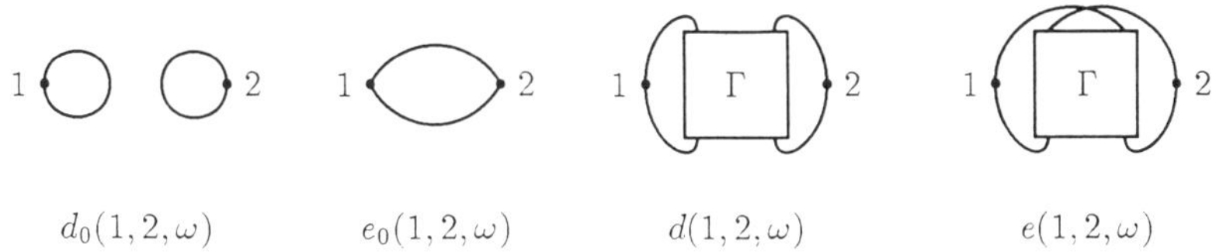

$$d_0(1,2,\omega) \qquad e_0(1,2,\omega) \qquad d(1,2,\omega) \qquad e(1,2,\omega)$$

Fig. 2. Contributions to the two-particle density and response function.

If the exact single-particle Green's function is used and if the complete set of irreducible diagrams is used as I, the parquet equations are an exact representation of the function Γ.

A number of useful physical quantities can be formed from the two-body vertex by closing off the left-hand Green's functions at $\mathbf{r}_i$ and the right-hand Green's functions at $\mathbf{r}_j$ as shown in fig. 2.

These diagrams are to be evaluated following the standard rules. The diagrams d_0 and e_0 come from the direct and exchange contributions of the free Green's function, while the diagrams d and e include all other contributions. While closing off the two-body vertex to form the response function removes some of the generality from the full two-body Green's function, the resulting quantities depend on only two spacial variables instead of four and on only one frequency variable. The polarization propagator χ denotes exactly the same object as the Π of Fetter and Walecka [6] and is related to the dynamic structure function S and the radial distribution function g.

The dynamic free polarization propagator χ_0 and full polarization propagator χ are given by

$$\chi(i,j,\omega) = -i\big[e_0(i,j,\omega) + d(i,j,\omega) + e(i,j,\omega)\big]. \tag{3.2}$$

A subscript 0 on χ indicates that only the free term (first on the right) is included. It is convenient to introduce *reduced* polarization propagators

$$\overline{\chi}(i,j,\omega) = \nu^{-2}\phi^{-2}(i)\phi^{-2}(j)\chi(i,j,\omega). \tag{3.3}$$

The dynamic free and full structure function S_0 and S are related to the retarded polarization propagators, which through analytic continuation can be expressed in terms of the imaginary parts of $\overline{\chi}_0$ and $\overline{\chi}$ as

$$S(i,j,\omega) = -\frac{1}{\pi}\,\mathrm{Im}\,\overline{\chi}(i,j,\omega). \tag{3.4}$$

The $S(i,j,\omega)$ is related to the experimentally measurable $S(\mathbf{k},\omega)$ by the double Fourier transform

$$S(\mathbf{k},\omega) = \frac{1}{N}\int d^3 r_i \int d^3 r_j e^{-i\mathbf{k}\cdot\mathbf{r}_i}e^{i\mathbf{k}\cdot\mathbf{r}_j}\rho(i)\rho(j)S(i,j,\omega), \tag{3.5}$$

which is normalized to 1.

The static functions are defined by

$$\chi(i,j) = i\int\frac{d\omega}{2\pi}\chi(i,j,\omega)$$
$$\overline{\chi}(i,j) = i\int\frac{d\omega}{2\pi}\overline{\chi}(i,j,\omega) \tag{3.6}$$
$$S(i,j) = \int_0^\infty d\omega\, S(i,j,\omega)$$

Since $\chi(i,j,\omega)$ is even in ω, one may also obtain the static structure function directly as

$$S(i,j) = \mathrm{Re}\,\overline{\chi}(i,j). \tag{3.7}$$

The radial distribution function $g(i,j)$ is given by

$$g(1,2) = \rho^{-1}(i)\rho^{-1}(j)\int\frac{d\omega}{2\pi}\big[d_0(i,j,\omega) + e_0(i,j,\omega) + d(i,j,\omega) + e(i,j,\omega)\big]$$
$$- \rho^{-1}(i)\delta(\mathbf{r}_i - \mathbf{r}_j). \tag{3.8}$$

This differs from the function g defined by Fetter and Walecka [6] by two factors of density and the inclusion of the uncorrelated part of the two-body density. For reference, the two-body density is $\rho_2(i,j) = \rho(i)\rho(j)g(i,j)$.

4. Parquet Summation for the Boson Γ

The *local approximate boson parquet* equations are defined by

$$\Gamma = V + L + C$$
$$L = W_L \, \mathbf{s} \, (W_L + L)$$
$$C = W_C \, \mathbf{c} \, (W_C + C) \tag{4.1}$$
$$W_L = V + C_{\mathrm{loc}}$$
$$W_C = V + L_{\mathrm{loc}}.$$

The L and C are nonlocal two-body vertex terms constructed from local W_C and W_L. The local approximations C_{loc} and L_{loc} are defined so that they give the same contribution to the static structure function as the full C or L driven by local functions, as illustrated in fig. 3.

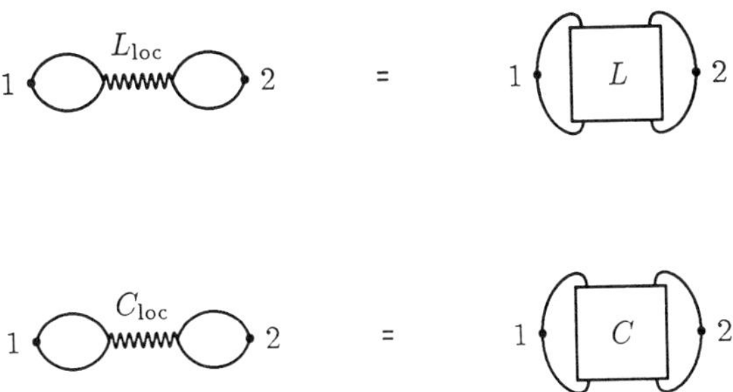

Fig. 3. The relationships defining C_{loc} and L_{loc}.

An important consequence of this choice is that the full radial distribution function and static structure function may be computed **either** from the RPA ring sum driven by W_L **or** from the ladder sum driven by W_C. Fig. 4 shows the contributions to the static structure function in the *local approximate boson parquet* theory.

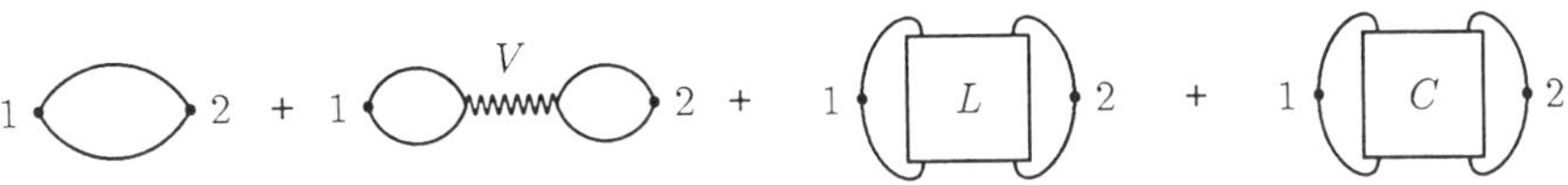

Fig. 4. Contributions to the static structure function.

Figs. 5 and 6 then show the interpretation of this static structure function as either a ladder sum driven by a local interaction $W_L = V + C_{\mathrm{loc}}$ (fig. 5) or as an RPA sum driven by an interaction $W_C = V + L_{\mathrm{loc}}$ (fig. 6).

The interpretation as a ladder sum ensures that the short-range behavior will be taken into account properly, while the interpretation as an RPA sum gives the

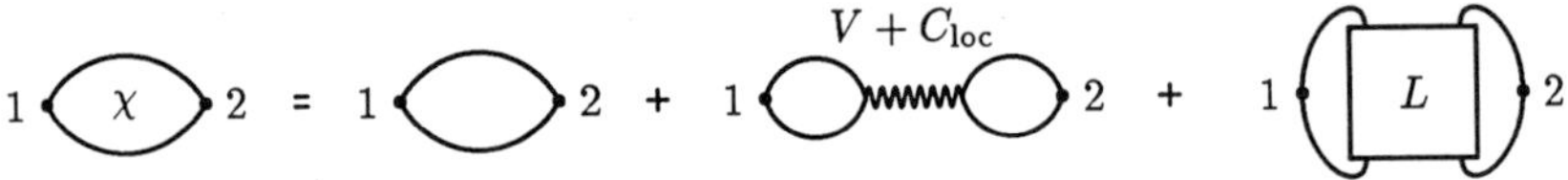

Fig. 5. The structure function viewed as coming from a ladder sum.

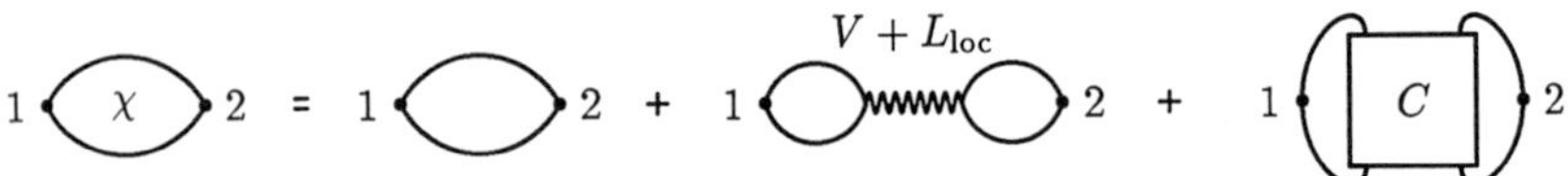

Fig. 6. The structure function viewed as coming from an RPA sum.

proper screening at long distances. This dual interpretation is a delightful result of the self-consistent sum of both ring and ladder diagrams. It also leads to an important simplification in deriving equations to be solved numerically. Although these may seem like strong truncations and approximations, the resulting theory is quite powerful.

For homogeneous systems, the corresponding theory turns out to be equivalent to the optimized boson hypernetted-chain variational theory[1]. Extensions of the theory have been made to deal with spin-dependent forces[7]. Briefly, the strong points of the theory are that it generates automatically the screening need for both strong short-range repulsions and long-range forces, that it tells you if you are trying to generate solutions in regions of physical unstability, and that is easy and efficient to implement.

For inhomogeneous systems functions $O(i,j)$ in the theory are no longer simply functions of the magnitude of the displacement $|\mathbf{r}_i - \mathbf{r}_j|$ and the non-interacting Green's function has to be replaced a suitable one-body Green's function. For homogeneous systems, a *local instantaneous* self-energy is just a constant, and so the one-body Green's function used is the non-interacting Green's function with the frequency offset by a constant, which means that in practical calculations, one can forget the self-energy and just use the non-interacting Green's function.

In discussing the nature of the approximations, we have already had to introduce most of the important functions which appear naturally in the parquet perturbation theory. The next order of business is to establish the formal methods by which they will actually be calculated. The next two sections establish a notation which relies on useful relationships between functions $O(i,j)$ and operators O acting in a one-particle or a two-particle space. With this notation, one finds relatively short expressions in deriving the theory. At the end, the notation also is very convenient for showing the correspondence between this theory, the inhomogeneous variational theory, and the homogeneous parquet theory.

5. One-Body Notation and Operators

The important single-particle states are the kets $|\alpha\rangle$ which are eigenstates of the operator H_0' with eigenvalues ϵ_α and the pseudo-kets $|i\rangle$ which denote pseudo-states localized at $\mathbf{r}_i$. The ket $|0\rangle$ always refers to the ground state of H_0'. The overlaps of these basis states give the single-particle wavefunctions

$$\phi_\alpha(i) = \langle i|\alpha\rangle. \tag{5.1}$$

The wavefunction $\phi(i)$ implicitly refers to the ground-state wavefunction $\langle i|0\rangle$. The operator

$$K = H_0' - \epsilon_0. \tag{5.2}$$

is useful because the difference $e_\alpha = \epsilon_\alpha - \epsilon_0$ is often what appears in energy denominators. The operator ϕ is defined by its action in the coordinate representation

$$\phi|i\rangle = \phi(i)|i\rangle. \tag{5.3}$$

If Σ and hence H_0' are local in coordinate space,

$$\begin{aligned}
\langle i|K|\psi\rangle &= -\frac{1}{2}\phi^{-1}(i)\nabla_i \cdot \phi^2(i)\nabla_i\phi^{-1}(i)\,\langle i|\psi\rangle \\
&= -\frac{1}{2}\phi(i)D(i)\phi^{-1}(i)\langle i|\psi\rangle,
\end{aligned} \tag{5.4}$$

where

$$D(i) = \phi^{-2}(i)\nabla_i \cdot \phi^2(i)\nabla_i \tag{5.5}$$

is the operator defined by KQK for the inhomogeneous variational theory. The hole operator h is defined by

$$h = |0\rangle\langle 0|. \tag{5.6}$$

The dimensions of the matrix element $\langle i|O|j\rangle$ are those of the operator divided by a volume. With each function $O(i,j)$, we associate a one-body operator O, writing

$$O \Leftrightarrow O(1,2) \tag{5.7}$$

to mean

$$O(i,j) = \nu^{-1}\langle i|\phi^{-1}O\phi^{-1}|j\rangle. \tag{5.8}$$

With this definition, the operator O and the function $O(i,j)$ have the same dimensions. Furthermore, a real symmetric function $O(i,j) = O(j,i)$ is associated with a Hermitian operator. Particular choices of the operator are of interest. For the unit operator,

$$1 \Leftrightarrow \nu^{-1}\phi^{-2}(i)\delta(\mathbf{r}_i - \mathbf{r}_j) = \rho^{-1}(i)\delta(\mathbf{r}_i - \mathbf{r}_j), \tag{5.9}$$

and for the hole operator,

$$\nu h \Leftrightarrow 1. \tag{5.10}$$

The operator $O = O_1 O_2$ is associated with the density-weighted convolution given by

$$O_1 O_2 \Leftrightarrow O(i,j) = \int d^3r_k\rho(k)O_1(i,k)\,O_2(k,j). \tag{5.11}$$

If O_1 and O_2 are Hermitian, then the product diagonal elements $O(i,i)$ are the same for the operator products O_1O_2 and O_2O_1. The product of K with an operator O has the useful association

$$KO \Leftrightarrow -\frac{1}{2}D(1)\,O(1,2). \tag{5.12}$$

6. Two-body Notation and Operators

The important two-particle states are the direct product states $|ij\rangle = |i\rangle \otimes |j\rangle$ or $|\alpha\beta\rangle = |\alpha\rangle \otimes |\beta\rangle$. No symmetrization is implied, so that $\langle ij|kl\rangle = \langle i|k\rangle\langle j|l\rangle$. The two-particle hole and particle projection operators are

$$h|\alpha\beta\rangle = (h|\alpha\rangle) \otimes (h|\beta\rangle)$$
$$p = 1 - h. \tag{6.1}$$

The two-particle energy operator K is defined by

$$K|\alpha\beta\rangle = (e_\alpha + e_\beta)|\alpha\beta\rangle, \tag{6.2}$$

and its effective inverse is

$$\Xi|\alpha\beta\rangle = -\frac{1}{e_\alpha + e_\beta}\, p|\alpha\beta\rangle. \tag{6.3}$$

With these definitions,

$$\langle ij|K|\psi\rangle = -\frac{1}{2}\left[\phi(i)D(i)\phi^{-1}(i) + \phi(j)D(j)\phi^{-1}(j)\right]\langle ij|\psi\rangle \tag{6.4}$$

and

$$\langle ij|K|00\rangle = 0. \tag{6.5}$$

Finally, a local interaction W is diagonal in the coordinate representation, with

$$W|ij\rangle = W(i,j)|ij\rangle. \tag{6.6}$$

A non-local interaction is not diagonal in the coordinate representation. The relationships between other functions of two variables and two-body operators will be defined as they arise.

In spite of the striking typographic similarity, there is little chance of confusing the one-body operator W with the two-body operator W. The context should make it clear which operator is meant. In general, the one-body operators will be used in deriving the RPA equations and expressing the energy functional, while the two-body operators will be used in deriving the ladder equations. The forms of the operators acting in the two spaces is generally quite distinctive.

7. Free and First-Order Diagrams

The derivation of the results in this section will be presented elsewhere[8]. The free polarizability operator is given by

$$\overline{\chi}_0(\omega) = \frac{2K}{\omega^2 - (K - i\eta)^2} \tag{7.1}$$
$$\overline{\chi}_0 = S_0 = 1.$$

The result of evaluating the first-order direct diagram is expressed in operator notation as

$$W = -\frac{1}{2}[K\overline{\chi}_W^{(1)} + \overline{\chi}_W^{(1)}K] = -\frac{1}{2}[KS_W^{(1)} + S_W^{(1)}K]. \tag{7.2}$$

The exchange diagram is ignored in the boson approximation.

8. The RPA Summation of Ring Diagrams

In the present notation, we obtain the RPA result for a driving term W_C

$$\overline{\chi}_{RPA}(\omega) = K^{\frac{1}{2}} \frac{2}{\omega^2 - \left[K^2 + 2K^{\frac{1}{2}} W_C K^{\frac{1}{2}} - i\eta\right]} K^{\frac{1}{2}}. \tag{8.1}$$

The frequency integral needed to compute the static polarization propagator $\overline{\chi}_{RPA}$ is easy to carry out in a basis given by the eigenstates of the operator in square brackets in eq. 8.1. The result is a Hermitian operator, and so

$$S_{RPA} = \overline{\chi}_{RPA} = K^{\frac{1}{2}} \left[K^2 + 2K^{\frac{1}{2}} W_C K^{\frac{1}{2}}\right]^{-\frac{1}{2}} K^{\frac{1}{2}}. \tag{8.2}$$

We define operators X and N by

$$\begin{aligned}
X &= 1 - S_{RPA}^{-1} \\
N &= S_{RPA} - X - 1.
\end{aligned} \tag{8.3}$$

Solving eq. 8.2 for W_C, we obtain the operator relation

$$W_C = -\frac{1}{2}\left[X\,K + K\,X - X\,K\,X\right]. \tag{8.4}$$

From fig. 6, it is apparent that

$$S_{RPA} = 1 + S_{W_C}^{(1)} + S_C, \tag{8.5}$$

where $S_{W_C}^{(1)}$ is the first-order contribution and S_C is the contribution to the static structure function from the chain sum C. The *local approximate boson parquet* theory is defined, as illustrated in fig. 3, by the relation

$$S_{C_{\text{loc}}}^{(1)} = S_C \tag{8.6}.$$

The full static structure function may then be expressed as

$$S_{RPA} = 1 + S_{W_C}^{(1)} + S_{C_{\text{loc}}}^{(1)} = 1 + S_{\Gamma_{\text{loc}}}^{(1)}, \tag{8.7}$$

with

$$\Gamma_{\text{loc}} = W_C + C_{\text{loc}}. \tag{8.8}$$

From eq. 7.2, Γ_{loc} is related to $S_{\Gamma_{\text{loc}}}^{(1)}$ by

$$\Gamma_{\text{loc}} = -\frac{1}{2}\left[K\,S_{\Gamma_{\text{loc}}}^{(1)} + S_{\Gamma_{\text{loc}}}^{(1)}\,K\right]. \tag{8.9}$$

Using eq. 8.7 to replace $S_{\Gamma_{\text{loc}}}^{(1)}$ by $S_{RPA} - 1$, we find

$$\begin{aligned}
\Gamma_{\text{loc}} &= -\frac{1}{2}\left[K\,(S_{RPA} - 1) + (S_{RPA} - 1)\,K\right] \\
&= -\frac{1}{2}\left[K\,(N + X) + (N + X)\,K\right].
\end{aligned} \tag{8.10}$$

Combining this with eq. 8.4, we find

$$C_{\text{loc}} = \Gamma_{\text{loc}} - W_C = -\frac{1}{2}[K\,N + N\,K + X\,K\,X]. \qquad (8.11)$$

The local driving term for the ladder equations is then

$$W_L = V + C_{\text{loc}}. \qquad (8.12)$$

Finally, the equation of motion for the radial distribution function $g(i,j)$ comes directly from the first line of eq. 8.10. The definitions and results from eq. 5.8 and eq. 5.12 enable it to be expressed in the form

$$\frac{1}{4}(D(i) + D(j))g_{RPA}(i,j) = W_C(i,j) + C_{\text{loc}}(i,j). \qquad (8.13)$$

This equation for $g_{RPA}(i,j)$ is valid in RPA sum for any driving term W_C. When used in the self-consistent parquet theory where $W_C = V + L_{\text{loc}}$, then $g_{RPA}(i,j)$ is also the full $g(i,j)$. Then $\Gamma_{\text{loc}} = V + L_{\text{loc}} + C_{\text{loc}}$ is the local approximation to the whole two-body vertex.

9. Summation of Ladders

The result of carrying out the ladder-diagram contribution to the distribution function for a driving term W_L is

$$g_L(i,j) = \phi^{-2}(i)\phi^{-2}(j)\langle ij|\frac{1}{1 - p\Xi W_L}|00\rangle\langle 00|\frac{1}{1 - W_L\Xi p}|ij\rangle. \qquad (9.1)$$

With the definitions

$$1 + \overline{\Gamma} = \frac{1}{1 - p\Xi W_L}$$

$$f(i,j) = \phi^{-1}(i)\phi^{-1}(j)\langle ij|1 + \overline{\Gamma}|00\rangle, \qquad (9.2)$$

we see that the ladder sum gives a distribution function

$$g_L(i,j) = f^2(i,j) \qquad (9.3)$$

which is a perfect square. For any W_L, the ladder distribution function is therefore non-negative. In the self-consistent parquet theory, where $W_L = V + C_{\text{loc}}$, the ladder distribution function is also the full distribution function.

The equation of motion for $f(1,2)$ is obtained by rewriting the first of eqs. 9.2 as

$$\overline{\Gamma} = p\Xi W_L(1 + \overline{\Gamma}). \qquad (9.4)$$

Multiplying this on both sides by $\Xi^{-1} = -K$ and taking the matrix element between states $\langle ij|$ and $|00\rangle$, one gets

$$-\langle ij|K(1 + \overline{\Gamma})|00\rangle = W_L(i,j)\langle ij|(1 + \overline{\Gamma})|00\rangle. \qquad (9.5)$$

In terms of $f(i,j)$, using eq. 5.5, this gives

$$\frac{1}{2}[D(i) + D(j)]f(i,j) = W_L(i,j)f(i,j). \qquad (9.6)$$

This is the ladder equation of motion for the square-root of the distribution function for any choice of W_L. When parquet self-consistency is satisfied by the particular choice $W_L = V + C_{\mathrm{loc}}$, it is also the equation of motion for the square-root of the full parquet distribution function.

10. The Parquet Equation of Motion

In the preceding two sections, the equations of motion for the RPA (eq. 8.13) and ladder (eq. 9.6) distribution functions have been derived. These two distribution functions are identical provided that the local approximations used satisfy the self-consistency conditions $W_C = V + L_{\mathrm{loc}}$ and $W_L = V + C_{\mathrm{loc}}$. Using

$$D(1)f^2(i,j) = 2\,f(i,j)D(i)f(i,j) + 2|\nabla_i f(i,j)|^2, \tag{10.1}$$

the RPA equation can be rewritten

$$\frac{1}{2}f(i,j)\big[D(i) + D(j)\big]f(i,j) = V(i,j) + L_{\mathrm{loc}}(i,j) + C_{\mathrm{loc}}(i,j)$$
$$- \frac{1}{2}\big(|\nabla_i f(i,j)|^2 + |\nabla_j f(1,2)|^2\big), \tag{10.2}$$

while the ladder equation multiplied by $f(i,j)$ becomes

$$\frac{1}{2}f(i,j)\big[D(i) + D(j)\big]f(i,j) = g(i,j)\big(V(i,j) + C_{\mathrm{loc}}(i,j)\big). \tag{10.3}$$

Equating the right-hand sides, one finds an alternative expression for the driving term of the RPA equation

$$W_C(i,j) = V + L_{\mathrm{loc}}(i,j)$$
$$= g(i,j)V(i,j) + \big(g(i,j) - 1\big)C_{\mathrm{loc}}(i,j) + \frac{1}{2}\big(|\nabla_i f(i,j)|^2 + |\nabla_j f(i,j|^2\big). \tag{10.4}$$

This is one of the major results of inhomogeneous variational theory and parquet perturbation theory. All of the terms on the right-hand side of this equation can be computed from just carrying out the RPA calculation. That is, starting with some guess for the RPA driving term W_C, one can solve RPA to find $g(1,2)$ and $f(1,2)$. From these, one can form a new W_C using eq. 10.4. This is the heart of the paired-phonon analysis iteration procedure described in the language of parquet perturbation theory. If this procedure converges, then it simultaneously generates a self-consistent sum of ladder diagrams driven by W_L. **The beauty of the process is that the self-consistent summation of rings and ladders is no harder than repeatedly solving a summation of rings; the ladders never need to be solved explicitly.** Actually, this conclusion is a little premature at this point, because we have not yet determined the right local instantaneous self-energy operator Σ to use.

11. The Energy Functional and the Self-Energy

In perturbation theory, the energy may be computed in several ways. The approach adopted here is biased by the previous work on the homogeneous parquet theory and the inhomogeneous variational theory. We assume the existence of an functional $F[\lambda, \phi, f]$ of the one-body function ϕ and the two-body function f with the following properties: (i) the coupling λ appears explicitly in F only as

$\frac{1}{2}\lambda \int d^3r_1 \int d^3r_2\rho_\lambda(1)\rho_\lambda(2)g_\lambda(1,2)V(1,2)$, (ii) stationarity of F with respect to $f(i,j)$ gives the parquet equation of motion for each λ, (iii) the variation of F with respect to a normalized $\phi(i)$ is zero for each λ, and (iv) F reduces to the energy in the non-interacting limit $\lambda = 0$ with $f(i,j) = 1$ and $\phi(i) = \phi_0(i)$, the exact solution of the non-interacting one-body problem. For such a functional, the Feynman-Hellmann theorem can be used to show that this functional is in fact the energy. Indeed, for such an F,

$$\frac{dF}{d\lambda} = \frac{1}{2}\int d^3r_1 \int d^3r_2\rho_\lambda(1)\rho_\lambda(2)g_\lambda(1,2)V(1,2)$$
$$+ \int d^3r_i \frac{\delta F}{\delta\phi}(i)\frac{d\phi(i)}{d\lambda} + \int d^3r_i \int d^3r_j \frac{\delta F}{\delta f}(i,j)\frac{df(i,j)}{d\lambda}. \tag{11.1}$$

The last two terms vanish because of the stationarity of F, and the remaining piece can be integrated to give

$$F[1,\phi,f] = F[0,\phi_0,1] + \int_0^1 d\lambda \frac{1}{2}\int d^3r_1 \int d^3r_2\rho_\lambda(1)\rho_\lambda(2)g_\lambda(1,2)V(1,2), \tag{11.2}$$

which is just the Feynman-Hellmann prescription for calculating the energy.

As a side benefit, setting the variation of F with respect to ϕ to zero also tells us what local instantaneous approximation is being used for the self-energy Σ. An alternative way of deriving the self-energy would be by an analysis of the diagrams for the self-energy; this approach is the subject of the next section.

We define

$$F = F_1 + F_2 + F_3 + F_4$$
$$F_1 = \nu \mathrm{Tr}\{\phi^{-1}(U_{\mathrm{ext}} - K)\phi h\}$$
$$F_2 = \frac{1}{2}\mathrm{Tr}\{gV\} \tag{11.3}$$
$$F_3 = \mathrm{Tr}\{fKf\}$$
$$F_4 = -\frac{1}{4}\mathrm{Tr}\{\frac{(S-1)^3}{S}K\}.$$

For $\lambda = 0$, it is the entire energy functional.

Some care is required in carrying out the variations, because eq. 5.8 involves the ϕ. Stationarity with respect to f gives the parquet equation of motion. Variation of ϕ gives the generalization of the Hartree equation

$$-\frac{1}{2}\nabla_i^2\phi(i) + U_{\mathrm{ext}}(i)\phi(i) + \Sigma(i)\phi(i) = \mu\phi(i), \tag{11.4}$$

where μ is the Lagrange multiplier which enforces normalization and represents the chemical potential. The self-energy term is given from the functional variations as

$$\Sigma(i) = \int d^3r_j\rho(j)\left\{g(i,j)V(i,j) + \frac{1}{2}|\nabla_i f(i,j)|^2 + \frac{1}{2}|\nabla_j f(i,j)|^2 + \frac{1}{2}C_{\mathrm{loc}}(i,j)(g(i,j) - 1)\right.$$
$$\left. - \frac{1}{16}D(i)[N(i,j)(g(i,j) - 1)]\right\}. \tag{11.5}$$

Since this functional satisfies the four requirements, it is the energy functional.

12. Diagrammatic Structure of the Parquet Self-Energy

How does the *local instantaneous* self-energy obtained in the last section compare with the self-energy one would compute in parquet theory? The parquet self-energy is constructed from the two-body vertex by putting a bare interaction line on top and closing off one side with a Green's function. The Hartree term comes from performing the same operation on the free Green's function. In the boson approximation scheme, the exchange diagrams are dropped. The resulting diagrammatic sum is shown in fig. 7.

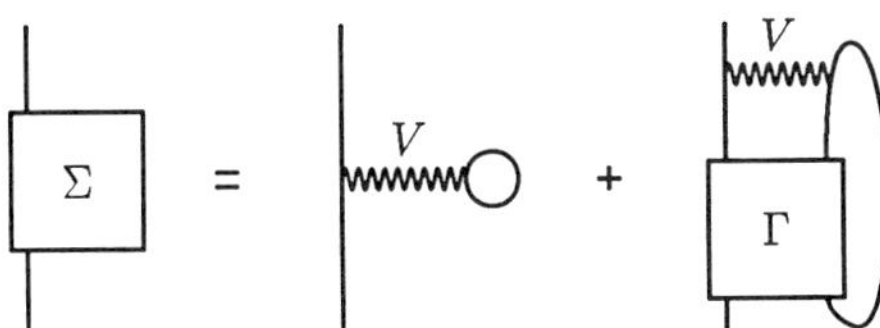

Fig. 7. The self-energy constructed from the two-body vertex and the Hartree term.

Substituting for Γ, one obtains the explicit representation of fig. 8. The disadvantage of this form is that it is expressed in terms of quantities which are ill-behaved.

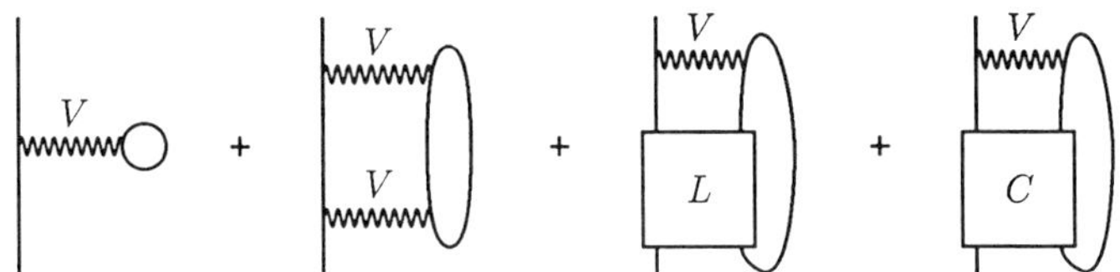

Fig. 8. An ill-behaved form for the self-energy.

The problem is that the bare interaction appears too many places. The first-order and second-order parts are both bad for short-range potentials. It is desirable to transform the diagrams so that they depend in a better-behaved fashion only on the better-behaved quantities $W_L = V + C_{\text{loc}}$ and $W_C = V + L_{\text{loc}}$. We note that in fig. 8, the L is constructed from local rungs W_L and the C is constructed from local functions W_C. Using the relation $\Gamma_{\text{loc}} = V + L_{\text{loc}} + C_{\text{loc}}$, the self-energy sum is rearranged to take the form of fig. 9. This form is exactly equivalent to that of fig. 8.

The first line of this figure gives a ladder contribution driven by W_L and the second line gives an RPA sum driven by W_C. The third line gives first-order and second-order contributions from a local potential Γ_{loc}. The only unattractive parts left are in the fourth line, where there are contributions to the self-energy depending on the difference $C - C_{\text{loc}}$ and $L - L_{\text{loc}}$. On closing these off with a Green's function and integrating over the frequency, these would vanish by the definitions of C_{loc} and L_{loc}. We make the obvious approximation of neglecting these contributions in all cases. The contribution to the self-energy from any of these diagrams is a factor of i times the value of the diagram.

The diagram sum of fig. 9 is the non-local frequency-dependent parquet self-energy. How is it related to the local instantaneous self-energy which was used in obtaining the two-body Green's function by the parquet summation? To see

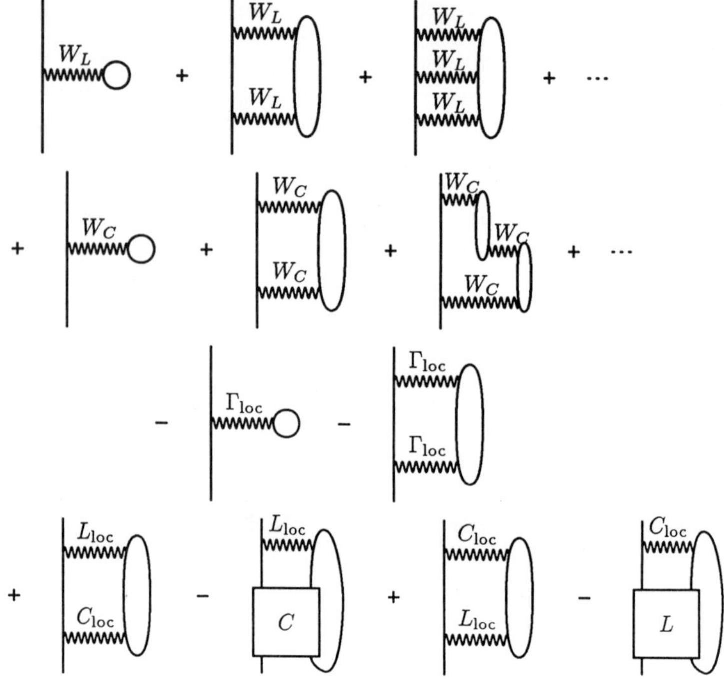

Fig. 9. A better-behaved form for the self-energy.

this, consider the self-energy Σ_λ calculated with the bare interaction V replaced everywhere by λV. From this, construct the canonical potential Φ illustrated in fig. 10 and given by

$$\Phi = \frac{i}{2} \int \frac{d\lambda}{\lambda} \left[\text{diagram value}\right]. \qquad (12.1)$$

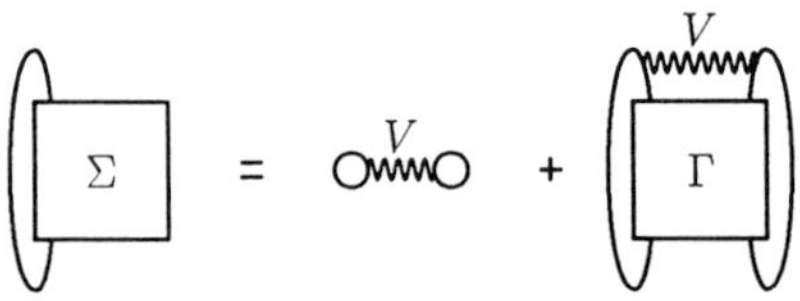

Fig. 10. The diagrammatic content of the canonical potential Φ.

This canonical potential has the property that its functional derivative with respect to the one-body Green's function gives the self-energy. In addition, comparison with the energy functional in sec. 11 shows that $\Phi = F_2 + F_3 + F_4$. The *local instantaneous self-energy* obtained in sec. 11 is therefore exactly the functional derivative of Φ with respect to the density. In performing the frequency integral in eq. 12.1, the error introduced by neglecting the differences $C - C_{\text{loc}}$ and $L - L_{\text{loc}}$ disappears.

It is then possible to understand just how the non-local frequency-dependent parquet self-energy is used to generate the *local instantaneous self-energy*. The frequency dependence of the parquet self-energy is first removed by closing it off with a Green's function and performing the coupling-constant integration. The nonlocality is them removed by integrating over both coordinates and then functionally differentiating with respect to the density to get the *local instantaneous self-energy.*

Several things are obviously lost in this process. In a better calculation, one would like to incorporate a self-consistent $\Sigma(k)$ from the outset for a homogeneous system. We have an algorithm for carrying this out which will be checked out for homogeneous systems elsewhere[9].

13. Discussion

We have given a self-contained derivation of the *local approximate boson parquet* equations solved with the *local instantaneous* self-energy approximation. The resultant theory is equivalent to the inhomogeneous boson variational theory of KQK. The present presentation provides an important alternative interpretation of their work in the framework of perturbation theory. A more complete presentation which includes details of the derivations and shows the relationship to the inhomogeneous variational theory and homogeneous parquet theory will be presented elsewhere.

The present context makes quite clear some points at which improvements could be made; *e.g.*, look at the italicized words in the first paragraph. We have some ideas on improving the treatment of the self-energy which are being tried out for homogeneous systems[9] but which could apply equally well to inhomogeneous systems.

The present context is also very useful for computing the parquet self-energy and dynamic structure function. These quantities are also being computed for homogeneous systems[9].

A more important use of the present theory is for looking at generalizations to fermion systems and finite temperature systems. Preliminary investigations of these theories suggest that the formalism and notation presented will enable these systems to be described in ways which are much more closely related to the ground-state boson theory than the corresponding variational approaches.

14. Acknowledgements

We are pleased to thank Eckhard Krotscheck for useful discussions of the inhomogeneous variational theory. This work was supported, in part, by the National Science Foundation under grant PHY-8806265. Hong-Wei He was supported, in part, by Texas Advanced Research Projects Grant 010366-012.

15. References

1. A. D. Jackson, A. Lande and R. A. Smith, Phys. Rev. Lett. **54**, 1469 (1985).
2. E. Krotscheck, G.-X. Qian and W. Kohn, Phys. Rev. **B31**, 4245 (1985).
3. B. Brandow, private communication.
4. R. A. Smith, Proceedings of the IX International Workshop on Condensed Matter Theories, San Francisco, Aug. 1985, ed. F. B. Malik, (Plenum Press, New York, 1986), 9.
5. A. Lande and R. A. Smith, Phys. Rev., in press.
6. A. L. Fetter and J. D. Walecka, *Quantum Theory of Many-Particle Systems,* (McGraw-Hill Book Company, New York, 1971)
7. R. A. Smith and A. D. Jackson, Nucl. Phys. **A476**, 448 (1988).
8. H.-W. He and R. A. Smith, submitted to Phys. Rev.
9. T. Ainsworth, H.-W. He and R. A. Smith, to be published.

QUANTUM PHASE TRANSITIONS IN BOSONIC SYSTEMS

Peter B. Weichman

Condensed Matter Physics
California Institute of Technology
Pasadena, CA 91125

The physics of bosons has found a fascinating rebirth in recent years, sometimes in systems far removed from superfluid helium. In this article I will review some of the theoretical advances and indicate some of the experiments that motivated them.

I. Helium in Porous Media

I will begin with superfluid ^{4}He since this boson system is the most familiar. The transition to superfluidity in bulk ^{4}He at temperatures of a few Kelvins (the exact temperature depending on the external pressure) is probably the most accurately studied phenomenon in condensed matter physics. Theoretically the transition is understood to be in the universality class of a three dimensional XY-model, exemplified by a lattice of two-component, unit-length classical spins, $\vec{s}_i$, interacting via a nearest neighbor coupling, $-J\vec{s}_i \cdot \vec{s}_j$, with ferromagnetic exchange $J > 0$ (how this comes about should become clearer below). Indeed, calculations based on this identification, especially using renormalization group methods, yield an amazingly accurate description of experiments close to the critical point.[1]

Although the basic physics of the *bulk* superfluid transition has been understood for several decades, it is only comparatively recently that we have begun to understand the superfluid transition under more exotic circumstances. Motivated to some extent by experiments on the Kosterlitz-Thouless transition in two-dimensional thin *films*,[2] experimentalists began to explore the behavior of ^{4}He adsorbed in various porous media, especially Vycor glass.[3] One interest was in very *low* coverages of helium, with the expectation that the resulting thermodynamic behavior might appear two-dimensional. If such a regime were achievable, the advantage of porous systems over the conventional rolled Mylar[2] would be the enormous surface area-to-volume ratio (over one-hundred square meters in a cubic centimeter sample of Vycor, as opposed to only a few tenths of a square meter in a Mylar roll of similar size).

The great surprise was that observations showed nothing of the kind! Thus over a wide range of coverages of helium in Vycor, from full pores down to fractions of a monolayer of surface adsorption, superfluid density data demonstrated clear bulk three-dimensional behavior—ρ_s vanishing with the characteristic three-dimensional 2/3-power law near the transition temperature T_λ,[3] *not* the discontinuous jump characteristic of two-dimensional Kosterlitz-Thouless behavior.[2] Even more surprising was the behavior at ultra-low coverages,[3] where the effective density, $\bar{\rho}$, of mobile helium is on the order of a few atoms per pore, i.e., interatomic separations, d_s, of order 50-100Å, 20-30 times the helium effective hard-core diameter, a. Note that there is always a localized/frozen "inert" monolayer or so—which will later be important, but at this stage is assumed to play no role—and we define $\bar{\rho}$ as the total density minus the density of this inert layer. In this regime the superfluid density profiles increasingly resemble those of an *ideal* Bose gas, vanishing nearly *linearly* as $T_\lambda(\bar{\rho})$ is approached. The interpretation, then, is that although the helium atoms are believed to reside on the pore surfaces, so that the film is *locally* two-dimensional, the pores are fully interconnected in three dimensions, and the atoms are in fact moving throughout a three-dimensional volume. Hence the three-dimensional nature of the phase transition. Furthermore, the characteristic size of an atomic wavepacket is set by the thermal de Broglie wavelength $\Lambda_T = h/(2\pi m k_B T)^{1/2}$, where m is the mass of a helium atom. The superfluid transition takes place when $d_s \sim \Lambda_T$, hence $T_\lambda(\bar{\rho}) \sim \bar{\rho}^{2/3}$. At the ultra-low coverages, T_λ is therefore strongly suppressed, and Λ_T becomes of order the pore size. One might imagine, then, that such a wavepacket sees the porous medium only in an *average* sense, with details below the scale Λ_T washed out. It is not a great leap of imagination to go from this observation to the picture of a dilute Bose gas in which the only role of the porous medium is to yield an effective mass m_{eff}, and some effective interatomic scattering potential. This last assumption is crucial since in pure bulk helium the atomic potential has a long-ranged attractive tail which causes it to condense into a dense strongly interacting fluid. A key effect of the porous medium must then be to screen out this attractive tail, leaving only the repulsive core. This allows the density to be continuously tuned without the usual first-order liquid-vapor transition intervening.

The above picture is intuitively appealing, but does it stand up to close theoretical scrutiny? The answer is both yes and no. On the one hand a detailed analysis,[4] based entirely on the above effective medium picture, of the *crossover* from strongly interacting to dilute weakly interacting Bose gas behavior as the density, $\bar{\rho}$, is reduced matches the experiments[3] remarkably well (the appropriate small parameter describing the strength of interactions is the ratio a/Λ_T). I will have no more to say about the details of this agreement, but refer the reader to the literature.[4] I will focus rather on the *no* part of the answer, which leads to even more interesting physics.

The problem with the above effective medium picture is that it does not account for the *disorder* or *randomness* inherent in porous media. The effects of disorder near a critical point can be very serious, sometimes completely changing the universality class of the transition. There is a simple criterion, called the Harris criterion,[5] which determines when this happens: If the specific heat exponent, α, of the pure nonrandom transition is positive (i.e., the specific heat actually *diverges* at the transition) then disorder is a *relevant* perturbation and will give rise to new critical behavior. If α is negative, disorder is irrelevant and the critical behavior should be unchanged. Now, the value of α for bulk ^{4}He is $\alpha \approx -.02 \pm .02$, very likely negative. This is consistent with the observation of bulk critical exponents for not too low coverages of helium in Vycor. However, as the coverage is reduced the critical behavior crosses over to that of an ideal Bose gas with specific heat exponent $\alpha = 1/2$, a strongly positive value. Thus we expect disorder to have an increasingly stronger effect at low coverages. This calls into question the validity of calculations based on the effective medium picture, despite their good agreement with the experimental data.

A partial way out is provided by understanding the detailed properties of Vycor glass.[6] It transpires that the method by which Vycor is made effectively ensures that the disorder on long length scales is *extremely weak*. It is therefore required that $\bar{\rho}$ be *extremely low*, probably below present experimental resolution, before the observed nearly ideal behavior is strongly perturbed.[6] Although this observation leaves one more comfortable with the effective medium theory, it merely sidesteps the really interesting question of what the behavior would be if the disorder was much stronger, or equivalently, if the experiments were to probe much lower coverages of helium in Vycor. Answering this question leads us into the realm of *localization effects* and *quantum phase transitions* in interacting boson systems.

II. Boson Localization at Zero Temperature

Quantum phase transitions are those that occur at zero temperature, where fluctuations are entirely due to the Heisenberg uncertainty principle: The fact that certain observables may not commute with the Hamiltonian. As observed previously, the superfluid transition temperature decreases with coverage, and *vanishes* at some critical coverage, ρ_c (the "inert" layer), below which superfluidity ceases to exist even at $T = 0$. Changing our perspective slightly, if we sit *at* $T = 0$ and consider the total density, ρ, as the independent variable (recall that $\bar{\rho} = \rho - \rho_c$), we see that the system passes from an insulating *localized* phase with superfluid density $\rho_s = 0$ for $\rho < \rho_c$ to a *superfluid* phase with $\rho_s > 0$ for $\rho > \rho_c$. This is precisely a quantum phase transition.

One may define critical exponents associated with this transition. For example we may ask how $\rho_s(T = 0)$ increases with $\bar{\rho} > 0$. We expect $\rho_s(0) \sim \bar{\rho}^{\zeta}$ for some exponent ζ. If, as in the effective medium picture, one views ρ_c as a totally inert substrate on which the excess density "skates", then we expect $\rho_s(0) \propto \bar{\rho}$, i.e., $\zeta = 1$. Indeed for a continuum dilute Bose gas with repulsive interactions one finds precisely $\rho_c = 0$ and $\rho_s(0) = \bar{\rho} = \rho$. In general we do not expect the precise equality $\rho_s(0) = \bar{\rho}$, even in cases where $\zeta = 1$ (for example in the case of regular *periodic* porous media), since it is a result of Galilean invariance, which will be broken by any residual interactions with the porous medium. As we have seen from the Harris criterion argument, the effective medium picture is incorrect in disordered systems when ρ is very close to ρ_c, and $\zeta \neq 1$ would not be unexpected. We can gain some intuition about the physics by thinking about Anderson localization in noninteracting Fermi systems. There, for a given external random potential, which may be thought of as a porous medium for our purposes, it is believed that one has low energy *localized* states separated from higher energy *extended* states by a *mobility edge*. Therefore if free fermions are added to such a system at $T = 0$, two per state, one first fills up the localized levels until a critical density ρ_c is reached, at which point all states below the mobility edge are filled; then any excess density $\bar{\rho} = \rho - \rho_c$ goes into the extended states and is therefore free to move about the entire system. This Anderson metal-insulator type transition is another example of a quantum phase transition. Furthermore it *violates* the effective medium picture. For example, although the analogue of ρ_s is absent, one may define a correlation length: The localization length $\xi_L \sim |\bar{\rho}|^{-\nu}$ for $\bar{\rho} < 0$, which measures the diverging linear extent of the localized states as the mobility edge is approached. The exponent ν, which turns out to take the value $\frac{1}{2}$ in the effective medium picture, is highly nontrivial, and its exact value is still controversial.

Building upon the above picture, we now consider bosons instead of fermions. It is immediately apparent that repulsive interactions are crucial: Without them the ground state would consist of all particles occupying the single lowest energy localized state. Clearly the particle density in the region of space occupied by this state would be infinite. Any kind of short ranged repulsion would therefore immediately preclude behavior of this type. A more accurate picture of what happens is the following: As particles are added to the system they more-or-less fill up the low-lying localized states until the particles' hard cores preclude further density increase. Further additions then have no choice but to occupy higher-lying localized states. In effect, the single particle states seen by the added particles are renormalized by interactions with the particles already present. Eventually, just as in the free fermion case, a critical density ρ_c is reached beyond which, in some self consistent sense to be discussed further below, subsequent added particles go into extended states. Since the particles are bosons, any occupying these states will Bose condense, forming a superfluid. This picture, then, provides the framework for a microscopic, quantum mechanical description of the formation of the frozen "inert" layer in Vycor, and an intuitive argument for the existence of a critical density ρ_c.

One can make this picture much more precise.[7] It is convenient to begin from a simple lattice boson model with nearest neighbor hopping and an on-site repulsion $U > 0$. Disorder is included via random site energies ϵ_i. Thus we consider the second-quantized Hamiltonian, sometimes called the boson Hubbard model,

$$H = J \sum_{<ij>} (a_i^{\dagger} a_j + h.c.) + \frac{1}{2} U \sum_i \hat{n}_i^2 + \sum_i (\epsilon_i - \mu) \hat{n}_i \tag{1}$$

where $\hat{n}_i = a_i^\dagger a_i$ is the site number operator and μ the overall chemical potential. The indices i are assumed to label the sites of a d-dimensional hypercubic lattice and $< ij >$ denotes nearest neighbor pairs. One may think of the lattice sites as idealized "pores", and the random ϵ_i as embodying the varying size, surface curvature, etc. of the pores. Clearly J and U ought also to have random components, but it turns out that all the important physics may be elucidated keeping only the ϵ_i.

As pointed out above, the random ideal Bose gas, represented by the first and last terms in (1), is the wrong starting point for the problem. Rather, one should begin in the *opposite* limit, where one throws out the *first* term in (1) by setting the hopping $J = 0$. In this limit there is no communication between different lattice sites and the Hamiltonian is diagonal in the basis of eigenstates of the $\hat{n}_j$: The ground state consists of an exact number, n_i, of bosons on each site, i, obtained by minimizing the single site energy

$$E_i = \frac{1}{2}Un_i^2 + (\epsilon_i - \mu)n_i, \quad n_i \geq 0 \ . \tag{2}$$

The solution is

$$n_i = \begin{cases} n_0 \geq 0 & \text{for} \quad n_0 - \frac{1}{2} < \frac{\mu - \epsilon_i}{U} < n_0 + \frac{1}{2} \\ 0 & \text{for} \quad \frac{\mu - \epsilon_i}{U} < \frac{1}{2} \ . \end{cases} \tag{3}$$

Thus as μ/U is increased by unity, one more particle is added to each site (except for those sites which remain empty). Note that if the ϵ_i are bounded such that $-\Delta \leq \frac{\epsilon_i}{U} \leq \Delta$, with $\Delta < \frac{1}{2}$ (by redefining the origin of μ if necessary, we assume the ϵ_i/U cover the symmetric interval $[-\Delta, \Delta]$), there are a discrete set of intervals $I_{n_0} = (n_0 - \frac{1}{2} + \Delta, n_0 + \frac{1}{2} - \Delta)$, for each $n_0 \geq 0$, such that when $\frac{\mu}{U} \in I_{n_0}$ all sites have precisely n_0 particles. We call these intervals *Mott insulating phases*, identified by the fact that the compressibility $\kappa = \partial n/\partial \mu$, where $n = \frac{1}{N}\sum_i n_i$ is the average number of particles per site, *vanishes* within each interval. For μ outside these intervals n varies continuously and $\kappa > 0$. We shall call these latter intervals *Bose glass* phases. Clearly both types of phase are insulating because $J = 0$. If $\Delta \geq \frac{1}{2}$ all Mott phases disappear, and only the Bose glass remains. Conversely, if $\Delta = 0$ (*no* disorder) the phase diagram consists entirely of Mott phases, pair-wise degenerate at half-integer values of μ/U. One may, of course, generate much richer phase diagrams with charge-density-wave-type Mott ground states (rational values of n) by allowing off-site interactions $\frac{1}{2}\sum_{i,j} U_{ij}n_in_j$, or by making the ϵ_i *periodic* (with, perhaps, a small random component added). None of this changes the nature of the superfluid transition so long as U_{ij} is short-ranged (see below), so we consider only the simplest case, $U_{ij} = U\delta_{ij}$.

Let us now introduce the possibility of superfluidity by adding back the hopping J. This adds a new dimension to our (so far) one-dimensional phase diagram. We begin by considering J as a perturbation, first on the Mott phases. One may construct the ground state perturbatively in J/U. It will consist primarily of the $J = 0$ ground state with precisely n_0 particles per site, however there will also be small admixtures of the excited states in which one, or more, of the sites have one or more extra particles, with corresponding depleted sites with one or more "holes" (note that the hopping term conserves overall particle number, so the number of particles equals the number of holes). However, each new particle-hole pair costs an extra energy of order U (the random potential, if present, will modulate this energy cost in the range $[U(1 - \Delta), U(1 + \Delta)]$, but this does not change the physics of the argument) and yields a component of the ground state with amplitude down by a factor J/U. Similarly, each time a particle is hopped one site further from the hole from which it originated, the amplitude goes down by a factor J/U. Roughly speaking, then, the probability that a particle will hop r steps varies as $(J/U)^r \sim e^{-r/\xi}$, where $\xi \sim 1/\ln(U/J)$ is a correlation length. From here it is not hard to argue that so long as ξ is finite one remains in the incompressible Mott phase. Thus, although there is a certain amount of local fluctuation in the particle number over regions of size ξ, the overall density remains precisely n_0. Correspondingly, there will remain an interval, $\mu_-(n_0, J) \leq \mu \leq \mu_+(n_0, J)$ on which $\kappa \equiv 0$, although the width $W(n_0, J) \equiv \mu_+(n_0, J) - \mu_-(n_0, J)$ will generally be less than the $J = 0$ value $W(n_0, 0) = U(1 - 2\Delta)$. The function $W(n_0, J)$ represents the effective energy required to create a particle-hole pair, and for positive J it will be reduced because the energy cost U is effectively shared among ξ^d particles through a net gain in kinetic (delocalization) energy. As J increases, there will come a point $J_c(n_0)$ at which $W(n_0, J_c(n_0))$ *vanishes*, i.e., it costs no energy to create particle-hole pairs. At this same point the correlation length ξ will diverge. In the absence of disorder, $\Delta \equiv 0$, this would signal the onset of superfluidity: Particles and holes are now free to hop throughout the system and will Bose condense. The phase diagram therefore consists of an infinite set of Mott *lobes*, one for each value of n_0, which meet pairwise only at $J = 0$ and μ/U half-integer. Everything outside the lobes is superfluid. Clearly the superfluid phase penetrates right to $J = 0$ at these half-integer points.

Consider now the transition to superfluidity for $J < J_c(n_0)$. For μ just above $\mu_+(n_0, j)$, so that n is slightly larger than n_0, it may be verified in detail by explicit calculation[7] that the system behaves like a dilute gas of bosons floating on top of the essentially inert Mott layer. Similarly, for μ just beneath $\mu_-(n_0, J)$ one has a dilute gas of holes. The transition from Mott to superfluid phase at fixed J is in every way identical to the bulk transition from an empty system to a dilute superfluid when μ increases through zero, and is described theoretically by the Bogoliubov model[8] (the same transition also occurs, of course, in the lattice system for $n_0 = 0$).

The point $J = J_c(n_0)$ is special. This is the point one goes through when the transition occurs at *fixed density* $n = n_0$, and is *particle-hole symmetric*. Since the density is precisely n_0 one can never view the system as a dilute addition to an otherwise inert background. It turns out[7] that this extra symmetry puts the transition in a different universality class: That of the $(d+1)$-dimensional XY-model. This is the same transition as occurs in helium at finite temperature, but in one higher dimension.

Let us now consider the effects of disorder. We expect, for bounded disorder $(\Delta < \frac{1}{2})$, that the phase diagram will consist of *three* phases in the $\mu - J$ plane, namely the superfluid phase in addition to the Mott and Bose glass insulating phases which exist at $J = 0$. We first discuss the perturbative effects of J on the $J = 0$ Bose glass phase. The fact that $n(\mu)$ is continuous follows simply from the fact that, due to the random nature of the ϵ_i, *somewhere* in the system there will be sites arbitrarily close in energy to a degenerate point $\frac{\mu - \epsilon_i}{U} =$ half integer. A slight increase in μ will then add a particle to these sites. Equivalently, there is no energy barrier to the creation of particle-hole pairs: Pick two sites, one with energy just below a degenerate point, the other with energy just above. At arbitrarily small cost in energy, $\Delta\epsilon$, one may then transfer a particle from the second site to the first. Why, then is the phase not superfluid? The argument that it is not is precisely along the lines of the original Anderson localization argument for free fermions.[7,9] Two sites with very small $\Delta\epsilon$ will generally be very far apart as $\Delta\epsilon \to 0$. But then the amplitude for hopping a particle from one site to the other will be of order $(J/U)^r$, where r is the distance between the two sites. For small J this factor will dominate and kill the net amplitude for the formation of such a particle-hole excitation. Thus for small J each individual Bose particle is localized in a self consistent way by a *combination* of interactions with other particles, and the residual hopping versus near-degeneracy effects described above.

Consider now the transitions between the three phases. Suppose one sits very close to the edge of a Mott lobe so that the average density, n, differs only slightly from n_0. One again expects the residual particles (or holes) to behave very much like a dilute fluid of density $\delta n = |n - n_0|$. However, in the presence of disorder this dilute fluid will still see a residual random potential due to the ϵ_i. For sufficiently small δn we therefore expect all particles to be localized [see the discussion above eq. (1)]. The conclusion is that at any fixed $J < J_c(n_0)$, the transition out of the Mott lobe must be into the Bose glass phase. This still leaves open the question of what happens at the special symmetric point $J = J_c(n_0)$, $n = n_0$. The argument is more shaky, but in essence we expect the system, for J slightly larger than $J_c(n_0)$, to look like a dilute "binary" fluid of particles and holes. Thus barring some exotic collective effect, we expect the residual random potential to localize both components of the fluid, and hence the Bose glass phase should completely surround the Mott lobes. Note that in any case, for strong randomness, $\Delta \geq \frac{1}{2}$, as certainly holds for all porous systems studied experimentally so far, the Mott lobes disappear entirely and Bose glass phase will exist for *all* values of μ, and sufficiently small values of J. For given μ we expect the onset of superfluidity to occur at some critical value $J = J_c(\mu)$.

III. The Bose Glass to Superfluid Transition

It is very likely, then, that the transition to superfluidity takes place *only* from the Bose glass phase,[7] and it is the physics of this transition which will concern us from now on. This is the transition which ought to describe the zero temperature insulator to superfluid transition in porous media.

It turns out that there are a number of conclusions one can draw about this transition simply on the basis of the expected *scaling* behavior of the thermodynamic functions. Central to scaling is the existence of a divergent correlation length, in this case the effective many body localization length, ξ. In quantum critical phenomena all fluctuations are driven by the quantum dynamics of the ground state wavefunction. This introduces, in addition to ξ, a correlation *time*, ξ_τ, into the equilibrium thermodynamics, with a corresponding frequency or energy scale $\Omega = 1/\xi_\tau$. Near the transition both ξ and ξ_τ diverge, and one defines a dynamical exponent, z, via $\xi_\tau \sim \xi^z \sim \delta^{-z\nu}$, where δ is any convenient measure of the deviation from the critical point, e.g., $\delta = \mu - \mu_c$, $J - J_c$, or $\rho - \rho_c$. The scaling hypothesis states that near the critical point all lengths scale with ξ and all times scale with ξ_τ. In addition the hypothesis of *hyperuniversality* states that the critical part of the free energy scales as $f_s \approx A\xi^{-d}(\hbar/\xi_\tau)$, where A is a universal constant. (This form is motivated by dimensional analysis – f_s is an energy per unit volume – and may be derived, with certain assumptions, from renormalization group theory.[10]) With the definition $f_s \sim |\delta|^{2-\alpha}$, where α is the $T = 0$ analogue of the specific heat exponent, this yields the generalized hyperscaling relation $2 - \alpha = (d + z)\nu$.

The superfluid density is derived by imposing a long wavelength *twist* with wavevector $k_0 \to 0$ on the superfluid order parameter,[11] then computing the derivative $\rho_s = \frac{m^2}{\hbar^2}(\partial^2 f/\partial k_0^2)_{k_0=0}$. Since k_0^{-1} is a length it must scale in the combination $k_0\xi$. We therefore have immediately[7] $\rho_s \sim |\delta|^\zeta$, with $\zeta = 2 - \alpha - 2\nu = (d + z - 2)\nu$. Using the Josephson relation between the time rate of change of the superfluid order parameter phase and the chemical potential, $\dot\phi = -\mu/\hbar$, one may write the total compressibility, $\kappa = -\frac{\partial^2 f}{\partial \mu^2}$, in the superfluid phase as a response to a slow twist in *time*, i.e. $\kappa = \hbar^{-2}\partial^2 f/\partial\omega_0^2$, where $\omega_0 \to 0$ is the frequency of the twist. We expect ω_0 to scale in the combination $\omega_0\xi_\tau$, therefore $\kappa \sim \xi_\tau\xi^{-d} \sim \delta^{(d-z)\nu}$ (note that in general the *total* compressibility κ includes, but is distinct from, the *singular* part of the compressibility $\kappa_s = -\partial^2 f_s/\partial\mu^2 \sim \delta^{-\alpha}$). We now come to the key observation: Since both the Bose glass and superfluid phases have a positive, non-zero compressibility (the argument fails for the Mott to superfluid transition in the pure case) we expect κ to be *finite* through the critical point. This clearly requires $\alpha < 0$, since κ includes κ_s, and immediately implies the equality $z = d$.

A further constraint on the exponents follows from a recent theorem[12] which states that in a random system one has the inequality $\nu \geq 2/d$. Using $z = d$ and $\alpha = 2 - (d+z)\nu$ this implies $\alpha < -2$, consistent with the finiteness of κ at $\delta = 0$. From the formula for ζ we also have $\zeta > 4(d-1)/d = 8/3$ for $d = 3$. This implies that ρ_s turns on very *slowly* as ρ passes through ρ_c, making it an extremely difficult exponent to measure.

One may define certain exponents which depend only on z and d, and which therefore may be evaluated explicitly. For example, for $\delta > 0$ there is a finite temperature transition with a T_c that should scale as some power, $T_c(\delta) \sim \delta^\theta$. What is θ? The quantity $\hbar/k_B T$ is a time scale, so T_c must scale with ξ_τ, implying $\theta = z\nu$. But $\rho_s(T=0) \sim \delta^{(d+z-2)\nu}$, so we may eliminate ν by writing $T_c \sim \rho_s(0)^x$ with $x = z/(d+z-2) = 3/4$ for $d=3$. This should be compared to $x = 2/d (= 2/3, d = 3)$ for the Mott to superfluid transition in the pure case.

Though we have come remarkably far using scaling ideas alone, we have yet to address the problem of calculating quantities, such as the exponent ν, explicitly. For the classical magnetic transitions, the primary analytic technique used for such calculations is the epsilon expansion about four dimensions.[13] This expansion relies upon the existence of a classical field theoretical representation for the problem—the φ^4 model in the classical magnetic case.[13] Such classical field theories can indeed be found for the dirty boson problem, but they are more complicated than their classical magnetic counterparts.[7,14] The crucial difference is the existence of the time scale ξ_τ: The quantum dynamics must be included in the effective classical theory, leading to complex fields $\varphi(\mathbf{x}, \tau)$ depending on both space *and* time. Furthermore space and time are not equivalent, so the effective field theory is anisotropic in time. The addition of disorder adds yet another complication: It couples $\varphi(\mathbf{x}, \tau)$ with an external *static* random potential $w(\mathbf{x})$. The fact that $w(\mathbf{x})$ does not depend on τ implies, in effect, that the disorder, instead of being interpreted as a set of bounded point-like impurities, must be interpreted, rather, as a set of one-dimensional rigid *rods*. This makes the problem even *more* anisotropic. To be more explicit, the effective classical Lagrangian is given by

$$\mathcal{L} = \int d^d x \int_0^\beta d\tau \left\{ \varphi^*(\mathbf{x}, \tau)(\frac{\partial}{\partial \tau} - J\nabla^2 - \mu)\varphi(\mathbf{x}, \tau) + w(\mathbf{x})|\varphi(\mathbf{x}, \tau)|^2 + u|\varphi(\mathbf{x}, \tau)|^4 \right\},$$

$$(4)$$

with $\beta = (k_B T)^{-1}$ and thermodynamics obtained from the partition function $\mathcal{Z} = tr^\varphi[e^{-\mathcal{L}}]$ in the usual way. If all τ-dependence is suppressed, the form reverts precisely to the classical φ^4-model for a random XY-magnet. Note that only for $T \to 0$ does the time dimension become infinite in extent. In this formulation, then, positive temperature represents a kind of finite size effect. As is standard in critical phenomena, any dimension of a system that is finite may be ignored when deciding the universality class of a transition. From this follows the irrelevance of quantum mechanics at $T > 0$: The behavior near $T_c(\delta)$ will be asymptotically governed by the classical φ^4-model in which all τ-dependence is suppressed.

Although the field theory defined by (4) is very complicated, there still exists an epsilon expansion type formalism within which one may derive renormalization group recursion relations.[14] The temporal anisotropy of the model, unfortunately, forces one to introduce a *second* expansion parameter, ϵ_d, the dimension of *time*, which along with $\epsilon = 4 - D$ (here $D = d + \epsilon_d$) must be presumed small in order to obtain a well defined expansion. Physically one has, of course, $\epsilon_d = 1$ and, for $d = 3$, $\epsilon = 0$. The argument leading to the exact exponent relation $z = d$ was very special to $\epsilon_d = 1$, and fails for general ϵ_d: One finds $z = 2 + a_1\epsilon + a_2\epsilon_d + \theta(\epsilon^2, \epsilon\epsilon_d, \epsilon_d^2)$ with nontrivial coefficients a_1, a_2.[14] The inequality $\nu > 2/d$ must still be obeyed, however, and the expansion for ν is consistent with this requirement.[14] This inequality also leads to very stringent requirements on how mean field theory becomes exact for $d > 4$ ($\epsilon + \epsilon_d < 0$), forcing the exponents to jump discontinuously as d increases through 4. Amazingly enough the ϵ, ϵ_d-expansion provides an explicit mechanism for this, whereby the nontrivial and mean field fixed points exchange stability without coalescing.[14]

This completes our general description of the dirty boson problem. We finally turn to applications outside helium, in particular to two-dimensional granular and amorphous superconductors, and, very briefly, to magnetic flux phases of high T_c compounds.

IV. Bosons and Superconductivity

The starting point for comparisons between superconductivity and superfluidity is the Josephson junction array Hamiltonian[15]

$$H_J = -\tilde{J} \sum_{<ij>} \cos(\hat{\phi}_i - \hat{\phi}_j) + \sum_i (\tilde{\epsilon}_i - \tilde{\mu}) + \frac{1}{2}\tilde{U} \sum_i \tilde{n}_i^2, \qquad (5)$$

where $\hat{\phi}_i$ is the *phase* operator on grain i, and $\tilde{n}_i$ is the conjugate number operator which measures the deviation of the number of Cooper pairs on grain i from some reference value N_0. The $\tilde{U}$ term is the so-called *charging energy* which disfavors large fluctuations in the $\tilde{n}_i$. The random site energies $\tilde{\epsilon}_i$ are precisely analogous to those in (1), and the chemical potential $\tilde{\mu}$ controls the average density. Note that if one were to accurately model the coulomb interactions between Cooper pairs the charging term should be replaced by $\frac{1}{2}\sum_{i,j} U_{ij}\tilde{n}_i\tilde{n}_j$, with $U_{ij} \propto e^2 |\mathbf{r}_i - \mathbf{r}_j|^{-1}$.

Quantum mechanics is completely defined by the commutation relations $[\hat{\phi}_i, \tilde{n}_j] = i\delta_{ij}$, all others vanishing. It is then easy to check that the eigenvalues of $\tilde{n}_i$ are integers (both positive and negative), and that if $\tilde{n}_i |n_i> = n_i |n_i>$, then $e^{i\hat{\phi}_i}|n_i> = |n_i + 1>$ and $e^{-i\hat{\phi}_i}|n_i> = |n_i - 1>$. Furthermore, the operators $a_i^\dagger = (N_0 + \tilde{n}_i)^{\frac{1}{2}} e^{i\hat{\phi}_i}$ and $a_i = e^{-i\hat{\phi}_i}(N_0 + \tilde{n}_i)^{\frac{1}{2}}$ obey *Bose* commutation relations, and $\hat{n}_i = a_i^\dagger a_i = N_0 + \tilde{n}_i$. Substituting these relations into (1) we find that the hopping term takes the form $J \sum_{<ij>} [(N_0 + \tilde{n}_i)^{\frac{1}{2}} e^{i(\hat{\phi}_i - \hat{\phi}_j)}(N_0 + \tilde{n}_j)^{\frac{1}{2}} + h.c.]$, which reduces precisely to the Josephson coupling term in (5) if the fluctuations $\tilde{n}_i$ are neglected relative to N_0, and if we identify $\tilde{J} = 2JN_0$. Modulo an N_0-dependent additive constant, the rest of the terms in (1) and (5) match exactly. We conclude that if N_0 is large (1) and (5) are quantitatively the same. For small N_0 the detailed physics of the two will differ, but will still possess precisely the same critical behavior. The general structure of the phase diagram, with Mott, Bose glass, and superfluid phases may be verified[7] in a straightforward way. Superfluidity and superconductivity, of course, correspond to long range or, perhaps, power law order in the phases $\hat{\phi}_i$, which for intuitive purposes may be thought of as classical angles in the interval $[0, 2\pi]$. The Josephson coupling is then precisely analogous to the usual XY-coupling between phases. For $T > 0$ this intuition is precisely correct, in some renormalized sense. All of quantum mechanics is buried in the operator character of the $\hat{\phi}_i$, and only at $T = 0$ does the classical intuition break down and more complicated behavior result.

The essence of the equivalence between granular superconductors and superfluids is the assumption of the existence of well defined Cooper pairs well *before* actual superconductivity occurs. Thus the operators $\hat{\phi}_i, \tilde{n}_i$ are well defined even above the critical temperature. In granular systems this assumption is valid because individual grains are usually sufficiently large that they behave like small pieces of bulk superconductor, and order at the bulk transition temperature, T_c^0. Ordering *between* grains, mediated by the Josephson coupling $\tilde{J}$, occurs at a much lower temperature $T_c(\tilde{J}) \ll T_c^0$. Thus well defined Cooper pairs exist within each grain and may be treated to good approximation as bosons. To the extent that all excitations of a fermionic character, such as pair breaking and residual interactions with normal electrons,[16] are separated from the bosonic excitations, embodied in (5), by a finite energy scale $\Delta E \gg T_c(\tilde{J})$ this treatment should be *exact* near the critical point.

For amorphous systems, without well defined grains, the validity of the boson model is less clear. However, one can argue[17] that in dirty systems, which are of main interest here, the role of grains is played by localized states in which it is favorable to put *pairs* of electrons. Nearby localized states are then assumed to interact via some effective Josephson coupling, eventually leading to bulk superconductivity. An experimental signal of this would again be the existence of a well defined energy gap between hopping of localized Cooper pairs and single electron-type excitations.

The experiments I wish to address are the two-dimensional thin film analogues of the zero temperature Bose glass to superfluid transition, or, more appropriately for electron systems, the Cooper-pair glass to superconducting transition. The transition can be accessed in various ways: For example, by changing the degree of disorder in the film; by varying the thickness of the film;[18] or by adjusting an external magnetic field.[19] The most interesting observation, which led to much of the interest in these systems, is that of *universal critical conductances*. Thus it was observed that the conductance $\sigma(T, \delta)$, where again δ is any of the above parameters which moves the system through the $T = 0$ transition located at $\delta = 0$, *diverges* at some $T_c(\delta)$ for $\delta > 0$, and *vanishes* as $T \to 0$ for $\delta < 0$, but approaches a *constant* of order unity (in units of $4e^2/h$, the inverse quantum of resistance) as $T \to 0$ for $\delta \equiv 0$. This constant, $\sigma^* \equiv \frac{h}{4e^2} \lim_{T \to 0} \sigma(T, 0)$, was seen to be remarkably close to unity, and a number of theories were proposed suggesting that $\sigma^* \equiv 1$.[20] Unfortunately, although σ^* is in fact a universal number[21] (see below) it is only unity for a very special set of *self-dual*[22] models, which unfortunately do not correspond to physical reality. One concludes then that the experiments which see $\sigma^* \approx 1$ are probably not yet in the asymptotic zero temperature limit.

The proof[21] that σ^* is universal follows from hyperuniversality (see Sec. 3) and the Kubo formula which relates the conductivity to the superfluid density

$$\sigma(T, \delta, \omega) = \frac{4e^2 \hbar}{m^2} \rho_s(-i\omega)/(-i\hbar\omega) \tag{6}$$

where m is the boson (i.e., Cooper pair) mass, and ω is the frequency. This formula holds only if the physics is described by the boson model. The detailed definition of the frequency dependent ρ_s is not important, only that it scales as $B\xi_\tau^{-1}\xi^{2-d}$, with a universal coefficient, B (this is equivalent to the hyperuniversality assumption for the free energy). Since ω and T scale with ξ_τ we have, as $\delta \to 0$, $\frac{h}{4e^2}\sigma \approx \xi^{2-d}R_\sigma(\omega\xi_\tau, T\xi_\tau)$. Once one sets the units of ω, τ, and σ, the function $R_\sigma(x, y)$ is *universal*.[21] In particular, for $d = 2$ the ξ^{2-d} prefactor drops out, and in the limit $T \to 0$ with $\omega = \delta \equiv 0$ we see that $\sigma^* = R_\sigma(0, \infty)$ is indeed a universal number. Note that this result does not depend on the values of the exponents z or ν. Thus it should hold even in the presence of long ranged Coulomb interactions, or applied magnetic fields, where the universality class of the transition will in general be different.

There are a number of other universal combinations one can define not involving the conductance.[21] For example, again in $d = 2$, the combination $\lim_{\delta \to 0} \rho_s(T = 0, \delta)/T_c(\delta)$ yields a universal number, which could in principle be measured in helium experiments. A wide-open problem is the actual calculation of some of these numbers for realistic models.

As a final example of boson physics in electronic systems, we mention the exotic magnetic flux phases of high temperature superconductors.[23,24] Using the well-known Feynman path-integral formulation of boson statistical mechanics,[25] one may view the flux lines in the mixed phase of high T_c compounds as boson *world lines*, with time progressing parallel to the applied field. The sample thickness then represents the effective *inverse temperature* of a two-dimensional system of interacting bosons. The proposed transitions between flux phases in bulk three-dimensional samples then correspond to zero temperature transitions between crystalline, superfluid, etc. phases of $2 - d$ bosons. A major complication, however, arises when one considers *disordered* materials. Physically the disorder comes from impurities spread randomly through the bulk sample. Thus the effective bosons will see a *time-varying* as well as spatially-varying random potential. This is very different from the "random-rod" problem discussed earlier for the conventional Bose glass. One has instead a "vortex glass" which is expected to display many novel properties similar to those of a spin glass.[24] Experiments now confirm quite unambiguously the existence of this transition,[26] but do not as yet give any reasonable estimates for the critical exponents, or any detailed properties of the glass phase. Much of the theory is of a phenomenological nature[24] but detailed calculations on reasonable models are also beginning to appear.[27]

References

1. For a relatively recent review of the state of the art see V. Dohm, J. Low Temp. Phys. **69**, 51 (1987).
2. D.J. Bishop and J.D. Reppy, Phys. Rev. **B22**, 5171 (1980).
3. See J.D. Reppy, Physica **B126**, 335 (1984) and references therein.
4. P.B. Weichman, M. Rasolt, M.E. Fisher and M.J. Stephen, Phys. Rev. **B33**, 4632 (1986); M. Rasolt, M.J. Stephen, M.E. Fisher and P.B. Weichman, Phys. Rev. Lett. **53**, 798 (1984). See also P.B. Weichman, Phys. Rev. **B38**, 8739 (1988).
5. A.B. Harris, J. Phys. C **7**, 1671 (1974).
6. P.B. Weichman and M.E. Fisher, Phys. Rev. **B34**, 7652 (1986).
7. M.P.A. Fisher, P.B. Weichman, G. Grinstein and D.S. Fisher, Phys. Rev. **B40**, 546 (1989). Recent Quantum Monte Carlo work has confirmed the basic structure of the phase diagram presented here. See G.G. Batrouni, R.T. Scalettar and G.T. Zimanyi, Phys. Rev. Lett. **65**, 1765 (1990); **66**, 3144 (1991).
8. See P.B. Weichman, Phys. Rev. **B38**, 8739 (1988) for a modern view of this transition.
9. P.W. Anderson, Phys. Rev. **109**, 1492 (1958).
10. K. Kim and P.B. Weichman, Phys. Rev. **B43**, 13583 (1991).
11. M.E. Fisher, M.N. Barber and D. Jasnow, Phys. Rev. **A8**, 1111 (1973).
12. J.T. Chayes, L. Chayes, D.S. Fisher and T. Spencer, Phys. Rev. Lett. **57**, 2999 (1986).
13. See, e.g., K.G. Wilson and J. Kogut, Physics Reports **12C**, 75 (1974).
14. P.B. Weichman and K. Kim. Phys. Rev. **B40**, 813 (1989).
15. S. Doniach, Phys. Rev. **B24**, 5063 (1981).
16. Such interactions are often treated by coupling harmonic oscillator baths to the superconducting grains: A.O. Caldeira and A.J. Leggett, Ann. Phys. (N.Y.) **149**, 374 (1983).
17. M. Ma, B.I. Halperin and P.A. Lee, Phys. Rev. **B34**, 3136 (1986).
18. D.B. Haviland, Y. Liu and A.M. Goldman, Phys. Rev. Lett. **62**, 2180 (1989); A.E. White, R.C. Dynes and J.P. Garno, Phys. Rev. **B33**, 3549 (1986).
19. A.F. Hebard and M.A. Paalanen, Phys. Rev. Lett. **65**, 927 (1990).
20. See, e.g., M.P.A. Fisher, Phys. Rev. Lett. **57**, 885 (1986); T. Pang, Phys. Rev. Lett. **62**, 2176 (1989).
21. K. Kim and P.B. Weichman, Phys. Rev. **B43**, 13583 (1991). Universality of σ^* was first proposed by M.P.A. Fisher, G. Grinstein and S.M. Girvin, Phys. Rev. Lett. **64**, 587 (1990), but the derivation given is incorrect.
22. See M.P.A. Fisher, *et al.*, Ref. 21.
23. D.R. Nelson and H.S. Seung, Phys. Rev. **B39**, 9153 (1989).
24. D.S. Fisher, M.P.A. Fisher and D.A. Huse, Phys. Rev. **B43**, 130 (1991).
25. R.P. Feynman and A.R. Hibbs, *Quantum Mechanics and Path Integrals* (McGraw Hill, 1965), Chap. 10.
26. See, e.g., P.L. Gammel, L.F. Schneemeyer and D.J. Bishop, Phys. Rev. Lett. **66**, 953 (1991).
27. See, e.g., D.R. Nelson and P. Le Doussal, Phys. Rev. **B42**, 10113 (1990); J.D. Reger, T.A. Tokuyasu, A.P. Young and M.P.A. Fisher (preprint, 1991).

COLLECTIVE COMPUTATION OF MANY-BODY PROPERTIES BY NEURAL NETWORKS

J. W. Clark,[a] S. Gazula,[a] K. A. Gernoth,[a]
J. Hasenbein,[b] and J. S. Prater[c]

[a]McDonnell Center for the Space Sciences
   and Department of Physics
[b]Department of Systems Science and Mathematics
[c]Department of Electrical Engineering
Washington University, St. Louis, MO 63130 USA

H. Bohr

School of Chemical Sciences
University of Illinois, Urbana, IL 61801 USA

INTRODUCTION

Artificial neural networks constitute a novel class of many-body systems in which the particles are neuron-like units and the interactions are weighted synapse-like connections between these units.[1,2] The most extraordinary feature of these systems is that the interactions are subject to modification, depending on the states recently visited by the system. Thus, as the network experiences varied stimuli, knowledge can be stored in the neuron-neuron interactions, for later retrieval in some information-processing task. Indeed, multilayered, feedforward networks of analog neurons can be taught by example to solve complex pattern-categorization problems using the backpropagation learning algorithm[3] or other procedures for modifying connection weights. During the learning process, inner neurons may evolve into useful feature detectors tailored to regularities or correlations inherent in the ensemble of input stimulus patterns and desired output response patterns used for training. The system builds an internal representation, or model, of its pattern environment, which may provide a good approximation to the actual rules determining the underlying input-output map. Accordingly, the artificial neural network may possess a useful generalization or predictive ability, as demonstrated by a high percentage of correct responses when presented with unfamiliar input patterns absent from the training set.

The application of neural networks to scientific problems raises intriguing possibilities and poses stimulating challenges: Can such artificially intelligent systems, when taught by example, develop economical rules for the correlations implicit in the data on a given class of physical systems, enabling them to make reliable predictions about cases for which experimental results are not available? Can neural networks

Recent Progress in Many-Body Theories, Vol. 3,
Edited by T.L. Ainsworth et al., Plenum Press, New York, 1992

discover illuminating new models of the physical systems under study, or even laws of nature? If so, how can we extract this new knowledge from the matrix of interactions that is created in the learning process?

Beginning with nets designed to predict protein secondary and tertiary structure from the primary amino-acid sequence,[4] there has been an explosive growth of neural network applications across many areas of science, including experimental high-energy physics,[5] astronomy,[6] chemistry,[7] and biology.[8]

The field of nuclear physics, with a wealth of data reflecting both the fundamental principles of quantum mechanics and the behavior of strong, electromagnetic, and weak interactions on the fermi scale of distances, provides ample opportunities for testing and exploiting new ideas for phenomenological analysis based on neural networks. Typically, the input patterns are just appropriately coded versions of the proton and neutron numbers Z and N defining particular nuclides; the output patterns represent associated nuclear properties such as ground-state mass, spin, parity, shape, deformation, etc. In this paper we shall report specific applications of neural nets to (i) discrimination between stable and unstable nuclides, (ii) fitting and prediction of atomic masses, (iii) analysis of the systematics of neutron separation energies, and (iv) assignment of ground-state spins and parities. Evidence will be presented that these novel adaptive computational systems can grasp essential regularities of nuclear physics including the valley of beta stability, the even-odd (pairing) effect, and the existence of shell structure. With suitable architecture and representation of input and output data, learning can be accomplished with high accuracy. Moreover, significant predictive ability is demonstrated, attesting to the potential practical value of the approach.

This pilot study has wider implications. Similar applications to problems in atomic and molecular physics as well as condensed matter and materials science can be envisioned. With further development, collective computation based on neural network models may offer a powerful new tool for the phenomenological description of complex many-body systems.

ELEMENTS OF NUCLEAR PHENOMENOLOGY WITH NEURAL NETS

A particular nuclear species, or nuclide, is identified by its proton number Z and its neutron number N. The Brookhaven National Nuclear Data Center (NNDC) provides on-line access to an extensive collection of experimental results for a wide range of properties associated with over 2000 nuclides, including mass, spin, parity, magnetic and electric moments, charge radius, decay modes, branching ratios, lifetimes, level schemes, etc. In principle, one should be able to reproduce all this data, and also to make valid predictions of nuclidic properties not yet determined experimentally, within the framework of fundamental theory as embodied in the Standard Model of the strong and electro-weak interactions. Less ambitiously, one should be able to calculate prescribed nuclear properties using the currently popular effective-interaction models based on nucleonic or hadronic degrees of freedom. The former alternative faces the intractability of quantum chromodynamics in the nonperturbative regime pertinent to nuclear physics (where the coupling constant is of order unity). While effective-interaction theories have generally been very useful for the formulation of tangible physical mechanisms, they are of limited accuracy and scope. Thus, the current state of nuclear theory leaves room for novel phenomenological approaches to nuclear physics.

We shall present here the results of a variety of computer experiments that demonstrate the ability of multilayered, feedforward neural networks to capture important regularities of the nuclear world embodied in the Brookhaven data base.

Success in these exercises requires suitable choices of (i) network architecture, (ii) coding schemes for representing input and output data, and (iii) learning algorithm.

Architecture and Dynamics

We consider layered feedforward networks containing an input layer, an output layer, and one or more intermediate or "hidden" layers. For input, hidden, and output neurons, we adopt the standard labels k, j, and i, respectively. Generic neurons are labeled m, m'. Every neuron extends a connection in the forward direction to every neuron in the next layer. The state of unit m is characterized by an analog activity variable a_m that ranges between 0 and 1. Thus, input and output patterns will be comprised of the corresponding sets of neuronal activities, $\{a_k\}$ and $\{a_i\}$. The activity of neuron m is a nonlinear function of its stimulus, which is a linear superposition $F_m = \sum_{m'} V_{mm'} a_{m'}$ of the signals the neuron receives from other neurons m' sending connections to it, where $V_{mm'}$ is the (signed) weight of the connection from m' to m. The threshold V_{mo} of neuron m is incorporated as a weight through the introduction of a "true unit"[3] that is always maximally active ($a_o = 1$) and extends a connection to m having weight $-V_{mo}$. The nonlinear "squashing function" $g(F_m)$ that transforms the stimulus F_m into the analog response $a_m \in [0, 1]$ is taken to be of the usual logistic form, $g(x) = [1 + \exp(-x)]^{-1}$. When a given activity pattern is imposed on the input layer, the system computes an output according to the following rules: (i) All units within a layer update their states in parallel. (ii) Successive layers are updated sequentially, starting at the input layer and proceeding forward until the output layer is reached.

Coding of Input and Output Data

In essentially all of our experiments, the pattern imposed on the neurons of the input layer represents the integers Z and N characterizing a given nuclide. The corresponding pattern of activity of the output neurons represents the value computed by the network for the desired property of that nuclide. For example, the output may code for the ground-state mass of the input nucleus, its shape and deformation, its neutron separation energy, its spin and parity, or some combination. The intermediate layers of neurons are called hidden layers because they do not communicate directly with the data environment, through the input or output interface. The hidden neurons are needed for the network to build an adequate model of the associations between input and output patterns, i.e. of the relation between the dependent (output) physical variables like mass, spin, etc., and the independent (input) variables Z and N. From a more mundane viewpoint, they are needed for an adequate parametrization of the data the system is trained to fit.

Quite generally, two basic problem types are seen in applications of feedforward neural networks involving supervised learning. In the first type – the standard *classification problem* – the output pattern serves to assign the input pattern to one of a discrete number of mutually exclusive classes. It will be useful to distinguish two subtypes of this kind of problem, namely (i) *detection problems*, in which the network is be asked to detect the presence of a *single* attribute (such as stability or even parity, in the nuclear context) and (ii) *sorting problems*, in which the net is required to assign the input pattern to one of *several* categories or "pigeonholes" (e.g. to the stable or the unstable set of nuclides, or to one of several values for ground-state spin). In the second general problem type, the output pattern is required to represent a real-number variable Y, which (like atomic mass and neutron separation energy) commonly will be regarded as taking values in a continuum $[Y_1, Y_2]$. For want of a better name, we shall use the term *real-function mapping problem* in this case. It is acknowledged that, to some extent, the distinctions between problem types may sometimes be blurred; however, they are important in choosing an effective representation of the output data.

We have tested a variety of coding schemes for representing the input and output data. Chief among these are:

(1) *Binary coding* of the discrete *input* variables Z and N. The first 8 of 16 input units are clamped "on" ($a_k = 1$) or "off" ($a_k = 0$) so as to express Z as a binary number; the remaining 8 units encode N in the same manner.

(2) *Analog coding* of the *input* variables Z and N, treated as numbers from respective continua $[Z_1, Z_2]$ and $[N_1, N_2]$. Two analog neurons k_1 and k_2 encode Z and N, respectively, as $a_{k_1} Z/(Z_2 - Z_1)$ and $a_{k_2} N/(N_2 - N_1)$.

(3) *Unary representation* of the *output*, for classification problems of detection and sorting subtypes. For detection problems, one analog output neuron i is used to decide the presence of the attribute according to the criterion $a_i \geq 0.5$, with $a_i < 0.5$ indicative of its absence. For sorting problems, each of the various possibilities, or categories, is represented by a dedicated output neuron, and the assignment made by the network is taken to be that category whose output neuron has the largest activity ("winner takes all").

(4) *Analog coding* of the *output* variable Y, for real-function mapping problems. The value of Y computed by the network is read out as $(Y_2 - Y_1)a_i$, i.e. as the activity of a single output neuron i multiplied by a suitable scaling factor.

(5) A special scheme for *real-number coding* of *input* and *output* variables. Intervals of the ranges of Z, N, and/or the real output variable Y are represented by a set of *several* analog input or output neurons, one unit being responsible for each of the disjoint, contiguous intervals. Technical details of this scheme may be found in Ref. 9, where a prescription is also given for reading out a computed value of the output variable from the set of output-neuron activities.

We have also considered other coding algorithms. For example, a unary representation of the input variables – with one "grandmother" neuron devoted to each possible Z value and one to each possible N – was tried and found to be inappropriate. On the other hand, there exist promising schemes that remain to be investigated, notably that involved in the Bayesian probabilistic formulation of Stolorz, Lapedes, and Xia,[10] designed for multi-option classification tasks like the spin-assignment problem.

The notation $^c(I + H_1 + H_2 + ... + H_L + O)_{c'}[P]$ permits an easy identification of particular choices of architecture and coding. In this expression, I and O are the numbers of input and output neurons and H_l is the number of neurons in the lth hidden layer; c and c' denote the input and output coding schemes, respectively; and P is the number of adaptive weights and thresholds. The thresholds of the input neurons do not contribute to P, since these units act simply as data registers. In the work to be described, the labels c and c' specialize to b for binary, a for analog, u for unary, and r for "real-number" coding.

Learning Algorithm

The artificial neural system is taught by example, using a supervised learning procedure designed to minimize the *cost function* $C = \sum_{\mu,i}[t_i^{(\mu)} - a_i^{(\mu)}]^2/2$, where $a_i^{(\mu)}$ is the actual activity level of output neuron i that results when the system is shown input pattern μ and $t_i^{(\mu)}$ is the corresponding target value. The input patterns of the training set are presented to the system in random order, errors in the responses of the output neurons are observed, and the weights and thresholds that parametrize the "education" of the network are incrementally adjusted in accordance with the standard backpropagation algorithm[3] including a momentum term. In one of the problems discussed below, we implemented an alternative procedure for minimizing the cost function in weight space, based on a conjugate-gradient algorithm[11] generally considered an improvement on the gradient-descent technique that underlies backpropagation. It should be noted that gradient-descent and conjugate-gradient methods can both get stuck in local minima.

Typically, some hundreds of passes through the training set are required before the cost function is effectively reduced to its asymptotic value. Since the initial values of the weights are chosen by random sampling from a uniform distribution, and since the order of presentation of the training patterns is random, different instances of the learning routine will yield different trained networks. We have executed a sufficient number of independent training sequences to ascertain that the behaviors described below are fairly representative.

Performance Measures

The two vital aspects of network performance are accuracy of learning and reliability of prediction. For satisfactory learning, one requires a high percentage of correct categorization decisions for the patterns on which the network has been taught, or computed values of the dependent output variable which closely approximate the target values. In prediction, the network is asked to generalize from its experience and produce valid assignments or computations for input patterns absent from the training set. Thus, learning performance relates to the accuracy of fit of the training data; whereas predictive performance bears on the validity of the model created by the adaptive network. We regard the latter aspect as essential, i.e., a neural network representation of a data set will only be deemed successful if an accurate fit of the training data is accompanied by significant predictive ability. (Conversely, given enough weight parameters, the system might make a perfectly accurate lookup table, but its "model" would then be empty and hence ineffectual when confronted with new examples.) Thus, in all of the applications described below, demonstrations of predictive power are sought by making runs in which a certain fraction of the database patterns are omitted, the network is trained on the remainder, and accuracy of response is subsequently tested on the unfamiliar patterns.

There is another practical issue in judging the merits of neural network phenomenology, namely, parametric efficiency. If the level of performance in learning and generalization is not sacrificed, networks with fewer neuronal units or fewer weight parameters are obviously preferred over larger systems. Indeed, excessively large nets tend to be poor at prediction, so economy in the number of hidden units is important. At this early stage of our investigations, we have not attempted to find optimal architectures. However, we have – with exceptions to be noted – worked with systems in which the number of modifiable weights (adjustable parameters P) is considerably smaller than the number of training examples (input data points n). Fault-tolerance studies indicate that substantial reductions can be made in the number of neurons and connections in our networks, without impairing performance.

STABILITY/INSTABILITY DISCRIMINATION

The most basic property of a given nucleus is whether or not it is stable. Thus, as a first exercise, we have constructed nets that can discriminate, with some reliability, between stable and unstable nuclides.

Data Base. There is patently some arbitrariness in the criterion for stability. On the one hand, no nuclei may be stable on a time scale of 10^{50} years; and on the other, it seems reasonable to include long-lived naturally occurring radioactive isotopes in the stable class. To be definite, we have taken as "stable" all those nuclides whose squares on the General Electric Chart of the Nuclides[12] contain gray patches. The full data base consists of 2226 nuclides, with unstables outnumbering stables by about 9 to 1. To test predictive performance, two reduced training sets (consisting of 1909 and 1689 patterns) were formed by deleting, at random, approximately 15% and approximately 25% of the unstable examples, and like percentages of the stables.

Architecture and Coding. Binary input coding is dictated by the importance of pairing and shell effects, which reflect the integral nature of Z and N. With regard to output coding, the stability/instability discrimination problem may be viewed either as a detection or a sorting task, the attribute in the former option being "stability" and the pigeonholes in the latter being "stable" and "unstable." A unary representation of the output is adopted in both cases (see previous specification). In the arrangement for detection, based on nets of type $^b(16+H+1)_a$, the activity $a_i \in [0,1]$ of the single output neuron can be interpreted as the confidence (or probability) with which the input nuclide (Z, N) is judged to be stable, and a "best-guess" decision is made using the simple rule that $a_i \geq 0.5$ means "stable" and $a_i < 0.5$, by default, means unstable. In the arrangement for sorting, based on nets of type $^b(16+H+2)_a$, there are two output neurons, for "stable" and "unstable" categories, with a "winner-takes-all" rule for reading out the decision of the network between these mutually exclusive categories.

Training Procedures. To equalize exposure to the two categories during the learning process, each stable example was presented 9 times and each unstable example once, in each pass through the training set. In different sets of runs, we employed (i) ordinary backpropagation (involving gradient descent) and (ii) the conjugate-gradient method to search for a minimum of the cost function in weight space. The former set of runs was carried out for detection networks, $^b(16 + H + 1)_a$, and included the choices $H = 0, 10, 19$, and 24 for the number of hidden units. The latter runs involved sorting networks, $^b(16 + H + 2)_a$, with $H = 10, 15$, and 20. For the backpropagation experiments, the learning rate η and momentum parameter α were respectively 0.5 (not 0.05 as stated in Refs. 9,13) and 0.9, initial weights were sampled from $[-0.5, 0.5]$, and weights were updated after each pattern. In the conjugate-gradient experiments, weights were updated after each pass through the training set.

Performance. We quote first some of the better results from the backpropagation runs. Learning performance was poor without a hidden layer ($H = 0$), improved for $H = 5$, and appeared to be saturating around $H = 19-24$. For $H = 19$ (thus $P = 343$ weight parameters), training on the full data base resulted (for a representative run) in an overall accuracy of 94% in the classification task, with a 75% accuracy in the identification of stables and 96% in the identification of unstables. It is useful to introduce the terms *efficiency* and *impurity*, to denote, respectively, the percentage of examples with a given attribute (here, stability) which are *correctly* classified and the percentage of input patterns without the attribute which are *incorrectly* categorized by the net. Thus, the net in the quoted example displays a 75% efficiency and 4% impurity. In predictive runs after training on the two reduced data sets, we found responses to the unfamiliar patterns characterized by 69% efficiency and 6% impurity for the smaller test set (317 patterns) and by 63% efficiency and 7% impurity for the larger test sample (537 patterns).

The conjugate-gradient runs show improved learning, particularly with respect to the stable nuclides, but similar predictive performance. Since the stability and instability decisions are complementary, we may still use efficiency and impurity as performance measures, where again these terms refer to the classification of nuclides as stable. To cite one example, a $^b(16 + 10 + 2)_a$ net (with $P = 192$ weights) learned the larger reduced data set (1909 patterns) with an efficiency of 100% and a impurity of 4.6% (meaning that the net decided correctly for 97.9% of the patterns). Testing prediction on the unfamiliar patterns, this net scored 77.1% in efficiency and 7.5% in impurity. The nets with larger numbers of hiddens did slightly better in learning the same training set: the $H = 20$ ($P = 382$) net learned perfectly and the $H = 15$ ($P = 287$) net mis-learned only a single pattern. On the other hand, predictive performance was somewhat worse, with the efficiency dropping to the 66% level but impurity remaining about the same.

Observations. A good representation of the data base, accompanied by useful predictive power, has been achieved with relatively simple networks. However, a simple yes-no formulation of the nuclear stability problem is unlikely to admit substantially better results including near-perfect prediction, since the network is given no information about degrees of stability or instability. This view is supported by the similarity of the results from the two independent sets of experiments described above.

Some remarkable features of the $^b(16 + H + 1)_a$ stability detection networks (which are presumably shared by the $^b(16 + H + 2)_b$ sorting nets) have been described in earlier reports.[13,14] In brief, examination of the learning dynamics of such networks and the receptive field patterns $\{V_{jk}\}$ of the hidden neurons j in mature nets shows that these systems are able to grasp the importance, for stability, of pairing and shell structure. They quickly learn to make a distinction between even-Z-even-N, even-Z-odd-N, odd-Z-even-N, and odd-Z-odd-N nuclei by inspection of the least significant input bit of Z and N, and certain hidden neurons develop into detectors for magic numbers (e.g. Z or $N = 2, 8, 20, 28, 50, 82$, and $Z = 126$) and tend to excite the stability output neuron. Other hidden neurons map out the unstable boundary regions of the valley of beta stability and inhibit activity of the stability output neuron for inputs in those regions. Of course, most of the receptive field patterns are highly complex and difficult to interpret. Nevertheless, there is clear evidence that the neural networks we have studied — although simple in architecture and coding — are capable of making physically sensible models of the stability data on which they are trained.

MODELING OF THE NUCLEAR MASS TABLE

From the yes-no stability-instability dichotomy, attention naturally turns to graded measures of nuclear stability, particularly nuclear masses or nuclear binding energies. A detailed neural-network analysis of the data base on nuclear masses (or, more properly, atomic masses) is now in progress, with the dual objectives of (i) accurate fits of the mass surface $M(Z, N)$ and (ii) reliable prediction of masses of novel nuclides. As is well known, nuclear binding energies and nucleon separation energies can be simply deduced from a table of mass values.

Data Base. The Brookhaven NNDC provides target values of the mass excess $\Delta = M - A$, in MeV, where $A = N + Z$ is the mass number and $\Delta(^{12}C) = 0$ in accordance with the usual convention. The table contains 2291 entries, from which training sets of various sizes may be chosen. In addition to the full data base, we have trained networks on a reduced set of 1719 nuclides, reserving a randomly selected set of 572 (some 25% of the data collection) for predictive tests. The empirical values of the mass excess range from $\Delta_{\min} = \Delta(^{118}\text{Sn}) = -91.6516$ MeV to $\Delta_{\max} = \Delta(^{266}109) = 128.21$ MeV.

Architecture, Learning Algorithm, and Coding Schemes. The existing experiments involve fully connected feedforward networks with one, two, and three hidden layers, trained using the standard backpropagation algorithm. For most runs, the learning rate and momentum parameter were again kept at the default values of 0.5

and 0.9, respectively, although the effect of smaller learning rates was investigated in some detail for several of the better architectures.

Since we are dealing with a real-function mapping problem (the pertinent map being $(Z, N) \rightarrow M(Z, N) - A$), the simple option of a one-neuron analog coding scheme is adopted for the output variable $Y = \Delta(Z, N) - \Delta_{\min}$. (Future studies will implement the more flexible real-number output coding scheme.) We have tried both analog and binary coding of the integers Z, N specifying the input patterns. Although performance measures will also be given for the nets with analog input coding, the binary representation is *qualitatively* superior. This is evident in global aspects of the corresponding error surface $D(Z, N) = M_{\exp}(Z, N) - M_{\mathrm{calc}}(Z, N)$ defined by deficiencies of the mass excess values calculated or learned by the trained network, relative to the experimental or database values. Projecting the error surface for the nets with analog input coding onto a plane of constant Z (or, alternatively, onto a plane of constant N), one finds exceptionally large negative errors at or near the magic numbers $N = 20, 28, 50, 82,$ and 126 (or $Z = 20, 28, 50,$ and 82). It is obvious that such networks are not able to recognize shell effects. Rather, the backpropagation training procedure creates a (generally) smooth mapping from (Z, N) to Δ, in accordance with the continuous nature conferred on the input and output variables by the analog coding prescription. In fact, the projected error plots for the pure analog nets bear a striking resemblance to the corresponding projections of the "experimental shell correction" determined by Myers and Swiatecki[15] and Möller and Nix[16] from the differences between experimental masses and the masses given by the liquid-droplet model. This model involves a semi-classical, continuum picture of nuclei which (like the analog input coding prescription) suppresses the discrete character of Z and N.

The global neural-network representation of the mass data is markedly improved when we revert to the binary input coding scheme used in the stability/instability problem. Binary coding of Z and N explicitly incorporates their integral nature and permits the system to make disjoint mappings for the different shell regions of the (Z, N) domain. Consequently, the large shell-edge errors and the associated scallop-shaped shell oscillations disappear from the error projection — just as they do when theoretical shell corrections are added to the liquid-droplet fits.[15,16] More details on this feature of our computer experiments may be found in Refs. 14,17. However, it is important to remark that the erroneous shell oscillations persist in nets with pure analog input coding even when large numbers of hidden neurons are introduced.

Although most trials have employed pure analog or pure binary input coding, we have also begun to investigate an apparently redundant "hybrid" scheme in which the 16-unit input layer of the standard binary representation of Z, N is supplemented by two additional neurons that code $A = N + Z$ and $N - Z$ as scaled analog variables. This arrangement is motivated by the fact that two of the leading non-trivial terms in the usual semi-empirical mass formula (the volume and asymmetry energies) are directly proportional to A and $(N - Z)^2$, considered as analog quantities. In our notation for network types, the hybrid scheme is specified by $c = h$.

Performance. To make quantitative assessments of performance in learning (fitting) and generalization (prediction), we adopt the conventional error measures of mean and rms deviations[18]:

$$\bar{D} = \Sigma_\mu (M_{\mathrm{calc}}^{(\mu)} - M_{\exp}^{(\mu)})/n \;, \qquad \sigma_{\mathrm{rms}} = [\Sigma_\mu (M_{\exp}^{(\mu)} - M_{\mathrm{calc}}^{(\mu)})^2/n]^{1/2} \;,$$

where the sum is performed over the n patterns or nuclides in the training or test set, as appropriate. For learning, σ_{rms} is directly proportional to the square root of the cost function C minimized by the backpropagation procedure (assuming single-unit

378

analog output coding), while $\bar{D}$ measures the size of any systematic overbinding or underbinding by the model that is developed. Selected results for various network types (with $c' = a$ and $c = a$ or b) are collected in Table 1. These results derive from training runs on the reduced data set formed by random omission of 25% of the nuclides. All training runs were long enough that the learning process had effectively reached completion. Nevertheless, substantial fluctuations in the cost function can still occur at about a 25% "noise" level due to the practical necessity of coursegraining the backpropagation search. The runs in question were stopped arbitrarily after about 2000 learning passes (in some cases more); with careful monitoring, appreciably better error figures can be obtained. The σ_{rms} and $\bar{D}$ values of the neural network models may be compared with those corresponding to conventional mass models, which generally involve 1500-1600 database nuclides in the fitting procedure. (For a compact survey of conventional results, see Ref. 18.) To provide some benchmarks, we have inserted error figures in Table 1 corresponding to one of the best of the Masson-Jänecke models[19] based on mass relations that generalize the Garvey-Kelson relations and to a late version of the Möller-Nix macroscopic-microscopic model.[16]

Table 1

Errors σ_{rms} and $\bar{D}$ in learning and prediction of atomic masses by neural networks of various types. Learning refers to the reduced data base of 1719 nuclei, prediction to the reserved test sample of 572. The last two lines give the error measures for two of the best traditional mass models,[19,16,18] which use respectively 1504 and 1593 database nuclei and 471 and 26 adjustable parameters. Units are MeV.

Net type	Learning error		Prediction error	
$^c(I + H_1 + \cdots + H_L + O)_{c'}[P]$	rms	mean	rms	mean
$^a(2 + 20 + 1)_a[81]$	5.254	1.165	5.100	1.001
$^a(2 + 60 + 1)_a[241]$	4.342	1.054	4.340	0.885
$^a(2 + 90 + 1)_a[361]$	10.219	5.086	10.231	5.041
$^a(2 + 10 + 10 + 1)_a[151]$	2.796	-1.192	2.929	-1.242
$^b(16 + 20 + 1)_a[361]$	2.013	1.156	2.278	-0.038
$^b(16 + 10 + 10 + 1)_a[291]$	1.499	-0.362	2.180	-0.278
$^b(16 + 10 + 10 + 10 + 1)_a[401]$	1.156	0.308	3.612	0.396
$^b(16 + 10 + 10 + 14 + 1)_a[449]$	1.569	-0.669	2.180	-0.559
Masson-Jänecke fit [471]	0.346	0.014	–	–
Möller-Nix fit [26]	0.849	0.013	–	–

The fitting problem represented by pure learning of the *full* data base is also of interest, although the performance measures for this case are less readily compared with those from traditional approaches. To date, the best results (stated in MeV) have been obtained with nets of types $^b(16 + 10 + 10 + 10 + 1)_a[401]$ ($\sigma_{\mathrm{rms}} = 1.008$, $\bar{D} = 0.005$), $^b(16+10+10+14+1)_a[449]$ ($\sigma_{\mathrm{rms}} = 0.932$, $\bar{D} = -0.049$), and $^h(18+10+$

$10+10+1)_a[421]$ ($\sigma_{\mathrm{rms}} = 0.697$, $\bar{D} = 0.010$). The second and third of these examples involved judicious choices of the termination points for weight changes. In the case of the net with hybrid input coding, the quoted results were arrived at by making three successive runs of 3000, 2000, and 2000 passes, with decreasing learning rates, respectively 0.5, 0.25, and 0.15. The weights corresponding to the smallest value of the cost function C found in the first run (which occurred at pass 2775) were used as a starting point for the second run; in turn the weights for the smallest C found in the second run (occurring at pass 1571) were used to start the third, which yielded the quoted minimum value of σ_{rms} (at pass 1939). The same strategy may lead to even smaller errors when the network is trained on the reduced data set.

Observations. The results in Table 1 clearly illustrate the superior quantitative performance of the nets with binary (as opposed to analog) input coding. Another general feature worth noting is the improvement in performance when the number of hidden layers is increased from one to two (which is seen to occur in spite of decreases in the number P of adjustable parameters); further improvement is possible upon going to three hidden layers. Perhaps the most remarkable feature, which attests to the validity of the underlying modeling process, is the rather modest decline in predictive accuracy relative to learning performance. With one exception, the networks with binary input coding display acceptable generalization ability. As expected, the nets with larger numbers of parameters, which learn more accurately but are susceptible to "overlearning," show greater relative differences between learning and prediction. (The error entries for the $^a(2 + 90 + 1)_a$ net are inconsistent with this observation. However, all three learning-prediction runs for this network type produced anomalous results; indeed, two independent backpropagation runs led to sets of weights such that the output unit remains "off." Such pathological behavior is presumably due to the excessively large number of units in the single hidden layer.)

It has yet to be demonstrated that this new approach to modeling the nuclear mass data can be competitive with mainstream approaches such as the macroscopic-microscopic model of Möller and Nix[16] or the sophisticated mass relations of Masson and Jänecke.[19] The required accuracy of fit (and of prediction) appears to be attainable; however, it is doubtful that the number of parameters can be reduced to the level of 30 or so typical of the models with the strongest physical motivation. However, our explorations of the potential of the neural network approach to this problem continue, with investigations of more advanced coding schemes, architectures, and training algorithms. Current effort focuses on various pruning schemes based on fault-simulation, skeletonization,[20] and optimal brain damage,[21] with a view to the enhancement of parametric efficiency and generalization ability.

ANALYSIS OF NEUTRON SEPARATION ENERGIES

Neutron separation energies $S(N; Z)$ provide another graded characterization of nuclear stability. Although the pertinent data base is more limited, they are more incisive than the masses themselves in revealing quantal features of the problem such as shell structure. We shall specialize to odd neutron numbers and even proton numbers, so as to suppress the even-odd periodicity due to pairing. When the data points for a common value of $N - Z$ are connected, a plot of $S(N; Z)$ versus N shows clear signs of neutron shell closures in the form of cusps, followed by sudden drops, near the magic numbers 8, 20, 28, 50, 82, and 126 (Ref. 22).

Data Base. We have extracted a set of 460 examples for $S(N; Z)$ that are especially well determined experimentally. These cases involve mass data for a large share of the stable nuclides. Training was performed using different subsets containing about 90% of these examples, the residual 10% being used to test prediction. One of the test samples included all of the data points for $N - Z = 19$.

Architecture, Coding, and Learning Procedure. Three-layer architectures have been studied, with "real-number" coding of input and output (thus $c = c' = r$, in our notation for network types). Training was by the standard backpropagation algorithm.

Performance. For this problem, we may again measure performance in terms of an rms deviation σ'_{rms} of the learned or predicted value from experiment, defined in a manner similar to σ_{rms} but of course with M replaced by S. The $S(N; Z)$ values in the data set are positive and typically around 6.5 MeV. Thus it also makes sense to consider the average value of $|S_{\text{exp}}(N; Z) - S_{\text{calc}}(N; Z)|/S_{\text{exp}}$ over the training or test set, which we denote by $\bar{d}'$ and express as a percentage. The results obtained in three of our experiments are presented in Table 2.

A strong test of predictive ability was made in the $^r(18 + 18 + 18)_r$ case, in a special run for which the test set contained all the examples with $N - Z = 19$. The net is remarkably successful in reproducing the missing line, especially the shell closure cusp.

Table 2

Errors in learning and prediction of even-Z, odd-N neutron separation energies by neural networks with real-number input and output coding. The quoted learning ("learn") errors refer to the accuracy of response on the 90% of the database used in training, and the prediction ("pred") errors, to the remaining 10%. Values of σ'_{rms} are in MeV.

Net type	σ'_{rms}(learn)	$\bar{d}'$ (learn)	σ'_{rms}(pred)	$\bar{d}'$(pred)
$^r(10 + 10 + 9)_r[209]$	0.143	1.63%	0.160	1.8%
$^r(18 + 18 + 18)_r[684]$	0.098	0.89%	0.117	1.6%
$^r(18 + 38 + 18)_r[1424]$	0.095	0.69%	0.197	1.9%

Observations. Networks of type $^r(I + H + O)_r$ acquit themselves very well in both learning and predictive aspects of the separation-energy problem. The results in Table 2 indicate that the larger nets are somewhat superior, but it is arguable whether or not the increase in precision is worth the greater parametric complexity. A tradeoff between accuracy of learning and reliability of prediction is also reflected in the table. The largest net learns with the highest accuracy, but its predictive performance is actually slightly worse than that of the smallest net. This set of experiments can be criticized for the use of an excessive number of parameters in comparison with the size of the training set; in defense it may be noted all three nets in the table show a respectable level of predictive accuracy.

In another study involving neutron separation energies, a network of type $^r(39 + 27 + 9)_r$ was used to extrapolate known semi-periodic trends in nuclear stability to the region of the hypothetical "magic island"[23] of stable, or nearly stable, super-heavy nuclei. Using a special training strategy, the network was taught to move systematically from one major stability region (major N-shell) to the next. Details are provided in Refs. 9,14,17. It is found that the next magic numbers in Z and N

beyond $Z = 82$ and $N = 126$ are shifted somewhat from the conventional shell model values of $Z_s = 126$ and $N_s = 184$ to 118 and 180, respectively.

LEARNING AND PREDICTION OF GROUND-STATE SPINS AND PARITIES

As an illustration of the application of neural networks to a highly nontrivial sorting problem in nuclear physics, we summarize the findings of some initial work on ground-state spins and parities. The network is asked to assign a total angular momentum quantum number J (or spin) and a parity π (even or odd) to the ground state of a given nuclide (Z, N). One goal of this work is to see if neural nets can apprehend certain key features of the addition of angular momenta in quantum mechanics, as well as those regularities of the data that are commonly described within the shell and collective models of nuclear structure. Another goal is to design networks that can predict J, π with some confidence for nuclides absent from the training set and possibly for exotic nuclei for which no measurements or empirical assignments are available.

Data Base. In the Brookhaven table, (J, π) values are given for 1889 nuclides, including assignments made from "weak" as well as "strong" arguments. This data base was broken down into data subsets corresponding to even-Z-even-N (EE), even-Z-odd-N (EO), odd-Z-even-N (OE), and odd-Z-odd-N (OO) nuclides, with populations of 575, 437, 442, and 435, respectively. Reduced data subsets of 523, 386, 401, and 392 nuclides were obtained by random deletion of tabulated nuclei with a probability 0.1, leaving test subsets of 52, 51, 41, and 43 in the respective classes. We refer to these reduced subsets and the corresponding test subsets as test sampling (a). Two additional test sampling, denoted (b) and (c), were generated in a similar way, but with the proviso that 5 "simple shell model cases" be included in each of the two odd-A classes (EO and OE). A "simple shell model nucleus" is defined as an odd-A nuclide in which the nucleonic subsystem containing the odd nucleon consists of a closed shell plus or minus one nucleon. The same 10 such special test nuclei were used in samplings (b) and (c). After removal of these nuclides from the data base, specification of the reduced training subsets of sampling (b) (and hence the corresponding test subsets) was completed by random deletion of individual database entries with probability 0.09; the same process was repeated independently in sampling (c).

Architecture, Coding, and Learning Procedure. Here we confine the discussion to results for three-layer feedforward nets with binary coding of the input variables (Z, N). The most obvious choice of output coding, adopted in our first round of experiments, is a unary representation in which one output neuron is associated with each quantized angular momentum value up to some reasonable cutoff, and an additional output neuron detects the presence of even parity. Setting the cutoff at $15/2$, the output layer then consists of 17 neurons: 15 to code for the spin values $J = 0, 1/2, 1, 3/2, ..., 7$, one to code for $J \geq 15/2$, and one for parity. According to our notational scheme, the nets studied are thus of type ${}^b(16 + H + 17)_u$. Standard backpropagation was used for training. We found rather poor performance in learning and prediction when a network (with $H = 20$) was trained on the *full* data base of 1889 examples or on a reduced set of approximately 90% of these patterns. Greatly improved results are obtained when individual nets are constructed for the four classes EE, EO, OE, and OO. Such nets (with $H = 20$, thus $P = 697$) learned the quantum mechanical restriction to integral spin values for EE and OO nuclides and to half-odd-integral spins for odd-A nuclei; these rules were *never* violated by the mature nets, either on the training set or the test sample. A further round of experiments was performed in which the output options were restricted to those permitted by quantum mechanics for the given nuclidic class, thus cutting down the number of output neurons by nearly one-half. Specifically, this work involved $^{b}(16+26+9)_{u}[685]$ nets for EO and OE classes and $^{b}(16 + 25 + 10)_{u}[685]$ nets for EE and OO, the last spin output neuron covering $J \geq 8$ and $J \geq 15/2$ for the even and odd-A nuclei, respectively. Another output coding scheme presently being examined maintains both integral and half-odd integral options, again with one unit for parity but also using one unit to signal half-odd-integral character for the spin; a further 10 output units serve to represent the integral part int(J) of the spin, 0 up to 8, with a "grab bag" unit to indicate int(J) ≥ 9. At this point we will only remark that, using this scheme, good results can be achieved with a *single net* trained on data sets containing all four classes of nuclides. A full account of such experiments will be given elsewhere.

Performance. In compiling performance figures, the angular momentum computed by the network for a specified input (Z, N) is decided by the "winner-takes-all" criterion among the output neurons dedicated to spin, and the parity assignment is taken as even [odd] when the parity neuron has activity $a_\pi \geq 0.5$ [$a_\pi < 0.5$]. An interpretation of the activities of the output neurons as probabilities attributed by the net to the corresponding outcomes requires a reformulation of the network design as described in Ref. 10; this more sophisticated treatment will be investigated in future work.

Results for individual EE, EO, OE, and OO networks trained on the reduced data subsets generated in test samplings (a)-(c) are shown in Table 3. It is to be noted that *all* mature networks constructed in our studies perform perfectly in both learning and prediction of the EE class. The simple fact that all even-even nuclei have spin-0, even-parity ground states is learned very quickly by these systems. Just as with human nuclear theorists, performance is rather poor on OO nuclides, so the most salient comparisons with established theory concern the odd-A examples.

Table 3

Percentages of successful assignment of spins and parities by networks of type $^{b}(16 + 20 + 17)_{u}[697]$. Separate networks are developed for even-Z-even-N (EE), even-Z-odd-N (EO), odd-Z-even-N (OE), and odd-Z-odd-N (OO) classes of nuclei. Respective training [test] samples involve 523 [52], 437 [51], 442 [41], and 435 [43] examples, from the full data base of 1889 nuclei. Learning performance refers to the reduced training sets.

Sampling	Spin-parity nets	Learning score		Prediction score	
	Nuclear class	Spin	Parity	Spin	Parity
(a)	EE	100%	100%	100%	100%
	EO	93	97	49	78
	OE	92	94	76	73
	OO	74	93	16	67
(b)	EO	92	99	64	91
	OE	95	95	60	81
(c)	EO	93	96	63	82
	OE	94	92	62	95

It may be remarked that learning performance of $^b(16 + 20 + 17)_u$ nets on the complete EE, EO, OO, and EE subsets is similar to that indicated in the table for the reduced subsets – better for parities, but about the same for spins. The predictive accuracy for the simple shell model examples in experiments based on test samplings (b) and (c) is of particular interest. It is found that in case (b) the spins and parities of these nuclides are all correctly predicted, and in case (c) the parities are all correctly reproduced and typical naive-shell-model errors are made on two of the spins, which are missed by one unit of angular momentum.

Information on performance for the nets of types $^b(16 + 25 + 9)_u$ and $^b(16 + 25 + 10)_u$ with unphysical spin outputs removed is limited to one run (on test sampling (a)). The learning scores are similar to those shown in Table 3 (slightly better for parities, slightly worse for spins). Prediction is somewhat worse for the odd-A nuclei, the accuracy dropping to around 50%. However, the main effect of the restrictive output coding appears to be a sharpening in the output decision, the "winning" spin neuron showing considerably greater dominance than in the experiments that admit both integral and half-odd-integral spin outputs for the same net.

Observations. State-of-the-art global nuclear structure calculations employing the macroscopic-microscopic approach[24] reproduce the ground-state spins of odd-A nuclei with an accuracy of 60% (agreement being cited in 428 cases out of 713). In this light, the results in Table 3 would appear to speak well for neural-network phenomenology. On the other hand, the network models involve an excessive number of parameters compared to the conventional physical models. Even so, it can be argued that neural nets show considerable promise for this problem, since many opportunities remain open for optimization of architecture, coding, and training. Indeed, some preliminary experiments suggest that the number of hidden neurons can be greatly reduced without adverse effect.

CONCLUDING REMARKS

The examples treated above should provide ample evidence that feedforward neural networks can be effective in learning and prediction – fitting and modeling – of the empirical properties of such a complex class of physical many-body systems as atomic nuclei. The overall strategy adopted has been to start with the simplest reasonable options for architecture, coding, and training procedure, i.e., the design of our networks has been guided only minimally by prior theoretical understanding of the class of systems under study. Having obtained significant and useful results under these conditions, it is obvious that we can do even better, either with more advanced forms of network connectivity, data representation, and weight modification, or with more specialized network designs incorporating elements of current nuclear models and tailored to specific phenomenological domains and the associated computational problems. We are now entering this second phase of the research, in which "nuclear" neural nets, which from the outset embody the broader aspects of established nuclear theory, can concentrate their resources upon the analysis of more subtle features of nuclear structure and dynamics. It is at this stage that we hope to learn something new from the artificial nuclear experts that have been created, by exploiting their predictive capabilities.

We anticipate that purely pragmatic implementation of neural network phenomenology along these lines can yield valuable new results, notably on nuclidic properties relevant to heavy-ion collisions and nuclear astrophysics. On the other hand, there is a clear need for a more fundamental theoretical understanding of the strengths and weaknesses of the neural-network approach in the nuclear context, and for complex many-body systems more generally. Along with other issues, this raises the problem of interpreting the parameters – the connection weights – of the neural models that are constructed. The most straightforward view of the present work is that the subset of training data has been fitted with rather complicated functions involving meshed sigmoids, by adjusting a rather *large* number of parameters (the weights). It will be an important goal of upcoming computer experiments to reduce the number of weights while maintaining accuracy of fit, with the expectation (or hope) that the corresponding increase in economy of the underlying model will entail an increase in its range of applicability.

There is, however, an alternative view in which the sterile notion of weights as fitting parameters recedes into the background. The neural network procedure is seen as automating what a model-building scientist does. Neural network phenomenology is a highly flexible, adaptive procedure for determining a model of the data environment, given only the *few* gross parameters characterizing the architecture, coding scheme, and learning rule (e.g. I, H, O, η, α) and an initial position in weight space. This view has merit to the extent that one can actually look inside the "black box" of the network, and determine what model, what rules, the system has arrived at through its experience. The deduction of these rules from the neuronic receptive fields is a highly nontrivial inverse problem which is beginning to receive serious attention (for example, see Ref. 25). In the solution of this challenging problem lies the prospect of discovering radically new models of physical systems.

Finally, we may offer still another interpretation of neural network analysis of complex many-body systems, a view that may strike a chord with the many-body theorist: the approach gives an operational prescription for transformation to an approximately equivalent many-body problem, one involving a system of "neurons" with highly specific interactions, a system which in turn carries out collective computation of salient properties of a given many-body system in the class under study. Among many other curious features of this transformation is the fact that *one* "equivalent" many-body system serves to generate the properties of *all* exemplars of the physical class – e.g., all nuclides, all atoms, all binary alloys, etc. "One size fits all."

This research was supported in part by the National Science Foundation under Grant No. PHY-9002863, and also partially by AFOSR Grant No. 89-0158 awarded to the Center for Optimization and Semantic Control at Washington University. We have benefited from useful discussions with S. Brunak, R. M. J. Cotterill, J. Jänecke, B. Lautrup, G. Senger, and E. B. Stockwell. J. W. C. thanks the Theoretical Physics Institute and the School of Physics and Astronomy, University of Minnesota, for hospitality and support during Summer 1991. K. A. G. gratefully acknowledges a postdoctoral fellowship award from the BASF Aktiengesellschaft and the Studienstiftung des deutschen Volkes.

REFERENCES

1. B. Müller and J. Reinhardt, *Neural Networks – an Introduction*, Springer, Heidelberg (1990).
2. J. W. Clark, "Neural network modelling," *Physics in Medicine and Biology* (in press).
3. D. E. Rumelhart, G. E. Hinton, and R. J. Williams, in *Parallel Distributed Processing: Explorations in the Microstructure of Cognition*, Vol. 1, D. E. Rumelhart, J. L. McClelland, *et al.*, eds., MIT Press, Cambridge, MA (1986).
4. H. Bohr, J. Bohr, S. Brunak, R. M. J. Cotterill, B. Lautrup, L. Nøskov, O. H. Olsen, and S. B. Petersen, *FEBS Letters* **B241**:223 (1988); N. Qian and T. J. Sejnowski, *J. Molec. Biol.* **202**:865 (1988); H. Bohr, J. Bohr, S. Brunak, R. M.

J. Cotterill, H. Fredholm, B. Lautrup, and S. B. Petersen, *FEBS Letters* **261**:43 (1990).

5. B. Denby, *Comput. Phys. Commun.* **49**:429 (1988); C. Peterson, *Nucl. Instr. Methods* **A279**:537 (1989); B. Denby and S. L. Linn, *Comput. Phys. Commun.* **56**:293 (1990); B. Humpert, *Comput. Phys. Commun.* **56**: 299 (1990).

6. J. R. P. Angel, P. Wizinowich, M. Lloyd-Hart, and D. Sandler, *Nature* **348**:221 (1990); D. Sandler, T. K. Barrett, D. A. Palmer, R. Q. Fugate, and W. J. Wild, *Nature* **351**:300 (1991); S. C. Odenwahn, E. B. Stockwell, R. L. Pennington, R. M. Humphreys, and W. A. Zumach, "Automated star/galaxy discrimination with neural networks," *Ap. J.* (in press).

7. B. Meyer, T. Hansen, D. Nute, P. Albersheim, A. Darvill, W. York, and J. Sellers, Science **251**: 542 (1991).

8. S. Brunak, J. Engelbrecht, and S. Knudsen, *J. Molec. Biol.* **220** (1991).

9. J. W. Clark, S. Gazula, and H. Bohr, "Teaching nuclear systematics to neural networks," in *Neural Networks: From Biology to High-Energy Physics*, O. Benhar, C. Bosio, P. del Giudice, and E. Tabet, eds., in press.

10. P. Stolorz, A. Lapedes, and Y. Xia, "Predicting protein secondary structure using neural net and statistical methods," *J. Molec. Biol.* (submitted).

11. D. G. Luenberger, *Linear and Nonlinear Programming*, Second Edition, Addison-Wesley, Reading, MA (1984).

12. F. W. Walker, D. G. Miller, and F. Feiner, *Chart of the Nuclides*, Thirteenth Edition, General Electric, San Jose, CA (1984).

13. J. W. Clark and S. Gazula, "Artificial neural networks that learn many-body physics," in *Condensed Matter Theories*, Vol. 6, S. Fantoni and S. Rosati, eds., Plenum, New York (1991).

14. S. Gazula, J. W. Clark, and H. Bohr, "Learning and prediction of nuclear stability by neural networks, *Nucl. Phys. A* (submitted).

15. W. D. Myers and W. J. Swiatecki, *Nucl. Phys.* **81**:1 (1966).

16. P. Möller and J. R. Nix, *Atomic Data and Nuclear Data Tables* **39**:213 (1988).

17. J. W. Clark, S. Gazula, and H. Bohr, "Nuclear phenomenology with neural nets," in *Complex Dynamics in Neural Networks*, E. R. Caianiello, J. W. Clark, R. M. J. Cotterill, and J. G. Taylor, eds., Springer, Heidelberg (1992), in press.

18. P. E. Haustein, *Atomic and Nuclear Data Tables* **39**:185 (1988).

19. P. J. Masson and J. Jänecke, *Atomic and Nuclear Data Tables* **39**:273 (1988).

20. M. C. Mozer and P. Smolensky, in *Neural Information Processing Systems*, Vol. 1, D. Touretzky, ed., Morgan Kaufmann, San Mateo (1989).

21. Y. Le Cun, J. S. Denker, and S. A. Solla, in *Neural Information Processing Systems*, Vol. 2, D. Touretzky, ed., Morgan Kaufmann, New York (1990).

22. A. Bohr and B. R. Mottelson, *Nuclear Structure*, Vol. I, W. A. Benjamin, New York (1969).

23. S. G. Thompson and C. F. Tsang, *Science* **178**:1047 (1972).

24. P. Möller and J. R. Nix, *Nucl. Phys.* **A520**:369c (1990).

25. C. McMillan, M. C. Mozer, and P. Smolensky, "The connectionist scientist game: rule extraction in a neural network," *Proceedings of the Thirteenth Annual Conference of the Cognitive Science Society*, Erlbaum, Hilsdale, NJ (1991).

THE COLLOIDAL MANY BODY PROBLEM:

COLLOIDAL SUSPENSIONS AS HARD SPHERE FLUIDS

E. G. D. Cohen

The Rockefeller University
New York, 10021, USA

I. M. de Schepper

IRI
University of Delft
Delft, the Netherlands

Abstract

In the last ten years computer simulations and modern kinetic theory have provided much physical insight into the microscopic properties of classical dense hard sphere fluids. In particular a quantitative understanding of the relaxation rate of microscopic structural deformations of the fluid has been obtained. These structural relaxations dominate the macroscopic diffusive and viscous behavior of the fluid. It appears that the same physics can be used to understand the much more complex system of a concentrated colloidal suspension of many Brownian particles in a solvent. In fact, explicit expressions for diffusion coefficients for concentrated colloidal suspensions will be given, obtained from those of hard sphere fluids. The physics of the origin of this connection will be discussed.

1   Introduction

Colloidal suspensions consist of particles large on the molecular scale but small on the macroscopic scale. i.e., particles of $10 - 1000$nm, suspended in a solvent, where they exert Brownian motion due to the irregular collisions with the molecules of the solvent. I will consider here concentrated colloidal suspensions of spherical particles, which constitute a complex strongly interacting system. That one can nevertheless say something about the behavior of this system is due to a simplification that will be discussed below. Two kinds of colloidal suspensions will be considered: those consisting of charged and those consisting of neutral Brownian particles. I will now summarize some of their properties.

1. <u>Charged Colloids</u>. An example is polystyrene spheres in water[1]. A negatively charged "pit" of about 200 electronic charges and diameter $\sigma_p \approx 50$nm is surrounded by a Debije cloud of positive counterions, such that the entire structure, pit and Debije cloud, has a diameter $\sigma_L \approx 300$nm. The diffusion coefficient D_0 of a single such Brownian particle in the surrounding solvent,for instance in water, is $D_0 \approx 0.86 \times 10^{-7}$cm^2/s. The interparticle interactions consist of a screened-Coulomb (Debije-Hückel) potential, van der Waals forces and, when moving, weak hydrodynamic interactions via the solvent.

2. <u>Neutral Colloids.</u> Examples are latex spheres in benzene[2] or silica spheres in cyclohexane[3]. The diameter of these hard-sphere-like particles is $\sigma_p \approx 220$nm, while their single particle diffusion coefficient D_0 is $D_0 \approx 0.80 \times 10^{-6}$cm^2/s. Their interparticle interactions are like those between hard spheres and, when moving, they exert strong hydrodynamic interactions on each other via the solvent.

Recent Progress in Many-Body Theories, Vol. 3,
Edited by T.L. Ainsworth et al., Plenum Press, New York, 1992

    The restriction to high concentrations means that the volume fraction of the particles,
i.e., that fraction of the total volume of the suspension occupied by the Brownian particles
alone, $\phi > 0.3$, where $\phi = \pi n\sigma^3/6$. Here n is the number density of the particles and σ,
the size of the particles, will be defined below. I note that at these high concentrations the
average surface to surface distance of two particles in the suspension is about $\sigma/4$, i.e., $< \sigma$.

    In the Brownian motion of the particles two time-scales can be distinguished[1]: 1) a
Brownian time scale $t_B \approx D_0/(k_BT/M)$, on which the particle forgets its initial velocity.
Here k_B is Boltzmann's constant, T the absolute temperature of the suspension, M the mass
of a Brownian particle and $t_B \approx 10^{-9}$s, typically; 2) an interaction time scale $t_I \approx \sigma^2/D_0$,
on which the particle diffuses a distance equal to its size. Typically $t_I \approx 10^{-3}$s and I am
interested here in times such that $t_B << t_I \leq t$. I note that D_0 is given by the Stokes-
Einstein relation. $D_0 = k_BT/3\pi\eta_0\sigma$, where η_0 is the viscosity of the solvent.

    I now summarize a few of the basic static properties of concentrated charged as well as
neutral colloidal suspensions. Because of the high concentrations, the particles exhibit a
short range ordering in their spatial arrangement. This is reflected in the behavior of the
(equilibrium) radial distribution function $g(r)$,which is proportional to the average density
of particles at a distance r from the origin, given that there is a particle at the origin.
It is defined such that it oscillates with decreasing amplitude as a function of increasing
r around the asymptotic value of 1 for $r \to \infty$. In concentrated suspensions and dense
atomic fluids $g(r)$ shows a pronounced maximum $\gg 1$ at $r \approx \sigma$. I note for future reference
that the $g(r)$ for a dense fluid of hard spheres with diameter σ has a similar oscillatory
behavior, but vanishes for $r < \sigma$ because of the impenetrability of the particles. It also
shows a very pronounced maximum at $r = \sigma$, such that $\chi = g(\sigma)$ can be $\gg 1$. In fact,
χ varies from a value of 1 for low densities to a value of about 6.7 at the high density
$n\sigma^3 = 1$ or $\phi = 0.52$. The sharp maxima of $g(r)$ at $r \cong \sigma$ imply that each particle in a
dense fluid or suspension finds itself on the average in a cage formed by its neighbors. A
Fourier transform of $g(r)$ yields the (equilibrium) static structure factor $S(k)$, which gives
the amplitude of finding a periodic particle density distribution in the colloidal suspension
or hard sphere fluid with wave number k, in the Fourier decomposition of $g(r)$. $S(k)$ exhibits
an oscillatory behavior with decreasing amplitude as a function of k around the ideal gas
value of 1 reached for $k \to \infty$ and it has a sharp maximum > 1 at a value k^* for which
$k^*\sigma = 2\pi$, i.e., $\lambda^* = 2\pi/k^* = \sigma$. This indicates that the ordering of the particles occurs
predominantly on a length scale λ^* determined by the size σ of the particles.

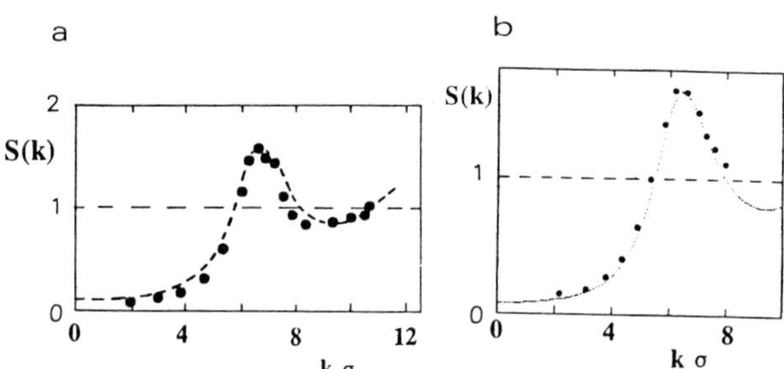

Fig. 1 Determination of effective hard sphere diameter of charged (a) and neutral (b) colloidal suspensions by plotting $S(k)$ as a function of $k\sigma$. (a) closed circles: experimental $S(k)$ for polystyrene
spheres in water for $\phi = 0.34$ (ref.1, p.129); - -,best fit with hard sphere $S(k)$ around maximum of
$S(k)$, for hard sphere diameter $\sigma = 310$nm, which is close to the diameter of the Debije cloud; (b)
same for latex in benzene for $\phi = 0.31$ (ref.2); closed circles: experiment; ....., best fit with hard
sphere $S(k)$, leading to $\sigma = 219$nm, close to the actual latex sphere diameter.

In the following we will assign the Brownian particles a size σ, or an equivalent hard sphere diameter σ, determined from the $S(k)$ of the suspension in the following way: we require σ to be such that the $S(k)$ of a fluid of hard spheres with diameter σ matches the $S(k)$ of the colloidal suspension, for k-values around k^* (cf. fig 1). Since such a matching is possible, those static properties, which are dominated by the behavior of $S(k)$ for k values near k^*, can be considered effectively to be like those of a (pure) hard sphere fluid, characterized by a single parameter σ. We will use this same σ to characterize dynamic properties of the concentrated suspensions, by mapping also those onto the corresponding ones of hard sphere fluids. To this end, I will now briefly discuss the diffusive behavior of dense hard sphere fluids.

2 Dense Hard Sphere Fluids

In the last ten years much progress has been made in the understanding of the dynamical (non-equilibrium) properties of dense hard sphere fluids. This is due to: 1) extensive computer simulations[4] and 2) the development of modern kinetic theory[5]. I will restrict myself here to the self-diffusion coefficient D_s. This coefficient characterizes the diffusion of one (tagged) particle through the (identical) other particles of the fluid. In this self-diffusion process, modern kinetic theory has identified three main physical processes that contribute to D_s.

1. Boltzmann diffusion, where uncorrelated binary collisions of the particles with each other are taken into account. Uncorrelated means here that the particles in the fluid (tagged as well as any of the others) collide with each other only once, or, equivalently, that no memory effects due to previous collisions are taken into account in recollisions of the same two particles. These uncorrelated particle collisions are already taken into account in the Boltzmann equation for dilute gases[5,6]. They contibute on a time-scale of about 3-5 collision times, i.e., for $0 < t < 3t_0$ to $5t_0$, where t_0 is the mean free time between successive collisions, given by $t_0 = (m/\pi k_B T)^{1/2}/4n\sigma^2\chi$, with m the mass of a hard sphere. The other two processes incorporate correlated binary collisions.

2. Cage (retarded) diffusion, where correlated binary collisions are taken into account associated with the "rattling" of the particles in the cages formed by their neighbors (cf. fig.2a)[7]. Clearly this contribution delays the diffusion process. The time scale of these contibutions is roughly is $t \geq 6t_0$. I note that the cage-diffusion is a "micro-diffusion" on the molecular scale, characterized by a length $\lambda^* \approx \sigma$ (or $k \approx k^*$), very different from the "macro-diffusion" on the hydrodynamic scale, characterized by $\lambda \approx L$ or $k \approx 0$, where L is a macroscopic length.

3. Vortex (enhanced) diffusion, where correlated binary collisions are taken into account, associated with a hydrodynamic-like backflow of fluid particles around a moving particle, as if this particle was a macroscopic sphere moving through the fluid (cf. fig.2b). Since the backflow ultimately returns to the particle and "kicks it in the back", it enhances the diffusion process. This backflow is referred to in the literature as "long time tails"[8] since the time scale of these contributions is long, viz. $t \geq 30t_0$. I note that the vortex diffusion is an orderly propagation of momentum via collisions through the fluid from particle to particle, where this orderliness is rooted in the conservation of momentum in particle collisions.

I will now give an approximate expression for the cage diffusion coefficient D_{cd}^{hs}, which characterizes the effectiveness of the particles to diffuse in and out of their cages. Before doing so, I remark, that while the actual diffusion process takes place in space and time, both theory and experiment are done in terms of Fourier components, i.e., in wave vector $\vec{k}$ and time language. This way the theory is simpler, since one is concerned with approximately independent $\vec{k}$-modes, while the experiments yield naturally results in $\vec{k}$-language, since they are performed by light scattering, which measures directly the Fourier components in a Fourier decomposition of the particle density fluctuations in space. Thus, when I speak of modes in the following, I mean either literally the (independent) eigenmodes of a linear kinetic operator in $\vec{k}$-language or figuratively a mode of motion in $\vec{k}$-language, i.e., a mode of motion of a periodic density profile.

The above mentioned expression for the cage-diffusion coefficient of a dense hard sphere fluid D_{cd}^{hs} is k-dependent and is derived from kinetic theory. The very complicated many particle problem of the motion of a central (tagged) particle moving in a (non-stationary) cage, consisting of many interacting particles, is replaced by an effective two particle problem. The equation which then takes into account the correlated motion of these two particle refers to the central particle and one "typical" cage-wall particle, where the presence of the cage is characterized by the occurrence in the equation of $S(k)$ and its sharp maximum at $k = k^*$. This equation is a generalization of the Boltzmann equation to high densities - sometimes called the generalized Enskog equation[5] - and $D_{cd}^{hs}(k)$ is then the lowest eigenvalue of a linear operator directly associated with this equation. In this linear operator, the correlated motions of all particles in their respective cages is taken into account in a self-consistent way in a two particle approximation, so that $D_{cd}^{hs}(k)$ is a collective diffusion coefficient. In other words, the operator recognizes that each particle is at the center of a cage as well as part of the walls of the cages of each of its neighbors (cf.fig.2a).

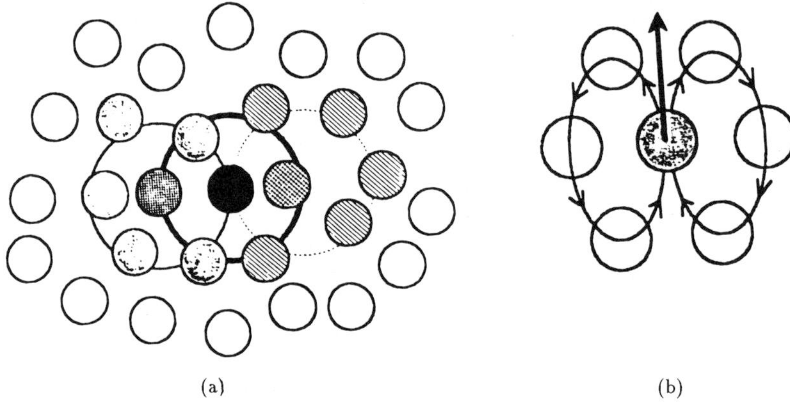

(a)   (b)

Fig. 2 (a) The central (tagged) particle (black) is in a cage whose wall is formed by the particles connected by the thick black line. Each wall particle is itself the center of a cage, of which the black particle is part of the wall. This is illustrated for two wall particles of the black particle, where cage walls are formed by particles connected by a thin line or a dotted line, respectively; (b) vortex motion induced into the neighboring (white) particles by a forward moving (dark) particle, with black arrow attached. Sitting on the forward moving particle, momentum transfer to the neighboring particles through collisions, induces a systematic backflow component in the velocities of these particles. This backflow ultimately returns as a forward flow to the forward moving particle, thus enhancing its diffusive motion.

The following expression for the short time cage diffusion coefficient $D_{cd}^{hs}(k)$ is then obtained:[7],[9],[10]

$$D_{cd}^{hs}(k) = \frac{D_B}{\chi} \cdot \frac{d(k)}{S(k)}, \tag{1}$$

where

$$d(k) = \frac{1}{1 - j_0(k\sigma) + 2j_2(k\sigma)}, \tag{2}$$

with $j_\ell(k\sigma)$ the spherical Bessel function of order ℓ. I note that the k-dependence of both $d(k)$ and $S(k)$ is via $k\sigma$. The eq.(1) is valid for cage-diffusion for short times of the order of $t \approx 6t_E$ and for $k \approx k^*$. It expresses the collective cage diffusion coefficient $D_{cd}^{hs}(k)$ in terms of the product of a cage-self-diffusion coefficient $D_{cd,s}^{hs} = D_B/\chi$[6] and a cage-factor terms of the product of a cage-self-diffusion coefficient $D_{\mathrm{cd,s}}^{\mathrm{hs}} = D_B/\chi$[6] and a cage-factor $d(k)/S(k)$. Here $D_B = 0.214(k_B T/m)^{1/2} n^{-1} \sigma^{-2}$ is the low density hard sphere gas self-diffusion coefficient, obtained from a linear Boltzmann equation and the other three factors represent corrections to this Boltzmann result: a) χ expresses the decrease of the diffusion coefficient in a dense fluid, as compared to that in a dilute gas, because of the increased probability to find two particles colliding, i.e., separated by a distance σ; b) $d(k)$ results from the instantaneous transfer of momentum between the central and a wall particle upon collision, i.e., the "knocking on the wall" and c) $1/S(k)$ expresses the ordering of the fluid particles, since $S(k \approx k^*) \gg 1$. This ordering increases the difficulty of diffusion and decreases therefore the cage-diffusion coefficient. The k-dependence of $D_{\mathrm{cd}}^{\mathrm{hs}}(k)$ is related to the finite size of the cage ($\sim 2\pi/k^*$).

$D_{\mathrm{cd}}^{\mathrm{hs}}(k)$ is the central quantity of the theory of the dynamical properties of dense hard sphere fluids. It expresses physically the structural relaxation rate in dense hard sphere fluids of a local structure or density disturbance via particle diffusion.

I will now use the results for $D_{\mathrm{cd}}^{\mathrm{hs}}(k)$ of a dense hard sphere fluid to obtain the corresponding cage-diffusion coefficient for a concentrated colloidal suspension, via a simple scaling relation.

3   Cage diffusion in concentrated colloidal suspensions

The scaling relation between dense hard sphere fluids and concentrated colloidal suspensions is based on the following comparison:

Hard Spheres	Colloids	Ratio
$\sigma \approx 10^{-1} \mathrm{nm}$	$\sigma \approx 10^{2} \mathrm{nm}$	10^{-3}
$D_B \approx 10^{-4} \mathrm{cm}^2/\mathrm{s}$	$D_0 \approx 10^{-7} \mathrm{cm}^2/\mathrm{sec}$	10^{3}
$t_{hs} \approx \sigma^2/D_B \approx 10\mathrm{ps}$	$t_I \approx \sigma^2/D_0 \approx 10^{10}\mathrm{ps}$	10^{-9}
Newtonian dynamics in vacuo: 5 conservation laws (mass, momentum, energy)	Brownian dynamics in solvent: 1 conservation law (mass)	

The scaling is to replace the low density Boltzmann diffusion coefficient of the hard sphere fluid D_B by the low density Stokes-Einstein diffusion coefficient of the colloidal suspension D_0. The corresponding time scales, taking into account the difference in particle size, differ then by a factor 10^9. Thus the idea is that the cage diffusion in the concentrated colloidal suspension is physically the same as that in a dense hard sphere fluid, except that it takes place very much slower, due to the cumbersome Brownian motion of the particles in the solvent rather than via free flight of hard spheres in vacuo. Using this analogy and eq.(1) we postulate for the cage diffusion coefficient $D_{\mathrm{cd}}^{\mathrm{coll}}(k)$ of a concentrated colloidal suspension[10]:

$$D_{\mathrm{cd}}^{\mathrm{coll}}(k) = \frac{D_0}{\chi} \cdot \frac{d(k)}{S(k)} \tag{3}$$

for time scales $t \approx 6t_I$ and $k \approx k^*$.

I will now discuss three experimental checks of eq.(3) for $k \approx k^*$ and $\phi > 0.3$.

1. The k-dependence of $D_{\mathrm{cd}}^{\mathrm{coll}}(k)$ for k near k^* is checked in fig.3. The experiments for both charged and neutral colloids are done with coherent light scattering. I note the sharp minimum of $D_{\mathrm{cd}}^{\mathrm{coll}}(k)$ due to the sharp maximum of $S(k)$ at $k = k^*$.

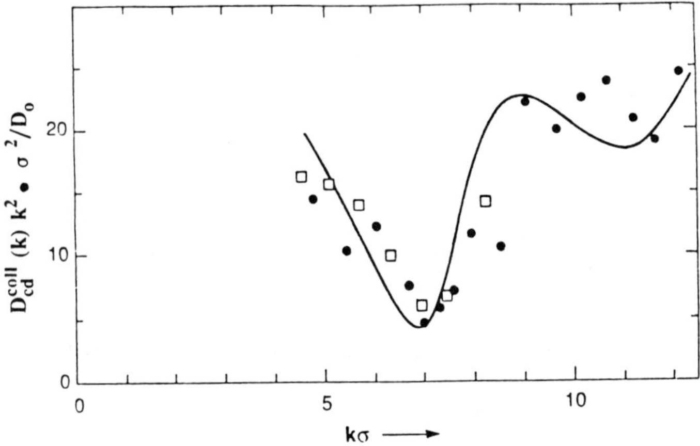

Fig. 3 Reduced decay rate $D_{cd}^{coll}(k)k^2\sigma^2/D_0$ as a function of $k\sigma$ for a charged colloid (circles), a neutral colloid (squares) and from theory (curve; eq.(3)).

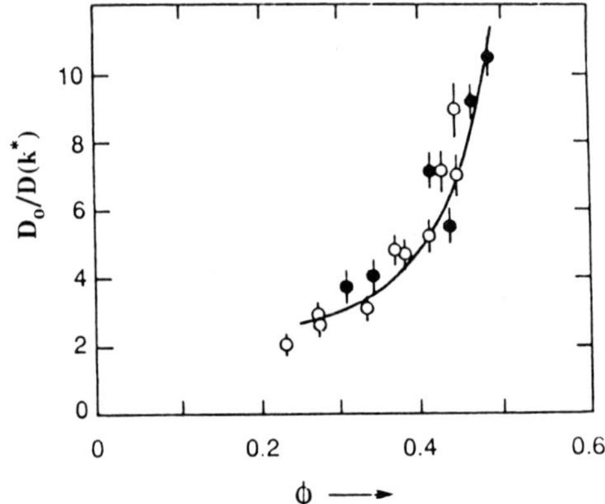

Fig. 4. Minimum value of $D_{cd}^{coll}(k^*) \equiv D(k^*)$, illustrated by $D_0/D(k^*)$, as a function of ϕ for charged colloids (closed circles), neutral colloids (open circles) and from theory (curve; eq.(3)).

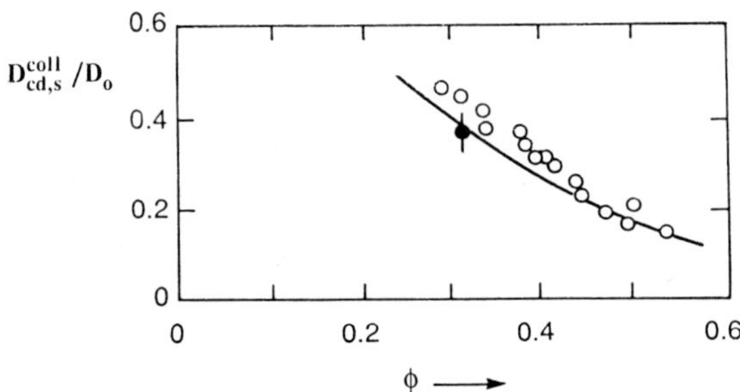

Fig. 5 Reduced cage self-diffusion coefficient $D_{cd,s}^{coll}/D_0$ as function of ϕ for a charged colloid (closed circle), for neutral colloids (open circles) and from theory (curve; eq.(3)).

2. $D_{\text{cd}}^{\text{coll}}(k^*)$ versus ϕ is checked in fig.4. The concentration, i.e., ϕ-dependence, of $D_{\text{cd}}^{\text{coll}}(k^*)$ comes via those of χ and $S(k)$.

3. For increasing $k > k^*$, $D_{\text{cd}}^{\text{coll}}(k)$ goes over in the cage-self-diffusion coefficient $D_{\text{cd,s}}^{\text{coll}} = D_0/\chi$, since $\lim_{k\to\infty} d(k) = \lim_{k\to\infty} S(k) = 1$. This can be understood physically by noting that $k \to \infty$, implies $\lambda \to 0$, and that in this limit only a single (tagged) particle is considered and all collective effects due to the other particles - incorporated in the two-particle functions $d(k)$ and $S(k)$ - disappear. In fig.5, $D_{\text{cd,s}}^{\text{coll}}$ is plotted versus ϕ.

The good agreement between eq.(3) and experiment gives support to the analogy of structural relaxation in dense hard sphere fluids and concentrated charged or neutral colloidal suspensions.

4 Self-diffusion in dense hard sphere fluids

The macroscopic self-diffusion coefficient D_s^{hs} of a tagged particle in a hard sphere fluid refers to the self-diffusion process on a much longer time scale than the cage diffusion process[10].

In fact, it incorporates the long time cumulative effects of many successive cage as well as vortex diffusions. Thus it requires a kinetic theory not just for short-times $t \approx 6t_0$, but for times $t \gg 6t_0$ as well. The result of such a theory is an expression for D_s^{hs}, which is a product of a short-time self-diffusion coefficient and the cumulative cage and vortex diffusion contributions that contain the long time contributions to D_s^{hs}:

$$D_s^{\text{hs}} = \frac{D_B}{\chi} \cdot \frac{1}{1 + \sum_{cd} + \sum_{vd}}, \tag{4}$$

with

$$\sum_{cd} \cong \frac{D_B}{6\pi n^2 \chi} \int_{6t_0}^{\infty} dt \int_0^{\infty} dk\, k^4 [S(k) - 1]^2 e^{-D_{\text{cd}}^{\text{hs}}(k)k^2 t} \cdot e^{-D_{\text{cd,s}}^{\text{hs}} k^2 t} \tag{5}$$

The first factor in (4), D_B/χ is a short-time contribution, since it is identical with the cage self-diffusion coefficient, $D_{\text{cd,s}}^{\text{hs}}$. The $\sum_{cd}$ and $\sum_{vd}$ in the second factor contain the cumulative contributions of cage diffusion and vortex diffusion respectively, integrated over $\vec{k}$ and t. $\sum_{cd}$ involves mode-mode coupling contributions from a cage-diffusion mode, represented by $D_{\text{cd}}^{\text{hs}}(k)$ and a self-diffusion mode, represented by $D_{\text{cd,s}}^{\text{hs}}$, respectively, in eq. (5). The exponentials in this equation are proportional to the amplitudes of a cage-diffusion mode and a cage-self-diffusion mode at time t, where $D_{\text{cd}}^{\text{hs}}k^2$ and $D_{\text{cd,s}}^{\text{hs}}k^2$ are the inverse relaxation times of these two modes, respectively. $\sum_{vd}$ involves similarly mode-mode coupling contributions due to two hydrodynamical modes: a viscous and a (self) diffusion mode. For a discussion of $\sum_{vd}$, I refer to the literature[8]. The appearance of the coupling of two modes in $\sum_{cd}$ and $\sum_{vd}$ is ultimately due to the fact that $\sum_{cd}$ and $\sum_{vd}$ involve the motion of two particles, which translates into integrals over $\vec{k}$ and time of two modes, each of which is associated with one of the two particles: e.g., for $\sum_{cd}$, $D_{\text{cd,s}}^{\text{hs}}$ with the central (tagged) particle and $D_{\text{cd}}^{\text{hs}}(k)$ with a wall particle of a cage. At the high densities we consider here, the time integral in eq.(5) is not very sensitive to the precise choice of the lower limit, $6t_0$.

5 Self-diffusion in concentrated colloidal suspensions

Using the same scaling as before, we postulate for the self-diffusion coefficient of a concentrated colloidal suspension:

$$D_s^{\text{coll}} = \frac{D_0}{\chi} \cdot \frac{1}{1 + \sum_{cd}}, \tag{6}$$

where $\sum_{cd}$ is given by (5). I note that there is no contribution here of $\sum_{vd}$, since no vortex enhanced diffusion can take place via colloidal particle collisions in colloidal suspensions, because of the absence of momentum conservation. For, unlike the mass, the momentum of the colloidal particles is not conserved in a collision, as it is dissipated in the surrounding solvent. The difference between (4) and (6) is clearly demonstrated in fig.6. Here we see good agreement between theory and experiment for both hard spheres and colloids, the difference between the two curves being entirely due to $\sum_{vd}$. The experimental data for the colloids follow from incoherent light scattering (or coherent light scattering for such large k, that $d(k)/S(k) \cong 1$), while for the hard spheres they were taken from computer simulations.

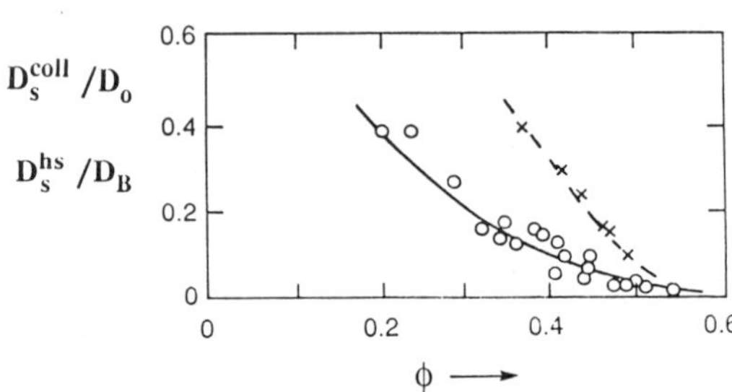

Fig. 6 Reduced self-diffusion coefficient D_s^{coll}/D_0 as function of ϕ for neutral colloids (circles) and solid curve, eq.(6); D_s^{hs}/D_B of hard spheres (crosses) and dashed curve, eq.(4).

6 Discussion

1. The simplification mentioned in the introduction that allows the derivation of explicit expressions for diffusion coefficients of colloidal suspensions from those of hard sphere fluids is due to the dominance of cage-diffusion at high densities in both systems. This, in turn, is due to the fact that at high densities on the single particle scale, i.e., for length scales $\lambda \approx \sigma$, the density, i.e., the number of particles, is the only relevant conserved quantity in collisions, both in hard sphere fluids and in colloidal suspensions[9]. In other words, for $\lambda \approx \sigma$, of the five conserved quantities in Newtonian dynamics only one is relevant.

2. A similar analogy as discussed here for self-diffusion also holds for the shear viscosity η^{hs} of dense hard sphere fluids and η^{coll} of concentrated colloidal suspensions [10,11]. In that case the fluid is subjected to an external shearing force, inducing a local fluid velocity gradient or shear rate γ in the fluid. Since the contribution of the vortex diffusion $\sum_{vd}$ to η^{hs} can be neglected at high densities $\phi > 0.3$, the viscosity η^{coll} can be obtained from η^{hs}, directly by an appropriate scaling, viz., by replacing in η^{hs}/η_B, the low density Boltzmann viscosity η_B, obtained from the Boltzmann equation, by the low density viscosity of the suspension η_0, the viscosity of the solvent, so that η^{coll}/η_0 is given by the same expression as η^{hs}/η_B. Also in this case one has good agreement between theory and experiment[10,11].

3. For the viscous case discussed under 2., a hint of a possible theoretical justification of the scaling procedure can perhaps be seen in the following formal analogy. Dhont[12] has derived from the Smoluchowski equation, the basic equation for colloidal suspensions, an equation for the non-equilibrium conditional probability $g(\vec{r}, \gamma)$ in a dilute charged colloidal suspension with a shear rate γ. For $\gamma = 0, g(\vec{r}, \gamma)$ reduces to the equilibrium $g(r)$ discussed in the introduction. Similarly, Kirkpatrick[13] has derived from the Liouville equation, the basic equation for hard sphere fluids, an equation for $g(\vec{r}, \gamma)$ of a dense hard sphere fluid with a shear rate γ. The two equations become formally identical, if one modifies Dhont's equation to that for a hard sphere potential and Kirkpatrick's equation to that for a dilute hard sphere gas and replaces then the Boltzmann diffusion coefficient D_B in his equation by the Stokes-Einstein diffusion coefficient D_0 in Dhont's equation.

References

1. P.N. Pusey and R.J.A. Tough in: "Dynamic Light Scattering", R. Pecora, ed., Plenum, New York (1985), p.85.
2. H.J.M.Fijnaut, C. Pathamananoharan, E.A. Nieuwenhuis and A. Vrij, Chem. Phys.Lett. $\underline{59}$, 351 (1978).
3. A.K. van Helden, J.W. Jansen and A. Vrij, J. Colloid Interface Sci.,$\underline{81}$, 354 (1981).
4. See, e.g., J.J. Erpenbeck and W.W. Wood, Phys. Rev. A $\underline{43}$, 4254 (1991).
5. See, e.g., J.R. Dorfman and H. van Beijeren, in: "Statistical Mechanics", Part B, B.J. Berne, ed., Plenum, New York (1977) p.65.
6. S. Chapman and T. G. Cowling, "The Mathematical Theory of Non-Uniform Gasas", Cambridge University Press (1953).
7. I.M. de Schepper, E.G.D. Cohen, P.N. Pusey and H.N.W. Lekkerkerker, J. Phys. Cond. Matter $\underline{1}$, 6503 (1989).
8. B. J. Alder and T. E. Wainwright, Phys. Rev. A $\underline{1}$, 18 (1970); Y. Pomeau and P. Résibois, Phys. Rep.$\underline{19C}$, 64 (1975).
9. E.G.D. Cohen, I.M. deSchepper and M.J.Zuilhof, Physica B $\underline{127}$, 282 (1984).
10. E.G.D. Cohen and I.M. de Schepper, J. Stat. Phys. $\underline{63}$, 241 (1991).
11. I.M. de Schepper and E.G.D. Cohen, Phys. Lett A $\underline{150}$, 308, (1990).
12. J.K.G. Dhont, J.C. van der Werff andC. G. Kruif, Physica A $\underline{160}$, 47 (1989).
13. T.R. Kirkpatrick, J. Non-Cryst. Solids $\underline{75}$, 437 (1985).

Acknowledgement: One of the authors (EGDC) is indebted to the Department of Energy for grant number DE-FGO2-88-13847 under which part of this work was performed.

APPLICATIONS OF MAXIMUM ENTROPY AND BAYESIAN METHODS

IN COMPUTATIONAL MANY-BODY PHYSICS

R. N. Silver, J. E. Gubernatis, D. S. Sivia†

Theoretical Division and Los Alamos Neutron Scattering Center
MS B262, Los Alamos National Laboratory, Los Alamos, NM 87545

M. Jarrell

Department of Physics, University of Cincinnatti, Cincinnatti, OH 45221

H. Roeder

Dept. of Physics, University of Bayreuth, W–8850 Bayreuth, Germany

INTRODUCTION

Computer simulations of many–body systems can present difficult data analysis problems. An example is the determination of dynamical properties from quantum Monte Carlo (QMC) simulations. Data on imaginary time Matsubara Greens' functions implicitly contain information about real time behavior. However, the spectral representation which relates imaginary time to real time is similar to a Laplace transform, and the numerical inversion of such transforms is notoriously unstable in the presence of statistical noise. Another example is the determination of the densities of states and the thermodynamic functions from imprecise knowledge of a finite number of moments. Data analysis problems in computer simulations are similar in many respects to those found in experimental research. They are often inverse problems, they can be ill–posed, the data may be incomplete, and they may be subject to statistical and systematic errors. Fortunately, the recent conceptual and algorithmic advances in maximum entropy (MaxEnt) and Bayesian data analysis methods[1] are equally applicable to computer simulations. We illustrate such applications by calculating the dynamical properties of the Anderson model for dilute magnetic alloys and the densities of states of the 2–D Heisenberg model.

The generic problem we address has the form

$$D(x) = \int dy \, R(x,y) \, O(y) \, + \, \delta D(x) \tag{1}$$

where $D(x)$ are the data, $O(y)$ is the object we wish to infer from the data, $R(x,y)$ is an integral

Recent Progress in Many-Body Theories, Vol. 3,
Edited by T.L. Ainsworth et al., Plenum Press, New York, 1992

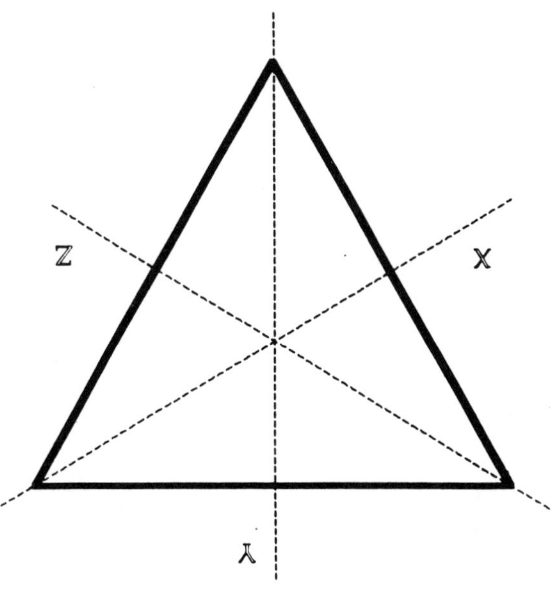

Fig. 1a Definitions of the projections of a two–dimensional distribution.

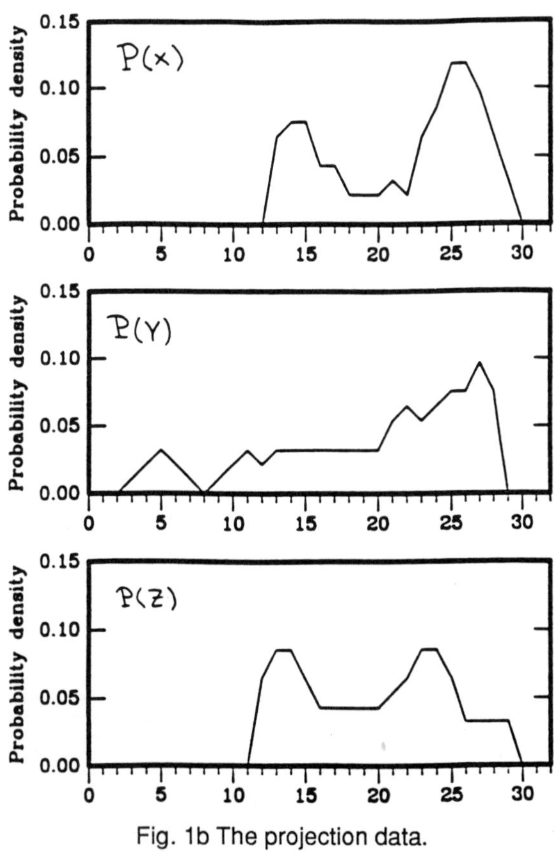

Fig. 1b The projection data.

transform, and $\delta D(x)$ represents the uncertainties in the data. They may be described by a covariance matrix

$$C(x,x') \;=\; <\delta D(x)\delta D(x')> \quad . \tag{2}$$

which must also be estimated from the simulation. One wishes to make a best estimate of the object from the available data along with estimates of the statistical errors on the object and any integrated function of it.

For example, for the analytic continuation of Quantum Monte Carlo data the spectral representation is

$$\overline{G}(\tau) \;=\; \int\limits_{-\infty}^{\infty} d\omega \; \frac{e^{-\omega\tau}}{1 \pm e^{-\beta\omega}} \; A(\omega) \;+\; \delta G(\tau) \quad , \tag{3}$$

where $\overline{G}(\tau)$ is the Greens' function data sampled at a discrete set of imaginary times τ_i, $A(\omega)$ is the spectral function we wish to infer, and $\delta G(\tau)$ is the uncertainty in our knowledge of $G(\tau)$ from the simulation. For the moment problem, the relation to the density of states is

$$\overline{M}_n \;=\; \int\limits_{-\infty}^{\infty} dE \; N(E) \; E^n + \delta M_n \quad , \tag{4}$$

where $\overline{M}_n$ is the moment data, $N(E)$ is the density of states we wish to infer, and δM_n is the uncertainty in our knowledge of the moments. Usually, it is feasible to calculate only a finite number of moments. One may be most interested in thermodynamic functions which are formed by weighted integrals over $N(E)$.

An illustrative example of such problems is that of inferring a two–dimensional distribution, $f(x,y)$, from data on a few projections at angles θ_i defined by

$$P_i(L) \;\equiv\; \int dxdy \; f(x,y) \; \delta(L - x\cos\theta_i - y\sin\theta_i) \quad . \tag{5}$$

Figure 1a,b shows such data.[2] The usual first response to such a problem is that the answer isn't unique, which is true. But it would be overly pessimistic to conclude that the problem is therefore impossible. The next response is to invent a model for $f(x,y)$ as, say, a sum of Gaussians (or Lorentzians, or whatever) with positions, widths and intensities as adjustable parameters to be fit to the data. This model fitting approach often works when one has prior knowledge which can justify the choice of model. But what if we can't get a good fit as a sum of Gaussians, or what if it requires too many Gaussians? Our response to such problems is to determine $f(x,y)$ from a variational principle. We find a functional, $S(f)$, to maximize subject to the constraints of the data. That is, maximize

$$Q \equiv S(f) - \int dL \sum_i \lambda_i(L)P_i(L) \quad , \tag{6}$$

as a functional of $f(x,y)$, where the $\lambda_i(L)$ are Lagrange multipliers. Fig 1c illustrates the resulting inference of $f(x,y)$ for our choice of $S(f)$.

How do we choose $S(f)$? This is illustrated by the Kangaroo[3] problem: *one third of Kangaroo's are left handed and two thirds are right handed; one third of kangaroo's are green eyed and two thirds are brown eyed; what fraction of kangaroo's are both left handed and green eyed?* A truth table, Fig. 2, illustrates the problem. In the squares are the possible values of f_i given the information. For example, if the number of left handed and green eyed kangaroo's is denoted by

Fig. 1c MEM reconstruction from the data in Fig. 1b.

	Left	Right	
Green	X	1/3 – X	1/3
Brown	1/3 – X	1/3 + X	2/3
	1/3	2/3	

Fig. 2 Truth table for the kangaroo problem.

$f_1 = x$, then the number of brown eyed right handed kangaroos must be $f_4 = 1/3 + x$. The only free parameter in the problem is x. The common sense answer to this problem is that most likely there is no correlation between handedness and eye color, and therefore the best estimate is the product of separate probabilities, i.e. $x=1/9$. But what do various choices for $S(f)$ produce? The results for some possible $S(f)$ are:

$S(f)$	x
$-\sum_i f_i^2$	1/12
$\sum_i \log f_i$	0.13013...
$\sum_i \sqrt{f_i}$	0.12176...
$-\sum_i f_i \log f_i$	1/9

The last $S(f)$ on this list is the *entropy*, and one can prove that only entropy gives the common sense (no correlations) answer of 1/9! The choice of entropy for the variational principle can be given a more rigorous mathematical justification[4,5,6] in terms of a few axioms which constitute desiderata for any statistical inference procedure.

Our examples of the projection problem and the kangaroo problem did not treat the possibility of statistical errors in the data, but this may be introduced via the law of conditional probabilities which is also known as *Bayes' theorem*. In our case, it reads[5]

$$P[f|D,I] \times P[D|I] = P[D|f,I] \times P[f|I] \quad . \tag{7}$$

Here, $P[f/D,I]$ for example is the conditional probability of the distribution, $\{f_i\}$, given the data, D, and any prior information we may have, I. The quantity $P[D/f,I]$ is termed the *Likelihood function*. For Gaussian independent errors it is given by

$$P[D|f,I] \propto \exp\left(-\frac{\chi^2}{2}\right) \quad , \tag{8}$$

where χ^2 is the usual statistical measure for the quality of a fit to data. It contains all the information about statistical errors and the integral transform between the object and the data. $P[f/I]$ is termed the *prior probability* (before we acquired the data), and for positive additive distribution functions, f, it is given by

$$P[f|I] \propto \exp(aS(f)) \quad , \tag{9}$$

where the Shannon–Jaynes entropy is given by

$$S(f) \equiv \sum_i \left[f_i - m_i - f_i \ln\left(\frac{f_i}{m_i}\right) \right] \quad . \tag{10}$$

$S(f)$ is measured relative to an initial default model m for f. The most likely f is the one which maximizes the *posterior probability*, $P[f/D,I]$, which is equivalent to maximizing

$$Q \equiv S(f) - \frac{1}{a}\frac{\chi^2}{2} \quad . \tag{11}$$

That is, our variational principle is to maximize the entropy subject to a constraint that f fits the data according to a χ^2 test, with $1/a$ the Lagrange multiplier (also termed the *statistical regular-*

ization parameter in the applied math). In the most recent versions of this procedure,[6] one chooses α by Bayesian arguments; i.e. one maximizes the conditional probability, $P[\alpha|D,I]$.

Generally, Bayes' theorem is a procedure for combining *prior knowledge* with the data. The prior knowledge includes the fact that f is a positive additive distribution, which is required to use the entropy functional. But prior knowledge can also include sum rules, symmetries, conservation laws, physical hypotheses, and the default model. The data includes the integral transform, statistical errors and backgrounds. The output of the procedure is an estimate of the most probable distribution, along with the statistical errors. This procedure is embodied in a variety of robust MaxEnt algorithms.[7] Of course, the ultimate goal is to develop physical insight into the problem which can lead to improved hypotheses and models for the data. In this way, the Bayesian approach can be iterative.

DYNAMICAL PROPERTIES OF THE ANDERSON AND KONDO MODELS

The dynamical properties of many–body systems one may wish to calculate include spectral functions which can be measured by photoemission, transport coeffecients such as resistivity and thermal conductivity, dynamical magnetic susceptibility which can be probed by neutron scattering and NMR, etc. Our approach[8,9,10] is to use Quantum Monte Carlo[11] to generate both the data on $G(\tau)$ and the covariance matrix (statistical errors) of the data. Special care in the data collection procedure is required to insure that the errors on the data are Gaussian distributed. We use MaxEnt to numerically invert the spectral representation, Eq. (3). And we use perturbation theory to generate informative default models as a starting point for the MaxEnt procedure, which can significantly reduce the variance of the result. In this way, MaxEnt provides the ability to combine analytic theory with Monte Carlo simulations.

The Anderson model[12] for dilute magnetic impurities in metals consists of a half–filled conduction band interacting with an impurity via a hybridization matrix element V, and Coulomb integral for two electrons on the impurity, U. The Hamiltonian can be written

$$\hat{H} = \sum_\sigma e_d \hat{d}_\sigma^+ \hat{d}_\sigma + \sum_{k\sigma} e_k \hat{a}_{k\sigma}^+ \hat{a}_{k\sigma} + \sum_{k\sigma} V\left(\hat{a}_{k\sigma}^+ \hat{d}_\sigma + \hat{d}_\sigma^+ \hat{a}_{k\sigma}\right) + U\left(\hat{n}_\uparrow - 1/2\right)\left(\hat{n}_\downarrow - 1/2\right). \quad (12)$$

Our calculations will be restricted to the symmetric Anderson model in which the impurity energy e_d is zero, and to the infinite bandwidth limit in which the only relevant property of the conduction band is the density of states at the Fermi surface, $N(0)$. We expect the single–particle spectral function, $A(\omega)$, to have peaks at energies $\omega = \pm U/2$ of hybridization width $\Gamma = N(0)\pi V^2$ corresponding to the addition or subtraction of an electron on the impurity. In the large U such charge fluctuations are suppressed, and the Anderson model reduces to the Kondo Hamiltonian[13] of a spin 1/2 impurity interacting with the conduction electrons with an antiferromagnetic coupling constant, $J = -8\Gamma/\pi N(0)U$. The characteristic energy scale is given by the Kondo temperature, $T_K \approx 0.515\Gamma\sqrt{u}\exp(-\pi^2 u/8)$, where $u \equiv U/\pi\Gamma$. Kondo behavior sets in for $u > 1.0$ resulting in a peak in the spectral function centered at $\omega = 0$ of width $\approx T_K$. Due to the absence of vertex corrections for this model,[14] the resistivity is related to $A(\omega)$ by

$$\frac{\varrho(0)}{\varrho(T)} = -\frac{1}{\pi\Gamma} \int_{-\infty}^{\infty} d\omega \, \frac{\partial f}{\partial \omega} A^{-1}(\omega) \quad (13)$$

Figure 3 shows the typical $G(\tau)$ data, which is nearly featureless. The errors are at the

10^{-4} level, and they are strongly covariant because the Metropolis algorithm for QMC is a Markov process. To handle this, we generalize[8] the χ^2 measure to

$$\chi^2 = \delta \vec{G}^T \cdot \overleftrightarrow{C}^{-1} \cdot \delta \vec{G} \tag{14}$$

where $\delta \vec{G}$ is the difference between the data and the fit to the data, and the i–Th component refers to τ_i. To make Gaussian independent data, one rotates the spectral representation by the orthogonal transformation required to diagonalize the covariance matrix, C. For a default model we use the Horvatic–Zlatic perturbation theory[15] for the Anderson model which is second order in the self–energy, even though it is valid only for $u < 1.0$. In contrast to the QMC data, the perturbation theory does not show Kondo scaling for $u > 1.0$.

Figure 4 shows the single–particle $A(\omega)$ obtained[9] at varying u and fixed T/T_K as indicated. The Kondo peaks for QMC – MaxEnt are *universal* (independent of u) functions for ω/T_K less than 10, while the perturbation theory is non–universal. The peaks at $\omega = \pm U/2$ are, of course, non–universal. Figure 5 shows the evolution of the Kondo peak at fixed u and varying T/T_K. The Kondo peak increases as T/T_K decreases and saturates at the Friedel sum rule value of 1.0 at $\omega = 0$ and $T = 0$. The dashed line is the comparison with the Doniach–Sunjic prediction[16] for the shape of the Kondo peak at $T=0$

$$\pi \Gamma A(\omega)_{T=0} = \operatorname{Re} \sqrt{\frac{i\Gamma_K}{\omega + i\Gamma_K}} \quad . \tag{15}$$

A best fit is obtained for $\Gamma_K = 2.5T_K$ which is approximately the Wilson temperature,[17] $T_K^o = \pi^2 T_K/4$, defined by $T_K^o \chi(0) = 1$.

Figure 6a shows our results for the resistivity of the Anderson model as a function T/T_K at a variety of u, calculated by inputting our results for $A(\omega)$ into Eq. (13). The error bars on the points represent statistical errors calculated by the MaxEnt code from the covariance matrix of the QMC data. We have found these error estimates to be reliable, in the sense of indicating the variance of the results when different QMC data sets are used. Since the resistivities agree within statistical errors at the same T/T_K and different u, the resistivity curve is universal. The solid line shows the prediction of Hamann et al. for the resistivity[18] at high temperatures

$$\begin{array}{c} \text{for} \\ T \gtrsim T_K^N \end{array} \quad \frac{\varrho(T)}{\varrho(0)} \simeq \frac{1}{2}\left[1 - \frac{\ln T/T_K^N}{\sqrt{\ln^2 T/T_K^N + 3\pi^2/4}} \right] \quad , \tag{16}$$

which is a function of a single parameter, T_K^N, termed the *Nagaoka temperature*. Remarkably, the best fit is obtained for $T_K^N = T_K^o$. The dashed line is the fit to the Fermi liquid behavior expected at low temperatures

$$\begin{array}{c} \lim \\ T \ll T_K \end{array} \quad \frac{\varrho(T)}{\varrho(0)} \simeq 1 - \alpha \left(\frac{T}{T_K} \right)^2 \quad , \tag{17}$$

with our best fit value of $\alpha = 0.83 \pm 0.06$ close to the Nozieres prediction[19] of $\alpha = 1.0$. The Fermi liquid behavior of the resistivity (linear in T^2) at low temperatures is further illustrated in the inset. Figure 6b shows that our QMC–MaxEnt resistivities[27] agree with experimental data[28] on $(La,Ce)B_6$.

One may also calculate two–particle spectral functions by the same procedure. The spectral representation for the dynamical magnetic susceptibility is given by

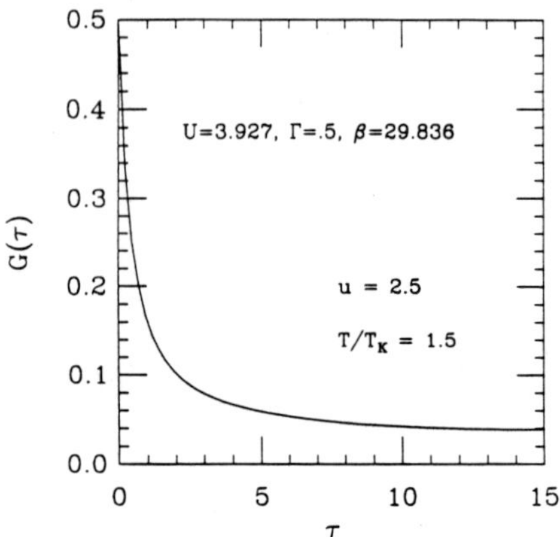

Fig. 3   Typical $G(\tau)$ QMC data for the Anderson model.

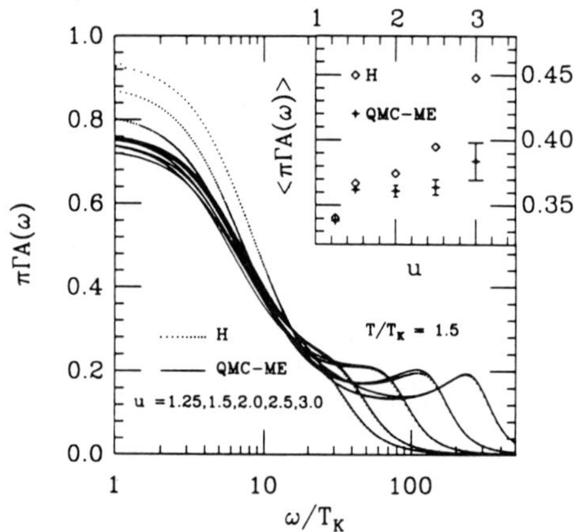

Fig. 4 $A(\omega)$ at fixed T/T_K and varying u. H is the prediction of Horvatic–Zlatic perturbation theory. QMC–ME are our results. Inset shows the average for $\omega/T_k \lesssim 10$.

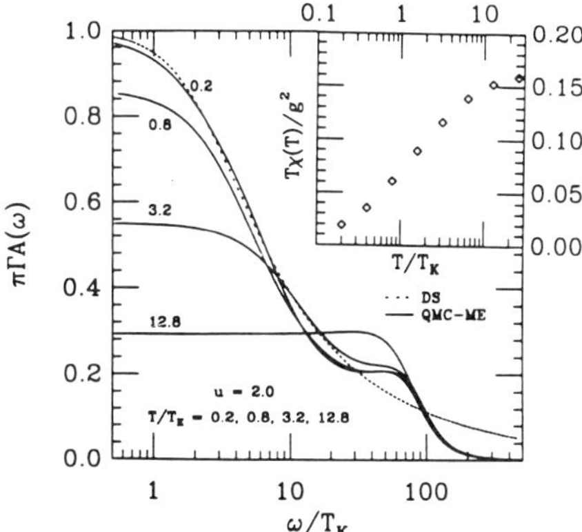

Fig. 5 Evolution of the Kondo peak as a function of T/T_K at fixed u. The dashed line is the prediction of Doniach–Sunjic theory for $\Gamma_K = T_K^o$. Inset shows the screened local moment.

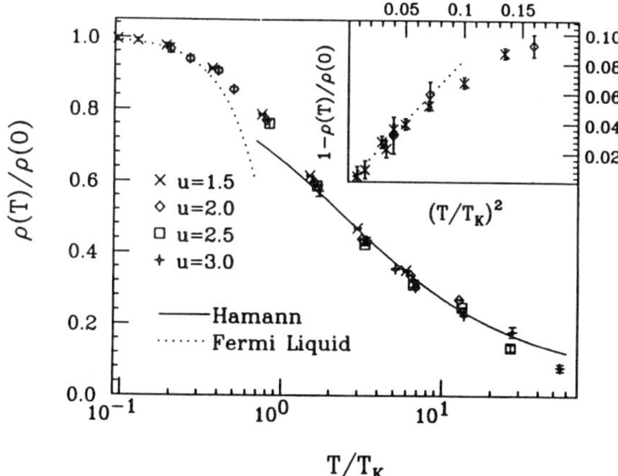

Fig. 6a Universal resistivity of the Anderson model. The solid line is the prediction of Hamann, et al. with $T_K^N = T_K^o$. The dashed line is the prediction of Fermi liquid theory.

$$\chi(\tau) = \int_{-\infty}^{\infty} \frac{d\omega}{\pi} \chi''(\omega) \frac{e^{-\tau\omega}}{1 - e^{-\beta\omega}} \quad . \tag{18}$$

Instead of $\chi''(\omega)$ we choose to plot

$$f(\omega) \equiv \frac{\pi\chi''(\omega)T_K}{2\omega\chi(T)} \quad , \tag{19}$$

which is normalized to unit integral over ω/T_K. Figure 7 shows the beautiful scaling we obtain[20] for $f(\omega)$ at fixed $T/T_K = 1.5$. In fact our lowest temperature results are nearly identical to recent numerical renormalization group results obtained by Costi,[21] which provides an independent check on our procedure. Our most interesting results are obtained for the zero–frequency intercept, which is related to the NMR relaxation rate by

$$\frac{1}{T_1 T} = A \frac{2\chi(T)}{\pi T_K} f(0) \quad . \tag{20}$$

Figure 8 plots $f(0)$ as a function of T/T_K. Not only are the results universal, but the $f(0)$ curve is identical to the universal resistivity curve within statistical error. The dynamical origin of this remarkable equivalence is not understood.

THE MOMENT PROBLEM

Another recurring interest in computational many–body physics is the diagonalization of Hamiltonians in as large a basis set as is feasible. The Lanczos method is ideally suited to obtaining the ground state energy and wave function. But one is often interested in thermodynamic functions which are formed by sums over the total eigenvalue spectrum. For large basis sets, exact knowledge of the eigenvalue spectrum may not be needed or even meaningful. One may be satisfied with approximate results for large systems provided the calculational method can also yield statistical error estimates and the computational time can be greatly reduced. There is a long history of attempts to estimate a spectrum from exact data on a limited number of moments, including use of the maximum entropy method.[22] However, recently Skilling[23] has shown how to estimate the eigenvalue spectra of mega–dimensional matrices from moment data which are subject to statistical errors. We adapt this procedure to many–body problems.

First we address the data generation process. Denote basis states by $|i>$ and eigenstates by $|n>$. One chooses an initial Gaussian random vector

$$|\psi_R> = \frac{1}{\sqrt{N_d}} \sum_i a_i |i> = \frac{1}{\sqrt{N_d}} \sum_n b_n |n> \quad , \tag{21}$$

by drawing a_i from a Gaussian random distribution of unit norm. Then the b_n will also be distributed in the same way. One generates the moment data by repeated application of the Hamiltonian to this random vector,

$$\overline{M}_k \equiv <\psi_R|H^k|\psi_R> \approx \frac{1}{\sqrt{N_d}} \sum_n (E_n)^k = \int dE \, N(E) \, E^k = M_k \quad . \tag{22}$$

The covariance matrix is given by

$$C_{k,l} = <\delta M_k \delta M_l> = \frac{2}{N_d} M_{k+l} \approx \frac{2}{N_d} \overline{M}_{k+l} \quad . \tag{23}$$

The inverse problem is to estimate $N(E)$ from a finite number of moment data subject to statistical errors described by the covariance matrix. Again, to create Gaussian independent data the

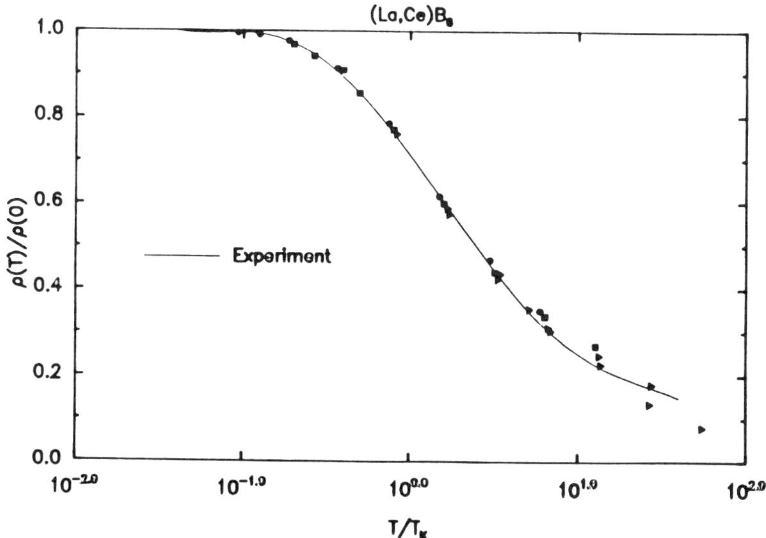

Fig. 6b Comparison of experimental resistivity data for *(La,Ce)B₆* with our QMC–ME results.

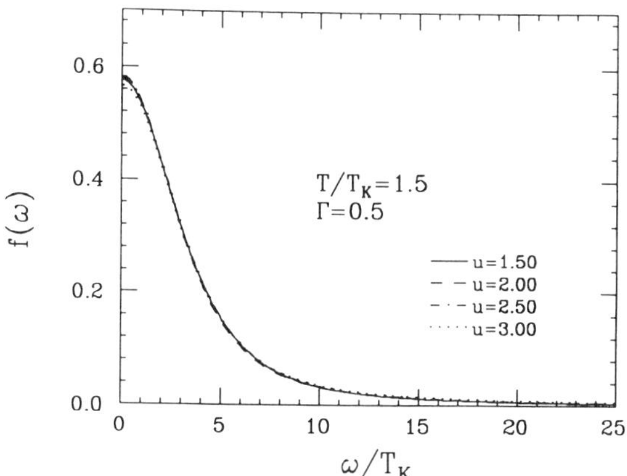

Fig. 7 Universal behavior of the dynamical magnetic susceptibility of the Anderson model at varying *u* and fixed $T/T_K = 1.5$.

covariance matrix must be diagonalized by an orthogonal transformation in the data space. Roundoff error in the data generation process results in some very small eigenvalues of the covariance matrix, which may be eliminated by singular value decomposition. Statistical errors may be systematically improved by sampling over several random vectors, but we have found that for large matrices the statistical errors are small.

In practice, the use of the power moments as described in Eqs. (21–23) is notoriously ill–conditioned, and it is better to use orthogonal polynomials defined such that the weight function is non–zero over the energy range of possible eigenvalues.[24] One can use a Lanczos procedure on the same Hamiltonian to determine the smallest and largest eigenvalues, and to set the energy scale for the polynomials. To generate the data, the recurrence relation for the polynomials is used in place of Eq. (22). The covariance matrix can then be shown to correspond to a sum over polynomials, with the largest polynomial required corresponding to twice the number of polynomials used in the data set. In his original work, Skilling used Chebyshev polynomials of the first kind, but we have found that this leads to an ill–conditioned covariance matrix because the polynomial weight function is singular. A better choice is Legendre polynomials because the weight function is uniform resulting in a better conditioned covariance matrix. The coeffecients in the expansion of the covariance matrix in terms of polynomials can be expressed in terms of the 3–J symbols in the quantum theory of angular momentum.

To illustrate the feasibility of the maximum entropy method for this problem, we consider[25] the periodic 4X4 spin 1/2 Heisenberg model on a square lattice

$$\hat{H} \equiv J \sum_{(ij)} \vec{S}_i \cdot \vec{S}_j \quad . \tag{24}$$

The problem is small enough (a maximum of 12870 basis states for total $S_z = 0$), that we can compare with exact diagonalization. Fig. 9 shows a superposition of the exact eigenvalue spectrum binned into 300 pixels, with the maximum entropy results obtained using 250 Legendre polynomials. The results are in essential agreement, although small details differ presumably due to statistical fluctuations. These differences are not important in constructing the thermodynamic functions, and the MaxEnt results for the energy per particle, entropy, specific heat, and susceptibility essentially agree with the exact diagonalization results.[26] We can also estimate statistical errors on such integrated functions of the energy spectrum.

CONCLUSIONS

Many problems in the analysis of computer simulations are similar to the analysis of experimental data. They may be addressed with maximum entropy and Bayesian methods for statistical inference. These provide a systematic approach to incorporating prior knowledge with the data, and they enable one to propagate statistical errors in the data into error estimates on any integrated function of the image. Using these methods, we have been able to analytically continue imaginary time quantum Monte Carlo data for the Anderson model to obtain real frequency spectral functions, dynamical magnetic susceptibility, and transport coeffecients. We have also been able to estimate the total many–body density of states of the Heisenberg model and thermodynamic functions constructed from it. We believe that these methods will have a much broader impact in future computational physics research.

ACKNOWLEDGEMENTS

This work was supported in part by the U. S. Dept. of Energy, Basic Energy Sciences, Division of Materials Research.

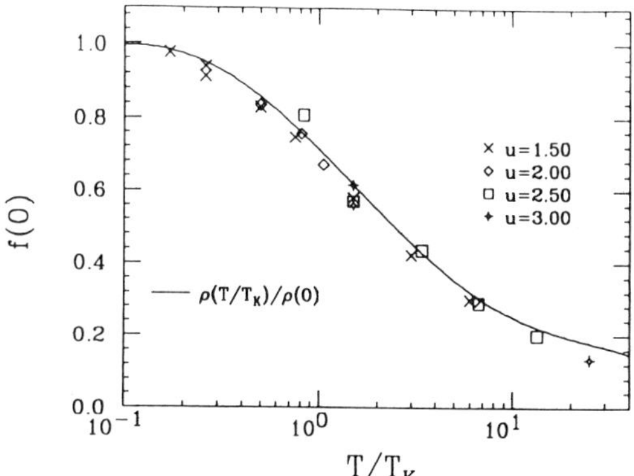

Fig. 8 Zero–frequency intercepts of the dynamical magnetic susceptibility of the Anderson model, compared to the universal resistivity curve.

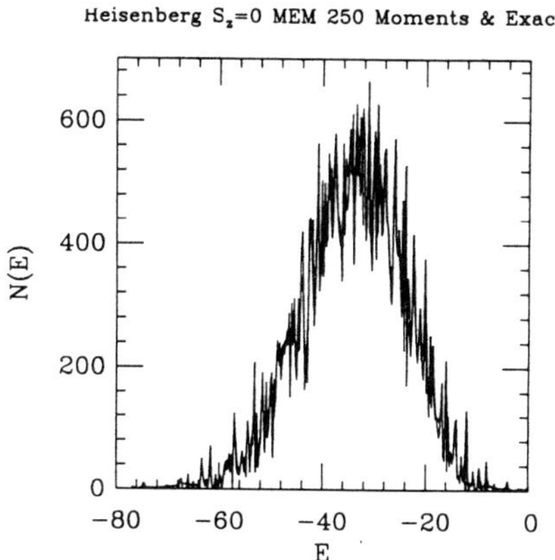

Fig. 9 Comparison of exact densities of states for the S$_z$ = 0 Heisenberg model (solid) with the MEM results obtained from 250 moments.

REFERENCES

[1] S. F. Gull, J. Skilling, IEE Proc. **131**(F), 646 (1984).

[2] C. E. M. Strauss, P. L. Houston, to be published in *Maximum Entropy and Bayesian Methods*, 1991, ed. G. Erickson, (Kluwer Academic, Dordrecht)..

[3] S. F. Gull, in *Maximum Entropy and Bayesian Methods in Science and Engineering*, ed. G. J. Erickson, C. R. Smith (Kluwer Academic, Dordrecht, 1988).

[4] E. T. Jaynes, in *Maximum Entropy and Bayesian Methods*, ed. J. H. Justice (Cambridge University, Cambridge, 1986).

[5] D. S. Sivia, Los Alamos Science **19**, 180 (1990) and references therein.

[6] J. Skilling, in *Maximum Entropy and Bayesian Methods*, ed. J. Skilling (Kluwer Academic, Dordrecht, 1989).

[7] MEMSYS3 produced by Maximum Entropy Data Consultants, Ltd. See also R. K. Bryan, Eur. Biophys. J. **18**, 165 (1990).

[8] R. N. Silver, D. S. Sivia, J. E. Gubernatis, Phys. Rev. B**41**, 2380 (1990).

[9] R. N. Silver, J. E. Gubernatis, D. S. Sivia, M. Jarrell, Phys. Rev. Lett. **65**, 496 (1990).

[10] J. E. Gubernatis, M. Jarrell, R. N. Silver, D. S. Sivia, in Sept 15, 1991 Phys. Rev. B. See also R. N. Silver, D. S. Sivia, J. E. Gubernatis, M. Jarrell, in *Maximum Entropy and Bayesian Methods*, ed. W. T. Grandy, L. H. Schick, Kluwer Academic Publishers, 1991, p. 109–126.

[11] J. E. Hirsch, R. M. Fye, Phys. Rev. Lett. **56**, 2521 (1986).

[12] P. W. Anderson, Phys. Rev. **124**, 41 (1961).

[13] J. Kondo, Prog. Theor. Phys. **32**, 37 (1964).

[14] N. E. Bickers, D. L. Cox, J. W. Wilkins, Phys. Rev. B**36**, 2036 (1987).

[15] B. Horvatic, D. Sokcevic, V. Zlatic, Phys. Rev. B**36**, 675 (1987).

[16] S. Doniach, M. Sunjic, J. Phys. C**3**, 285 (1970).

[17] K. G. Wilson, Rev. Mod. Phys. **47**, 773 (1975).

[18] D. R. Hamann, Phys. Rev. **158**, 570 (1967); Y. Nagaoka, Phys. Rev. **138**, A1112 (1965); H. Suhl, Phys. Rev. **138**, A515 (1965).

[19] P. Nozieres, J. Low Temp. Phys. **17**, 31 (1974).

[20] M. Jarrell, J. E. Gubernatis, R. N. Silver, Phys. Rev. B Rap. Comm., to be published.

[21] T. Costi, private communication.

[22] See, e.g., L. R. Mead, N. Papanicolaou, J. Math. Phys. **25**, 2404 (1984).

[23] J. Skilling, "The Eigenvalues of Mega–Dimensional Matrices", in ref. [6], p. 455–466.

[24] J. C. Wheeler, M. G. Prais, C. Blumstein, Phys. Rev. B**10**, 2429 (1974).

[25] H. Roeder, R. N. Silver, to be published.

[26] J. Shi, T. Rezayi, Phys. Rev. B**43**, 13618 (1991).

[27] M. Jarrell, J. E. Gubernatis, R. N. Silver, to be published.

[28] K. Winzer, Z. Physik **256**, 139 (1973).

† Present Address, Rutherford Appleton Laboratory, Chilton, Didcot, UK

GREEN'S FUNCTION MONTE CARLO FOR FERMIONS

M.H. Kalos and Shiwei Zhang

Laboratory of Atomic and Solid State Physics
and Center for Theory and Simulation in Science and Engineering
Cornell University, Ithaca, New York 14853, USA

ABSTRACT

We present an exact and stable Green's function Monte Carlo algorithm for few-fermion systems. Results are given for low-Z atoms. Possible improvements in constructing the importance functions are discussed. We also present a bilinear algorithm for sampling the product of two wavefunctions.

1. INTRODUCTION

Quantum Monte Carlo has been successfully applied to many different systems and has proved an extremely powerful method in studying many-body problems, especially of bosons. The approach allows one to treat explicitly many-dimensional integral equations that result from the Schroedinger equation. Based on an iterative random walk process, it can project out the state with the lowest energy, in principle, for any given symmetry. But for fermion systems, due to the requirement of antisymmetry on the wavefunctions and the nature of Monte Carlo (MC) methods, a sign problem arises. In the MC process, a random walker is a point in configuration space and the collection of such walkers (possibly with weights) samples the wavefunction when the random walk has converged. It is straightforward to represent an antisymmetric wavefunction in a fermion problem by two types of random walkers carrying positive and negative weights respectively. But in the usual algorithm, the random walkers are independent of each other, and therefore the two populations have the same asymptotic form, a symmetric one. This happens with or without the existence of a symmetric state with lower energy. In the former case, both types of walkers converge to this symmetric state. In the latter, each becomes $\max\{\psi, -\psi\}$,

Recent Progress in Many-Body Theories, Vol. 3,
Edited by T.L. Ainsworth et al., Plenum Press, New York, 1992

since $-\psi$ is also a correct antisymmetric solution to the Schroedinger equation. So the asymptotic overlap with any antisymmetric test function is zero and thus the algorithm does not provide a long-term stable solution. The fixed-node MC method[1,2] imposes nodes from some trial wavefunction as boundary conditions and solves the Schroedinger equation under such conditions. This avoids the instability but yields only an upper bound to the energy. In many cases this energy is very close to the correct answer. But it relies on the trial wavefunction and it is often hard to know *a priori* or even *a posteriori* how accurate the calculation is.

We have recently proposed an algorithm[3–5] with the Green's function Monte Carlo (GFMC) to treat the sign problem for small systems of fermions. Rather than using independent walkers, we use equal numbers of interacting positive and negative walkers. This interaction, in the dynamical process of sampling, has the effect of a repulsion between the two types of walkers in configuration space. It employs the fact that the antisymmetry is a global property of the system and can be satisfied exactly only if the random walk is carried out also in a "global" fashion. We use different importance functions for positive and negative walkers. This along with the repulsion, breaks the symmetry that otherwise would exist between an ensemble of walkers and an ensemble derived by interchanging positive and negative walkers. These permit a stable and exact method whose results are independent of other approaches.

In section 4 we propose a bilinear sampling algorithm in the GFMC for evaluating energy differences and expectation values of operators that do not commute with the Hamiltonian. It provides the possibility of sampling directly the product of two wavefunctions by GFMC and will enhance its applicability to various problems.

2. DESCRIPTION OF THE FERMION ALGORITHM

2.1. Some Basics of GFMC

In GFMC[6–8] , the Schroedinger equation for a many-body system is transformed into an integral equation with the following form:

$$\psi(\mathbf{R}) \propto \int d\mathbf{R}' G(\mathbf{R}, \mathbf{R}') W(\mathbf{R}') \psi(\mathbf{R}'), \tag{1}$$

where $\mathbf{R}$ is a vector denoting all degrees of freedom of the system. We assume the Green's function (or an approximation to the actual Green's function) is $G(\mathbf{R}, \mathbf{R}')$ and has an explicitly known form. The wavefunction ψ is represented by a finite ensemble of random walkers that live in configuration space. Thus, each walker corresponds to a vector $\mathbf{R}$ or a certain configuration of the many-body system. This ensemble can have any initial distribution provided that the distribution function is not orthogonal to the desired lowest energy eigenfunction with a particular symmetry. The random walk evolves from one generation to the next with the walkers moving from one position to another according to the kernel $G(\mathbf{R}, \mathbf{R}')$. The known function $W(\mathbf{R})$ affects the weights of walkers. Asymptotically the random walk yields a distribution of walkers that samples the wavefunction ψ with the lowest energy of a given symmetry. With sufficiently many walkers drawn from the asymptotic distribution, information can be extracted on the properties of the state, such as the energy.

Importance sampling is a powerful method in reducing the fluctuations in the random walk. The integral equation (1) can be transformed to obtain

$$\tilde{\psi}(\mathbf{R}) \propto \int d\mathbf{R}' \left[i(\mathbf{R})G(\mathbf{R},\mathbf{R}')/i(\mathbf{R}') \right] W(\mathbf{R}')\tilde{\psi}(\mathbf{R}') \tag{2}$$

with $\tilde{\psi}(\mathbf{R}) = i(\mathbf{R})\psi(\mathbf{R})$. The random walk is now carried out according to Eq (2) to sample a different distribution $\tilde{\psi}$. The importance function $i(\mathbf{R})$ is the asymptotic contribution of a walker starting at $\mathbf{R}$. For a system of bosons, that is just the desired ground state wavefunction and therefore a trial wavefunction can be used for i.

2.2. Features of the New Algorithm

For a system of fermions, the wavefunction must change sign under interchange of particles with like spins. To reflect the antisymmetry of ψ, we can write it as:

$$\psi(\mathbf{R}) = \psi^+(\mathbf{R}) - \psi^-(\mathbf{R}), \tag{3}$$

where $\psi^\pm(\mathbf{R})$ map into each other by any odd permutation of like spins. We note that the above decomposition is not unique. In the usual MC approach as outlined in 2.1, $\psi^\pm(\mathbf{R})$ separately satisfy Eq (1), and when found by solving that equation, the two populations yield the same symmetric asymptotic distribution. This leads to a zero MC estimate of ψ coupled with random fluctuations and thus the decay of signal to noise characteristic of the sign problem. Moreover, as noted in the introduction, even with the positive and negative populations separated at any instant in the random walk, the asymptotic representation of ψ could still be zero due to the "super-symmetry" between ψ and $-\psi$. Therefore, the essence of the new algorithm is to: (i) introduce a "repulsion" between positive and negative walkers to separate the two populations; and (ii) introduce different importance functions for positive and negative walkers to break the "super-symmetry".

We choose for ψ the form $\psi^\pm = \max\{\pm\psi, 0\}$ and rewrite (1) in the following coupled way:

$$\psi^\pm(\mathbf{R}) \propto \max\left\{ \pm\int d\mathbf{R}' G(\mathbf{R},\mathbf{R}')W(\mathbf{R}')[\psi^+(\mathbf{R}') - \psi^-(\mathbf{R}')], 0 \right\}. \tag{4}$$

We generalize an idea of correlating pairs of positive and negative walkers by Arnow, et $al.$[9]. The whole ensemble of walkers are taken into account to sample the "difference" on the right-hand side in Eq (4) for the next generation. Again we can introduce importance sampling. But instead of a single function, we use different importance functions $i^\pm(\mathbf{R})$ for $\psi^\pm(\mathbf{R})$ respectively. We shall see below that this follows from the meaning of the importance function in such an interacting ensemble. Now Eq (4) can be written as

$$\tilde{\psi}^{\pm}(\mathbf{R}) \propto i^{\pm}(\mathbf{R}) \max\left\{ \pm \int d\mathbf{R}' G(\mathbf{R}, \mathbf{R}') W(\mathbf{R}') \left[\frac{\tilde{\psi}^{+}(\mathbf{R}')}{i^{+}(\mathbf{R}')} - \frac{\tilde{\psi}^{-}(\mathbf{R}')}{i^{-}(\mathbf{R}')} \right], 0 \right\}, \quad (5)$$

where $\tilde{\psi}^{\pm} = i^{\pm}\psi^{\pm}$ is the distribution function to be sampled.

2.2.1 <u>Interaction Between Walkers.</u> In the MC process, $\tilde{\psi}^{\pm}$ are represented by two populations of walkers that carry positive/negative signs:

$$\tilde{\psi}(\mathbf{R}) = \sum_k s_k \delta(\mathbf{R} - \mathbf{R}_k), \quad (6)$$

with k being the label of walkers and s_k the sign of walker k, i.e., $s_k = \pm 1$ for positive/negative walkers. According to Eqs (5) and (6), we will sample positions for new walkers from the density distribution

$$K(\mathbf{R}, \{\mathbf{R}'\}) = \sum_k s'_k G(\mathbf{R}, \mathbf{R}'_k) W(\mathbf{R}'_k) / i(\mathbf{R}'_k, s'_k), \quad (7)$$

where $\{\mathbf{R}'\}$ denote the current ensemble of walkers and $i(\mathbf{R}, \pm 1) = i^{\pm}(\mathbf{R})$. The density K is as often negative as positive for the ground states of fermion systems. Technically, the sampling can be accomplished by a rejection method: First sample from $K'(\mathbf{R}, \{\mathbf{R}'\}) = \sum_k |G(\mathbf{R}, \mathbf{R}'_k) W(\mathbf{R}'_k) / i(\mathbf{R}'_k, s'_k)|$ and then use $|K|/K'$ to accept the new walker. The walker is assigned a minus sign if K is negative.

2.2.2 <u>Differing Importance Functions.</u> The importance of a walker at $\mathbf{R}$ is the overlap with a trial function of the asymptotic distribution that arises from it. That is, let

$$\eta_1(\mathbf{R}; \mathbf{R}') = \lambda \int G(\mathbf{R}, \mathbf{R}'') W(\mathbf{R}'') \delta(\mathbf{R}'' - \mathbf{R}') d\mathbf{R}'',$$

$$. \qquad .$$

$$. \qquad .$$

$$\eta_{\ell+1}(\mathbf{R}; \mathbf{R}') = \lambda \int G(\mathbf{R}, \mathbf{R}'') W(\mathbf{R}'') \eta_{\ell}(\mathbf{R}''; \mathbf{R}') d\mathbf{R}''.$$

The importance $i(\mathbf{R}')$ is calculated as

$$i(\mathbf{R}') = \lim_{\ell \to \infty} \int \phi_T(\mathbf{R}) \eta_{\ell}(\mathbf{R}; \mathbf{R}') d\mathbf{R},$$

where ϕ_T is a trial function. For the antisymmetric ground state we are dealing with, the trial wavefunction must also be antisymmetric. With the coupling of all walkers as described in 2.2.1, the importance function for a positive walker is no longer symmetric: a positive walker originated from the neighborhood of positive walkers has a high probability of surviving the walk and thus making more contribution; while a positive walker near a dense distribution of negative walkers is more likely to be "canceled" in a few steps, although with a small probability it can give birth

414

to walkers in the positive region and contribute asymptotically. So the importance function $i^+(\mathbf{R})$ (> 0) for positive walkers will have the effect of biasing the walker toward regions in configuration space where the wavefunction ψ is positive. Where the wavefunction is likely to be negative, it falls off because of the repulsion by negative walkers. Thus the following form seems a reasonable description of i^+,

$$i^+(\mathbf{R}) = \int d\mathbf{R}' G(\mathbf{R}, \mathbf{R}')\psi^+(\mathbf{R}'). \tag{8}$$

Of course, this form relies on our knowledge of ψ, which is very likely to come from the test function ϕ_T. But we emphasize that, as with boson systems, the method yields exact results even when used with a less than optimal importance function. From the antisymmetry of the problem, the importance function for negative walkers is also positive and is related to i^+ by $i^-(\mathbf{R}) = i^+(P\mathbf{R})$, where P denotes an odd permutation of like spins. After choosing an analytical form (such as one mapped from the test function) for the importance functions, $i^\pm$ in Eq (5) can be dealt with, e.g., by an additional rejection.

    2.2.3. <u>"Shadow" Importance Functions.</u> With a reasonable test function $\phi_T(\mathbf{R})$, it is possible to replace ψ^+ by $\phi_T^\pm$ in Eq (8) and use it directly as the importance function. We note that this resembles a class of functions called "shadow" wavefunctions developed by Vitiello, Runge, and Kalos[10] in describing systems of ^{4}He, which may be used as importance functions for such systems[11]. As we will discuss below, in many cases it is easy to sample the product of Green's functions. Also we can take advantage of the antisymmetry in fermion systems and just keep an ensemble of walkers consisting of positive walkers and their "images" derived from odd permutations. We can then generalize their method[11,12] of alternating GFMC steps with Metroplis steps so as to sample

$$\tilde{\psi}^\pm(\mathbf{R}, \mathbf{S}) = \iota^\pm(\mathbf{R}, \mathbf{S})\psi^\pm(\mathbf{R}) \equiv G(\mathbf{R}, \mathbf{S})\phi_T^\pm(\mathbf{S})\psi^\pm(\mathbf{R}), \tag{9}$$

where $\mathbf{S}$ is an ensemble of positive shadows $(\mathbf{S}^+)$ and their images $(\mathbf{S}^-)$ to "anchor" the walkers to desired regions as given by ϕ_T. Each walker is paired with a shadow and the object of the random walk is $\{\mathbf{R}_k, \mathbf{S}_k\}$. The shadows are sampled solely by Metroplis, which in effect does the integral in the importance function. We denote the Metroplis step on the shadows (conditional on the walkers) by a kernel Γ, i.e.,

$$\Gamma[\iota^\pm(\mathbf{R}', \mathbf{S}')] = \iota^\pm(\mathbf{R}', \mathbf{S}), \tag{10}$$

where the primed coordinates denote the old generation and unprimed the new generation. Each sampling step is now a Metroplis step from $\{\mathbf{R}', \mathbf{S}'\}$ to $\{\mathbf{R}', \mathbf{S}\}$ followed by a GFMC step. From Eqs (4) and (5), in the GFMC step a positive walker $\mathbf{R}$ is sampled from the density distribution

$$K(\mathbf{R}) = \sum_k \left[\frac{G(\mathbf{R}, \mathbf{R}_k'^+)G(\mathbf{R}, \mathbf{S}_k^+)}{G(\mathbf{R}_k'^+, \mathbf{S}_k^+)} W(\mathbf{R}_k'^+) - \frac{G(\mathbf{R}, \mathbf{R}_k'^-)G(\mathbf{R}, \mathbf{S}_k^+)}{G(\mathbf{R}_k'^-, \mathbf{S}_k^-)} W(\mathbf{R}_k'^-) \right] \tag{11}$$

when it is greater than zero; and a positive pair is then formed with the shadow $\mathbf{S}^+$ inherited from its parent. (If the parent is negative, an odd permutation is performed on $\mathbf{S}$ to obtain the shadow for the new positive walker.) The parent pair is chosen from a distribution K' defined as the sum of the absolute value of every term in K, in a similar fashion as in 2.2.1.

3. APPLICATION OF THE FERMION ALGORITHM

We have applied the algorithm to compute energies in small atoms[5] including those of the He 1s2s triplet state and of the ground states of Li, Be and N. The method is stable and yields answers to the energies that agree with experiment. Except for technical issues, application of the algorithm to small Coulombic systems (including molecules) is straightforward. We mention in passing that for systems with interaction other than the Coulomb potential, the product of importance function by potential is integrable; if, instead of $\tilde{\psi}^\pm = i^\pm \psi^\pm$, we use $\tilde{\psi}^\pm = i^\pm W\psi^\pm$, then the potential term W appears only when multiplied by i and a density can be defined and sampled. Thus our method can be extended to non-integrable repulsive potentials. In our calculation, we kept a constant number of positive and negative walkers. This population control introduces a bias in the answer for any finite number of walkers. The bias is corrected in the final answer by doing separate runs with different numbers of walkers and extrapolating to infinite population. The results of our calculations are shown in Table 1 together with the exact answers. The test wavefunctions used consisted of products of a simple determinant and a simple Jastrow factor[13,14].

3.1. The Green's Function

We use Green's function for the time-integrated diffusion operator with constant absorption[6]. In atomic units, the many-body Schroedinger equation can be written as

$$[-\frac{1}{2}\nabla_{\mathbf{X}}^2 + V(\mathbf{X})]\phi(\mathbf{X}) = E\phi(\mathbf{X}), \tag{12}$$

where $\mathbf{X}$ is a $3N$ dimensional vector denoting the coordinates of all N electrons in 3D real space. The energy E is negative. We can change the length scale of the system

Table 1. Energies by the GFMC for low-Z atoms compared with exact values. The expected statistical errors in the GFMC results are in the last digits and are indicated by parentheses. The exact results are estimates of the non-relativistic energies with infinite nuclear mass. Their sources are indicated and only relevant digits in these results are kept. All values are in Hartrees.

	He 1s2s triplet	Li ground	Be ground	N ground
exact	-2.175229[17]	-7.4781[18]	-14.667[18]	-54.59[18]
GFMC	$-2.17524(7)$	$-7.4776(6)$	$-14.664(4)$	$-54.57(5)$*

* preliminary result.

by $\mathbf{R} = \sqrt{2|E|}\mathbf{X}$. Noticing that the Coulomb potential V scales as 1/length, we can transform Eq (12) to

$$(-\nabla_\mathbf{R}^2 + 1)\psi(\mathbf{R}) = -\sqrt{\frac{2}{|E|}}V(\mathbf{R})\psi(\mathbf{R}). \tag{13}$$

We note that the rescaling allows us to work with $\mathbf{R}$ in solving for the ground state wavefunction without knowing E and conveniently evaluate the energy at the end with samples whose distribution is the rescaled wavefunction ψ. We have the following integral equation

$$\psi(\mathbf{R}) = \sqrt{\frac{2}{|E|}}\int d\mathbf{R}' G(\mathbf{R}, \mathbf{R}')[-V(\mathbf{R}')]\psi(\mathbf{R}'). \tag{14}$$

The Green's function G above has a known analytical form. It also has an integral representation[15] that is very advantageous:

$$G(\mathbf{R}, \mathbf{R}') = (4\pi)^{-3N/2}\int_0^\infty t^{-3N/2}\exp[-t - |\mathbf{R} - \mathbf{R}'|^2/(4t)]dt. \tag{15}$$

This is easy to sample either as is, or when multiplied by other superpositions of gaussians, for example, other Green's functions of this class.

3.2. Regulating the Coulomb Potential

The Coulomb potential term in Eq (14) poses some difficulties: 1) It is sometimes negative; and 2) It is singular. Since we have to use walkers that carry signs to account for the antisymmetry, the first point is effectively taken care of, except that the importance function is affected to some extent. But this term is rarely negative, so the effect is small. It is possible to add in a small correction term in the importance function when two electrons are close. The singularity of $-V$ can in principle cause large fluctuations in the sampling since it affects the weights of walkers. In ref 5, we used another similarity transformation with V' constructed from the sum of the absolute values of all the terms in V. Since V/V' is well behaved and both $\frac{1}{|\mathbf{r}_i|}G(\mathbf{R}, \mathbf{R}')$ and $\frac{1}{|\mathbf{r}_i - \mathbf{r}_j|}G(\mathbf{R}, \mathbf{R}')$ can be sampled exactly[16], the second issue can be resolved nicely.

3.3. Obtaining the Importance Functions

The importance functions used were derived by qualitative application of the spirit of 2.2.2. They can be written as $i^+(\mathbf{R}) = a(\mathbf{R})b(\mathbf{R})$, where a is a function that accounts for the antisymmetry and b reflects the orbital properties of the electrons and prevents them from drifting off to infinity. The function a falls off exponentially as $\mathbf{R}$ departs from the nodes into the negative region as given by $\phi_T(\mathbf{R})$ and it increases and approaches 1 if $\mathbf{R}$ goes in the opposite direction.

The "shadow" importance functions seem to provide a more direct link between the test function and the importance functions. Since it is straightforward to sample a product of two Green's functions in this class, this type of importance function also avoids the additional rejection with i and is likely to be more efficient. We are now testing this idea.

4. BILINEAR SAMPLING

In many cases it is desirable to be able to sample directly the product of two wavefunctions, since the expression for expectation values of quantum operators is quadratic in ψ. We employ the fact that it is sometimes possible to sample easily the product of Green's functions, such as with the choice of the Green's function in 3.1, and propose here a bilinear sampling method.

From Eq (1), consider two boson systems denoted by subscripts α and β respectively. We can obtain the following pair of equations

$$\psi_\alpha(\mathbf{x})G_\alpha(\mathbf{x},\mathbf{y})\psi_\beta(\mathbf{y}) = \lambda_\alpha \int \frac{G_\alpha(\mathbf{x},\mathbf{y})G_\alpha(\mathbf{x},\mathbf{z})}{G_\alpha(\mathbf{z},\mathbf{y})} W_\alpha(\mathbf{z})\psi_\alpha(\mathbf{z})G_\alpha(\mathbf{z},\mathbf{y})\psi_\beta(\mathbf{y})d\mathbf{z} \quad (16a)$$

and

$$\psi_\alpha(\mathbf{x})G_\alpha(\mathbf{x},\mathbf{y})\psi_\beta(\mathbf{y}) = \lambda_\beta \int \frac{G_\alpha(\mathbf{x},\mathbf{y})G_\beta(\mathbf{y},\mathbf{z})}{G_\alpha(\mathbf{x},\mathbf{z})} W_\beta(\mathbf{z})\psi_\alpha(\mathbf{x})G_\alpha(\mathbf{x},\mathbf{z})\psi_\beta(\mathbf{z})d\mathbf{z}. \quad (16b)$$

Here $\mathbf{x}$, $\mathbf{y}$, and $\mathbf{z}$ are all vectors denoting the collection of all particle coordinates. Note that both equations yield $\psi_\alpha(\mathbf{x})G_\alpha(\mathbf{x},\mathbf{y})\psi_\beta(\mathbf{y})$, bilinear in ψ_α and ψ_β as asymptotic density.

To solve these equations by Monte Carlo, we introduce random walks on the enlarged configuration space $\mathbf{x} \otimes \mathbf{y}$ as follows:

For a walker at $(\mathbf{u}, \mathbf{v})$, compute

$$p_1 \propto \lambda_\alpha W_\alpha(\mathbf{u}) \int \frac{G_\alpha(\mathbf{x},\mathbf{u})G_\alpha(\mathbf{x},\mathbf{v})}{G_\alpha(\mathbf{u},\mathbf{v})} d\mathbf{x} \qquad (17a)$$

$$p_2 \propto \lambda_\beta W_\beta(\mathbf{v}) \int \frac{G_\alpha(\mathbf{y},\mathbf{u})G_\beta(\mathbf{y},\mathbf{v})}{G_\alpha(\mathbf{u},\mathbf{v})} d\mathbf{y}. \qquad (17b)$$

With probability $p_1, (\mathbf{u}, \mathbf{v}) \to (\mathbf{x}, \mathbf{y})$ using the kernel

$$G_\alpha(\mathbf{x},\mathbf{v})G_\alpha(\mathbf{x},\mathbf{u})\delta(\mathbf{y} - \mathbf{v})$$

and with probability $p_2, (\mathbf{u}, \mathbf{v}) \to (\mathbf{x}, \mathbf{y})$ using the kernel

$$G_\alpha(\mathbf{y},\mathbf{u})G_\beta(\mathbf{y},\mathbf{v})\delta(\mathbf{x} - \mathbf{u}).$$

In each case one new configuration ($\mathbf{x}$ or $\mathbf{y}$) is selected and the complementary configuration is kept the same. With the Green's function given by Eq (15), the above calculations and sampling are straightforward. This is also true with certain quantum spin systems where the operation of the Green's function is simply flipping a pair of nearest-neighbor spins.

If $\{\mathbf{x}_k, \mathbf{y}_k\}$ are pairs of configurations in the asymptotic regime, then $W_\alpha(\mathbf{x}_k)$ is an estimator for $\psi_\alpha(\mathbf{y}_k)\psi_\beta(\mathbf{y}_k)$ and $W_\beta(\mathbf{y}_k)\frac{G_\beta(\mathbf{x}_k,\mathbf{y}_k)}{G_\alpha(\mathbf{x}_k,\mathbf{y}_k)}$ is an estimator for $\psi_\alpha(\mathbf{x}_k)\psi_\beta(\mathbf{x}_k)$. If $H_\beta - H_\alpha = V_\beta - V_\alpha$, then

$$(E_\beta - E_\alpha) = \left\langle \frac{\sum_k W_\alpha(\mathbf{x}_k)[V_\beta(\mathbf{y}_k) - V_\alpha(\mathbf{y}_k)]}{\sum W_\alpha(\mathbf{x}_k)} \right\rangle. \tag{18}$$

A similar expression involves an average of $V_\beta(\mathbf{x}_k) - V_\alpha(\mathbf{x}_k)$. Since both of these are direct averages of $\triangle V(\mathbf{x})$ or $\triangle V(\mathbf{y})$ they give small variances for $\triangle E$ when $\triangle V$ is small. In other words, it is now possible to compute energy differences directly rather than the difference between two independent estimates of energies.

When $\alpha = \beta$, we get from the above estimators for $\psi_\alpha^2(\mathbf{x})$ that do not require asymptotic "side walks[19]" or "extrapolations[8]". Therefore it is possible to compute directly with the GFMC method expectation values of quantum operators that do not commute with the Hamiltonian.

ACKNOWLEDGMENTS

We thank C.J. Umrigar and S.A. Vitiello for helpful discussions. The Cornell Theory Center is funded by the U.S. National Science Foundation, by New York State, by IBM, and by Cornell University.

REFERENCES

1. K.E. Schmidt and M.H. Kalos, in *Applications of the Monte Carlo Method in Statistical Physics*, ed. by K. Binder (Springer Verlag 1984).

2. J.B. Anderson, J. Chem. Phys. <u>63</u>, 1499 (1975); J. Chem. Phys. <u>65</u>, 4121 (1976); J. Chem. Phys. <u>73</u>, 3897 (1980).

3. M.H. Kalos, in *Computational Atomic and Nuclear Physics*, ed. by C. Bottcher, M.R. Strayer and J.B. McGrory (World Scientific 1989).

4. M.H. Kalos, J. Stat. Phys. <u>63</u>, 1269, (1991).

5. S. Zhang and M.H. Kalos, Phys. Rev. Lett. <u>67</u>, 3074, (1991).

6. M.H. Kalos, Phys. Rev. <u>128</u>, 1791, (1962).

7. M.H. Kalos, D. Levesque, and L. Verlet, Phys. Rev. <u>A9</u>, 2178 (1974).

8. D.M. Ceperley and M.H. Kalos, in *Monte Carlo Methods in Statistical Physics*, ed. by K. Binder (Springer Verlag 1979).

9. D.M. Arnow, M.H. Kalos, M.A. Lee and K.E. Schmidt, J. Chem. Phys. <u>77</u>, 5562 (1982).

10. S.A. Vitiello, K.J. Runge, and M.H. Kalos, Phys. Rev. Lett., <u>60</u> ,1970 (1988).

11. S.A. Vitiello, K.J. Runge, and M.H. Kalos, in *Nuclear and Atomic Physics at One Gigaflop*, ed. by C. Bottcher, M.R. Strayer and J.B. McGrory, Nuclear Science Research Conference Series, Vol. 16 (Harwood Academic, Chur, 1989).

12. S.A. Vitiello and P.A. Whitlock, Phys. Rev. B. <u>44</u>, 7373, (1991).

13. C.J. Umrigar, K.G. Wilson and J.W. Wilkins in *Computer Simulation Studies in Condensed Matter Physics: Recent Developments*, ed. by D.P. Landau, K.K. Mon and H.B. Schuttler (Springer Verlag 1988).

14. C.J. Umrigar, private communication.

15. G.N. Watson, "Treatise on the Theory of Bessel Functions", 2nd edition, p. 183. Cambridge Univ. Press, New York (1945).

16. M.H. Kalos, J. Comp. Phys. $\underline{2}$, 257, (1967).

17. Y. Accad, C.L. Pekeris and B. Schiff, Phys. Rev. A $\underline{4}$, 516 (1971).

18. A. Veillard and E. Clementi, J. Chem. Phys. $\underline{49}$, 2415 (1968).

19. K. J. Runge and R. J. Runge, in *Quantum Simulations of Condensed Matter Phenomena*, ed. by J. D. Doll and J. E. Gubernatis (World Scientific 1990); R.N. Barnett, P.J. Reynolds, and W.A. Lester, Jr., J. Comput. Phys. (to be published).

MONTE CARLO CALCULATION OF THE GROUND AND EXCITED STATE STRUCTURES OF ^{4}He CLUSTERS

S. A. Chin and E. Krotscheck

Center for Theoretical Physics and Department of Physics

Texas A&M University, College Station, TX 77843, USA

I. INTRODUCTION

The study of Helium clusters is interesting from many perspectives. A current theme is to understand the emergence of bulk physical properties from finite systems, *e.g.*, the onset of superfluidity as a function of cluster size. Another theme, inherited from nuclear physics, is to understand the extent to which bulk properties can be inferred from that of a finite system, *e.g.*, the relation between bulk liquid compressibility and the cluster giant resonance energies. Finally, Helium clusters are unique *bosonic*, but fully quantum, finite systems. Because of the simplicity of the ^{4}He-^{4}He potential, these clusters provide an exceptionally clean laboratory for the study of intrinsic many-body effects unhampered by the complexity of the interaction. The scattering of neutron off ^{4}He clusters would provide a unique opportunity of directly confronting experimental data with first-principle calculations.

In this work, we describe our efforts to understand both the ground and excited state properties of ^{4}He clusters through the use of Monte Carlo simulations[1,2]. We evolve the ground state wave function via a second order Diffusion Monte Carlo algorithm with Jastrow and triplet trial functions for importance sampling. We then obtain the collective excitation spectrum, transition densities, and the dynamic structure function by solving a generalized Feynman eigenvalue equation. We will focus primarily on presenting results of our investigation, interested readers can consult Refs.3-5 for more details and references.

II. GROUND STATE RESULTS

We evolve the product state $\Psi_0\Phi_0$ of a N-atom Helium cluster interacting through the Aziz[6] potential by a second order Diffusion Monte Carlo (DMC) algorithm[2]. The trial function Φ_0 used are a McMillian-type two-body trial function

$$\Phi_0^{\mathrm{I}} = \prod_{i<j}\exp\{-\frac{1}{2}(a/r_{ij})^5\}\prod_i\exp\{-\frac{1}{2}b^{-2}(\mathbf{r}_i - \mathbf{r}_{cm})^2\}, \tag{1}$$

and a triplet trial function of the form

$$\Phi_0^{\mathrm{II}} = \Phi_0^{\mathrm{I}}\exp\{-\frac{1}{4}\lambda\sum_k\mathbf{G}_k\cdot\mathbf{G}_k + \frac{1}{2}\lambda\sum_{i<j}g^2(r_{ij})r_{ij}^2\}, \tag{2}$$

Recent Progress in Many-Body Theories, Vol. 3,
Edited by T.L. Ainsworth et al., Plenum Press, New York, 1992

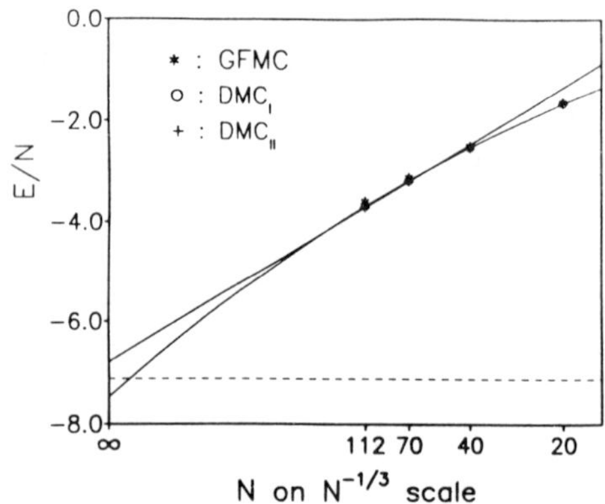

Fig. 1 The ground state energy per particle as a function of cluster
size. The two lines give the linear and the quadratic fits.
The broken line indicates the bulk liquid value.

where $\mathbf{G}_k = \sum_{l \neq k} g(r_{kl})\mathbf{r}_{kl}$ and $g(r) = \exp[-(r - r_t)^2/\omega^2]$. We shall refer to DMC
results obtained by using these two trial functions for importance sampling as DMC_I
and DMC_II respectively. Results obtained by using them only as variational wave
functions will be denote by VMC_I and VMC_II respectively. The values of variational
parameter used can be found in Ref. 5. Fig. 1 shows the resulting ground state energy
per particle as compared with a previous GFMC calculation[7]. The two DMC results
are in excellent agreement with one another, but are both slightly below that of GFMC.
Also shown are a linear (through the two larger cluster sizes) and a quadratic fits to
the energy. They bracket the experimental bulk value of -7.15 K. In contrast, the
same two fits would *both* under-estimate the bulk binding energy in the case of GFMC
results.

To obtain the ground state expectation value of operators other than the Hamiltonian, such as the one- and two-body densities,

$$\rho_1(\mathbf{r}) = \langle\Psi_0| \sum_i \delta(\mathbf{r}_i - \mathbf{r}_{cm} - \mathbf{r})|\Psi_0\rangle,$$

$$\rho_2(\mathbf{r}, \mathbf{r}') = \langle\Psi_0| \sum_{i \neq j} \delta(\mathbf{r}_i - \mathbf{r}_{cm} - \mathbf{r})\delta(\mathbf{r}_j - \mathbf{r}_{cm} - \mathbf{r}')|\Psi_0\rangle, \tag{3}$$

we use the perturbative estimate

$$\langle\Psi_0\,|\,O\,|\,\Psi_0\rangle = 2\langle\Psi_0\,|\,O\,|\,\Phi_0\rangle - \langle\Phi_0\,|\,O\,|\,\Phi_0\rangle. \tag{4}$$

Fig. 2 shows the one-body ground state densities for clusters sizes $N = 20, 40, 70,$
112. Despite differences in the VMC densities, there are excellent agreements among
the two DMC_I and one DMC_II results using the perturbative estimate (4). While all
VMC densities are generally smooth, DMC densities show persistent oscillations.

Fig. 3 compares our VMC and DMC results for the pair-distance distribution
$\rho_2(r)$ defined by

$$\rho_2(r) = \int d^3r_1\, d^3r_2 \delta(r - |\mathbf{r}_1 - \mathbf{r}_2|)\rho_2(\mathbf{r}_1, \mathbf{r}_2). \tag{5}$$

422

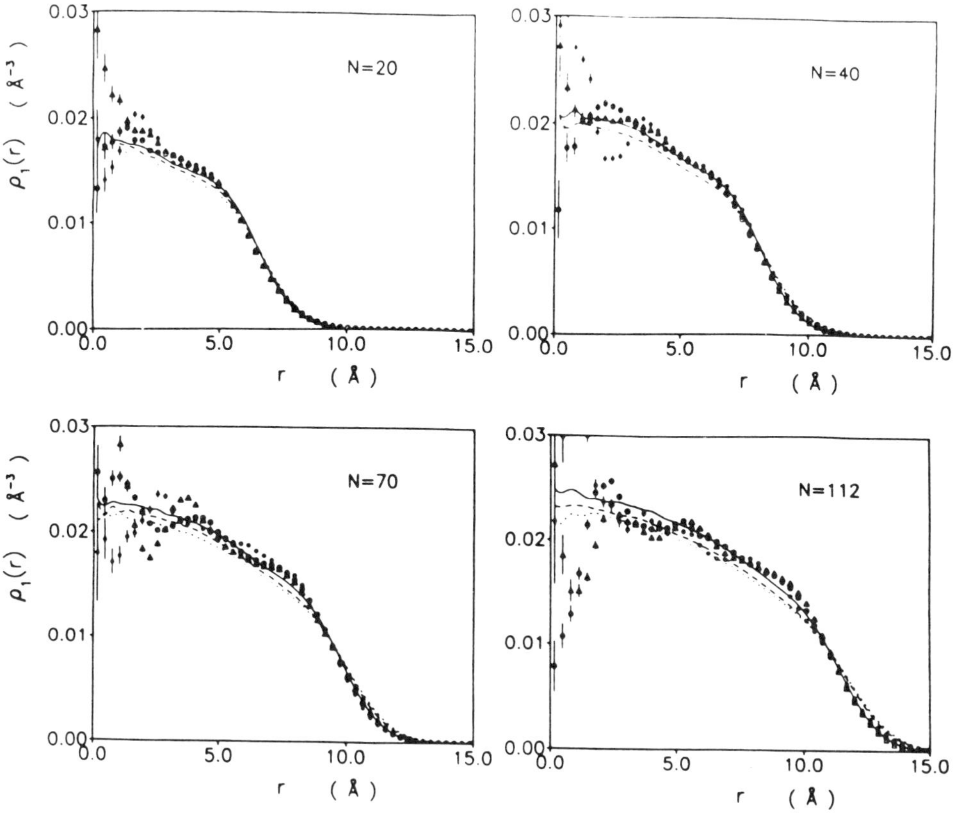

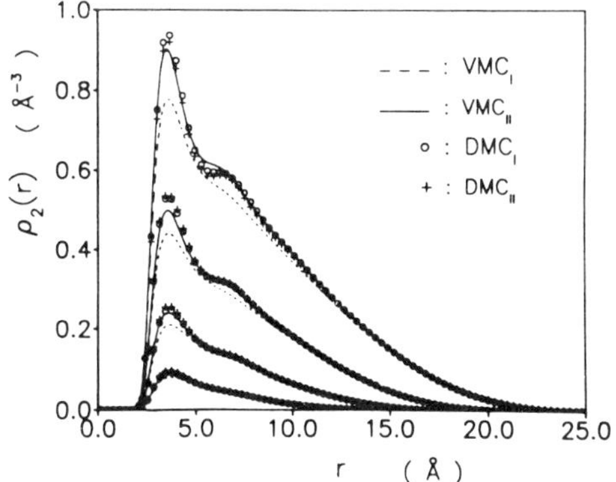

Fig. 2 The one-body density for four cluster sizes. The solid line and solid circles are VMC_{II} and DMC_{II} results. The two broken lines and open circles and triangles are two VMC_I and DMC_I results.

Fig. 3 The pair-distance distribution function as defined by (5) for cluster sizes, from bottom to top, $N = 20$, 40, 70 and 112.

Again the agreement between the two extrapolated DMC results is excellent. Rather than directly binning the two-body density $\rho_2(\mathbf{r}_1, \mathbf{r}_2)$, it is useful to bin its partial wave amplitudes $\rho_2^{(\ell)}(r, r')$ defined by

$$\rho_2^{(\ell)}(r, r') \equiv \int d\Omega \rho_2(\mathbf{r}, \mathbf{r}') P_\ell(\hat{\mathbf{r}} \cdot \hat{\mathbf{r}}'). \tag{6}$$

We have carried out such binnings for $\ell = 0, 1, 2, 4, 5$. Fig. 4 shows the $\ell = 0$ and $\ell = 2$ partial wave amplitude for $N = 112$ in the normalized form of

$$g_\ell(r, r') = \frac{(2\ell + 1)}{4\pi} \frac{\rho_2^{(\ell)}(r, r')}{\rho(r)\rho(r')}. \tag{7}$$

One observes that $g_0(r, 0)$ has the expected behavior of a pair correlation function familiar from bulk liquid helium.

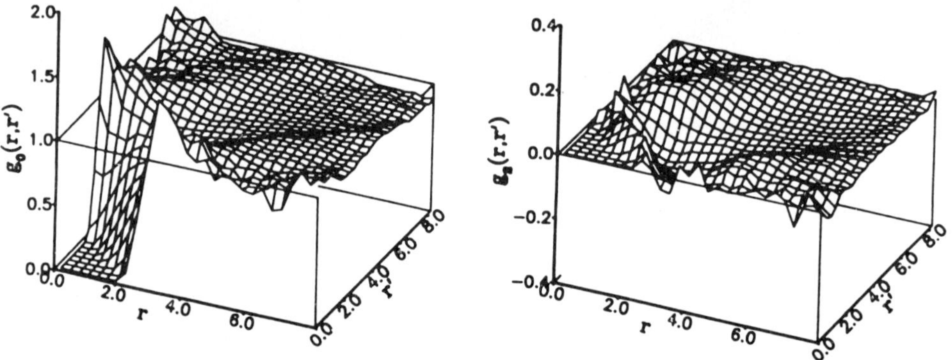

Fig. 4 The ground state energy per particle as a function of cluster size. The two lines give the linear and the quadratic fits. The broken line indicates the bulk liquid value.

III. COLLECTIVE EXCITATIONS

We generalizing Feynman's original theory[8] of collective excitation for bulk liquid ^{4}He to the case of finite clusters by writing the trial excited state as

$$\Psi_F(\mathbf{r}_1, \ldots \mathbf{r}_N) = F(\mathbf{r}_1, \ldots \mathbf{r}_N)\Psi_0(\mathbf{r}_1, \ldots \mathbf{r}_N), \tag{8}$$

where $F(\mathbf{r}_1, \ldots \mathbf{r}_N) = \sum_{i=1}^{N}\left[f(\mathbf{r}_i - \mathbf{r}_{cm}) - \langle f_i \rangle \right]$, $\langle f_i \rangle = \langle \Psi_0 \,|\, f(\mathbf{r}_i - \mathbf{r}_{cm}) \,|\, \Psi_0 \rangle$. A rigorous upper bound to the excitation energy $E_F - E_0 = \hbar\omega$ can be obtained by minimizing

$$\hbar\omega = \frac{1}{2} \frac{\langle \Psi_0 \,|\, [F, [H, F]] \,|\, \Psi_0 \rangle}{\langle \Psi_0 \,|\, F^2 \,|\, \Psi_0 \rangle} \tag{9}$$

with respect to the excitation function $f(\mathbf{r}_i - \mathbf{r}_{cm})$. This yields the generalized Feynman eigenvalue equation in terms of $u(\mathbf{r}) = \sqrt{\rho_1(\mathbf{r})}f(\mathbf{r})$,

$$\int d^3r' H_1(\mathbf{r}, \mathbf{r}')u(\mathbf{r}') = \hbar\omega \int d^3r' S(\mathbf{r}, \mathbf{r}')u(\mathbf{r}'), \tag{10}$$

where

$$H_1(\mathbf{r}, \mathbf{r}') = - \left[1 - \frac{1}{N}\right] \delta(\mathbf{r} - \mathbf{r}') \frac{\hbar^2}{2m} \frac{1}{\sqrt{\rho_1(\mathbf{r})}} \nabla \rho_1(\mathbf{r}) \cdot \nabla \frac{1}{\sqrt{\rho_1(\mathbf{r})}}$$
$$- \frac{\hbar^2}{2mN} \frac{\nabla_\mathbf{r} \cdot \nabla'_\mathbf{r} \rho_2(\mathbf{r}, \mathbf{r}')}{\sqrt{\rho_1(\mathbf{r})}\sqrt{\rho_1(\mathbf{r}')}}, \tag{11}$$

and $S(\mathbf{r}, \mathbf{r}') = \delta(\mathbf{r} - \mathbf{r}') + [\rho_2(\mathbf{r}, \mathbf{r}') - \rho_1(\mathbf{r})\rho_1(\mathbf{r}')]/\sqrt{\rho_1(\mathbf{r})\rho_1(\mathbf{r}')}$. With inputs of one- and two-body ground state densities, Eq. (10) can be solved by partial-wave expansion to determine the *entire* excitation spectrum within the Feynman ansatz. The bound state energies for the collective monopole ($\ell = 0$) and the quadrupole ($\ell = 2$) mode are shown in Figs. 5. The VMC results are obtained by solving the eigenvalue equation (12) with corresponding variational one- and two-body densities. As we improve upon the Jastrow trial function to the triplet trial function, our VMC monopole energy moves up, in agreement with our two DMC calculations. The emerging trend from VMC to DMC results gives an excellent impression of how the exact monopole energy must behave: It must initially increases with N and then comes down around $N = 112$ and approaches eventually the liquid drop limit. This is physically plausible in that for small N, an increase in cluster size increases the interior density, raises the zero sound speed and hence the monopole energy. For large N, the interior density is already constant, an increase in cluster size simply increases the wavelength of the excitation and therefore lowers the monopole energy.

For the quadrupole mode, the first surprise is that all VMC results cross over the liquid drop line. Our DMC calculations tend to bring their respective VMC values back down below the liquid drop line, but eventually fail at larger N's. The second surprise is that our triplet wave function not only did not close the gap between VMC and DMC results, it widens them further. This is very discomforting, even though our DMC$_{II}$ results seemed reasonable except for the last, $N = 112$ data point. This suggests that although Φ_0^{II} is a clear improvement over Φ_0^{I} in accounting for cluster ground state energies and densities, at large N it may still be inadequate for describing surface structures which are important for the quadrupole mode. It is also possible

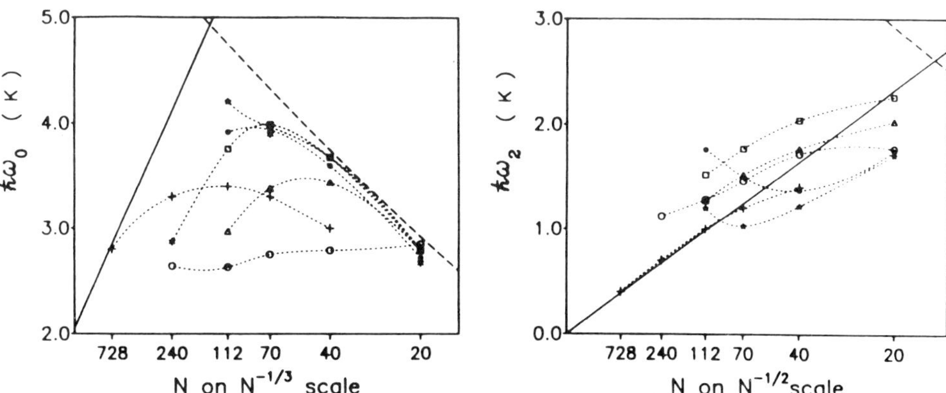

Fig. 5 The monopole and quadrupole collective excitation energy as a function of cluster size. The straight and broken lines denote the liquid drop limit and the chemical potential respectively. Crosses are the RPA results of Casas and Stringari[9]. Circles and asterisks are triplet VMC calculations of Wiringa *et al.*[10] and Krishna and Whaley[11] respectively. VMC$_I$ and VMC$_{II}$ results are denoted by triangles and squares and DMC$_I$ and DMC$_{II}$ results by dots and stars.

that the perturbative estimate (4) is inadequate for extrapolating the $\ell = 2$ component of the ground state two-body densities.

Once the excitation function is known, the corresponding transition density can be obtained via

$$\delta\rho(\mathbf{r}) = \langle \Psi_F \,|\, \sum_i \delta(\mathbf{r}_i - \mathbf{r}_{cm} - \mathbf{r}) \,|\, \Psi_0 \rangle,$$

$$= \rho_1(\mathbf{r})f(\mathbf{r}) + \int d^3 r'[\rho_2(\mathbf{r},\mathbf{r}') - \rho_1(\mathbf{r})\rho_1(\mathbf{r}')]f(\mathbf{r}'). \tag{12}$$

Figs. 6 show the resulting monopole transition densities. Marked oscillations with wavelengths approximately equal to the average particle separation suggest that they are connected with the geometric, hard-sphere like, shell structure of the droplets. By contrast, the variational monopole transition density shows no such oscillation. Similar oscillations are also evident in the quadrupole transition density.

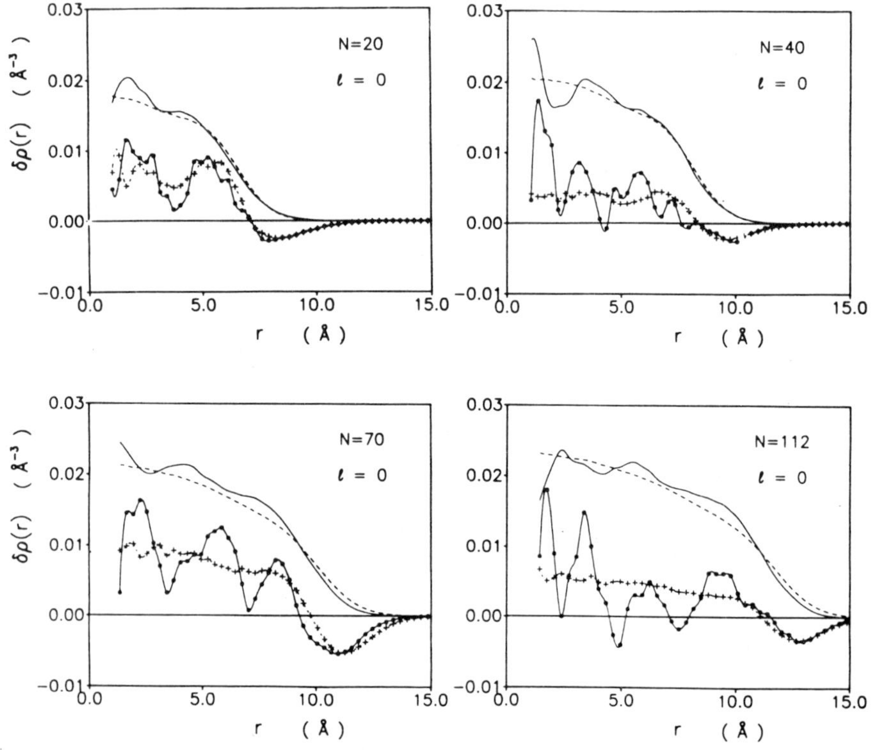

Fig. 6 The monopole transition density for four cluster sizes. The dots and crosses are DMC$_{\mathrm{II}}$ and VMC$_{\mathrm{II}}$ results, respectively. They are connected by smooth and broken lines to guide the eye. For comparison, their corresponding one-body densities are indicated by a solid and a dashed line. The scale for the transition density is arbitrary.

IV. DYNAMIC STRUCTURE FUNCTION

In scattering a weak probe from a composite system, the scattering cross-section is given in the Born approximation by[12]

$$\frac{d^2\sigma}{d\Omega d\omega} = \frac{p_f}{p_i}\left(\frac{d\sigma}{d\Omega}\right)_0 S(\mathbf{k},\omega),\tag{13}$$

where $(d\sigma/d\Omega)_0$ is the differential scattering cross-section for scattering from a single constitutent, $\mathbf{p}_i$ and $\mathbf{p}_f$ are the initial and final momenta, $\mathbf{k} = \mathbf{p}_f - \mathbf{p}_i$ is the momentum transfer, and

$$S(\mathbf{k},\omega) = \sum_{n\neq 0} |\langle\Psi_n|\rho(\mathbf{k})|\Psi_0\rangle|^2\delta(E_n - E_0 - \omega)\tag{14}$$

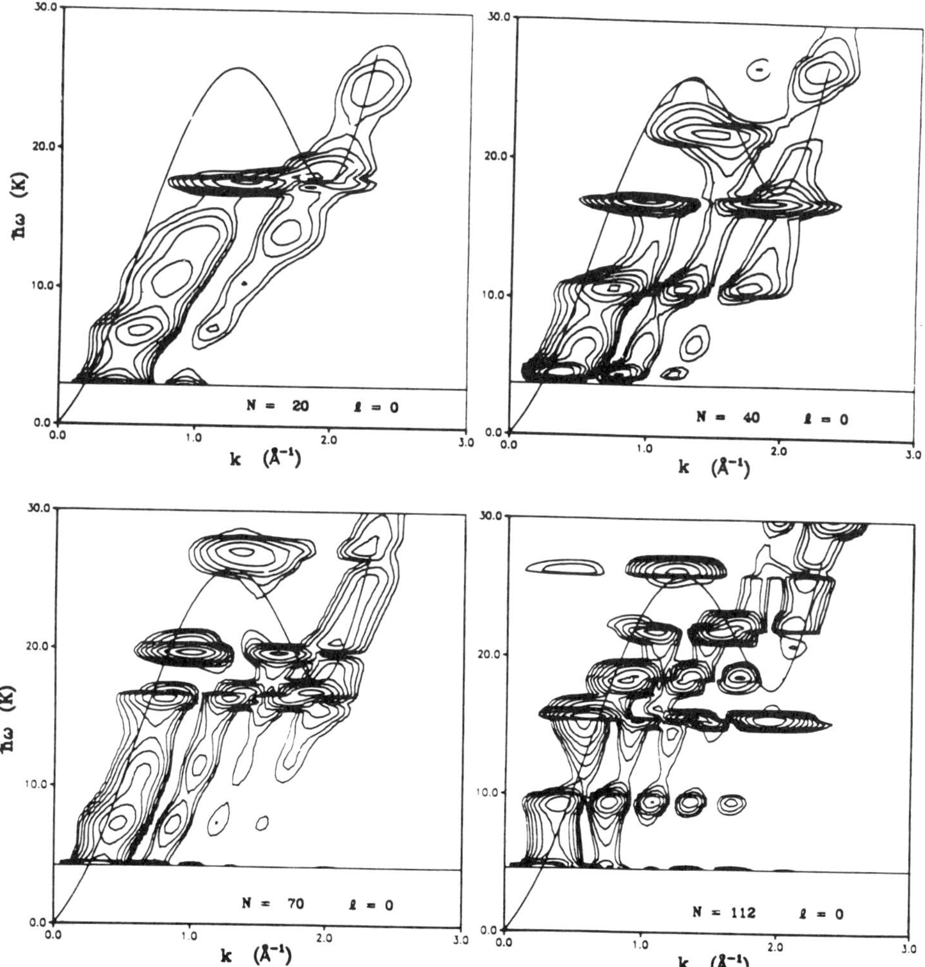

Fig. 7 Contour plot of the $\ell=0$ component of the dynamic structure function $S(\mathbf{k},\omega)$ for four cluster sizes. The contour lines depict levels of exponentially varying strength.

is the dynamic structure function. Within the Feynman ansatz, since for a given excitation energy $\hbar\omega = E_n - E_0$, one can solve (10) to obtained the corresponding transition density $\delta\rho_\omega(\mathbf{r}) = \langle \Psi_n|\rho(\mathbf{k})|\Psi_0\rangle$ via (12), one automatically determines a parameter-free two-point dynamical structure function,

$$S(\mathbf{r},\mathbf{r}';\omega) = \delta\rho_\omega(\mathbf{r})\delta\rho_\omega(\mathbf{r}'). \tag{15}$$

What is measured in inclusive scattering experiments is its diagonal Fourier transform,

$$S(\mathbf{k},\omega) = \int d^3r d^3r' e^{i\mathbf{k}(\mathbf{r}-\mathbf{r}')} S(\mathbf{r},\mathbf{r}';\omega) = \sum_\ell S_\ell(k,\omega), \tag{16}$$

where

$$S_\ell(k,\omega) = |\delta\rho_{\ell\omega}(k)|^2, \tag{17}$$

and $\delta\rho_{\ell,\omega}(k) = \int d^3r \delta\rho_{\ell\omega}(r)j_\ell(kr)$.

Figs. 7 show contour plots of $S_0(k,\omega)$ for four cluster sizes studied. Also plotted is the bulk liquid Feynman dispersion relation. One observes a series of "ridges" running roughly parallel to the Feynman curve, punctuated by discrete sets of peaks. As the cluster size increases, these peaks become denser along the Feynman curve, following it with remarkable faithfulness. The sharply defined peaks that space out at discrete values of $\hbar\omega$ clearly reflect the quantized character of excitations appropriate for a finite droplet. One can thus directly read off the droplet's continuum excitation spectrum by simply noting the locations of these peaks. The regular repetition of peaks and ridges along k with diminishing strength appears to be due mostly to "diffractive" effects, $i.e.$, from squaring the Fourier transforms of a transition densitiy which is non-zero only within the droplet. The ridges have the correct spacing in k of roughly π/R, where R is the droplet radius. Thus the dynamic structure function directly maps the cluster excitation spectrum and provides a graphic demonstration of how the continuous bulk dispersion relation emerges out of the discrete cluster excitation spectrum.

V. SUM RULES

The static structure function $S(k)$ can either be expressed directly as

$$
\begin{aligned}
S(k) &= \frac{1}{N} \int d^3r d^3r' \sqrt{\rho_1(\mathbf{r})\rho_1(\mathbf{r}')}e^{i\mathbf{k}\cdot(\mathbf{r}-\mathbf{r}')} S(\mathbf{r},\mathbf{r}'), \\
&= 1 + \frac{1}{N}\left\{\tilde\rho_2(k) - |\tilde\rho_1(k)|^2\right\},
\end{aligned}
\tag{18}
$$

where $\tilde\rho_2(k)$ and $\tilde\rho_1(k)$ are Fourier transforms of the pair-distance and one-body densities respectively, or as a sum over integrated contribution of each angular momentum modes,

$$S(k) = \sum_\ell \frac{1}{N} \int_0^\infty d(\hbar\omega) S_\ell(k,\omega). \tag{19}$$

Comparing the latter with the former then allows us to determine the specific contribution of each excitation to the full static structure function $S(k)$. This is shown in Figs. 8. Aside from the reduced overshoot of the cluster $S(k)$ around $k \approx 2\text{Å}^{-1}$, there appears a shoulder around $k \approx 0.5\text{Å}^{-1}$ which is absent in the bulk $S(k)$. This is primarily due the quadrupole mode. For cluster size considered, this mode accounts

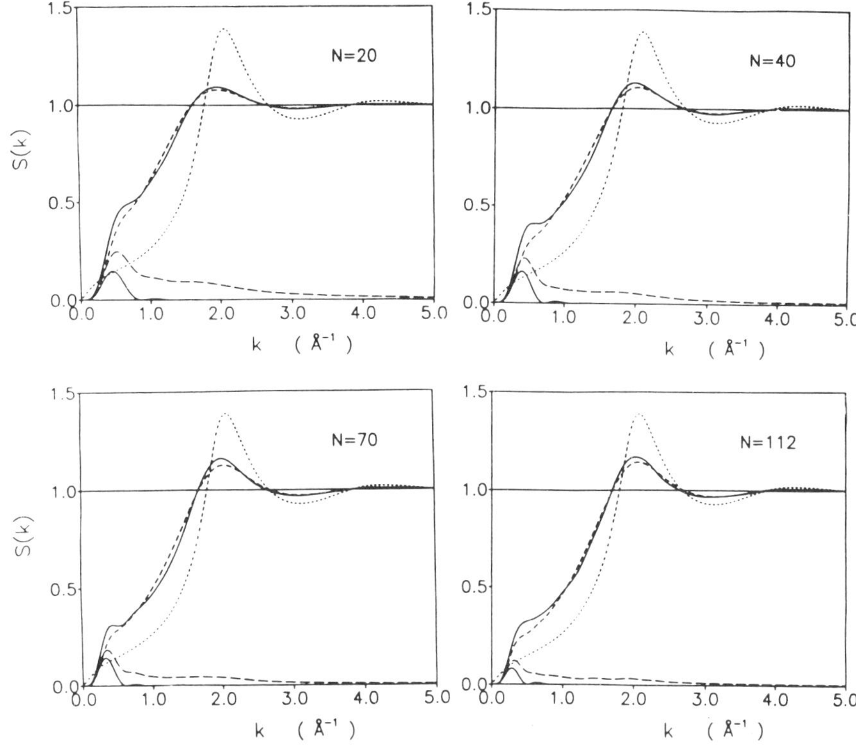

Fig. 8 Sum rule results for four cluster size. The solid and short dash line are the
DMC$_{\mathrm{II}}$ and the VMC$_{\mathrm{II}}$ static structure function respectively. The dotted
line is the experimental bulk liquid structure function. The contribution to
$S(k)$ from the quadrupole collective state is given by the solid hump. The
sum of contributions from all states in the three lowest angular momentum
mode is given by the long dash line.

for 25%-30% of the overall excitation strength. By contrast, the monopole's contribution is negligible ($< 5\%$). In the limit of $N \to \infty$, the quadrupole (which is mainly a
surface mode) ceases to be important, and the shoulder goes away.

VI. CONCLUSIONS

In this work, we have demonstrated that by knowing just the ground state densities, a tremendous amount of physics about the excited states of a finite quantum
system can be extracted. In solving the generalized Feynman eigenvalue equation, we
have determined, 1) the systematic variation of collective energies with cluster size, 2)
shell structure related oscillations in the transition densities, and 3) the cluster excitation spectrum as mapped out by the dynamic structure function. Once the one- and
two-body densities of any finite system are known, the same theory can be applied.

It is therefore of great interest to apply this generalized Feynman theory to Fermi systems as well as to improve the theory itself by including higher order ground state densities.

ACKNOWLEDGMENT

This work was supported, in part, by the National Science Foundation grants PHY89-07986 (to SAC), PHY-8806265 (to EK), and the Texas Advanced Research Program under Grant 010366-012 (to EK). The simulations of the ground-state structure were performed on the CRAY-YMP of the Texas A&M University Supercomputer Center.

REFERENCES

1. D. Ceperly and M. Kalos, in *Monte Carlo Methods in Statistical Mechanics*, edited by K. Binder (Springer, New York, 1979).
2. S. A. Chin, Phys. Rev. **A42**, (1990) 6991.
3. S. A. Chin and E. Krotscheck, Phys. Rev. Lett. **65**, 2658 (1990).
4. S. A. Chin and E. Krotscheck, Chem. Phys. Lett. **178**, 435 (1991).
5. S. A. Chin and E. Krotscheck, Texas A&M University CTP preprint 51/91.
6. R. A. Aziz, V. P. S. Nain, J. C. Carley, W. L. Taylor, and G. T. McConville, J. Chem. Phys. **70**,4330 (1979).
7. V. R. Pandharipande, J. G. Zabolitzky, S. C. Pieper, R. B. Wiringa, and U. Helmbrecht, Phys. Rev. Lett. **50**, 1676 (1983).
8. R. P. Feynman, Phys. Rev. **94**, 262 (1954).
9. M. Casas and S. Stringari, J. Low. Temp. Phys. **79** (1990) 135.
10 R. B. Wiringa, S. C. Pieper, and V. R. Pandharipande (Argonne preprint Phy-4676-TH-85) (1985).
11 M. V. Rama Krishna and K. B. Whaley, J. Chem. Phys. **93**, (1990) 6738.
12. L. Van Hove, Phys. Rev. **95**, (1954) 249.

MONTE CARLO APPROACHES TO EFFECTIVE FIELD THEORIES

J. Carlson

T-Division, Los Alamos National Laboratory
Los Alamos, NM 87545

K. E. Schmidt

Department of Physics and Astronomy
Arizona State University, Tempe, AZ 85287

INTRODUCTION

In this paper, we explore the application of continuum Monte Carlo methods to effective field theory models. Effective field theories, in this context, are those in which a Fock space decomposition of the states is useful. These problems arise both in nuclear and condensed matter physics. In nuclear physics, much work has been done on effective field theories of mesons and baryons. While the theories are not fundamental, they should be able to describe nuclear properties at low energy and momentum. After describing the methods, we solve two simple scalar field theory problems; the polaron and two nucleons interacting through scalar meson exchange.

The methods presented here are rather straightforward extensions of methods used to solve quantum mechanics problems. Monte Carlo methods are used to avoid the truncation inherent in a Tamm-Dancoff approach and its associated difficulties (such as cluster separability). Nevertheless, the methods will be most valuable when the Fock space decomposition of the states is useful. Hence, while they are not intended for ab initio studies of QCD, they may prove valuable in studies of light nuclei, or for systems of interacting electrons and phonons. In these problems a Fock space decomposition can be used to reduce the number of degrees of freedom and to retain the rotational symmetries exactly. These symmetries can be recovered in a lattice approach, but only at the cost of a very large number of degrees of freedom, most of which may be physically irrelevant.

The problems we address here are comparatively simple, but offer useful initial tests of the method. We present results for the polaron[1] (a slow electron in a polar crystal) and two non-relativistic nucleons interacting through scalar meson exchange. In each case, it is possible to integrate out the boson degrees of freedom exactly, and obtain a retarded form of the action that depends only upon the fermion paths. Here we keep the bosons explicitly, though, since (1) we would like to retain information about the boson components of the states and (2) it will be necessary to keep these components in order to treat non-scalar or interacting bosonic fields. We conclude with a few remarks on future applications of the method.

Recent Progress in Many-Body Theories, Vol. 3,
Edited by T.L. Ainsworth et al., Plenum Press, New York, 1992

HAMILTONIAN AND METHODS

We consider a simple Hamiltonian consisting of free-particle energies for mesons and baryons plus a simple interaction term:

$$H = \int d\mathbf{k}\, \omega_f(k)\, a^\dagger(\mathbf{k})a(\mathbf{k}) + \int d\mathbf{k}\, \omega_b(k)\, b^\dagger(\mathbf{k})b(\mathbf{k}) + \left[\int d\mathbf{k}d\mathbf{q}\, \beta(q)\, a^\dagger(\mathbf{k}-\mathbf{q})a(\mathbf{k})b^\dagger(\mathbf{q}) + h.c. \right],$$

$$(1)$$

where the $a(\mathbf{k})$ $(b(\mathbf{k}))$ are the fermion (boson) destruction operators, and $a^\dagger(\mathbf{k})$ $(b^\dagger(\mathbf{k}))$ the creation operators. The function $\omega_f(k)$ describes the free fermion propagator, and $\omega_b(k)$ the boson propagator. The coupling term involving $\beta(k)$ is linear here, creating or destroying one boson at a time. While this feature may not be essential, it is necessary that the theory not include fermion loops. Relativistic extensions incorporating only Z-graphs do seem feasible, however. These 'quenched' theories are used to describe the polaron, and may be useful for hadronic effective field theories due to the composite nature of the nucleons.

The problems discussed in this paper involve only scalar bosons and one or two fermions, hence the sign problem does not arise. Of course, it will appear for systems with more fermions, in which case transient estimation or fixed node methods can be employed. It is also possible that the methods proposed elsewhere in this volume by Kalos for handling the fermion sign problem for few-body systems will prove useful here.

We employ a momentum space description here, although conversion to coordinate space is straightforward. The expressions in momentum space are highly non-local, but in coordinate space the non-locality is governed by the form factors at the vertices. Each representation has advantages and disadvantages. For example, momentum space is useful in determining the self-energy and effective mass of one fermion, while coordinate space may prove more useful in studying few-fermion systems. This would be particularly true if, for example, a contact nucleon-nucleon term were added to the Hamiltonian to represent part of the repulsive core.

A simple variational (trial) wave function for one fermion and many bosons is:

$$\Psi_t = P(\mathbf{K}) \int d\mathbf{k_f}\, g(\mathbf{k_f},\mathbf{K})\, a^\dagger(\mathbf{k_f})\, \exp\left[\int d\mathbf{k}\, f(k)\, b^\dagger(\mathbf{k})\,\right] \mid 0 \,\rangle. \tag{2}$$

The operator $P(\mathbf{K})$ is a projection on total momentum $\mathbf{K}$, $g(\mathbf{k_f},\mathbf{K})$ describes the fermion wave function, and $f(\mathbf{k})$ describes the bosons. This wave function describes independent mesons multiplied by a fermion wave function, which is appropriate to the problems studied here.

Variational Monte Carlo can be performed as a random walk in both coordinate and Fock space, using the Metropolis method[2] to obtain points distributed with a density distribution proportional to the square of a trial wave function. Three types of moves are employed in the algorithm; creation, destruction, and diagonal moves. The diagonal moves are those that do not create or destroy mesons; they can be treated in the standard fashion. The creation and destruction moves must be constructed to satisfy detailed balance. The condition may be written generically as:

$$|\Psi(\mathbf{K}_{n+1})|^2 T(\mathbf{K}_{n+1} \to \mathbf{K}_n)A(\mathbf{K}_{n+1} \to \mathbf{K}_n) = |\Psi(\mathbf{K}_n)|^2 T(\mathbf{K}_n \to \mathbf{K}_{n+1})A(\mathbf{K}_n \to \mathbf{K}_{n+1}), \tag{3}$$

where $T(a \to b)$ represents the probability of proposing a move from configuration a to configuration b, and A represents the acceptance probability for that move. The arguments $\mathbf{K}_n$ indicate the momenta of each of the particles, n of which are bosons. For a wave function of the form of Eq. 2,

$$|\Psi(\mathbf{K}_n)|^2 = g^2(\mathbf{k_f}) \prod_{i=1}^{n} f^2(k_i)/n!. \tag{4}$$

Many valid choices of the transition and acceptance probabilities exist. For simple wave functions, it is possible to obtain very efficient Metropolis algorithms. Of course, the wave function must be symmetric under the exchange of any two boson momenta. Also, some care must be undertaken when constructing the detailed balance condition, since in going from the n+1 to the n boson state there are n+1 possible choices of the meson to destroy. For other than 1-body problems, there is also the choice of which fermion absorbs the destroyed meson's momentum. Given this algorithm, variational methods can be applied to determine approximate ground state properties.

The Green's function Monte Carlo method[3,4] projects out the true ground state by applying $\exp(-H\tau)$ to the trial wave function. This method can be thought of as corresponding to time-ordered perturbation theory, where the 'time' variable is a euclidean time. For Hamiltonians of the form of equation 1, no time-step error need be associated with the calculation.

The method is most easily described algorithmically. We assume some familiarity with standard Green's function or Diffusion Monte Carlo methods.[3,4] The variational Monte Carlo algorithm can be used to generate a set of configurations drawn with probability proportional to the square of the trial wave function. Each configuration consists of the coordinates of the fermions plus a set of coordinates for the bosons. In order to propagate the configurations in imaginary time, we first calculate the following quantities for each configuration:

$$E_{diag} = \omega_f(k_f) + \sum_i \omega_b(k_i), \tag{5}$$

$$E_{cre} = \frac{\int d\mathbf{q}\langle\Psi_t(\mathbf{K}_{n+1})|\beta(q)b^\dagger(\mathbf{q})a^\dagger(k-q)a(k)|\Psi_t(\mathbf{K}_n)\rangle}{\langle\Psi_t(\mathbf{K}_n)|\Psi_t(\mathbf{K}_n)\rangle}, \tag{6}$$

$$E_{des} = \frac{\sum_i\langle\Psi_t(\mathbf{K}_{n-1})|\beta(q)b(\mathbf{q})a^\dagger(k+q)a(k)|\Psi_t(\mathbf{K}_n)\rangle}{\langle\Psi_t(\mathbf{K}_n)|\Psi_t(\mathbf{K}_n)\rangle}. \tag{7}$$

The first quantity E_{diag} is just the sum of boson and meson energies. The 'creation energy' E_{cre} involves an integral over all momenta for a created meson. The vector $\Psi_t(\mathbf{K}_{n+1})$ indicates a state with n boson momenta equal to the original configuration, and one additional boson of momentum q. The fermion must lose the created boson's momentum. For the simple wave functions used here, this integral can be performed analytically. The 'destruction energy' E_{des} sums over all possible destroyed mesons. Generalizations of these expressions to systems with more than one fermion are straightforward, and involve sums over the fermions that absorb or destroy mesons. We assume everywhere that the occupation probability for bosons of any specific momentum $\mathbf{q}$ goes as the inverse of the volume.

Once we have calculated these quantities, a configuration can be propagated without any time step error. The probability per unit time of creating or destroying a meson is given by $|E_{cre}| + |E_{des}|$. The time until the next creation or destruction can then be sampled from an exponential distribution

$$\delta\tau = \frac{-1}{|E_{cre}| + |E_{des}|}\log(\zeta), \tag{8}$$

where ζ is a random number distributed uniformly between 0 and 1. The growth per unit time of the population is

$$R = |E_{cre}| + |E_{des}| - E_{diag} - E_t, \tag{9}$$

where E_t is an arbitrary constant added to the Hamiltonian to keep the size of the population stable. After choosing $\delta\tau$, the weight of the configuration is multiplied by $R\delta\tau$.

At this point either a creation or destruction move must be made, with probabilities:

$$P_{cre} = |E_{cre}|/[|E_{cre}| + |E_{des}|] \tag{10}$$

$$P_{des} = |E_{des}|/[|E_{cre}| + |E_{des}|]. \tag{11}$$

If a creation move is chosen, the weight is multiplied by the sign of $-E_{cre}$ and a momentum for the new boson is picked with relative probability proportional to the coefficient of the creation term in H for a boson of momentum q. If a destruction move is chosen, one of the bosons is destroyed with a relative probability governed by the destruction energy for that boson. In this case the weight of the configuration is multiplied by the sign of $-E_{des}$. In the problems discussed here, these signs are always positive. Each configuration can be propagated until it has reached a specified imaginary time. Note that different numbers of steps will be required for each configuration. At this point, the branching technique is used to retain equal weights for all the configurations.

Alternative formulations of the Green's function method are possible. The method outlined above requires that the creation energy E_{cre} be evaluated exactly for each configuration in order to eliminate time step errors. If this is not feasible, one can employ iteration schemes where the interaction term of the Hamiltonian enters linearly, allowing one to sample them directly. Such a method would be essentially a Monte Carlo implementation of Rayleigh-Schroedinger perturbation theory.

THE POLARON

In this section we briefly review the polaron,[1] and discuss some of the results we have obtained. The polaron has been studied extensively; a review is given in reference 5. For this reason, it offers a valuable testing ground for our method. The Hamiltonian is of the form of equation 1 for a slow electron interacting with a field of phonons. The free-particle energies and coupling are:

$$\omega_f(k) = k^2/2 \quad \omega_b(k) = 1 \quad \beta(k) \propto \sqrt{\alpha}/k. \tag{12}$$

The electron mass and phonon energy have been absorbed into the coupling constant. Note that the phonon energy is independent of momentum in this Hamiltonian.

A variety of variational schemes have been employed to study this problem.[5,6] While I cannot list them all, I will try to illustrate the most important features. The variational wave functions presented here are very simple, but they serve to illustrate the two limits.

At weak coupling (small α), the ground state wave function can be written in the form of equation 2, with

$$g(\mathbf{k}) = 1 \qquad f(\mathbf{k}) \propto \frac{\sqrt{\alpha}}{k}\left(\frac{1}{2+k^2}\right) \tag{13}$$

A wave function of this form gives an $E(\alpha) = -\alpha$, and an expectation value of $\alpha/2$ for the number of bosons. Better wave functions give corrections of order α^2.

At strong coupling this wave function is inadequate because the large number of uncorrelated phonons in the trial wave function leads to large fermion kinetic energies. The simplest, but by no means most accurate, strong coupling wave function assumes that g and f are both gaussians.[7] Neglecting momentum conservation, the parameters of the gaussians can be optimized to yield an energy of $E = -\alpha^2/(3\pi)$ and an average value of $2\alpha^2/(3\pi)$ bosons. The variational method described above can be used to project out a state of zero total momentum, which lowers the energy. Note that before the projection on zero momentum this wave function confines the electron. Once the projection is introduced, though, the electron is not confined since the no-phonon component of the state is a delta function in momentum space.

As mentioned previously, the phonons can be integrated out exactly. The result, due to Feynman,[8] is an effective action for the electron which contains a retarded interaction:

$$S_{eff} = \int_0^\beta dt\frac{\dot{\mathbf{x}}^2}{2} + \frac{\alpha}{2}\int_0^\beta dt \int_0^t d\tau \frac{\exp(-[t-\tau])}{|\mathbf{x}(t) - \mathbf{x}(\tau)|} \tag{14}$$

From the action, the ground state energy can be obtained through the path integral:

$$Z(\beta) \equiv \int d\mathbf{x}_0 \int_0^{\mathbf{x}(\beta)=\mathbf{x}_0} \mathcal{D}[\mathbf{x}] \exp(-S[\mathbf{x}]) \overset{\beta\to\infty}{\to} \exp[-\beta E_0] \tag{15}$$

The integral over the end-point $\mathbf{x}_0$ is a projection onto the zero-momentum state. Of course, in principle this is not necessary in the infinite β limit, but it greatly improves the convergence in numerical calculations.

434

This retarded interaction cannot be treated analytically, so various approximations and numerical methods have been employed. Feynman constructed a gaussian trial action which can be explicitly evaluated to determine an upper bound to the ground state energy:

$$S_T = \int_0^\beta dt \frac{\dot{\mathbf{x}}^2}{2} + C \int_0^\beta dt \int_0^t d\tau \ \exp(-\omega[t - \tau]) \ [\mathbf{x}(t) - \mathbf{x}(\tau)]^2 \tag{16}$$

This form contains two variational parameters, w and C. These can be optimized numerically for any value of the coupling constant; the results are given in table 1.

The second-order corrections to this trial action have also been evaluated,[9] essentially through calculating moments of the path integral. The results of this calculation are also given in the table. Note that these corrections do not necessarily preserve the upper bound property for the estimated ground state energy.

Finally, the effective action can be evaluated numerically to achieve an 'exact' result. Alexandrou, Fleischer, and Rosenfelder[10] have recently done this by breaking the path integral into fourier components. The lowest M components (with M as large as 19) are treated by Monte Carlo integration, while Feynman's trial action is used for the higher components. This approach requires one to extrapolate to the $\beta \to \infty, \mathrm{M} \to \infty$ limit, while keeping $M >> \beta$.

In order to compare the methods, we have computed the ground state energy at the same coupling constants as Alexandrou, et al.[10] Note that the Feynman trial action (column 1) gives a result correct to within a few per cent for all values of the coupling constant. The next column adds the second order corrections to Feynman's trial action. The third and fourth columns indicate the results of the Fourier decomposition and our GFMC results, respectively. The sum of statistical and systematic errors are indicated in brackets. At small coupling constants these two numerical methods give results identical within their statistical errors. As the coupling constant increases, though, the results begin to disagree.

Table 1. Polaron Ground-State Energy

α	S_T	+pert	Fourier	GFMC
1.0	-1.013	-1.0135	-1.0166[20]	-1.0167[2]
3.0	-3.133	-3.161	-3.153[9]	-3.168[2]
5.0	-5.440	-5.523	-5.484[24]	-5.555[15]

The two methods have somewhat different weaknesses. The effective action technique has the equivalent of a finite time-step error, and requires an extrapolation to infinite β limit. The extrapolations become more difficult as the coupling constant increases; these extrapolations are described in their paper.

Any errors in the GFMC results are likely due to the bias caused by fluctuating populations. This bias increases with coupling constant due to the fairly primitive trial wave functions used here. We have checked our results by varying the population size and by using different trial functions which are linear superpositions of the weak and strong coupling wave functions. In the future, we will use more accurate trial wave functions, which will reduce both the bias and the statistical error of the results.

We now examine some properties of the phonons, as that was one of the reasons for keeping these variables explicitly in the first place. Figure 1 shows the probability of different numbers of bosons for the case $\alpha = 3$. The solid line is the result of calculations with the weak coupling wave function, and the circles are mixed estimates obtained in the GFMC calculations, using the weak coupling wave function as the importance function. Mixed estimates are defined as $\langle O \rangle_M = \langle \Psi_T | O | \Psi_0 \rangle$. Linear extrapolation can be used to approximate the true ground state expectation value. I have not done that in this case since the differences in probabilities can be quite large when the number of phonons is large, and the extrapolation procedure is likely to be inadequate. Undoubtedly this could be cured with more accurate trial wave functions.

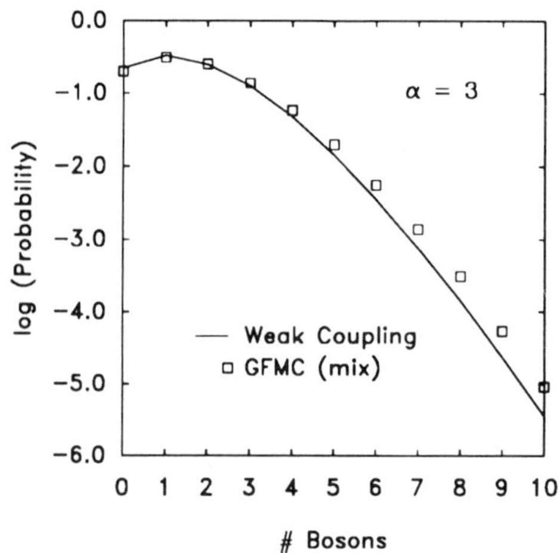

Figure 1. Probability for different numbers of phonons

In the second figure, we plot the expectation value of the square of the momentum vs. the number of phonons, again for $\alpha = 3$. For the bosons, the sum of the squares of the individual momenta are plotted. The weak coupling results are a straight line since the phonons are uncorrelated in this approximation. In our GFMC results (again mixed estimates are plotted), the fermion has on average a smaller momenta than the weak coupling limit, while the bosons momenta are larger. These results are due to correlations induced between the phonons by the fermion kinetic energy; the system prefers phonons with anti-correlated momenta.

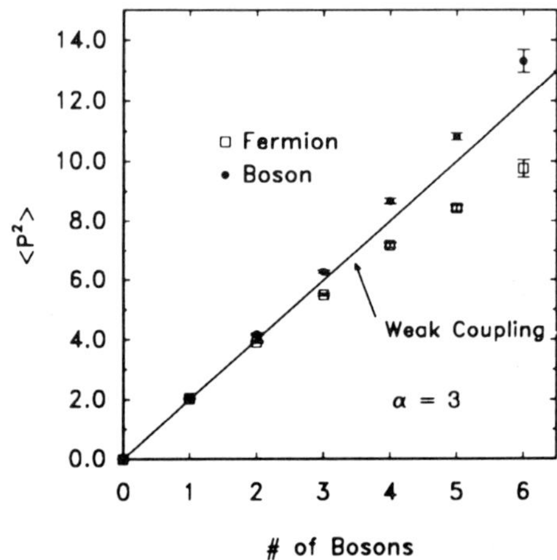

Figure 2. Expectation value of P^2 vs. the number of phonons

In the near future, these calculations will be extended to larger coupling constants and to finite momentum. To improve the accuracy of the calculations, the trial wave functions should be improved, but the methods are basically unchanged. A variety of polaron problems have been studied in the literature (for example, polarons bound in a well); exact GFMC calculations are also feasible for these problems.

NUCLEONS AND SCALAR MESONS

In this section, we look at solutions for one and two nucleons interacting through scalar meson exchange. This problem is similar to that studied initially by Serot, Koonin, and Negele.[11]
They treated a one-dimensional system of nucleons interacting with scalar and vector meson
fields using a lattice approach. Eventually one would like to include interacting scalar, vector, and pion fields (and perhaps a contact interaction) in three dimensions. The hope is
that eventually one will be able to treat effects such as exchange currents, three-body forces,
meson-nucleon scattering and meson production in a more consistent manner.

Although we are still a long way from that goal, perhaps we can gain some insight in even
these very simple models. Here we treat a Hamiltonian in the form of Equation 1, with:

$$\omega_f(k) = k^2/(2M) \quad \omega_b(k) = \sqrt{m^2 + k^2} \quad \beta(k) = \alpha N(\alpha)\frac{\exp(-\mathbf{k}^2/k_0^2)}{\sqrt{\omega_b(k)}}, \tag{17}$$

where we take a 'nucleon' mass of 1 GeV, a scalar mass of 0.1 GeV, and two different sets of
values for the coupling:

$$\text{set}[A] \quad g = 0.04 \quad k_0 = 1.0\,fm^{-}1 \tag{18}$$

$$\text{set}[B] \quad g = 0.35 \quad k_0 = 4.0\,fm^{-}1. \tag{19}$$

In each case, the factor N is a normalization factor: $N = [1/(k_0\sqrt{\pi})]^{3/2}$. Parameter set A
corresponds to a weak-coupling situation and gives total energies at a nuclear physics scale
(tens of MeV). Set B is an intermediate coupling, but due to the lack of short range repulsion
in this model, gives very large total energies (hundreds of MeV).

We compare the nucleon self-energies in the static approximation to the GFMC results in
table 2. In the static approximation, one solves for the state in the $M \to \infty$ limit, and adds
the kinetic energy of the fermion as a perturbation. As one can see from the table, the static
approximation is very accurate at weak coupling but somewhat less accurate at intermediate
coupling.

Table 2. Nucleon Self-Energy

Weak Coupling			
Self Energy (GeV)		$\langle b^\dagger b \rangle$	
Static	GFMC	Static	GFMC
-0.018807	-0.018865[3]	0.1297	0.1247[3]
Intermediate Coupling			
Self Energy (GeV)		$\langle b^\dagger b \rangle$	
Static	GFMC	Static	GFMC
-0.1672	-0.1751[6]	0.681	0.599[1]

In the static approximation, one can also construct a potential from the relation:

$$V_{NN}(static) = -\int d\mathbf{k} \; \frac{|\beta(k)\rho(k)|^2}{\omega_b(k)} + \Delta E_{recoil} - 2\Delta E_1. \tag{20}$$

The nucleon density is indicated by ρ, and the last two terms are the recoil kinetic energy in the
two-body system and twice the recoil energy of an isolated nucleon, respectively. The resulting
potential and recoil correction are shown for the weak-coupling case in figure 3.

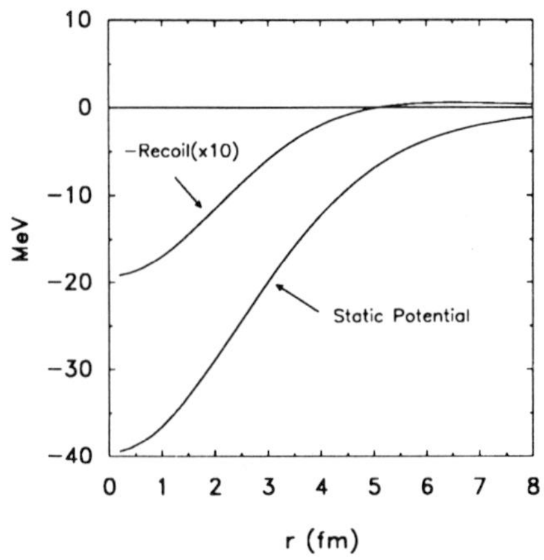

Figure 3. Nucleon-Nucleon potential in the static approximation.

We can compute the solution of the Schroedinger equation with this static potential and compare to the solution of the full two-nucleon problem. For weak coupling (case A), the Schroedinger equation gives an energy of -12.3 MeV, with a contribution of approximately +0.8 MeV from the recoil terms. We have used a somewhat simplified variational form for the two-nucleon state in the full calculation; it has the solution of the Schroedinger equation built into the fermion sector but the bosonic wave function is simplified. Treating the full static approximation to the wave function is computationally expensive in momentum space. Nevertheless, the Variational Monte Carlo result for the simplified wave function is -9.6 MeV. The Green's function Monte Carlo method gives a slightly lower energy of -13 $\pm$ 0.1 MeV.

For the intermediate coupling problem, the variational result for the difference between two interacting nucleons and two isolated nucleons is 145 MeV, while the GFMC result is 150 $\pm$ 2 MeV. Thus the variational result is surprisingly close to the correct answer. Of course the difference is larger (approximately 20 MeV) in terms of absolute energies, but it appears that the error is largely in the one-body self-energy.

The errors made in the static approximation here are 5 to 10 %, similar to those found in reference 11. When making detailed comparisons of particular expectation values the error may be larger, however. Of course, this picture is much too simplified for even primitive comparisons with nuclear physics. At a minimum, a repulsive core must be added, either through a vector meson exchange or a contact interaction. In addition, the full field theory should be compared with a potential model that gives the same two-nucleon phase shifts, not simply with the static approximation, since conventional NN interactions are obtained through fits to experimental data.

CONCLUSION

We have introduced Monte Carlo methods applicable to effective field theories and used them to study the polaron and nucleon plus scalar meson problems. The GFMC method gives the exact ground state energy for these field theories, and also the self-energy as a function of the momentum. There are several interesting yet straightforward applications of the method that remain to be carried out. These include further studies of polaron problems (higher coupling, bound polarons, ...) and nuclear systems with repulsive cores.

More interesting, though, are solutions of effective field theories with pion degrees of freedom. These are more complicated due to the nature of the pi-nucleon coupling. It is not yet clear if these approaches will be successful in dealing with this problem.

If a method can be developed, though, a great deal of interesting physics could be studied in this context. The pion provides a very large fraction of the interaction in potential models and is essential for a realistic description of light nuclei. Calculations of π-N scattering are the first natural candidates for this approach. Others include comparisons of deuteron form factors in the field theory picture with traditional potential plus exchange current models. Of course, one would hope eventually to be able to tie together much of medium-energy physics in this way, including general models of exchange currents, and three nucleon interactions that arise in potential models due to the resonance structure of the nucleon. We are currently pursuing work along these lines.

Acknowledgements

This work was supported by the National Science Foundation under grant CHE-9015337 and the U. S. Department of Energy.

References

1. H. Frölich, Philos. Mag. Suppl **3**, 325 (1954).

2. N. Metropolis, A. W. Rosenbluth, M. N. Rosenbluth, A. H. Teller, and E. Teller, J. Chem. Phys. **21**, 1087 (1953).

3. M. H. Kalos, Phys. Rev. **128** 1791 (1962).

4. M. H. Kalos, D. Levesque, and L. Verlet, Phys. Rev. **A9** 2178 (1974).

5. T. K. Mitra, A. Chatterjee, and S. Mukhopadhyay, Phys. Rep. **153**, 91 (1987).

6. T. D. Lee, F. E. Low, and D. Pines, Phys. Rev. **90**, 297 (1953).

7. V. M. Buimistrov and S. I. Pekar, Sov. Phys. JETP **5**, 970 (1957), and Sov. Phys. JETP **6**, 977 (1958).

8. R. P. Feynman, Phys. Rev., **97**, 660 (1955).

9. J. T. Marshall and L. R. Mills, Phys. Rev. **B8**, 3143 (1970).

10. C. Alexandrou, W. Fleischer, and R. Rosenfelder, Phys. Rev. Lett. **65**, 2615 (1990).

11. Brian D. Serot, S. E. Koonin, and J. W. Negele, Phys. Rev. **C28**, 1679 (1983).

QUANTUM MONTE CARLO STUDY OF SYMMETRY BREAKING
IN A DOUBLE-WELL CHAIN

J. E. Gubernatis

Theoretical Division
Los Alamos National Laboratory
Los Alamos, NM 87545

D. K. Campbell and Xidi Wang

Center for Nonlinear Studies
Los Alamos National Laboratory
Los Alamos, NM 87545

INTRODUCTION

We report the results of a quantum Monte Carlo simulation of a double-well chain. This chain is a system of particles that move on a lattice of symmetric, double-well potentials which are coupled harmonically to one another. The physical properties of this system are invariant, like those of the the Ising model, under the symmetry operations of the Z_2 group. In this case, changing the sign of the displacement variables leaves the energy unchanged and leads to a doubly-degenerate ground-state. Classically, this symmetry is always broken, and the particles all sit in the left- or the right-hand side of their wells. Quantum mechanically, however, we find that below a critical value of the double-well coupling constant the symmetry is restored by quantum fluctuations.

Our interest in this model was motivated by a series of quantum Monte Carlo simulations we are performing on one-dimensional models of conducting polymers and synthetic metals. The properties of these materials are described by a system of interacting electrons coupled to a system of phonons. Several years ago, for similar models, Fradkin and Hirsch[1] investigated how the electron motion can generate an effective double-well potential for the phonons and thereby cause the lattice to dimerize. They also argued, based on continuum renormalization group considerations and quantum Monte Carlo simulations, that for certain models quantum fluctuations at low temperatures restore symmetry (i.e., destroy the dimerization). We were attracted to the quantum double-well chain because it is a simpler problem than the electron-phonon models on which to test new numerical methods and to study similar issues.

The model, however, is also interesting on its own. It is a discretized version of a $1 + 1$ dimensional quantum ϕ^4 field theory. Although considerable numerical work has been done for these models in higher dimensions, little has been done for one spatial dimension. Analytically, some interesting results are known about the continuum version of the one-dimensional model.[2] For instance, kink/anti-kink pairs are the elementary excitations above the ground state. One of our objectives was to learn more about the nature and consequence of these excitations.

Recent Progress in Many-Body Theories, Vol. 3,
Edited by T.L. Ainsworth et al., Plenum Press, New York, 1992

The model is also relevant to the study of the structural properties of such hydrogen-bonded materials as the hydrogen halides.[3] In many cases, the energy as a function of lattice spacing for these chain-like materials has been obtained by local density approximation calculations of their electronic structure and fitted to the physical parameters in the double-well chain. Thus, with the phase diagram of the model, we can suggest whether these materials exist in the broken symmetry phase.

MODEL HAMILTONIAN

The Hamiltonian we are considering is

$$\hat{H} = : \sum_{n=1}^{N} \left[\frac{\hat{\pi}_n^2}{2m} + \frac{w}{2}(\hat{\phi}_n - \hat{\phi}_{n+1})^2 - \frac{k}{2}\hat{\phi}_n^2 + \frac{g}{4}\hat{\phi}_n^4 \right] : \tag{1}$$

where $\hat{\phi}_n$ and $\hat{\pi}_n$ are the coordinate and momentum operators of a particle with mass m on n^{th} site of a chain of length N. At each site, the particle moves within a potential well whose harmonic and anharmonic parts are characterized by the constants k and g. These wells are coupled to their nearest neighbors with coupling strength w. The constants m, w, k, and g are all non-negative, and the bracketing colons denote normal ordering of the quantum operators. Periodic boundary conditions are assumed. We also assumed that the quantum fluctuations do not cause the neighboring particles to interchange and that the lattice structure does not melt.

The first two terms on the right-hand side of (1) are those of a harmonic oscillator. The remaining two terms are those for the symmetric, on-site double-well potentials. These potentials have two absolute minima displaced $\pm\sqrt{k/g}$ relative to each lattice site and have an energy barrier $E_b = k^2/4g$ between these minima. Using the frequency $\omega = \sqrt{w/m}$, the energy $\hbar\omega$, and the displaced distance of minima to scale time, energies, and lengths, we rewrite (1) in the following dimensionless form

$$\hat{H} = : \epsilon \sum_{n=1}^{N} \left[\frac{\gamma}{2}\hat{\pi}_n^2 + \frac{\gamma}{2}(\hat{\phi}_n - \hat{\phi}_{n+1})^2 + \frac{1}{4}(\hat{\phi}_n^4 - 1)^2 \right] : \tag{2}$$

where $\epsilon = 4E_b/\hbar\omega$ and $\gamma = w/k$, and a constant term has been added. In (1), the parameter γ measures the relative strengths of the on-site part of the harmonic potential to the harmonic part implicit in the double-well. In (2), it is the harmonic frequency Ω associated with the coupling of the double-wells.

We will calculate the ground-state properties of (2) by doing quantum Monte Carlo simulations at successively lower and lower temperatures. The formalism on which the simulations are based requires the partition function of the model in terms of the Feynman path integral.[4,5] In general,

$$Z = \int \mathcal{D}\phi \, e^{-S\{\phi\}}$$

where $S\{\phi\}$ is the action associated with the scalar field ϕ. In terms of (2), we find that

$$S\{\phi\} = \epsilon \sum_n \int_0^\beta d\tau \left[\frac{\gamma}{2}\left(\frac{\partial\phi_n}{\partial\tau}\right)^2 + \frac{\gamma}{2}(\phi_n - \phi_{n+1})^2 + \frac{1}{4}(\phi_n^2 - 1)^2 \right] \tag{3}$$

where τ is the imaginary-time variable and β is the inverse of the temperature T.

An implicit parameter in (1) is the lattice constant a. If we let a vanish while keeping wa constant, we can show that in the ground state ($\beta \to \infty$)

$$S\{\phi\} = \epsilon\gamma \int_{-\infty}^{\infty} dx \int_{0}^{\infty} dt \left[\frac{1}{2}\left(\frac{\partial\phi}{\partial\tau}\right)^2 + \frac{1}{2}\left(\frac{\partial\phi}{\partial x}\right)^2 + \frac{1}{4}(\phi^2 - 1)^2 \right] \qquad (4)$$

This expression is the continuum version of the ϕ^4 field theory we are investigating. If treated classically, this field theory has the well known ground states of $\phi = \pm 1$ with energy $E_g = 0$ and has soliton solutions representing excitations from the ground state. If treated quantum mechanically, the ground state properties are renormalized. The lowest order correction (at the one-loop level) has been calculated,[2] and in this approximation

$$E_g = \frac{2\sqrt{2}}{3}\epsilon\sqrt{\gamma} + \frac{1}{\sqrt{\gamma}}\left(\frac{-3}{\pi\sqrt{2}} + \frac{1}{2\sqrt{2}}\right) \qquad (5)$$

When

$$\epsilon < \frac{18 - \sqrt{3}\pi}{8\pi\gamma} = \frac{0.49969}{\gamma} \qquad (6)$$

E_g becomes negative, then the broken symmetry ground state is unstable to the formation of kink/anti-kink pairs. This instability is an indication of the restoration of symmetry. We will find that the above estimate, while not very useful quantitatively, does point to the correct physical picture.

NUMERICAL METHODS

For the simulations, we discretize the integration in imaginary time into L steps of size Δ defined by $L\Delta = \beta$ and express the action $S\{\phi\}$ in terms of variable $\phi_{i,j}$,

$$S\{\phi\} = \Delta\epsilon \sum_{i,j}^{L,N} \left[\frac{\gamma}{2}\left(\frac{\phi_{i+1,j} - \phi_{i,j}}{\Delta}\right)^2 + \frac{\gamma}{2}(\phi_{i,j} - \phi_{i,j+1})^2 + \frac{1}{4}(\phi_{i,j}^4 - 1)^2 \right] \qquad (7)$$

Here, the subscripts i and j are the labels for the imaginary time and space dimensions.

Formally, this action is similar to the discrete, classical, two-dimensional ϕ^4 system.[6,7] In the present case, however, to insure we are describing quantum behavior, we have to require Δ to be much smaller than the reciprocal of Ω. In contrast to the classical ϕ^4 system, this requirement poses an extreme anisotropic nature on the problem. It also makes measurements of quantum quantities, such as the energy, specific heat, *etc.*, different from those of their classical counterparts. We will now describe how we performed the simulations and how we made our measurements.

Hybrid Monte Carlo Algorithm

The conventional, path integral Monte Carlo procedure evaluates the expectation value of a physical quantity $A = A\{\phi\}$,

$$\langle A \rangle = \int \mathcal{D}\phi \, A\{\phi\} \frac{e^{-\beta S\{\phi\}}}{Z}$$

by generating a sequence ϕ_i of independent configuration of the ϕ-fields with weight $e^{-\beta S}/Z$. The procedure reduces the computation of the expectation values to the simple summation

$$\langle A \rangle = \frac{1}{M}\sum_{i=1}^{M} A\{\phi_i\}$$

where M is the number of independent configurations generated.

With the commonly-used Metropolis algorithm, the configurations are generated by making local, trial changes of ϕ-fields, thereby changing the action from $S\{\phi\}$ to $S\{\phi'\}$, and then deciding to accept or reject the proposed configuration with the probability of $\min(1, e^{\delta S})$ where $\delta S = S\{\phi'\} - S\{\phi\}$. This movement of the $\phi_{i,j}$ is diffusive, and strong correlations can exist between consecutive configurations. Related simulation methods based on Lanegvin or molecular dynamics motions are also used, but often they share similar difficulties in efficiently generating independent configurations. Hybrid methods, which combine two or more different simulation methods, can greatly reduce this problem. In our simulations, we used the Hybrid Method proposed by Duane et al.[8]

In the Hybrid Method, a fictitious time t is added to the problem to allow the $\phi_{i,j}$ to evolve globally. Instead of the particles drifting in the phase space by diffusion, fictitious momenta (usually called pseudo-momenta) are introduced to guide their motion. The equations of motion are given by the Hamilton-Jacobi equations derived from the pseudo-Hamiltonian H_p. In our case, we assign each $\phi_{i,j}$ a pseudo-mass μ, such that the pseudo-Hamiltonian of the new system can be written as

$$H_p\{\phi, p\} = \sum_{i,j} \frac{p_{i,j}^2}{2\mu} + S\{\phi\}$$

The partition function of this system is

$$Z_p = \int \prod_{i,j} d\phi_{i,j}\, dp_{i,j}\, e^{-H_p\{\phi,p\}}$$

and since the integrand for the $p_{i,j}$ integration is Gaussian, the integration can be carried out easily to give to $Z_p = (2\pi\mu)^{NL/2} Z$. Thus, the pseudo-partition function Z_p differs from the true partition function Z only by a multiplicative constant so the physical observables will have the same expectation values as in the original system.

In our simulations, we first generate a Gaussian distribution for momentum $p_{i,j}$, and then let we let $\phi_{i,j}$ and $p_{i,j}$ evolve in a fictitious time t under the following equations of motion

$$\frac{d\phi_{i,j}}{dt} = \frac{p_{i,j}}{\mu}$$

$$\frac{dp_{i,j}}{dt} = -\epsilon\Delta\left[\frac{\gamma}{\Delta^2}(2\phi_{i,j} - \phi_{i+1,j} - \phi_{i-1,j})\right.$$
$$\left. - \gamma(2\phi_{i,j} - \phi_{i,j+1} - \phi_{i,j-1}) - \phi_{i,j} - \phi_{i,j}^3\right]$$

This propagation should be energy conserving, but after the system has evolved for a while in t with time-step δt, the energy will begin to deviate from its starting value because of the numerical errors in the integration procedure. We stop the evolution and then calculate the difference between the current and initial energies, $\delta H_p\{\phi\} = H_p\{\phi'\} - H_p\{\phi\}$, and accept the new configuration of the ϕ-fields according to Metropolis Algorithm, i.e., accept with a probability of $\min(1, e^{-\delta H_p})$. This step removes the cumulative errors that arise from the numerical integration because of the finite-size of the integration step. We carry out this procedure repeatedly until the statistical errors of the measured quantities are as small as required.

Because of the integration error is removed by the Metropolis step, the simple leap-frog algorithm[8,9] is more than adequate for evolving the configurations of the

$\phi_{i,j}$. This algorithm evolves the system with accuracy second order in δt, with the same computational efficiency as first order integrations. Since we can allow the $\phi_{i,j}$ to drift far away from their starting configuration, this method also has the advantage of reducing correlations among consecutive measurements. Because of parallel nature of the equations of motion, the hybrid method is very efficient on parallel (and vector) computers. This efficiency allows us to study long chains at low temperatures.

Measurements

To study whether the symmetry of the ground state is broken, we adopt the following strategy: for successively lower values of the temperature T, we study the behavior of the order parameter, the energy, and their mean-squared fluctuations as a function of ϵ and γ. In terms of the discretized action (7), we define a "quasi" inverse temperature $\beta_q \equiv \epsilon$ and a "quasi" (classical) two-dimensional Hamiltonian

$$H_q \equiv \sum_{i,j} \left[\frac{\gamma}{2\Delta}(\phi_{i+1,j} - \phi_{i,j})^2 + \frac{\Delta\gamma}{2}(\phi_{i,j+1} - \phi_{i,j})^2 + \frac{\Delta}{4}(\phi_{i,j}^2 - 1)^2 \right] \qquad (8)$$

such that $\beta_q H_q\{\phi\} = S\{\phi\}$. At each value of the physical temperature, we fix γ and then use standard methods to study the energy, specific heat, and their mean-squared fluctuations to determine whether the system defined by H_q undergoes a transition from the broken to the restored symmetry state at some critical value of β_q, $\beta_{qc} \equiv \epsilon_c$.

Within this strategy, finding the condition for the broken symmetry at a fixed physical temperature for our one-dimensional quantum model is equivalent to finding the critical inverse quasi-temperature ϵ_c for the two-dimensional Hamiltonian H_q. In the absence of infinite-ranged interactions, however, a true phase transition in one-dimension can only occur for an infinite-sized system at zero temperature (i.e. for N and $L \to \infty$). Within our strategy, we search for a phase transition in a two-dimensional system whose inverse temperature is ϵ. Again, a true transition will only occur in an infinite system (i.e., for $N \to \infty$), but it can occur at a finite value of inverse quasi-temperature ϵ. We seek to determine if such a transition is indicated and if these indications remain as we increase L and N. What will distinguish our quantum simulations from those for the classical system H_q is the need to require that the physical quantities we compute to be independent of Δ to within the accuracy of our calculation.

In our simulations, we chose $L = N$ or $L = 2N$. These choices were a compromised concession to the intuitive and empirical fact that high aspect-ratio rectangular space-time lattices inhibit efficient propagation of the $\phi_{i,j}$ configurations. As we lower the temperature, we also reduce finite-size effects by making L increasingly larger.

Order Parameter and Susceptibility. A criterion for the symmetry state of the system is the expectation value of $\Phi = \sum_{i,j}^{L,N} \phi_{i,j}$. Since only finite-sized systems can be simulated, this definition of an order parameter will change its sign as the simulation progresses and in general averages to zero independent of whether the ground-state exhibits broken symmetry or not. Thus, as an indicator of the symmetry state of the system, we take for the order parameter

$$X(\epsilon) = \frac{1}{LN}\langle |\Phi| \rangle \qquad (9)$$

As the size of the system N becomes very large and the true temperature $T = 1/\Delta L$ approaches zero, the value of the order parameter will change from a finite positive value to zero if quantum fluctuations are restoring the symmetry.

Although the order parameter $X(\epsilon)$ gives direct evidence of the quantum symmetry, simply measuring it is not always a very accurate way of finding the critical value ϵ_c. For determining a critical value, the susceptibility is more useful since it will diverge near ϵ_c. This susceptibility is defined as

$$\chi(\epsilon) = NL\beta(\langle \Phi^2 \rangle - \langle |\Phi| \rangle^2) \tag{10}$$

Energy and Specific Heat. The expectation value $\langle S\{\phi\} \rangle$ is not a meaningful estimator for the energy $E(\beta)$ of the quantum system, since it diverges as $\Delta \to 0$. The difficulty lies in using

$$K_q(\epsilon) \equiv \frac{\gamma}{2\Delta} \sum_{i,j} (\phi_{i+1,j} - \phi_{i,j})^2$$

as an estimator for the kinetic energy.[4,5] To estimate the kinetic energy $K(\beta)$ of the quantum system correctly, we used the following estimator based on the Virial Principle[5]

$$K(\beta) = \frac{1}{2}\langle \phi V'\{\phi\} \rangle$$

where in the present case

$$V\{\phi\} = \Delta \sum_{i,j} \left[\frac{\gamma}{2}(\phi_{i,j+1} - \phi_{i,j})^2 + \frac{1}{4}(\phi_{i,j}^2 - 1)^2 \right] \tag{11}$$

Hence, $E(\beta) \equiv K(\beta) + V(\beta)$ where $V(\beta)$ is given by (11). The specific heat $C(\beta)$ for the quantum system is simply

$$C(\beta) = \beta^2(\langle E(\beta)^2 \rangle - \langle E(\beta) \rangle^2) \tag{12}$$

To find ϵ_c, on the other hand, we take $E_q(\epsilon) = \langle H_q \rangle$ and

$$C_q(\epsilon) = \epsilon^2(\langle E_q(\epsilon)^2 \rangle - \langle E_q(\epsilon) \rangle^2) \tag{13}$$

It is important to note that a divergence of $C_q(\epsilon)$ does *not* necessarily imply divergent behavior in actual energy $E(\beta)$ or the specific heat $C(\beta)$. The actual and quasi energies are computed differently. This difference again underscores the intrinsic difference between a two-dimensional classical system and a $1 + 1$ dimensional quantum system, notwithstanding their formal similarity.

RESULTS

We simulated a variety of system sizes, ranging from $N = 7$ ($\beta = 4.95$) to $N = 128$ ($\beta = 45.25$). The $N = 128$ calculations were done on a Thinking Machines CM-2 computer. The remainder were done on a Sun Sparcstation 1, a Convex 230, and a Cray X-MP. The Cray computer ran our programs about up to 20 times faster than the Sun computer and about twice as fast as the Convex computer. The Cray computer was, however, at least 10 times slower, and sometimes 20 times slower, than the Thinking Machines computer. The speed of the latter provided us the opportunity to simulate efficiently a rather large system at a very low temperature. The simulations on the CM-2 took about 65 minutes of computation time for 40,000 measurements and about 120 minutes for 80,000.

We found that Δ satisfying $\Delta\Omega \approx \frac{1}{4}$ was sufficient to reduce the Δ dependence of our results to within our statistical error. This choice is a comprise between the need to have Δ small to be in the quantum regime, at the cost of increased computation time, and to have Δ sufficiently large to control computation time, at the risk of being in the classical regime.

For the pseudo-time-step δt, we typically chose $\delta t \Omega \approx 3.5$. With this choice our Monte Carlo acceptance rate was 90% to 95%. This rate seemed reasonable. In the Hybrid Method, one of the things that we are trying to achieved is a global updating of the configurational variables $\phi_{i,j}$. We accomplish this by integrating the equations of motion and using the Monte Carlo step to eliminate the need to monitor and adjust the integration step size δt. The idea is to choose this step-size to keep things close and to use the Metropolis algorithm to cull out cases that deviate a bit too far from the initial energy. The average, absolute value of the relative deviation from energy conservation was approximately 1.5×10^{-4}.

To promote decorrelated measured values, we usually made measurements after every second Monte Carlo step. For the smaller chain lengths, we made 20,000 measurements, while for the larger ones, 100,000 measurements. The measurements were grouped into bins, and the average of each bin was computed. The desired expectation values were the average of the bin averages. Our error estimates were based on the sample estimate of the variance of the bin averages.[10]

To estimate ϵ_c, we used the Cumulant Intersection Method.[6,10,11] Here, one computes

$$U_N = 1 - \frac{\langle \phi^4 \rangle_N}{3 \langle \phi^2 \rangle_N} \tag{14}$$

for each chain length (and hence T) as a function of the inverse quasi-temperature ϵ. This quantity measures the size of the system relative to the coherence length ξ. To find ϵ_c, one plots the ratio $R(\epsilon) = U_{N'}/U_N$ for various pairs of lattice sizes as a function of ϵ. At the critical point, $R(\epsilon_c) = 1$. Accordingly, ϵ_c is found by searching for the point of common intersection of the curves . Since the point of intersection is independent of system size, this method returns an estimate for ϵ_c extrapolated to the thermodynamic limit and to zero temperature.

In practice, the curves never intersect precisely at a point and care must be taken in interpolation between measure values of $R(\epsilon)$. To reduce the computation time near the critical point, we used the Histogram Method of Ferrenberg and Swendsen.[12] In this method, at a given inverse quasi-temperature ϵ' (and actual inverse temperature β), one collects a histogram of the measured values of the energy to construct an estimate of the density of states $\rho(E)$ and the dependence of ϕ on E. With these, one then estimates the moments of $\phi(E)$ as a function of ϵ by using

$$\langle \phi^n \rangle_N = \frac{\int \rho(E) \phi(E)^n e^{(\epsilon - \epsilon')E} dE}{\int \rho(E) e^{(\epsilon - \epsilon')E} dE}$$

and then obtains (14) and hence $R(\epsilon)$ for the different pairs of N and N'. This estimate for (14) is accurate in a narrow region around ϵ', but this region is wide enough so that the overlap of ϵ dependence from adjacent values of ϵ' will allow a smooth accurate curve for $R(\epsilon)$ to be produced with fewer simulations than needed to compute the same curve with another sequence of necessarily more closely-spaced ϵ' values.

Our principal result is shown in Fig. 1 where we plot the phase diagram as a function of the model parameters ϵ and $1/\gamma$. The straight line with a slope of approximately 2 is the phase boundary estimated from the continuum theory results of (5) and (6). It predicts the existence of a restored symmetry phase above this line. The markers in this figure are the results from the quantum Monte Carlo simulation. Above the curve represented by these points lies the restored symmetry phase. We

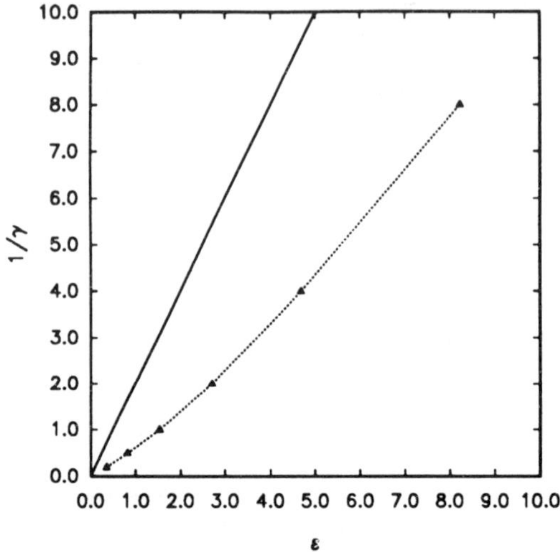

Fig. 1. The phase diagram. The straight line is the phase boundary predicted from the perturbation analysis of the continuum limit; the makers are the results obtained from the simulations. The restored symmetry phase lies above the curve represented by these markers.

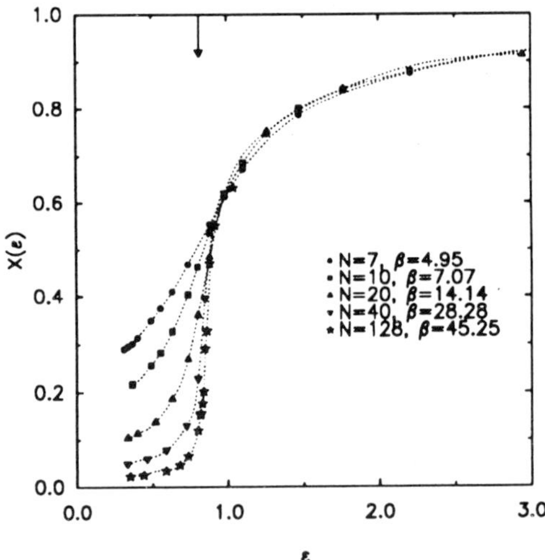

Fig. 2. The order parameter as a function of ϵ for different system sizes. Its value for a classical system is unity. The down-ward arrow marks $\epsilon_c \approx 0.82$. Here, $\gamma = 2$.

see that the continuum theory qualitatively predicts the correct physics, but quantitatively the the transition actually occurs more easily.

The behavior of the order parameter from different chain lengths as a function of ϵ is shown in Fig. 2. Out normalizations are such that the classical value of $X(\epsilon)$ is unity, independent of size and ϵ. From this figure, we see that quantum effects always reduce the order parameter relative to the classical value and that above an $\epsilon_c \approx 0.82$ the symmetry is broken. Clearly, as N (and hence β becomes large) the quantum order parameter curves are tending toward a definite value of ϵ_c.

The susceptibility (10) for different chain lengths as a function of ϵ is shown in

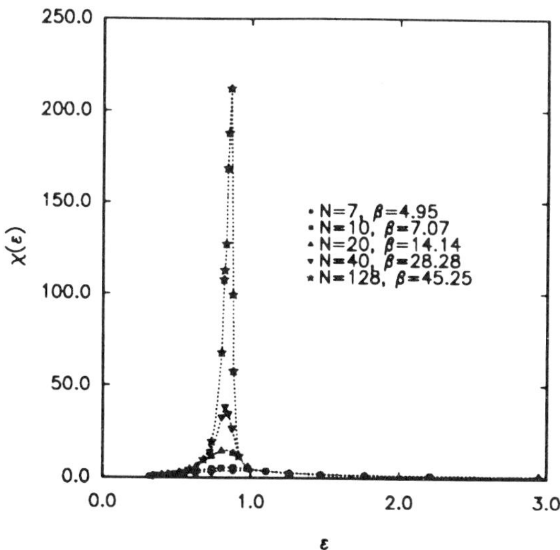

Fig. 3. The susceptibility $\chi(\epsilon)$ as a function of ϵ for different sizes. $\gamma = 2$. The peak in $\chi(\epsilon)$ occurs at approximately the same value of ϵ at which the order parameter, shown in Fig. 2, is vanishing.

Fig. 3. As N increases, the peak in $\chi(\epsilon)$ is clearly diverging around the same value of ϵ (≈ 0.82) at which the order parameter was disappearing in Fig. 2.

CONCLUDING REMARKS

We have presented numerical evidence that quantum fluctuations can produce a symmetric ground-state in the double-well chain, restoring the symmetry that is broken classically. We presented the phase diagram for this model that shows the symmetry restoration occurs more easily than predicted by a perturbation theory calculation of the continuum limit of the model. In another paper will report the full details of our analysis and results.[13]

We are currently trying to determine the universality class for the model by a combination of standard finite-size scaling methods[10] and more novel conformal-charge methods.[14,15] The identity of the universality class of related, two-dimensional, classical versions of the model has been controversial.[16,17]

ACKNOWLEDGMENTS

We thank W. R. Somsky and H. Q. Lin for helpful conversations. The work was supported by the U. S. Department of Energy. We also thank the Advanced Computing Laboratory at the Los Alamos National Laboratory for the use of its facilities.

REFERENCES

1. E. Fradkin and J. E. Hirsch, Phys. Rev B27: 1680 (1982); J. E. Hirsch and E. Fradkin, Phys. Rev. B27: 4032 (1983).
2. R. Dashen, B. Hasslacher, and A. Neveu, Phy. Rev. D10: 4114, 4139 (1974).
3. For example, R. W. Jansen, R. Bertoncini, D. A. Pinnick, A. I. Katz, R. C. Hanson, O. F. Sankey and M. O'Keeffe, Phys. Rev. B35: 9830 (1987).
4. R. P. Feynman and A. R. Hibbs, *Quantum Mechanics and Path Integrals*, McGraw-Hill, New York (1965).

5. M. Creutz and B. Freedman, Ann. Phys. 132: 427 (1981).
6. A. Milchev, D. W. Heermann and K. Binder, J. Stat.Phys. 44: 749 (1986).
7. R. Toral and A. Chakrabari, Phys. Rev. B42: 2445 (1990).
8. S. Duane, Nucl. Phys. B257: 652 (1985); S. Duane and J. B. Kogut, Nucl. Phys. B275: 398 (1986).
9. W. H. Press, B. R. Flannery, S. A. Teukolsky, and W. T. Vetterling, *Numerical Recipes*, Cambridge Press, Cambridge (1986), p. 631.
10. K. Binder, in *Applications of the Monte Carlo Method in Statistical Physics*, Springer-Verlag, Berlin (1984), Chap. 1.
11. K. Binder, Z. Phys. B 43: 119 (1981); Phys. Rev. Lett. 47: 693 (1981).
12. A. M. Ferrenberg and R. Swendsen, Phys. Rev. Lett. 61: 2635 (1988); Phys. Rev. Lett. 63: 1195 (1989).
13. Xidi Wang, D. K. Campbell, J. E. Gubernatis, "Symmmetry breaking in a quantum double-well chain," unpublished.
14. Rajiv R. P. Singh and G. A. Baker, Jr., Phys. Rev. Lett. 61: 1 (1991).
15. Xidi Wang, D. K. Campbell, J. E. Gubernatis. G. A. Baker, Jr., "Conformal charge of the two-dimensional ϕ^4 field theory," unpublished.
16. G. A. Baker, Jr., and J. D. Johnson, J. Phys. A17: L275 (1984).
17. A. D. Bruce, J. Phys. A18: L873 (1985).

QUANTUM MONTE CARLO CALCULATIONS ON MATERIALS:
TESTS ON CRYSTALLINE SILICON AND THE SODIUM DIMER

Richard M. Martin,[a,b] X.-P. Li,[a,b] E. L. Shirley,[a,b]
L. Mitáš,[a] and D. M. Ceperley[a,c]

Physics Department,[a] Material Research Laboratory[b] and

National Center for Supercomputer Applications[c]

University of Illinois at Urbana-Champaign

1110 W. Green Street, Urbana, IL 61801

INTRODUCTION

Quantum Monte Carlo (QMC) is a general method to calculate the exact (or nearly exact) ground state energy and correlation functions of a many-body quantum system.[1] Green's Function Monte Carlo (GFMC) can find exact results for Boson systems; however, for Fermions the famous "sign problem" has prevented the formulation of an exact method that is feasible for more than a few particles. Nevertheless, the fixed node approximation is extremely accurate if one has a trial function with appropriate nodes. The accuracy has been established by carrying out "release node" calculations for many systems, including the homogeneous electron gas, small molecules, and solid hydrogen. [2,3,4]

The work described here is part of our efforts to make it possible to carry out such nearly exact calculations on general condensed matter systems. The obstacles to be overcome are caused by the presence of core electrons. Although they are relatively inert, core electrons have crucial effects upon the active valence electrons. (The ideal would be to devise a theory that involves only valence electrons, yet takes into account all effects of the cores and respects the fact that all electrons of the same spin - core and valence - are in fact identical.) The core states are problematic because their characteristic energy scales are so large and time scales so small; in GFMC this causes an increase in computational time to achieve a given accuracy for the total energy which scales as $Z^{6.5}$, where Z is the atomic number.[5] For this reason, direct calculations are not feasible for heavy atoms, let alone solids! Our approach involves full many-body calculations on the valence electrons only, with effects of the core electrons replaced by a pseudopotential[6] (PP) or pseudohamiltonian[5] (PH). One of our primary results is that for a given PP or PH, it is indeed possible to calculate the total valence energy to a precision of order 0.05eV per atom.[7] This is sufficient for many real problems and leads to properties of Si in good agreement with experiment, including improvement of the well-known errors in the cohesive energy found in the local density approximation (LDA).[8] Fahy, Wang, and Louie[9] have also done VMC pseudopotential calculations using the same types of trial functions as considered here.

Recent Progress in Many-Body Theories, Vol. 3,
Edited by T.L. Ainsworth et al., Plenum Press, New York, 1992

The second part of our work addresses the question: how can one generate a pseudopotential or pseudohamiltonian that treats the effects of the cores with sufficient accuracy that they are worthy to be used in such accurate valence calculations? One point is that the pseudohamiltonian approach cannot be applied in all cases[5,7,10] and the non-local potentials, which are more general, are not consistent with the fixed node approach in GFMC. Progress in this direction has been made in recent work,[11] which has shown how to estimate the non-local term from a variational function while treating the local terms in the full fixed-node GFMC. We also must face the fact that up to now pseudopotentials have been generated only in approximate one-electron methods such as LDA or Hartree Fock.[6] Any errors made in generating the potentials propagate directly into the final answers. To overcome this defect, we have recently developed methods for a more rigorous many-body "core-valence partitioning" that incorporates core-valence exchange and correlations into a self-energy for the valence states.[12] The most convenient form of the results are "quasiparticle pseudopotentials"[12,13] which can be used as readily any non-local pseudopotential (with only a change requiring a simple modification of the electron-electron interaction). We have tested our pseudopotentials by essentially exact valence-only CI and QMC calculations on many atoms and on the Na_2 dimer.[12,13] Comparison with experiment shows excellent agreement in many cases and significant improvements over previous one-electron type pseudopotentials in essentially all cases.

CALCULATIONS ON SOLIDS

In VMC one minimizes the energy with respect to a trial function, which we choose to have the Jastrow-Slater form:

$$\Psi(\mathbf{r}_1,\ldots,\mathbf{r}_N) = \exp\left[\sum_{i=1}^{N}\chi(\mathbf{r}_i) - \sum_{i \le j} u(r_{ij})\right] D(\mathbf{r}_1,\ldots,\mathbf{r}_N) \tag{1}$$

The trial function is also the starting point for GFMC[14] where the the operator $\exp(-tH)$ projects out the ground state from the starting trial function. Here $u(r)$ is a two body correlation function obtained from random phase approximation for a homogeneous electron gas[2]; $\chi(\mathbf{r})$ is a one body term which modifies the VMC charge density[9] and D is a Slater determinant of single particle states. In our work to make a practical algorithm, LDA calculations are used in two important ways - construction of good trial orbitals and in the extrapolation to large cell size. The orbitals are taken from LDA calculations with a plane-wave basis set, including all reciprocal lattice vectors with an energy less than the "energy cutoff." We have used two energy cutoffs, 7 Ry and 15 Ry to test the influence on energy. In VMC, the energy obtained with orbitals cutoff at 15 Ry is 0.44 eV lower than the energy with a 7 Ry cutoff, while the two GFMC energies are the same within statistical errors (0.04 eV/atom.) This test suggests that the error in the nodal locations caused by the truncation of the LDA trial function is small. We use the larger cutoff in our calculations here which is more efficient because fluctuations are reduced.

The present results are calculated with a cubic supercell containing 64 Si atoms with periodic boundary conditions. The difference in an LDA calculation between a 64 atom and infinite system is 0.11 eV, and our QMC results are corrected assuming that they have the same size dependence as LDA. The number of walkers in the GFMC ensemble is chosen to be 200 and the initial distribution was obtained from VMC. A time step of 0.015 in atomic units was used. (This gives an acceptance ratio of 98% in the Metropolis portion of the time evolution.) A test calculation using half the time step gave identical results, showing that the time step error is less than 0.03eV/atom. A typical run with 3×10^4 steps, took 20 hours of CRAY-XMP time.

Fig. 1 shows the energy as a function of lattice constant from the LDA, VMC and GFMC calculations. The curves are least square fits to Murnaghan equation of state.[15] The total energy dropped 0.21(3) eV in the atom and 0.34(3) eV/atom in the solid at zero pressure (with the most accurate variational function) in going from the VMC to GFMC. This difference reflects the fact that it is easier to construct a good trial function in the atom than in the solid. In VMC it is important to construct equally good trial functions at all the lattice constants, otherwise there will be a systematic bias in the results. To achieve the same error bars, the GFMC calculation takes only 2.6 as much computer time as VMC but does not require systematic search of trial functions.

Also shown in Fig. 1 are the LDA total energies using the same PH. For semiconductors like silicon, LDA is known to work very well, and indeed the total energies from LDA are very close (~ 0.2 eV) to those from GFMC (even closer to those of VMC). Although these differences are not negligible, the largest change is in the atom where the spin polarized LDA energy is about 0.8 eV higher than GFMC.

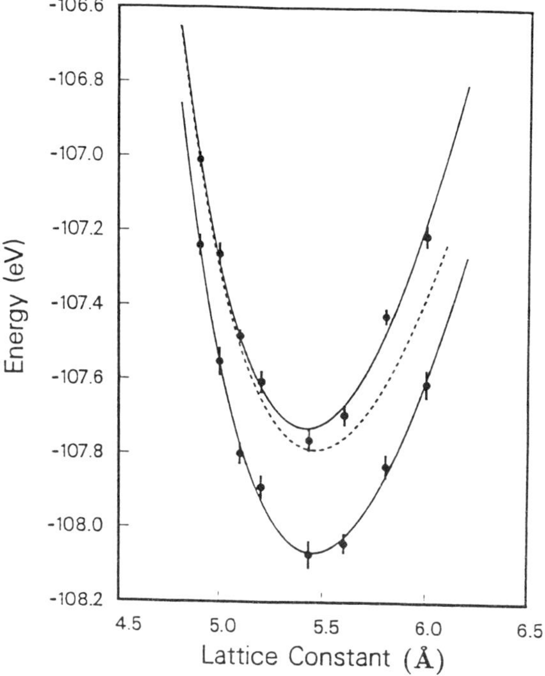

Fig. 1. Total energy of silicon versus lattice constant. The upper curve with error bars is from VMC and the lower from GFMC. Both have been corrected for the finite size of the system using LDA. The error bars show the estimated statistical errors. The solid lines are the fits to the Murnaghan equation of state. The dashed line represents the results of an LDA calculation with the same PH.

Table I. Comparison of the GFMC and VMC results with experiment and other calculations. All quantities have been corrected for finite size effects, zero point motion of the Si ions but here are no corrections applied to the difference between the pseudo- Hamiltonian and the pseudo-potential. Footnotes: [a] Our calculation using the pseudopotential from Ref. 8 with an energy cutoff of 30 Ry. [b] Fahy, *et. al.*, Ref. 9. [c] At T=0 K, Y. Okada and Y, Tokumaru, J. Appl. Phys. **56**, 314(1984). [d] At T=0 K, J.J. Hall, Phys. Rev. **161**, 756(1967). [e] Landolt-Börnstein: Numerical Data and Functional Relationships Science, New Series, Vol.3, 17:a, (Springer, New York, 1982). [f] JANAF Thermodynamical Tables, 3rd ed., J. Phys. Chem. Ref. Data **14**, Suppl. 1, 1795(1985); and B. Farid and R. W. Godby, Phys. Rev. **B43**, 14248(1991).

	a_0 (eV)	B_0 (Å)	dB_0/dP	$E_{at.}$ (eV)	$E_{sol.}$ (eV)	E_{coh} (eV)
LDA PP[a]	5.373	0.946	4.00	-102.71	-108.01	5.30
LDA PH	5.455	0.916	3.93	-102.71	-107.79	5.08
VMC-PP[b]	5.40(4)	1.08(10)	—	-103.42(3)	-108.23(6)	4.81(7)
VMC-PH	5.42(2)	1.08(5)	3.5(6)	-103.35(3)	-107.73(2)	4.38(4)
GFMC-PH	5.45(2)	1.03(7)	3.8(8)	-103.56(2)	-108.07(2)	4.51(3)
Exp.	5.430[c]	0.992[d]	3.20–4.68[e]	—	—	4.63(8)[f]

The final comparison with experiment and other calculations is given in Table I. For solid silicon, at least, LDA is working very well, and the LDA error in the cohesive energy comes mainly from using the LDA value for the energy of the atom. Our VMC energy is 0.43±0.08 eV smaller than that of Fahy, et al. After correcting for the difference between the PH and PP as calculated by LDA, there is still a 0.2 eV difference. We performed additional VMC calculations with their non-local pseudopotential, and reproduced their results. This implies that the transferability of the pseudopotentials is different between LDA and a many-body calculation such as VMC or GFMC (by 0.2 eV/atom). Together with evidence discussed below, this shows that accurate construction of the pseudopotentials from a many-body theory is necessary for the full accuracy of the QMC approach to be reached.

The structural properties from GFMC are in slightly better agreement with experiment than the VMC results of Fahy, et al. Our GFMC cohesive energy, bearing in mind the unknown transferability of the PH, should be between 4.51 (assuming no correction) and 4.73 eV (assuming the LDA gives correctly the difference between the PH and PP), in general agreement with the most quoted experimental value, 4.63(8) eV (see Table caption). In future work, we plan GFMC calculations with a non-local "quasiparticle" pseudopotential (described below) using a new method,[11] which treats the non-local parts in a variational manner.

QUASIPARTICLE PSEUDOPOTENTIALS

As the work above has demonstrated, the accuracy of QMC calculations applied to real solids is now limited by the quality of the pseudopotentials used to represent the effects of the cores. However, essentially all work to derive theoretical pseudopotentials has been in the context of a one-electron method such as Hartree-Fock or LDA.[6] In a one-electron method it is straightforward to carry out this operation because core and valence states are separate eigenfunctions and a pseudopotential can readily be derived which reproduces the valence function outside the core. In a many- body theory it is not clear *a priori* how to make such a separation. We have devised a way to do this "core/valence partitioning" taking into account the exchange and correlation between the core and valence electrons.[12]

The essence of our core/valence partitioning method is to treat the valence electrons as "quasiparticles" with self-energies which reflect the effects of the cores, i.e., exchange, dynamical correlation, and relaxation. These effects are calculating in the atom (or ion) using the Greens function method[16] which is a variant of Hedin's GW approximation, based upon the generalized RPA. We call our way of including vertex corrections the generalized GW (GGW) which is fully conserving.[17] The basic reason that summing diagrams is successful in describing the effects of cores is that the large gap for core excitations makes the sums converge rapidly.[12] The result is that the Green's function for electron addition and removal in the range of valence energies can be described by valence electrons moving in the presence of frozen core orbitals plus a self-energy described by a "core polarization potential" V_{cp} which takes into account core relaxation and dynamic correlation. The form of V_{cp} was chosen following the work of Muller, et. al.[18] In addition, core polarization modifies the electron-electron interaction in a simple way near the ion cores. Finally, usual methods[6,13] can be used to transform this to a non-local pseudopotential valence-only problem.

The first test of the "quasiparticle" pseudopotential is for one electron outside a core. In that case the valence problem is a simple one-electron problem and the eigenvalue of the "quasiparticle" pseudopotential should agree with the experimental binding energy of the electron in the exact many-body atom. Representative examples of results on atoms showing accuracy of our potential are given in Table II. The row labelled GGW shows the results of our full atom GGW calculation; the next row (GGW/PP) shows the small errors introduced in the pseudopotential transformation. Both agree very well with experiment compared to the single body methods listed below in the table. This is a necessary, but not sufficient test for any pseudopotential, and we see the the new quasiparticle pseudopotential has substantial improvements. Conversely, the errors shown in table II for the one electron methods will propagate in any many-body valence calculation which uses them.

Table II. Removal energies (in eV) for one electron bound to a core from experiment, the full atom generalized GW (GGW), the quasiparticle pseudopotential (GGW/PP) derived from the GGW, Koopman's eigenvalue (HF), self-consistent HF, and pseudopotential local spin density (LDA PP).

	Be 2s	Be 2p	Na 3s	Na 3p	Sc 4s	Sc 4p	Sc 3d	Sc 4f
Expt.	18.21	14.25	5.14	3.04	21.58	16.99	24.73	7.76
GGW	18.21	14.25	5.11	3.03	21.56	17.00	24.53	7.77
GGW/PP	18.21	14.25	5.12	3.03	21.58	17.01	24.54	7.77
HF	18.13	14.14	4.96	2.98	20.90	16.57	23.09	7.69
HF-SCF	18.13	14.14	4.96	2.98	20.97	16.60	23.76	7.69
LDF PP	18.30	14.61	5.30	3.20	21.34	16.91	25.12	—

Our goal, however, is to apply these potentials to many-valence-electron systems such as molecules and solids. This involves also the changes in the e-e interactions and static core relaxations. We have carried out one nearly exact Monte Carlo simulation to test the quality of the results: the binding curve for the Na dimer. This is a two-valence-electron system which can be solved exactly by QMC. Parallel calculations have been done with different pseudopotentials[12] - ones derived from the atom in the Hartree-Fock approximation, our GGW "quasiparticle" pseudopotentials plus e-e and static core

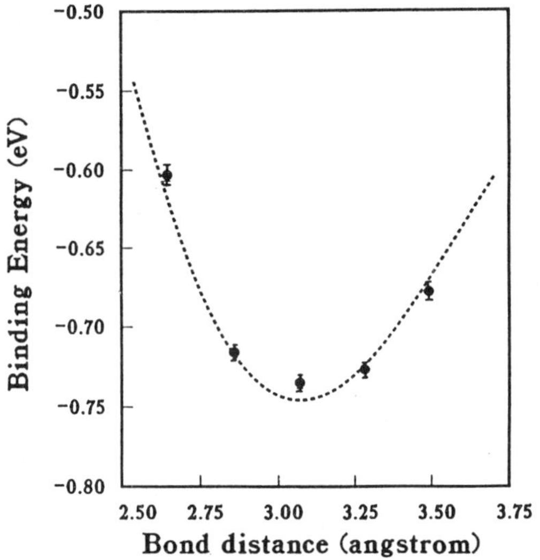

Fig. 2. GFMC calculations of energy vs. distance for the two-valence electron Na dimer and the GGW quasiparticle pseudopotential. Points with error bars are the theoretical results compared to the experimental curve derived from spectroscopic data. Comparison with HF and LDA pseudopotentials is described in the text.

polarization terms, and LDA potentials. Figure 2 shows the nearly exact agreement of the QMC calculation using the GGW potential with the experimental binding curve for the dimer. In contrast, the curves with the HF generated pseudopotential are too weakly bound (0.02eV) with equilibrium distance about 5% too large. Results with an LDA generated potential[12] are overbound by a much larger amount, 0.1 eV.[13]

CONCLUSIONS

There are two primary conclusions of our work. The first is that given the pseudopotential, the valence electron calculation can be carried out with an absolute accuracy of order 0.05 eV per atom in a real solid.[7] The precision can be better and the fundamental limitation is the fixed node approximation, for which the error has been estimated to be of this order in atomic and molecular calculations. However, there are still developments required to have effective methods to carry out fixed node calculations with non-local potentials.[11] The second conclusion is that promising results have been found using a new approach to "core/valence partitioning" that leads to "quasiparticle pseudopotentials."[13,12] These have been derived for atoms using many-body Green's function methods to determine the self-energy of a valence electron due to exchange and correlation with the core. Tests have shown the new potentials are transferable and predict many different states of atoms and binding of the Na dimer more accurately than potentials derived from single body methods like Hartree-Fock or LDA. Together these results are promising for accurate *an initio* calculations on molecules and solids.

ACKNOWLEDGEMENTS

This project was supported by the NSF grants DMR-88 08126 and DMR- 89 20538 and the U. S. Department of Energy under grant DE FG02- 91ER4539. The computations were done at the National Center for Supercomputer Applications at the University of Illinois.

REFERENCES

1. D. M. Ceperley and M. H. Kalos, in Monte Carlo Methods in Statistical Physics, edited by K. Binder (Springer, Berlin, 1979) p. 145.

2. D. M. Ceperley, Phys. Rev. **B 18**, 3126(1978); D. M. Ceperley and B. J. Alder, Phys. Rev. Lett. **45**, 566(1980).

3. P. J. Reynolds, D. M. Ceperley, B. J. Alder and W. A. Lester, J. Chem. Phys. **77**, 5593(1982); D. M. Ceperley and B. J. Alder, ibid. **81**, 5833(1984).

4. D. M. Ceperley and B. J. Alder, Phys. Rev. **B36**, 2092(1987).

5. G. B. Bachelet, D. M. Ceperley and M. G. B. Chiochetti, Phys. Rev. Lett. **62**, 2088(1989).

6. D. R. Hamann, M. Schlüter and C. Chiang, Phys. Rev. Lett. **43**, 1494(1979); G. B. Bachelet, D. R. Hamann and M. Schlüter, Phys. Rev. **B26**, 4199(1982).

7. X. P. Li, D. M. Ceperley, and R. M. Martin, submitted to Phys. Rev.

8. P. Hohenberg and W. Kohn, Phys. Rev. **136**, B864(1964); W. Kohn and L. J. Sham, ibid. **140**, A1133(1965).

9. S. Fahy, X. W. Wang and S. G. Louie, Phys. Rev. Lett. **61**, 1631(1988); Phys. Rev. **B42**, 3503(1990).

10. W. M. C. Foulkes and M Schlüter, Phys. Rev. **B42**, 11505(1990).

11. L. Mitáš, E. L. Shirley and D. M. Ceperley, J. Chem. Phys. Aug. 1991.

12. E. L. Shirley,thesis, University of Illinois, 1991.

13. E. L. Shirley, L. Mitáš, and R. M. Martin, to be published in Phys. Rev.

14. Some authors distinguish between GFMC and DMC or diffusion Monte Carlo. Our calculations use a short time approximation to the Green's function (see Ref. [3]), and we have tested that the time step is small enough that this is not an essential approximation.

15. O. L. Anderson, J. Phys. Chem. Solids, **27**, 547(1966).

16. L. Hedin and S. Lundquist, Solid State Physics, vol. 23, p.1 (Academic Press, New York, 1969).

17. G. Baym and L. P. Kadanoff, Phys. Rev. **124**, 287 (1961); G. Baym, Phys. Rev. **127**, 1391 (1962).

18. W. Muller, J. Flesch, and W. Meyer, J. Chem. Phys. **80**, 3297 (1982); W. Muller, and W. Meyer, J. Chem. Phys. **80**, 3311 (1982).

INCOMMENSURATE SOLID MONOLAYER OF ^{3}He ADSORBED ON GRAFOIL:
GENERALIZED HEISENBERG MODEL WITH EXCHANGE FREQUENCIES
EVALUATED BY PATH INTEGRAL TECHNIQUES

B. Bernu[†], D. Ceperley[‡], C. Lhuillier[†] and L. Pierre[†]

[†]Laboratoire de Physique Théorique des liquides.
U.P.M.C., Boite 121, 75252 Paris Cedex 05, France
équipe associée au CNRS

[‡]National Center for Supercomputer Applications
Dept. of Physics, University of Illinois at Urbana-Champaign
1110 W. Green. St., Champaign, Il 61820

ABSTRACT

Recent experiments on ^{3}He adsorbed on grafoil have shown a very rich phase diagram: commensurate and incommensurate solid phases, coexistence of first solid layers with liquid second or third layers. We focus here on the study of one incommensurate solid layer, for densities between 0.08 and 0.1 atom/Å^2, at temperatures less than $1K$. The solid forms a triangular lattice and its Debye temperature is between $20K$ and $30K$. Below $1K$, the phonons vanish and the physics is governed only by multiple spin exchanges: the full hamiltonian can be mapped on to a generalized Heisenberg model.

Up to now, experimental data have been analyzed using approximate solutions of the Heisenberg model with an effective pair exchange energy J: however measurements of the specific heat lead to values of J different from those extracted from magnetic succeptibility measurements. In order to go beyond these approximations, we compute *ab initio* various exchange frequencies (2, 3 and 4 body exchanges) at two densities by path integral techniques by evaluating the probability of tunneling from one configuration to its permuted one. These exchange frequencies are then introduced in a generalized Heisenberg model for which we calculate the full spectrum for small periodic systems and derive the thermodynamics.

INTRODUCTION

Helium 3 adsorbed on surfaces presents a large variety of phases depending on the area density and on the strength of the attractive potential with the substrate. The potential between an helium atom and the graphite is very attractive; its depth falls down to $-200K$. On the contrary, it varies smoothly with the x, y coordinates, with an amplitude of the order of $10K$. At low coverage, ^{3}He is well described as a two-dimensional

Fermi liquid. The solid phase begins for a densitie larger than $\rho = 0.041$ atom/Å^2. First, the adsorbed helium atoms sit on the minimum of the potential surface to give a commensurate phase. Then for densities greater than 0.07 atom/Å^2, the system evolves to an incommensurate phase[1, 2]. For densities larger than 0.11 atom/Å^2, the second layer is promoted, which leads in the second layer to a similar scenario as the coverage increases: first, it is a Fermi liquid, then a commensurate solid phase, and finally an incommensurate solid phase.

Here, we focus on the study of an incommensurate solid monolayer, and hence neglect the x, y-dependence of the interaction potential between helium atoms and the substrate. We believe that, in this phase, the corrugation effects are weak. Including them would complicate considerably the calculations. Then, in the incommensurate solid phase, and without the corrugation, the helium atoms form a triangular lattice. The lower collective density-density excitations are the phonons whose energy scale is the Debye temperature θ_D, which is of the order of 20 K, and whose contributions to the specific heat C_V vanish as $(T/\theta_D)^2$. The picture of this system is that helium atoms vibrate randomly in their local ground state. As the temperature decreases, the DeBroglie wavelength increases and very rarely, as the random zero point motions of atoms permit, a few helium atoms permute their position and therefore their spin. These spin exchanges (P) break down the degeneracy of the distinguishable particle ground state into a spectrum of 2^N states, whose energy scale is now given the exchange frequencies J_P, which are in the range of μK to mK. The physics is therefore well described with a generalized Heisenberg hamiltonian for triangular lattice in two dimensions. The sign of the effective pair exchange fixes the ferro or antiferromagnetic character of the system. For dense solids, the cyclic triple exchange is the largest exchange, which leads to a ferromagnetic phase, whereas for less dense solids the two body and the cyclic four body exchanges become comparable to the cyclic triple exchange, which may lead to an antiferromagnetic phase. A possible ferro/antiferromagnetic transition is then possible in the first or in the second layer[1].

Specific heat[2] and magnetization[3] measurements are now available as functions of the coverage and the temperature. Effective pair exchanges are evaluated from these two kinds of experiments, but lead to somewhat different values. Among the possible explanations one can propose: the importance of other multiple exchanges which can lead to different temperature dependencies, the role of defects or the presence of vacancies which would imply more complex effective hamiltonian.

We assume that the lattice is perfect and evaluate its "exact" thermodynamical properties. Exchange frequencies are computed with path integral techniques and, in a second step, the generalized Heisenberg model is solved for small periodic clusters. For simplicity, we focus on the first layer.

THE MODEL

First, we check if our potential model recovers thermodynamical experimental data: solid phase and Debye temperature. The hamiltonian is written down as:

$$\mathbf{H}_{\text{layer}} = \mathbf{H}_{\text{bulk}} + \sum_i V_S(z_i) \tag{1}$$

where z_i is the distance from the surface, $\mathbf{H}_{\text{bulk}}$ is the hamiltonian for bulk helium, where we use the Aziz potential[4]. V_S is the potential between an helium atom and the surface of graphite. This potential is fitted to reproduce the excitation spectrum of one

---

[1] Because the J's vary very rapidly with the density and the layer, we can assume in a first approximation that the first and the second layers are uncoupled.

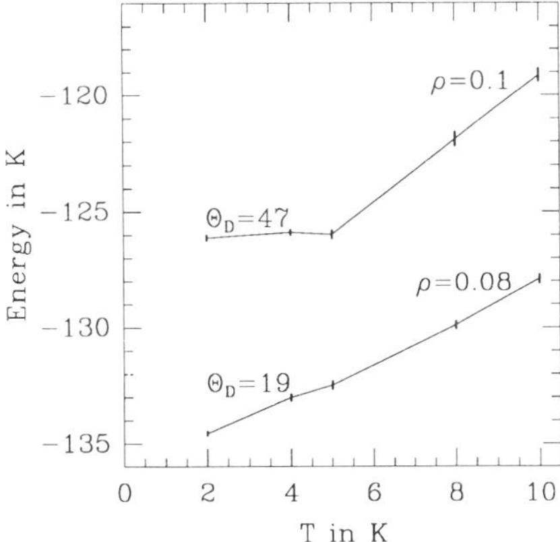

Figure 1. Energy versus temperature for a ^{3}He monolayer .

atom at the surface:

$$V_S(z) = V_0 \exp(-z/z_s) - C_3/(z - z_0) \tag{2}$$

where $V_0 = 3.7\ 10^6$, $z_s = 0.281$, $C_3 = 2130$ and $z_0 = 1$, the distances are expressed in angstroms and energies in Kelvin. Its depth is about $-200K$.

The thermodynamical properties are evaluated from path integral techniques, using an analytical expression for the approximate many-body density matrix at high temperature:

$$\rho_T(R, R'; \tau) = \prod_{i=1}^{N} \rho_0(r_i, r_i'; \tau) \exp(u_i^s(z_i, z_i'; \tau)) \atop \prod_{1 \le i < j \le N} \exp(u_{ij}^b(r_{ij}, r_{ij}'; \tau)) \tag{3}$$

where ρ_0 is the free particle density matrix, $\exp(u^s)$ is the contribution of the exact density matrix of one helium at the surface of the grafoil and $\exp(u^b)$ is the contribution the exact density matrix of two isolated helium atoms[2]. In Eq. 3, $R \equiv \{r_1, \ldots, r_N\}$ stands for $3N$-dimensional vectors and r for 3-dimensional vectors. The correction to this trial density matrix is evaluated by calculating $\partial\rho/\partial\tau - \mathbf{H}_{\mathrm{layer}}\rho$ [5]. We believe that the error on the energy due to this approximate density matrix is less than $0.5K$. In figure 1 are reported the total energies for the two densities 0.08 and 0.1 atom per $\AA^2$. In the simulations, we check the stability of the solid phase by computing the average one body density. The density plots show a tendency for melting at temperatures greater than $5K$ in agreement with experimental results[6]. The Debye temperature is determined by fitting the energy at the lowest temperatures using[2, 7]:

$$E(T) = E(T = 0) + \frac{28.848}{3} \frac{T^3}{\Theta^2} \tag{4}$$

---

[2]Corrugation would lead to a 6-dimensional potential tabulation of u^s instead of a 2-dimensional in our case .

At the lowest density the Debye temperature is close to the experimental evaluation of $20K$, and the variations with the density agree with a ρ^3 dependence[8].

The one body density profile $n(z)$ is very close to $\Psi_0^2(z)$, where Ψ_0 is the ground state of an helium atom in the external potential $V_S(z)$. $n(z)$ does not depend on the density, but as the temperature increases some particles jump to the second layer and one can see a second maximum around the equilibrium position of a second layer. It is important to notice that the second excited state of the one boby problem has an energy $60K$ above the ground state[9].

PATH INTEGRAL EVALUATION OF EXCHANGE FREQUENCIES

We evaluate the exchange frequencies with the same method as was used in solid ^{3}He by Ceperley et al.[10]. The equilibrium lattice sites are denoted by $Z = \{z_1, z_2, \ldots, z_N\}$, and a permutation by P. If the barriers between the $N!$ equilibrium positions PZ are infinite, all configurations have the same energy. This degeneracy is broken by the exchange of atoms.

Permutations are very rare and occur roughly every 10^5 atomic vibrations. So it is not practical to wait for the system to jump by itself from one configuration to another. But this property allows us to focus on one permutation at a time.

To define the exchange frequency J_P associated to a permutation P, we assume that the ground state is split into two states ϕ_0 and ϕ_1 with even and odd symmetry, and energies E_0 and E_1. If the system is localized around Z (resp. PZ), it is described by $\Psi_Z = \phi_0 + \phi_1$ ($\Psi_{PZ} = \phi_0 - \phi_1$). It oscillates from Z to PZ at the frequency $\hbar\omega_P = E_1 - E_0 = 2J_P$.

For solid ^{3}He, it is the repulsive part of the potential and not the Fermi statistics, which creates a cage effect, restricting the motion in the phase space around Z. Therefore the density matrix around an equilibrium position Z or PZ is given by the distinguishable particle density matrix:

$$\rho(R, R'; \beta) = \sum_n e^{-\beta E_n} \phi_n(R)\phi_n(R')$$

and for the unpermuted and permuted reference positions, we have:

$$\rho(Z, Z; \beta) \sim e^{-\beta E_0} \left[\phi_0^2(Z) + e^{-\beta(E_1-E_0)}\phi_1^2(Z)\right]$$
$$\rho(Z, PZ; \beta) \sim e^{-\beta E_0} \left[\phi_0(Z)\phi_0(PZ) + e^{-\beta(E_1-E_0)}\phi_1(Z)\phi_1(PZ)\right]$$

From the ratio of these two matrix elements, we deduce:

$$F_P(\beta) = \frac{\rho(Z, PZ; \beta)}{\rho(Z, Z; \beta)} = \tanh\left(J_P(\beta - \beta_P)\right) \tag{5}$$

where $J_P\beta_P = \ln\left\{\phi_1^2(Z)/\phi_0^2(Z)\right\}$.

Inserting (M-1) intermediate points $R_1, R_2, \ldots, R_{M-1}$ into Eq. 5 we convert it into a path integral:

$$F_P(\beta) = \frac{\int dR_1 dR_2 \ldots dR_{M-1}\rho(Z, R_1; \tau) \ldots \rho(R_{M-1}, PZ; \tau)}{\int dR_1 dR_2 \ldots dR_{M-1}\rho(Z, R_1; \tau) \ldots \rho(R_{M-1}, Z; \tau)} \tag{6}$$

with $\tau = \beta/M$. If M is large enough, an accurate expression for $\rho(R, R'; \tau)$ can be written down, and we use Eq. 3.

Since the potential between helium atoms is hard-core like, most of the points R_i are close to either Z or PZ. This means that only a few points of the path in the numerator of Eq. 6 go from Z to PZ. This picture implies that the numerator of Eq. 6 can be written as:

$$\rho(Z, PZ, \beta) = \int dR_1 dR_2 \ldots dR_{M-1}\rho(Z, R_1; \tau) \ldots \rho(R_{M-1}, PZ; \tau) \tag{7}$$

$$= \int dR_i dR_{i+k}\rho(Z, R_i; i\tau)\rho(R_i, PR_{i+k}; k\tau)\rho(PR_{i+k}, PZ; {\scriptstyle (M\text{-}k\text{-}i)\tau}) \tag{8}$$

$$= \int dR_i dR_{i+k}\rho(Z, R_i; i\tau)\rho(R_i, PR_{i+k}; k\tau)\rho(R_{i+k}, Z; {\scriptstyle (M\text{-}k\text{-}i)\tau}) \tag{9}$$

$$\begin{aligned}
= \int & dR_1 \ldots dR_i dR'_{i+1} \ldots dR'_{i+k-1} dR_{i+k} \ldots dR_{M-1} \\
& \times \rho(Z, R_1; \tau) \ldots \rho(R_{i-1}, R_i; \tau) \\
& \times \rho(R_i, R'_{i+1}; \tau) \ldots \rho(R'_{k+i-1}, PR_{i+k}; \tau) \\
& \times \rho(R_{i+k}, R_{i+k+1}; \tau) \ldots \rho(R_{M-1}, Z; \tau)
\end{aligned} \tag{10}$$

$$\begin{aligned}
= \int & dR_1 \ldots dR_i dR_{i+1} \ldots dR_{i+k-1} dR_{i+k} \ldots dR_{M-1} \\
& \times \rho(Z, R_1; \tau) \ldots \rho(R_{M-1}, Z; \tau) \\
& \times \frac{\int dR'_{i+1} \ldots dR'_{i+k-1}\rho(R_i, R'_{i+1}; \tau) \ldots \rho(R'_{k+i-1}, PR_{i+k}; \tau)}{\rho(R_i, R_{i+1}; \tau) \ldots \rho(R_{k+i-1}, R_{i+k}; \tau)}
\end{aligned} \tag{11}$$

where we use $\rho(PR, PR'; \tau) = \rho(R, R'; \tau)$ to go from Eq. 8 to Eq. 9. Eq. 10 is found by introducing again elementary paths with the time step τ and by labeling R' the dummy variables for the exchange part of the path. The last step is obtained by introducing new variables $R_{i+1} \ldots R_{i+k-1}$ and multiplying and dividing by the corresponding ρ's. The points R are near Z, and the points R' form a path which goes from R_i to PR_{i+k} in a time $k\tau$. This means that during an equilibrium walk of distinguishable particles, we try to map a path between R_i and PR_{i+k}. $F_P(\beta)$ is then the mean value of the last line of Eq. 11. For a given path $\{R_1, \ldots, R_{M-1}\}$, we choose all initial positions R_i and try all permutations P' equivalent to P (for example, all first neighbors pair permutations): the permutation P' fixes the end point of the partial path at $P'R_{i+k}$. Then the multisampling of $R'_{i+1}, \ldots, R'_{i+k-1}$ is done and the histogram of path contribution ratio is accumulated.

Computed directly by this algorithm, the convergence of the method will be poor. The reason is, that during an equilibrium configuration only the points close to Z (permutation identity) are sampled and there is a small probability of finding configurations R_i and R_{i+k} so that $\rho(R_i, PR_{i+k}; \tau)$ is not small. Much faster convergence is achieved by splitting this computation in two parts: the first one is described above and in the second one evaluate the inverse of Eq. 5: starting from a permuted configuration, one tries to map on to unpermuted configurations. Optimization of the computational time spent in these two parts is done as in classical statistical mechanics by Bennett[11].

The advantage of this method combining classical statistical mechanics and quantum techniques is that the convergence is independent of the magnitude of J_P, the tunneling frequency, since we evaluate the ratio of two probabilities and not the differences between eigenvalues.

The following exchange frequencies have been obtained from simulations at $T = 1K$, with $M = 40$ and $k = 4$. We do not find a ferromagnetic/antiferromagnetic transition in the first layer. The comparison with a pure 2D model (see the table I) shows that the additional motions of atoms in the z-direction enhance the exchange frequencies, specially for high densities and for the two and four body exchanges.

Table 1. Exchange frequencies in μK. ρ is in atom/Å^2. α is fitted from $J_P \sim \rho^{-\alpha}$. J_{eff} is defined by Eq. 16. The upper box is the 3D simulation with the substrate potential, while the lower box corresponds to pure 2D exchange frequencies.

3D	$\rho = .0785$	$\rho = .09$	$\rho = .1$	α
●—●	29 ± 2	1.3 ± 0.2	0.157 ± 0.014	22
△	48 ± 5	3.5 ± 0.3	0.33 ± 0.03	21
◇	25 ± 4	0.78 ± 0.14	0.066 ± 0.009	25
J_{eff}	36	4	0.34	19

2D	$\rho = .081$	$\rho = .105$	
●—●	4.6 ± 0.6	0.0026 ± 0.0003	29
△	28 ± 2	0.024 ± 0.002	27
◇	2.6 ± 0.2	0.00055 ± 00005	33
J_{eff}	45	0.044	26

GENERALIZED HEISENBERG MODEL, COMPARISONS WITH EXPERIMENTS

Here, we evaluate the thermodynamical properties for the generalized Heisenberg model, where the hamiltonian is defined by:

$$\mathbf{H} = -\sum_P (-1)^P J_P P \tag{12}$$

where the summation includes all cyclic permutations involving up to four nearest neighbor particles and P is the spin permutation operator (for a transposition $P_{ij} = 1/2 + 2\vec{S}_i.\vec{S}_j$). The system is an infinite periodic triangular lattice where the separation between two neighbors is $a = \sqrt{2/\rho\sqrt{3}}$. On this lattice, we superpose a lattice of diamond cells with a side length d (see figure 2), and acute angles of $\pi/3$. The side of a diamond can be any line between two sites and is of the form $l\vec{u} + m\vec{v}$, where $\vec{u}$ and $\vec{v}$ are vectors of length a and with an angle of $\pi/3$. The length d of this vector fixes the number N of atoms in the cell: $N = (d/a)^2 = l^2 + m^2 + lm$. The cases where N is less or equal to 16 have been studied.

The full spectrum of the generalized Heisenberg hamiltonian is calculated by the Lanczos method. The group G of symmetries of the hamiltonian is the direct product

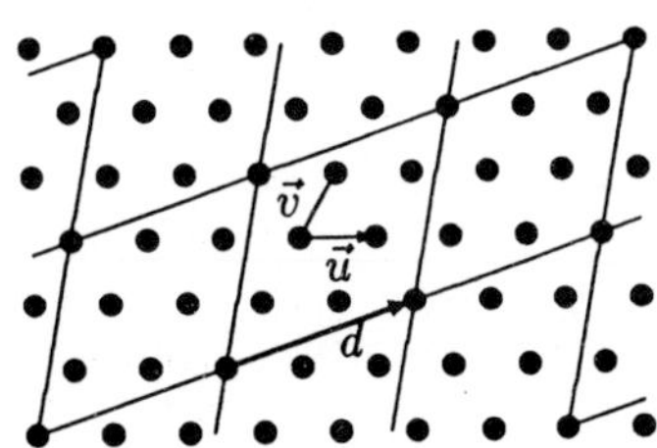

Figure 2. Example of a periodic system: $N = 7$, $l = 2$ and $m = 1$. The diamond side length d is $\sqrt{7}a$, where a is the separations between sites.

$G_s \times G_l$ of the spin symmetry group G_s ($\equiv SU_2$) and the lattice symmetry group G_l, which contains N translations $\times$ 6 rotations $= 6N$ permutations of the sites. Casually, G_l may also contain axial reflexions and then it has $12N$ elements. These symmetries are responsible for the energy degeneracy. Each eigenvalue is associated with an irreducible representation (I.R.) of G which characterizes the symmetry of its eigenvectors. Its multiplicity is equal to the dimension of this irreducible representation. For each relevant I.R., i, the Lanczos method is applied in a space, $\mathcal{E}_i$, where the degeneracy is removed. Because the dimension d_i of $\mathcal{E}_i$ are small (the maximum is 116 for $N = 16$), the full spectrum may be computed. In fact, we use only the I.R. of the group G_l to reduce the dimensionality. The I.R. of G_s (i.e. the total spin) are taken into account in another way. At each iteration of the Lanczos method, the new vector is projected on the eigen subspace of S^2 under consideration and then orthogonalized to all the previous ones. This last precaution, even if expensive in computational time, is necessary to prevent wrong degeneracies from appearing.

The general hamiltonian involving up to four body exchanges is written from the hamiltonian of one diamond. Labeling particles as follows:

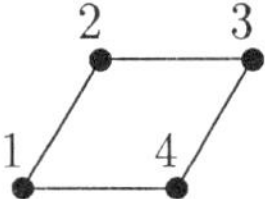

the most general hamiltonian for a diamond involves 8 different types of permutations:

$$
\begin{aligned}
-\mathbf{H}_D = {} & -J_{11}[P_{12} + P_{23} + P_{14} + P_{24} + P_{34}] - J_{13}P_{13} \\
& + K_{111}[P_{124} + P_{142} + P_{234} + P_{243}] \\
& + K_{113}[P_{123} + P_{132} + P_{134} + P_{143}] \\
& - L_{1111}[P_{1234} + P_{1432}] \\
& - L_{1113}[P_{1243} + P_{1342} + P_{1423} + P_{1324}] \\
& + L'_{1111}[P_{12}P_{34} + P_{14}P_{32}] + L'_{1113}P_{13}P_{24}
\end{aligned}
$$

All cyclic permutations are then grouped using the properties for cyclic permutations [12, 13]:

$$
\begin{aligned}
P_{123} + P_{132} &= P_{12}P_{13} + P_{13}P_{12} = P_{12} + P_{23} + P_{13} - 1 \\
P_{1234} + P_{1432} &= P_{12}P_{13}P_{14} + P_{14}P_{13}P_{12} \\
&= -1 + P_{13} + P_{24} + P_{12}P_{34} + P_{14}P_{23} - P_{13}P_{24}.
\end{aligned}
$$

and when summing over all diamonds, one finally gets:

$$
\mathbf{H} = \sum_{<11>} \tilde{J}_{11}P_{ij} + \sum_{<13>} \tilde{J}_{13}P_{ij} \tag{13}
$$
$$
+ \sum_{<1111>} \tilde{J}_{1111}P_{ij}P_{kl} + \sum_{<1113>} \tilde{J}_{1113}P_{ij}P_{kl} + CN
$$

where $< 11 >$ means first neighbors and $< 13 >$ means second neighbors ($1-3$ means a distance of $\sqrt{3}a$) and the $\tilde{J}$'s are defined as:

$$
\begin{aligned}
\tilde{J}_{11} &= J_{11} - 2K_{111} - 4K_{113} + L_{1111} + 4L_{1113} \\
\tilde{J}_{13} &= J_{13} - 2K_{113} + L_{1111} \\
\tilde{J}_{1111} &= -L'_{1111} + L_{1111} \\
\tilde{J}_{1113} &= -L'_{1113} - L_{1111} + 2L_{1113} \\
C &= 2K_{111} + 6K_{113} - 3L_{1111} - 6L_{1113}
\end{aligned} \tag{14}
$$

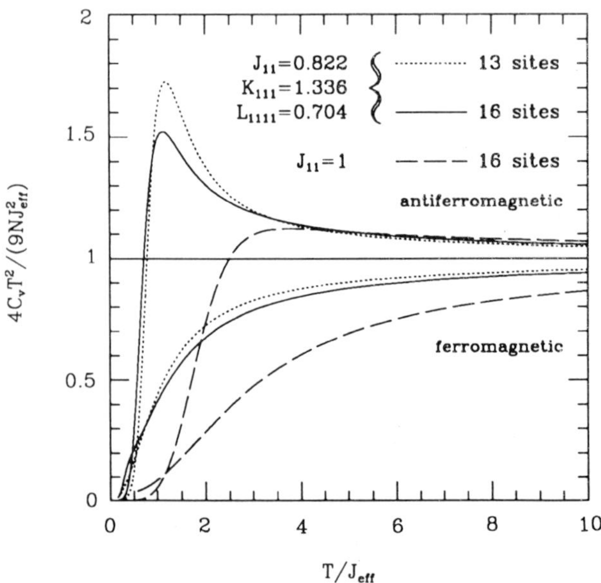

Figure 3. Deviations of C_V from its high temperature limit (Eq. 15) for two different sets of exchange frequencies. Down (up) is the (anti)ferromagnetic case.

Note that with these definitions, the $\tilde{J}$'s are not necessarily positive. The phase diagram for this hamiltonian depends only on four independent parameters[14]. The specific heat at high temperature for this model is then given by:

$$C_V/(N\beta^2) = \frac{9}{4}J_{eff}^2 \tag{15}$$

where

$$\begin{aligned}
J_{eff}^2 = \ &\tilde{J}_{11}^2 + 4\tilde{J}_{11}\tilde{J}_{1111} + \tilde{J}_{11}\tilde{J}_{1113} \\
&+\tilde{J}_{13}^2 + \tilde{J}_{13}\tilde{J}_{1113} \\
&+6\tilde{J}_{1111}^2 + 3\tilde{J}_{1111}\tilde{J}_{1113} + \frac{5}{4}\tilde{J}_{1113}^2
\end{aligned} \tag{16}$$

When we retain the three most important exchange (J_{11}, K_{111} and L_{1111}) in J_{eff}, we obtain $J_{eff}^2 = (J_{11} - 2K_{111} + \frac{5}{2}L_{1111})^2 + 2L_{1111}^2$. In table are given the values of J_{eff} for three densities.

On figure 3 are plotted the specific heat deviations from the high temperature limit for the case of a pure Heisenberg model (only pair exchange) and for the case of multi-exchange frequencies (of a ^{3}He monolayer at the density 0.08 atom/Å^2. For comparison the data are scaled with J_{eff}. Important changes in C_V are obtained between these two cases, where the effect are more dramatic in the antiferromagnetic case (reverse all the signs of $\tilde{J}$).

By comparison with the experiments of Greywall[2], it appears that our values are one order of magnitude smaller than his last evaluations of J_{eff}. Our data seem to be in better agreement with magnetization data[15], even if the complete analysis of this comparison is not yet achieved.

REFERENCES

[1] D.S. Greywall and P.A. Busch, *Phys. Rev. Lett.* **65** 2788 (1990);

[2] D.S. Greywall *Phys. Rev. B* **41** 1842 (1990);

[3] H. Godfrin, R.R. Ruel and D.D. Osheroff, *Phys. Rev. Lett.* **60** 305 (1988);
H. Godfrin, *Can. J. Phys.* **65** 1430 (1987);
L.J. Friedman, S.N. Ytterboe, H.M. Bozier, A.L. Thomson, M.C. Cross,
Can. J. Phys. **65** 1351 (1987);
C.P. Lusher, J. Saunders, B.P. Cowan, *Europhys. Lett.* **14** 809 (1991);
J. Saunders, C.P. Lusher, B.P. Cowan, *Phys. Rev. Lett.* **21** 2523 (1990);

[4] R.A.Aziz, V.P.S. Nain, J.S.Carley, W.L. Taylor and G.T. McConville,
J. Chem. Phys. **70** 4330 (1979);

[5] D.M. Ceperley and E. L. Pollock, Proceedings of the Elba Conference on Monte
Carlo Methods for Physics, 1990.

[6] S.H. Hering, S.W. Van Sciver and O.E. Vilches *J. Low Temp. Phys.* **25** 793 (1976);

[7] J.G. Dash, *Films on Solid Surfaces* Academic, New-York (1975);

[8] S.V. Hering and O.E. Vilches in *Monolayer and Submonolayer Helium films*
edited by J.G. Daunt and E. Lerner (Plenum, New-York, 1873);

[9] F. Joly, C. Lhuillier, B. Brami submited to *Surface Sciences*;

[10] D.M. Ceperley and G. Jacucci, *Phys. Rev. Lett.* **58** 1648 (1987);

[11] C.H. Bennett *J. Comput. Phys.* **22** 245 (1976);

[12] D.J. Thouless, *Proc. R. Phys. Soc. London* **86** 893 (1965);

[13] M. Roger, J.H. Hetherington, and J.M. Delrieu, *Rev. Mod. Phys.* **55** 1 (1983);

[14] see also for another hamiltonian model : M. Roger, *Phys. Rev. Lett.* **64** 297 (1990);

[15] H. Godfrin, private communication;

MASSIVELY-PARALLEL REALIZATIONS OF SELF-CONSISTENT PERTURBATION THEORIES

J.W. Serene and D.W. Hess

Complex Systems Theory Branch
Naval Research Laboratory
Washington, D.C. 20375-5000

Self-Consistent Perturbation Theory

Propagator-renormalized perturbation theory for the temperature Green's functions and thermodynamic potential was introduced in 1960 by Luttinger and Ward [1], who showed that the thermodynamic potential can be expressed in terms of the Green's function and self-energy as

$$\Omega(T,\mu) = -2\mathrm{Tr}\left[\Sigma G + \ln(-G_0^{-1} + \Sigma)\right] + \Phi[G], \tag{1}$$

where 'Tr' denotes a generalized trace over all arguments of the Green's function; for example, in the momentum-frequency representation, $\mathrm{Tr}[A] = (T/N)\sum_{\mathbf{k}}\sum_n A(\mathbf{k},\varepsilon_n)$, with N the number of lattice sites (for simplicity we will consider only paramagnetic systems). Here $\Phi[G]$ is a set of Feynman diagrams whose functional derivative with respect to G generates the skeleton-diagram expansion for Σ,

$$\Sigma(\mathbf{k},\varepsilon_n) = \frac{1}{2}\frac{\delta\Phi[G]}{\delta G(\mathbf{k},\varepsilon_n)}, \tag{2}$$

and G and Σ satisfy Dyson's equation,

$$G(\mathbf{k},\varepsilon_n) = [G_0^{-1}(\mathbf{k},\varepsilon_n) - \Sigma(\mathbf{k},\varepsilon_n)]^{-1}. \tag{3}$$

Viewed as a functional of G and Σ, the right-hand side of Eq. (1) has the attractive property of being stationary under independent variations of G and Σ whenever Eqs. (2) and (3) are satisfied. Baym completed the formal structure of the theory for normal (i.e., without long-range order) systems by establishing the connection between conservation laws and the condition expressed by Eq. (2), which he named Φ-derivability [2].

Luttinger used this formalism (particularly the stationary property of the free-energy functional) to establish a number of the central formal results of the microscopic Fermi liquid theory, including the famous 'Luttinger theorem' on the volume enclosed by the true Fermi surface of a Fermi liquid, and the relation between the quasiparticle effective mass and the low temperature specific heat [3]. Four years later, De Dominicis and Martin generalized the propagator-renormalized theory of Luttinger and Ward to cover systems with both diagonal and off-diagonal long-range order (including, for example, Bose superfluids and both singlet and triplet superconductors) [4]. They also pointed out that re-expressing the perturbation theory in terms of the interacting Green's function can be viewed as a generalized Legendre transformation to eliminate the bare one-body potential, and extended the formalism by eliminating the bare two-body interaction in favor of the full two-particle Green's function (or,

Recent Progress in Many-Body Theories, Vol. 3,
Edited by T.L. Ainsworth et al., Plenum Press, New York, 1992

equivalently, the two-particle vertex function). In this case the natural stationary thermodynamic function is the entropy, and the stationary conditions for this functional generate the parquet equations (again including all types of long-range order).

During the next twenty-five years, the propagator-renormalized formalism was used in two different types of applications. One line of work followed the spirit of the Luttinger-Ward approach to Fermi liquid theory, and exploited the stationary properties of the free-energy functional to evaluate leading-order corrections to Landau theory. Examples of this approach include the calculation of $T^3 \ln(T)$ corrections to the specific heat by Pethick and Carneiro [5], and the theory of strong-coupling corrections for superfluid ^{3}He [6]. The other line of work involved full solutions of the self-consistent renormalized theory. On account of computational limitations, this work was limited to generalized molecular-field approximations. The most important of these applications were to the BCS theory of superconductivity and its generalization to include the true electron-phonon interaction, for which the propagator-renormalized perturbation theory and the idea of eliminating an external 'pair-potential' in favor of the off-diagonal Gor'kov Green's function provide the correct formal setting (a point obscured in almost all textbook treatments) [7, 8].

In 1989 Bickers, Scalapino, and White applied the propagator-renormalized formalism to the 2D Hubbard Hamiltonian [9],

$$H = -t \sum_{<i,j>,\sigma} (c_{i\sigma}^\dagger c_{j\sigma} + c_{j\sigma}^\dagger c_{i\sigma}) + U \sum_i n_{i\uparrow} n_{i\downarrow}, \tag{4}$$

using what they call the fluctuation-exchange approximation, in which the functional $\Phi[G]$ is the sum of a single second-order graph, plus particle-hole and particle–particle bubble chains describing exchanged density, spin-density, and (singlet) pair fluctuations,

$$\Phi \quad = \quad \Phi_2 + \Phi_{ph}^{df} + \Phi_{ph}^{sf} + \Phi_{pp} \tag{5}$$

$$\Phi_2 \quad = \quad -\tfrac{1}{2}\mathrm{Tr}\left[\chi_{ph}^2\right] \tag{6}$$

$$\Phi_{ph}^{df} \quad = \quad \tfrac{1}{2}\mathrm{Tr}\left[\ln(1+\chi_{ph}) - \chi_{ph} + \tfrac{1}{2}\chi_{ph}^2\right] \tag{7}$$

$$\Phi_{ph}^{sf} \quad = \quad \tfrac{3}{2}\mathrm{Tr}\left[\ln(1-\chi_{ph}) + \chi_{ph} + \tfrac{1}{2}\chi_{ph}^2\right] \tag{8}$$

$$\Phi_{pp} \quad = \quad \mathrm{Tr}\left[\ln(1+\chi_{pp}) - \chi_{pp} + \tfrac{1}{2}\chi_{pp}^2\right]. \tag{9}$$

The particle-hole and particle-particle bubbles are expressed most simply in the position-imaginary time $(\mathbf{r}\text{-}\tau)$ representation,

$$\chi_{pp}(\mathbf{r},\tau) \quad = \quad U G(\mathbf{r},\tau) G(\mathbf{r},\tau) \tag{10}$$

$$\chi_{ph}(\mathbf{r},\tau) \quad = \quad -U G(\mathbf{r},\tau) G(-\mathbf{r},-\tau), \tag{11}$$

as is the self-energy,

$$\Sigma(\mathbf{r},\tau) = U(\chi_{ph}(\mathbf{r},\tau) + T_{ph}(\mathbf{r},\tau))G(\mathbf{r},\tau) + U T_{pp}(\mathbf{r},\tau)G(-\mathbf{r},-\tau). \tag{12}$$

The particle-hole and particle-particle T-matrices appearing in Σ, on the other hand, are simplest in the momentum-frequency $(\mathbf{k}\text{-}\varepsilon)$ representation,

$$T_{ph}(\mathbf{q},\omega_m) \quad = \quad \frac{U}{2}\left[-\frac{\chi_{ph}(\mathbf{q},\omega_m)^2}{1+\chi_{ph}(\mathbf{q},\omega_m)} + \frac{3\chi_{ph}(\mathbf{q},\omega_m)^2}{1-\chi_{ph}(\mathbf{q},\omega_m)}\right] \tag{13}$$

$$T_{pp}(\mathbf{q},\omega_m) \quad = \quad \frac{U\chi_{pp}(\mathbf{q},\omega_m)^2}{1+\chi_{pp}(\mathbf{q},\omega_m)}. \tag{14}$$

This approximation can also be thought of as a fully-symmetrized (in the sense of Ref. [4]) and self-consistent generalization of the familiar random phase approximation.

470

The fluctuation-exchange approximation for the Hubbard model (and related lattice models such as the Anderson lattice model) has a special feature that makes it especially well-suited for a fine-grained SIMD parallel computer such as the Connection Machine. The crucial observation is that the bare interaction is completely local in space and time, and that the approximation does not introduce any nonlocal (i.e., momentum or frequency dependent) effective interaction vertices. As a result, all equations of the theory can be solved completely in parallel at each point in either $\mathbf{k}$-ε space (Dyson's equation and the T-matrix equations) or $\mathbf{r}$-τ space (susceptibilities and self-energy). The necessary transformations between $\mathbf{k}$-ε and $\mathbf{r}$-τ representations can be carried out efficiently by fast Fourier transforms. Using this algorithm on the NRL Connection Machine, we routinely treat 128×128 lattices with up to 1024 Matsubara frequencies [10, 11, 12].

Our discussion so far has glossed over an important issue, which is the need to impose some type of cutoff on the (infinite) set of Matsubara frequencies $\varepsilon_n = (2n+1)\pi T$ and $\omega_m = 2m\pi T$. This issue is closely related to the use of FFTs, as outlined above, to transform between $\mathbf{k}$-ε and $\mathbf{r}$-τ representations for the Green's functions. Periodic boundary conditions are the standard way to terminate a finite spatial lattice, and it is well known that the corresponding functions in momentum space are periodic in the reciprocal lattice. Hence there are no additional approximations in our treatment of the lattice (or momentum) variables beyond those implicit in the use of a finite lattice (and a discrete lattice Hamiltonian). On the other hand, the (imaginary) time τ is a continuous variable and the Green's functions, susceptibilities, etc. can all be taken as periodic in τ with period $1/T$ (for convenience we include a factor of $\exp(i\pi T\tau)$ in Fermionic Green's functions). By performing a Fourier transform using only M Matsubara frequencies ε_n we obtain a (properly periodic) function evaluated at M equally-spaced points, $\tau_k = k/(MT)$ in the interval $[0, 1/T]$. When we use this discrete Green's function to evaluate a susceptibility bubble $\chi(\tau_k)$ and then compute $\chi(\omega_m)$ by a discrete inverse Fourier transform, we have effectively used the trapezoid rule to approximate the integral that defines the exact $\chi(\omega_m)$.

Described in this way, our procedure sounds perfectly natural. It may appear more unusual when viewed as a method for approximating the exact expression for the (particle-hole) susceptibility bubble expressed in the frequency representation,

$$\chi_{ph}(\mathbf{q},\omega_m) = -U(T/N)\sum_{\mathbf{k}}\sum_{n} G(\mathbf{k}+\mathbf{q}, \varepsilon_n + \omega_m)G(\mathbf{k}, \varepsilon_n). \tag{15}$$

The problem with approximating convolution sums such as this using only M frequencies is that $\varepsilon_n + \omega_m$ can fall outside the chosen frequency interval even when ε_n and ω_m do not, and the procedure described above is equivalent to handing this problem by periodically extending the Green's functions in frequency with period M. On first sight a more natural scheme might seem to be to set to zero any Green's function whose frequency argument falls beyond the cutoff, but this method clearly underestimates the susceptibility bubbles and the self-energy at high frequencies even more seriously than at low frequencies, while the Fourier transform scheme presumably overestimates these functions at frequencies near the cutoff. We see no fundamental reason for preferring one over the other.

We have explored this question by comparing results using the two cutoff schemes, since we can also impose a sharp cutoff by using twice as many frequencies as desired and then setting all functions to zero beyond the true cutoff. Fortunately the low-frequency self-energies appear to be insensitive to the cutoff procedure, although (not surprisingly) integrated thermodynamic quantities such as the free energy and its derivatives are relatively more sensitive. When we need accurate results for the free energy, we can significantly reduce the cutoff sensitivity by calculating exactly the noninteracting susceptibility bubbles, for which the cutoff-free frequency sums can be carried out analytically, and using the self-consistent interacting Green's functions to calculate only the difference between the interacting and noninteracting susceptibilities. (Because the FFT-based algorithm cannot be used to evaluate the noninteracting susceptibilities, this is the most time-consuming step in our calculations.)

In almost all of our calculations, and also in the work of Bickers and collaborators [9, 13], the self-consistent renormalized equations were solved by iteration, starting from the exact noninteracting Green's function in the frequency representation, subject to one cutoff scheme or the other. For stability of this iterative algorithm, it is necessary to use a mixture of the new self-energy and the previous self-energy as input to the next iteration. We generally found good convergence using an equal mixture of the old and new self-energies. Our normal procedure is to stop iterating when the all elements of the new and old self-energies differ by less than 1 part in 10^4, but we have frequently iterated to 1 part in 10^6 precision (using the NRL Connection Machine's double-precision floating-point hardware) to verify that our standard precision is adequate.

The discussion above suggests an alternative approach, beginning from the exact noninteracting Green's functions in the tau representation at the discrete points τ_k. This approach has two attractive features: (1) the noninteracting susceptibility bubbles are calculated exactly at the points τ_k, in contrast to the schemes starting from the cut off (or periodically continued) frequency-representation Green's functions, in which the susceptibilities are never reproduced exactly in either tau or frequency representation; and (2) the Green's functions are smooth in both tau and frequency representation, as opposed to the alternative schemes, which impose either a sharp cutoff or a kink at the maximum frequency, and lead to corresponding end-point oscillations in the tau representation. Further insight into the structure of this approximation can be obtained by calculating analytically the discrete Fourier transform of the exact tau-representation noninteracting Green's function and (particle-hole) susceptibility. Using M discrete points τ_k, these are

$$G_0(\mathbf{k}, \varepsilon_n | M) \;=\; \frac{1}{MT} \left\{ \frac{1}{\exp[G_0^{-1}(\mathbf{k}, \varepsilon_n)/MT] - 1} + \frac{1}{2} \right\} \tag{16}$$

$$\chi_0^{ph}(\mathbf{q}, \omega_m | M) \;=\; -\frac{1}{NMT} \sum_{\mathbf{k}} \frac{f(\xi_{\mathbf{k+q}}) - f(\xi_{\mathbf{k}})}{\exp[(i\omega_m - \xi_{\mathbf{k+q}} + \xi_{\mathbf{k}})/MT] - 1}. \tag{17}$$

We have begun to test this approximation against our usual approach, with some encouraging results. Fig. 1 shows essentially identical results from the two approaches for $\mathrm{Im}\Sigma(\mathbf{k}, \varepsilon_0)$ for the 1D Hubbard model at half filling with $U = t$ and $T = 0.03t$.

Application to The Hubbard Model in 2D

Varma and coworkers [14] proposed that a unified description of the anomalous temperature-dependent resistivity and tunneling data in cuprate superconductors follows from an electronic self-energy of the general form

$$\Sigma(\varepsilon) = \frac{2\alpha}{\pi} \left[\varepsilon \ln \frac{x}{\omega_c} - i\frac{\pi}{2}x \right], \tag{18}$$

where $x = \max(|\varepsilon|, T)$ and ω_c is a cutoff of order the bandwidth W. They called systems with this type of self-energy marginal Fermi liquids (MFLs), because the weight of the Fermi-liquid quasiparticle pole, $a_k = (1 - d\Sigma(\varepsilon)/d\varepsilon)^{-1}$, vanishes (weakly) as an inverse logarithm when the "quasiparticle" approaches the Fermi surface. A number of authors have suggested that a MFL-like self-energy could result from special features of 2D bandstructure, at least over some nontrivial range of temperature and energy (but not extending to $T = 0$ or $\varepsilon = 0$). Read and Lee [15] considered the 2D Hubbard model near half-filling, and argued that the combination of Fermi surface nesting and saddle points in the dispersion relation (which lead to van Hove singularities in the electronic density of states) result in a quasiparticle lifetime that goes like $1/T$, and might account for observed linear temperature dependence of the resistivity. Virosztek and Ruvalds [16] later used an approximate self-consistent calculation to show that the enhanced phase space for quasiparticle-quasiparticle scattering due to Fermi surface nesting can produce a temperature and frequency dependent MFL-like self-energy. In this calculation,

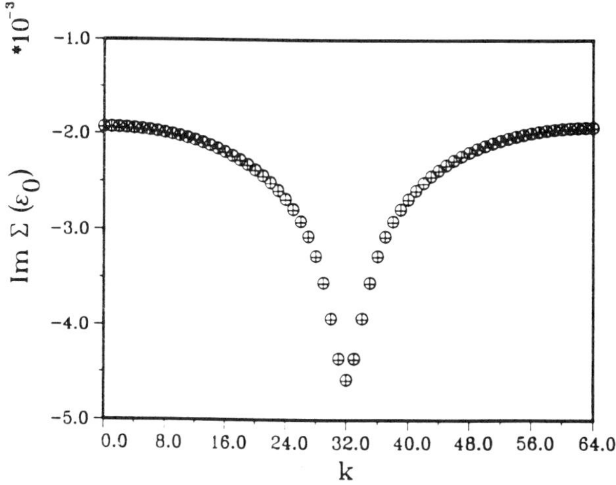

Figure 1. Comparison of Im $\Sigma(\mathbf{k}, \varepsilon_0)$ calculated beginning from the exact noninteracting Green's function in the τ representation ($\circ$) and ε representation ($+$), for the 1D Hubbard model at half filling with $U = t$ and $T = 0.03t$.

as in the original phenomenological MFL model, the self-energy was assumed to be independent of momentum. Newns and coworkers [17] have demonstrated that even without Fermi surface nesting, saddle points in the quasiparticle dispersion relation enhance the phase space for scattering of two quasiparticles and lead to a quasiparticle lifetime that goes like $1/\varepsilon$, at least to leading order in perturbation theory. All of these proposals argue for MFL-like behavior from conventional Green's-function perturbation theory in weak to moderate coupling, where the fluctuation-exchange approximation should be qualitatively accurate. Furthermore, a self-consistent calculation is essential to validate any proposal that argues for MFL-like behavior from perturbation theory using conventional Fermi liquid quasiparticles. In this section we discuss some of our results for 128×128-site Hubbard models in 2D, with particular attention to the possibility of marginal-Fermi-liquid behavior.

Our self-consistent solution of Eqs. (2) and (3) yields the self-energy on the imaginary axis, while quasiparticle properties are most naturally expressed in terms of the self-energy for real frequencies. We use two methods to extract quasiparticle properties from our imaginary-axis calculations. The first employs the temperature-dependent ratio Im $\Sigma(\mathbf{k}, \varepsilon_0)/\varepsilon_0$ as an imaginary-axis approximation to the frequency derivative of the real part of the real-frequency self-energy. For a Fermi liquid, this quantity is weakly temperature dependent and yields the quasiparticle renormalization factor in the limit of zero temperature.

$$\lim_{T \to 0} \left[1 - \frac{\mathrm{Im}\ \Sigma(\mathbf{k}, \varepsilon_0)}{\varepsilon_0} \right] = a_{\mathbf{k}}^{-1}. \tag{19}$$

We observe in passing that the quantity in square brackets is the renormalization factor $Z(\mathbf{k}, i\varepsilon_0)$ of Eliashberg theory [18]. The second method is to obtain the retarded self-energy on the real axis by analytic continuation from the imaginary axis using N-point Padé approximants [19]. Although this procedure is known to be sensitive to the accuracy of the imaginary axis data, it has been used extensively for strong-coupling superconductors, where it gives satisfactory results for the gap function and renormalization factor at low temperatures, even with highly-structured phonon densities of states. We expect the self-energy of the Hubbard model to have a less structured frequency dependence, so that the Padé approximant approach should be quite accurate. To test this notion, we used the imaginary part of Eq. (18) as a spectral

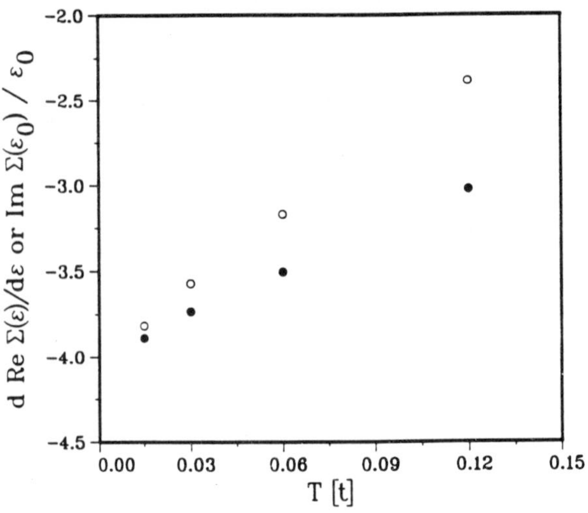

Figure 2. Im $\Sigma(\mathbf{k}_F, \varepsilon_0)/\varepsilon_0$ compared with $d\,\mathrm{Re}\,\Sigma(\mathbf{k}_F, \varepsilon)/d\varepsilon$ (obtained by analytic continuation using an N-point Padé approximant), for the 2D Hubbard model with $n = 0.53$ and $U = 8t$.

function to generate a model MFL self-energy on the imaginary axis, and then continued back to the real axis using Padé approximants. The linear part of the imaginary self-energy was reproduced accurately; not surprisingly, the (unphysical) sharp high-energy cutoff was always distorted. The slope of the real part of the self-energy could be determined to within 10% accuracy; the kink in the model spectral function at $\varepsilon = T$ apparently precludes obtaining this quantity with higher accuracy.

We will first discuss the standard Hubbard model with nearest-neighbor hopping only, for which the dispersion relation is

$$\epsilon_{\mathbf{k}} = -2t\,(\cos k_x + \cos k_y). \tag{20}$$

We have previously reported that in the fluctuation exchange approximation, the self-energy and thermodynamic potential of this model near quarter filling are apparently consistent with Fermi liquid theory [12]. In Fig. 2 we compare the imaginary-frequency and N-point Padé estimates for $d\mathrm{Re}\,\Sigma(\mathbf{k}_F, \varepsilon)/d\varepsilon$ with $U = 8t$, at four temperatures. As expected for a Fermi liquid, these quantities are weakly temperature dependent and extrapolate to the same value in the limit of zero temperature. This result can be viewed both as confirming the Fermi-liquid character of the fluctuation-exchange approximation for this case and also as demonstrating that our two estimators for the renormalization factor behave as expected for a Fermi liquid. Because the Fermi surface of the quarter-filled Hubbard model is neither nested nor near a van Hove singularity, from a conventional perspective the correctness of the Fermi liquid picture is expected. On the other hand, Anderson has argued that the 2D positive-U Hubbard model is never a Fermi liquid for any density or interaction strength [20]. From this perspective our results imply that any universal breakdown of Fermi liquid theory must have a very subtle origin.

At half filling the Fermi surface is perfectly nested, as assumed by Virosztek and Ruvalds, and in addition passes through the van Hove singularities, as noted by Lee and Read. In agreement with these authors, we find clear evidence for the failure of the Fermi liquid picture at half filling, but our results also differ substantially from the MFL model. For example, as shown in the top panel of Fig. 3, Im $\Sigma(\mathbf{k}, \varepsilon_0)/\varepsilon_0$ shows strong frequency, momentum and temperature dependence [11]. As a function of $\mathbf{k}$ this quantity displays a sharp 'notch' precisely at the

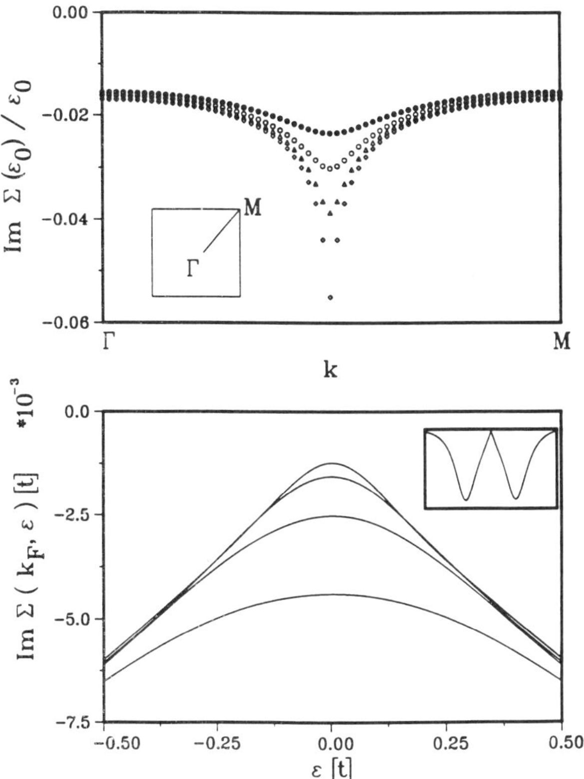

Figure 3. *Upper Panel:* The momentum dependence of Im $\Sigma(\mathbf{k}, \varepsilon_0)/\varepsilon_0$ for the Hubbard model at half filling with $U = 0.75t$ and $\mathbf{k}$ along the (11) direction. The calculations were performed at temperatures $T = 0.12$ ($\bullet$), 0.06 ($\circ$), 0.03 ($\triangle$) and 0.015 ($\diamond$). *Lower Panel:* $\mathrm{Im}\Sigma(\mathbf{k}_F, \varepsilon)$ obtained from an analytic continuation of the imaginary axis calculations at the temperatures above. The inset shows $\mathrm{Im}\Sigma(\mathbf{k}_F, \varepsilon)$ over the range $-12.0 < \varepsilon < 12.0$.

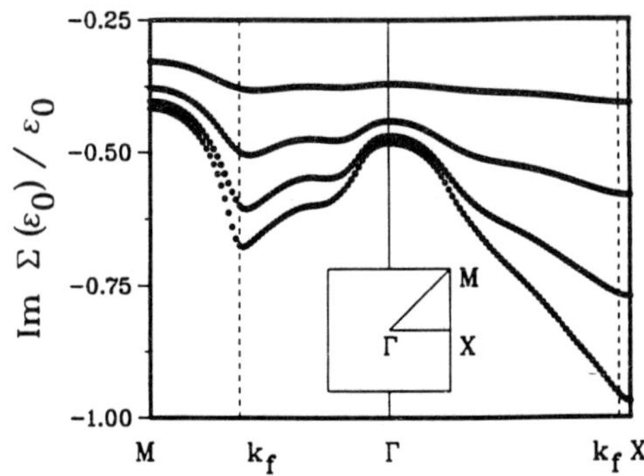

Figure 4. Momentum dependence of Im $\Sigma(\mathbf{k}, \varepsilon_0)/\varepsilon_0$ along the (10) and (11) directions for a 2D Hubbard model with next-nearest-neighbor hopping and $U = 2t$, at the same four temperatures as in Fig. 3 (descending, top to bottom) and $n = 1.30$. The Fermi surface is just below the van Hove singularities in the (10) directions.

Fermi surface. As we have discussed elsewhere, a similar feature occurs in the fluctuation-exchange approximation for the 1D Hubbard model; in both cases this structure reflects the strong momentum dependence of the particle-hole susceptibility bubble near $q = 2k_F$ [10, 21]. The depth of the notch also has a substantial momentum dependence in the direction parallel to the Fermi surface, presumably due to the van Hove singularities at the corners of the diamond-shaped Fermi surface, where the notch is deepest. The depth of the notch grows with decreasing temperature and for $U = 0.5t$ and $U = 0.75t$ seems to be well described by a power law, Im $\Sigma(\mathbf{k}_F, \varepsilon_0)/\varepsilon_0 = bT^{-\beta(\mathbf{k})}$ with $\beta(\mathbf{k}) < 1$; for $U/t = 0.5$ we find $\beta = 0.23[0.35]$ for $\mathbf{k}_F$ in the (11)[(10)] direction. The quality of the power-law fits degrades with increasing U and for $U = t$ neither a power law nor the MFL model appears adequate. In the bottom panel of Fig. 3, we show Padé results for Im $\Sigma(\mathbf{k}, \varepsilon)$ for $\mathbf{k} = (\pi/2, \pi/2)$ at temperatures of $T/t = 0.12, 0.06, 0.03$, and 0.015. The inset shows Im $\Sigma(\mathbf{k}_F, \varepsilon)$ for frequencies from $\varepsilon = -12.0$ to 12.0; as expected at half filling, Im $\Sigma(\mathbf{k}_F, \varepsilon)$ is symmetric in energy. Over a wide energy range Im $\Sigma(\mathbf{k}_F, \varepsilon)$ is roughly linear in ε and independent of temperature, as in the MFL model. However, the (roughly) linear frequency-dependence does not extrapolate to zero at zero frequency, and the minimum of the scattering rate (at $\varepsilon = 0$) decreases less rapidly than T at low temperatures and does not appear to extrapolate to zero at $T = 0$. At low energy, the linear energy dependence of Im $\Sigma(\mathbf{k}_F, \varepsilon)$ is terminated by a roughly parabolic region at an energy of $\sim \pi T$ with a crossover region that displays a an unusual negative curvature. These characteristics (which are even more pronounced at half-filling in 1D) are stable features of the N-point Padé approximants for varying N, which leads us to believe that they are either unexpected properties of the self-consistent fluctuation-exchange approximation exactly on half filling, or else signals that the Padé approach is unable to handle some idiosyncratic analytic structure of the self-energy in this case. It may be relevant to note that for $U \geq 1$ and $T = 0.015t$, we were unable to avoid a pole of the particle-hole T-matrix at $q = 2k_F$ (presumably signaling an SDW instability), even with very gradual ramps of U, T, or density, beginning from converged solutions. We emphasize, however, that all of the results for which solutions could be obtained are effectively in the weak-coupling limit, in the sense that the

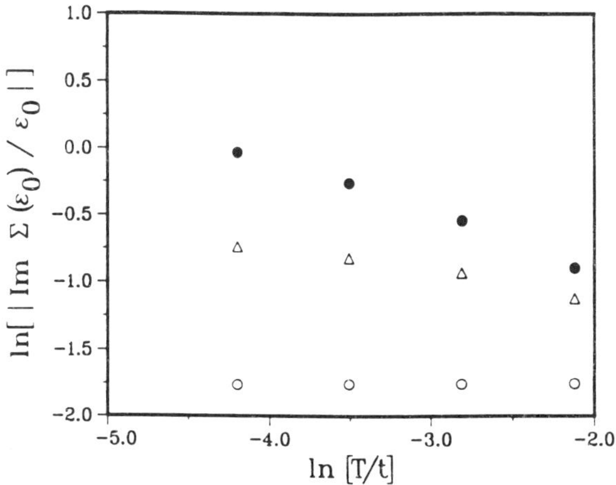

Figure 5. The temperature dependence of $\ln|\mathrm{Im}\ \Sigma(\mathbf{k}_F,\varepsilon_0)/\varepsilon_0|$ for a 2D Hubbard model with next-nearest-neighbor hopping and $U = 2t$, for $n = 0.89(\circ)$, $1.30(\bullet)$, and $1.54(\triangle)$, plotted vs. $\ln T/t$.

renormalization factor, as estimated by either the Padé or imaginary-frequency method, is never much less than one.

In order to study the effect of van Hove singularities alone (without the perfect nesting that occurs on the same energy surfaces in the standard Hubbard model), we used a dispersion relation that includes next-nearest-neighbor hopping [22],

$$\epsilon_k = -2t\ (\cos k_x + \cos k_y) - 4t'\cos k_x \cos k_y, \qquad (21)$$

with $t'/t = 0.35$. For this choice of t'/t, Fermi-surface nesting occurs at a filling substantially larger than that for which the Fermi surface passes through the van Hove singularities. Fig. 4 shows $\mathrm{Im}\ \Sigma(\mathbf{k},\varepsilon_0)/\varepsilon_0$ for $\mathbf{k}$ along the (10) and (11) directions, with $U = 2t$ and filling $n = 1.30$, for which the Fermi surface falls between two and three mesh points below the van Hove singularities. Along both directions, $\mathrm{Im}\ \Sigma(\mathbf{k},\varepsilon_0)/\varepsilon_0$ shows a strongly temperature-dependent feature in the vicinity of the Fermi surface. These features are broad in comparison with the notches found for the perfectly nested Fermi surface, with the most gradual momentum dependence along the cut passing directly through the van Hove singularity. Nonetheless, as shown in Fig. 5, when the Fermi surface is near the van Hove singularities, $\mathrm{Im}\ \Sigma(\mathbf{k},\varepsilon_0)/\varepsilon_0$ increases strongly with decreasing temperature in contrast to the nearly temperature independent $\mathrm{Im}\ \Sigma(\mathbf{k},\varepsilon_0)/\varepsilon_0$ at low filling where the Fermi surface is far from the van Hove singularities. Since a van Hove singularity is a critical *point*, it is not surprising that the the temperature dependence of $\mathrm{Im}\ \Sigma(\mathbf{k}_F,\varepsilon_0)/\varepsilon_0$ is somewhat weaker than the power law behavior observed for Fermi surfaces with large nested regions; a slight downward curvature is apparent in the log-log plot in Fig. 5. The anomalous temperature and momentum dependence of the self-energy is still apparent for $n = 1.54$, even though the Fermi surface is again far from the van Hove singularities. This probably reflects the fact that the Fermi surface is becoming nested at this density. Fig. 6 shows $\mathrm{Im}\ \Sigma(\mathbf{k},\varepsilon)$ for $T = 0.03t$ and $U = 2t$, with $n = 0.89$ and $n = 1.30$. The low-density results provide a useful point of reference: the imaginary part is small for $\varepsilon = 0$, and has an approximately parabolic energy dependence; and the real part is smooth near $\varepsilon = 0$. When the Fermi surface passes near to the van Hove singularities, however, a region nearly linear in frequency is apparent in the imaginary part of the self-energy, and the real part develops a

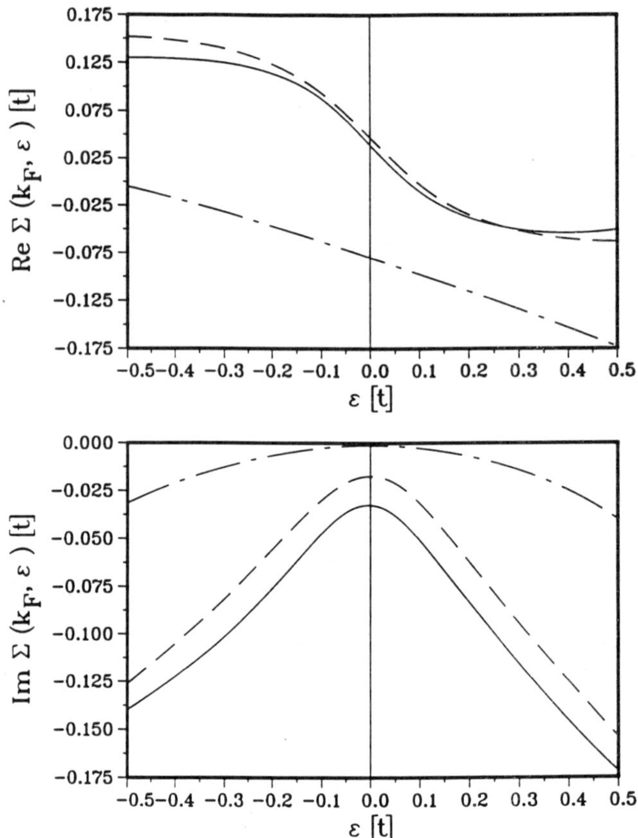

Figure 6. *Upper Panel:* Re $\Sigma(\mathbf{k}_F, \varepsilon)$ for a 2D Hubbard model with next-nearest-neighbor hopping and $U = 2t$. The filling is $n = 1.30$, where the Fermi surface passes just below the van Hove singularities. The solid (dashed) line refers to $\mathbf{k}_F$ along the (10) $[(11)]$ direction. The dash-dot line is for for $n = 0.89$, where the Fermi surface is far below the van Hove singularities. *Lower Panel:* Im$\Sigma(\mathbf{k}_F, \varepsilon)$ for the same calculations as in the upper panel. The results for $n = 0.89$ have been scaled relative to those for $n = 1.30$ by the inverse ratio of the densities.

low-frequency kink, as expected from Fig. 5; this feature is qualitatively similar to what we observe for the 1D Hubbard model and for the 2D Hubbard model at half filling [21].

Summary

We have shown that a massively-parallel algorithm allows one to solve the self-consistent fluctuation-exchange approximation on a sufficiently fine computational mesh to obtain meaningful information on the low-energy quasiparticle spectrum. Application of this approach to the Hubbard model shows non-Fermi-liquid behavior can occur in weak to moderate coupling when the Fermi surface exhibits nesting or passes near a van Hove singularity in two dimensions, and is apparent for all fillings in one dimension.

Acknowledgments

We thank D.M. Newns for stimulating conversations, and J. Thorp and R. Whaley for invaluable assistance with the Connection Machine.

References

[1] J.M. Luttinger and J.C. Ward, Phys. Rev. **118**, 1417 (1960).

[2] G. Baym, Phys. Rev. **127**, 1391 (1962).

[3] J.M. Luttinger, Phys. Rev. **119**, 1153 (1960).

[4] C. De Dominicis and P.C. Martin, J. Math. Phys. **5**, 14 (1964); **5**, 31 (1964).

[5] G.M. Carneiro and C.J. Pethick, Phys. Rev. B **11**, 1106 (1975).

[6] J.W. Serene and D. Rainer, Phys. Rep. **101**, 221 (1983).

[7] G.M. Eliashberg, Sov. Phys. JETP **16**, 780 (1963).

[8] J. Bardeen and M. Stephen, Phys. Rev. **136**, 1485 (1964).

[9] N.E. Bickers, D.J. Scalapino and S.R. White, Phys. Rev. Lett. **62**, 961 (1989).

[10] J.W. Serene and D.W. Hess, in *Electronic Structure and Mechanisms for High Temperature Superconductivity*, edited by J. Ashkenazi, S.E. Barnes, F. Zuo, G. Vezzoli and B.M. Klein (Plenum, New York, in press).

[11] D.W. Hess and J.W. Serene, J. Phys. Chem. Solids, to be published, (1991).

[12] J.W. Serene and D.W. Hess, Phys. Rev., **B44** , (1991).

[13] N.E. Bickers and S.R. White, Phys. Rev. B **43**, 8044 (1991).

[14] C.M. Varma, P.B. Littlewood, S. Schmitt-Rink, E. Abrahams, and A.E. Ruckenstein, Phys. Rev. Lett. **63**, 1996 (1989).

[15] P.A. Lee and N. Read, Phys. Rev. Lett. **58**, 2691 (1987).

[16] A. Virosztek and J. Ruvalds, Phys. Rev. B **42**, 4064 (1990).

[17] C.L. Kane, D.M. Newns, P.C. Pattnaik, C.C. Tsuei and C.C. Chi, in *Electronic Structure and Mechanisms for High Temperature Superconductivity*, edited by J. Ashkenazi, S.E. Barnes, F. Zuo, G. Vezzoli and B.M. Klein (Plenum, New York, in press).

[18] P.B. Allen and B. Mitrović, in *Solid State Physics*, edited by H. Ehrenreich, F. Seitz and D. Turnbull, Vol 37 (Academic Press, New York, 1982).

[19] H.J. Vidberg and J.W. Serene, J. Low Temp. Phys. **19**, 179(1977).

[20] P.W. Anderson, Phys. Rev. Lett. **64**, 1839 (1990); **65**, 2306 (1990).

[21] D.W. Hess and J.W. Serene, in preparation.

[22] This dispersion relation was suggested by D.M. Newns, who encouraged us to investigate the effect of the van Hove singularity in a fully self-consistent calculation.

COMPUTER SIMULATION OF NETWORK GLASSES AND MOLECULAR DYNAMICS ALGORITHM ON SIMD AND MIMD MACHINES

Priya Vashishta, Donald L. Greenwell, Rajiv K. Kalia, and Aiichiro Nakano

Concurrent Computing Laboratory for Materials Simulations,
Department of Physics and Astronomy,
and Department of Computer Science

Louisiana State University, Baton Rouge, Louisiana 70803-4001, USA

ABSTRACT

The nature of intermediate-range order manifested in the form of first sharp diffraction peak (FSDP) in the structure factor, $S(q)$, in the glassy states of SiO_2, GeO_2, $GeSe_2$, and $SiSe_2$ is investigated using the molecular dynamics (MD) method. The origin of anomalous temperature dependence of the FSDP is discussed. We also investigate the nature of tetrahedral-molecular fragments in a-$SiSe_2$. MD simulations reveal that a-$SiSe_2$ consists of half the tetrahedra in corner-sharing (C-S) configurations and the remaining tetrahedra share edges. Edge-sharing (E-S) tetrahedra are the building blocks of tetrahedral-molecular fragments where two-edge-sharing tetrahedra are the nucleus and C-S configurations provide the connecting hinges between the molecular fragments.

Recent development of the quantum molecular dynamics (QMD) technique has made it feasible to study the electron transport in disordered systems. We have developed a nonequilibrium QMD method which yields a good signal-to-noise ratio in the calculation of transport coefficients. Simulation results for the excess electron mobility in amorphous silicon and in gaseous helium are presented.

Implementation of the molecular dynamics algorithm on the Connection Machine, a single instruction multiple data (SIMD) architecture, and multiple time step MD algorithm on the Intel iPSC/860, a multiple instruction multiple data (MIMD) machine, is discussed.

1. INTRODUCTION

Neutron and x-ray diffraction has provided a wealth of structural data on binary chalcogenide glasses, e.g. SiO_2, GeO_2, $GeSe_2$, $SiSe_2$, As_2Se_3, B_2O_3, etc.[1,2] The most celebrated structural feature in these glasses is the first sharp diffraction peak (FSDP) observed between q $\cong$ 1.0 and 1.5 Å^{-1}, which is believed to be a signature of intermediate order correlations.[3] These experiments reveal many similarities in the static structure factor, $S(q)$, when expressed in terms of of the dimensionless quantity $Q = qd_s$ ($d_s = (6/\pi\,\rho)^{1/3}$, ρ is the number density), the data for all these systems show the first sharp diffraction peak at $Q \sim$ 4.5. Although the chemical constituents, structure in crystalline phases, and topologies of these glasses are widely different, the dimensionless quantity is nearly constant in all these systems.[3] The FSDP also displays its characteristics anomalous temperature dependence, i.e. the height of the FSDP increases as the temperature is increased - a feature common to $GeSe_2$ and As_2Se_3 and many other binary chalcogenide glasses.[3,4]

In amorphous SiO_2 and $GeSe_2$ the FSDP is observed at 1.5 Å^{-1} and 1.0 Å^{-1}, respectively. Neutron and X-ray experiments reveal that the peak persists in the molten state and if fact its height in molten $GeSe_2$ at 1100K is about the same as in the low temperature glass.[1] The other peaks in $S(q)$, as expected, show normal behavior, i.e. their height in the molten phase is smaller than in the low temperature glass. So the key structural questions are: (i) What is the nature of the correlations that give rise to the FSDP at 1.0 Å^{-1} in a-$GeSe_2$, and (ii) Why does the FSDP show anomalous temperature dependence in some glasses and not in others, and what is the origin of the anomalous behavior. Related to these structural issues is the question of the connectivity of the glassy network: In other words, what is the population of corner-sharing (CS) and edge-sharing (ES) tetrahedra, and what is the statistics of rings. Can some form of crystalline fragments be identified in the amorphous state? And if the answer is yes, then what is the nature of these tetrahedral-molecular fragments. We will investigate the nature of structural and dynamical correlations in network glasses using the molecular dynamics method.

Recent development of the quantum molecular dynamics (QMD) method has made it feasible to study the real time dynamics of the coupled quantum and classical particles at finite temperatures.[5-8] We have studied the transport of an electron in dense helium gas and amorphous silicon at finite temperatures using the QMD method.

Molecular dynamics algorithm has a very high degree of inherent parallelisms. It is therefore desirable to run MD simulations on massively parallel architectures. We have investigated a variety of the MD algorithms on single instruction multiple data (SIMD) and multiple instruction multiple data (MIMD) architectures. We have implemented MD with long range interactions on the Connection machine, and the multiple time step method in conjunction with a spatial subdivision scheme has been implemented on the Intel iPSC/860.

2. INTERMEDIATE RANGE ORDER IN GLASSES

We have investigated structural correlations and their temperature and density dependence using the MD method.[9-17] In MD simulations, the motion of atoms or molecules is followed by numerically integrating Newton's equations. The knowledge of trajectories of particles allows one to investigate microscopically various material properties. In a typical MD simulation the system consists of a few hundred to a thousand particles, although with present day parallel supercomputers much larger systems can be simulated. Interparticle potentials used in our MD simulations include of two-body (v_2) and three-body (v_3) terms,

$$V = \sum_{i,j} v_2(r_{ij}) + \sum_{i,j,k} v_3(\vec{r}_{ij}, \vec{r}_{ik}) \; .$$

In network glasses - namely, SiO_2, GeO_2, $GeSe_2$, $SiSe_2$, As_2Se_3, B_2O_3 - the important two-body interactions are long-range Coulomb potentials due to charge-transfer effects, charge-dipole interactions due to the large electronic polarizability of anions, and the steric repulsion due to the size of atomic cores, and three-body bond bending and bond stretching forces resulting from covalent interactions. Details of the interaction potential for SiO_2,[9] $GeSe_2$,[12,13] $SiSe_2$[15,16] have been discussed elsewhere. MD calculations were performed on a system of 648 particles in a cubic cell with periodic boundary conditions. Long-range Coulomb interactions were treated with Ewald's summation. To assess finite-size effects, we also simulated a system of 5184 particles in a cell whose edge for $GeSe_2$ is 55.02 Å in the liquid and 53.75 Å in the glass corresponding to experimental number densities $\rho_L=3.114 \times 10^{22}$ and $\rho_G=3.338 \times 10^{22}$ cm^{-3}.[12,13] MD simulation for SiO_2,[9] GeO_2,[11] and $SiSe_2$,[15,16] were also carried out at their experimental glass densities. To prepare the glassy state, the system in the molten state at a temperature of about $1.25 \times T_M$, where T_M is the melting temperature, is quenched and thermalized at a rate of 0.1% per $10\Delta t$, leading to a state at about $0.7 \times T_M$. At this temperature, long range diffusion ceases and the system undergoes thermal arrest. However, local rearrangements continue to take place because of considerable thermal energy in the system. The system is thermalized for 30,000 time steps and then the averages are accumulated for additional 30,000 time steps. A detailed discussion of cooling and thermalization schedule is given in Vashishta et al.[9]

Several structural and dynamical quantities - static structure factor, bond-angle distributions, temperature dependence of the first sharp diffraction peak, and phonon density-of-states - have been calculated for network glasses mentioned above. Figure 1 shows a comparison between our MD results and neutron diffraction data for molten and a-$GeSe_2$,[12,13] a-$SiSe_2$,[15,16] and a-SiO_2.[9] The simulation results for the static structure factor are in good agreement with experiments. This is also true for other quantities such as bond-angle distributions and phonon density-of-states. Recently we have also determined[11] the structural properties a-GeO_2 at normal pressure and the results are in good agreement with neutron scattering experiments.

MD partial structure factors, $S_{\alpha\beta}(q)$, and results from differential anomalous x-ray scattering experiment for a-$GeSe_2$ are shown in Figure 2. From the MD results we find that the first peak in the total $S(q)$ arises from Ge-Ge and Ge-Se correlations. X-ray experiments on $GeTe_2$-$GeSe_2$ glasses also reveal that the FSDP grows as Te is replaced by the weaker Se so as to enhance the Ge contributions. However, differential anomalous x-ray scattering experiment clearly indicates that Ge correlations are responsible for the first peak, in agreement with the MD results for the partial structure factors. We have also determined that intermediate-range correlations between 4 and 8 Å are responsible for the FSDP.

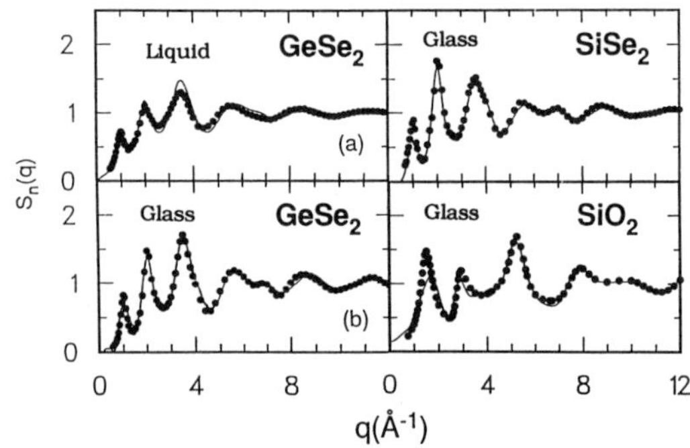

Figure 1. Neutron static structure factors, $S_n(q)$, of molten and amorphous $GeSe_2$, a-$SiSe_2$, and a-SiO_2. Liquid $GeSe_2$ is at 1100K and three glassy states are at 300K. Results of MD calculations and neutron experiments are represented by continuous curves and solid circles, respectively.

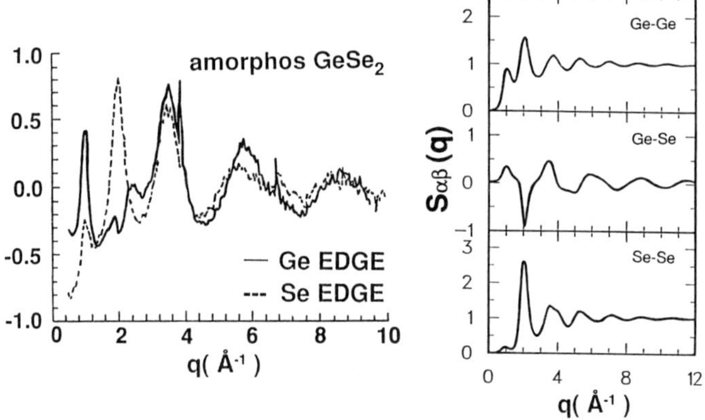

Figure 2. A comparison between the differential anomalous x-ray scattering experiment,[2] left, and the MD results for partial static structure factors, $S_{\alpha\beta}(q)$, of a-$GeSe_2$ at 300K, right.

Let us now discuss the anomalous temperature dependence of the FSDP in a-$GeSe_2$. As Fig. 1 shows, in both experiment and MD simulation the height of the FSDP for $GeSe_2$ is nearly the same in the liquid (1100K) and in the glass (300K). In contrast, the other peaks in $S(q)$ show the expected broadening and a decrease in their height on heating. To understand this behavior, let us remember that cooling from 1100 to 300K results in a 7% increase in the density of the system (ρ_L=3.114x10^{22}, ρ_G=3.338x10^{22} cm^{-3}, and ρ_G/ρ_L=1.07). To illustrate the effect of temperature alone, we form a glass at 300 K from the melt at 1100 K while keeping the density fixed at ρ_L. The structure factors for this lower density glass (ρ_L), as shown in the Figure 3 (b), and the molten state (ρ_L) at 1100 K, as given in the Figure 1 (a), show that the height of all the peaks decrease upon heating. On the other hand, Fig. 3 indicates that the FSDP is higher in the low density glass (ρ_L) than in the glass with the correct experimental density(ρ_G). The reason for this behavior is that an increase in the density as a result of cooling enhances the frustration associated with the packing of tetrahedral units. Opposite happens when the system expands as a result of heating. The decrease in the height of the FSDP due to the increase in the temperature is off set by the increase in the height due to the expansion of the volume resulting into a net increase in the height of the FSDP upon heating in a-GeSe$_2$.

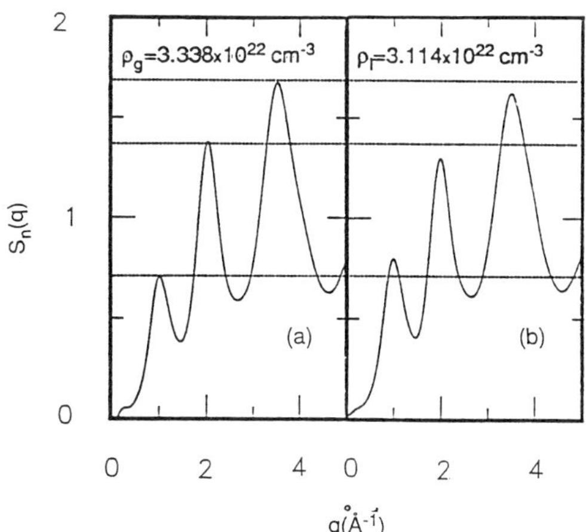

Figure 3. The neutron structure factor $S_n(q)$ for a-GeSe$_2$ at 300 K (a) the glass density ρ_G and (b) at the liquid density ρ_L.

Recently, we have studied the structural properties of permanently densified a-SiO$_2$.[10] These MD simulations were carried out in conjunction with neutron scattering experiments[10] on samples of a-SiO$_2$ compressed to 16 GPa at room temperature. The density of recovered samples is 20% higher than that of normal a-SiO$_2$. MD calculations show substantial changes in the intermediate-range order of densified a-SiO$_2$, which is in agreement with neutron scattering measurements. These changes are due to increased frustration caused by the decrease in the Si-O-Si bond angle.

3. CRYSTALLINE FRAGMENTS IN GLASSES

Silicon diselenide has an unusual crystal structure. The local structural units are $Si(Se_{1/2})_4$ tetrahedra, each of which is bonded at two edges. Such configurations form molecular entities which are non-intersecting infinite chains of edge-sharing (E-S) tetrahedra. In contrast, crystalline $GeSe_2$ consists of two-dimensional planes which are made of both corner-sharing (C-S) and edge-sharing tetrahedra, and crystalline forms of SiO_2 and GeO_2 are three-dimensional networks of only corner-sharing tetrahedra. Infrared,[18] Raman,[19] and NMR measurements,[20] when combined with diffraction data, can provide structural information about larger molecular entities and crystalline fragments in glasses.

We have investigated the nature of tetrahedral-molecular fragments in a-$SiSe_2$ using the molecular dynamics method.[21] MD simulations reveal that unlike the crystalline phase where every tetrahedron shares two edges, a-$SiSe_2$ consists of half the tetrahedra in C-S configurations and the remaining tetrahedra share edges. Among the E-S tetrahedra, 89% share one edge and the remaining 11% share two edges. E-S tetrahedra are the building blocks of tetrahedral-molecular fragments where two-edge-sharing tetrahedra are the nucleus and C-S configurations provide the connecting hinges between the molecular fragments. Analysis of rings and fragments in a-$SiSe_2$ reveals that: (1) 3- and 8-fold rings are most abundant; (2) chain-like fragments occur predominantly in 8-fold rings; and (3) the longest fragments occur in 11-fold rings. These chain-like fragments are typically 10-15 Å long.

MD simulations were performed[21] on 648 and 5,184 particle systems in a cubic box of 27.380 Å (28.102 Å) at the experimental density of 3.25 g/cm^3 (3.01 g/cm^3) for the glass (liquid). A snapshot of a 9 Å thick slice of a computer glass at T = 293 K is shown in Figure 4. The figure shows that in the glass each Si has four nearest-neighbor Se atoms and so the local structural unit is a $Si(Se_{1/2})_4$ tetrahedron. These local structural units are connected to one another to form rings of various sizes. MD results are in good agreement with the diffraction data. From the radial distribution functions, we determine that the Si-Se and Se-Se bond lengths are 2.28 Å and 3.73 Å, respectively, which are in good agreement with the corresponding experimental values of 2.30 Å and 3.80 Å. This establishes that our simulation of a-$SiSe_2$ is a satisfactory representation of the experimental system.

To examine distortions of tetrahedra and their connectivity, we have calculated Se-Si-Se and Si-Se-Si bond angle distributions in the glass. For an ideal tetrahedron Se-Si-Se angle is 109°. In crystalline $SiSe_2$, c-$SiSe_2$, the Se-Si-Se angle has three values 100.0°, 112.2°, and 116.7° indicating distortions of tetrahedra each of which shares two edges. Nearest-neighbor connectivity of tetrahedra is described by the Si-Se-Si bond angle distribution. This angle has a value of 80° in c-$SiSe_2$, indicating large distortions the tetrahedra have to undergo while sharing two edges. In c-$GeSe_2$, which has edge- and corner-sharing tetrahedra, Ge-Se-Ge angle distribution has a peak at 94° and individual tetrahedra have smaller distortions compared to those in c-$SiSe_2$. Figure 5 shows Se-Si-Se and Si-Se-Si distributions in a-$SiSe_2$. The two peak structure of Se-Si-Se at 93° and 108° is a reflection of crystalline angles. The Si-Se-Si distribution in a-SiSe$_2$ also shows two peaks, the first at 86° has an analog in c-SiSe$_2$ and indicates E-S tetrahedra, whereas the second peak at 111° is indicative of C-S configurations in the glass. The E-S configurations are also manifested in an interesting way in the Si-Si pair-distribution.

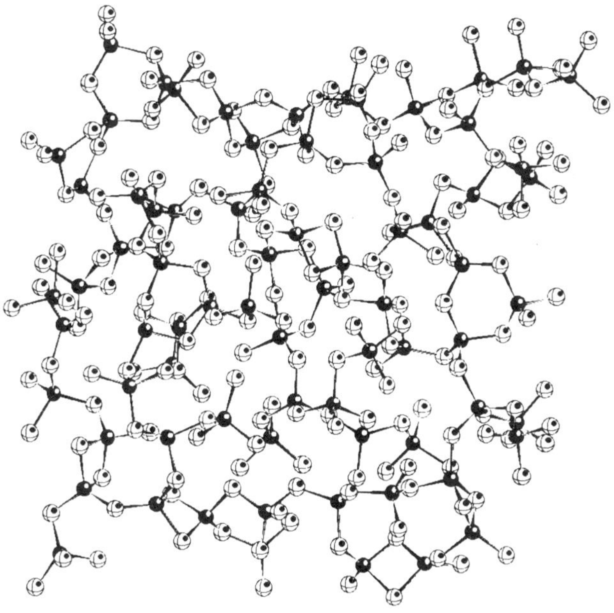

Figure 4. Snapshot of a 9Å thick slice of a SiSe$_2$ MD glass at 293K. Si atoms are indicated by dark circles with white dot and Se atoms by circles with black dot.

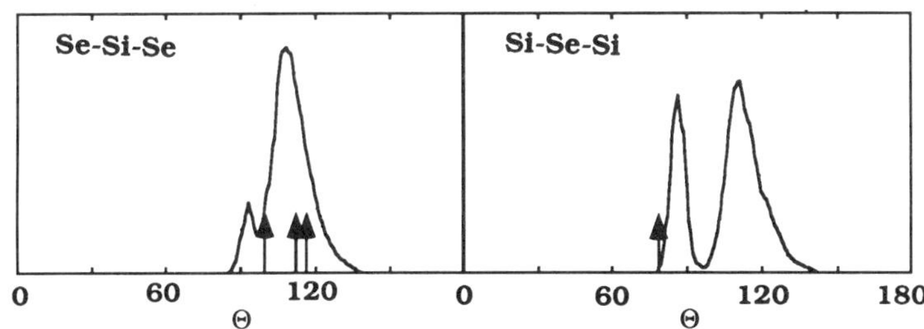

Figure 5. Se-Si-Se and Si-Se-Si bond angle distributions in a-SiSe2 at 293K. Crystalline angles are denoted by arrows.

The topology of covalent glasses is commonly discussed in terms of rings. We have calculated the distribution of 2- to 12-fold rings from MD configurations. In the glass the environment of Si atoms can be characterized into three categories: 1) Si atoms which participate only in C-S configurations. These Si atoms with no 2-fold rings are labeled as Si(0). 2) Si atoms which have only one 2-fold ring (tetrahedron sharing one edge). These are denoted by Si(1). 3) Si atoms which have two 2-fold rings, as in c-SiSe$_2$, are labeled as Si(2). Having analyzed the glass in terms of rings, let us now examine if there are chain-like molecular fragments in the glass. A molecular fragment is defined as an entity with three tetrahedra. The central tetrahedron contains a Si(2) atom whereas the remaining two tetrahedra have either Si(0), Si(1) or Si(2) atoms. Such fragments come in six flavors: 0-2-0, 0-2-2, 1-2-0, 1-2-1, 1-2-2, 2-2-2, where 0-2-0 denotes an Si(0)-Si(2)-Si(0) fragment. Statistics of these fragments in 3- to 12-fold rings are given in Ref. 21. No fragments of type 0-2-0, 0-2-2 or 2-2-2 are found. The most commonly occurring fragments are 1-2-0 and 1-2-1, see Fig. 6. The latter occurs mostly in 8-fold rings. The average length of these fragments is about 10-12 Å.

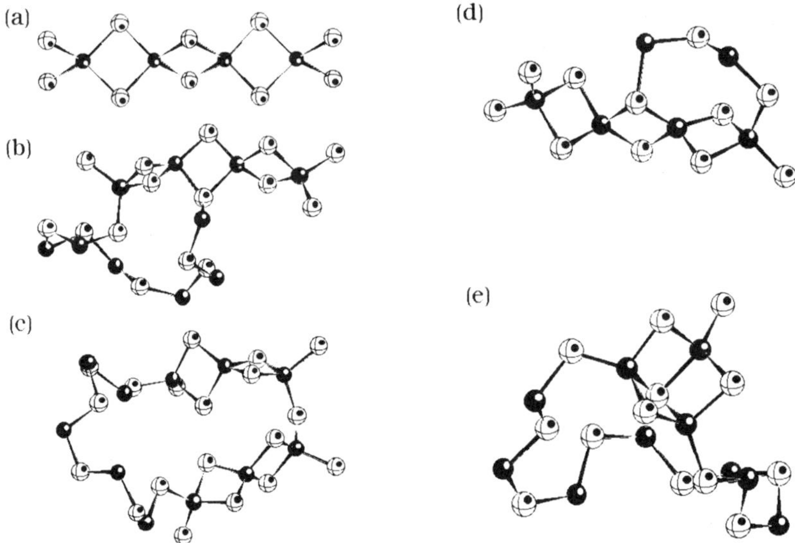

Figure 6. (a) shows the tetrahedral chain in c-SiSe$_2$. (b) and (c) show snapshots of fragments and rings taken from an MD glass at 293 K, while (d) and (e) are those in an MD liquid at 1,293 K.

4. QUANTUM TRANSPORT IN AMORPHOUS SYSTEMS

An important extension of the MD technique, applicable to mixed systems of quantum and classical particles, was developed three years ago. This so called quantum molecular dynamics (QMD) technique[5-8] involves the numerical solutions of the time-dependent Schrödinger and Newton's equations for quantum and classical particles, respectively. This approach allows the investigation of quantum transport at finite temperatures.

We have carried out nonequilibrium quantum simulations to investigate electron transport in disordered materials. In these simulations, one common source of difficulty is that large fluctuations in the electron current can mask the small current induced by an external electric field. Recently, we have developed an algorithm that overcomes this problem and yields a good signal-to-noise ratio.[22] Implementing this algorithm with the QMD approach,

488

we have calculated the mobility of an excess electron in gaseous helium[22] and amorphous Si.[23] In both cases the simulation results are in good agreement with time-of-flight mobility measurements, see, Fig. 7. The new algorithm makes it feasible to calculate the mobility in quasi free as well as localized states.

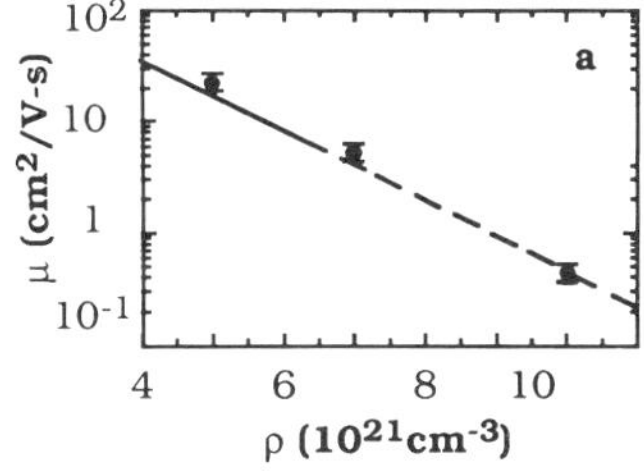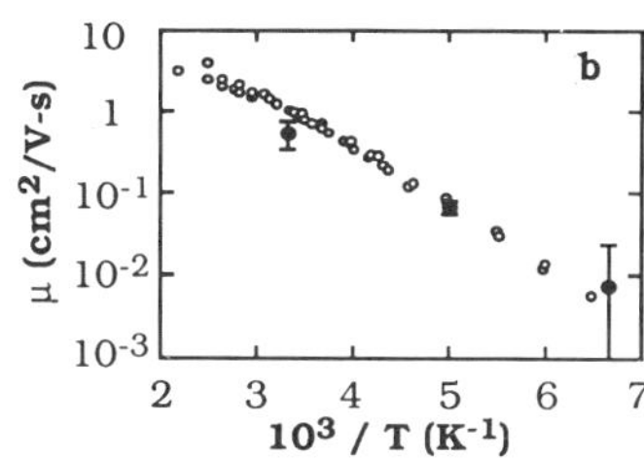

Figure 7. (a) Excess electron mobility in gaseous helium. Solid circles are nonequilibrium QMD results. Solid and dashed lines are the experimental result and its extrapolation [A. Bartels, Appl. Phys. **8**, 59 (1975)]. (b) Excess electron mobility in amorphous silicon. Solid circles are nonequilibrium QMD results. Open circles are from time-of-flight experiments [A. C. Hourd and W. E. Spear, Phil. Mag. B **5 1**, L13 (1985)].

5. MOLECULAR DYNAMICS ALGORITHM ON THE CONNECTION MACHINE

Molecular dynamics simulations have considerable inherent parallelism and therein lies the reason for doing these simulations on parallel machines. Three years ago, we performed MD simulations on the Argonne-Caltech Connection Machine.[24] These simulations were implemented for an ensemble of particles interacting via an inverse power law interaction.[24] In this implementation, each processor has the position and velocity of a given particle and it receives the position of each of the other particles at some point in time during the execution of the MD algorithm. Since the local memory is too small to allow all the particles to reside in a single processor, it is necessary to do a large amount of communication.

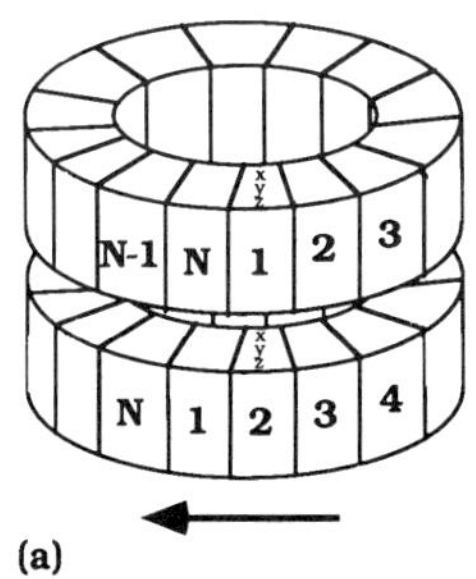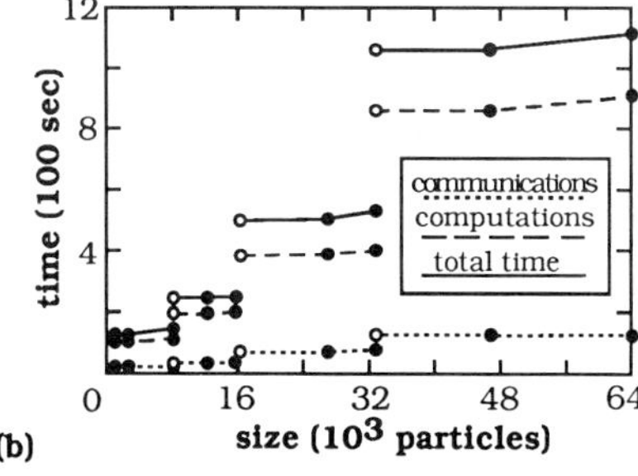

Figure 8. (a) Orrery Shift with grid communication on Connection Machine. (b) Timing of an MD simulation with NEWS grid on Connection Machine.

The Orrery Shift Algorithm was used where each processor is associated with a unique particle and the processors are lined up in a circle, see Fig. 8 (a). The processors move a copy of their particle coordinates into the processor on their left. The appropriate computations are made and the shifted data are again passed to the left and the computations repeated (in parallel). This process continues until the interaction between every pair of particles has been calculated. The critical part of this code is the shift operation. By organizing the particle processors in a one-dimensional NEWS grid, good use is made of fast nearest neighbor communication, see Fig. 8 (b). The performance in floating point operations per second for a 65,536 particle problem run on a 16,384-node machine (with virtual-processor ratio of 4 to 1) was 399 Mflops (or 1.6 Gflops scaled to a full machine).

6. MOLECULAR DYNAMICS ALGORITHM ON THE INTEL iPSC/860

On the distributed-memory Intel iPSC/860 MIMD machine, we have recently implemented[25] molecular dynamics simulations for semiconducting glasses in which the particles interact through a combination of two- and three-body potentials. The algorithm combines the multiple time-step method[26] with a spatial subdivision scheme. This efficient combination is applicable to a variety of realistic materials simulations. Decomposing the three-body potential into a separable form further enhances the performance.[27] First, the MD simulations were implemented in a conventional way by dividing the total system into N subsystems of equal volume. The data associated with particles in each subsystem resides on one of the N processors. In this implementation the total cpu time for each MD step is too large to perform any realistic large-scale simulations, even though the computation dominates the communication time. To this spatial subdivision implementation, we add a multiple time-step (MTS) approach which is based on the division of the total force on each particle into a rapidly varying primary component due to nearest neighbors and a slowly varying component due to secondary neighbors.[26] The secondary force varies much more slowly than the primary force and it is updated after every 10-20 time steps. On the 8-node Intel iPSC/860 machine in the Concurrent Computing Laboratory for Materials Simulations, the MTS approach gives a speedup of more than 10 over the conventional MD algorithm, see Fig. 9.

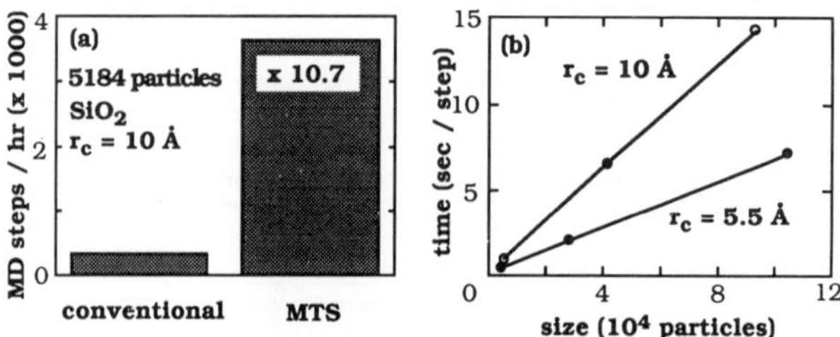

Figure 9. (a) Timing of an MD simulation and (b) cpu time versus the size of the MD system using the MTS algorithm on the 8-node iPSC/860 system for two different values of the interaction ranges, r_c.

7. CONCLUDING REMARKS

In this manuscript we have reviewed our investigations of the structural properties of network materials SiO_2, GeO_2, $GeSe_2$, $SiSe_2$ in molten and glassy states using the molecular dynamics technique. The calculations are based on an effective potential consisting of two- and three-body terms. The MD results reveal that a-SiO_2 consists entirely of corner sharing tetrahedra with Si-O-Si bond angle distribution peaking around 142° and having a FWHM of 25°. Recent NMR measurements determine the peak in Si-O-Si distribution to be at 142° with a FWHM of 26°. In MD results for the S(q), the FSDP occurs around 1.55 Å^{-1}, which is in good agreement with diffraction measurements. We have also investigated structural correlations in molten SiO_2. The proposed interaction potential for $GeSe_2$ also satisfactorily describes the structural features associated with short- and intermediate-range order including the anomalous temperature dependence of the FSDP. The investigation of tetrahedral-molecular fragments in a-$SiSe_2$ reveals that the material does indeed have chain-like fragments where two-edge-sharing tetrahedra are the nucleus and corner-sharing configurations provide connecting hinges between fragments.

The quantum molecular dynamics simulations for an excess electron in a-Si reveal that defects such as voids tend to localize excess electrons. Applying the multiple-trapping model to the simulation results, the values for the activation energy and the density-of states ratio are in good agreement with the corresponding experimental values. We have shown the viability of the non-equilibrium QMD simulations based on the subtraction technique. The generalization to many-electron systems is also, in principle, relatively straightforward.

From our work on the implementation of MD algorithm on SIMD and MIMD computer architectures, it is clear that very large scale MD simulations ($N>10^5$) can be performed on massively parallel computers with out too much difficulty. There are a variety of physical problems where reliable simulations require large systems and long-time behavior.

ACKNOWLEDGEMENT

This work was supported in part by a grant from Louisiana Education Quality Support Fund Grant No. LEQSF-(1991-1992)-RD-A-05. A. Nakano was supported by National Science Foundation Grant No. ASC-9109906.

REFERENCES

1.   O. Uemura et al., Phys. Status Solidi (a) **32**, K91 (1975); O. Uemura et al., J. Non-Cryst. Solids **30**, 155 (1978).

2.   F. H. Fuoss et al., Phys. Rev. Lett. **46**, 1537 (1981); P. H. Fuoss and A. Fisher-Colbrie, Phys. Rev. B **38**, 1875 (1988).

3.   S. C. Moss and D. L. Price, in *Physics of Disordered Materials*, edited by D. Adler, H. Fritzsche, and S. R. Ovshinsky (Plenum, New York, 1985), p. 77.

4.   L. E. Busse and S. R. Nagel, Phys. Rev. Lett. **47**, 1848 (1981); L. E. Busse, Phys. Rev. B **29**, 3639 (1981).

5.   A. Selloni, P. Carnevali, R. Car, and M. Parrinello, Phys. Rev. Lett. **59**, 823 (1987).

6.   R. N. Barnett, U. Landman, and A. Nitzan, J. Chem. Phys. **89**, 2242 (1988).

7.   R. K. Kalia, P. Vashishta, and S. W. de Leeuw, J. Chem. Phys. **90**, 6802 (1989).

8.   R. K. Kalia and J. Harris, Solid State Commun. **73**, 839 (1990).

9.   P. Vashishta, R. K. Kalia, J. P. Rino, and I. Ebbsjö, Phys. Rev. B **41**, 12197 (1990).

10. S. Susman, K. J. Volin, D. L. Price, M. Grimsditch, J. P. Rino, R. K. Kalia, P. Vashishta, G. Gwanmesia, Y. Wang, and R. C. Liebermann, Phys. Rev. B **43**, 1194 (1991).

11. P. Vashishta, J. P. Rino, G. A. Farias, and R. K. Kalia, to be published.

12. P. Vashishta, R. K. Kalia, G. A. Antonio, and I. Ebbsjö, Phys. Rev. Lett. **62**, 1651 (1989).

13. P. Vashishta, R. K. Kalia, and I. Ebbsjö, Phys. Rev. B **39**, 6034 (1989).

14. H. Iyetomi, P. Vashishta, and R. K. Kalia, Phys. Rev. B **43**, 1726 (1991).

15. G. A. Antonio, R. K. Kalia and P. Vashishta, J. Non-Cryst. Solids **106**, 305 (1988).

16. G. A. Antonio, R. K. Kalia, P. Vashishta, and I. Ebbsjö, Phys. Rev. B - in press.

17. H. Iyetomi, R. K. Kalia, and P. Vashishta - to be published.

18. M. Tenhover et al., Solid State Commun. **51**, 455 (1984).

19. M. Tenhover, M. A. Hazel, and R. K.. Grasselli, Phys. Rev. Lett. **51**, 404 (1983).

20. M. Tenhover et al., Solid State Commun. **65**, 1517 (1988).

21. G. A. Antonio, R. K. Kalia, A. Nakano, and P. Vashishta, Phys. Rev. B **45**, 7455 (1992).

22. A. Nakano, P. Vashishta and R. K. Kalia, Phys. Rev. B **43**, 10928 (1991).

23. A. Nakano, P. Vashishta, R. K. Kalia, and L. H. Yang, Phys. Rev. B **45**, 8363 (1992).

24. D. L. Greenwell, R. K. Kalia, J. C. Patterson, and P. Vashishta, *Scientific Applications of the Connection Machine*, Edited by H. D. Simon, (World Scientific, Singapore, 1989), p. 252; Int. J. High Speed Computing **1**, 321 (1989).

25. A. Nakano, R. K. Kalia, and P. Vashishta, to be published.

26. W. B. Streett, D. J. Tildesley and G. Saville, Mol. Phys. **35**, 639 (1978).

27. D. Frenkel, in *Simple Molecular Systems at High Density*, Edited by A. Polian, P. Loubeyre and N. Boccara, (Plenum, New York, 1989), p. 411.

ACHIEVEMENTS IN SOLID STATE PHYSICS AND
DENSITY FUNCTIONAL THEORY

Presentation of the Fourth Eugene Feenberg Memorial Medal
in Many-Body Physics to Walter Kohn

Malvin H. Kalos

Cornell Theory Center
Cornell University
Ithaca, NY 14853-3801, USA

Priya Vashishta

Concurrent Computing Laboratory for Materials Simulations,
Departments of Physics & Astronomy and Computer Science
Louisiana State University
Baton Rouge, LA 70803-4001, USA

It is a great privilege and a high honor to present the fourth *Feenberg Memorial Medal* on the first day of this International Conference on Recent Progress in Many-Body Theories. The Medal was established to commemorate the outstanding contributions made by Eugene Feenberg to the advancement of many-body theories based on first principles, as well as his intellectual leadership of our discipline.

Eugene Feenberg's death in St. Louis on November 7, 1977 ended a brilliant career in theoretical physics spanning nearly five decades. Almost all of his work in many-body theory and related areas was carried out at Washington University, where he guided the graduate research of a number of very good physicists. These former students include John Clark, H. W. Jackson, F. Y. Wu, Charles Campbell (co-organizer of this conference along with Eckhard Krotscheck), Chia-Wei Woo (who became president of San Francisco State University at a young age and is the Founding President of the Hong Kong University of Science and Technology), and Walter Massey, the current director of the National Science Foundation.

Feenberg received B.S. and M.A. degrees from the University of Texas in 1929 and was awarded a Ph.D. in theoretical physics at Harvard in 1933. After spending a year visiting the European centers of quantum physics, he returned to Harvard, where he was made a Junior Fellow. Feenberg held positions at the University of Wisconsin, the Institute for Advanced Study, and New York University before joining the faculty of Washington University in 1946. He became Wayman Crow Professor of Physics in 1970. Feenberg's profound and original contributions to theoretical nuclear physics, to perturbation theory and approximation methods, and to the correlated-basis-functions theory of quantum fluids have had an important and lasting impact on diverse subfields

Recent Progress in Many-Body Theories, Vol. 3,
Edited by T.L. Ainsworth et al., Plenum Press, New York, 1992

of many-body physics.[1] His pioneering investigations in nuclear physics were fundamental to the development of modern nuclear shell theory. This work culminated in his 1955 book *Shell Theory of the Nucleus*. His path-breaking research on strongly interacting quantum systems was summarized in 1969 in the now-classic monograph *Theory of Quantum Fluids*. Feenberg's work on the microscopic theory of the ground and low-lying excited states of quantum many-particle systems bears the stamp of his independent and penetrating mind. As a result of the development of variational, Green's function, and path-integral Monte Carlo procedures, we have been witnessing an impressive growth of theoretical activities in which the correlated-basis-functions approach launched by Feenberg has proven itself as a powerful tool for analyzing strongly interacting many-body systems on a quantitatively accurate level.

The first Feenberg Medal was awarded to David Pines of the University of Illinois, at the 1985 meeting of this conference series in San Francisco. John W. Clark of Washington University received the second medal in 1987 at Oulu, Finland.[2] The third medal was awarded to Malvin H. Kalos of Cornell University in 1989 at Arad, Israel.[3] Nominations for the 1991 Feenberg Medal were submitted to a selection committee consisting of Christopher Pethick (Chairman), John Clark, and Malvin Kalos.

The fourth Eugene Feenberg Medal is awarded to Walter Kohn for his seminal contributions to theoretical solid-state physics and to the development of density-functional theory. Before attempting to convey the scope of Kohn's achievements in solid-state physics and many-body theory, a brief review of his education and scientific career is in order.

Walter Kohn was born in Vienna, Austria in 1923. He obtained his early education in Austria, England, and Canada. He earned a B.A. in mathematics and physics and an M.A. in applied mathematics from the University of Toronto. During World War II, he served in the Canadian infantry. In 1948, Kohn received a Ph.D. in theoretical physics from Harvard University under Julian Schwinger (Thesis: Theory of Nuclear Collisions) and remained at Harvard for two more years as an instructor. Kohn joined the faculty of the Carnegie Institute of Technology (now Carnegie-Mellon University) in 1950 and moved to the University of California at San Diego in 1960, where he concentrated on research in theoretical solid-state physics and density-functional theory. In 1960, Kohn was awarded the prestigious Oliver Buckley Prize by the American Physical Society for his pivotal work in solid-state physics. He was elected to the National Academy of Sciences in 1969. Kohn received the 1977 Davison-Germer Prize for his contributions to surface physics. From its inception in 1979, through 1984, Kohn was the director of the National Science Foundation's Institute for Theoretical Physics at UC-Santa Barbara. At the end of his tenure as director, the institute was cited by an NSF panel as "a leading world-wide institution" playing a "key role in expanding the scope of theoretical physics." In 1988, President Reagan presented the National Medal of Science to Walter Kohn. It is noteworthy that Kohn has been an opponent of the "Star Wars" space-based missile defense system proposed by the Reagan administration.

The current award of the Feenberg Medal to Professor Kohn recognizes his outstanding accomplishments in solid-state physics and density-functional theory. These achievements include

- Development of a new procedure, the KKR method or Green's function method, for calculating the electronic structure of solids, in a collaboration with Norman Rostoker of the University of California at Irvine.[4]
- Studies of the electronic properties of superconductors conducted with Joaquin Luttinger of Columbia University in the mid-50's — elucidating the shallow impurity states and issues relating to the insulator-metal transition.
- Discovery of Kohn anomalies – In 1959, Kohn pointed out that when the wave

vector $\mathbf{q}$ of a phonon disturbance in a metal has magnitude $2k_F$, the $2k_F$ singularity of the dielectric function should be conveyed via the screened ion-ion interaction into the phonon spectrum itself. He predicted that this will result in weak but discernible "kinks" in the spectrum, arising from infinities in $\frac{d\omega}{dq}$ at values of q corresponding to the extremal diameters of the Fermi surface.[6] We have always suspected that Walter must have been involved in some sort of kinky stuff. The phenomenon was indeed observed in 1967 by R. Stedman, a Swedish neutron scatterer, through highly accurate measurements of the phonon dispersion.[7]

- Proposal of a non-phonon mechanism for superconductivity. – Kohn and Luttinger showed in 1965 that one could obtain a superconducting state without the conventional electron-phonon mechanism.[8] The main point of their work was to demonstrate that a system of electrons with purely repulsive interactions can give rise to the formation of Cooper pairs and the occurrence of superconductivity. The origin of the attractive pairing interaction lies in the Friedel oscillations, which are a manifestation of the singularity in the dielectric function of an electron gas at $q = 2k_F$.

- Foundations of density-functional theory. – In 1964-65 Kohn, in collaboration with Pierre Hohenberg of AT&T Bell Laboratories and Lu Sham of UC-San Diego proved two basic theorems, (i) expressing the ground state of a many-electron system in terms of the particle-density distribution, and (ii) providing a variational solution in terms of a single-particle Schrödinger equation (the Kohn-Sham equation).[9,10] In effect, they validated the historical fascination with the density as a descriptor of many-particle systems (a function $\rho(\mathbf{r})$ of the three coordinates x, y, z of a point in 3-dimensional space acting as proxy for the wave function, which depends on the $3N$ spatial variables of the electrons). Simultaneously, they offered a formal basis for the local-density approximation.[11,12] Kohn and Norton Lang, in 1970, reported the first self-consistent calculation of surface properties within the framework of density-functional theory.[13,14] This theory has truly revolutionized the calculation of electronic structure for atoms, molecules, surfaces, and solids in physics, chemistry, and materials science. In fact, to date the density-functional theory, applied within the local-density approximation, is the only viable method for computing the electronic structure of very large systems.[15] The theory has been extended to take into account time-dependent effects.[16] Extensions have also been proposed for treating excited states.[17,18] In 1988 Kohn, together with L. N. Oliveira and E. K. Gross, proposed a density-functional theory for superconductors.[19] It is fair to say that the development of density-functional theory has brought us into a new era in the quantitative description of electronic structure.

Walter Kohn has also been an exemplary scientist-as-citizen. He has long been concerned about the involvement of the University of California in the management of the Department of Energy's Livermore and Los Alamos weapons laboratories. At UC-Santa Barbara he has been on the steering committee of the Institute on Global Conflict and Cooperation. To be thorough and gain an appreciation of all sides of these complex issues, Kohn has had discussions with His Holiness the Dalai Lama of Tibet on "Buddhism and Global Security."

ACKNOWLEDGMENTS

We would like to thank John Clark, Charles Campbell and Eckhard Krotscheck for their help in preparing this article.

REFERENCES

1. Eugene Feenberg Obituary, Nucl. Phys., **A317** (1979).
2. R. F. Bishop, in *Recent Progress in Many-Body Theories*, eds. A. J. Kallio, E. Pajanne and R. F. Bishop (Plenum, New York, 1988) Vol. I, p. 385.
3. M. L. Ristig, in *Recent Progress in Many-Body Theories*, ed. Y. Avishai (Plenum, New York, 1990) Vol. II, p. 347.
4. W. Kohn and N. Rostoker, Phys. Rev., **94** (1954) 111.
5. W. Kohn and J. M. Luttinger, Phys. Rev., **97** (1955) 1721; Phys. Rev., **98** (1955) 915; W. Kohn, in *Solid State Physics*, (Academic Press, New York, 1957) Vol. V, p. 257.
6. W. Kohn, Phys. Rev. Lett., **2** (1959) 393.
7. R. Stedman *et al.*, Phys. Rev., **162** (1967) 545.
8. W. Kohn and J. M. Luttinger, Phys. Rev. Lett., **15** (1965) 524.
9. P. C. Hohenberg and W. Kohn, Phys. Rev., **B136** (1964) 864.
10. W. Kohn and L. J. Sham, Phys. Rev., **A140** (1965) 1133.
11. W. Kohn and P. Vashishta, in *Theory of Inhomogeneous Electron Gas*, eds. S. Lundqvist and N. H. March (Plenum, New York, 1983) p. 79.
12. P. C. Hohenberg, W. Kohn and L. J. Sham, in *Advances in Quantum Chemistry*, ed. S. B. Trickey (Academic Press, New York, 1990) Vol. 21, p. 7.
13. N. D. Lang and W. Kohn, Phys. Rev., **B1** (1970) 4555.
14. N. D. Lang and W. Kohn, Phys. Rev., **B3** (1971) 1215.
15. R. O. Jones and O. Gunnarsson, Rev. Mod. Phys., **61** (1989) 689; in *Advances in Quantum Chemistry*, ed. S. B. Trickey (Academic Press, New York, 1990) Vol. 21.
16. E. K. U. Gross and W. Kohn, in *Advances in Quantum Chemistry*, ed. S. B. Trickey (Academic Press, New York, 1990) Vol. 21, p. 255.
17. A. K. Theophilou, J. Phys., **C12** (1978) 5419; N. Hadjisavvas and A. K. Theophilou, Phys. Rev., **A32** (1985) 720.
18. W. Kohn, Phys. Rev., **A34** (1986) 737; Phys. Rev. Lett., **56** (1986) 2219.
19. L. N. Oliveira, E. K. U. Gross and W. Kohn, Phys. Rev. Lett., **60** (1988) 2430.

CONFERENCE SUMMARY

Alexander L. Fetter

Department of Physics, Stanford University

Stanford, CA 94305

Many months ago, Chuck Campbell asked me if I would be willing to serve as the Summary Speaker for the conference MBVII. Having agreed to do so, I have been worrying about this rash promise for the past several weeks. Clearly, it is an impossible task to provide a complete summary of a conference as large and diverse as this one, with more than 40 invited papers, and I shall necessarily give only an incomplete picture.

Let me start by making a few observations. I have attended only one other conference in this series—that in Oaxtapec, Mexico in 1981. Compared to that one, the present conference has a remarkable mix of fields, interests, and topics; it incorporates most, if not all, of many-body physics in its true sense, with much more than just Feynman diagrams, variational wave functions, and model hamiltonians.

To give an idea of the scope of the conference, I have made a brief list of the major topics:

Nuclei and nuclear matter
    bulk *vs.* finite systems
    nonrelativistic *vs.* relativistic formalism

4He and 3He
    liquid, solid, clusters
    variational methods
    quantum fluids

Condensed matter
    density functional theory
    spin systems
    Fermi liquids, marginal Fermi liquids, Luttinger liquids
    statistical transmutation in two dimensions
    Hubbard model
    high-T_c superconductivity

Atoms and molecules
    variational *vs.* many-body methods
    nonrelativistic *vs.* relativistic

Recent Progress in Many-Body Theories, Vol. 3,
Edited by T.L. Ainsworth et al., Plenum Press, New York, 1992

QED corrections

bound states *vs.* collisions

I personally have heard *every* invited talk, apart from the double sessions that were used for the shorter seminars. It has been a rigorous but rewarding experience. Obviously, I cannot mention every talk, or even every topic. Instead, I plan to concentrate on a few intriguing areas.

First, however, I would like to award a few symbolic prizes for talks that seemed particularly meritorious:

1. Bravos for Academician A. Abrikosov's warm and moving account of the scientific life and personal reminiscences of Landau: "L. D. Landau—his life, achievements, and my own memories". In common with many of those here in the audience, I have long considered Landau a personal hero, so that this presentation was truly memorable.

2. Congratulations to Professor W. Kohn on his receiving the Eugene Feenberg Medal for Many-Body Physics. His talk on "density functional theory in physics and chemistry" was an inspiration to us all.

One of the major difficulties at a meeting like this is the disparate nature of the participants, with few common threads in their background. We all appreciated the efforts that everyone made to communicate to the nonspecialist. To encourage us to do even better, I am going out on a limb to mention three talks that I found especially clear, compelling, and accessible. In order of presentation, they are

1. M. H. Kalos on "exact Monte Carlo for fermion systems".
2. V. Korepin on "complete solution to the one-dimensional Hubbard model".
3. E. G. D. Cohen on "the colloidal many body problem".

Telling a coherent story is never easy, and these talks struck me as models that we can all emulate.

Having gone this far, I shall now go a bit farther and focus on three areas that I found particularly interesting — in increasing order of my own knowledge.

I. FUTURE CALCULATIONS ON NUCLEAR MATTER (A. D. Jackson, B. Serot, and J. D. Walecka)

We heard that studies of the energy of bulk nuclear matter involve a very delicate cancelation between the kinetic energy and the potential energy (much more so than in bulk liquid 3He). As a result, a basic question remains unresolved. The conventional nonrelativistic theory of nuclear matter uses two-body potentials V_{12} that are fit to nucleon-nucleon scattering data. Is this approach adequate to provide an accurate description? The calculated energy per particle gives a reasonable value compared to E/A extrapolated from large nuclei, but the corresponding calculated value of the bulk equilibrium density ρ is off by roughly a factor of 2. This latter discrepancy suggests that the nonrelativistic theory is incomplete, and it is natural to consider what must be included in an improved theory.

One approach has been to introduce the hadronic degrees of freedom, which, in essence, can be considered a more fundamental picture of the two-body potentials. This procedure relies on relativistic quantum-field theory, which incorporates Lorentz invariance and causality. One specific model is known as quantum hadrodynamics (QHD); it introduces a baryon, a scalar meson, and a vector meson, with parameters fit to the properties of nuclear matter. Although such a theory may be unnecessary for bulk nuclear matter, it is certainly needed for an understanding of the properties under extreme conditions of high density and temperature, such as those in astrophysical applications and in the new heavy-ion accelerators.

Another important question involves nuclear compositeness—based on quarks and gluons as described in quantum chromodynamics (QCD). Although there has been considerable discussion, the precise relation between QHD and QCD remains unclear (at least to me).

Still another topic of interest is chiral symmetry, which has not been included fully in the previous descriptions. This symmetry is broken in our world, but it is assumed to be restored at sufficiently high temperatures and pressures. One very intriguing question is the nature of this phase transition—first order? second order? Various versions of the σ model have been used to characterize the chiral symmetry, but much work remains to be done on deciding whether it can provide a complete picture.

II. VARIATIONAL DESCRIPTIONS OF ^{4}He (S. Fantoni, L. Reatto, and S. Vitiello)

Over the past few years, the "shadow" wave function seems to have provided an intrinsically new approach to variational studies of liquid ^{4}He. This procedure introduces an extra set of "shadow" variables that are coupled to the "physical" variables by gaussian integral kernels, allowing a smearing of the original localized variables. Apparently, it can encompass both a liquid phase and a crystal phase, opening up the exciting possibility of a uniform theoretical description that interpolates between the two distinct phases. It has the important advantage that it uses only two-body correlation functions (as in the simpler Jastrow trial functions), yet it effectively incorporates three-body effects. We saw that it gives a ground-state energy that rivals more elaborate calculations with explicit three-body correlations. It also seems to give good values for the excitation energy of rotons. One interesting question is whether the shadow wave function can help provide a more physical picture of a roton than that provided by the old Feynman-Cohen trial function. Years ago, Onsager characterized a roton as "a ghost of a vanishing vortex ring". Despite Onsager's legendary intuition, I have always considered a roton as continuously connected to the phonon branch of the excitation spectrum, and hence without circulation. Can the shadow wave function provide a definitive answer to this uncertainty?

One exciting new application of this formalism is to a singly quantized vortex line in superfluid ^{4}He. The old Feynman model wave function1 involves a product of one-body factors $\Pi_{j=1}^{N} \left[f(r_j) \, e^{i\phi_j} \right]$ multiplying the exact ground-state wave function $\Phi_0(\mathbf{r}_1, \mathbf{r}_2, \ldots, \mathbf{r}_N)$. The centrifugal barrier associated with the exponential phase factor $e^{i\phi}$ implies that the radial function $f(r)$ vanishes linearly at $r = 0$, which eliminates the singular kinetic energy near the origin. As a result, this Feynman trial function automatically yields an empty core with a hole in the center of the vortex. Such behavior is almost certainly not present in the real system, and the shadow wave function provides an opportunity to include the spreading out of the core, which can be considered to reflect the zero-point motion.

To justify this claim, it is helpful to recall the behavior of the weakly interacting Bose gas at zero temperature. In the uniform bulk system, the total particle density n contains two contributions: the condensate density n_0 and the noncondensate density n'. As shown by Bogoliubov's celebrated calculation, most of the particles are in the condensate, so that n_0 is of order n; nevertheless, n' is definitely positive, although small ($n' \ll n$ or n_0). For a singly quantized vortex line in such a system, all the excited states have wave functions that vanish at the vortex core, except for the *one* set of states with appropriate angular momentum to cancel that of the condensate2 (this situation is analogous to the s states of the hydrogen atom). These special states contribute a nonzero density at the center of the vortex, which is comparable with the noncondensate value n' at infinity.

In real bulk ^{4}He, the condensate density n_0 is believed to constitute only about 10% of the total density, so that $n'/n \approx 0.9$. Based on the preceding discussion, we may conjecture that the particle density in the center of a real vortex line will be of order n', and it could even exceed the asymptotic total density n. The shadow wave function may well provide an answer to this long-standing question.

III. ANYONS IN TWO DIMENSIONS

Earlier in the conference, we heard from Yu Lu about charge fractionation in the one-dimensional Hubbard model. In that case, the electron effectively separates into a "spinon"—a neutral spin-$\frac{1}{2}$ object, and a "holon"—a charged spin-0 object. Similar behavior has been proposed by Anderson for the copper-oxide planes of high-T_c superconductors, and it has been suggested that these two-dimensional spinons and holons both obey fractional statistics. Specifically, if the coordinates of two such particles are interchanged, the corresponding two-body wave function $\Psi(1,2)$ changes into $\Psi(2,1)$, and the fractional statistics means that $\Psi(2,1) = e^{i\nu\pi}\Psi(1,2)$, where ν is the "fractional-statistics" parameter. Generically, particles obeying fractional statistics have been called "anyons"; in the special case of $\nu = \pm\frac{1}{2}$, they are known as "semions". Note that there are two types of semions associated with the $\pm$ sign in the exponent, so that the ground state of such a system will break time-reversal symmetry. It is not difficult to see that *pairs* of semions act like bosons, and they might therefore be expected to condense, leading to superfluidity and superconductivity.

The talk of S. M. Girvin pointed out that semionic holons can be modeled a spinless charged fermions plus a flux tube (a solenoid) carrying $\frac{1}{2}hc/e$ units of magnetic flux. Over the past three years, I have worked with C. Hanna and R. Laughlin (and, more recently, also with Q. Dai and J. Levy) to understand the properties of a gas of these exotic particles. In particular, we have demonstrated that the ground state exhibits two crucial and characteristic features of superfluidity and superconductivity:[3]

1. There is a low-lying undamped collective mode with a linear dispersion relation: $\omega = v_s q$ as $q \to 0$; this behavior is analogous to that of phonons in superfluid ^{4}He, and Feynman's beautiful arguments for superfluidity[4] apply to semions as well.

2. The linear response to an *external* electromagnetic field is explicitly gauge invariant. In the special case of a static magnetic field, the system displays a Meissner effect, with full flux expulsion at zero temperature. Thus all the particles participate in the superconductivity, with the superfluid density ρ_s equal to the total density ρ.

Having listened to all the other talks at the conference, I hope that you will allow me to describe this work briefly, with particular attention to its many-body aspects. The first and most basic question is: what is the hamiltonian? As mentioned by Girvin, we consider spinless fermions with charge $-e$, each carrying a solenoid with flux $\frac{1}{2}hc/e$ [see the top part of Fig. 1, where the parameter ν describes general fractional statistics (with $\nu = \frac{1}{2}$ for semions)]. Thus, the basic hamiltonian is just

$$H = \frac{1}{2m} \sum_{j=1}^{N} \left(\mathbf{p}_j + \frac{e}{c}\mathbf{A}_j\right)^2,$$

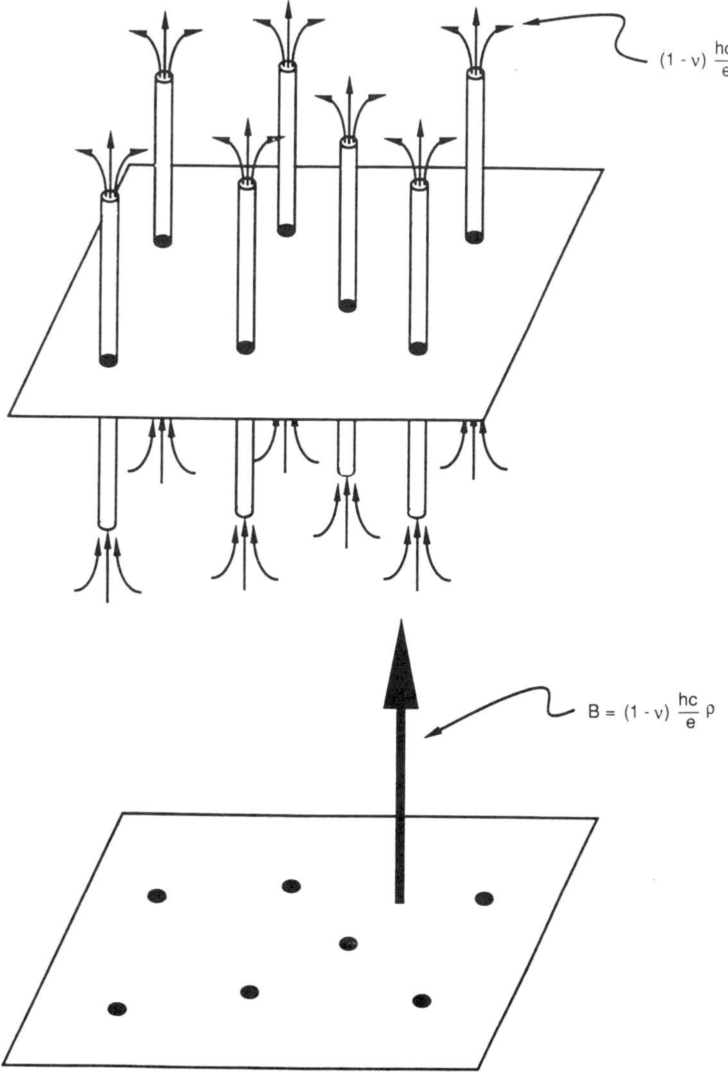

Fig. 1 Top: The ν-fractional-statistics gas consists of spinless fermions at density ρ, carrying solenoids with a fraction $1 - \nu$ of the basic flux quantum hc/e. For semions, we take $\nu = \frac{1}{2}$. Bottom: In the Hartree average-field approximation, each particle sees a uniform magnetic field $B = (1 - \nu)(hc/e)\rho$.

and all the physics is contained in the vector potential $\mathbf{A}_j$ seen by the jth particle. It has the form of a sum of the contributions from all the *other* particles

$$\mathbf{A}_j = \sum_{k \neq j}^{N} \mathbf{A}_{jk},$$

where $\mathbf{A}_{jk}$ is the vector potential at $\mathbf{r}_j$ generated by the solenoid at $\mathbf{r}_k$:

$$\mathbf{A}_{jk} = \tfrac{1}{2} \frac{\hbar c}{e} \frac{\hat{z} \times \mathbf{r}_{jk}}{\left| \mathbf{r}_{jk} \right|^2}.$$

This vector interaction potential is transverse to the line joining the two particles and varies inversely with the separation; it is analogous to the velocity field around a quantized rectilinear vortex filament. We note several unusual features of the hamiltonian.

1. The interaction is long-ranged.
2. The cross terms like $\mathbf{p}_1 \cdot \mathbf{A}_1 = \mathbf{p}_1 \cdot \sum_{k \neq 1}^{N} \mathbf{A}_{1k}$ introduce momentum-dependent terms in the interactions.
3. The squared terms like $\mathbf{A}_1 \cdot \mathbf{A}_1 = \sum_{j \neq 1}^{N} \sum_{k \neq 1}^{N} \mathbf{A}_{1j} \cdot \mathbf{A}_{1k}$ introduce *three-body* interactions when $j \neq k$. To develop a set of Feynman rules for this hamiltonian, it is necessary to devise a new symbol for these three-body potentials, as seen in Figs. 2 and 3 below.

Given this simple hamiltonian, how can one proceed? As a first step, we used the analogy with the electron gas and developed a random phase approximation (RPA) that led to the two results mentioned previously: the phonon-like collective mode and the Meissner effect. That calculation left one basic question unanswered—why is the linear response automatically gauge invariant? In the BCS theory of superconductivity, only the transverse response is gauge invariant, and the demonstration of gauge invariance for the longitudinal response required the introduction of a collective density mode through a generalized RPA[5] (this mode is analogous to that found here for the semion gas). Recently, we have developed a new way to understand this RPA result and to improve the theory, based on the self-consistent and conserving formalism developed by Baym and Kadanoff.[6] Their approach ensures that an approximation will satisfy the condition of particle conservation, which here implies gauge invariance as well.

The simplest such description is the Hartree approximation, in which the self energy Σ_H has the form shown in Fig. 2; it contains both two-body and three-body contributions. Taken together, they generate an average uniform magnetic field (mentioned in Girvin's talk—see the bottom part of Fig. 1), with

$$B = \tfrac{1}{2} \frac{hc}{e} \rho,$$

where ρ is the two-dimensional particle density. This relation is analogous to that for flux lines in a type-II superconductor. Correspondingly, there is an average vector potential

$$\bar{\mathbf{A}} = \tfrac{1}{2} B \hat{z} \times \mathbf{r}.$$

A similar result arises in rotating superfluid helium, where the uniform array of vortices produces an effective solid-body rotation Ω with average velocity $\bar{\mathbf{v}} = \Omega \hat{z} \times \mathbf{r}$. It is not surprising that the self-consistent Hartree eigenstates are the Landau levels in this uniform magnetic field B.

Fig. 2. The self-consistent Hartree self energy Σ_H, with both two-body and three-body contributions. The heavy solid line is the self-consistent one-body Green's function.

The calculation proceeds by finding the density and current correlation functions, which determine the linear response to an external electromagnetic field. They are obtained from a particle-hole Bethe-Salpeter equation, and the Baym-Kadanoff formalism provides a precise prescription for the kernel of this equation. In the Hartree approximation, this integral equation is exactly soluble, and the results reproduce those found previously in the RPA. Here, the gauge invariance of the response is ensured by the conserving formalism itself.

To obtain an improved conserving approximation, we merely add the exchange contributions to the self energy, leading to the Hartree-Fock approximation. In addition to the familiar two-body exchange diagram, there are several additional three-body exchange diagrams, as shown in Fig. 3. Remarkably, the resulting Hartree-Fock equations7 can still be solved exactly, despite the nonlocal exchange potential and the linear spatial dependence of the average vector potential $\bar{\mathbf{A}}$. To appreciate this result, recall that the Hartree-Fock equations involve a self-consistent solution for the one-body eigenvalues *and* one-body eigenfunctions. In a translationally invariant medium, plane waves automatically serve as the eigenfunctions, leaving only the eigenvalues to be determined self-consistently. For other cases, such as an atom, the eigenfunctions must be found as part of an iterative solution. In the present problem, however, the set of Landau eigenfunctions remains the exact self-consistent solution, and the associated one-body eigenvalues can be found explicitly and analytically. They have a logarithmically divergent contribution (negative for filled states and positive for empty ones). The resulting logarithmically divergent gap at the Fermi surface reflects the vortex character of the single-particle excitations in the Hartree-Fock picture.[7]

The Baym-Kadanoff formalism again provides the corresponding particle-hole Bethe-Salpeter equation for the Hartree-Fock approximation. The effective attraction between the particles and holes precisely cancels the logarithmic divergencies in the single-particle energies, leading to collective modes with finite excitation energies. The calculation involves significant numerical analysis, but the final answers in essence reproduce those found in the much simpler RPA, with only minor modifications. In particular, the lowest collective density wave has a linear spectrum at long wavelengths, with the speed of sound given by the bulk Hartree-Fock compressibility (as expected for a conserving approximation). At short wavelengths, however, the spectrum varies logarithmically, which reflects the vortex character of the particle-hole states. The linear response kernel is again gauge-invariant and displays a Meissner effect.

Several questions remain to be answered. First is the behavior at finite temperature. Girvin already argued that the system should display a Kosterlitz-Thouless type
of transition to the normal state,[8] but it is not yet clear how to extract it from the
present many-body formalism. More generally, it is still unclear whether semions have
anything to do with real high-temperature superconductors.

In closing, let me say how much we all enjoyed the Thursday evening river cruise.
The trip through the locks was perfect recreation for weary physicists, who acted more
like children than adults as the water level dropped 50 feet! Finally, on behalf of all
the participants, I would like to thank the organizing committee and all their helpers
(especially S. Smith and F. Krotscheck) for a wonderful week. We know well how
much work is involved, and it is much appreciated. I am confident that we can look
forward to an equally successful MBVIII.

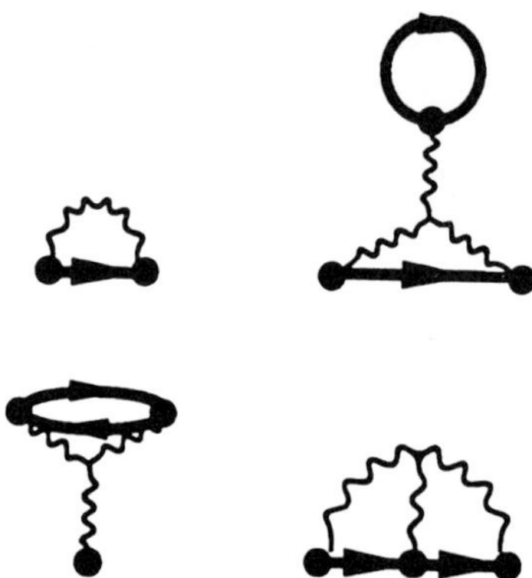

Fig. 3. The self-consistent exchange contributions to the Hartree-Fock self energy Σ_{HF}. Note the presence of several three-body contributions, in
addition to the usual two-body one.

ACKNOWLEDGMENT

This work has been supported in part by the National Science Foundation under
Grant No. DMR 88-19585.

References

1. R. P. Feynman, Prog. Low Temp. Phys. **I**, 17 (1955).
2. A. L. Fetter, Phys. Rev. Lett. **27**, 986 (1971); Ann. Phys. (N. Y.) **70**, 67 (1972).
3. A. L. Fetter, C. B. Hanna, and R. B. Laughlin, Phys. Rev. **B39**, 9679 (1989);
   Int. J. Mod. Phys., to be published.
4. R. P. Feynman, Phys. Rev. **91**, 1301 (1953); **94**, 262 (1954).

5. P. W. Anderson, Phys. Rev. **112**, 1900 (1958).
6. G. Baym and L. P. Kadanoff, Phys. Rev. **124**, 287 (1961); G. Baym, Phys. Rev. **127**, 1391 (1962).
7. R. B. Laughlin, Phys. Rev. Lett. **60**, 2677 (1988); C. B. Hanna, R. B. Laughlin, and A. L. Fetter, Phys. Rev. **B40**, 8745 (1989).
8. See also Y. Kitazawa and H. Murayama, Phys. Rev. **B41**, 11101 (1990).